Communications and Control Engineering

Series Editors

E.D. Sontag • M. Thoma • A. Isidori • J.H. van Schuppen

Published titles include:

Stability and Stabilization of Infinite Dimensional Systems with Applications
Zheng-Hua Luo, Bao-Zhu Guo and Omer Morgul

Nonsmooth Mechanics (Second edition)
Bernard Brogliato

Nonlinear Control Systems II
Alberto Isidori

L_2-Gain and Passivity Techniques in Nonlinear Control
Arjan van der Schaft

Control of Linear Systems with Regulation and Input Constraints
Ali Saberi, Anton A. Stoorvogel and Peddapullaiah Sannuti

Robust and H_∞ Control
Ben M. Chen

Computer Controlled Systems
Efim N. Rosenwasser and Bernhard P. Lampe

Control of Complex and Uncertain Systems
Stanislav V. Emelyanov and Sergey K. Korovin

Robust Control Design Using H_∞ Methods
Ian R. Petersen, Valery A. Ugrinovski and Andrey V. Savkin

Model Reduction for Control System Design
Goro Obinata and Brian D.O. Anderson

Control Theory for Linear Systems
Harry L. Trentelman, Anton Stoorvogel and Malo Hautus

Functional Adaptive Control
Simon G. Fabri and Visakan Kadirkamanathan

Positive 1D and 2D Systems
Tadeusz Kaczorek

Identification and Control Using Volterra Models
Francis J. Doyle III, Ronald K. Pearson and Babatunde A. Ogunnaike

Non-linear Control for Underactuated Mechanical Systems
Isabelle Fantoni and Rogelio Lozano

Robust Control (Second edition)
Jürgen Ackermann

Flow Control by Feedback
Ole Morten Aamo and Miroslav Krstić

Learning and Generalization (Second edition)
Mathukumalli Vidyasagar

Constrained Control and Estimation
Graham C. Goodwin, María M. Seron and José A. De Doná

Randomized Algorithms for Analysis and Control of Uncertain Systems
Roberto Tempo, Giuseppe Calafiore and Fabrizio Dabbene

Switched Linear Systems
Zhendong Sun and Shuzhi S. Ge

Subspace Methods for System Identification
Tohru Katayama

Digital Control Systems
Ioan D. Landau and Gianluca Zito

Multivariable Computer-controlled Systems
Efim N. Rosenwasser and Bernhard P. Lampe

Dissipative Systems Analysis and Control (2nd Edition)
Bernard Brogliato, Rogelio Lozano, Bernhard Maschke and Olav Egeland

Algebraic Methods for Nonlinear Control Systems
Giuseppe Conte, Claude H. Moog and Anna M. Perdon

Polynomial and Rational Matrices
Tadeusz Kaczorek

Simulation-based Algorithms for Markov Decision Processes
Hyeong Soo Chang, Michael C. Fu, Jiaqiao Hu and Steven I. Marcus

Iterative Learning Control
Hyo-Sung Ahn, Kevin L. Moore and YangQuan Chen

Distributed Consensus in Multi-vehicle Cooperative Control
Wei Ren and Randal W. Beard

Control of Singular Systems with Random Abrupt Changes
El-Kébir Boukas

Nonlinear and Adaptive Control with Applications
Alessandro Astolfi, Dimitrios Karagiannis and Romeo Ortega

Aziz Belmiloudi

Stabilization, Optimal and Robust Control

Theory and Applications in Biological and Physical Sciences

Springer

Aziz Belmiloudi, PhD
Institut de Recherche Mathématique
de Rennes (IRMAR)
Centre de Mathématiques
Institut National des Sciences Appliquées
(INSA) de Rennes
20 Avenue des buttes de Coesmes
35043 Rennes Cedex
France

ISBN 978-1-84800-343-9 e-ISBN 978-1-84800-344-6

DOI 10.1007/978-1-84800-344-6

Communications and Control Engineering ISSN 0178-5354

British Library Cataloguing in Publication Data
Belmiloudi, Aziz
Stabilization, optimal and robust control : theory and applications in biological and physical sciences. - (Communications and control engineering)
1. Robust control 2. Automatic control - Mathematical models 3. Differential equations, Partial 4. Differential equations, Nonlinear 5. Game theory
I. Title
629.8'312
ISBN-13: 9781848003439

Library of Congress Control Number: 2008928175

Cover design: le-tex publishing services oHG, Leipzig, Germany

Printed on acid-free paper

9 8 7 6 5 4 3 2 1

springer.com

To my mother Fatima Elyoubi

To my wife Géraldine
and to our children
Nora, Gwenda and Yann

To the memory of my grandmother

Preface

This book focuses on stabilization, control and fluctuation of systems governed by non-linear partial differential equations (PDEs), which arise in many applications. The developed approach is based on robust control theory in the dynamic non-cooperative optimization framework (*i.e.*, game theory).

Robust control problems are the subject of a huge and varied range of research worldwide. Research into the dynamical systems governed by PDEs is a relatively new area of study, but exciting and vitally important. This theory deals with optimization, identification, stability, robustness and regulation. It is impossible to mention all the applications of this theory, but they include: fluid flow in domains of variable configuration; advanced, composite and smart materials; aerospace and mechanical systems; electronic and optical devices; economic models; and biological, medical and chemical processes (for example, to maintain the various constituents at their appropriate levels). We can also mention the so-called delay systems, which are mathematical models used for many diffusion processes, in which time-delayed feedback signals are used to describe propagation phenomena, with applications in population dynamics, plasma physics, ecology, epidemiology, immunology, neural networks, *etc.*

The objective of robust control theory, which generalizes optimal control theory, is to compensate for the undesirable effects of system uncertainties through control actions so that a cost function achieves its minimum for the worst uncertainties. In other words, the goal is to find the best control which stabilizes the fluctuations of the dynamic system with a limited control effort, by taking into account the worst-case disturbances which destabilize the dynamic behavior of the system. The existence, uniqueness, qualitative properties and good behavior under perturbation of solutions of the model are becoming an important prerequisite and are a research domain of their own, especially in view of the possibility of modeling by states and controls.

The techniques developed in this book concern the robust control of infinite-dimensional dynamical systems. All these systems are derived from time-dependent coupled PDEs associated with boundary-valued problems that arise in the physical and biological sciences. It is clear that to make a rigorous analysis of these systems, it is necessary to take into account not only the non-linear dynamics of the problems but also the evolutionary and coupled behavior of the PDEs governing such systems. It is thus important to make a good choice of function spaces, both in terms of solvability (in general, a suitable treatment of such systems requires the use of several function spaces) and of the correct modeling of realistic problems.

Therefore, our approach in this book is to combine the general theory of control, the optimization theory, the modeling process and the theory of time-dependent coupled PDEs into one complete unified theory.

Mathematical foundations essential to the analysis of stabilization, robust control and fluctuation in the context of control systems described by dynamical coupled non-linear PDEs are provided, while remaining accessible to the non-specialist.

Therefore, this book will be useful to researchers in mathematics, physics, biology and chemistry, and to professionals involved in complex problems in fluid mechanics, biological systems and material sciences. Most of the topics developed in this book are new or have been published recently.

The book is divided roughly into three parts. In the first part, mathematical results necessary to control theory are presented. Proofs are only provided for those results that either cannot be easily found in the standard textbooks, or are useful in order to understand related problems or concern new results. Some essential results for convex functions are given in Chapter 2, and the basic features of Sobolev spaces with useful compactness results in Chapter 3. In Chapter 4, the convex conjugate duality theory (with the Legendre–Fenchel transformation) is developed. Chapter 5 discusses very important tools used in the study of non-linear systems: critical point, Lagrange duality theory and minimax principles. The minimax theorems have many useful applications and play a central role in the notion of stability and robust control theory. The non-convex parametric variational problem for a geometrically non-linear system, by introducing a new gap function, is also studied. This part is illustrated with different applications including the Navier–Stokes equations for fluid mechanics, Maxwell equations for electric and magnetic fields, Ginzburg–Landau equations for ferroelectric models, and the elasticity problem for deformation processes.

In the second part of the book, classical optimal control theory and, the heart of this book, robust control theory (on PDEs) are developed. For both, several different realistic cases of observations and controls are analyzed. In the robust control approach, different cases of disturbances are also considered. Linear, bilinear and non-linear control problems for dynamical systems with or without time-varying delays are discussed. In Chapter 6, some elements of functional analysis are introduced: function spaces and linear evolution problems of first order in time; it is considered as a reference chapter. Chapter 7 contains the general results and concepts for the optimal control problem of several time-dependent and differential systems, in different contexts. A very basic problem is first given to explain the theory as simply as possible. In each section, the existence, uniqueness and optimality conditions for the optimal solution (for different observation and control situations) are studied. Chapter 8 is devoted to stabilization and robust regulation problems, using some of the new mathematical objects that have been recently introduced in relation to the stabilization of dynamical systems. This chapter contains the essential and fundamental developments of the robust control theory of distributed parameter systems. This area concerns investigation into the control, stability and adjoint control optimization of infinite-dimensional dynamical systems. Moreover, we are interested in the robust regulation of the deviation of the systems from the desired target, by analyzing the *full non-linear* systems, which models large perturbations to the desired target. Several different mathematical applications are given in these two main chapters to illustrate optimal and robust control theory, such as a biochemical pollutant model (a non-linear problem) and a nuclear fission reactor model (a bilinear problem), the last model also describing cancer chemotherapy. A general case of time-varying delay in non-linear parabolic systems is also studied: the delays occur naturally in biological and chemical systems, in population dynamics, *etc.* Finally, the last chapter of this part briefly presents some numerical approaches.

In the last part of the book, some applications to biological and physical sciences are given. Chapter 10 is devoted to vortex dynamics in superconducting films. In Chapter 11, the multiscale modeling solidification of binary alloys is studied. Chapter 12 concerns the large-scale ocean in the climate system. In Chapter 13, the impact of heat transfer laws on temperature distribution in biological systems with directional blood flow (with application in the cancer treatment) is analyzed. Chapter 14 concerns resource management problems and the stabilization of uncertain species resources (*i.e.*, population dynamics). Chapter 15 presents two other interesting models, namely micropolar fluids (*e.g.*, animal blood) and semiconductor melts.

I would like to thank the editors of Springer, especially A. Doyle who proposed that I should write this book, and O. Jackson for his kind suggestions on the layout of the manuscript, and for support and patience. I thank the le-tex publishing services oHG and my PhD student, A. Rasheed, for their help in improving the English language. I am grateful to the commission of INSA of Rennes who allowed me a half-year of CRTC to write this book. I wish to thank the librarians of IRMAR (Institute of Mathematical Research of Rennes), V. Cohoner, A. Guillemer and D. Hervé, who have provided me with all the books and articles necessary for my research, and many friends and colleagues who encouraged me to complete this work. I thank also all the anonymous reviewers (who will recognize themselves, I hope) who, during their reading of my different articles, provided me with very helpful suggestions.

Last, but not least, I would like to express my acknowledgment and thanks to my wife Géraldine for her patience, support and encouragement during this project and over the years.

Rennes (France),
March 2008

Aziz Belmiloudi

Contents

Part III Applications in the Biological and Physical Sciences: Modeling and Stabilization

Notation and Symbols

a.e.	Almost everywhere
■	End of definition
♣	End of example
□	End of proof
◇	End of remark
$\mathbb{N} = \{0, 1, 2, \ldots\}$	Set of integers
$\mathbb{N}^* = \{1, 2, \ldots\}$	Set of non-null integers
$\mathbb{R}$	Field of real numbers
$\mathbb{R}^+$	Field of non-negative real numbers
$\mathbb{C}$	Field of complex numbers
$\overline{\mathbb{R}} = \mathbb{R} \cup \{-\infty, +\infty\}$	$-\infty$ and $+\infty$ are allowed to the functions under consideration
$\in$	Belongs to
$\emptyset$	Empty set
$A \subset B$	A is a subset of B
$A \cap B$	Intersection of subsets A and B
$A \cup B$	Union of subsets A and B
$\mid \alpha \mid$	Absolute value of $\alpha \in \mathbb{C}$
$\overline{\alpha}$	Complex conjugate of $\alpha \in \mathbb{C}$
$\mathcal{I}(\alpha)$	Imaginary part of $\alpha \in \mathbb{C}$
$\mathcal{R}(\alpha)$	Real part of $\alpha \in \mathbb{C}$
$\max(a, b)$	The greater of a and b
$\min(a, b)$	The lesser of a and b
$\mathrm{clco}S$	Closed convex hull of a set S
$\mathrm{co}S$	Convex hull of a set S
$\mathrm{int}(C)$	Interior of a set C

$\mathcal{X}_S$	Indicator function of a set S
$F : V \longrightarrow \overline{\mathbb{R}}$	Extended real-valued mapping on V
$\mathrm{cl}F,\ \overline{F}$	Closure of a mapping F
$\mathrm{cv}F$	Convex envelope of a mapping F
$\mathrm{dom}(F),\ D(F)$	Effective domain of a mapping F
$\mathrm{epi}F$	Epigraph of a mapping F
$\mathrm{hypo}F$	Hypograph of a mapping F
$\mathrm{ls}F$	Lower semi-continuous envelope of a mapping F
2^V	Set of all subsets of V (power set of V)
$\mathcal{L}(V;W)$	Space of continuous linear functionals from V to W
$\Gamma(V)$	Set of extended real-valued mappings on V which are pointwise supremum of a family of continuous affine functions
Λ^*	Adjoint of the linear operator Λ
V'	Dual of a space V
V''	Bidual of a space V
$\sigma(V',V)$	Weak-star topology defined on V'
$\sigma(V,V')$	Weak topology defined on V
$F^* : V' \longrightarrow \overline{\mathbb{R}}$	Convex conjugate function of $F : V \longrightarrow \overline{\mathbb{R}}$, defined as, $F^*(f) = \sup\limits_{u \in V}(\langle f, u\rangle_{V',V} - F(u)),\ \forall f \in V'$
$\liminf$	Limit infimum
$\limsup$	Limit supremum
$\longrightarrow$	Strong convergence in V
$\rightharpoonup$	Weak convergence in V
$\rightharpoonup^*$	Weak star convergence in V'
$\partial\Omega$	Boundary of a domain Ω
$\mathcal{C}(\Omega)$	Space of continuous functions on a domain Ω
$\mathcal{C}^k(\Omega)$	Space of k times continuously differentiable functions on Ω, $k \in \mathbb{N}$ or $k = +\infty$
$\mathcal{C}_0^\infty(\Omega),\ \mathcal{D}(\Omega)$	Space of $\mathcal{C}^\infty$ functions on Ω with a compact support in Ω
$\mathcal{C}^{0,\alpha}(\Omega)$	Hölder space of order $\alpha \in (0,1]$ on $\overline{\Omega}$ *i.e.*, the space of continuous functions u on $\overline{\Omega}$ such that $\sup\limits_{\substack{x,y\in\Omega \\ x\neq y}} \frac{\mid u(x) - u(y) \mid}{\mid x - y \mid^\alpha} < \infty$
$\mathcal{C}^k([a,b];U)$	Space of k times continuously differentiable functions v from $[a,b]$ into space U, $k \in \mathbb{N}$ or $k = +\infty$
$\mathcal{D}'(\Omega)$	Space of distributions on Ω

$L^p(\Omega)$	Space of (class of) measurable functions v on Ω such that $x \longrightarrow \|v(x)\|^p$ is integrable on Ω, $p \in [1, +\infty[$
$L^\infty(\Omega)$	Space of (class of) measurable functions v on Ω such that $x \longrightarrow \|v(x)\|$ is essentially bounded on Ω
$L^p_{\text{loc}}(\Omega)$	Space of functions which are L^p on any bounded sub-domain of Ω, $p \in [1, +\infty[$
$L^p(a, b; U)$	Space of (class of) L^p functions v from (a, b) into space U, with $a, b \in \overline{\mathbb{R}}$, $1 \leq p \leq +\infty$
$\text{grad}(f)$, ∇f	Gradient of a function $f : \Omega \subset \mathbb{R}^n \longrightarrow \mathbb{R}$ or X (X is a Banach space): $\nabla f = (\frac{\partial f}{\partial x_1}, \ldots, \frac{\partial f}{\partial x_n})$
$\text{div}(v)$, $\nabla . v$	Divergence of a function $v : \Omega \subset \mathbb{R}^n \longrightarrow \mathbb{R}^n$: $\nabla . v = \sum_{i=1,n} \frac{\partial v_i}{\partial x_i}$, where $v = (v_i)_{i=1,n}$
$\text{curl} f$, $\nabla \times f$	Curl of a function $f : \Omega \subset \mathbb{R}^n \longrightarrow \mathbb{R}$: $\text{curl} f = (\frac{\partial f_i}{\partial x_j} - \frac{\partial f_j}{\partial x_i})_{i,j=1,\ldots n}$
$W^{m,p}(\Omega)$	Sobolev space of order (m, p), $m \in \mathbb{N}^*$, $p \in [1, +\infty]$
$H^m(\Omega)$	Sobolev space $W^{m,2}(\Omega)$ of order m, $m \in \mathbb{N}^*$
$W_0^{m,p}(\Omega)$	Closure of $\mathcal{D}(\Omega)$ for the $W^{m,p}(\Omega)$ norm, $m \in \mathbb{N}^*$, $p \in [1, +\infty]$
$H_0^m(\Omega)$	Closure of $\mathcal{D}(\Omega)$ for the $H^m(\Omega)$ norm, $m \in \mathbb{N}^*$
$W^{-m,p^*}(\Omega)$	Dual of $W_0^{m,p}(\Omega)$, $m \in \mathbb{N}^*$, $p \in]1, +\infty[$, $\frac{1}{p} + \frac{1}{p^*} = 1$
$H^{-m}(\Omega)$	Dual of $H_0^m(\Omega)$, $m \in \mathbb{N}^*$
$H(\text{div}; \Omega)$	The space $\{v \in L^2(\Omega) : \text{ div}(v) \in L^2(\Omega)\}$
$H_0(\text{div}; \Omega)$	Closure of $\mathcal{D}(\Omega)$ for the $H(\text{div}; \Omega)$ norm
$[X, Y]_\theta$	Interpolation space between X and Y, $X \subset Y$, which is endowed with the norm $\| \cdot \|_{[X,Y]_\theta}$, $\theta \in [0, 1]$ such that $\forall u \in X$, $\| u \|_{[X,Y]_\theta} \leq C_\Theta \| u \|_X^{1-\theta} \| u \|_Y^\theta$

1

General Introduction

"As far as the laws of mathematics refer to reality, they are not certain; as far as they are certain, they do not refer to reality."

Albert Einstein

In the analysis of contemporary dynamical systems, scientists and engineers are often confronted with increasingly complex models that can simultaneously include terms taking into account non-linear dynamics, time delays and hysteresis effects, and uncertainties in parameters. Classical optimal control techniques have allowed them to optimize the control systems they build for cost and performance, but these techniques are not always tolerant of fluctuations in the dynamical system or in the real world. The goal of robust control theory is to estimate the performance changes of a dynamical system with changing system parameters and functions, and to develop alternatives that are insensitive to changes in the system in order to maintain the stability and the performance. In a broad sense, the goal of the robust control is to maintain the transformation from the desired state to the output state as close to unity as possible, despite these fluctuations.

In this chapter, we explain the motivations and our general ideas for using the robust control theory to study non-linear dynamical systems. Then, we present the general process for our robust control approach and finally, in order to explain our theoretical proposals on practical cases, we briefly give various applications, that exhibit graceful degradation subject to many disturbances, as well as more fluctuations which affect considerably the model of the dynamical system, where the use of robust control theory is extremely important.

1.1 Motivations and Objectives

The differential systems that we want to study are spatially and temporally distributed and they are governed by partial differential equations (PDEs) which are mostly non-linear. They may represent many fields of physics or bio-technological processes, which are modeled by systems with parameters distributed and governed by time-dependent PDEs (with or without time delays), in which it is of interest to prescribe a suitable dynamical behavior.

The mathematical robust control theory is a part of applied mathematics serving perhaps the most important link between modeling, mathematics (either from a theoretical or computational point of view), industrial processes and technology. We note that the three main steps in the area of research in robust control of dynamical systems are inextricably linked, as shown below:

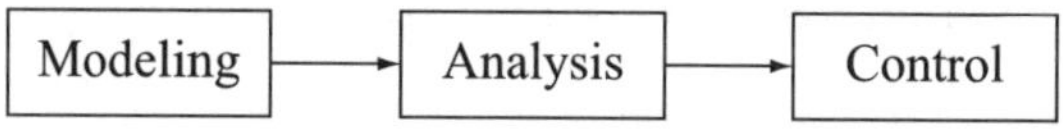

The investigated problems are of various nature and deal both with the analysis of the structural properties of parametrically dependent differential equations and with their regulation according to some task or cost. Three types of problems arise then naturally:

(*i*) *Identification*: Certain parameters or functions intervening in these models are unknown, or rather badly known (for example, coefficients of diffusion, non-linear source, initial conditions or boundary conditions, *etc.*). We propose to identify these parameters or functions starting from experimental observations: these problems are called "inverse problems" (in opposition to the resolution of equations themselves which constitutes the direct problem). Indeed, certain parameters or functions can influence considerably the material behavior or modify phenomena in environmental, bio-economic, biological or medical matter; then their knowledge is an invaluable help for the physicists, biologists or chemists who, in general, use a mathematical model for their problem, but with a great uncertainty on its parameters. The resolution of the inverse problems thus provides them essential informations which are necessary to the comprehension of the various processes which can intervene in these models.

(*ii*) *Regulation*: The most real physical, biological or chemical systems can only be described by means of an uncertain model may induce instability. Moreover, the systems are destabilizing by unmeasured noises and disturbances. Consequently, even if some "well-posedness" property is verified, the systems become often unstable. The idea is to regulate the response of systems by modifying the dynamical nature of the system.

(*iii*) *Optimization*: The physicists, biologists and chemists control, in general, their experimental devices by using a certain number of functions of control which enable them to optimize and/or to stabilize the system. The work of the mathematician consists in determining these functions in an optimal way. The methods consist in designing a trajectory for the control inputs and are normally based on optimization of the performance of the system relative to some performance functional.

The optimal control methods are used to determine the unknown parameters or control certain functions for problems where uncertainties (disturbances, noises, fluctuations, *etc.*) are neglected. But it is well known that many uncertainties occur in more realistic studies of physical, biological or chemical problems. The presence of these uncertainties may induce complex behaviors, *e.g.*, oscillations, instability, bad performances, *etc.* Problems with uncertainties are the most challenging and difficult in control theory but their analysis are necessary and important for applications.

The fundament of robust control theory, which is a generalization of the optimal control theory, is to take into account these uncertain behaviours and to analyze how the control system can deal with this problem. From Chandrasekharan [70], "*Robust control refers to the control of unknown plants with unknown dynamics subject to unknown disturbances.*"

The uncertainty can be of two types: first, the errors (or imperfections) coming from the model (difference between the reality and the mathematical model, in particular if some parameters are badly known) and, second, the unmeasured noises and fluctuations that act on the physical, biological or chemical systems. These uncertainty terms can have additive and/or multiplicative components. They often lead to great instability: for example, the *El Niño* phenomenon, tropical instability waves, which are essentially the consequence of a small perturbation of ocean-surface temperature (these equatorial waves are the precursors for hurricanes and typhoons). The goal of robust control theory is to control these instabilities, either by acting on some parameters to maintain the system in a desired state (target), or by calculating the limit of these parameters before the system becomes unstable ("predict to act"). In other words, the robust control allows engineers to analyze instabilities and their consequences and helps them to determine the most acceptable conditions for which a system remains stable. The goal is then to define the maximum of noises and fluctuations that can be accepted if we want to keep the system stable. Therefore, we can predict that if the disturbances exceed this threshold, the system becomes unstable (for example, we can predict the fluctuation of ocean-surface temperature from which the *El Niño* phenomenon can occur). It also allows us, in a system where we can control the perturbations, to provide the threshold at which the system becomes unstable. The robust control theory for dynamical systems can be described by the block diagram Figure 1.1, where the function ψ corresponds to the disturbance, ϕ is

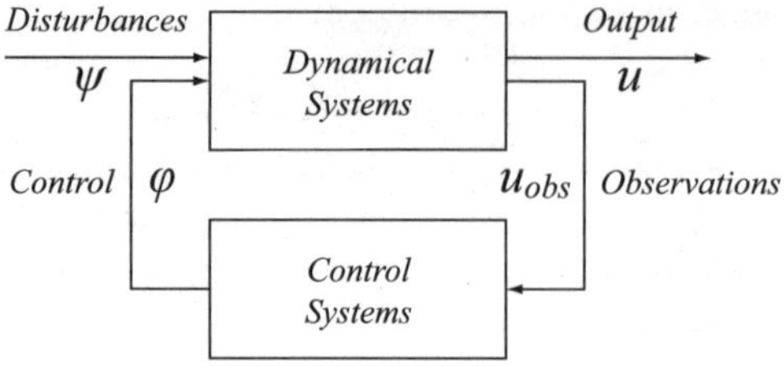

Figure 1.1. Process of robust control

the control function, u_{obs} is the observation (*e.g.*, measurement), u represents the perturbation of the desired target (that it is desired to keep small).

The fundamental idea of our approach is the connection between the game theory approach and the problem of stabilizing uncertain non-linear distributed parameter systems[1] where the fluctuation dynamics and noises are *deterministic*. This is motivated, by the fact that the robust control theory can be represented as a *differential game*[2] between an engineer seeking the best control which stabilizes the system perturbations with limited control efforts, and simultaneously plant (during physical, biological or chemical experiences) or unexpected events (during the physical, biological or chemical dynamics) seeking the maximally malevolent disturbance which destabilizes the system perturbations with limited disturbance magnitude.

In other words, the idea of our approach is to transform robust stability and performance problems into constrained game-type minimax optimization ones (of infinite-dimensional dynamical systems). The objective of a robust control is then to compensate the undesirable effects of system disturbances through control actions such that a performance functional achieves its minimum for the worst disturbances, *i.e.*, to find the best control which takes into account the worst-case disturbance. This area concerns investigation of the control, stability and adjoint control optimization of infinite-dimensional dynamical systems.

Since the late 1970s, a great variety of techniques have been developed for this new area of research. Many different models describing physical, biological or chemical systems (from chemical to mechanical engineering, from biological systems to population dynamics, *etc.*) with taking into account some uncertainties are giving raise to different research directions. In the framework of many robust control problems studied in the literature, the considered problems mainly correspond to the construction of controllers for linear (or linearized) plants with additive disturbances (because by assuming these considerations, the analysis is greatly simplified). But it is well-known that in the real world, the systems have non-linear dynamical behaviors and dis-

[1] The goal of the game is to find a controller in the presence of an adversary that changes the process.

[2] In the sense of two-person zero-sum game, see Section 5.3.2.

turbances may act additively and/or multiplicatively. Moreover, the systems may present hard control, disturbance or state constraints (*e.g.*, pointwise constraints).

Therefore, we are interested in the robust regulation of the deviation of the systems from the desired target, by analyzing the *full non-linear and time varying systems* (with or without time delays), which models large perturbations to the desired target, and by considering different actions of disturbances and controls with various constraints.

In our approach it is not assumed that the system is stabilizable or detectable, as opposed to other books in the treatment of robust control problems to some classes of finite (or infinite)-dimensional systems which are considered in terms of a general Riccati operator (see, *e.g.*, Başar and Bernhard, [26], Chen [76], Foias *et al.* [125], Dullerud and Paganini [232], Petersen *et al.* [241], Sanchez-Pena and Sznaier [257], Van Keulen [288], Whittle [300], Zhou *et al.* [311, 312], and references therein), a hypothesis that is difficult to verify in practice. Moreover, in the more general case, numerical realizations based on the adjoint control optimization are preferred over techniques based on Riccati approaches. To the best of our knowledge, the approaches and results developed in the previous and below references are not applicable to the applications analyzed in this book.

The previous references give an interesting background though most of these references consider linear systems (of finite or infinite dimensions) and optimizations over the infinite time horizon, and are referred to differential games and $\mathcal{H}_\infty$-control theory[3] (that is well described in Green and Limebeer [140] and Zhou *et al.* [311]) or its stochastic counterpart, *i.e.*, risk-sensitive control theory. For the robust design where the $\mathcal{H}_\infty$-control problem is regarded as loop-shaping problem, the reader can refer to Vinnicombe [290], in which the author introduces a new metric for systems. Other methods based on Lyapunov design method were proposed for robust stabilization of non-linear uncertain systems see, *e.g.*, Qu [246], in which the author considers the robust stabilization of systems described by ordinary differential equations. For practical examples, we can cite Ackermann *et al.* [3], in which the authors present stability analysis for problems described by linear time-invariant, based on Kharitonov-type criteria. Finally, for robust control of time-delay systems, the reader may refer to Zhong [310], in which used tools are chain-scattering approach and J-spectral factorizations and Niculescu [230], in which the author treats the stability of finite-dimensional delay differential equations (we can also refer to Mahmoud [213]). The reader can find other references which complete this survey in the introduction of each chapter.

The approach described in this book, which is based on minimax theorems in relation with non-linear PDEs in finite time horizon and motivated by practical application, has been highlighted from the beginning of the 2000s. For

[3] $\mathcal{H}_\infty$-control problem is worked and posed in Hardy spaces (spaces of all stable transfer functions).

these developments, we can mention for the robust control of a class of nonlinear parabolic systems with time-varying delays, Belmiloudi [38, 39]; for robust control of the incompressible Navier–Stokes equations, Bewley *et al.* [52]; for the Kuramoto–Sivashinsky model, Hu and Temam [162]; for the stability of solidification processes, Belmiloudi *et al.* [40, 41]; and for the Ginzburg–Landau system and superconductivity, Belmiloudi [45].

We shall now present the process of our control robust approach.

1.2 General Process of the Robust Control Theory

In contrast to the optimal control problems,[4] the relation between the problems of identification, regulation and optimization, lies in the fact that it acts, in these cases, to find a saddle point of a functional calculus depending on the control, the disturbance and the solution of the perturbed PDEs. Indeed, the problems of control can be formulated as the robust regulation of the deviation of the systems from the desired target; the considered control and disturbance variables, in this case, can be in the parameters or in the functions to be identified. This optimization problem (a minimax problem), depending on the solution of PDEs, with respect to control and disturbance variables (intervening either in the initial conditions, or boundary conditions or equation itself), is the base of the robust control theory of PDEs.

The essential data used in our robust control problem are the following:

1. A "control" variable φ in a set U_{ad} (known as set of "admissible controls") and a "disturbance" variable ψ in a set V_{ad} (known as set of "admissible disturbances").
2. The state $u(\varphi, \psi)$ of the system to be controlled, which is given, for a chosen control-disturbance (φ, ψ), by the resolution of a perturbed equation
$$\tilde{\mathcal{F}}(t)(u(\varphi, \psi)) = \text{"given function of } (\varphi, \psi)\text{"}$$
where $\tilde{\mathcal{F}}(.)$ is an operator (supposed to be known) which represents the system to be controlled and u is the perturbation of the desired target U. The operator $\tilde{\mathcal{F}}(.)$, which depends on U, is the perturbation of the model $\mathcal{F}(.)$ of the studied system.
3. An "observation" u_{obs} which is supposed to be known exactly (for example, the desired tolerance for the perturbation or the offset given by measurements).
4. A "cost" functional (or "objective" functional) $J(\varphi, \psi)$ which is defined from a real-valued and positive function $G(X, Y)$ by
$$J(\varphi, \psi) = G((\varphi, \psi), u(\varphi, \psi)).$$

[4] The optimal control problem corresponds to minimize or maximize a calculus function depending on the control and the solution of PDEs.

The goal is to find a saddle point of J, *i.e.*, a solution $(\varphi^*, \psi^*) \in U_{\rm ad} \times V_{\rm ad}$ of

$$J(\varphi^*, \psi) \leq J(\varphi^*, \psi^*) \leq J(\varphi, \psi^*) \quad \forall \varphi \in U_{\rm ad}, \psi \in V_{\rm ad}.$$

It should be noted that there is no general method to analyze the problems of robust control (it is necessary to adapt it in each situation). Moreover, in non-linear systems or bilinear systems, the analysis is more complicated than in the case of optimal control problems, because we are interested in the robust regulation of the deviation of the systems from the desired target, by analyzing the *full non-linear* systems which model large perturbations to the desired target. Consequently the perturbations of the initial models, governed by PDEs, which show additional operators (and then difficulties) generate new primal problem and then new dual problem which, often, seem of a new type.

On the other hand, we can define the process to be followed for each situation:

(*i*) Solve the initial problem (analysis of PDEs, existence of solutions, stability according to the data, regularity, *etc.*).

(*ii*) Define the function or the parameter to be identified and the type of disturbance to be controlled.

(*iii*) Introduce and solve the perturbed problem which plays the role of the primal problem (analysis of PDEs, existence of solutions, stability according to the data, regularity, differentiability of the operator solution, *etc.*).

(*iv*) Define the cost (or objective) functional, which depends on control and disturbance functions.

(*v*) Obtain the existence of an optimal solution (as a saddle point of the cost functional) and analyze the necessary (and if possible the sufficient) conditions of optimality (which require to obtain before a very fine regularity on the state functions).

(*vi*) Characterize the optimal solutions.

(*vii*) Define an algorithm allowing to solve numerically the robust control problem (which requires sometimes the development of new methods of numerical resolution).

We now present some applications to biological and physical sciences (which are studied and analyzed in the third part of this book).

1.3 Applications to Biological and Physical Sciences

The mathematical, physical and biological systems studied in the third part of the book include the following:

- reaction-diffusion equations from population growth (*e.g.*, Lotka–Volterra model)
- bioheat transfer and Pennes-type model
- Ginzburg–Landau system and superconductivity
- Warren–Boettinger-type model and solidification
- incompressible Navier–Stokes equations coupled with transport-diffusion equations and other equations of fluid mechanics
- micropolar fluid model.

For all these problems, the following questions will be addressed:

(a) modeling and governing system
(b) existence, uniqueness, regularity and continuous dependence on the data of the model
(c) perturbation model and same points as in (b)
(d) motivation and formulation of the robust control problem in different situations
(e) existence of an optimal solution (a saddle point)
(f) optimality conditions and identification of gradients
(g) uniqueness of the optimal solution.

Applications are of three main types and are described next.

1.3.1 Material Sciences

Material sciences concern with the synthesis and manufacture of new materials, the modification of materials, the understanding and prediction of material properties, and the evolution and control of these properties over a time period. Today it is a vast growing body of knowledge based on physical sciences, engineering, and mathematics.

The goal is to present some mathematical treatments for the non-linear evolution systems which arise in material sciences. During the manufacture of the material, small perturbations caused by the introduction of noises terms in the data (which are regarded as impurities) give rise to surface and convective instabilities. To manufacture the materials free from impurities, it is essential to control both surface and convective instabilities.

• *Vortex dynamics in superconducting films with Ginzburg–Landau systems*: The phase transitions taking place in superconductor films with variable thickness is modeled by a two-dimensional, time-dependent Ginzburg–Landau type model with Robin boundary conditions on a phase-field parameter. To take into account thermal fluctuations and material impurities (which affect the motion of vortices in superconductors), we use a variant of Ginzburg–Landau type model containing additive noise (this work is a generalization of the recent research developed by Belmiloudi in [45]); we can note that the unknown phase-field parameters are complex valued. The objective is to control and stabilize the motion of vortices in superconductors.

• *Multi-scale modeling solidification of binary alloys and phase-field model*: The isothermal solidification of a binary alloy (*i.e.*, a mixture of two elements) is modeled by a two-dimensional, time-dependent and solutal phase field model of Warren–Boettinger type. To take into account thermal fluctuations and material impurities, we use a variant of the Warren–Boettinger model containing additive noise due to thermal fluctuations (this work is a generalization of the recent research developed by Belmiloudi in [40]). This studied model involves dentrite growth of highly supersatured binary melts. The objective is to predict and stabilize the microstructure dynamics by taking into account thermal fluctuations and material impureties. The developed technique can be used to study different general physical models concerning the solidification process, for example the problems presented recently by Granasy *et al.* [139], and Warren *et al.* [295].

1.3.2 Fluid Mechanics

Our aim is to describe some non-linear time-dependent PDEs which arise in fluid mechanics. In each case, we present briefly the physical model and the governing equations. Then we present the mathematical setting of the obtained equation and we study the robust control problem in order to control the fluctuations of the system. We also discuss the mechanisms of control of these instabilities. Different techniques and methods, used to investigate these instabilities, are developed.

• *Large-scale ocean in the climate system*: The phenomenon of long waves in tropical ocean is modeled by equations of non-linear Navier–Stokes type for the velocity and pressure, and of transport-diffusion type for the temperature and the salinity. The oceanic currents, which play a key role in the regulation of the climate, are characterized, in the tropical zone, by steady zonal currents and by long waves propagating westward along the equator and superimposed to the mean currents. The equatorial waves can be connected with strong vertical currents, which are very sensitive to small changes in temperature (for example the *El Niño* phenomenon begins with a temperature elevation of 2 or 3 °C of surface waters). The goal is to predict the deviation of circulation from the mean circulation caused by these small variations of the surface temperature. Our work therefore complements and generalizes research works developed for several years by Belmiloudi (by using the optimal control techniques) on the analysis of fluctuations and perturbations in the equatorial zone. We study this problem with two types of hypothesis: the Boussinesq approximation and the Hydrostatic approximation with vertical viscosity.

1.3.3 Biological Models

The goal is to present some mathematical treatments for the non-linear evolution systems which arise in life sciences. The questions that we address are the same as developed in the previous applications. We provide a brief review of various models in mathematical biology, and describes how these models arise. Then we study the properties of solutions of non-linear time-dependent partial differential systems (with and without time delays). In particular, the existence and the uniqueness of these solutions are discussed. Various mechanisms of control of the perturbations of the system by using different techniques under different boundary conditions, are also presented.

• *Impact of heat transfer laws on temperature distribution in biological tissues*: The temperature distribution in living tissues is modeled by some generalized transient bioheat transfer type models with directional blood flow and Robin boundary conditions. The model equation depends on the blood perfusion rate, the heat transfer parameter, the distributed energy source terms and the heat flux due to the evaporation, which affect the effects of thermal and physical properties on the transient temperature of biological tissues. The knowledge about the behavior of the temperature in tissues can be very beneficial for thermal diagnostics and treatments in medical practices, for example thermotherapy for regional hyperthermia, often used in treatment of cancer (the aim of the thermal therapy is to destroy the pathological tissues by rising the temperature with minimal damage to the surrounding tissues). The goal of our study is to control and stabilize the desired online temperature.

• *Parabolic Lotka–Volterra type systems with logistic time-varying delays*: The studied systems are governed by parabolic equations governing diffusive biological species with logistic growth terms and multiple time-varying delays. A very important ecological and economical problem is resource management, *i.e.*, the stabilization of uncertain biological species taking into account the influence of the population at earlier times on the regulatory effect. In population dynamics, this includes the multiple time-varying delay model, for example the birth rate, which does not act instantaneously (time to reach maturity), the finished period of gestation, *etc.* The applications are varied and include: forest or agriculture (trapping animals, damage cost to environment), fisheries (resource stock to prevent overfishing), *etc.*

1.3.4 Other Systems

We also develop two very interesting systems: first, we analyze the motion of animal blood which is described by micropolar fluid models and, second, we present the semiconductor melts in zone-melting and Czochralski growth configuration.

Most of the topics developed here are new or have recently appeared. Moreover, the methods developed in this book can be extended to the well-posedness non-linear hyperbolic systems. Relevant questions which are not developed here include those of hybrid systems involving a mixture of continuous and discrete dynamics. These aspects are currently being investigated and will be the object of publications in the near future.

Part I

Convex Analysis and Duality Principles

2

Convexity and Topology

This chapter is devoted to the presentation of several fundamental and practical aspects of the theory of real-valued convex functions. Convex functions play an important role in many fields of mathematics such as optimization, control theory, operational (or operations) research, geometry, differential equations, functional analysis *etc.*, as well as in applied sciences, *e.g.*, in physics, biology, medicine, *etc.*

Within this chapter we introduce some essential results for convex functions which will be used frequently later in the book. The detailed treatment of this topic can be found in the references listed at the end of the book. Hence, we provide proofs only for those results that either cannot be easily found in the standard textbooks or are insightful to the understanding of some related problems.

2.1 Convex Sets

2.1.1 Definitions

Definition 2.1. *Let V be a vector space over $\mathbb{R}$. If u and v are two vectors of V, u and v are called the endpoints of the line-segment denoted by $[u,v]$, where*

$$[u,v] = \{\alpha u + (1-\alpha)v : \; 0 \leq \alpha \leq 1\}.$$

■

Definition 2.2. *(Convex set) The subset C of V is said to be convex if and only if the segment $[u,v]$ is entirely included in C, whenever its endpoints u and v are in C. As convention, the empty set is considered to be convex.* ■

In the following proposition we give some properties of convex sets.

Proposition 2.3. *(i) The set $C \subset V$ is convex if and only if for every finite subset of elements $(u_i, \alpha_i)_{i=1,n}$ of $C \times I\!R^+$ such that $\sum_{i=1,n} \alpha_i = 1$, we have $\sum_{i=1,n} \alpha_i u_i \in C$.*

(ii) The intersection of any arbitrary family of convex sets is convex, but the union of convex sets is usually not convex.

(iii) If $C \subset V$ is convex then the interior of C denoted by $\text{int}(C)$ is convex. Moreover, if $\text{int}(C)$ is not empty then the closure of C is convex and is equal to the closure of $\text{int}(C)$.

Proof. For the proof see, for example, Schwartz [264]. □

Definition 2.4. *(Convex hull) Let S be any non-empty subset of V. The smallest convex set containing S is said to be the convex hull of the subset S and is denoted by $\text{co}S$ (sometimes $\text{conv}S$).* ■

The following proposition states that the convex hull of any set exists.

Proposition 2.5. *(i) Let S be any non-empty subset of V, then the convex hull of S is the intersection of all the convex subsets of V containing S.*

(ii) The subset $\text{co}S$ can also be described as the set of all convex combinations of the elements of S:

$$\text{co}S = \{ \sum_{i=1,n} \alpha_i u_i : \ n \in I\!N, \sum_{i=1,n} \alpha_i = 1, \alpha_i \geq 0, u_i \in S, 1 \leq i \leq n \}.$$

□

2.1.2 Topological Spaces and Properties

We recall that topological vector space (t.v.s) is defined as a vector space V equipped with a topology for which the operations of addition and scalar multiplication:

$$(u, v) \in V \times V \longrightarrow u + v \in V \text{ and } (\alpha, u) \in I\!R \times V \longrightarrow \alpha u \in V$$

are continuous. In this case, we say that the topology is linear.

The neighborhoods of any point $u \in V$ may then be deduced from those of the origin, because by translation we can obtain a base of neighborhoods of u: each neighborhood of u is of the form $u + \mathcal{X}_0$, where $\mathcal{X}_0$ is a neighborhood of origin. Consequently, we have the following lemma.

Lemma 2.6. *A linear mapping between two topological vector spaces is continuous if and only if it is continuous at the origin.* □

Let us now give the definition of compact subset.

Definition 2.7. *(Open cover) Let V be a topological vector space. An open cover of a subset D of V is an indexed family of open subsets $\{\mathcal{U}_i, i \in J\}$ of V such that*

$$D \subset \bigcup_{i \in J} \mathcal{U}_i.$$

■

Definition 2.8. *(Compacity) Let V be a topological vector space. A subset K of V is said to be a compact set if every open cover of K has a finite subcover, i.e., from any open cover of K, it is possible to extract a finite subcover.* ■

A particular type of topological vector spaces, which is very important and has extensive properties, is the locally convex space.

Definition 2.9. *(Locally convex space) A topological vector space is said to be a locally convex space if the origin possesses a fundamental system of convex neighborhoods. Otherwise, if for each vector there exists a base of neighborhoods consisting of convex sets.* ■

Remark 2.10. A linear mapping on a topological vector space is continuous if and only if there exists a neighborhood of origin on which the mapping is bounded. ◇

Definition 2.11. *(Separated points) Let V be a topological vector space. Let u and v be two points in V. We say that u and v can be separated by neighborhoods if there exist two open neighborhoods $\mathcal{U}_u$ (of u) and $\mathcal{U}_v$ (of v) such that $\mathcal{U}_u$ and $\mathcal{U}_v$ are disjoint, i.e., $\mathcal{U}_u \cap \mathcal{U}_v = \emptyset$.* ■

Definition 2.12. *(Separated (Hausdorff) space) Let V be a topological vector space. V is said to be a Hausdorff space (or separated space) if any two distinct points of V can be separated by neighborhoods.* ■

The following results give examples of the role that can play the Hausdorff assumption. More precisely:

Lemma 2.13. *In Hausdorff topological spaces, each compact subset is closed.*

Proof. Let V be a Hausdorff topological space, K be a compact subset of V and u_0 be any element in the complement of K. Then there exists an open neighborhood of u_0 which doesn't intersect K.

Since V is Hausdorff then for each $u \in K$ there exist two open neighborhoods $\mathcal{U}_u$ (of u) and $\mathcal{B}_u$ (of u_0) such that $\mathcal{U}_u \cap \mathcal{B}_u = \emptyset$. As K is compact and $K \subset \bigcup_{u \in K} \mathcal{U}_u$, there exists a finite subset $\{u_i : i \in J\}$ of K such that

$$K \subset \bigcup_{i \in J} \mathcal{U}_{u_i}.$$

Since $\bigcup_{i \in J} \mathcal{U}_{u_i}$ and $\bigcap_{i \in J} \mathcal{B}_{u_i}$ are disjoint sets, then

$$K \cap \left(\bigcap_{i \in J} \mathcal{B}_{u_i}\right) = \emptyset.$$

As a finite intersection of open neighborhoods is an open neighborhood, then $\bigcap_{i \in J} \mathcal{B}_{u_i}$ is an open neighborhood of u_0. This achieves the proof of the lemma. □

The immediate consequence of this lemma is given in Proposition 2.14.

Proposition 2.14. *Let V be a Hausdorff topological space. For any non-increasing sequence of compact subsets $(K_n)_{n\geq 1}$ such that $\bigcap_{n\geq 1} K_n = \emptyset$ (is empty), there exists $m \geq 1$ such that $K_n = \emptyset$ for all $n \geq m$. Otherwise, for any non-increasing sequence of non-empty compact subsets $(K_n)_{n\geq 1}$, we have that $\bigcap_{n\geq 1} K_n$ is non-empty.*

Proof. The proof is left to the reader as an exercise. □

Remark 2.15. Proposition 2.14 is used to derive the well-known Weierstrass theorem (see below), which corresponds to the existence and attainment of an optimum of a function in compact subsets. ♢

Let us give now some interesting families of topological spaces that are Hausdorff and locally convex spaces, namely the normed vector spaces.

Definition 2.16. *(Normed vector space) A topological vector space V is said to be a normed vector space if V is endowed with a norm $\| . \|$.* ■

Remark 2.17. (*i*) Normed spaces are locally convex spaces: it is sufficient to take the set of neighborhoods formed by the balls centred at the origin to obtain the result.

(*ii*) The normed spaces are also separate Hausdorff spaces.

(*iii*) The topology of a normed space V can be generated by the distance function $d(u, v) = \| u - v \|, \forall (u, v) \in V$.

(*iv*) One of the advantages of the metrizability of a topological vector space is that many of topological properties can be simply characterized by the analysis of sequences. ♢

Definition 2.18. *(Banach space) A topological vector space V is said to be a Banach vector space if V is endowed with a norm $\| . \|$ and is complete or Cauchy,[1] with respect to the metric $d(u, v) = \| u - v \|$, $\forall (u, v) \in V$.* ■

Example 2.19. Let Ω be a domain in $\mathbb{R}^d$, $d \in \mathbb{N}^*$. Then:

(*i*) The space $\mathcal{C}(\Omega)$ of real-valued and continuous functions on the domain Ω is a Banach space with the norm

$$\| u \|_{\mathcal{C}(\Omega)} := \sup_{x\in\Omega} | u(x) | \quad \forall u \in \mathcal{C}(\Omega). \tag{2.1}$$

(*ii*) The spaces $L^p(\Omega)$, $1 \leq p < \infty$, of real-valued and p-integrable, in the Lebesgue sense, functions on the domain Ω are Banach spaces with the norm

$$\| u \|_{L^p(\Omega)} := \left(\int_\Omega | u(x) |^p \, dx \right)^{1/p} \quad \forall u \in L^p(\Omega). \tag{2.2}$$

[1] If every Cauchy sequence of points in V has a limit that is also in V.

(*iii*) The space $L^\infty(\Omega)$ of real-valued and essentially bounded functions on the domain Ω is a Banach space with the norm

$$\| u \|_{L^\infty(\Omega)} := \sup_{x\in\Omega} \text{ess} \mid u(x) \mid \quad \forall u \in L^\infty(\Omega). \tag{2.3}$$

♣

We finish with Hilbert vector spaces.

Definition 2.20. *(Hilbert space) A topological vector space V is said to be a Hilbert vector space if V is a Banach space under the norm defined by the inner (or scalar) product: $\langle .,. \rangle : V \times V \longrightarrow I\!R$, i.e., with respect to the norm $\| u \| = (\langle u, u \rangle)^{1/2}$, $\forall u \in V$.* ■

Example 2.21. Let Ω be a domain in $\mathbb{R}^d$, $d \in \mathbb{N}^*$. Then the space $L^2(\Omega)$, of real-valued and 2-integrable (square integrable), in the Lebesgue sense, functions on the domain Ω, is a Hilbert space under the norm

$$\| u \|_{L^2(\Omega)} := \left(\int_\Omega \mid u(x) \mid^2 \mathrm{d}x\right)^{1/2}, \tag{2.4}$$

defined by the inner product $\langle .,. \rangle_{L^2(\Omega)} : L^2(\Omega) \times L^2(\Omega) \longrightarrow \mathbb{R}$ such that

$$\langle u, v \rangle_{L^2(\Omega)} := \int_\Omega u(x)v(x)\mathrm{d}x \quad \forall (u, v) \in (L^2(\Omega))^2. \tag{2.5}$$

♣

2.1.3 Hahn–Banach and Separation Between Convex Sets

Before presenting the results of Hahn–Banach, we give some definitions.

Definition 2.22. *(Half-spaces bounded by hyperplane) Let H be an affine hyperplane with equation $L(u) = \lambda$, where L is a non-zero linear form on V and $\lambda \in I\!R$. The following sets*

$$\{u \in V : \; L(u) < \lambda\}, \qquad \{u \in V : \; L(u) > \lambda\}$$

are called open half-spaces bounded by H and

$$\{u \in V : \; L(u) \leq \lambda\}, \qquad \{u \in V : \; L(u) \geq \lambda\}$$

are called closed half-spaces bounded by H. All these sets are convex. ■

Remark 2.23. (*i*) The hyperplane H is topologically closed if and only if the linear form L is continuous.

(*ii*) The half-spaces bounded by H do not depend on the choice of the equation of the hyperplane H.

(*iii*) If the function L is continuous then the open (respectively closed) half-space bounded by H will be topologically open (respectively topologically closed). ◇

Definition 2.24. *(Internal point) Let $C \subset V$ be a convex set and u be in C. The point u is called internal if every line passing through u meets C in a segment $[u_1, u_2]$ such that $u \in]u_1, u_2[$.* ■

Remark 2.25. Every interior point is internal and if the interior of C is nonempty then every internal point is interior. ♢

Definition 2.26. *(Closed convex hull) For S be any subspace of V, the intersection of all the closed convex subsets containing S is the smallest closed convex subset containing S. It is also the closure of the convex hull of S and it is called the closed convex hull of S denoted by* $\text{clco}S$. ■

Remark 2.27. In general the closed convex hull of a subspace S of V is not the convex hull of the closure of S, *i.e.*, $\text{clco}S \neq \text{co}\overline{S}$ (where $\overline{S}$ is the closure of S). ♢

The proofs of all the following results of this section can be found, for example, in Trèves [283].

Next we recall the Hahn–Banach theorem in its analytical form (based on the well-known Zorn's lemma argument).

Theorem 2.28. *(Hahn–Banach, analytical form) Let $p: V \longrightarrow I\!R$ be a sublinear mapping such that*

$$p(\alpha u) = \alpha p(u), \forall u \in V \text{ and } \forall \alpha > 0,$$

$$p(u+v) \leq p(u) + p(v), \forall (u,v) \in V \times V.$$

Let also S be any vector subspace of V and let $g: S \longrightarrow I\!R$ be a linear mapping such that

$$g(u) \leq p(u), \forall u \in S.$$

Then there exists a linear form f on V which extends g such that

$$g(u) = f(u), \forall u \in S \text{ and } f(u) \leq p(u), \forall u \in V.$$

□

Before giving the geometric form of the Hahn–Banach theorem, we introduce the following separate definition.

Definition 2.29. *(Separate sets by hyperplane) Let H be an affine hyperplane with equation $L(u) = \lambda$ and let S_1, S_2 be two subsets of V.*
(i) The hyperplane H is said to separate (strictly separate) S_1 and S_2 if S_1 lies in one of the closed (open) half-spaces bounded by H and S_2 in the other, i.e., (for example)

$$L(u) \leq \lambda, \forall u \in S_1 \quad \text{and} \quad L(u) \geq \lambda, \forall u \in S_2, \quad \text{(for separation)},$$

$$L(u) < \lambda, \forall u \in S_1 \quad \text{and} \quad L(u) > \lambda, \forall u \in S_2, \quad \text{(for strict separation)}.$$

(ii) The hyperplane H is said to separate properly S_1 and S_2 if S_1 lies in one of the closed half-spaces bounded by H and S_2 in the other and both are not contained in H. ■

Definition 2.30. *(Supporting hyperplane) Let S be a subset of V and H be a closed hyperplane which contains at least one point $u \in S$.*
The hyperplane H is said to be a supporting hyperplane of the subset S when S is completely contained in one of the two closed half-spaces bounded by H, and u is said to be a supporting point of S. ■

We now recall the Hahn–Banach theorem in its geometrical form and its consequences for the separation of convex sets.

Theorem 2.31. *(Hahn–Banach, first geometric form) Let V be a topological vector space, $C_1 \subset V$ be an open non-empty convex subset and $C_2 \subset V$ be a non-empty convex subset which does not intersect C_1 (i.e., $C_1 \cap C_2 = \emptyset$). Then there exists a closed hyperplane H which separates C_1 and C_2.* □

Theorem 2.32. *(Hahn–Banach, second geometric form) Let V be a locally convex topological vector space, $C_1 \subset V$ be a closed non-empty convex subset and $C_2 \subset V$ be a compact non-empty convex subset which does not intersect C_1. Then there exists a closed hyperplane H which strictly separates C_1 and C_2.* □

We can now give some useful corollaries (the first is an application of Theorem 2.31 and the second is an application of Theorem 2.32).

Corollary 2.33. *Let V be a topological vector space, C be a convex subset of V with non-empty interior and H be a supporting hyperplane of C. Then every boundary point of C is a supporting point of C.* □

Corollary 2.34. *Let V be a locally convex topological vector space, C be a closed convex subset of V and H be a supporting hyperplane of C. Then C is the intersection of the closed half-spaces which contain it.* □

Let us give finally a very useful corollary that one seeks to prove the density of a topological subspace (an application of Theorem 2.32).

Corollary 2.35. *Let V be a Hausdorff locally convex space and S be a topological subspace of V such that $\overline{S} \neq V$ ($\overline{S}$ denotes the closure of S). Then there exists a non-zero continuous linear form f over V such that*

$$f(u) = 0, \quad \forall u \in S.$$

□

2.2 Convex Functions

2.2.1 Definitions

In the sequel we will work with the extended real-valued functions, that is $\overline{\mathbb{R}}$-valued functions[2] (*i.e.*, $-\infty$ and $+\infty$ are allowed to the functions under consideration), except mentioned contrarily.

---

[2] $\overline{\mathbb{R}} = \mathbb{R} \cup \{-\infty, +\infty\}$.

Definition 2.36. *(Convex and concave functions) Let C be a non-empty convex subset of V and F be a mapping of C into $\overline{\mathbb{R}}$.*
(i) The mapping F is said to be convex on C if, for all pair (u, v) in $C \times C$ and all α in $]0, 1[$, the inequality (if the right-hand side is well defined)

$$F(\alpha u + (1-\alpha)v) \leq \alpha F(u) + (1-\alpha)F(v) \quad (Jensen's\ inequality) \tag{2.6}$$

holds. The function F is called strictly convex on C when (2.6) holds as a strict inequality if $u \neq v$ with $F(u) < \infty$ and $F(v) < \infty$.
(ii) The function F is said to be (strictly) concave on C if the function $-F$ is (strictly) convex on C. ■

Remark 2.37. (*i*) If the mapping $F : V \longrightarrow \overline{\mathbb{R}}$ is convex, we obtain easily that, for every finite subset of elements $(u_i, \alpha_i)_{i=1,n}$ of $V \times \mathbb{R}^+$ such that $\sum_{i=1,n} \alpha_i = 1$ (if the right-hand side is well defined)

$$F(\sum_{i=1,n} \alpha_i u_i) \leq \sum_{i=1,n} \alpha_i F(u_i). \tag{2.7}$$

The inequality (2.7) is called the general inequality of *Jensen.*
(*ii*) If the mapping $F : V \longrightarrow \overline{\mathbb{R}}$ is convex, the following level sets (for each $\lambda \in \overline{\mathbb{R}}$)

$$\{u \in V : F(u) \leq \lambda\} \text{ and } \{u \in V : F(u) < \lambda\}$$

are convex (but the converse is false and such functions are called *quasi-convex*). ◇

Remark 2.38. (*i*) Let S be a non-empty subset of V and h be a mapping of S into $\mathbb{R}$. We can associate with it the extended mapping F on V by

$$F(u) = h(u) \text{ if } u \in S \text{ and } F(u) = +\infty \text{ if } u \notin S. \tag{2.8}$$

The function F is convex if and only if the subset S is convex and the function h is convex. Therefore, we need only to consider functions defined on V everywhere.
(*ii*) The study of a convex set is naturally reduced to the study of a convex function. Because for a subset S of V it is clear enough that the function $\mathcal{X}_S : S \longrightarrow \overline{\mathbb{R}}$ defined by

$$\mathcal{X}_S(u) = 0 \text{ if } u \in S \text{ and } \mathcal{X}_S(u) = +\infty \text{ if } u \notin S$$

(called the *indicator function* of S) is a convex function if and only if S is a convex set.
(*iii*) The notion of strict convexity is very important to derive the uniqueness of the minimum of a function (see below). ◇

Example 2.39. Let $(V, \| . \|)$ be a real normed vector space then the functions

$$F: V \longrightarrow \mathbb{R} \text{ defined by } F(u) := \| u \|^p, \quad 1 \leq p < \infty, \forall u \in V$$

are convex functions. ♣

Let us now introduce the useful notions of effective domain, epigraph and hypograph of a function.

Definition 2.40. *(Effective domain of a function) Let F be a mapping of V into $\overline{\mathbb{R}}$ (extended real-valued functions). The domain (or also the effective domain) of F,* dom$F$ *(sometimes $D(F)$), is the non-empty set*

$$\text{dom}F = \{u \in V : \ F(u) < +\infty\}.$$

It is clear that if the mapping F is convex then the domain dom$F$ *is a convex set. Moreover, if F is given by (2.8),* dom$F$ *is a subset of S.* ■

Definition 2.41. *(Epigraph and hypograph of a function) Given a mapping $F: V \longrightarrow \overline{\mathbb{R}}$ (extended real-valued function):*

(i) The epigraph of F is the non-empty set

$$\text{epi}F = \{(u, \lambda) \in V \times \mathbb{R} : \ F(u) \leq \lambda\}.$$

Its strict epigraph $\text{epi}_s F$ *is defined by*

$$\text{epi}_s F = \{(u, \lambda) \in V \times \mathbb{R} : \ F(u) < \lambda\}.$$

(ii) The hypograph of F is the non-empty set

$$\text{hypo}F = \{(u, \lambda) \in V \times \mathbb{R} : \ F(u) \geq \lambda\}.$$

Its strict hypograph $\text{hypo}_s F$ *is defined by*

$$\text{hypo}_s F = \{(u, \lambda) \in V \times \mathbb{R} : \ F(u) > \lambda\}.$$

■

The following proposition shows that the epigraph of the pointwise supremum of a family of functions is the intersection of the epigraphs of these functions.

Proposition 2.42. *Let $(F_i)_{i \in J}$ be an arbitrary family of functions of V into $\overline{\mathbb{R}}$ and F their pointwise supremum $F := \sup_{i \in J} F_i$. Then*

$$\bigcap_{i \in J} \text{epi}F_i = \text{epi}F.$$

□

Remark 2.43. (*i*) The epigraph or hypograph of a function is a subset of $V \times \mathbb{R}$ but not of $V \times \overline{\mathbb{R}}$.

(*ii*) If $(u, \lambda) \in V \times \mathbb{R}$ is in the epigraph of a function then $u \in V$ is in the effective domain of the function under consideration. More precisely the projection of the epigraph of a function on V is its effective domain. ◇

Let us now give some characterizations and properties of convex functions.

Proposition 2.44. *Given a mapping $F: V \longrightarrow \overline{I\!R}$. The following properties are equivalent:*

(i) F is a convex function on V
(ii) its epigraph is a convex subset of $V \times I\!R$
(iii) its strict epigraph is a convex subset of $V \times I\!R$. □

Remark 2.45. (*i*) A similar characterization of concave function can be given in terms of its hypograph.

(*ii*) The function which is identically the constant function $+\infty$ is convex since $\text{epi}(+\infty) = \emptyset$ is a convex set, and the constant function $-\infty$ is also convex since its epigraph is $V \times \mathbb{R}$. These functions are also concave. ◇

Proposition 2.46. (*i*) *If F is a convex function from V into $\overline{I\!R}$ then λF is a convex function, $\forall \lambda \in I\!R^+$.*

(*ii*) *If F and G are convex functions from V into $\overline{I\!R}$ then $F + G$ is a convex function.*

(*iii*) *Let $(F_i)_{i \in J}$ be an arbitrary family of convex functions of V into $\overline{I\!R}$, then their pointwise supremum $F := \sup_{i \in J} F_i$ is a convex function.* □

We can now introduce the following definition.

Definition 2.47. *(Convex envelope) Given a mapping $F: V \longrightarrow \overline{I\!R}$. The convex envelope of F is the greatest convex function which is less than F, and is denoted by $\text{cv}F$. Otherwise*

$$\text{cv}F = \sup\{\textit{convex functions},\ G: V \longrightarrow \overline{I\!R}:\ G \leq F\}.^3$$

■

Let us now introduce an essential notion of regularity for an optimization problem, *i.e.*, the semi-continuity.

2.2.2 Closure and Semi-continuous Functions

Definition 2.48. *(Semi-continuous function) Let us consider the mapping $F: V \longrightarrow \overline{I\!R}$:*

(i) The function F is said to be lower semi-continuous (l.s.c), if for each $u \in V$

$$\liminf_{v \longrightarrow u} F(v) \geq F(u), \tag{2.9}$$

where $\liminf_{v \longrightarrow u} F(v) = \sup_{\mathcal{O} \in \mathcal{N}(u)} \inf_{v \in \mathcal{O}} F(v)$ and $\mathcal{N}(u)$ is the family of all open neighborhoods of u.

[3] That is $G(u) \leq F(u), \forall u \in V$.

(ii) The function F is said to be upper semi-continuous (u.s.c), if $-F$ is a lower semi-continuous function. In other words, if

$$\limsup_{v \longrightarrow u} F(v) = \inf_{\mathcal{O} \in \mathcal{N}(u)} \sup_{v \in \mathcal{O}} F(v) \leq F(u).$$

■

Remark 2.49. (*i*) Since the converse inequality of (2.9), *i.e.*, $\liminf_{v \longrightarrow u} F(v) \leq F(u)$ always holds, then the inequality (2.9) is equivalent to the equality

$$F(u) = \liminf_{v \longrightarrow u} F(v).$$

(*ii*) A function F is continuous at point u if and only if F is both lower and upper semi-continuous at point u. ◇

Let us now give some geometrical characterizations and properties of semi-continuous functions.

Proposition 2.50. *For a function $F : V \longrightarrow \overline{I\!R}$, the following three properties are equivalent:*

(i) F is lower semi-continuous on V
(ii) the epigraph of the function F, $\mathrm{epi}F$ is a closed subset of $V \times I\!R$
(iii) the level sets $\{u \in V : F(u) \leq \lambda\} \subset V$ are closed, for each $\lambda \in I\!R$. □

In particular we have

Corollary 2.51. *A subset C of V is closed if and only if its indicator function $\mathcal{X}_C$ is lower semi-continuous.* □

Proposition 2.52. *Let $(F_i)_{i \in J}$ be an arbitrary family of lower semi-continuous functions of V into $\overline{I\!R}$, then their pointwise supremum $F := \sup_{i \in J} F_i$ is a lower semi-continuous function.* □

We can now introduce the following definitions.

Definition 2.53. *(Lower semi-continuous envelope) Given a mapping $F : V \longrightarrow \overline{I\!R}$. The lower semi-continuous envelope of F is the greatest lower semi-continuous function which is less than F, and is denoted by $\mathrm{ls}F$. Otherwise*

$$\mathrm{ls}F = \sup\{\textit{lower semi-continuous functions},\ G : V \longrightarrow \overline{I\!R} : G \leq F\}.$$

■

Definition 2.54. *(Closure of function) Given a mapping $F : V \longrightarrow \overline{I\!R}$. The closure (or the lower semi-continuous regularization) of F is the function $\overline{F} : V \longrightarrow \overline{I\!R}$ (sometimes $\mathrm{cl}F$) defined by*

$$\overline{F}(u) = \liminf_{v \longrightarrow u} F(v), \qquad \textit{for every } u \in V,$$

or equivalently

$$\mathrm{epi}\overline{F} = \overline{\mathrm{epi}F} \quad (\textit{the closure of } \mathrm{epi}F).$$

■

An important property of lower semi-continuous functions is given by the following well-known theorem of Weierstrass, which corresponds to the attainment of the minimum on a compact set (based on Propositions 2.14 and 2.50).

Theorem 2.55. *(Weierstrass) Let V be a Hausdorff topological space, K be a compact subset of V and $F : V \longrightarrow \overline{I\!R}$ be a given lower semi-continuous function. Then the function F takes a minimum value on K, i.e., there exists at least one point $u \in K$ such that*

$$F(u) = \inf_{v \in K} F(v).$$

Moreover, if F takes only finite values then F is bounded from below. □

As a consequence of the last result, we have (according to Lemma 2.13 and Proposition 2.50) the following corollary.

Corollary 2.56. *Let V be a Hausdorff topological space and $F : V \longrightarrow \overline{I\!R}$ be a given function such that all its level sets are compact (F is said to be inf-compact). Then the function F is lower semi-continuous and its infimum is attained.* □

Let us now introduce the notion of weak topologies.

2.2.3 Weak Topologies and Dual Spaces

Let V be a topological vector space.

Definition 2.57. *(Dual topological space) The vector space V' (or V^*) of continuous linear functionals over V is said to be the (topological) dual of V. For f in V' and u in V we will denote, in general, the value at u of the linear functional f by $\langle f, u\rangle_{V',V}$ ($\langle .,.\rangle_{V',V}$ is said to be the scalar product between V' and V).* ■

We can thus introduce over V' (respectively over V) the topology of weak convergence over V (respectively over V'). This will be termed weak topology of V' (respectively of V) associated with the duality between V and V' and will be denoted by $\sigma(V', V)$ (respectively by $\sigma(V, V')$).

Definition 2.58. *(Weak topology) The topology $\sigma(V, V')$ (respectively $\sigma(V', V)$) is the weakest topology on V (respectively on V') which makes the linear forms $\phi_f : u \in V \longrightarrow \langle f, u\rangle_{V',V} \in I\!R$, continuous for all $f \in V'$ (respectively the linear forms $\psi_u : f \in V' \longrightarrow \langle f, u\rangle_{V',V} \in I\!R$, continuous for all $u \in V$). It is called the topology of V (respectively of V') weakned by V' (respectively by V).* ■

Remark 2.59. (*i*) The functions $p_f : u \in V \longrightarrow \mid \langle f, u\rangle_{V',V} \mid \in \mathrm{I\!R}^+$ are continuous seminorms on V, for all $f \in V'$. Moreover, if we assume that the family

$\mathcal{P}$ of all seminorms defined by: $\mathcal{P} = \{p_f : f \in V'\}$ is separating, then the topology $\sigma(V, V')$, generated by the family $\mathcal{P}$, makes V into a locally convex space: a fundamental system of neighborhoods of any u_0 in V is given by $(\{u \in V : | \langle f_i, u - u_0 \rangle_{V',V} | < \epsilon, \ \forall i \in J\}$, where J is a finite set, $f_i \in V'$ and $\epsilon > 0)$.

(*ii*) If V is a locally convex space, then $\mathcal{P}$ is separating by the Hahn–Banach theorem (second geometric form) and consequently, $(V, \sigma(V, V'))$ (V with the topology $\sigma(V, V')$) is a locally convex space.

(*iii*) A sequence $(u_n)_{n \in \mathbb{N}}$ converges to $u \in V$ in the topology $\sigma(V, V')$ if and only if for all $f \in V'$, $(\langle f, u_n - u \rangle_{V',V})_{n \in \mathbb{N}}$ converges to 0. ♢

In general the dual of V', denoted by V'' and called the *bidual* of V, and V are different. In the context of normed space, it is observed that, this last property characterizes a special class of normed spaces called reflexive. Moreover, in general the weak topologies $\sigma(V, V')$ and $\sigma(V', V'')$ are different. For more clearness, we preserve the name of *weak topology* only for $\sigma(V, V')$ and the weak topology for $\sigma(V', V)$ will be called *weak star topology* (or weak* topology).

In contrast to these topologies, the initial topologies on V (respectively on V'), will be called *strong topology* on V (respectively on V').

We denote by $\longrightarrow$, $\rightharpoonup$ and $\rightharpoonup^*$ the *strong convergence* in V, the *weak convergence* in V and the *weak star convergence* in V'. Similarly we will speak about strong neighborhood, strongly closed, strongly bounded, *etc.*, and weak (or weak*) neighborhood, weakly (or weakly*) closed, weakly (or weakly*) bounded, *etc.*

Weak Topology

Proposition 2.60. *If V is a (Hausdorff) locally convex space, then:*

(*i*) *The topology $\sigma(V, V')$ (respectively $\sigma(V', V)$) is a (Hausdorff) locally convex topology on V (respectively on V').*

(*ii*) *The weak topology $\sigma(V, V')$ (respectively the weak star topology $\sigma(V', V)$) is always weaker than the strong topology on V (respectively on V'). The equality holds if and only if V is a finite dimensional space.* □

Remark 2.61. (*i*) Every strongly convergent sequence is weakly convergent (the converse is, in general, false).

(*ii*) The closed (respectively open) subsets of the weak topology $\sigma(V, V')$ are closed (respectively open) subsets of the strong topology (the converse is, in general, false).

(*iii*) If V is a finite dimensional space, then every weakly convergent sequence is strongly convergent (and vice versa). ♢

Example 2.62. Suppose that $(V, \| \cdot \|)$ is a Banach space, equipped with the norm $\| \cdot \|$, then we can prove easily that:

(*i*) The subset $B = \{u \in V : \| u \|= 1\}$ is never a closed subset for the weak topology $\sigma(V, V')$ and the weak closure of B is the set

$$\overline{D} = \{u \in V : \| u \| \leq 1\}.$$

(*ii*) The subset $D = \{u \in V : \| u \| < 1\}$ is never an open subset for the weak topology $\sigma(V, V')$. ♣

It is known that the closed subsets of the weak topology are closed subsets of the strong topology and the converse is, in general, false. The next result shows that these two concepts coincide, in the case of convex subsets (by applying Hahn–Banach theorem).

Theorem 2.63. *If V is a locally convex space and $C \subset V$ is a convex subset, then C is weakly closed if and only if it is strongly closed.*

Proof. Let C be strongly closed and u_0 be any point in the complement of C. We have to prove that there exists an open neighborhood of u_0, in the weak topology $\sigma(V, V')$, which doesn't intersect C.
According to the Hahn–Banach theorem (second geometric form), there exists a closed hyperplane which strictly separates $\{u_0\}$ and C. So, there exists $(f, \lambda) \in V' \times \mathbb{R}$ such that

$$\langle f, u_0 \rangle_{V',V} < \lambda < \langle f, v \rangle_{V',V} \quad \text{for all } v \in C.$$

Consider the following open set (in the weak topology $\sigma(V, V')$):

$$\mathcal{B} = \{v \in V : \ \ \langle f, v \rangle_{V',V} < \lambda\},$$

we have that $u_0 \in \mathcal{B}$, $\mathcal{B} \cap C = \emptyset$ and then the result of Theorem 2.63 follows. □

In particular, we have that:

Corollary 2.64. *(Mazur theorem) If V is a metrizable locally convex space, and the sequence $(u_n)_{n \in \mathbb{N}}$ converges weakly to u, then there exists a sequence of convex combinations $(v_k)_{k \in \mathbb{N}}$:*

$$v_k = \sum_{n \in J} \lambda_{k,n} u_n, \ \lambda_{k,n} \geq 0, \ \sum_{n \in J} \lambda_{k,n} = 1, \ \textit{with } J \textit{ a finite set}$$

which converges strongly to u.

Proof. Let H be the convex hull of $(u_n)_{n \in \mathbb{N}}$ and C its weak closure, the weak limit of the sequence $(u_n)_{n \in \mathbb{N}}$ is in C. Because of Theorem 2.63, we can deduce that u is in the strong closure of H. Since V is metrizable then there is some sequence $(v_k)_{k \in \mathbb{N}}$ converging strongly to u. □

Moreover, we have an interesting result pertaining to lower semi-continuous convex functions when the topology of V is weakened.

Proposition 2.65. *Let V be a locally convex space. Then any convex function $F: V \longrightarrow \overline{I\!R}$ is weakly lower semi-continuous (in the weak topology $\sigma(V,V')$) if and only if it is (strongly) lower semi-continuous. In particular, if V is metrizable and the sequence $(u_n)_{n\in I\!N}$ converges weakly to u then*

$$F(u) \leq \liminf_n F(u_n).$$

Proof. ($\Rightarrow$) A simple consequence of the fact that $\sigma(V,V')$ is a subset of the original (strong) vector topology of V.

($\Leftarrow$) Suppose that F is lower semi-continuous and prove that F is weakly lower semi-continuous. For this it is sufficient to prove that the following subsets:

$$H_\lambda = \{v \in V: \; F(u) \leq \lambda\},$$

are closed in the weak topology $\sigma(V,V')$. Since F is lower semi-continuous and convex then H_λ is convex and strongly closed. According to Theorem 2.63, we have that H_λ is weakly closed. □

As an immediate corollary of the last proposition we have Corollary 2.66.

Corollary 2.66. *Let $(V, \|\;.\;\|)$ be a normed space, so that V' is a Banach space with the dual norm $\|\;.\;\|_{V'}$ defined by*

$$\| f \|_{V'} = \sup_{\|u\|=1} \mid \langle f, u\rangle_{V',V} \mid .$$

If a sequence $(u_n)_{n\in I\!N}$ converges weakly to u, then $(u_n)_{n\in I\!N}$ is uniformly bounded in V. Moreover,

$$\| u \| \leq \liminf_n \| u_n \| .$$

Proof. The first result is a consequence of the classical Banach Steinhaus's theorem (or principle of uniform boundedness theorem). The second result is a simple consequence of the fact that the norm is a convex and continuous function. □

Weak* Topology

On the dual space V', the family of seminorms $\mathcal{P}^* = \{q_u = \mid \psi_u \mid: \; u \in V\}$ is separating, consequently the topology $\sigma(V',V)$, which is generated by $\mathcal{P}^*$, makes V' into locally convex space: a fundamental system of neighborhoods of any f_0 in V' is given by $(\{f \in V' : \mid \langle f - f_0, u_i\rangle_{V',V} \mid < \epsilon, \; \forall i \in J\}$, where J is a finite set, $u_i \in V$ and $\epsilon > 0)$. Moreover (as in the topology $\sigma(V,V')$), a sequence $(f_n)_{n\in \mathbb{N}}$ converges to f in the topology $\sigma(V',V)$ if and only if $((\langle f_n - f, u\rangle_{V',V})_{n\in\mathbb{N}}$ converges to 0, for all $u \in V$.

A priori, we can introduce the second dual Z of the locally convex space V with respect to (*w.r.t* in short) the topology $\sigma(V',V)$ (*i.e.*, $(V, \sigma(V',V))$) as follows:

$$Z = \{\Lambda : V' \longrightarrow \mathbb{R};\ \Lambda \text{ linear and continuous w.r.t the topology } \sigma(V', V)\}.$$

The next proposition shows that the dual of V' endowed with the topology $\sigma(V', V)$ can be identified with V.

Proposition 2.67. *Let Λ be an element of Z. Then there exists $u \in V$ such that*

$$\Lambda(f) = \langle f, u\rangle_{V',V}, \quad \forall f \in V'.$$

□

As a consequence, we have Corollary 2.68.

Corollary 2.68. *Let H be an affine hyperplane of V', closed with respect to the topology $\sigma(V', V)$. Then*

$$H = \{f \in V' :\ \langle f, u\rangle_{V',V} = \lambda\},$$

where u is a non-zero element of V and $\lambda \in \mathbb{R}$. □

Remark 2.69. Weak topologies are interesting in regards to compactness. Indeed, weak topologies have less open sets and have on the other hand more compact sets than the finer (original) topology. Moreover, the compact sets play a fundamental role when we seek to analyze the existence results. ◇

Using Tychonoff's theorem, which states that the product of an arbitrary number of compact spaces is still compact with respect to the corresponding product topology, we have the following fundamental result.

Theorem 2.70. *(Banach–Alaoglu) Let $\mathcal{U}$ be a neighborhood of $0 \in V$. Then the set $K = \{f \in V' :\ \mid \langle f, u\rangle_{V',V} \mid \leq 1, \forall u \in \mathcal{U}\}$ is a compact subset of V' with respect to the weak* topology $\sigma(V', V)$.* □

As an immediate consequence, we have the following Banach–Alaoglu–Bourbaki's theorem (in the case of normed spaces).

Corollary 2.71. *(Banach–Alaoglu–Bourbaki) If V is a normed space equipped with the norm $\| \cdot \|_{V'}$, then the unit ball in V':*

$$\mathcal{B}_{V'}(0,1) = \{f \in V' :\ \| f \|_{V'} \leq 1\}$$

is a compact subset of V' with respect to the weak topology $\sigma(V', V)$.* □

It is known that in general in a compact set there are sequences without converging subsequences. The next results show that, in a particular type of topological vector space, each sequence in a compact set has a converging subsequence.

2.2.4 Separable Spaces

Definition 2.72. *(Separability) Let V be a topological vector space. We say that V is separable if there exists a subset D of V countable and dense.* ■

Proposition 2.73. *Let V be a separable metric space and U be a subspace of V. Then U is separable.*

Proof. Let $(u_n)_{n \in \mathbb{N}}$ be a countable and dense sequence in V and let $(\eta_m)_{m \in \mathbb{N}}$ be a real positive sequence converging to 0. If, arbitrarily, we choose the values $d_{m,n} \in \mathcal{B}(u_n, \eta_m) \cap U$, it is clear that the sequence $(d_{m,n})$ constitute a countable and dense subset in U. □

Proposition 2.74. *Let V be a Banach space such that its dual V' is separable. Then V is separable.*

The converse is false in general.

Proof. Let $(f_n)_{n \in \mathbb{N}}$ be a countable and dense sequence in V' and consider the sequence $(u_n)_{n \in \mathbb{N}}$ in V such that

$$\| u_n \|_V = 1 \quad \text{and} \quad \langle f_n, u_n \rangle_{V',V} \geq \frac{\| f_n \|_{V'}}{2}.$$

It is clear that W_0 the vector space on $\mathbb{Q}$ generated by the sequence $(u_n)_{n \in \mathbb{N}}$ is countable and dense in W the vector space on $\mathbb{R}$ generated by the sequence $(u_n)_{n \in \mathbb{N}}$. Prove now that W is dense in V. For this, let f be in V' such that $\langle f, u \rangle_{V',V} = 0$, for all $u \in W$ and prove that $f = 0$ (see Corollary 2.35). Let $\epsilon > 0$ be given then there exists $n > 0$ such that $\| f - f_n \|_{V'} < \frac{\epsilon}{3}$. Then (since $u_n \in W$, $\| u_n \|_V = 1$ and $\langle f, u_n \rangle_{V',V} = 0$)

$$\frac{\| f_n \|_{V'}}{2} \leq \langle f_n, u_n \rangle_{V',V} = \langle\, f_n - f, u_n \,\rangle_{V',V} + \langle f, u_n \rangle_{V',V} \leq \frac{\epsilon}{3}.$$

Consequently, $\| f \|_{V'} \leq \| f_n - f \|_{V'} + \| f_n \|_{V'} \leq \epsilon$ and then $f = 0$. So, W is dense in V and then W_0 is dense in V. We can conclude that the space V is separable. □

The following results show that the properties of separability are closely dependent on the metrizability of the weak topology.

Definition 2.75. *(Metrizability) A set D is said to be metrizable in the weak* topology $\sigma(V', V)$ if there exists a metric defined in D such that the associate topology coincides with $\sigma(V', V)$.* ■

Theorem 2.76. *Let V be a Banach space and V' be its dual. Then the closed unit ball of V', $\mathcal{B}_{V'}(0, 1)$, is metrizable in the weak* topology $\sigma(V', V)$ if and only if V is separable.*

Proof. Let $(u_n)_{n \in \mathbb{N}}$ be a subset countable and dense in $\mathcal{B}_V(0, 1)$. For a couple (f, g) in $\mathcal{B}_{V'}(0, 1)$ we define the following function:

$$d(f, g) = \sum_{n=1}^{\infty} \frac{1}{2^n} \mid \langle f - g, u_n \rangle_{V',V} \mid .$$

It is clear that the function d is a metric, since it satisfies all the metric axioms.

($\Rightarrow$) Suppose that $\mathcal{B}_{V'}(0,1)$ is metrizable in the topology $\sigma(V',V)$. Let $U_n = \{f \in \mathcal{B}_{V'}(0,1) : \ d(f,0) < 1/n\}$ and let H_n be a neighborhood of 0 for the topology $\sigma(V',V)$ in $\mathcal{B}_{V'}(0,1)$ such that $H_n \subset U_n$. We can assume that

$$H_n = \{f \in \mathcal{B}_{V'}(0,1) : \ | \langle f, u\rangle_{V',V} | < \epsilon_n \text{ for all } u \in D_n\},$$

where D_n is a finite subset of V and $\epsilon_n > 0$.

We have that $D := \bigcup_{i=1}^{\infty} D_n$ is a countable set and $\bigcap_{i=1}^{\infty} H_n = \{0\}$, and then

$$\text{If } \langle f, u\rangle_{V',V} = 0 \ \text{ for all } u \in D \quad \text{then} \quad f = 0.$$

Consequently, the vector space generated by D is dense in V and then the space V is separable.

($\Leftarrow$) Suppose now that the space V is separable. Prove now that the topology generated by the metric d coincides on $\mathcal{B}_{V'}(0,1)$ with $\sigma(V',V)$. In order to deal with this issue, we will proceed in two stages:

Stage 1: Let $f_0 \in \mathcal{B}_{V'}(0,1)$ and take a neighborhood H of f_0 for the topology $\sigma(V',V)$ by

$$H = \{f \in \mathcal{B}_{V'}(0,1) : \ | \langle f - f_0, v_i\rangle_{V',V} | < \epsilon \text{ for all } i = 1, \ldots, k\},$$

where $v_i \in \mathcal{B}_V(0,1)$ (*i.e.*, $\| v_i \| \leq 1$), for all $i = 1, \ldots, k$, and $\epsilon > 0$.

Prove that there exists $\eta > 0$ such that

$$U_\eta = \{f \in \mathcal{B}_{V'}(0,1) : \ \ d(f, f_0) < \eta\} \subset H.$$

Since the sequence $(u_n)_{n\in\mathbb{N}}$ is dense in $\mathcal{B}_V(0,1)$, then, for each i, we can find an integer n_i such that $\| u_{n_i} - v_i \| < \epsilon/4$. Let $\eta > 0$ such that $\eta < 2^{-(n_i+1)}\epsilon$, for all $i = 1, \ldots, k$ and prove that $U_\eta \subset H$.

Let f be in U_η, we have $f \in \mathcal{B}_{V'}(0,1)$ and $d(f, f_0) < \eta$, and then

$$| \langle \ f - f_0, u_{n_i}\rangle_{V',V} | < 2^{n_i}\eta < \frac{\epsilon}{2} \text{ for all } i = 1, \ldots, k.$$

According to the previous inequality, we can deduce that, for all $i = 1, \ldots, k$ (since f, f_0 are in $\mathcal{B}_{V'}(0,1)$ and $\| u_{n_i} - v_i \| < \epsilon/4$)

$$| \langle f - f_0, v_i\rangle_{V',V} | = | \langle f - f_0, v_i - u_{n_i}\rangle_{V',V} + \langle f - f_0, u_{n_i}\rangle_{V',V} | \leq \frac{\epsilon}{2} + \frac{\epsilon}{2}$$

and then $f \in H$.

Stage 2: Let $f_0 \in \mathcal{B}_{V'}(0,1)$. For a given $\eta > 0$, prove that there exists a neighborhood H of f_0 for the topology $\sigma(V',V)$ in $\mathcal{B}_{V'}(0,1)$ such that

$$H \subset U = \{f \in \mathcal{B}_{V'}(0,1) : \ \ d(f, f_0) < \eta\}.$$

For this, we take H as

$$H = \{f \in \mathcal{B}_{V'}(0,1) : \ | \langle f - f_0, u_i\rangle_{V',V} | < \epsilon \text{ for all } i = 1, \ldots, k\},$$

where $u_i \in \mathcal{B}_V(0,1)$, and we prove the existence of $\epsilon > 0$ and $k \in \mathbb{N}$ such that $H \subset U$.

If f is in H, we have then

$$d(f, f_0) = \sum_{n=1}^{\infty} \frac{1}{2^n} \mid \langle f - f_0, u_n \rangle_{V',V} \mid$$
$$= \sum_{n=1}^{k} \frac{1}{2^n} \mid \langle f - f_0, u_n \rangle_{V',V} \mid + \sum_{n=k+1}^{\infty} \frac{1}{2^n} \mid \langle f - f_0, u_n \rangle_{V',V} \mid .$$

Since $\mid \langle f - f_0, u_n \rangle_{V',V} \mid < \epsilon$ for all $n = 1, \ldots, k$, $u_n \in \mathcal{B}_V(0,1)$, $f \in \mathcal{B}_{V'}(0,1)$ and $f_0 \in \mathcal{B}_{V'}(0,1)$ we can deduce that

$$d(f, f_0) \leq \epsilon \sum_{n=1}^{k} \frac{1}{2^n} + 2 \sum_{n=k+1}^{\infty} \frac{1}{2^n} < \epsilon + \frac{1}{2^{k-1}}.$$

We can choose ϵ such that $0 < \epsilon < \eta/2$ and k sufficiently large such that $1/2^{k-1} < \eta/2$, and we obtain the existence of H. □

We can obtain a generalization of the above result.

Theorem 2.77. *If the topological vector space V is separable, and K a subset of the dual space V' is weakly* compact, then K is metrizable in the weak* topology $\sigma(V', V)$.*
As a consequence, each sequence in K has a convergent subsequence.

Proof. The proof is left as an exercise (by replacing the distance d in V' by $d(f,g) = \sum_{n=1}^{\infty}(1/2^n) \min (\mid \langle f - g, u_n \rangle_{V',V} \mid, 1)$, by proving the uniform continuity of any element $f \in K$ and by using the same technique as to obtain the second result of the above theorem). □

Theorem 2.78. *Let V be a Banach space and V' be its dual. Then the closed unit ball of V', $\mathcal{B}_{V'}(0,1)$, is metrizable in the weak* topology $\sigma(V, V')$ if and only if V' is separable.*

Proof. To prove the implication: V' separable $\Rightarrow \mathcal{B}_{V'}(0,1)$ is metrizable in $\sigma(V', V)$, we can use the proof of Theorem 2.76, by exchanging the roles of V and V'. The converse of this theorem is more delicate and we can find the proof, *e.g.*, in Dunford and Schwartz [108]. □

As a corollary, we have (according to Banach–Alaoglu–Bourbaki's theorem) the following.

Corollary 2.79. *Let V be a Banach separable space and (f_n) be a bounded sequence in V'. Then there exists a subsequence denoted also by (f_n) converging for the topology $\sigma(V', V)$.* □

2.2.5 Dual of Banach Spaces and Reflexivity

A particular case is when V is a normed topological space, equipped with the norm $\| \, . \, \|$. In this case the dual space V' is a Banach space with the dual norm $\| \, . \, \|_{V'}$ (that can be denoted by $\| \, . \, \|_*$) defined by

$$\| f \|_{V'} = \sup_{\|u\|=1} | \langle f, u \rangle_{V',V} | \, .$$

Example 2.80. Let Ω be a domain in $\mathbb{R}^d$, $d \in \mathbb{N}^*$. Then:

(*i*) The dual of the space $V := \mathcal{C}(\Omega)$ (of real-valued and continuous functions on the domain Ω) is the space $V' := \mathcal{M}(\Omega)$ of Radon measures μ with

$$\langle \mu, u \rangle_{V',V} := \int_\Omega u d\mu \quad \forall u \in V.$$

(*ii*) The dual of the space $L^p(\Omega)$, $1 \le p < \infty$ (of real-valued and p-integrable, in the Lebesgue sense, functions on the domain Ω) is the space $L^{p^*}(\Omega)$, where p^* is the conjugate to p, *i.e.*, $(1/p) + (1/p^*) = 1$.

(*iii*) The dual of the space $L^1(\Omega)$ is the space $L^\infty(\Omega)$ (of real-valued and essentially bounded functions on the domain Ω).

(*iv*) The dual of the space $L^\infty(\Omega)$ is the space of Borel measures and contains the space $L^1(\Omega)$. ♣

We can introduce the bidual of V, *i.e.*, the dual of V', denoted by V'' with the bidual norm

$$\| \xi \|_{V''} = \sup_{\|f\|_{V'}=1} | \langle \xi, f \rangle_{V'',V'} | \, .$$

We have a canonical immersion (or a natural embedding) $J : V \longrightarrow V''$ defined by: let $u \in V$ be given, the application $f \in V' \longrightarrow \langle f, u \rangle_{V',V} \in \mathbb{R}$ constitute a linear and continuous form, *i.e.*, is an element of V'', denoted by Ju. We then have

$$\langle Ju, f \rangle_{V'',V'} = \langle f, u \rangle_{V',V}, \ \forall u \in V, \ \forall f \in V'.$$

It is clear that the immersion J is linear and continuous, and $J(V)$ is a closed subset of V''. Moreover, J is an isometric injection, *i.e.*, $\| Ju \|_{V''} = \| u \|$, $\forall u \in V$.

Remark 2.81. (*i*) The function J is not in general surjective, *i.e.*, $J(V) \neq V''$.
(*ii*) By using the function J, we can always identify the space V to a subspace of its bidual V''. ◇

Definition 2.82. *(Reflexive space) Let V be a Banach space and J a canonical immersion of V into V''. The space V is said to be reflexive if $J(V) = V''$ (J is surjective), i.e., it can be identified under J with its bidual V''.* ■

The next result shows an important characterization of the reflexive spaces.

Theorem 2.83. *(Characterization of Kakutani) Let $(V, \| \, . \, \|)$ be a Banach space, with norm $\| \, . \, \|$. The space V is said to be reflexive if and only if its closed unit ball,*

$$\mathcal{B}_V(0,1) = \{u \in V; \ \| u \| \leq 1\},$$

is compact in the weak topology $\sigma(V, V')$, where V' is its dual.

This implies that every bounded sequence admits a weakly converging subsequence.

Proof. If J is reflexive then J is a linear, continuous and bijective function. Consequently, J^{-1} is linear and continuous with respect to the strong topologies V and V'' (then J^{-1} is also an isometric function).

It is clear that $(\forall f \in V'$, and $\forall \eta > 0)$

$$J(\{u \in V; \ | \langle f, u \rangle_{V',V} | \leq \eta\}) = J(\{\xi \in V''; \ | \langle \xi, f \rangle_{V'',V'} | \leq \eta\}),$$

so that the topology $J^{-1}(\sigma(V'', V'))$ coincides with the topology $\sigma(V, V')$. Since the closed unit ball in V'' is weak* compact, then it is the unit closed ball of V.

Conversely, if the closed unit ball in V is compact (for the weak topology $\sigma(V, V')$), then as $J(\mathcal{B}_V(0,1))$ is closed, and by using the result of Goldstine, based on the result of Helly (see below), it coincides with $\mathcal{B}_{V''}(0,1)$ and then $J(V) = V''$. □

Now we give the results previously used in the proof of the theorem (that can be proved easily).

Lemma 2.84. *(Helley) Let $(V, \| \, . \, \|)$ be a Banach space, f_i, $i = 1, n$, n elements in V' and λ_i, $i = 1, n$, n real values. Then the following properties are equivalent:*

(i) for all $\eta > 0$, there exists u_η in the closed unit ball of V $(\| u_\eta \| \leq 1)$ such that

$$| \langle f_i, u_\eta \rangle_{V',V} - \lambda_i | \leq \eta, \ \forall i = 1, n$$

(ii) for all $\beta_i \in I\!R$, $i = 1, n$, we have

$$| \sum_{i=1,n} \beta_i \lambda_i | \leq \| \sum_{i=1,n} \beta_i f_i \|_{V'} .$$

Proof. The proof is left to the reader as an exercise. □

Lemma 2.85. *(Goldstine) Let V be a Banach space. Then $J(\mathcal{B}_V(0,1))$ is dense in $\mathcal{B}_{V''}(0,1)$ for the weak topology $\sigma(V'', V')$.*

Proof. Let $\xi \in \mathcal{B}_{V''}(0,1)$ and take a neighborhood of ξ (for the topology $\sigma(V'', V')$) of the form

$$\Omega = \{\psi \in V'' : | \langle \psi - \xi, f_i \rangle_{V'',V'} | < \eta, \ \forall i = 1, n\},$$

where f_i, $i = 1, n$ are n elements in V'.

To prove the result of the lemma, it is sufficient to prove that $J(\mathcal{B}_V(0,1)) \cap \Omega \neq \emptyset$. Otherwise, we need only to find $u \in \mathcal{B}_V(0,1)$ such that

$$| \langle f_i, u \rangle_{V',V} - \langle \xi, f_i \rangle_{V'',V'} | < \eta, \forall i = 1, n.$$

Let then λ_i be the values $\langle \xi, f_i \rangle_{V'',V'}$, for $i = 1, n$, then for $\beta_i \in \mathbb{R}$, $i = 1, n$, we have (since $\| \xi \|_{V''} \leq 1$)

$$| \sum_{i=1,n} \beta_i \lambda_i | = | \sum_{i=1,n} \langle \xi, \beta_i f_i \rangle_{V'',V'} | \leq \| \sum_{i=1,n} \beta_i f_i \|_{V'} .$$

By using Helley's lemma, it follows that there exists an element u_η of $\mathcal{B}_V(0,1)$ such that $| \langle f_i, u_\eta \rangle_{V',V} - \lambda_i | < \eta, \forall i = 1, n$, and then $J(u_\eta) \in J(\mathcal{B}_V(0,1)) \cap \Omega$. □

Now we give some elementary properties of the reflexive spaces (by using the characterization of Kakutani).

Proposition 2.86. *Let V be a reflexive Banach space and U be a closed subspace of V (with the norm induced by V). Then U is reflexive.*

Proof. The proof follows by proving that the topology $\sigma(U, U')$ coincides with the trace on U of the topology $\sigma(V, V')$ (by using the restrictions and the prolongations of the linear forms), and by using Theorem 2.63 (closed for strong topology and convex), in order to obtain that $\mathcal{B}_U(0,1)$ is compact for the topology $\sigma(V, V')$ and then $\mathcal{B}_U(0,1)$ is compact for the topology $\sigma(U, U')$. □

As corollaries, we have the following.

Corollary 2.87. *Let V be a Banach space and let V' be its dual. Then V is a reflexive space if and only if V' is a reflexive space.*

Proof. ($\Rightarrow$) According to Banach–Alaoglu–Bourbaki's theorem (Corollary 2.71), we have that $\mathcal{B}_{V'}(0,1)$ is a compact subset of V' with respect to the topology $\sigma(V', V)$. Moreover, as V is reflexive, $\sigma(V', V) = \sigma(V', V")$. Consequently, $\mathcal{B}_{V'}(0,1)$ is a compact subset with respect to the topology $\sigma(V', V'')$ and then V' is reflexive (according to Kakutani's theorem (Theorem 2.83)).

($\Leftarrow$) According to ($\Rightarrow$) we have that V'' is reflexive (since V' is reflexive) and then, because of Proposition 2.86, $J(V)$ is reflexive (since $J(V)$ is a closed subset of V''). Hence, V is reflexive. □

Corollary 2.88. *Let V be a reflexive Banach space and K be a closed, convex and bounded subset of V. Then K is compact for the topology $\sigma(V, V')$.*

Proof. Because of Theorem 2.63 and Theorem 2.83, we have first that K is closed for the topology $\sigma(V, V')$ and then K is compact for the topology $\sigma(V, V')$. □

Corollary 2.89. *Let V be a reflexive Banach space and let V' be its dual. Then V is separable if and only if V' is separable.*

Proof. ($\Rightarrow$) If V is reflexive and separable then $V'' = J(V)$ is reflexive and separable. Consequently, V' is separable (according to Corollary 2.87).

($\Leftarrow$) Because of Proposition 2.74 (the reflexivity assumption is not necessary). □

Now we give the interesting Eberlein–Smulian's theorem, that shows the characterization of the reflexive space by sequences.

Theorem 2.90. *(Eberlein–Smulian) Let V be a Banach space. Then V is reflexive if and only if for any bounded sequence (u_n) in V, there exists a subsequence, denoted also by (u_k), converging for the topology $\sigma(V, V')$.*

Proof. ($\Rightarrow$) Let U_0 be the span of $u_1, u_2, \ldots, u_n, \ldots$, and U the closure of U_0. Then the subspace U of V is closed, separable and reflexive (according to Proposition 2.86), so U' is separable (according to Corollary 2.89). Consequently, $\mathcal{B}_{U''}(0,1) = \mathcal{B}_U(0,1)$ is metrizable in the topology $\sigma(U'', U') = \sigma(U, U')$ (according to Theorem 2.78) and then $\mathcal{B}_U(0,1)$ is compact (characterization of Kakutani) and metrizable in the topology $\sigma(U, U')$. We can then extract a subsequence of (u_n) (denoted also by (u_n)) which converges in the topology $\sigma(U, U')$ and then (u_n) converges in $\sigma(V, V')$.

($\Leftarrow$) This result (the veritable result of Eberlein–Smulian) is more delicate to prove; see, *e.g.*, Dunford and Schwartz [108] and Holmes [160] for details. □

Definition 2.91. *(Uniformly convex Banach spaces) A Banach space $(V, \| . \|)$ is said to be uniformly convex, for the norm $\| . \|$, if for all $\eta > 0$, there exists $\delta > 0$, such that*

$$\text{if } u \text{ and } v \text{ are in } \mathcal{B}_V(0,1) : \| u - v \| > \eta \text{ then } \left\| \frac{u+v}{2} \right\| < 1 - \delta.$$

■

The uniform convexity is a geometrical property of a Banach space. In the following theorem, we prove that this property involves a topological property.

Theorem 2.92. *(Milman–Pettis) If a Banach space $(V, \| . \|)$ is uniformly convex, then V is reflexive.*

Proof. Let ξ be in V'' such that $\| \xi \|_{V''} = 1$. We want to prove that $\xi \in J(\mathcal{B}_V(0,1))$ or otherwise (since $J(\mathcal{B}_V(0,1))$ is strongly closed in V''), to prove that

$$\forall \eta > 0, \ \exists u \in \mathcal{B}_V(0,1) : \ \| \xi - Ju \|_{V''} < \eta.$$

For $\eta > 0$ fixed, consider the value $\delta > 0$ which is chosen by the uniform convexity estimate (corresponding to η). Let $f \in V'$ be given such that $\| f \|_{V'} = 1$ and satisfies $\langle \xi, f \rangle_{V'',V'} > 1 - \delta/2$ (this inequality is valid, since $\| \xi \|_{V''} = 1$ and $\| f \|_{V'} = 1$). We consider the neighborhood of the value ξ by

$$\Omega = \{ \psi \in V'' : \ | \langle \xi - \psi, f \rangle_{V'',V'} | < \delta/2 \}.$$

According to Goldstine's lemma, we have that $J(\mathcal{B}_V(0,1)) \cap \Omega \neq \emptyset$ and then there exists an element $u \in \mathcal{B}_V(0,1)$ such that $Ju \in \Omega$, *i.e.*, by the definition of J, $| \langle \xi, f \rangle_{V'',V'} - \langle f, u \rangle_{V',V} | < \delta/2$.

Prove now that $\| \xi - Ju \|_{V''} \leq \eta$. Suppose that the result is false, then we can obtain a new neighborhood of ξ for the topology $\sigma(V'', V')$ which does not contain u. By using again Goldstine's lemma, we have the existence of $v \in \mathcal{B}_V(0,1)$, such that $\| \xi - Jv \|_{V''} > \eta$ and $| \langle \xi, f \rangle_{V'',V'} - \langle f, v \rangle_{V',V} | < \delta/2$. So, we have (by adding the two previous inequalities)

$$2\langle \xi, f \rangle_{V'',V'} \leq \langle f, u + v \rangle_{V',V} + \delta \leq \| u + v \| + \delta \ (\text{since } \ \| f \|_{V'} = 1).$$

Since $\langle \xi, f \rangle_{V'',V'} > 1 - \delta/2$, then $\| (u+v)/2 \| \geq 1 - \delta$ and, consequently, according to the uniform convexity of V, $\| u - v \| \leq \eta$, which is a contradiction.
□

As a corollary, we have the following proposition.

Proposition 2.93. *If V is a Hilbert space, then V is a reflexive space.*

Proof. Clearly, a Hilbert space is a uniformly convex Banach space (by using the *parallelogram law*: $2(\| u \|^2 + \| v \|^2) = \| u + v \|^2 + \| u - v \|^2$, for all $(u, v) \in V \times V$). Consequently, because of Milman–Pettis's theorem (Theorem 2.92), a Hilbert space is reflexive. □

Let us terminate this section with the following proposition.

Proposition 2.94. *Let V be an uniformly convex Banach space, and (u_n) be a sequence of V converging weakly, for the topology $\sigma(V, V')$, into u such that*

$$\limsup_{n \longrightarrow \infty} \| u_n \| \leq \| u \| .$$

Then the sequence (u_n) converges strongly to u.

Proof. Suppose that the limit u is not null (else the proof is immediate). Let us introduce the following sequence (v_n) by

$$v_n = \frac{u_n}{\max(\| u_n \|, \| u \|)} \in \mathcal{B}_V(0,1),$$

then (v_n) converges weakly, for the topology $\sigma(V, V')$, into $v = u/\| u \|$ (we have $\| v \|= 1$). Moreover, according to Corollary 2.66, we can deduce that

$$1 = \| v \| < \liminf_{n \longrightarrow \infty} \| \frac{v_n + v}{2} \| \leq 1.$$

So, $\| (v_n + v)/2 \|$ converges into 1. Consequently, according to the uniform convexity of the space V, $\| v_n - v \|$ converges into 0.

This completes the proof. □

Example 2.95. In view of Examples 2.80, the spaces $L^p(\Omega)$, $1 < p < \infty$, are reflexive, but the spaces $L^1(\Omega)$ and $L^\infty(\Omega)$ are not reflexive. ♣

2.2.6 Closure and Continuity of Convex Functions

The convex functions appreciate the remarkable properties of continuity: they are locally Lipschitzian on the relative interior of its effective domain.

Before studying the continuity of convex functions, we introduce some definitions and give some results concerning the lower semi-continuous convex function.

Definition 2.96. *(Proper function) A function* $F : V \longrightarrow \overline{I\!R}$ *is said to be proper if it takes nowhere the value* $-\infty$ *(i.e.,* $F(u) > -\infty$*, for every* $u \in V$*, and then* $F : V \longrightarrow]-\infty, +\infty]$*) and not identically equal to* $+\infty$ *(i.e.,* $F \not\equiv +\infty$ *or also* $\text{dom}F \neq \emptyset$*).* ■

Definition 2.97. *(Closed function) A function* $F : V \longrightarrow \overline{I\!R}$ *is said to be closed if* $\overline{F} = F$ *(where* $\overline{F}$ *is the closure of* F *or the lower semi-continuous regularization).* ■

In particular we have (according to Proposition 2.50) the following corollary.

Corollary 2.98. *A proper function is closed if and only if it is lower semi-continuous, or if and only if its epigraph is closed, or if and only if its level sets are closed.* □

The following proposition shows that if a convex function takes the value $-\infty$, then the points where the function is finite are not numerous. More precisely:

Proposition 2.99. *If* $F : V \longrightarrow \overline{I\!R}$ *is a convex and lower semi-continuous function and takes the value* $-\infty$ *(i.e., there exists a point* $u_0 \in V$ *such that* $F(u_0) = -\infty$*), then* F *cannot take any finite value.*

Proof. The proof is trivial and is left to the reader as an exercise. □

Next we give an interesting result pertaining to lower semi-continuous, convex and proper functions for a minimization problem (according to Proposition 2.65 and Theorem 2.90).

Definition 2.100. *(Minimizing sequence) A sequence $(u_n)_{n \in I\!N}$ in G (no structure is assumed on G) is said to be a minimizing sequence of some mapping F if*

$$\lim_{n \longrightarrow \infty} F(u_n) = \inf_{v \in G} F(v).$$

■

Theorem 2.101. *Let $(V, \| \, . \, \|)$ be a reflexive Banach space, $F : V \longrightarrow \overline{I\!R}$ be a proper, convex and lower semi-continuous function on V, and C a convex and closed subset of V. If we assume that C is bounded or that the functional F is coercive over C, i.e.,*

$$F(v) \longrightarrow +\infty \text{ if } v \in C, \quad \| v \| \longrightarrow \infty, \tag{2.10}$$

then the function F assumes a minimum value on C. Otherwise there exists a point $u \in C$ such that

$$F(u) = \inf_{v \in C} F(v). \tag{2.11}$$

Proof. Let (u_n) be a minimizing sequence in C of F, then

$$\lim_{n \longrightarrow \infty} F(u_n) = \inf_{v \in C} F(v).$$

The sequence (u_n) is uniformly bounded in V (since C is bounded or since $F(u_n)$ is bounded according to (2.10)), then we can extract from (u_n) a subsequence, denoted also by (u_n) which converges weakly in V (in the topology $\sigma(V, V')$) to a value $u \in C$ (according to Eberlein–Smulian's theorem (Theorem 2.90)). According to Proposition 2.65, we have that F is lower semi-continuous with respect to the weak topology $\sigma(V, V')$ and so,

$$F(u) \leq \liminf_{n \longrightarrow \infty} F(u_n).$$

Consequently, u is a solution of problem (2.11). □

The next results of this section concern the continuity properties of convex functions. The proofs of these results can be found, for example, in Ekeland *et al.* [112].

The main result is contained in the following proposition.

Proposition 2.102. *Let V be a locally convex topological vector space and $F : V \longrightarrow \overline{I\!R}$ be a given convex function on V. If in the neighborhood of a point $u \in V$ such that $F(u) > -\infty$, the function F is bounded above by a finite constant, then F is continuous at point u.* □

More generally we have the following proposition.

Proposition 2.103. *Let V be a locally convex topological vector space and $F : V \longrightarrow \overline{I\!R}$ be a given proper and convex function. The following statements are equivalent:*

(*i*) *there exists a non-empty open set* $\mathcal{V}$ *on which* F *is bounded from above by a finite constant*

(*ii*) F *is continuous and locally Lipschitz over interior of its effective domain* dom$F$, *which is non-empty.* □

As a consequence we have with more precision, in many special situations, some results which clarify this principal result, as follows.

Corollary 2.104. *Let* V *be a locally convex topological vector space and* $F : V \longrightarrow \overline{I\!R}$ *be a proper and convex function. Then, if* F *is upper semi-continuous at a point which is interior to its effective domain* dom$F$, *then* F *is continuous over interior of* dom$F$. □

Corollary 2.105. *Let* V *be a Banach space and* $F : V \longrightarrow \overline{I\!R}$ *be a lower semi-continuous, proper and convex function. Then* F *is continuous over the interior of its effective domain* dom$F$. □

Corollary 2.106. *Let* V *be a finite dimensional and separated topological space and* $F : V \longrightarrow \overline{I\!R}$ *be a proper and convex function. Then* F *is continuous over the interior of its effective domain* dom$F$. □

2.3 Γ-Regularization and Continuous Affine Functions

Definition 2.107. *(Continuous affine function) A function* $f : V \longrightarrow I\!R$ *on a vector space is a continuous affine function if it is of the form* $f(x) = l(x)+\lambda$, *for some linear continuous function* l *and some real* λ. ■

It is clear that every linear functional is affine, and every affine function is both convex and concave.

Definition 2.108. *(Pointwise supremum of continuous affine functions) The set of functions* $F : V \longrightarrow \overline{I\!R}$ *which are the pointwise supremum of a family of continuous affine functions is denoted by* $\Gamma(V)$ *and the subset of* $\Gamma(V)$ *other than the constant functions* $-\infty$ *and* $+\infty$ *is denoted by* $\Gamma_0(V)$. ■

From this definition we observe that all functions of $\Gamma(V)$ are convex (according to Proposition 2.46) and lower semi-continuous functions on V (according to Proposition 2.52). Moreover, if a function of $\Gamma(V)$ attains the value $-\infty$ then this function is exactly the constant function $-\infty$ (because the pointwise supremum of an empty family is $-\infty$ and if the family under consideration is not empty, the pointwise supremum can not take the value $-\infty$). More precisely we have the following characterization of $\Gamma(V)$.

Proposition 2.109. *(Characterization of* $\Gamma(V)$*) The function* $F : V \longrightarrow \overline{I\!R}$ *is in* $\Gamma(V)$ *if and only if* F *is a lower semi-continuous convex function, and if* F *takes the value* $-\infty$ *then* F *is identically equal to* $-\infty$. *Otherwise,* $\Gamma(V)$ *is the set of all closed convex functions on* V.

Proof. Suppose that F is a lower semi-continuous convex function where F is different from the constant function $-\infty$ and from the constant function $+\infty$. We show that for every $(u_0, \lambda_0) \in V \times \mathbb{R}$ such that $F(u_0) > \lambda_0$, there exists a continuous affine function L such that

$$\lambda_0 \leq L(u_0) \leq F(u_0). \tag{2.12}$$

According to Propositions 2.44 and 2.50 we have that epi$F$ is a closed convex set with $(u_0, \lambda_0) \notin \text{epi}F$. Then (because of Hahn–Banach's second geometric form) there exists a closed hyperplane H which strictly separates (u_0, λ_0) and epi$F$, *i.e.*, there exists a non-zero continuous linear function l and $(\alpha, \beta) \in \mathbb{R}^2$ such that $H = \{(u, \lambda) \in V \times \mathbb{R} : \ l(u) + \alpha\lambda = \beta\}$ and

$$\begin{aligned} &l(u_0) + \alpha\lambda_0 < \beta, \\ &l(u) + \alpha\lambda > \beta \ \ \forall (u, \lambda) \in \text{epi}F. \end{aligned} \tag{2.13}$$

To finish the proof we can distinguish the two following cases:
• If $F(u_0)$ is finite, $(u_0, F(u_0)) \in \text{epi}F$ and then $\alpha(\lambda_0 - F(u_0)) < 0$ (according to (2.13)). Since $\lambda_0 - F(u_0) < 0$, we can conclude that $\alpha > 0$ and then

$$\lambda_0 < \frac{\beta - l(u_0)}{\alpha} < F(u_0).$$

So, we can take the continuous affine function L as follows: $L(.) := (\beta - l(.))/\alpha$ (which satisfies the relation (2.12)).
• If $F(u_0)$ is infinite: either $\alpha \neq 0$ and then we find the preceding case, or $\alpha = 0$ and then the continuous affine function $\beta - l(.)$ satisfies (according to (2.13))

$$\beta - l(u_0) > 0 \ \text{ and } \ \beta - l(u) < 0 \ \ \forall u \in \text{dom}F.$$

Therefore, there exists a continuous linear function r and $\gamma \in \mathbb{R}$ such that

$$\gamma - r(u) < F(u) \ \ \forall u \in V$$

and then, for all $\eta > 0$, the continuous affine functions $L_\eta(.) := \gamma - r(.) + \eta(\beta - l(.))$ satisfy

$$L_\eta(u) < F(u) \ \ \forall u \in V.$$

By taking η sufficiently large such $\lambda_0 < L_\eta(u_0)$, we can deduce that the affine function L_η satisfies the relation (2.12). This completes the proof. □

A similar definition can be given for a concave function, as follows.

Definition 2.110. *(Closed concave function) The function $F : V \longrightarrow \overline{\mathbb{R}}$ is said to be a closed concave function if $-F$ is a closed convex function and we denote the set of all closed concave functions on V by*

$$-\Gamma(V) = \{F : \ -F \in \Gamma(V)\}.$$

■

Before giving a regularization of an extended real-valued function in $\Gamma(V)$, we introduce the following definition.

Definition 2.111. *We say that the extended real-valued function G dominates (or is greater than) the extended real-valued function F on V, written $G \geq F$, if we have $G(u) \geq F(u)$, for every $u \in V$. In the same way we say that G is less than F, written $G \leq F$, if we have $G(u) \leq F(u)$, for every $u \in V$.* ■

Definition 2.112. *(Γ-regularization of a function) Let F and G be two functions of V into $\overline{I\!R}$. The function G is said to be the Γ-regularization of F if G is the largest minorant of F in $\Gamma(V)$, or if G is the pointwise supremum of all the continuous affine functions less than F.* ■

In particular we have the following corollary.

Corollary 2.113. *If $F : V \longrightarrow \overline{I\!R}$ is in $\Gamma(V)$ then F coincides with its Γ-regularization.* □

In general, we have the following properties of the *Γ-regularization.*

Proposition 2.114. *Let $F : V \longrightarrow \overline{I\!R}$ and G be its Γ-regularization. If there exists a continuous affine function less than F, we have that the epigraph of G is exactly the closed convex hull of the epigraph of F:*

$$\text{epi}G = \text{clco}(\text{epi}F).$$

Proof. The proof is left to the reader as an exercise. □

Proposition 2.115. *Let $F : V \longrightarrow \overline{I\!R}$, $\overline{F}$ be its closure (or its lower semi-continuous regularization) and G be its Γ-regularization. Then the following statements are true:*

(i) $G \leq \overline{F} \leq F$.

(ii) If F is convex and there exists a continuous affine function less than F then $\overline{F} = G$.

Proof. This is a direct consequence of Proposition 2.114 and the definition of the closure of functions. □

3

A Brief Overview of Sobolev Spaces

The aim of this chapter is to recall the basic features of Sobolev spaces and useful compactness results. For the proofs and more details, the reader is referred, for instance, to Adams [4], Lions and Magenes [204] and Maz'ja [221].

3.1 Tools and Definitions

3.1.1 Definitions and Notations

We denote by Ω an open domain in $\mathbb{R}^n$, $n \geq 1$, with a smooth boundary $\Gamma = \partial\Omega$. In general, some regularity of Ω will be assumed. We will suppose that either

$$\Omega \text{ is Lipschitz,} \tag{3.1}$$

i.e., the boundary Γ is locally the graph of a Lipschitz function, or

$$\Omega \text{ is of class } \mathcal{C}^r,\ r \geq 1, \tag{3.2}$$

i.e., the boundary Γ is a manifold of dimension $n-1$ of class $\mathcal{C}^r$ (see below). In both cases we assume that Ω is totally on one side of Γ. These definitions mean that locally the domain Ω is below the graph of some function ψ, the boundary Γ is represented by the graph of ψ and its regularity is determined by that of the function ψ. Moreover, it is necessary to note that a domain with a continuous boundary is never on both sides of its boundary at any point of this boundary and that a Lipschitz boundary has almost everywhere a unit normal vector $\mathbf{n}$.

We will also use the following multi-index notation for partial differential derivatives of a function:

$$
\begin{aligned}
&\partial_i^k u := \frac{\partial^k u}{\partial x_i^k} \text{ for all } k \in \mathbb{N} \text{ and } i = 1, \ldots, n, \\
&D^\alpha u := \partial_1^{\alpha_1} \partial_2^{\alpha_2} \cdots \partial_j^{\alpha_j} \cdots \partial_n^{\alpha_n} u = \frac{\partial^{\alpha_1 + \cdots + \alpha_n} u}{\partial x_1^{\alpha_1} \cdots \partial x_n^{\alpha_n}}, \\
&\alpha = (\alpha_1, \ldots \alpha_j, \ldots, \alpha_n) \in \mathbb{N}^n, \quad [\alpha] := \sum_{i=1}^n \alpha_i.
\end{aligned} \tag{3.3}
$$

We denote by $\mathcal{C}(D)$ (respectively $\mathcal{C}^k(D)$, $k \in \mathbb{N}$ or $k = +\infty$) the space of real continuous functions on D (respectively the space of k times continuously differentiable functions on D), where D plays the role of Ω or its closure $\overline{\Omega}$. The space of real $\mathcal{C}^\infty$ functions on Ω with a compact support in Ω is denoted by $\mathcal{C}_0^\infty(\Omega)$ or $\mathcal{D}(\Omega)$ as in the theory of distributions of Schwartz [262] and the space of distributions on Ω is denoted by $\mathcal{D}'(\Omega)$, *i.e.*, the space of continuous linear form over $\mathcal{D}(\Omega)$ (for more details on the theory of distribution, see Schwartz [262, 263]).

For $p \in [1, +\infty]$, we denote by $L^p(\Omega)$ the space of (class of) measurable functions v on Ω such that

$$
\begin{aligned}
\| v \|_{L^p(\Omega)} &= \left(\int_\Omega | v(x) |^p \, \mathrm{d}x\right)^{1/p} < \infty \quad \text{if } p < \infty, \\
\| v \|_{L^\infty(\Omega)} &= \sup_{x \in \Omega} \text{ess} \, | v(x) | < \infty \quad \text{if } p = +\infty.
\end{aligned}
$$

The space $L^p(\Omega)$ equipped with the norm $\| \, . \, \|_{L^p(\Omega)}$ (for $1 \leq p \leq +\infty$) is a Banach space: it is reflexive and separable for $1 < p < \infty$ (its dual is $L^{\frac{p}{p-1}}(\Omega)$), separable but not reflexive for $p = 1$ (its dual is $L^\infty(\Omega)$), and not separable, not reflexive for $p = \infty$ (its dual contains strictly $L^1(\Omega)$). In particular the space $L^2(\Omega)$ is a Hilbert space for the scalar product

$$
(u, v) = (u, v)_{L^2(\Omega)} = \int_\Omega u(x).v(x) \mathrm{d}x.
$$

We denote by $L^p_{\text{loc}}(\Omega)$ the space of functions which are L^p on any bounded subdomain of Ω.

Of course, similar space can be defined on any open set other than Ω, in particular, on the cylinder set $\Omega \times]a, b[$ or on the set $\Gamma \times]a, b[$, where $(a, b) \in \mathbb{R}$ and $a < b$.

Let U be a Banach space, $1 \leq p \leq +\infty$ and $-\infty \leq a < b \leq +\infty$, then $L^p(a, b; U)$ is the space of (class of) L^p functions v from (a,b) into U which is a Banach space for the norm

$$
\| v \|_{L^p(a,b;U)} = \left(\int_a^b \| v(t) \|_U^p \, \mathrm{d}t \right)^{1/p} \quad \text{if } p < \infty
$$

and for the norm

$$
\| v \|_{L^\infty(a,b;U)} = \sup_{t \in (a,b)} \text{ess} \, \| v(t) \|_U \quad \text{if } p = +\infty.
$$

Similarly, for a Banach space U, $k \in \mathbb{N}$ and $-\infty < a < b < +\infty$, we denote by $\mathcal{C}([a,b];U)$ (respectively $\mathcal{C}^k([a,b];U)$) the space of continuous functions (respectively the space of k times continuously differentiable functions) v from $[a,b]$ into U, which are Banach spaces, respectively, for the norms $\| v \|_{\mathcal{C}([a,b];U)} = \sup_{t\in[a,b]} \| v(t) \|_U$ and $\| v \|_{\mathcal{C}^k([a,b];U)} = \sum_{i=0,k} \| \frac{\partial^i v}{\partial t^i} \|_{\mathcal{C}([a,b];U)}$.

Finally, we define the Hölder space $\mathcal{C}^{k,\alpha}(\Omega)$, for $k \in \mathbb{N}$ and $\alpha \in (0,1]$, as the space of functions in $\mathcal{C}^k(\overline{\Omega})$ whose partial derivatives of order k, $\partial_j u$, $[j] = k$, are Hölder uniformly continuous with exponent α in Ω, *i.e.*, if

$$\sup_{\substack{x,y\in\Omega \\ x\neq y}} \frac{| \partial_j u(x) - \partial_j u(y) |}{| x - y |^\alpha} < \infty, \quad \forall j \text{ such that } [j] = k.$$

It is a Banach space for the norm given by

$$\| v \|_{\mathcal{C}^{k,\alpha}(\Omega)} := \| v \|_{\mathcal{C}^k(\overline{\Omega})} + \sup_{\substack{x,y\in\Omega \\ x\neq y \\ [j]=k}} \frac{| \partial_j u(x) - \partial_j u(y) |}{| x - y |^\alpha}.$$

3.1.2 Some Fundamental Inequalities and Convergence Criteria

Some Fundamental Inequalities

Our study involves the following fundamental inequalities, which are repeated here for review:

(*i*) *Hölder's inequality* :

$$\int_D \Pi_{i=1,k} f_i \mathrm{d}x \leq \Pi_{i=1,k} \| f_i \|_{L^{q_i}(D)}, \text{ where}$$
$$\| f_i \|_{L^{q_i}(D)} = \left(\int_D | f_i |^{q_i} \mathrm{d}x \right)^{1/q_i} \quad \text{and} \quad \sum_{i=1,k} \frac{1}{q_i} = 1.$$

(*ii*) *Young's inequality* ($\forall a, b > 0$ and $\lambda > 0$):

$$ab \leq \frac{\lambda}{p} a^p + \frac{\lambda^{-q/p}}{q} b^q, \text{ for } p, q \in]1, +\infty[\text{ and } \frac{1}{p} + \frac{1}{q} = 1.$$

(*iii*) *Gronwall's lemma:*

$$\text{If } \frac{\mathrm{d}\Phi}{\mathrm{d}t} \leq g(t)\Phi(t) + h(t), \ \forall t \geq 0 \text{ then}$$
$$\Phi(t) \leq \Phi(0) \exp\left(\int_0^t g(s)\mathrm{d}s \right) + \int_0^t h(s) \exp\left(\int_s^t g(\tau)\mathrm{d}\tau \right) \mathrm{d}s, \ \forall t \geq 0.$$

Some Convergence Criteria

The aim of this subsection is to recall some notions of the convergence of sequences on the L^p-spaces. For the proofs and more details, the reader is referred, for instance, to Hewitt and Stromberg [155].

Lemma 3.1. *(Fatou's lemma) Let $(u_n)_{n\geq 1}$ be a sequence of measurable and positive functions. Then*

$$\int_\Omega \liminf_{n\longrightarrow\infty} u_n(x)\mathrm{d}x \leq \liminf_{n\longrightarrow\infty} \int_\Omega u_n(x)\mathrm{d}x.$$

□

Theorem 3.2. *(Monotone convergence) Let $(u_n)_{n\geq 1}$ be a decreasing sequence of measurable and positive functions. Then*

$$\lim_{n\longrightarrow\infty} \int_\Omega u_n(x)\mathrm{d}x = \int_\Omega \lim_{n\longrightarrow\infty} u_n(x)\mathrm{d}x.$$

□

Theorem 3.3. *(Lebesgue's dominated convergence theorem) Let $(u_n)_{n\geq 1}$ be a sequence of L^1-functions, which is almost everywhere (a.e.) convergent to a measurable function u. If there exists a positive function $v \in L^1(\Omega)$ such that $| u_n |\leq v$ a.e. in Ω, then $u \in L^1(\Omega)$ and we have*

$$\lim_{n\longrightarrow\infty} \| u_n - u \|_{L^1(\Omega)}= 0, \quad \lim_{n\longrightarrow\infty} \int_\Omega u_n(x)\mathrm{d}x = \int_\Omega u(x)\mathrm{d}x.$$

□

We give now some convergence results in L^p-spaces (see Brezis and Lieb [61]).

Lemma 3.4. *(Brezis–Lieb's lemma) Let $(u_n)_{n\geq 1}$ be a bounded sequence of L^p-functions, $1 \leq p < \infty$, which is almost everywhere convergent to a measurable function u. Then $u \in L^p(\Omega)$ and*

$$\lim_{n\longrightarrow\infty} (\| u_n \|^p_{L^p(\Omega)} - \| u_n - u \|^p_{L^p(\Omega)}) =\| u \|^p_{L^p(\Omega)} .$$

□

A direct corollary of Brezis–Lieb's lemma is the following result.

Corollary 3.5. *Let $u \in L^p(\Omega)$, $1 \leq p < \infty$, and $(u_n)_{n\geq 1}$ be a sequence of L^p-functions which is almost everywhere convergent to u.*

$$\text{If } \lim_{n\longrightarrow\infty} \| u_n \|_{L^p(\Omega)}=\| u \|_{L^p(\Omega)} \text{ then } \lim_{n\longrightarrow\infty} \| u_n - u \|_{L^p(\Omega)}= 0.$$

□

Lemma 3.6. *Let $u \in L^p(\Omega)$, $1 \leq p < \infty$, and $(u_n)_{n\geq 1}$ be a sequence of L^p-functions. If $\lim_{n\longrightarrow\infty} \| u_n - u \|_{L^p(\Omega)}= 0$ then there exist $v \in L^p(\Omega)$ and a subsequence of $(u_n)_{n\geq 1}$, denoted also by $(u_n)_{n\geq 1}$, such that $| u_n |\leq v$ a.e. in Ω and $(u_n)_{n\geq 1}$ is almost everywhere convergent to u.* □

We finish this subsection by a proposition which gives a relation between the almost everywhere convergence and the weak-convergence results on L^p-spaces, for $1 < p < \infty$ (which are reflexive spaces).

Proposition 3.7. *Let $(u_n)_{n\geq 1}$ be a bounded sequence of L^p-functions, $1 < p < \infty$, which is almost everywhere convergent to u. Then*

$$u_n \rightharpoonup u \quad \textit{weakly in} \quad L^p(\Omega).$$

□

Remark 3.8. Since the space $L^p(\Omega)$, for $1 < p < \infty$, is reflexive then, according to Eberlein–Smulian's theorem (Theorem 2.90), for $(u_n)_{n\geq 1}$ being a bounded sequence of L^p-functions, we can extract from (u_n) a subsequence, denoted also by (u_n) which converges weakly in $L^p(\Omega)$ to a function $v \in L^p(\Omega)$. If, moreover, (u_n) is almost everywhere convergent to u then $v = u$. ♢

3.1.3 Definition of Sobolev Spaces

We introduce now the *Sobolev spaces*, which will be considered in more details in Adams [4]. Let $\Omega \subset \mathbb{R}^n$ be an open domain (not necessarily bounded). For an integer $m > 0$ and $1 \leq p \leq \infty$, the Sobolev space of order (m, p), denoted by $W^{m,p}(\Omega)$, is defined as the space of functions in the space $L^p(\Omega)$ whose (distribution) derivatives of order $\leq m$ are also in $L^p(\Omega)$, *i.e.*,

$$W^{m,p}(\Omega) := \{v \in L^p(\Omega) : \; D^\alpha v \in L^p(\Omega) \;\; \forall \alpha \in \mathbb{N}^n \text{ such that } [\alpha] \leq m\}.$$

This is a Banach space for the Sobolev norm given by

$$\| v \|_{W^{m,p}(\Omega)} := \sum_{[\alpha]\leq m} \| D^\alpha v \|_{L^p(\Omega)} \qquad 1 \leq p < \infty$$

(which is equivalent to the norm $(\sum_{[\alpha]\leq m} \| D^\alpha v \|^p_{L^p(\Omega)})^{1/p}$)

and, in the case of $p = \infty$,

$$\| v \|_{W^{m,\infty}(\Omega)} := \max_{[\alpha]\leq m} \| D^\alpha v \|_{L^\infty(\Omega)} \, .$$

Moreover, $W^{m,p}(\Omega)$ is a reflexive space for $1 < p < \infty$ and a separable space for $1 \leq p < \infty$. In particular, if $p = 2$, the Sobolev space $W^{m,2}(\Omega)$ denoted by $H^m(\Omega)$ is a Hilbert and separable space for the following scalar product:

$$(u, v)_{H^m(\Omega)} := \sum_{[\alpha]\leq m} (D^\alpha u, D^\alpha v)_{L^2(\Omega)}.$$

The space $H^m(\Omega)$ and $W^{m,p}(\Omega)$ contain $\mathcal{C}^\infty(\overline{\Omega})$ and $\mathcal{C}^m(\overline{\Omega})$. The closure of $\mathcal{D}(\Omega)$ for the $H^m(\Omega)$ norm (respectively $W^{m,p}(\Omega)$ norm) is denoted by $H^m_0(\Omega)$ (respectively $W^{m,p}_0(\Omega)$). In particular, the H^1 type spaces and the corresponding dual spaces, will be frequently used. The space $H^1(\Omega)$ denotes the space of L^2 functions u on Ω such that $\partial_i u \in L^2(\Omega)$, for $i = 1, \dots, n$, and $H_0^1(\Omega)$ is the closure of $\mathcal{D}(\Omega)$ for the $H^1(\Omega)$ norm. They are both Hilbert spaces for the following scalar product:

$$(u, v)_{H^1(\Omega)} := \int_\Omega u(x)v(x)\mathrm{d}x + \int_\Omega \nabla u(x).\nabla v(x)\mathrm{d}x,$$

where $(\nabla u)_i := \partial_i u$.

The space $H^{-1}(\Omega)$ is defined as the dual space of $H_0^1(\Omega)$ for the duality between $\mathcal{D}'(\Omega)$ and $\mathcal{D}(\Omega)$ and its norm is

$$\| f \|_{H^{-1}(\Omega)} := \sup_{\substack{v \in H_0^1(\Omega) \\ \| v \|_{H^1(\Omega)} \leq 1}} | \langle f, v \rangle_{\mathcal{D}'(\Omega), \mathcal{D}(\Omega)} | .$$

Let us recall the Poincaré inequality, where Ω is a bounded domain. Let v be in $\mathcal{D}(\Omega)$ and write $v(x) = v(x_1, x')$, for $x = (x_1, x')$ and $x' = (x_2, \dots, x_n)$ as the form

$$v(x_1, x') = \int_{-\infty}^{x_1} \partial_1 v(t, x')\mathrm{d}t.$$

Then, by using the Cauchy–Schwartz inequality, we can deduce that

$$| v(x_1, x') |^2 \leq C(\Omega) \int_{\mathbb{R}} | \nabla v(t, x') |^2 \, \mathrm{d}t$$

and then (by integrating)

$$\| v \|_{L^2(\Omega)}^2 \leq C(\Omega) \| \nabla v \|_{L^2(\Omega)}^2 .$$

By density (by definition) of $\mathcal{D}(\Omega)$ on $H_0^1(\Omega)$, we can deduce that for all $v \in H_0^1(\Omega)$ we have $\| v \|_{L^2(\Omega)}^2 \leq C(\Omega) \| \nabla v \|_{L^2(\Omega)}^2$ and then we can define the norm in $H_0^1(\Omega)$ by

$$\| v \|_{H_0^1(\Omega)} := \| \nabla v \|_{L^2(\Omega)},$$

which is equivalent to the $H^1(\Omega)$ norm (if Ω is a bounded domain). Consequently, $H_0^1(\Omega)$ is a Hilbert space for the following scalar product:

$$(u, v)_{H_0^1(\Omega)} := \int_\Omega \nabla u(x).\nabla v(x)\mathrm{d}x.$$

If we identify $L^2(\Omega)$ to its dual, but if *we do not identify* $H_0^1(\Omega)$ and its dual $H^{-1}(\Omega)$, we have

$$H_0^1(\Omega) \subset L^2(\Omega) \equiv (L^2(\Omega))' \subset H^{-1}(\Omega) = (H_0^1(\Omega))',$$

where the injections are dense and continuous.

3.2 Some Properties of Sobolev Spaces

Let us recall some important properties of Sobolev spaces, namely the density, embedding, compactness and trace theorems.

Let p be a real value such that $p \geq 1$ and $m \geq 1$ be an integer value. Assume that the open domain $\Omega \subset \mathbb{R}^n$ (not necessarily bounded) is sufficiently regular.

It is clear that, if Ω is a bounded domain then

$$\begin{aligned} &L^p(\Omega) \subset L^q(\Omega) \quad \text{for all } p \geq q, \\ &W^{m,p}(\Omega) \subset W^{k,q}(\Omega) \quad \text{for all } m \geq k,\ p \geq q, \\ &H^m(\Omega) \subset H^k(\Omega) \quad \text{for all } m \geq k. \end{aligned} \tag{3.4}$$

with continuous embedding.

3.2.1 Density Results

If the open domain Ω is sufficiently regular, for example of class $\mathcal{C}^m$, then

$$\mathcal{C}^m(\overline{\Omega}) \text{ is dense in } W^{m,p}(\Omega) \ \forall 1 \leq p < \infty.$$

This implies that (if Ω is sufficiently regular)

$$\begin{aligned} &W^{m,p}(\Omega) \text{ is dense in } W^{m-1,p}(\Omega) \ \forall 1 \leq p < \infty, \\ &H^m(\Omega) \text{ is dense in } H^{m-1}(\Omega). \end{aligned} \tag{3.5}$$

3.2.2 Embedding Results

Assume that the open domain Ω is sufficiently regular, for example of class $\mathcal{C}^m$, we have

$$\begin{aligned} &\text{if } \frac{1}{p} - \frac{m}{n} > 0 \text{ then } W^{m,p}(\Omega) \subset L^q(\Omega) \text{ where } \frac{1}{p} - \frac{1}{q} = \frac{m}{n}, \text{ and} \\ &\qquad \| u \|_{L^q(\Omega)} \leq C \| u \|_{W^{m,p}(\Omega)} \text{ (the embedding is continuous),} \\ &\text{if } \frac{1}{p} - \frac{m}{n} = 0 \text{ then } W^{m,p}(\Omega) \subset L^q_{\text{loc}}(\Omega) \text{ where } 1 \leq q < \infty, \text{ and} \\ &\qquad \| u \|_{L^q(\omega)} \leq C_\omega \| u \|_{W^{m,p}(\Omega)} \text{ for all bounded } \omega \subset \Omega, \\ &\text{if } \frac{1}{p} - \frac{m}{n} < 0 \text{ then } W^{m,p}(\Omega) \subset \mathcal{C}^{k,\alpha}(\omega) \text{ where } m - \frac{n}{p} = k + \alpha, \\ &\quad k \text{ is the integer part of } m - \frac{n}{p} \text{ and } \forall j \text{ such that } [j] = k \\ &\qquad | \partial_j u(x) - \partial_j u(y) | \leq C_\omega | x - y |^\alpha \| u \|_{W^{m,p}(\Omega)} \\ &\qquad \text{for all bounded } \omega \subset \Omega \text{ and } x, y \in \omega. \end{aligned} \tag{3.6}$$

In particular, for $m = 1$, if the boundary Γ of Ω is bounded then

$$\begin{aligned}
&\text{if } 1 \le p < n \text{ then } W^{1,p}(\Omega) \subset L^q(\Omega) \text{ where } \frac{1}{p} - \frac{1}{q} = \frac{1}{n},\\
&\text{if } p = n \text{ then } W^{1,p}(\Omega) \subset L^q(\Omega) \text{ for all } p \le q < \infty,\\
&\text{if } p > n \text{ then } W^{1,p}(\Omega) \subset L^\infty(\Omega),
\end{aligned} \tag{3.7}$$

with continuous embeddings.

If the domain Ω is bounded, we have (because of (3.4))

$$\begin{aligned}
&\text{if } \frac{1}{p} - \frac{m}{n} < 0, \text{ then } W^{m,p}(\Omega) \subset \mathcal{C}^{k,\alpha}(\Omega) \text{ where } m - \frac{n}{p} = k + \alpha,\\
&\text{if } \frac{1}{p} - \frac{m}{n} > 0, \text{ then } W^{m,p}(\Omega) \subset L^q(\Omega) \text{ where } q \in [1, \frac{np}{n - mp}],\\
&\text{if } \frac{1}{p} - \frac{m}{n} = 0, \text{ then } W^{m,p}(\Omega) \subset L^q(\Omega) \text{ where } q \in [1, +\infty[,
\end{aligned} \tag{3.8}$$

with continuous embeddings.

3.2.3 Compactness Results

Assume now that $m = 1$ and the domain Ω is bounded and of class $\mathcal{C}^1$. Then we have the following results (due to Rellich–Kondrachov)

$$\begin{aligned}
&\text{if } 1 \le p < n \text{ then } W^{1,p}(\Omega) \subset L^{\tilde{q}}(\Omega)\ \forall 1 \le \tilde{q} < q \ \text{ where } \frac{1}{p} - \frac{1}{q} = \frac{1}{n},\\
&\text{if } p \ge n \text{ then } W^{1,p}(\Omega) \subset L^q(\Omega) \text{ for all } 1 \le q < \infty,\\
&\text{if } p > n \text{ then } W^{1,p}(\Omega) \subset \mathcal{C}^{0,\tilde{\alpha}}(\Omega) \text{ for all } \tilde{\alpha} < \alpha = 1 - \frac{n}{p},
\end{aligned} \tag{3.9}$$

with compactness embeddings.

3.2.4 Trace Results and Green's Formula

Trace Results

Suppose now that the domain Ω is of class $\mathcal{C}^{m+1}$, then for a given function $u \in W^{m,p}(\Omega)$, we can define its trace on Γ, $u|_\Gamma$, which coincides with the value of u on Γ. More generally, we can define the linear and continuous operator trace $\gamma :\ W^{m,p}(\Omega) \longrightarrow (L^p(\Gamma))^m$ by

$$\begin{aligned}
&\gamma := \{\gamma_0, \gamma_1, \ldots, \gamma_{m-1}\} \text{ and}\\
&\gamma_i u = \frac{\partial^i u}{\partial \mathbf{n}^i}|_\Gamma \quad i = 1, \ldots, m-1,
\end{aligned} \tag{3.10}$$

where $\mathbf{n} = (n_1, \ldots, n_n)$ is the unit vector outward normal on Γ.

The space $W_0^{m,p}(\Omega)$ can be written as

$$W_0^{m,p}(\Omega) = \ker\gamma \quad (i.e., \text{ the kernel of } \gamma). \tag{3.11}$$

So,

$$\text{if } \; u \in W_0^{m,p}(\Omega) \; \text{ then } \; \frac{\partial^i u}{\partial \mathbf{n}^i}|_\Gamma = 0 \text{ for all } \; i = 1, \ldots, m-1.$$

The dual spaces of $W_0^{m,p}(\Omega)$ are the following spaces:

$$\begin{aligned} W^{-m,p^*}(\Omega) &:= (W_0^{m,p}(\Omega))' \; \text{ where } \; \frac{1}{p} + \frac{1}{p^*} = 1, \\ H^{-m}(\Omega) &:= (H_0^m(\Omega))' \; \text{ (the case of } p = 2), \end{aligned} \tag{3.12}$$

where

$$\| f \|_{W^{-m,p^*}(\Omega)} := \sup_{\substack{v \in W_0^{m,p}(\Omega) \\ \|v\|_{W^{m,p}(\Omega)} \leq 1}} | \langle f, v \rangle_{W^{-m,p^*}(\Omega), W^{m,p}(\Omega)} | .$$

In particular:

- For $p = 2$, $\gamma_i(H^m(\Omega))$, which is not the whole space $L^2(\Gamma)$, is denoted by $H^{m-i-1/2}(\Gamma)$ and is endowed with the following quotient norm

$$\| w \|_{H^{m-i-1/2}(\Gamma)} := \inf_{\substack{v \in H^m(\Omega) \\ \gamma_i v = w}} \| v \|_{H^m(\Omega)}$$

  and its dual is denoted by $H^{-m+i+1/2}(\Gamma)$. For example, for $m = 1$, we have that $H_0^1(\Omega) = \{v \in H^1(\Omega) : \; \gamma_0 u = u|_\Gamma = 0\}$, $\gamma_0(H^1(\Omega)) = H^{1/2}(\Gamma)$ and $(H^{1/2}(\Gamma))' := H^{-1/2}(\Gamma)$.
- For $m = 1$, if Ω is bounded then

$$W_0^{1,p}(\Omega) \subset L^2(\Omega) \subset W^{-1,p^*}(\Omega) \; \text{ if } \; \frac{2n}{n+2} \leq p < \infty,$$

  with continuous and dense embeddings. Moreover, because of Poincaré's inequality, we have that for all $u \in W_0^{1,p}(\Omega)$ $(1 \leq p < \infty)$, $\| u \|_{L^p(\Omega)} \leq C \| \nabla u \|_{L^p(\Omega)}$ and then we can endowed the space $W_0^{1,p}(\Omega)$ with the norm $\| \nabla u \|_{L^p(\Omega)}$, which is equivalent to the norm $\| u \|_{W^{m,p}(\Omega)}$. If Ω is not bounded then

$$W_0^{1,p}(\Omega) \subset L^2(\Omega) \subset W^{-1,p^*}(\Omega) \; \text{ if } \; \frac{2n}{n+2} \leq p \leq 2.$$

Some Properties of Spaces Related to Divergence Operator

We assume in this subsection that the domain Ω is a bounded subset of $\mathbb{R}^n$ with Lipschitz boundary Γ. For a vector function $v := (v_i)_{i=1,n}$, we define the divergence operator div by

$$\mathrm{div}(v) = \sum_{i=1,n} \frac{\partial v_i}{\partial x_i}.$$

Introducing now the following subspace of $H^1(\Omega)$ by

$$H(\mathrm{div};\Omega) := \{v \in L^2(\Omega) : \ \mathrm{div}(v) \in L^2(\Omega)\},$$

which is a Hilbert space for the norm

$$\| v \|_{H(\mathrm{div};\Omega)} = \left(\| v \|^2_{L^2(\Omega)} + \| \mathrm{div}(v) \|^2_{L^2(\Omega)} \right)^{1/2}.$$

The closure of $\mathcal{D}(\Omega)$ for the $H(\mathrm{div};\Omega)$ norm is denoted by $H_0(\mathrm{div};\Omega)$.

The following theorem concerns the normal component of boundary values of functions of the space $H(\mathrm{div};\Omega)$.

Theorem 3.9. *Let Ω be a bounded subset of $I\!R^n$ with Lipschitz boundary Γ, we have the following properties:*

$$\mathcal{D}(\overline{\Omega}) \text{ is dense in } H(\mathrm{div};\Omega) \text{ (the boundedness of } \Omega \text{ is not necessary)}, \quad (3.13)$$

there exists a mapping $\gamma_n : \ v \longrightarrow v.\mathbf{n}|_\Gamma$ defined on $\mathcal{D}(\overline{\Omega})$ which can be extended by continuity to a linear and continuous mapping, denoted also by γ_n, from $H(\mathrm{div};\Omega)$ into $H^{-1/2}(\Gamma)$. (3.14) □

A consequence of the previous theorem is the following *Green's formula*: for all vector functions $v \in H(\mathrm{div};\Omega)$ and scalar functions $\phi \in H^1(\Omega)$,

$$\int_\Omega v.\nabla\phi \mathrm{d}x + \int_\Omega \phi \mathrm{div}(v) \mathrm{d}x = \langle v.\mathbf{n}, \phi \rangle_{H^{-1/2}(\Gamma), H^{1/2}(\Gamma)}. \quad (3.15)$$

The next results give some properties of $H_0(\mathrm{div};\Omega)$.

Theorem 3.10. *Let Ω be a bounded subset of $I\!R^n$ with Lipschitz boundary Γ, we have the following properties:*

$$H_0(\mathrm{div};\Omega) = \ker\gamma_n = \{v \in H(\mathrm{div};\Omega) : \ \gamma_n v = v.\mathbf{n} = 0 \ \text{on } \Gamma\}, \quad (3.16)$$

$$\mathcal{V} \text{ is dense in } \mathcal{H}, \quad (3.17)$$

where $\mathcal{V} = \{v \in \mathcal{D}(\Omega) : \ \mathrm{div}(v) = 0\}$, $\mathcal{H} = \{v \in H_0(\mathrm{div};\Omega) : \ \mathrm{div}(v) = 0\}$. □

Green's Formula

Let us now give the interesting result of the integrating by parts for the Laplace operator, due to a trace theorem, namely the extended *Green's formula* (which is a direct consequence of (3.15)).

If $u \in H^1(\Omega)$ and $\Delta u \in L^2(\Omega)$, then the traces on Γ, $\gamma_0 u = u|_\Gamma$ is in $H^{1/2}(\Gamma)$ and $\gamma_1 u = (\partial u/\partial \mathbf{n})|_\Gamma$ is in $H^{-1/2}(\Gamma)$. Moreover, we have

$$(-\Delta u, v)_{L^2(\Omega)} = (\nabla u, \nabla v)_{L^2(\Omega)} - \langle \gamma_1 u, \gamma_0 u \rangle_{H^{-1/2}(\Gamma), H^{1/2}(\Gamma)}. \tag{3.18}$$

If we assume that $\gamma_1 u \in L^2(\Omega)$ then the Green's formula becomes

$$(-\Delta u, v)_{L^2(\Omega)} = (\nabla u, \nabla v)_{L^2(\Omega)} - (\gamma_1 u, \gamma_0 u)_{L^2(\Gamma)}. \tag{3.19}$$

More generaly, we can define the generalized form of Green's theorem (see, for instance, Lions and Magenes [204]). For this, let $a_{ij} : \Omega \longrightarrow \mathbb{R}$, for $i, j = 1, \ldots, n$, be functions in $L^\infty(\Omega)$ and satisfy the ellipticity conditions

$$\sum_{i,j=1,n} a_{ij}(x)\xi_i\xi_j \geq \alpha \sum_{k=1,n} \xi_k^2 \quad \forall(\xi)_{i=1,n} \in \mathbb{R}^n, \quad \text{a.e. in } \Omega. \tag{3.20}$$

Consider the operator A by

$$Au = -\sum_{i,j=1,n} \frac{\partial}{\partial x_i}(a_{ij}(x)\frac{\partial u}{\partial x_j}). \tag{3.21}$$

The normal derivative $\partial u/\partial \eta_A$ at Γ, directed towards the exterior of Ω, can be written in the following form:

$$\frac{\partial u}{\partial \eta_A} = \sum_{i,j=1,n} a_{ij}(x)\cos(\eta, x_i)\partial_j u, \tag{3.22}$$

where η is the unit normal vector at Γ exterior to Ω and $\cos(\eta, x_i)$ is an ith direction cosine of η. Finally, we introduce the following bilinear form a:

$$a(u, v) = \sum_{i,j=1,n} \int_\Omega a_{ij}(x)\partial_i u \partial_j v dx, \quad \forall u, v \in H^1(\Omega). \tag{3.23}$$

We can now give the *generalized form of Green's theorem*:

If $u, v \in H^1(\Omega)$ and $Au \in L^2(\Omega)$, then the traces on Γ, $\gamma_0 v = v|_\Gamma$ is in $H^{1/2}(\Gamma)$ and $\gamma_1 u = (\partial u/\partial \eta_A)|_\Gamma$ is in $H^{-1/2}(\Gamma)$. Moreover, we have

$$(Au, v)_{L^2(\Omega)} = a(u, v) - \langle \gamma_1 u, \gamma_0 v \rangle_{H^{-1/2}(\Gamma), H^{1/2}(\Gamma)}. \tag{3.24}$$

3.2.5 Truncation Operations

Let us consider the following notations:

- For $r \in \mathbb{R}$, we write classically $r^+ = \max(r, 0)$, $r^- = (-r)^+$ and then

$$r = r^+ - r^- \quad \text{and} \quad | r | = r^+ + r^-.$$

- For u a real function on Ω, we define the functions u_+ and u_- by

$$u_+(x) = (u(x))^+ \quad \text{and} \quad u_-(x) = (u(x))^-, \ \forall x \in \Omega, \text{ respectively.}$$

According, for instance, to Gilbarg and Trudinger [132], we have that:

(*i*) If u is in $L^p(\Omega)$, $1 \leq p \leq +\infty$, then u_+ and u_- are also in $L^p(\Omega)$ with

$$\| u_+ \|_{L^p(\Omega)} \leq \| u \|_{L^p(\Omega)} \quad \text{and} \quad \| u_- \|_{L^p(\Omega)} \leq \| u \|_{L^p(\Omega)} .$$

(*ii*) If u is in $W^{1,p}(\Omega)$, $1 \leq p < +\infty$, then u_+ and u_- are also in $W^{1,p}(\Omega)$. Moreover, for a.e. $x \in \Omega$, we have that (for $i = 1, \ldots, n$)

$$\begin{aligned} \partial_i u_+(x) &= \partial_i u(x) \ \text{ if } \ u(x) > 0 \ \text{ and } \ \partial_i u_+(x) = 0 \ \text{ if } \ u(x) \leq 0, \\ \partial_i u_-(x) &= \partial_i u(x) \ \text{ if } \ u(x) < 0 \ \text{ and } \ \partial_i u_-(x) = 0 \ \text{ if } \ u(x) \geq 0, \end{aligned} \tag{3.25}$$

and

$$\begin{aligned} &\| \partial_i u_+ \|_{L^p(\Omega)} \leq \| \partial_i u \|_{L^p(\Omega)}, \quad \| u_+ \|_{W^{1,p}(\Omega)} \leq \| u \|_{W^{1,p}(\Omega)}, \\ &\| \partial_i u_- \|_{L^p(\Omega)} \leq \| \partial_i u \|_{L^p(\Omega)}, \quad \| u_- \|_{W^{1,p}(\Omega)} \leq \| u \|_{W^{1,p}(\Omega)} . \end{aligned} \tag{3.26}$$

3.2.6 Interpolation Theory

Let X and Y be two Hilbert spaces, endowed respectively by the norm $\| \,.\, \|_X$ and $\| \,.\, \|_Y$, such that $X \subset Y$, with the continuous and dense embedding. The interpolation theory gives a family of Hilbert spaces denoted by $[X, Y]_\theta$, $\theta \in [0, 1]$, such that

$$[X, Y]_0 := X, \ [X, Y]_1 := Y \text{ and } X \subset [X, Y]_\theta \subset Y, \tag{3.27}$$

with continuous and dense embeddings.

The spaces $[X, Y]_\theta$, $\theta \in [0, 1]$, are endowed with the norm $\| \,.\, \|_{[X,Y]_\theta}$ such that

$$\| u \|_{[X,Y]_\theta} \leq C_\Theta \| u \|_X^{1-\theta} \| u \|_Y^\theta \quad \forall u \in X, \ \forall \theta \in [0, 1]. \tag{3.28}$$

By using the interpolation between $H^m(\Omega)$, $0 < m \in \mathbb{N}$, and $H^0(\Omega) = L^2(\Omega)$ we can define, the following space by

$$H^{\theta m}(\Omega) := [H^m(\Omega), H^0(\Omega)]_{1-\theta}, \text{ for all } \theta \in (0, 1), \tag{3.29}$$

which gives the definition of $H^s(\Omega)$, for a real $s \geq 0$.

The density, embedding, compactness and trace results given above, can be extended, without modifications to Sobolev spaces $H^s(\Omega) = W^{s,2}(\Omega)$, for a real $s \geq 0$. Moreover, if Ω is bounded and sufficiently regular domain, we can complete the previous results by (for all s_1, s_2 such that $0 \leq s_2 < s_1$)

$$\begin{aligned} &H^{s_1}(\Omega) \subset H^{s_2}(\Omega), \text{ with compactness embedding,} \\ &H^{(1-\theta)s_1 + \theta s_2}(\Omega) := [H^{s_1}(\Omega), H^{s_2}(\Omega)]_\theta, \text{ for all } \theta \in]0, 1[, \end{aligned} \tag{3.30}$$

with equivalent norms.

In particular, if the domain Ω is bounded, we have (because of (3.8)) the following useful estimates, namely *the Gagliardo–Nirenberg inequalities*:

$$\| v \|_{L^p(\Omega)} \leq C \| v \|^{\theta}_{H^q(\Omega)} \| v \|^{1-\theta}_{L^2(\Omega)} \quad \text{for all } v \in H^q(\Omega), \tag{3.31}$$

where q is a non-negative integer, $\theta \in [0,1[$ and $p = 2m/(m - 2\theta q)$ (with the exception that if $q - m/2$ is a non-negative integer, then θ is restricted to 0).

Remark 3.11. (*i*) We can define $H^s_0(\Omega)$ as the closure of $\mathcal{D}(\Omega)$ for the $H^s(\Omega)$ norm and $H^{-s}(\Omega) := (H^s_0(\Omega))'$.

(*ii*) We can also define the family of Sobolev spaces $W^{s,p}(\Omega)$, $H^s(\Gamma)$ and $W^{s,p}(\Omega)$, for $s \in \mathbb{R}$.

(*iii*) In the whole space $\mathbb{R}^n$ and $p = 2$ (in (*ii*)), Sobolev spaces can be defined also in terms of integrability properties in frequency space by using the Fourier transformation. For all $u \subset \mathcal{D}'(\mathbb{R}^n)$, the Fourier transform $\widehat{u}$ (denoted also by $\mathcal{F}u$) is defined by

$$\widehat{u}(y) := \int_{\mathbb{R}^n} \exp(-ix.y)u(x)\mathrm{d}x \quad \text{for} \quad y \in \mathbb{R}^n$$

and the inverse Fourier transform (which allows us to recover u from $\widehat{u}$) is defined by

$$u(x) = \mathcal{F}^{-1}\widehat{u}(x) := \frac{1}{(2\pi)^n} \int_{\mathbb{R}^n} \exp(ix.y)\widehat{u}(y)\mathrm{d}y \quad \text{for} \quad y \in \mathbb{R}^n.$$

For all $s \in \mathbb{R}$, we can introduce the Sobolev spaces

$$H^s(\mathbb{R}^n) := \{u \in \mathcal{S}'(\mathbb{R}^n) : \| u \|^2_{H^s(\mathbb{R}^n)} := \int_{\mathbb{R}^n} (1+ | y |^2)^s | \widehat{u}(y) |^2 \mathrm{d}y < \infty\},$$

and similarily the homogeneous Sobolev spaces

$$\dot{H}^s(\mathbb{R}^n) := \{u \in \mathcal{S}'(\mathbb{R}^n) : \; u \in L^1_{\text{loc}}(\mathbb{R}^n), \\ \| u \|^2_{\dot{H}^s(\mathbb{R}^n)} := \int_{\mathbb{R}^n} | y |^{2s} | \widehat{u}(y) |^2 \mathrm{d}y < \infty\},$$

where $\mathcal{S}'(\mathbb{R}^n)$ is the space of tempered distributions, *i.e.*, the dual of the Schwartz space $\mathcal{S}(\mathbb{R}^n)$ (the space of rapidly decreasing functions):

$$\mathcal{S}(\mathbb{R}^n) := \{f \in \mathcal{C}^\infty(\mathbb{R}^n) : \| x^\alpha D^\beta f \|_\infty < \infty \quad \text{for all multi-indices } \alpha, \beta\}.$$

Here, $\| \, . \, \|_\infty$ is the supremum norm.

We notice that $H^s(\mathbb{R}^n)$ is a Hilbert space for all $s \in \mathbb{R}$, but $\dot{H}^s(\mathbb{R}^n)$ is not a Hilbert space if $s \geq n/2$. ♢

These last results, and others, will be recalled only when needed.

4

Legendre–Fenchel Transformation and Duality

For any physical or biological system, at least two types of dependent variables can be considered. The *source variables*, which represent the *input variables* of the system (for example the external force in mechanics, the distributed source energy or drug terms in medical treatment, ..., *etc.*); the *configuration variables*, which represent the *output variables* of the system and describe the state of the system, are also said to be the state variables. These two types of variables, in most systems, usually appear in pairs. For each given configuration variable $u \in V$, where V is said to be the space of configuration or state variables, there exists a variable $f \in V'$ which is dual to u; the dual f is said to be the dual configuration variable and V' is said to be the dual configuration space. Moreover, usually, the space of source variables is a subspace of V'.

In order to study the mathematical theory of duality in natural phenomena, we will study in this chapter the main part of convex conjugate duality theory, and we will present different applications.

4.1 Fenchel Conjugate Functions

4.1.1 Definitions and Properties

Let V be a topological vector space and let V' be its topological dual space (or conjugate space). We denote by $\langle .,. \rangle_{V',V} : V \times V' \longrightarrow \mathbb{R}$ the bilinear form with respect to the duality between V' and V. We say that paired spaces V and V' are placed in duality by the bilinear form $\langle .,. \rangle_{V',V}$.

Definition 4.1. *(Conjugate function) Let $F : V \longrightarrow \overline{\mathbb{R}}$ be an extended real-valued function. The function $F^* : V' \longrightarrow \overline{\mathbb{R}}$ defined by*

$$F^*(f) = \sup_{u \in V} (\langle f, u \rangle_{V',V} - F(u)), \forall f \in V' \tag{4.1}$$

is said to be the Fenchel (convex) conjugate, or simply conjugate (or also the polar) function of F.

The mapping $F \longrightarrow F^*$ *is called the Legendre–Fenchel transformation.* ■

A direct consequence of the definition of F^* is the following result.

Proposition 4.2. *Let* $F : V \longrightarrow \overline{\mathbb{R}}$ *be a given extended real-valued function, the following statements are true:*

(*i*) *(Fenchel's inequality):* $F^*(f) + F(u) \geq \langle f, u \rangle_{V',V}, \forall f \in V', \forall u \in V.$

(*ii*) *Let* f *be in the dual* V' *of* V *and* $\lambda \in \mathbb{R}$, *the continuous affine function* $u \longrightarrow \langle f, u \rangle_{V',V} - \lambda$ *is less than* F *if and only if*

$$F^*(f) \leq \lambda.$$

(*iii*) *If* F *is identically equal to* $+\infty$ *then* F^* *is identically equal to* $-\infty$. *Moreover, if the function* F *is proper, then the relation (4.1) may be restricted to the points* u *in the effective domain of* F, $\text{dom}F$.

(*iv*) *The function* F^* *is always in* $\Gamma(V')$ *(since* F^* *is the pointwise supremum of a family of affine continuous functions on* V'*). Therefore,* F^* *is always a lower semi-continuous convex function on* V'. *Moreover, if* F^* *takes the value* $-\infty$ *then* F^* *is identically equal to* $-\infty$. □

We can also deduce immediately the following properties.

Proposition 4.3. (*i*) *Let* F *and* G *be two given extended real-valued functions of* V *into* $\overline{\mathbb{R}}$, *the following properties hold:*

(*a*) $F^*(0) = -\inf_{u \in V} F(u)$.

(*b*) *If* F *is less than* G ($F \leq G$) *then* G^* *is less than* F^* ($G^* \leq F^*$).

(*c*) *If* $G(u) = F(\alpha u), \forall u \in V$, *with* $\alpha \neq 0$ *then* $G^*(f) = F^*(f/\alpha),\ \forall f \in V'$.

(*d*) $(\alpha F)^*(f) = \alpha F^*(f/\alpha),\ \forall f \in V',\ \forall \alpha > 0$.

(*e*) $(F + \beta)^* = F^* - \beta,\ \ \forall \beta \in \mathbb{R}$.

(*ii*) *Given a family* $(F_i)_{i \in J}$ *of functions from* V *into* $\overline{\mathbb{R}}$, *we have*

$$(\inf_{i \in J} F_i)^* = \sup_{i \in J} F_i^*,$$
$$(\sup_{i \in J} F_i)^* \leq \inf_{i \in J} F_i^*.$$

(*iii*) *For every* $a \in V$, *we denote by* F_a *the translated function (i.e.,* $F_a(u) = F(u - a), \forall u \in V$*). Then*

$$F_a^*(f) = F^*(f) + \langle f, a \rangle_{V',V},\ \ \forall f \in V'.$$

□

We can also define the biconjugate or bipolar function of F.

Definition 4.4. *(Biconjugate function) Let* $F : V \longrightarrow \overline{\mathbb{R}}$. *The function* $F^{**} : V \longrightarrow \overline{\mathbb{R}}$ *defined by*

$$F^{**}(u) = \sup_{f \in V'} (\langle f, u \rangle_{V',V} - F^*(f)), \forall u \in V \tag{4.2}$$

is said to be the biconjugate (or the bipolar) function of F. ■

Now, we recall some geometric characterizations of a biconjugate function.

Theorem 4.5. *(Fenchel–Moreau) Let* F *be a function of* V *into* $\overline{\mathbb{R}}$. *Then its biconjugate* $F^{**} \in \Gamma(V)$ *is its* Γ*-regularization (i.e., the pointwise supremum of all affine functions which are less than* F*). In particular the following results hold:*

(*i*) epiF^{**} = clco(epi$F$) *(the closed convex hull of* epi$F$*).*

(*ii*) $F^{**} \leq F$.

(*iii*) *If the function* F *is convex then* F *coincides with its closure, i.e.,*

$$F^{**} = \overline{F}.$$

Proof. By the definition of F^{**} we have easily that $F^{**} \in \Gamma(V)$.
Let now u be any element of V, then

$$F^{**} = \sup_{f \in V'} (\langle f, u \rangle_{V',V} - F^*(f)) = \sup_{f \in V'} \sup_{\{\lambda : F^*(f) \leq \lambda\}} (\langle f, u \rangle_{V',V} - \lambda).$$

Since for all $\lambda \in \mathbb{R}$, $F^*(f) \leq \lambda$ if and only if the continuous affine functions $u \longrightarrow \langle f, u \rangle_{V',V} - \lambda$ are less than F (see Proposition 4.2), we can conclude that F^{**} is supremum of all the affine functions which are less than F and then F^{**} is the *Γ-regularization* of F.

The last statements can be deduced directly from Proposition 2.115. □

A direct consequence is given by Corollary 4.6.

Corollary 4.6. *Let* F *be a function of* V *into* $\overline{\mathbb{R}}$. *If the function* F *is in* $\Gamma(V)$, *then* $F^{**} = F$. □

According to Theorem 4.5 and Proposition 4.3, we can deduce immediately that the repetition of the conjugation operation is limited.

Corollary 4.7. *Let* F *be a function of* V *into* $\overline{\mathbb{R}}$. *Then*

$$F^{***} = F^*.$$

In general manner, if we denote by $F^{(n)*} = (F^{(n-1)*})^*$ *the n-th repetition of the conjugation operation* ($n \geq 2$)*, we have*

$$\begin{aligned} F^{(n)*} &= F^{**}, \text{ if } n \text{ is even}, \\ F^{(n)*} &= F^{*}, \text{ if } n \text{ is odd}. \end{aligned} \tag{4.3}$$

□

Let us end this section with another property of the conjugate functions

Theorem 4.8. *(Fenchel–Rockafellar or Fenchel duality) Let V be a locally convex Hausdorff topological vector space over $\mathbb{R}$ with its dual V'. Let F and G be two proper convex functions of V into $\overline{\mathbb{R}}$. Suppose that there exists $u_0 \in \mathrm{dom}F \cap \mathrm{dom}G$ such that F is continuous in u_0. Then*

$$\inf_{u \in V}(F(u) + G(u)) = \sup_{f \in V'}(-F^*(-f) - G^*(f)).$$

Proof. It follows from Fenchels inequality that for any function H

$$H^*(f) + H(u) \geq \langle f, u \rangle_{V',V} \quad \forall (f, u) \in V' \times V.$$

Consequently, we have that

$$\sup_{f \in V'}(-F^*(-f) - G^*(f)) \leq \inf_{u \in V}(F(u) + G(u))$$

(this fact is usually referred to as *weak duality*).

Denote $p := \inf_{u \in V}(F(u) + G(u))$, $q := \sup_{f \in V'}(-F^*(-f) - G^*(f))$ and $C := \mathrm{epi}F$. To complete the proof, we show that $p \leq q$.

If $p = -\infty$ there is nothing to prove. Suppose now that $p \neq -\infty$.

It is clear that the interior of C: $\mathrm{int}C$ is not empty (because F is continuous in u_0). We introduce now the following sets:

$$A := \mathrm{int}C,$$
$$B := \{(u, \lambda) \in V \times \mathbb{R} : \quad \lambda \leq p - G(u)\}.$$

The set A and B are convex (since F and G are convex, Proposition 2.44) and disjoint (according to the definition of p), therefore, (because of Hahn–Banach's first geometric form) there exists a closed hyperplane H which separates A and B and then separates $\overline{C} = \overline{A}$ (see Proposition 2.3) and B, *i.e.*, there exists a non-zero continuous linear function $f \in V'$ and $(\alpha, \beta) \in \mathbb{R}^2$ such that $H = \{(u, \lambda) \in V \times \mathbb{R} : \quad \langle f, u \rangle_{V',V} + \alpha\lambda = \beta\}$ and

$$\begin{aligned} \langle f, u \rangle_{V',V} + \alpha\lambda \geq \beta \ \ \forall (u, \lambda) \in C, \\ \langle f, u \rangle_{V',V} + \alpha\lambda \leq \beta \ \ \forall (u, \lambda) \in B. \end{aligned} \tag{4.4}$$

By taking $u = u_0$ in the first part of (4.4) and by passing to the limit on λ ($\lambda \longrightarrow +\infty$), we can deduce that $\alpha \geq 0$.

Prove now that $\alpha \neq 0$; for this we proceed by contradiction. Suppose that $\alpha = 0$, then according to (4.4), we arrive at

$$\langle f, u \rangle_{V',V} \geq \beta, \quad \forall u \in \mathrm{dom}F, \quad \text{and} \quad \langle f, u \rangle_{V',V} \leq \beta, \quad \forall u \in \mathrm{dom}G.$$

In particular $\langle f, u_0 \rangle_{V',V} = \beta$ (since $u_0 \in \mathrm{dom}F \cap \mathrm{dom}G$) and then $\langle f, u - u_0 \rangle_{V',V} \geq 0$ for all u in $\mathrm{dom}F$. Consequently, $f = 0$ since $\mathrm{dom}F$ is a neighborhood of u_0. We thus have $\alpha > 0$.

According to (4.4) and dividing by $\alpha > 0$, we obtain easily that

$$\begin{aligned} F^*(-f_\alpha) &\leq -\beta_\alpha, \\ G^*(f_\alpha) &\leq \beta_\alpha - p \end{aligned}$$

and then

$$-F^*(-f_\alpha) - G^*(f_\alpha) \geq p$$

where $f_\alpha = f/\alpha$ and $\beta_\alpha = \beta/\alpha$.

Therefore, $p \leq q$. This completes the proof. □

4.1.2 Examples

1. Let C be a non-empty subset of a topological vector space V and $\mathcal{X}_C$ be its indicator function. Then the conjugate function $\mathcal{X}_C^*$ of $\mathcal{X}_C$ is defined by

$$\mathcal{X}_C^*(f) = \sup_{u \in C} \langle f, u \rangle_{V',V}$$

and is called the *support function* of C. Moreover, if C is a closed and convex set, $\mathcal{X}_C$ is closed and convex, and by the conjugacy theorem (Theorem 4.5) the conjugate of its support function is its indicator function.

2. Let $(V, \| \cdot \|)$ be a Banach space, $(V', \| \cdot \|_*)$ its dual, $\psi_\alpha : t \in \mathbb{R} \longrightarrow | t |^\alpha / \alpha$ and $F_\alpha : V \longrightarrow \mathbb{R}$ such that $F_\alpha(u) = \psi_\alpha(\| u \|)$, where $1 < \alpha < \infty$. Then

$$\begin{aligned} F_\alpha^*(f) &= \sup_{u \in V} \left(\langle f, u \rangle_{V',V} - F_\alpha(u) \right) \\ &= \sup_{u \in V} \left(\langle f, u \rangle_{V',V} - \frac{\| u \|^\alpha}{\alpha} \right) \\ &= \sup_{\lambda \geq 0} \sup_{\|u\| = \lambda} \left(\langle f, u \rangle_{V',V} - \frac{\| u \|^\alpha}{\alpha} \right) \\ &= \sup_{\lambda \geq 0} \sup_{\|u\| = \lambda} \left(\langle f, u \rangle_{V',V} - \frac{\lambda^\alpha}{\alpha} \right) \\ &= \sup_{\lambda \geq 0} \left(\sup_{\|u\| = 1} \left(\langle f, u \rangle_{V',V} \right) \lambda - \frac{\lambda^\alpha}{\alpha} \right) \\ &= \sup_{\lambda \geq 0} \left(\| f \|_* \lambda - \frac{\lambda^\alpha}{\alpha} \right) \quad \text{(by definition of } \| \cdot \|_*). \end{aligned}$$

Hence (by analyzing the function $r(\lambda) := \theta\lambda - \lambda^\alpha/\alpha$, where $\theta := \| f \|_*$ and $\lambda \in [0, +\infty[$), $F_\alpha^*(f) = (\| f \|_*^{\alpha^*})/\alpha^*$, where $1/\alpha^* + 1/\alpha = 1$. Consequently,

$$F_\alpha^*(f) = \psi_{\alpha^*}(\| f \|_*), \quad \text{with } \frac{1}{\alpha^*} + \frac{1}{\alpha} = 1.$$

In the same way we have that $F_1^*(f) = \mathcal{X}_{\mathcal{B}_*}(f)$ where $\mathcal{B}_*$ is the unit closed ball $\mathcal{B}_{V'}(0,1)$.

3. We finish with an interesting example for the boundary-valued problems in a lemma form.

Lemma 4.9. *Let* $(V, \| \,.\, \|)$ *be a Banach space,* $(V', \| \,.\, \|_*)$ *its dual and* C *be a non-empty closed and convex subset of* V*. Consider the convex and lower semi-continuous real-valued function* F *on* V *given by*

$$F(v) := \langle f, v\rangle_{V',V} + \mathcal{X}_C(v-u) \quad \forall v \in V,$$

where $u \in V$ *and* $f \in V'$ *are given elements.*

Then the conjugate function of F *is defined by*

$$F^*(g) = \langle g-f, u\rangle_{V',V} + \mathcal{X}_{C^*}(g-f) \quad \forall g \in V',$$

where $C^* = \{g \in V' : \ \langle g, v\rangle_{V',V} = 0 \ \ \forall v \in C\}$ *(which is said to be the polar set of* C*).*

Proof. Let $g \in V'$, we have that

$$\begin{aligned} F^*(g) &= \sup_{v\in V}(\langle g, v\rangle_{V',V} - \langle f, v\rangle_{V',V} - \mathcal{X}_C(v-u)) \\ &= \sup_{w\in C}\langle g-f, w+u\rangle_{V',V} \\ &= \langle g-f, u\rangle_{V',V} + \sup_{w\in C}\langle g-f, w\rangle_{V',V}. \end{aligned}$$

This completes the proof (since $\sup_{w\in C}\langle g-f, w\rangle_{V',V} = \mathcal{X}_C^*(g-f) = \mathcal{X}_{C^*}(g-f)$). □

4.2 Subdifferentials and Superdifferentials of Extended-value Functions

Let V be a topological vector space and let V' be its topological dual space. We denote by $\langle .,. \rangle_{V',V}$ the bilinear form with respect to the duality between V' and V.

4.2.1 Definition and Characterization

Definition 4.10. *Let* F *be a function of* V *into* $\overline{I\!R}$*. The continuous affine function* L *everywhere less than* F *(minorizing* F *on* V*) is said to be exact at the point* $u \in V$ *if* $L(u) = F(u)$*.* ■

It is clear that we have the following proposition.

Proposition 4.11. *Let* F *be a function of* V *into* $\overline{I\!R}$ *and* L *a continuous affine function that is everywhere less than* F*. If* L *is exact at point* $u \in V$ *then the value* $F(u)$ *is finite and the affine function* L *will have the form:*

there exists $f \in V'$ such that

$$L(v) = \langle f, v-u \rangle_{V',V} + F(u), \forall v \in V. \tag{4.5}$$

Moreover, the function L is maximal and its constant term

$$\beta = -\langle f, u \rangle_{V',V} + F(u)$$

is the greatest $\lambda \in I\!R$ such that $F(w) \geq \langle f, w \rangle_{V',V} + \lambda$, i.e.,

$$\beta = -\langle f, u \rangle_{V',V} + F(u) = -F^*(f), \tag{4.6}$$

where F^* is the conjugate of F. □

Definition 4.12. *(Subdifferential and subgradient) Let F be a function of V into $\overline{I\!R}$ (possibly non convex):*

(i) The function F is said to be subdifferentiable at the point $u \in V$ if it has a continuous affine minorant L which is exact at the point u (an affine function minorizing F and coinciding with F at u).

(ii) A linear form $f \in V'$ defined by (4.5) is called a subgradient of F at u, and the set of all these subgradients of F at u is called the subdifferential of F at u and is denoted by $\partial F(u)$. ■

A direct corollary is the following characterization.

Corollary 4.13. *Let F be a function of V into $\overline{I\!R}$. The function F is subdifferentiable at a point $u \in V$ if $\partial F(u) \neq \emptyset$.* □

Remark 4.14. (*i*) The subdifferential ∂F is a set-valued mapping.

(*ii*) If $F(u) = +\infty$ and F is not identically equal to $+\infty$ then $\partial F(u)$ is empty. On the other hand, if F is identically equal to $+\infty$, the set $\partial F(u)$ is non-empty, for all $u \in V$.

(*iii*) Compared to the classical notion of the differentiability which is local and requires some regularity, the subdifferentiability concept is global and the function F may not be necessarily continuous at the point u so that $\partial F(u)$ is non-empty.

(*iv*) For $u \in V$, consider the function R by $R(v) := F(u) + \langle f, v-u \rangle_{V',V}$, for all $v \in V$, then F is subdifferentiable at u and $f \in \partial F(u)$ if and only if the graph of R (*i.e.*, graph$(R) = \{(v, R(v)) : \ v \in V\}$) is a supporting hyperplane of the epigraph of F, epi(F) in $(u, F(u))$. ◇

We shall give now the characterization of the subdifferential of F at u, $\partial F(u)$.

Theorem 4.15. *Let $F : V \longrightarrow \overline{I\!R}$ be a subdifferentiable function at point $u \in V$. Then*

$$f \in \partial F(u) \text{ if and only if } F(u) \text{ is finite and} \\ \langle f, v-u\rangle_{V',V} + F(u) \leq F(v), \quad \forall v \in V. \tag{4.7}$$

Otherwise,

$$\partial F(u) = \{f \in V' : \quad F(v) \geq F(u) + \langle f, v-u\rangle_{V',V}, \quad \text{for all } v \in V\}. \tag{4.8}$$

Moreover,

$$\text{if } \partial F(u) \text{ is non-empty then } F(u) = F^{**}(u), \tag{4.9}$$

$$\text{if } F(u) = F^{**}(u) \text{ then } \partial F(u) = \partial F^{**}(u). \tag{4.10}$$

□

By using the definition of the conjugate function of F and the relation (4.6) we have also the following characterization of $\partial F(u)$.

Proposition 4.16. *Let F be a function of V into $\overline{I\!R}$ and F^* its conjugate. Then*

$$f \in \partial F(u) \text{ if and only if } F(u) + F^*(f) = \langle f, u\rangle_{V',V}. \tag{4.11}$$

Proof. Suppose that, for $u \in V$ and $f \in V'$ we have that $F(u) + F^*(f) = \langle f, u\rangle_{V',V}$. Then, by Fenchel's inequality, *i.e.*, $F(v)+F^*(f) \geq \langle f, v\rangle_{V',V} \;\; \forall v \in V$, we can deduce that $F(u) - F(v) \leq \langle f, u-v\rangle_{V',V} \;\; \forall v \in V$. Consequently, according to Theorem 4.15, $f \in \partial F(u)$.

Conversely, consider $u \in V$ and $f \in \partial F(u)$ then, according to Theorem 4.15, $F(u) - F(v) \leq \langle f, u-v\rangle_{V',V} \;\; \forall v \in V$. So, $F(u) + (\langle f, v\rangle_{V',V} - F(v)) \leq \langle f, u\rangle_{V',V} \;\; \forall v \in V$ and then

$$F(u) + F^*(f) = F(u) + \sup_{v \in V}(\langle f, v\rangle_{V',V} - F(v)) \leq \langle f, u\rangle_{V',V}.$$

Because of Fenchel's inequality again we have that $F(u)+F^*(f) = \langle f, u\rangle_{V',V}$.

□

Example 4.17. Let $(V, \|\ .\ \|)$ be a Banach space with its dual $(V', \|\ .\ \|_*)$ and $F : V \longrightarrow I\!R$ such that $F(u) :=\| u \|$ for all $u \in V$. Then

$$\partial F(0) = \{f \in V' : \quad \| f \|_* \leq 1\}, \\ \partial F(u) = \{f \in V' : \quad \| f \|_* = 1, \ \langle f, u\rangle_{V',V} = \| u \|\}, \quad \text{if } u \neq 0. \tag{4.12}$$

The proof is left as an exercise. ♣

It is clear that the following statements hold.

Corollary 4.18. *Let $F : V \longrightarrow \overline{I\!R}$ be a subdifferentiable function at point $u \in V$. Then the set $\partial F(u)$ is convex and closed in V' with respect to the topology $\sigma(V', V)$.* □

Corollary 4.19. *Let $F : V \longrightarrow \overline{I\!R}$ be any extended real-valued function on V. A point $u \in V$ is a minimizer of F, i.e., $F(u) = \inf_{v \in V} F(v)$ if and only if $0 \in \partial F(u)$.* □

Corollary 4.20. *Let F be a proper function of V into $\overline{I\!R}$ and subdifferentiable at point $u \in V$. Then the following properties hold:*

(i) If $f \in \partial F(u)$ then $u \in \partial F^*(f)$.

(ii) If, furthermore, $F \in \Gamma(V)$ then: $f \in \partial F(u)$ if and only if $u \in \partial F^*(f)$. □

Remark 4.21. The subgradient funtional ∂F is monotone, *i.e.*,

$$\langle f_1 - f_2, u_1 - u_2 \rangle_{V',V} \geq 0 \ \ \forall u_i \in \mathrm{dom}(\partial F), f_i \in \partial F(u_i), i = 1, 2.$$

Indeed, by the characterization of ∂F, we have that

$$F(u_1) - F(u_2) \leq \langle f_1, u_1 - u_2 \rangle_{V',V} \text{ and } F(u_2) - F(u_1) \leq \langle f_2, u_2 - u_1 \rangle_{V',V}.$$

Consequently, by adding the previous relations, we obtain that

$$0 \leq \langle f_1 - f_2, u_1 - u_2 \rangle_{V',V}.$$

This implies the result. ◊

In the case of convex functions we have the following criterion for subdifferentiability.

Proposition 4.22. *Let F be a proper and convex function of V into $\overline{I\!R}$, finite valued and continuous at the point $u \in V$. Then $\partial F(v) \neq \emptyset$ for all v in the interior of the effective domain of F dom$F$ (in particular $\partial F(u) \neq \emptyset$).*

Proof. The main idea of the proof is based essentially on the usage of Hahn–Banach separation in order to separate the point $(u, F(u))$ (element of the boundary of epi$F$) and the interior of epi$F$. The proof is left to the reader as an exercise (he can be inspired by the proof of Theorem 4.8). □

We can also introduce the notion of superdifferential and supergradient that occurs naturally in the case of concave functions and maximization problems.

Definition 4.23. *(Superdifferential and supergradient) Let G be a function of V into $\overline{I\!R}$ (possibly non concave). The function G is said to be superdifferentiable at the point $u \in V$ if there exists a linear form $f \in V'$ such that*

$$G(v) \leq G(u) + \langle f, v - u \rangle_{V',V}, \quad \text{for all } v \in V. \tag{4.13}$$

A linear form $f \in V'$ defined by (4.13) is called a supergradient of G at u, and the set of all these supergradients of G at u is called the superdifferential of G at u and is denoted by $\hat{\partial} G(u)$. ■

Definition 4.24. *(Concave conjugate) Let G be a function of V into $\overline{\mathbb{R}}$. The function $G^{*c} : V' \longrightarrow \overline{\mathbb{R}}$ defined by*

$$G^{*c}(f) = \sup_{u \in V}(G(u) - \langle f, u\rangle_{V',V}) \text{ for all } f \in V', \tag{4.14}$$

is said to be the concave conjugate function of G. ■

Remark 4.25. A similar result, at the subdifferentiability case, can be given for the superdifferentiability case in terms of its concave conjugate because

$$(-G)^*(f) = -G^{*c}(-f), \forall f \in V' \text{ and } \partial(-G)(u) = -\widehat{\partial}G(u), \forall u \in V.$$

In particular, a point $u \in V$ is a maximizer of an extended real-valued function G of V into $\overline{\mathbb{R}}$, *i.e.*, $G(u) = \sup_{v \in V} G(v)$ if and only if $0 \in \widehat{\partial}G(u)$. ♢

4.2.2 General Case

In this section we give some generalization of the explicit results in Section 4.2.1 to the case of general functions $c(.,.)$. For the results we refer, for example, to Dietrich [103]. This notion plays, for example, an important role in the study of the Monge–Kantorovich mass transportation problem (see, for instance, Rachev and Ruschendorf [248]).

Let V and X be two topological spaces and $c : V \times X \longrightarrow \overline{\mathbb{R}}$ be an extended real-valued function.

Example 4.26. A particular case is $X = V'$ the dual of V and

$$c(u, f) = \langle f, u\rangle_{V',V}.$$

Another particular interest is the case where $(X, \| . \|) = (V, \| . \|)$ is a Banach space and

$$c(u, v) = \frac{1}{p} \| u - v \|^p, \quad p \geq 1.$$

♣

Definition 4.27. *(c-convex function) Let C be a non-empty convex subset of V and F be a mapping of C into $\overline{\mathbb{R}}$.*

The mapping F is said to be c-convex on C if F can be written as

$$F(u) := \sup_{i \in J}(c(u, f_i) + \alpha_i), \quad \forall u \in V \tag{4.15}$$

for some index set J, $f_i \in X$ and $\alpha_i \in \overline{\mathbb{R}}$ for $i \in J$. ■

Definition 4.28. *The set of functions $F : V \longrightarrow \overline{\mathbb{R}}$ which are c-convex and lower semi-continuous, such that if F takes the value $-\infty$ then F is identically equal to $-\infty$, is denoted by $\Gamma^c(V)$ and the subset of $\Gamma^c(V)$ other than the constant functions $-\infty$ and $+\infty$ is denoted by $\Gamma_0^c(V)$.* ■

Definition 4.29. *(c-conjugate and double c-conjugate function) Let us consider an extended real-valued function* $F: V \longrightarrow \overline{I\!R}$*:*

(i) *The function* $F^c: X \longrightarrow \overline{I\!R}$ *defined by*

$$F^c(f) := \sup_{v \in V}(c(v, f) - F(v)) \quad \text{for all } f \in X, \tag{4.16}$$

is said to be the c-conjugate or the c-polar of F.

(ii) *The function* $F^{cc}: V \longrightarrow \overline{I\!R}$ *defined by*

$$F^{cc}(u) := \sup_{g \in X}(c(u, g) - F^c(g)) \quad \text{for all } u \in V, \tag{4.17}$$

is said to be the double c-conjugate of F. ■

We can now introduce the Fenchel-type inequality for the general case.

Proposition 4.30. *(Fenchel-type inequality) Let* $F: V \longrightarrow \overline{I\!R}$ *be a given extended real-valued function. Then*

$$F^c(f) + F(u) \geq c(u, f), \forall f \in X, \forall u \in V.$$

□

The following results are analogous to the results obtained in the case of the classical conjugate (which do not depend on the bilinearity of form $\langle ., . \rangle$).

Proposition 4.31. *Let* F *and* G *be two given extended real-valued functions of* V *onto* $\overline{I\!R}$*. Then the following properties hold:*

(i) *If* F *is less than* G *then* G^c *is less than* F^c.

(ii) $(F + \beta)^c = F^c - \beta, \quad \forall \beta \in I\!R$.

(iii) $F^c \in \Gamma^c(X)$ *and* $F^{cc} \in \Gamma^c(V)$.

(iv) F^{cc} *is less than* F.

(v) $\begin{cases} F^{(n)c} = F^{cc}, & \text{if } n \text{ is even}, \\ F^{(n)c} = F^c, & \text{if } n \text{ is odd}. \end{cases}$ □

Proposition 4.32. *Let* $F: V \longrightarrow \overline{I\!R}$ *be an extended real-valued function. Then the following statements are true:*

(i) F^{cc} *is the largest c-convex minorant of* F.

(ii) $F \in \Gamma^c(V)$ *if and only if* $F = F^{cc}$. □

Definition 4.33. *(c-subdifferentiability of function) Let* $F: V \longrightarrow \overline{I\!R}$ *be an extended real-valued function. The c-subdifferential of* F *at point* u*, the set-valued function* $\partial_c F: V \longrightarrow X$*, is defined by*

$$\partial_c F(u) = \{f \in X : F(u) - F(v) \leq c(u, f) - c(v, f), \text{ for all } v \in V\}. \tag{4.18}$$

Any element $f \in \partial_c F(u)$ *is called a c-subgradient of* F *at* u *and if* $\partial_c F(u) \neq \emptyset$ *then* F *is c-subdifferentiable at* u. ■

It is obvious that the following statement holds.

Proposition 4.34. *Let* $F : V \longrightarrow \overline{I\!R}$ *be an extended real-valued function. Then the following properties hold:*

(i) $f \in \partial_c F(u)$ *if and only if* $F(u) + F^c(f) = c(u, f)$.

(ii) If, furthermore, $F \in \Gamma^c(V)$, *then* $f \in \partial_c F(u)$ *if and only if* $u \in \partial_c F^c(f)$.

□

4.2.3 Calculus Rules with Subdifferentials

In this section, we give some results of calculus with subdifferentials which are the extension of the usual differential calculus.

Proposition 4.35. *(i) Let* $F : V \longrightarrow \overline{I\!R}$ *be a subdifferentiable function at point* $u \in V$ *and* α *be positive. Then*

$$\partial(\alpha F)(u) = \alpha \partial F(u).$$

(ii) Let $F_i : V \longrightarrow \overline{I\!R}, i = 1, 2$ *be two subdifferentiable functions at point* $u \in V$. *Then*

$$(\partial F_1(u) + \partial F_2(u)) \subset \partial(F_1 + F_2)(u).$$

□

Proposition 4.36. *(Moreau–Rockafellar) Let* $F_i : V \longrightarrow \overline{I\!R}, i = 1, 2$ *be two functions of* $\Gamma(V)$ *and* $\alpha_i, i = 1, 2$ *be positive. If there is a point* $u_0 \in \mathrm{dom}F_1 \cap \mathrm{dom}F_2$ *where* F_1 *is continuous, then*

$$\partial(\alpha_1 F_1 + \alpha_2 F_2)(u) = \alpha_1 \partial F_1(u) + \alpha_2 \partial F_2(u), \quad \forall u \in V.$$

Proof. We prove first that $\partial(F_1 + F_2)(u) = \partial F_1(u) + \partial F_2(u)$, $\forall u \in V$. According to the result *(ii)* of Proposition 4.35 we have that $\partial F_1(u) + \partial F_2(u) \subset \partial(F_1 + F_2)(u)$, $\forall u \in V$. We have now to show that the inverse inclusion holds, *i.e.*, to prove that for each $g \in \partial(F_1 + F_2)(u)$, there exists $(g_1, g_2) \in \partial F_1(u) \times \partial F_2(u)$ such that $g = g_1 + g_2$.

According to Theorem 4.15, we have that $F_1(u)$ and $F_2(u)$ are finite and that

$$F_1(v) + F_2(v) \geq F_1(u) + F_2(u) + \langle g, v - u \rangle_{V',V}, \quad \forall v \in V. \tag{4.19}$$

Consider the convex and continuous function at u_0

$$G(.) := F_1(.) - \langle g, . \rangle_{V',V} - F_1(u) + \langle g, u \rangle_{V',V}$$

and the convex set with a non-empty interior $C := \mathrm{epi}G$. We now introduce the following sets

$$A := \mathrm{int}C,$$

$$B := \{(v, \lambda) \in V \times \mathrm{I\!R} : \ \lambda \leq F_2(u) - F_2(v)\}.$$

The set A and B are convex (since F and G are convex, see Proposition 2.44) and disjoints. To end the proof, we can use (again!) Hahn–Banach's separation (which is left to the reader as an exercise).

According to the result (*i*) of Proposition 4.35 and to the previous result, we can obtain easily that, for all positive constants $\alpha_i, i = 1, 2$ and for all $u \in V$,

$$\partial(\alpha_1 F_1 + \alpha_2 F_2)(u) = \partial(\alpha_1 F_1)(u) + \partial(\alpha_2 F_2)(u) = \alpha_1 \partial F_1(u) + \alpha_2 \partial F_2(u).$$

This completes the proof. □

We shall now examine the subdifferential of a composition function with an affine mapping. For this, let us consider two pairs of topological vector spaces V, V' and W, W' in duality with respect to certain bilinear forms $\langle .,.\rangle_{V',V}$ and $\langle .,.\rangle_{W',W}$, respectively. Let $\Lambda : V \longrightarrow W$ be a continuous linear mapping (*geometrical operator*) with the transpose (adjoint) $\Lambda^* : W' \longrightarrow V'$ defined by

$$\langle g, \Lambda u\rangle_{W',W} = \langle \Lambda^* g, u\rangle_{V',V} \quad \forall (g,u) \in W' \times V. \tag{4.20}$$

Remark 4.37. (*i*) The two paired dual spaces V, V' and W, W' are linked respectively by a so-called *geometrical equation*

$$p = \Lambda u \tag{4.21}$$

and a so-called *equilibrium equation*

$$g = \Lambda^* f. \tag{4.22}$$

(*ii*) In calculus of variations, if Λ is a gradient-like operator "grad" then Λ^* should be a divergence-like operator "div". ♢

Proposition 4.38. *Let $F : W \longrightarrow \overline{\mathbb{R}}$ be a function of $\Gamma(W)$ and suppose that there exists a point $u_0 \in V$ such that F is continuous and finite at point Λu_0. Then the function $F \circ \Lambda : V \longrightarrow \mathbb{R}$ (belong to $\Gamma(V)$) is a subdifferential function on V and satisfies*

$$\partial(F \circ \Lambda)(u) = \Lambda^* \partial F(\Lambda u), \quad \textit{for all } u \subset V. \tag{4.23}$$

Proof. The first inclusion, *i.e.*, $\Lambda^* \partial F(\Lambda u) \subset \partial(F \circ \Lambda)(u)$ is based on the fact that for $g \in \Lambda^* \partial F(\Lambda u)$, we have (by the definition of the subdifferential)

$$\langle \Lambda^* g, v - u\rangle_{V',V} + F \circ \Lambda(u) = \langle g, \Lambda(v-u)\rangle_{W',W} + F \circ \Lambda(u) \leq F \circ \Lambda(v) \ \forall v \in V.$$

The second inclusion $\partial(F \circ \Lambda)(u) \subset \Lambda^* \partial F(\Lambda u)$ is based on the Hahn–Banach's separation (which is left to the reader as an exercise). □

Remark 4.39. The result of Proposition 4.38 holds in the case of affine functions. Precisely, let $F: W \longrightarrow \overline{\mathbb{R}}$ be a function of $\Gamma(W)$ and A be an affine mapping, *i.e.*, $A: V \longrightarrow W$ such that $Au = \Lambda u - \mu$, where $\mu \in W$; if there exists a point $u_0 \in V$ such that F is continuous and finite at point Au_0, then the function $F \circ A: V \longrightarrow \mathbb{R}$ (belong to $\Gamma(V)$) is a subdifferential function on V and satisfies

$$\partial(F \circ A)(u) = \Lambda^* \partial F(Au), \quad \text{for all } u \in V.$$

◇

The subdifferentiability of a convex function F at a given point is connected with the directional derivative of F at this point. Precisely, it is closely connected with the other classical differentiability concepts such as Gâteaux or Fréchet derivative.

4.2.4 Connection with Directional Derivative

Differentiablility of Extended Real-valued Functions

Definition 4.40. *(Gâteaux-differentiability or G-differentiability) Let F be a function of V into $\overline{\mathbb{R}}$:*

(i) The limit as $t \longrightarrow 0_+$, if it exists, of the difference quotient of F

$$\frac{F(u+tv)-F(u)}{t} \quad \textit{for } u, v \in V, \tag{4.24}$$

is said to be the directional derivative (or Gâteaux variation) of F at u in the direction v and is denoted by $\delta F(u, v)$.

(ii) If furthermore, there exists $f \in V'$ such that

$$\delta F(u, v) = \langle f, v \rangle_{V',V} \quad \textit{for all } v \in V,$$

we say that F is Gâteaux(or G)-differentiable at u, and the Gâteaux(or G)-differential at u of F, is denoted by $F'(u)$ or by $DF(u)$ (in finite dimensional space $F'(u)$ is denoted by ∇F or by grad $F(u)$*) and is characterized by*

$$\langle F'(u), v \rangle_{V',V} := \delta F(u, v) = \lim_{t \longrightarrow 0_+} \frac{F(u+tv)-F(u)}{t} \quad \forall v \in V. \tag{4.25}$$

■

Remark 4.41. If the function F is convex, then the difference quotient (4.24) is a monotonically increasing function of the stepsize $t \in]0, +\infty[$ and its limit for $t \longrightarrow 0_+$ exists, which, however, can be $-\infty$ or $+\infty$. ◇

Definition 4.42. *(Fréchet-differentiability) Let F be a function of V into $\overline{\mathbb{R}}$. The function F is said to be Fréchet(or F)-differentiable at a given point $u \in V$ if the difference quotient (4.24) as a function of t converges uniformly on every bounded set.* ■

We will note that the F-differentiability of a mapping F implies the G-differentiability of F, but the converse is not true. Moreover, the F-differentiability of a mapping F implies the continuity of F, but the G-differentiability of F does not imply necessarily the continuity of F. However, there are some cases where G-differentiability implies F-differentiability.

Suppose now that $(V, \| \,.\, \|)$ is a Banach space and let F be a function of K (an open subset of V) into $\mathbb{R}$. We say that F is a class $\mathcal{C}^1$ functional on K if the F-derivative F' exists at every point $u \in K$ and the mapping $u \longrightarrow F'(u)$ is continuous from K into V', *i.e.*, if $\lim_{n \longrightarrow \infty} u_n = u \in K$ then

$$\lim_{n \longrightarrow \infty} \langle F'(u_n) - F'(u), v \rangle_{V',V} = 0 \text{ uniformly on } \mathcal{B}_V(0,1).$$

We have the following result:

If F has a continuous G-derivative on K then F is F-differentiable on K and F is of class $\mathcal{C}^1$ on K.

This result follows by using the following simple manipulation for $u, v \in K$:

$$F(u) - F(v) - \langle F'(u), u - v \rangle_{V',V} := \int_0^1 \langle F'(tu + (1-t)v) - F'(u), u - v \rangle_{V',V} \mathrm{d}t.$$

Example 4.43. Let Ω be an open subset of $\mathbb{R}^d$, $d \geq 1$, with finite measure, *i.e.*, $| \, \Omega \, | < \infty$, and let a functional $\phi : \; V = L^{p+1}(\Omega) \longrightarrow \mathbb{R}$, $p \in]1, \infty[$, is defined by $(\forall u \in L^{p+1}(\Omega))$

$$\phi(u) = \frac{1}{p+1} \int_\Omega | \, u(x) \, |^{p+1} \, \mathrm{d}x.$$

Then the functional ϕ is of class $\mathcal{C}^1$ on $L^{p+1}(\Omega)$ and we have

$$\langle \phi'(u), h \rangle_{V',V} = \int_\Omega u(x) \, | \, u(x) \, |^{p-1} \, h(x) \mathrm{d}x \quad \forall h \in L^{p+1}(\Omega),$$

where $V' = L^{\frac{p+1}{p}}(\Omega)$.

Indeed, let u and h be in $L^{p+1}(\Omega)$ and $t \in (0,1)$, we have

$$\begin{aligned} &\frac{| \, u(x) + th(x) \, |^{p+1} - | \, u(x) \, |^{p+1}}{t(p+1)} \\ &\quad = \int_0^1 | \, s(u(x) + th(x)) + (1-s)u(x) \, |^p \\ &\qquad\qquad \mathrm{sign}(s(u(x) + th(x)) + (1-s)u(x)) h(x) \mathrm{d}s \\ &\quad = \int_0^1 | \, u(x) + sth(x) \, |^p \, \mathrm{sign}(u(x) + sth(x)) h(x) \mathrm{d}s \end{aligned}$$

and then

$$\left| \frac{| u(x) + th(x) |^{p+1} - | u(x) |^{p+1}}{t(p+1)} \right| \leq || u(x) | + | h(x) ||^p | h(x) | .$$

Using Hölder's inequality, we can deduce that

$$\begin{aligned}
&\int_\Omega | (| u(x) | + | h(x) |) |^p | h(x) | \, \mathrm{d}x \\
&\quad \leq \left(\int_\Omega | (| u(x) | + | h(x) |) |^{p+1} \, \mathrm{d}x \right)^{\frac{p}{p+1}} \left(\int_\Omega | h(x) |^{p+1} \, \mathrm{d}x \right)^{\frac{1}{p+1}} \\
&\quad \leq C \left(\| u \|^{p+1}_{L^{p+1}(\Omega)} + \| h \|^{p+1}_{L^{p+1}(\Omega)} \right)^{\frac{p}{p+1}} \| h \|^{p+1}_{L^{p+1}(\Omega)} < \infty
\end{aligned}$$

and then the function $| (| u(.) | + | h(.) |) |^p | h(.) |$ is in $L^1(\Omega)$.

Because of Lebesgue's theorem (the dominated convergence theorem), it follows that

$$\begin{aligned}
\langle \phi'(u), h \rangle_{V',V} &= \lim_{t \longrightarrow +} \frac{\phi(u+th) - \phi(t)}{t} \\
&= \lim_{t \longrightarrow +} \int_\Omega \int_0^1 | u(x) + sth(x) |^p \operatorname{sign}(u(x) + sth(x)) h(x) \mathrm{d}s \mathrm{d}x \\
&= \int_\Omega | u(x) |^p \operatorname{sign}(u(x)) h(x) \mathrm{d}x \\
&= \int_\Omega | u(x) |^{p-1} u(x) h(x) \mathrm{d}x.
\end{aligned}$$

The continuity of the mapping $u \longrightarrow \phi'(u)$ is a direct consequence of the continuity of the well-known *Nemytskii operator*

$$N_f : L^{p+1}(\Omega) \longrightarrow L^{\frac{p+1}{p}}(\Omega)$$

defined by

$$N_f u(x) := f(x, u(x)),$$

which is generated by the Carathéodory function $f : (x, s) \in \Omega \times \mathrm{I\!R} \longrightarrow s \mid s \mid^p$ (see below).

This completes the proof. ♣

Now we give a relation between subdifferentiability and G-differentiability of a given convex function.

Proposition 4.44. *Let F be a convex function of V into $\overline{\mathrm{I\!R}}$. If F is Gâteaux-differentiable at the point $u \in V$, it is subdifferentiable at u and $\partial F(u) = \{F'(u)\}$ ($\partial F(u)$ has exactly one element). Conversely, if at the point $u \in V$, F is continuous and finite and has only one subgradient, then F is Gâteaux-differentiable at the point u and $\partial F(u) = \{F'(u)\}$.*

Proof. For the proof see Rockafellar [252]. □

We shall now give some characterizations of the convexity of the G-differentiable function F.

Proposition 4.45. *(Inequality of the G-differential of a convex function) Let C be a non-empty convex subset of V and F be a G-differentiable function of C into $I\!R$. Then F is convex over C if and only if*

$$F(v) \geq F(u) + \langle F'(u), v-u\rangle_{V',V} \quad \text{for all } (u,v) \in C \times C \tag{4.26}$$

and F is strictly convex over C if and only if (4.26) holds as a strict inequality if $u \neq v$. □

Proposition 4.46. *(Monotonicity of the G-differential of a convex function) Let C be a non-empty convex subset of V and F be a G-differentiable function of C into $I\!R$. Then:*

(i) F is a convex function over C if and only if

$$\langle F'(u) - F'(v), v-u\rangle_{V',V} \geq 0 \quad \text{for all } (u,v) \in C \times C, \tag{4.27}$$

i.e., if F' is a monotone mapping of V into V'.

(ii) F is a strictly convex function over C if and only if (4.27) holds as a strict inequality if $u \neq v$, i.e., if F' is a strictly monotone mapping of V into V'. □

In the same way we can give some characterizations of the concavity of a G-differentiable function F by considering the functional $(-F)$.

Some Differentiable Functionals

We consider, in this subsection some basic results on the Nemytskii operator and certain differentiable functionals. The proofs of these results can be found, for instance, in Berger[49], Figueiredo [123], Kavian [171], Krasnoselskii [182] or Vainberg [286].

Definition 4.47. *(Carathéodory function) Let D be an open domain of $I\!R^m$, $m \in I\!N^*$, V and W be two Banach spaces and F be a function from $D \times V$ into W.*
F is said to be a Carathéodory mapping if and only if the function F satisfies the following conditions:

(i) for each u in V, the function $\eta \longrightarrow F(\eta, u)$ is measurable in D

(ii) for the almost everywhere element η in D, the function $u \longrightarrow F(\eta, u)$ is continuous in V. ■

Let Ω be an open subset of $\mathrm{I\!R}^d$, $d \in \mathrm{I\!N}^*$, with finite measure, *i.e.*, $|\,\Omega\,| < \infty$, and $\mathcal{M}$ be the space of real-valued measurable functions on Ω. We then have the following results.

Proposition 4.48. *Let $f : \Omega \times I\!R \longrightarrow I\!R$ be a Carathéodory function. Then for each function $u \in \mathcal{M}$, the function $N_f u : \Omega \longrightarrow I\!R$, which is generated by the function f in the sense (for a.e. $x \in \Omega$)*

$$N_f u(x) := f(x, u(x)),$$

is measurable in Ω.

This operator is said to be a Nemytskii operator (or superposition operator). □

The next result shows sufficient conditions on the continuity when a Nemytskii operator is a mapping from L^p space into L^q space, for $1 \leq p, q < \infty$.

Proposition 4.49. *Let Ω be an open and bounded domain in $I\!R^d$, $d \in I\!N^*$, and let $f : \Omega \times I\!R \longrightarrow I\!R$ be a Carathéodory function. Assume that the function f satisfies the following growth condition:*

$$| f(x, s) | \leq C | s |^r + a(x) \quad a.e\ x \in \Omega \quad and \quad for\ all\ s \in I\!R, \tag{4.28}$$

where $C \geq 0$, $r > 0$ are real constants and a is a function in $L^q(\Omega)$ $(1 \leq q < \infty)$. Then N_f is a continuous function from $L^p(\Omega)$ into $L^q(\Omega)$, where $p = rq$, and maps bounded subsets into bounded subsets. □

Conversely, we have the following proposition.

Proposition 4.50. *Let Ω be an open and bounded domain in $I\!R^d$, $d \in I\!N^*$, and let $f : \Omega \times I\!R \longrightarrow I\!R$ be a Carathéodory function.*

If the function N_f maps $L^p(\Omega)$ into $L^q(\Omega)$ $(1 \leq p, q < \infty)$, then N_f is a continuous function. □

The previous result is a particular case of the more general results of Krasnoselskii [182]

Theorem 4.51. *(Krasnoselskii's theorem) Let V and W be two Banach spaces, D be a Borel subset of $I\!R^m$, $m \in I\!N^*$, and F be a Carathéodory mapping from $D \times V$ into W. For each function $u \in \mathcal{M}$, let $\mathcal{G}u : D \longrightarrow W$ be the measurable function, which is generated by the function F in the sense (for a.e. $\eta \in \mathcal{D}$)*

$$\mathcal{G}u(\eta) := F(\eta, u(\eta)). \tag{4.29}$$

If $\mathcal{G}$ maps $L^p(D; E)$ into $L^q(D; E)$ $(1 \leq p, q < \infty)$, then $\mathcal{G}u$ is a continuous function. □

We present now some results concerning differentiable functionals (defined on L^p-spaces), which are important in the calculus of variations. A direct corollary of Krasnoselskii's theorem is the following result.

Corollary 4.52. *Let D be a bounded open subset of $I\!R^m$, $m \in I\!N^*$, and F be a Carathéodory mapping from $D \times I\!R^l$ into $I\!R$, $l \in I\!N^*$. If $\mathcal{G}$, defined by (4.29),* *maps* $(L^p(D))^l$ *into* $L^1(D)$ $(1 \leq p < \infty)$*, then the function* $P_{\mathcal{G}} : (L^p(D))^l \longrightarrow$ $I\!R$ *defined by*

$$P_{\mathcal{G}}(u) = \int_{\Omega} F(x, u(x))\mathrm{d}x, \tag{4.30}$$

is continuous. □

Concerning the potentiality of the Nemitskii operator, the following results hold.

Proposition 4.53. *Let* Ω *be an open and bounded domain in* $I\!R^d$, $d \in I\!N^*$ *and let* $F : \Omega \times I\!R \longrightarrow I\!R$ *be a Carathéodory function such that*

$$| F(x,s) | \leq C | s |^p + a(x) \quad a.e.\ x \in \Omega \ \text{ and } \ for\ all\ s \in I\!R, \tag{4.31}$$

where $C \geq 0$, $p \geq 1$ *are real constants and* a *is a function in* $L^1(\Omega)$*. Then the functional* $P_F : L^p(\Omega) \longrightarrow I\!R$ *defined by*

$$P_F(u) = \int_{\Omega} F(x, u(x))\mathrm{d}x, \tag{4.32}$$

is continuous.

In particular, for a Carathéodory function $f : \Omega \times I\!R \longrightarrow I\!R$ *such that*

$$| f(x,s) | \leq C | s |^{p-1} + b(x) \quad a.e.\ x \in \Omega \ \text{ and } \ for\ all\ s \in I\!R, \tag{4.33}$$

where $C \geq 0$, $p > 1$ *are constants and* b *is a function in* $L^q(\Omega)$, $1/p + 1/q = 1$, *we have that the function* $F : \Omega \times I\!R \longrightarrow I\!R$ *defined by* $F(x,s) := \int_0^s f(x,\tau)\mathrm{d}\tau$ *is a Carathéodory function and satisfies the condition (4.31). Moreover, the functional* P_F *is continuously F-differentiable from* $L^p(\Omega)$ *into* $I\!R$ *and*

$$P_F'(u) = f(., u(.)) = N_f u \in L^q(\Omega).$$

□

According to Proposition 4.50 and Proposition 4.53, we can deduce the following proposition.

Proposition 4.54. *Let* Ω *be an open and bounded domain in* $I\!R^d$, $d \in I\!N^*$*, then the following results hold:*

(i) If $F : \Omega \times I\!R \longrightarrow I\!R$ *is a Carathéodory function such that* N_F *maps* $L^p(\Omega)$ *into* $L^1(\Omega)$, $1 \leq p < \infty$*, then the functional* P_F *defined by (4.32), is continuously on* $L^p(\Omega)$*.*

(ii) If $f : \Omega \times I\!R \longrightarrow I\!R$ *is a Carathéodory function such that the function* N_f *maps* $L^p(\Omega)$ *into* $L^q(\Omega)$, $1 < p < \infty$, $1/p + 1/q = 1$*, then the function* $F : \Omega \times I\!R \longrightarrow I\!R$ *defined by* $F(x,s) := \int_0^s f(x,\tau)\mathrm{d}\tau$ *is a Carathéodory function and* N_F *maps* $L^p(\Omega)$ *into* $L^1(\Omega)$*. Moreover, the functional* P_F *is continuously F-differentiable on* $L^p(\Omega)$ *and*

$$P_F'(u) = f(., u(.)) = N_f u \in L^q(\Omega).$$

□

Differentiability of Operators and Second-order Derivative

The G (or F)-differentiability also works for operators.

Definition 4.55. *(G-differentiability of an operator) Let V and W be locally convex topological vector spaces and $K \subset V$ be an open subset. Let $\mathcal{D}$ be a function of V into W.*

(i) The limit as $t \longrightarrow 0_+$, if it exists, of the difference quotient of the operator $\mathcal{D}$

$$\frac{\mathcal{D}(u + t\psi) - \mathcal{D}(u)}{t} \tag{4.34}$$

is said to be the directional derivative (or Gâteaux variation) of the operator $\mathcal{D}$ at point $u \in K$ in the direction ψ and is denoted by $\delta\mathcal{D}(u, \psi)$.

(ii) If the previous limit exists for all $\psi \in V$, we say that the operator $\mathcal{D}$ is Gâteaux(or G)-differentiable operator at u. The Gâteaux(or G)-differential at u of the operator $\mathcal{D}$ is an operator from V to W denoted by $\mathcal{D}'(u)$ (or by $D\mathcal{D}(u)$) and defined by

$$\begin{aligned} &\mathcal{D}'(u): \ V \longrightarrow W \ \textit{such that} \\ &\mathcal{D}'(u)\psi := \delta\mathcal{D}(u, \psi) = \lim_{t \longrightarrow 0_+} \frac{\mathcal{D}(u + t\psi) - \mathcal{D}(u)}{t} \quad \forall \psi \in V. \end{aligned} \tag{4.35}$$

■

The operator $\mathcal{D}'(u)$ is 1-homogeneous (*i.e.*, $\mathcal{D}'(u)(\alpha\psi) = \alpha\mathcal{D}'(u)(\psi)$ for all $(\alpha, \psi) \in \mathbb{R} \times V$), but is not additive in the general case and the linearity of G-derivative is a nontrivial requirement in the general case.

Definition 4.56. *(F-differentiability of operators) Let $\mathcal{D}$ be a function of V into W. The operator $\mathcal{D}$ is said to be Fréchet(or F)-differentiable at a given point $u \in V$ if the difference quotient $(\mathcal{D}(u + t\psi) - \mathcal{D}(u))/t$ (as a function of t) converges uniformly on every bounded set. In this case the F-derivative of $\mathcal{D}$ at point u, denoted also by $\mathcal{D}'(u)$, is a linear continuous operator on V.* ■

Remark 4.57. The F-differentiability of $\mathcal{D}$ implies the G-differentiability of $\mathcal{D}$ and the G-derivative $\mathcal{D}'(u)$ of $\mathcal{D}$ at point u is a linear continuous operator on V. Conversely, if $\mathcal{D}$ is G-differentiable in an open subset K of V, if the G-derivative $\mathcal{D}'(u)$ at point u is bounded and linear for all $u \in \mathcal{K}$ and if $\mathcal{D}' : \ v \in V \longrightarrow \mathcal{D}'(v) \in \mathcal{L}(V; W)$ [1] is continuous then $\mathcal{D}$ is F-differentiable (see, *e.g.*, Schwartz [261]). ♢

Example 4.58. (*i*) Let $\Lambda : \ V \longrightarrow W$ be a G-differentiable and linear operator, where V and W are two locally convex topological vector spaces. Then the G-derivative of Λ at point u is exactly the operator Λ, *i.e.*,

$$\Lambda'(u) = \Lambda. \tag{4.36}$$

[1] $\mathcal{L}(V; W)$ denotes the space of continuous linear functionals from V to W.

In this case the operator $\Lambda'(u)$ is a linear operator.

(*ii*) Let $\Lambda : V \longrightarrow W$ be a G-differentiable and quadratic operator (*i.e.*, there exists a bilinear operator $\mathcal{B} : V \times V \longrightarrow W$ such that $\Lambda(u) := \mathcal{B}(u,u)$ for all $u \in V$) then the G-derivative Λ at point u in the direction ψ is

$$\Lambda'(u)\psi := \mathcal{B}(u,\psi) + \mathcal{B}(\psi,u). \tag{4.37}$$

Here also the operator $\Lambda'(u)$ is a linear operator. ♣

We can now give the second-order G-derivative of a real valued on V. Let $F : V \longrightarrow \overline{\mathbb{R}}$ be a given G-differentiable functional. Then its G-derivative $F'(u)$ at a point $u \in V$ is an operator from V into V'. If now the operator $F'(u)$ is also G-derivative, then the second-order G-variation of F at point $u \in V$ in the direction $v \in V$ and $\psi \in V$ is defined by

$$\begin{aligned} \delta^2 F(u,v,\psi) &:= \lim_{t \longrightarrow 0_+} \frac{F'(u+t\psi).v - F'(u).v}{t} \\ &= \langle D^2F(u)\psi, v\rangle_{V',V}, \end{aligned} \tag{4.38}$$

where $D^2F(u)$ (or $F^{(2)}(u)$) is said to be the *second-order G-derivative* of F at point $u \in V$.

It is clear that $\delta^2 F(u,.,.) : V \times V \longrightarrow \overline{\mathbb{R}}$ is a symmetric bilinear functional. As a direct consequence (because of Proposition 4.59), we have a characterization of the convexity of the second G-differentiable function F.

Proposition 4.59. *(Monotonicity of the second G-differential of a convex function) Let C be a non-empty convex subset of V and F be a G-differentiable function of C into $\mathbb{R}$. Then F is convex over C if and only if*

$$\langle D^2F(u_\alpha)(v-u), v-u\rangle_{V',V} \geq 0 \quad \textit{for all } (u,v) \in C \times C, \tag{4.39}$$

where $u_\alpha = v + \alpha(u-v)$, $\alpha \in (0,1)$.

Moreover, F is strictly convex over C if and only if (4.39) holds as a strict inequality if $u \neq v$. □

We finish this subsection with the following definitions:

(*i*) The second F-derivative $\delta^2\mathcal{D}(u,.,.) : V \times V \longrightarrow W$ of the F-differentiable operator $\mathcal{D} : V \longrightarrow W$ (*i.e.*, the F-derivative of the function $\mathcal{D}' : v \in V \longrightarrow \mathcal{D}'(v) \in \mathcal{L}(V;W)$) at point $u \in V$ belongs to the space of symmetric bilinear functionals and is denoted by $D^2\mathcal{D}(u)$ (or $\mathcal{D}^{(2)}(u)$).

(*ii*) Similarly as above, the d-th F-derivative $\mathcal{D}^{(d)}(u, \underbrace{.,\ldots,.}_{d\text{-times}})$ is a symmetric continuous d-linear function on V.

4.3 Applications of the Duality

In this section we shall apply the duality results to different problems of calculus of variations arising from mechanics and physics.

Before presenting some applications of the duality, we need to make some classical definitions for general non-linear systems.

4.3.1 Fundamental Equations

Assume that the paired topological vector space V and its dual V' are placed in duality by the bilinear form $\langle .,. \rangle_{V',V}$. The bilinear form $\langle .,. \rangle_{V',V}$ is defined on a space domain $\Omega \subset \mathrm{I\!R}^n$, $n \geq 1$ in static systems, and on a space-time domain $\mathcal{Q} := \Omega \times (0,T) \subset \mathrm{I\!R}^{n+1}$ (T is the final time), in dynamical systems. Consider an operator $\mathcal{A}: \ V \longrightarrow V'$, where its effective domain dom(A) is non-empty. The set $\{A(u) : \ u \in V\} \subset V'$ is said to be the *range* of the operator $\mathcal{A}$ and denoted by $\mathcal{R}(A)$. Then the operator $\mathcal{A}$ is a mapping from dom$(\mathcal{A})$ into $\mathcal{R}(A)$ and is said to invertible, if there exists an inverse operator $\mathcal{A}^{-1}$ from $\mathcal{R}(A)$ into dom$(\mathcal{A})$ such that

$$\mathcal{A}(u) = u^* \text{ if and only if } \mathcal{A}^{-1}(u^*) = u. \tag{4.40}$$

Let $\mathcal{S}$ be a source space, assumed to be a subspace of V'. For a given source $f \in \mathcal{S}$ the equation

$$\mathcal{A}(u) = f, \tag{4.41}$$

is called the *fundamental equation* and the operator $\mathcal{A}$ is said to be the *fundamental operator.*

Generally, between the configuration and the source variables, there exist intermediate variables, which always appear in pairs placed in duality by other types of bilinear forms. These duality relations, between intermediate variables, describe different physical or biological interior properties or characteristics of the system (for example, the environmental, the physical or biological parameters). Consequently, the fundamental operator (respectively equation) is depending on certain intermediate equations (respectively operators).

Assume now that the paired topological vector space W and its dual W' are placed in duality by the bilinear form $\langle .,. \rangle_{W',W}$. The paired spaces V, V' and W, W' are linked by the following operators:

(*i*) *The geometrical operator* $\Lambda : U \longrightarrow W$: this operator describes the geometrical transformation of the system (does not concern the physical or biological properties of the system). The associated *geometrical equation* (also called the *kinematic equation*, see Oden and Reddy [231]) is given by

$$p = \Lambda u. \tag{4.42}$$

(*ii*) *The constitutive operator* $\mathfrak{C} : \ W \longrightarrow W'$: this operator describes the physical or biological properties of the system. The associated *constitutive equation* (which gives the duality relation between the spaces W and W') is given by

$$p^* = \mathfrak{C}p. \tag{4.43}$$

(*iii*) *The balance operator* $\mathfrak{B} : W' \longrightarrow V'$: this operator describes the local state of the system. The associated *balance equation* (also called the *equilibrium equation*) is given by

$$f = \mathfrak{B}p^*. \tag{4.44}$$

Then, by taking account of (*i*)–(*iii*) we can deduce easily that the fundamental operator $\mathcal{A} = \mathfrak{B} \circ \mathfrak{C} \circ \Lambda : V \longrightarrow V'$ and the fundamental equation $\mathcal{A}u = f$ can be written as

$$\begin{aligned} p &= \Lambda u, \\ p^* &= \mathfrak{C}p, \\ f &= \mathfrak{B}p^*. \end{aligned} \tag{4.45}$$

4.3.2 Duality Mapping in Banach Spaces

Let $(V, \| \, . \, \|)$ be a real Banach space and its dual $(V', \| \, . \, \|_*)$. Before introducing the notion of duality mapping, we give a smoothness definition and a result concerning the G-differentiablity of a norm.

Definition 4.60. *(Smooth space) The real Banach space* $(V, \| \, . \, \|)$ *is said to be smooth, if for each non-null element* u *in* V*, there exists a unique element* f_u *in* V' *such that* $\| f_u \|_* = 1$ *and* $\langle f_u, u \rangle_{V',V} = \| u \|$ *.* ■

Definition 4.61. *(Strictly convex space) The real Banach space* $(V, \| \, . \, \|)$ *is said to be strictly convex if for all* u, v *in* V *with* $\| u \| = \| v \| = 1$ *and* $u \neq v$ *we have* $\| u + v \| < 2$.

It is obvious that a uniform convex Banach space is a strictly convex Banach space. ■

Theorem 4.62. *(G-differentiability of a norm) Let* $(V, \| \, . \, \|)$ *be a real Banach space. The norm* $\| \, . \, \|$ *of* V *is G-differentiable if and only if* $(V, \| \, . \, \|)$ *is a smooth Banach space.*

Proof. The proof can be found, *e.g.*, in Diestel [102]. □

We can now study the duality mapping on a real Banach space.

Definition 4.63. *(Gauge duality mapping) Let* $\varphi : I\!R \longrightarrow I\!R$ *be a so-called gauge function, i.e.,*

$$\varphi \text{ is continuous, strictly increasing }, \varphi(0) = 0 \text{ and } \lim_{t \longrightarrow \infty} \varphi(t) = \infty$$

and let $2^{V'}$ *be the so-called power set of* V'*, i.e., the set of all subsets of* V'*. Then, the multivalued mapping* $\mathcal{J}_\varphi : V \longrightarrow 2^{V'}$*, defined as follows:*

$$\begin{aligned} &\mathcal{J}_\varphi(0) := \{0\}, \\ &\mathcal{J}_\varphi(u) := \{f \in V' : \| f \|_* = \varphi(\| u \|), \ \langle f, u \rangle_{V',V} = \varphi(\| u \|) \, \| u \|\} \\ &\qquad\qquad \text{if } u \neq 0 \end{aligned} \tag{4.46}$$

is called the gauged (or weighted) duality mapping corresponding to the gauge function φ*.* ■

Remark 4.64. (*i*) Because of the Hahn–Banach theorem, it is easy to see that the domain of $\mathcal{J}_\varphi$ is the whole space, *i.e.*, $\mathrm{dom}(\mathcal{J}_\varphi) = V$.

(*ii*) By the expression (4.46) we have that, for each $u \in V$, $\mathcal{J}_\varphi(u)$ is bounded, closed and convex subset of V'.

(*iii*) According to (4.12) we have that

$$\mathcal{J}_\varphi(u) := \varphi(\aleph u)\partial\aleph(u) \quad \forall u \in V, \tag{4.47}$$

where $\aleph: \; u \in V \longrightarrow \aleph u := \parallel u \parallel$ and $\partial\aleph$ is the subdifferential of $\aleph$.

The mapping $\mathcal{J}_\varphi$ can be also written as

$$\mathcal{J}_\varphi := \partial F, \;\text{ with }\; F(u) := \int_0^{\aleph u} \varphi(t)dt \;\; \forall u \in V.$$

◇

Proposition 4.65. *Let* $(V, \parallel . \parallel)$ *be a real Banach space, then* $\mathcal{J}_\varphi$ *is monotone, i.e.,*

$$\langle f_1 - f_2, u_1 - u_2\rangle_{V',V} \geq 0,$$

for all $u_i \in V, f_i \in \mathcal{J}_\varphi(u_i),\; i = 1, 2.$

Moreover, if V *is also strictly convex then* $\mathcal{J}_\varphi$ *is strictly monotone and, in particular,*

$$\mathcal{J}_\varphi(u) \cap \mathcal{J}_\varphi(v) = \emptyset \;\text{ for } u, v \text{ in } V \text{ with } u \neq v.$$

Proof. Because of the expression (4.46) and the fact that φ is increasing, we obtain

$$\langle f_1 - f_2, u_1 - u_2\rangle_{V',V} \geq (\varphi(\parallel u_1 \parallel) - \varphi(\parallel u_2 \parallel))(\parallel u_1 \parallel - \parallel u_2 \parallel) \geq 0,$$

for all $u_i \in V, f_i \in \mathcal{J}_\varphi(u_i),\; i = 1, 2$.

Prove now the strict monotonicity of $\mathcal{J}_\varphi$. Suppose then, by contradiction, that there exist two elements u_1, u_2 in V such that $u_1 \neq u_2$ and $f_i \in \mathcal{J}_\varphi(u_i)$, $i = 1, 2$ satisfying

$$\langle f_1 - f_2, u_1 - u_2\rangle_{V',V} = 0.$$

According to the first result we have that

$$0 = \langle f_1 - f_2, u_1 - u_2\rangle_{V',V} \geq (\varphi(\parallel u_1 \parallel) - \varphi(\parallel u_2 \parallel))(\parallel u_1 \parallel - \parallel u_2 \parallel) \geq 0$$

and then $\parallel u_1 \parallel = \parallel u_2 \parallel = \alpha \neq 0$ (since $u_1 \neq u_2$). Now, putting $v_i = u_i/\alpha$, $i = 1, 2$, then $\parallel v_1 \parallel = \parallel v_2 \parallel = 1$ and $\langle f_1 - f_2, v_1 - v_2\rangle_{V',V} = 0$. Consequently (since, by definition, $\langle f_i, v_i\rangle_{V',V} = \varphi(\parallel u_i \parallel), i = 1, 2$),

$$0 = [\varphi(\parallel u_1 \parallel) - \langle f_1, v_2\rangle_{V',V}] + [\varphi(\parallel u_2 \parallel) - \langle f_2, v_1\rangle_{V',V}].$$

So (since, by definition, $\varphi(\parallel u_i \parallel) = \parallel f_i \parallel_* \geq \langle f_i, v_j\rangle_{V',V}$, $i, j = 1, 2$ with $i \neq j$) we have

$$\parallel f_1 \parallel_* = \parallel f_2 \parallel_* = \langle f_2, v_1\rangle_{V',V} = \langle f_1, v_2\rangle_{V',V}$$

Consequently, $\langle f_1, v_1 + v_2\rangle_{V',V} = 2 \parallel f_1 \parallel_*$ and then $\parallel v_1 + v_2 \parallel \geq 2$, which is a contradiction with the fact that the space V is strictly convex. □

It is clear that the following results hold (according to the definition of $\mathcal{J}_\varphi$, the expression (4.47), the convexity of $\aleph$, Proposition 4.44 and Theorem 4.62).

Proposition 4.66. *Let $(V, \| \cdot \|)$ be a real Banach space. Then:*

(*i*) *V is a smooth Banach space if and only if $\mathcal{J}_\varphi$ is a singleton (i.e., $\mathcal{J}_\varphi(u)$ is a singleton for all u in V).*

(*ii*) *If V is a smooth Banach space then $\mathcal{J}_\varphi(u) = F'(u)$. Otherwise*

$$\mathcal{J}_\varphi(0) = 0 \text{ and } \mathcal{J}_\varphi(u) = \varphi(\aleph u)\aleph'(u) \quad \text{if } u \neq 0,$$

where $\aleph'(u)$ is the G-differential of $\aleph$ at u. □

Remark 4.67. If $(V, \| \cdot \|)$ is a smooth Banach space then (since $\aleph'(u)$ satisfies $\| \aleph'(u) \|_* = 1$, $\langle \aleph'(u), u \rangle_{V',V} = \| u \|$, for all $u \neq 0$ in V)

$$\| \mathcal{J}_\varphi(u) \|_* = \varphi(\aleph u) \text{ and } \langle \mathcal{J}_\varphi(u), u \rangle_{V',V} = \varphi(\aleph u) \| u \| \quad \forall u \in V.$$

◊

Proposition 4.68. *(Demi-continuous result) Let $(V, \| \cdot \|)$ be a Banach space. If V is smooth and reflexive then $\mathcal{J}_\varphi$ is demi-continuous, i.e.,*

$$\text{if } u_n \longrightarrow u \text{ strongly on } V \quad \text{then} \quad \mathcal{J}_\varphi(u_n) \rightharpoonup \mathcal{J}_\varphi(u) \text{ weakly on } V'.$$

Proof. Since (u_n) is uniformly bounded then $\varphi(\| u_n \|)$ and $\| \mathcal{J}_\varphi(u_n) \|_* = \varphi(\aleph u_n) = \varphi(\| u_n \|)$ is also bounded. Consequently, $\mathcal{J}_\varphi(u_n)$ is uniformly bounded on a reflexive Banach space V'. Then, we can extract from $\mathcal{J}_\varphi(u_n)$ a subsequence, denoted also by $\mathcal{J}_\varphi(u_n)$, converging to f.

Prove now that $f = \mathcal{J}_\varphi(u)$. By the lower semi-continuity of the norm $\| \cdot \|$ we have that (according to the continuity of φ)

$$\| f \|_* \leq \liminf_n \| \mathcal{J}_\varphi(u_n) \|_* = \lim_n \varphi(\| u_n \|) = \varphi(\| u \|). \tag{4.48}$$

Moreover, since

$$\langle \mathcal{J}_\varphi(u_n), u_n \rangle_{V',V} - \langle f, u \rangle_{V',V} = \langle \mathcal{J}_\varphi(u_n) - f, u \rangle_{V',V} + \langle \mathcal{J}_\varphi(u_n), u_n - u \rangle_{V',V},$$

then

$$\begin{aligned} &| \langle \mathcal{J}_\varphi(u_n), u_n \rangle_{V',V} - \langle f, u \rangle_{V',V} | \\ &\qquad \leq | \langle \mathcal{J}_\varphi(u_n), u_n - u \rangle_{V',V} | + \| \mathcal{J}_\varphi(u_n) \|_* \| u_n - u \| . \end{aligned}$$

From $u_n \longrightarrow u$ strongly in V, $\mathcal{J}_\varphi(u_n) \rightharpoonup f$ weakly in V' and the boundedness of $\mathcal{J}_\varphi u_n$, it follows easily that $\langle \mathcal{J}_\varphi(u_n), u_n \rangle_{V',V} \longrightarrow \langle f, u \rangle_{V',V}$.

On the other hand, $\langle \mathcal{J}_\varphi(u_n), u_n \rangle_{V',V} = \varphi(\| u_n \|) \| u_n \| \longrightarrow \varphi(\| u \|) \| u \|$ so, $\varphi(\| u \|) \| u \| = \langle f, u \rangle_{V',V}$ and then $\varphi(\| u \|) \leq \| f \|_*$. Consequently, by (4.48) we can deduce that

$$\varphi(\| u \|) = \| f \|_* .$$

Since $\varphi(\| u \|) = \| f \|_*$ and $\langle f, u \rangle_{V',V} = \varphi(\| u \|) \| u \|$, then, by definition, this implies that $\mathcal{J}_\varphi(u) = f$. □

Remark 4.69. It is clear that $\mathcal{J}_\varphi$ is coercive, because $\lim_{t \longrightarrow \infty} \varphi(t) = \infty$ and so,

$$\frac{\langle \mathcal{J}_\varphi(u), u \rangle_{V',V}}{\| u \|} = \varphi(\| u \|) \longrightarrow \infty \text{ for } u \in V, \ \| u \| \longrightarrow \infty.$$

◊

The following theorem shows an important result namely the surjectivity of the gauge duality mapping $\mathcal{J}_\varphi$.

Theorem 4.70. *Let $(V, \| \, . \, \|)$ be a Banach space. If V is smooth and reflexive then $\mathcal{J}_\varphi$ is surjective. Moreover, if V is also a strictly convex space, then $\mathcal{J}_\varphi$ is a bijection function.*

Proof. The surjectivity result follows from the well-known theorem of Browder [64]:

If V is a real Banach and reflexive space, then any monotone, coercive and demi-continuous operator $\mathcal{T} : V \longrightarrow V'$ is surjective.

Indeed, according to Proposition 4.65, Remark 4.69 and Proposition 4.68, $\mathcal{J}_\varphi$ is monotone, coercive and demi-continuous. Consequently, by Browder's theorem $\mathcal{J}_\varphi$ is surjective.

The injectivity of the functional $\mathcal{J}_\varphi$ is a direct consequence of Proposition 4.65. □

Example 4.71. Let Ω be a domain in $\mathbb{R}^d$ $(d \in \mathbb{N}^*)$ and $V = L^p(\Omega)$ (for $1 < p < \infty$). The space V is a reflexive, smooth and uniformly convex Banach space and its dual is $V' = L^{p^*}(\Omega)$, where $1/p + 1/p^* = 1$. It is clear that the function $\varphi(t) = t^{p-1}$ is a gauge function. The gauge duality mapping $\mathcal{J}_\varphi$ corresponding to φ is $\mathcal{J}_\varphi(u)(x) = | u(x) |^{p-1} \operatorname{sign}(u(x)), \quad \text{a.e. } x \in \Omega.$ ♣

4.3.3 Duality and Fundamental Equations

Let $\mathcal{F}$ and $\mathcal{S}$ be two real valued functions on two reflexive Banach spaces V and W respectively which are either convex or concave, and Gâteaux-differentiable on the convex sets $K \subset V$ and $M \subset W$, respectively. Then the two duality equations between the paired spaces V, V' and W, W' can be given by

$$\begin{aligned} \mathcal{F}'(u) &= g \in V', \\ \mathcal{S}'(p) &= f \in W', \end{aligned} \tag{4.49}$$

where $\mathcal{F}' : \ K \subset V \longrightarrow V'$ and $\mathcal{S}' : \ M \subset W \longrightarrow W'$ stand for the G-derivative of $\mathcal{F}$ and $\mathcal{S}$, respectively.

In mathematical modeling, the duality equation $f = \mathcal{S}'(p)$ is known as the *constitutive equation* and the duality equation $g = \mathcal{F}'(u)$ usually gives *natural boundary conditions* or *external energy* in variational boundary value problems.

Let now a so-called *feasible set* or *admissible set* (the set of possible solutions to a given problem (4.49))

$$K_{\text{ad}} = \{u \in V : \ u \in K \text{ and } \Lambda u \in M\},$$

where Λ is a continuous linear mapping $\Lambda : \ V \longrightarrow W$ with the adjoint $\Lambda^* : \ W' \longrightarrow V'$ defined by (4.20).

According to (4.21) and (4.22), the system (4.49) can be written in a so-called *fundamental equation*

$$\mathcal{F}'(u) = \Lambda^* \mathcal{S}'(\Lambda u). \tag{4.50}$$

If, for a given $\xi \in V'$ (external source) and a linear operator $\mathcal{C} : \ W \longrightarrow W'$, the functional $\mathcal{F}$ is written as $\mathcal{F}(u) := \langle \xi, u \rangle_{V',V}$ for all $u \in V$ and the functional $\mathcal{S}$ is written as $\mathcal{S}(p) := (1/2)\langle \mathcal{C}p, p \rangle_{W',W}$ for all $p \in W$ (*i.e.*, $\mathcal{S}$ is a quadratic function), then Equation (4.50) becomes

$$\Lambda^* \mathcal{C}_{\text{s}} \Lambda u = \xi,$$

where $\mathcal{C}_{\text{s}} := (\mathcal{C} + \mathcal{C}^*)/2$ (the symmetric part of $\mathcal{C}$) and $\mathcal{C}^* : \ W \longrightarrow W'$ the adjoint operator of $\mathcal{C}$ is defined by

$$\langle \mathcal{C}p, q \rangle_{W',W} = \langle \mathcal{C}^* q, p \rangle_{W',W}.$$

It is clear that the operator $\mathcal{A} := \Lambda^* \mathcal{C}_{\text{s}} \Lambda$ is self-adjoint, *i.e.*, $\mathcal{A}^* = \mathcal{A}$.

Example 4.72. Consider Maxwell's system, which describes electric and magnetic fields in a homogeneous and isotropic medium, by (a.e. on $[0, T]$)

$$\begin{aligned} &-\text{div}(\nu E) = f \quad \text{on } \Omega, \\ &\frac{\partial E}{\partial t} + \text{curl}(\frac{1}{\mu}\text{curl} A) = g \quad \text{on } \Omega, \\ &E = \nabla \phi + \frac{\partial A}{\partial t} \quad \text{on } \Omega, \end{aligned} \tag{4.51}$$

where Ω is a bounded open subset of $\mathbb{R}^d$ ($d \in \mathbb{N}^*$), sufficiently regular with $\Gamma = \partial\Omega$ its boundary, T is the final time, ϕ is the scalar potential, B is the vector magnetic potential, E is the electric field intensity, f is the charge density, g is the current density, ν is the induction capacity of the medium and μ is the permeability of free space.

We introduce now the following state variables:

$$\mathbf{u} = \begin{pmatrix} \phi \\ A \end{pmatrix}, \ \mathbf{p} = \begin{pmatrix} E \\ B \end{pmatrix}, \ \mathbf{u}^* = \begin{pmatrix} f \\ g \end{pmatrix}, \ \mathbf{p}^* = \begin{pmatrix} D \\ H \end{pmatrix},$$

where D is the electric flux density, B is the magnetic flux density and H is the magnetic field intensity.

We can prove easily that the previous system (4.51) can be written in the form

$$\mathbf{p} = \begin{pmatrix} \nabla & \frac{\partial}{\partial t} \\ 0 & \text{curl} \end{pmatrix} \mathbf{u} \ \text{ on } \Omega,$$

$$\mathbf{p}^* = \begin{pmatrix} \nu & 0 \\ 0 & \frac{1}{\mu} \end{pmatrix} \mathbf{p} \ \text{ on } \Omega, \tag{4.52}$$

$$\mathbf{u}^* = \begin{pmatrix} -\text{div} & 0 \\ -\frac{\partial}{\partial t} & \text{curl} \end{pmatrix} \mathbf{p}^* \ \text{ on } \Omega.$$

Let V, V' and W, W' be two pairs of real topological vector spaces, in duality with respect to the bilinear forms $\langle .,. \rangle_{V',V}$ and $\langle .,. \rangle_{W',W}$, respectively defined by

$$\begin{aligned} \langle \mathbf{u}^*, \mathbf{u} \rangle_{V',V} &:= \int_0^T \int_\Omega (\mathbf{u}, \mathbf{u}^*)_2 \mathrm{d}x\mathrm{d}t, \\ \langle \mathbf{p}^*, \mathbf{p} \rangle_{W',W} &:= \int_0^T \int_\Omega (\mathbf{p}, \mathbf{p}^*)_2 \mathrm{d}x\mathrm{d}t, \end{aligned} \tag{4.53}$$

where $(.,.)_2$ is the euclidian scalar product in $\mathbb{R}^4$. Moreover, we suppose that W is a Hilbert space and can be identified to its dual W'.

By using Green's formula and the identity

$$H.\text{curl}(A) = \text{div}(A \times H) + A.\text{curl}(H),$$

we have that

$$\int_\Omega H.\text{curl}(A)\mathrm{d}x = \int_\Gamma (A \times H).\mathbf{n}\mathrm{d}x + \int_\Omega A.\text{curl}(H)\mathrm{d}x, \tag{4.54}$$

where $\mathbf{n}$ is the unit outward normal on $\partial\Omega$.

If $(A \times H).\mathbf{n} = 0$ on Γ then

$$\int_\Omega H.\text{curl}(A)\mathrm{d}x = \int_\Omega A.\text{curl}(H)\mathrm{d}x. \tag{4.55}$$

Let now

$$\mathcal{K} = \{\mathbf{u} \in V : \mathbf{u} = 0 \ \text{ on } \ \Gamma, \ \ \mathbf{u}(.,t=0) = \mathbf{u}(.,t=T)\}$$

and the admissible set

$$\mathcal{K}_{\text{ad}} := \{\mathbf{u} \in \mathcal{K} : \ \Lambda\mathbf{u} \in W\},$$

where $\Lambda : \ V \longrightarrow W$ is defined by

$$\Lambda\mathbf{u} = \begin{pmatrix} \nabla & \frac{\partial}{\partial t} \\ 0 & \text{curl} \end{pmatrix} \mathbf{u}.$$

Then for a given $\mathbf{u} \in \mathcal{K}_{\rm ad}$ and $\mathbf{p}^* \in W'$ such that $\mathbf{p}^*(.,t=0) = \mathbf{p}^*(.,t=T)$ we have that (since $\mathbf{u} = 0$ on Γ)

$$\langle \mathbf{p}^*, \Lambda \mathbf{u} \rangle_{W',W} = \int_0^T \int_\Omega ((\nabla\phi + \frac{\partial A}{\partial t}).D + H.\mathrm{curl}(A))dxdt.$$

By using Green's formula and by integrating by parts in time, we can deduce that (according to (4.55))

$$\langle \mathbf{p}^*, \Lambda \mathbf{u} \rangle_{W',W} = \int_0^T \int_\Omega (-\mathrm{div}(D)\phi + A.(-\frac{\partial D}{\partial t} + \mathrm{curl}(H)))dxdt = \langle \Lambda^* \mathbf{p}^*, \mathbf{u} \rangle_{V',V},$$

where $\Lambda^* : W' \longrightarrow V'$ is defined by

$$\Lambda^* \mathbf{p}^* := \begin{pmatrix} -\mathrm{div} & 0 \\ -\frac{\partial}{\partial t} & \mathrm{curl} \end{pmatrix} \mathbf{p}^*.$$

Let now C be the following linear and symmetric matrix

$$\mathcal{C} := \begin{pmatrix} \nu & 0 \\ 0 & 1/\mu \end{pmatrix}.$$

For a given external source $\mathbf{u}^* \in V'$, we introduce the functionals $\mathcal{F} : V \longrightarrow \overline{\mathrm{IR}}$ and $\mathcal{S} : W \longrightarrow \overline{\mathrm{IR}}$ defined respectively by

$$\mathcal{F}(\mathbf{u}) := \langle \mathbf{u}^*, \mathbf{u} \rangle_{V',V} \text{ and } \mathcal{S}(\mathbf{p}) := \frac{1}{2} \langle C\mathbf{p}, \mathbf{p} \rangle_{W',W}.$$

The functionals $\mathcal{F}$ and $\mathcal{S}$ are G-differentiable and their G-differentials satisfy $\mathcal{F}'(u) = \mathbf{u}^*$, $\mathcal{S}'(p) = \mathcal{C}\mathbf{p}$ respectively. Consequently, according to (4.49) and (4.50), we can deduce that the fundamental system is $\Lambda^* \mathcal{C} \Lambda \mathbf{u} = \mathbf{u}^*$ (which is an abstract form of System (4.51)), the constitutive system is $\mathcal{C}\mathbf{p} = \mathbf{p}^*$, and finally the equilibrium system is $\Lambda^* \mathbf{p}^* = \mathbf{u}^*$. ♣

Example 4.73. Let us consider the following mixed boundary problem in electrostatics

$$\begin{aligned} -\mathrm{div}(\nu \nabla u) = f \quad \text{on } \Omega, \\ u = 0 \quad \text{on } \Gamma_0, \quad \nu \nabla u.\mathbf{n} = g \quad \text{on } \Gamma_1, \end{aligned} \tag{4.56}$$

where Ω is a bounded open subset of IR^d $(d \in \mathrm{IN}^*)$, sufficiently regular with $\partial\Omega = \Gamma_0 \cup \Gamma_1$ its boundary, $\mathbf{n}$ is the unit outward normal on $\partial\Omega$, and $\nu > 0$ is the dielectric constant.

Let V, V' and W, W' be two pairs of real topological vector spaces, in duality with respect to the bilinear forms $\langle .,. \rangle_{V',V}$ and $\langle .,. \rangle_{W',W}$, respectively.

Let $\Lambda := -\nabla$, where $p := \Lambda u$ is the electric field intensity. The variational functionals are

$$\begin{aligned} \mathcal{S}(p) &:= \int_\Omega \frac{\nu}{2} \mid p(x) \mid_2^2 dx + \mathcal{X}_{\mathcal{A}}(p), \\ F(u) &:= \int_\Omega f u dx + \int_{\Gamma_1} g u dx - \mathcal{X}_{\mathcal{K}}(u), \end{aligned} \tag{4.57}$$

where $\mathcal{A} := H(\mathrm{div};\Omega) \subset W$, $\mathcal{K} := \{u \in H^1(\Omega) : \; u = 0 \text{ on } \Gamma_0\} \subset V$, $|\,.\,|_2$ is the euclidian norm and $\mathcal{X}_{\mathcal{K}}$ (respectively $\mathcal{X}_{\mathcal{A}}$) is the indicator function of $\mathcal{K}$ (respectively of $\mathcal{A}$).

Then it is easy to prove that the functionals $\mathcal{S}$ and F are finite and G-differentiable on $\mathcal{A}$ and $\mathcal{K}$ respectively. Moreover,

$$p^* = \mathcal{S}'(p) = \nu p, \quad u^* = F'(u) = f \;\text{ on } \Omega \;\text{ and }\; u^* = F'(u) = g \;\text{ on } \Gamma_1.$$

By using Green's formula we can deduce that

$$\begin{aligned}\langle \mathcal{S}'(p), \Lambda u\rangle_{W',W} &= \langle p^*, \Lambda u\rangle_{W',W}\\ &= \int_\Omega -p^* \nabla u \mathrm{d}x\\ &= \int_\Omega u \mathrm{div}(p^*)\mathrm{d}x - \langle p^*.\mathbf{n}, u\rangle_{H^{-1/2}(\Gamma),H^{1/2}(\Gamma)}\\ &= \langle \Lambda^* p^*, u\rangle_{V',V}.\end{aligned}$$

Then we have the following abstract equilibrium equation:

$$u^* = \Lambda^* p^* = \mathrm{div}(p^*) = f \text{ on } \;\Omega \text{ and } u^* = \Lambda^* p^* = -p^*.\mathbf{n} = g \;\text{ on } \Gamma_1,$$

and the fundamental operator in this problem is the operator $-\nu\Delta$. ♣

Remark 4.74. More examples can be found, *e.g.*, in Oden and Reddy [231] and Strang [275]. ♢

4.3.4 Euler–Lagrange Equation and the Non-linear Operator

Many non-linear boundary value problems can be written as

$$\mathcal{A}u = 0,$$

where $\mathcal{A} : \; V \longrightarrow W$ is a non-linear operator between two topological vector spaces V and W.

If the previous problem is variational then there exists a real-valued and G-differentiable function $\mathcal{F}$ on V, such that $\mathcal{A} := \mathcal{F}'$, where the space W is a subset of V' (the dual of V). The operator $\mathcal{A}$ is called the *potential operator* and $\mathcal{F}$ is called the potential of $\mathcal{A}$ or *the variational functional.* Then the previous problem is equivalent to solving the so-called *Euler–Lagrange equation*

$$\mathcal{F}'(u) = 0 \text{ or } \langle \mathcal{F}'(u), v\rangle_{V',V} = 0 \qquad \forall v \in V. \tag{4.58}$$

Definition 4.75. *(Critical point of a function) A point $u_c \in V$ is said to be a critical point (or stationary point) of $\mathcal{F}$ if u_c is a solution of (4.58), i.e., $\mathcal{F}'(u_c) = 0$ or $\langle \mathcal{F}'(u_c), v\rangle_{V',V} = 0 \;\; \forall v \in U$. Its value $\alpha := \mathcal{F}(u_c)$ is called a critical value of $\mathcal{F}$ and the set $\mathcal{F}^{-1}(\alpha)$ is called a critical level of $\mathcal{F}$.* ■

Example 4.76. Consider the following semilinear elliptic problem

$$\begin{aligned} -\Delta u + \lambda u &= f(.,u) \quad \text{on } \Omega, \\ \alpha u + \beta \nabla u.\mathbf{n} &= 0 \quad \text{on } \partial\Omega. \end{aligned} \tag{4.59}$$

where Ω is a bounded open subset of $\mathbb{R}^d$ ($d \in \mathbb{N}^*$), sufficiently regular with $\partial\Omega$ its boundary, $\mathbf{n}$ is the unit outward normal on $\partial\Omega$, the constants α, β are such that $\alpha + \beta \neq 0$, the constant λ is positive, $f(.,u)$ is a non-linear operator satisfying $f(.,0) = 0$ and other standard conditions.

Let $F(.,v) := \int_0^v f(.,w)dw$. The variational functional is

$$\mathcal{F}(u) := \int_\Omega (-\frac{1}{2}\Delta u(x).u(x) + \frac{\lambda}{2} \mid u(x) \mid^2 - F(x,u(x)))dx, \tag{4.60}$$

where $V := H_0^1(\Omega)$ if $\beta = 0$ and $V := H^1(\Omega)$ if $\beta \neq 0$. Then it is easy to prove that the weak solutions of problem (4.59) coincide with critical points of $\mathcal{F}$. ♣

There exist two classes of critical points, the first corresponds to the classical local extrema (infimum or supremum) of $\mathcal{F}$ and the second class corresponds to critical points which are not local extrema.

Definition 4.77. *(Saddle point of a function) A critical point u_c of a real-valued G-differentiable function $\mathcal{F}$ that is not a local extremum is said to be a saddle point of $\mathcal{F}$, i.e., for any neighborhood $\mathcal{K}_c$ of u_c, there exist v, w of $\mathcal{K}_c$ such that*

$$\mathcal{F}(v) < \mathcal{F}(u_c) < \mathcal{F}(w). \tag{4.61}$$

■

Remark 4.78. In theoretical chemistry and physics, saddle points appear as unstable equilibria or transient excited states (as they correspond to the transition states and lead to the minimum energy paths between reactant molecules and product molecules). A vast literature can be found in the theoretical and computational chemistry and physics. ◊

To study the critical point problem, which is generally motivated by the search of solutions to non-linear partial differential equations, to an infinite dimensional setting, requires certain compactness conditions in order to express the compacity of sequences (*e.g.*, minimizing sequences) which converge to a point when we hope to be the critical point. Moreover, the compactness condition is a fundamental tool to carry out the deformation and then the well-known *Mountain–Pass* theorem of Ambrosetti-Rabinowitz, which is a minimax characterization of a critical point of a functional $\mathcal{F}$ in the sense of a minimax over a suitable non-empty class $\mathcal{S}$ of non-empty subsets of U

$$c = \inf_{A \in \mathcal{S}} \sup_{u \in A} \mathcal{F}(u). \tag{4.62}$$

The one most frequently used is due to Palais and Small (see Palais [233]).

Definition 4.79. *(Palais–Small condition) Let* $(U, \| \cdot \|)$ *be a Banach space,* $(U', \| \cdot \|_*)$ *be its dual space and* $\mathcal{F}$ *be a real-valued and G-differentiable function on a subset* $\mathcal{K}$ *of* U*. The function* $\mathcal{F}$ *is said to satisfy the so-called Palais–Small condition (denoted by* (PS) *for short), if every sequence* (u_n) *in* $\mathcal{K}$*, for which* $\mathcal{F}(u_n)$ *is bounded and* $\lim_{n \longrightarrow \infty} \| \mathcal{F}'(u_n) \|_* = 0$*, has a convergent subsequence in* $\mathcal{K}$*.*

A variant of Palais–Small condition is the following (see Brezis *et al.* [59]).

Definition 4.80. *(Palais–Small condition at level c) Let* $(U, \| \cdot \|)$ *be a Banach space,* $(U', \| \cdot \|_*)$ *be its dual space and* $\mathcal{F}$ *be a real-valued and G-differentiable function on a subset* $\mathcal{K}$ *of* U*. For a given* $c \in I\!R$*, the function* $\mathcal{F}$ *is said to satisfy the so-called Palais–Small condition at the level c (denoted by* $(PS)_c$ *for short), if every sequence* (u_n) *in* $\mathcal{K}$*, for which* $\lim_{n \longrightarrow \infty} \mathcal{F}(u_n) = c$ *and* $\lim_{n \longrightarrow \infty} \| \mathcal{F}'(u_n) \|_* = 0$*, has a convergent subsequence in* $\mathcal{K}$*.* ■

Remark 4.81. (*i*) If a bounded function $\mathcal{F}$ satisfies the condition (PS), the set of critical points is compact.

(*ii*) If a function $\mathcal{F}$ satisfies the condition $(PS)_c$ at a fixed level c, we have that the set of critical points at level c

$$Z(c) := \{u \in U : \quad \mathcal{F}(u) = c \text{ and } \mathcal{F}'(u) = 0\},$$

is compact.

(*iii*) It is clear that the condition (PS) implies the condition $(PS)_c$ at a fixed level c. ◇

The following theorem shows the existence of the infimum of a function that satisfies the Palais–Small condition.

Theorem 4.82. *Let* $(U, \| \cdot \|)$ *be a Banach space and* $\mathcal{F}$ *be a real valued, lower semi-continuous and G-differentiable function on a subset* $\mathcal{K}$ *of* U*. If the function* $\mathcal{F}$ *is bounded from below (i.e., the set of its values has a lower bound) and satisfies the Palais–Small condition* (PS)*, then there exists at least one critical point* $u_c \in \mathcal{K}$ *of* $\mathcal{F}$ *(i.e.,* $\mathcal{F}'(u_c) = 0$*) such that*

$$\mathcal{F}(u_c) = \inf_{v \in \mathcal{K}} \mathcal{F}(v).$$

□

This theorem is a consequence of Ekeland's lemma in complete metric spaces (see Ekeland [111]), see below.

Lemma 4.83. *(Ekeland's lemma) Let* (X, d) *be a complete metric space equipped with a metric* d*,* $\phi : X \longrightarrow I\!R$ *be a real-valued, lower semi-continuous function which is bounded from below on* X *and set by* $c := \inf_{u \in X} \phi(u)$*. Then for all* $\epsilon > 0$*, there exists* $u_\epsilon \in X$ *such that*

$$
\begin{aligned}
& c \le \phi(u_\epsilon) \le c + \epsilon, \\
& \frac{\phi(u_\epsilon) - \phi(v)}{\epsilon} < d(v, u_\epsilon) \quad \forall v \in X, \quad v \neq u_\epsilon \; .
\end{aligned} \tag{4.63}
$$

□

We are now presenting a minimax characterization of a critical point of the type (4.62), namely the well-known Mountain–Pass theorem which has been proposed by Ambrosetti and Rabinowitz. This type of minimax problems, in which the basic ideas go back to Lusternik and Schnirelman [211], has been studied extensively in critical point theory since last few years (see Chang [71], Ekeland [113], Ghoussoub [131], Mawhin and Willem [220], Struwe [277]).

Before giving this theorem, we introduce some definitions and a variant of the so-called deformation lemma.

Definition 4.84. *(Pseudo-gradient) Let $(U, \| \cdot \|)$ be a Banach space and $(U', \| \cdot \|_*)$ be its dual space. Let ϕ be a real-valued and continuously F-differentiable function on U and $K_r(\phi) := \{u \in U : \phi'(u) \neq 0\}$ be the set of non critical points of ϕ:*

(i) Let $u \in U$. An element $v \in U$ is said to be a pseudo-gradient point of ϕ on u, if we have

$$
\| v \| \le 2 \| \phi'(u) \|_* \quad \text{and} \quad \langle \phi'(u), v \rangle \ge \| \phi'(u) \|_*^2 \; . \tag{4.64}
$$

(ii) The function $V_\phi : \; K_r(\phi) \longrightarrow U$ is said to be a pseudo-gradient field for ϕ on $K_r(\phi)$ if V_ϕ is a locally Lipschitz continuous mapping on $K_r(\phi)$. For all $u \in K_r(\phi)$, $V_\phi(u)$ is a pseudo-gradient point of ϕ on u. ■

We have the following lemma.

Lemma 4.85. *Under the assumption of Definition 4.84, there exists a pseudo-gradient field for ϕ on $K_r(\phi)$.*

Proof. For the proof, we can refer, for example, to Rabinowitz [247] or Willem [301]. □

We can deduce easily the following corollary.

Corollary 4.86. *Let $(U, \| \cdot \|)$ be a Banach space and $(U', \| \cdot \|_*)$ be its dual space and ϕ be a real-valued, even and continuously F-differentiable function on U. Then there exists a pseudo-gradient field for ϕ on $K_r(\phi)$, which is an odd function.*

Proof. By using Lemma 4.85, we have the existence of a pseudo-gradient field V_ϕ for ϕ on $K_r(\phi)$. We verify easily that the mapping $\tilde{V}_\phi$ defined by, for all $u \in U$,

$$
\tilde{V}_\phi(u) := \frac{V_\phi(u) - V_\phi(-u)}{2},
$$

is a pseudo-gradient field for ϕ on $K_r(\phi)$ and an odd function.

This completes the proof. □

Definition 4.87. *(Homotopy) Let $(U, \| \cdot \|)$ be a Banach space and $\eta : [0,1] \times U \longrightarrow U$ be a continuous function:*

(i) The function η is called a homotopy function, if

$$\eta(0, u) = u \quad \text{for all } u \in U. \tag{4.65}$$

(ii) The homotopy function η is said to be a homotopy of homeomorphisms if, for all $t \in [0,1]$,

$$\text{each map } \eta(t, .) : [0,1] \times U \longrightarrow U \text{ is an homeomorphism function.} \tag{4.66}$$

(iii) Let f be a real-valued and continuously F-differentiable function on U. The homotopy function η is called f-increasing (respectively f-decreasing) if for $0 \leq s \leq t \leq 1$ we have, for all $u \in U$,

$$f(\eta(s, u)) \leq f(\eta(t, u)) \quad (\text{respectively } f(\eta(s, u)) \geq f(\eta(t, u))). \tag{4.67}$$

■

We can now give the well-known deformation lemma (see Rabinowitz [247]).

Lemma 4.88. *(Deformation lemma) Let $(U, \| \cdot \|)$ be a Banach space and $(U', \| \cdot \|_*)$ be its dual space. Let ϕ be a real-valued and continuously F-differentiable function on U and $c \in I\!R$ be a given value. If the function ϕ satisfies the Palais–Small (PS) condition at level c, then:*

(i) for a given $\epsilon_0 > 0$, there exists a homotopy of homeomorphisms and ϕ-decreasing function η_{ϵ_0} such that:

(a) for all $t \in [0,1]$ and $u \in U$ such that $| \phi(u) - c | > \epsilon_0$, we have

$$\eta_{\epsilon_0}(t, u) = u$$

(b) if, moreover, ϕ is even then, for all $t \in [0,1]$, each map $\eta_{\epsilon_0}(t, .)$ is odd

(ii) if the set of critical points of ϕ at level c, $Z(c)$ is empty then we can find $\epsilon_0 > 0$ such that the corresponding homotopy of homeomorphisms and ϕ-decreasing function η_{ϵ_0} satisfies the following condition:

(c) for all $\epsilon \in]0, \epsilon_0[$ and $u \in U$ such that $\phi(u) \leq c + \epsilon$, we have that

$$\phi(\eta_{\epsilon_0}(1, u)) \leq c - \epsilon.$$

Proof. The proof is based essentially on the existence of a pseudo gradient field V_ϕ, because of Lemma 4.85 (which can be chosen odd if the functional ϕ is even, by Corollary 4.86) and finally by considering the function η_{ϵ_0} (for a given $\epsilon_0 > 0$), the unique solution of the following Cauchy problem:[2]

[2] For the existence and uniqueness theorem see the general theory of ordinary differential equations in Banach spaces given, for example, in Cartan [69].

$$\begin{aligned}&\frac{d\eta(t,u)}{dt} = -W_\phi(\eta(t,u)),\\ &\text{with the initial condition}\\ &\eta(0,u) = u,\end{aligned} \tag{4.68}$$

where $W_\phi := \xi \min\left(1, \frac{1}{\| V_\phi \|}\right) V_\phi$ and $\xi : U \longrightarrow [0,1]$ is given by

$$\begin{aligned}\xi(u) &= \frac{d(u,P)}{d(u,P)+d(u,Q)},\\ P &= \{u \in U :\ \phi(u) \leq c - \epsilon_0\} \cup \{u \in U :\ \phi(u) \geq c + \epsilon_0\},\\ Q &= \{u \in U :\ c - \epsilon_0 \leq \phi(u) \leq c + \epsilon_0\},\end{aligned} \tag{4.69}$$

and where $d(u,v) :=\| u - v \|$ is the distance in U.

Since P and Q are closed disjoint non-empty subsets then the map ξ is locally Lipschistz continuous. Moreover, we have that $\xi = 0$ on P, $\xi = 1$ on Q and if the function ϕ is even, ξ is even also and then W_ϕ is odd (by the choice of V_ϕ, odd map if ϕ is even). We note also that the solution η_{ϵ_0} is locally Lipschistz continuous (since W_ϕ is locally Lipschistz continuous, see *e.g.*, Cartan [69]) and satisfies $\eta_{\epsilon_0}(t, \eta_{\epsilon_0}(s,u)) = \eta_{\epsilon_0}(t+s,u)$, for all $t,s \in \mathbb{R}$, $u \in U$ and then by the uniqueness of Cauchy problem (4.68) we can deduce that, for every t, each map $\eta_{\epsilon_0}(t,.)$ is a homeomorphism from U into U and its inverse $\eta_{\epsilon_0}^{-1}(t,.)$ is $\eta_{\epsilon_0}(-t,.)$. Consequently, the function η_{ϵ_0} is a homotopy of homoemorphisms function. To obtain the property ϕ-decreasing of η_{ϵ_0}, we calculate the derivative by time of the function $\phi(\eta_{\epsilon_0}(t,u))$. Indeed (for simplicity we denote $\eta_{\epsilon_0}(.,u)$ by $\eta_{\epsilon_0}(.)$)

$$\begin{aligned}&\frac{d}{dt}\phi(\eta_{\epsilon_0}(t))\\ &= \langle \phi'(\eta_{\epsilon_0}(t)), \frac{d\eta_{\epsilon_0}(t)}{dt} \rangle_{U',U}\\ &= -\xi(\eta_{\epsilon_0}(t)) \min\left(1, \frac{1}{\| V_\phi(\eta_{\epsilon_0}(t)) \|}\right) \langle \phi'(\eta_{\epsilon_0}(t)), V_\phi(\eta_{\epsilon_0}(t)) \rangle_{U',U}\\ &\leq -\xi(\eta_{\epsilon_0}(t)) \min\left(1, \frac{1}{\| V_\phi(\eta_{\epsilon_0}(t)) \|}\right) \| \phi'(\eta_{\epsilon_0}(t)) \|_*^2 \ \text{(by (4.64))}\\ &\leq 0\end{aligned} \tag{4.70}$$

and then the ϕ-decreasing result. The point (b) is immediate and the point (a) is a direct consequence of the uniqueness of the Cauchy problem (4.68) and the fact that if u is such that $| \phi(u) - c |> \epsilon_0$, $W_\phi(u) = 0$ ($\eta_{\epsilon_0}(.) - u$ is a solution of (4.68) with null initial condition).

For the point (c), since the set of critical points of ϕ at level c is empty, according to the Palais–Small condition, we have the existence of $\beta > 0$ and $0 < \delta \leq 1$ such that

$$\forall u \in \{u \in U : | \phi(u) - c | \leq \beta\}, \ \| \phi'(u) \|_* \geq \delta \tag{4.71}$$

and we can take $\epsilon_0 := \min(\beta, \delta^2/8)$.

Let now ϵ be in $]0, \epsilon_0]$ and u be in U such that $\phi(u) \leq c + \epsilon$. We remark that if there exists $t \in [0, 1[$ such that $\phi(\eta_{\epsilon_0}(t)) \leq c - \epsilon$ then $\phi(\eta_{\epsilon_0}(1)) \leq c - \epsilon$ (since η_{ϵ_0} is a ϕ-decreasing function). Assume then, for all $t \in [0, 1[$, $| \phi(\eta_{\epsilon_0}(t)) - c | \leq \epsilon \leq \epsilon_0 \leq \beta$. So, according to relations (4.70), (4.71) and (4.64), and the fact that $\delta \leq 1$ and $\xi \leq 1$, we can deduce that

$$\frac{\mathrm{d}}{\mathrm{d}t}\phi(\eta_{\epsilon_0}(t)) \leq \frac{-\delta^2}{4} \leq -2\epsilon_0 \quad (\text{since } \epsilon_0 = \min(\beta, \delta^2/8))$$

and then $\phi(\eta_{\epsilon_0}(1)) \leq \phi(\eta_{\epsilon_0}(0)) - 2\epsilon_0 \leq \phi(u) - 2\epsilon_0 \leq c + \epsilon - 2\epsilon_0 \leq c - \epsilon_0$.

This completes the proof. □

Let us now recall the well-known Mountain–Pass theorem in a useful and popular form.

Theorem 4.89. *(Mountain–Pass theorem) Let $(U, \| \, . \, \|)$ be a Banach space and $(U', \| \, . \, \|_*)$ be its dual space. Let ϕ be a real-valued and continuously F-differentiable function on U satisfying the Palais–Small (PS) condition. Assume that $\phi(0) = 0$ and that:*

(i) there exist $\rho > 0$ and $a > 0$ such that if $u \in U : \| u \| = \rho$ then $\phi(u) \geq a$

(ii) there exists an element $u_0 \in U$ such that $\| u_0 \| > \rho$ and $\phi(u_0) \leq a$.

Then ϕ admits a critical value $c \in I\!R$, which is charaterized by the following minimax condition:

$$a \leq c := \inf_{A \in \mathcal{S}} \sup_{u \in A} \phi(u),$$

where $\mathcal{S} := \{g([0, 1]) : \; g \in \mathcal{C}([0, 1]; U), \; g(0) = 0$ and $g(1) = u_0\}$.

Proof. The proof is based on the deformation principle. It is clear that $\mathcal{S} \neq \emptyset$ and, by convexity, for all $A \in \mathcal{S}$, $A \cap \{u \in U : \| u \| = \rho\} \neq \emptyset$. Consequently,

$$\max_{w \in A} \phi(w) \geq a \text{ and } c = \inf_{A \in \mathcal{S}} \sup_{u \in A} \phi(u) \geq a.$$

Prove now that $c = \inf_{A \in \mathcal{S}} \sup_{u \in A} \phi(u)$ is a critical value of ϕ. For this suppose, by contradiction, that the set of critical points of ϕ at level c, $Z(c)$ is empty.

Let $0 < \epsilon \leq a/2$ then, from (i) and (ii), we can find $A \in \mathcal{S}$ such that

$$A = g([0, 1]) \text{ and } a \leq c \leq \max_{w \in A} \phi(w) \leq c + \epsilon.$$

Consequently, according to (PS) condition, the Palais–Small condition at level c, $(PS)_c$, holds.

By using the deformation lemma, we can find $\epsilon_0 > 0$ and a homotopy of homeomorphisms and ϕ-decreasing function η_{ϵ_0} such that the points (a)–(c) of Lemma 4.88 hold.

By taking $f(\tau) := \eta_{\epsilon_0}(1, g(\tau))$ we have that $f(0) = \eta_{\epsilon_0}(1, g(0)) = 0$ and $f(1) = \eta_{\epsilon_0}(1, g(1)) = \eta_{\epsilon_0}(1, u_0) = u_0$ since $\mid \phi(0) - c \mid > \epsilon_0$, $\mid \phi(u_0) - c \mid > \epsilon_0$. Then $f([0,1]) \in \mathcal{S}$. By the deformation lemma (point (c)), we obtain that

$$f([0,1]) \subset \{u \in U . \quad \phi(u) \leq c - \epsilon\},$$

which is a contradiction with the fact that $\sup_{u \in f([0,1])} \phi(u) \geq c$ (according to the definition of c). Therefore, $Z(c)$ is non-empty.

This completes the proof. □

4.3.5 Minimization of Convex Functions

In this section, we give some well-known results concerning the minimization of convex functionals in a reflexive Banach space – the existence of the minimum, the uniqueness and the characterization of the optimal solution – and some results concerning the variational inequalities.

Existence, Uniqueness and Characterization of the Optimal Solution

Let us consider a functional F of $K \subset V$ into $\mathbb{R}$, where

$$\begin{aligned} &(V, \|\ .\ \|) \text{ is a reflexive Banach space on } \mathbb{R} \text{ with norm } \|\ .\ \|_*, \\ &(V', \|\ .\ \|_*) \text{ be the dual space of } V \text{ with norm } \|\ .\ \|_*, \\ &K \text{ is a non-empty closed convex subset of } V, \\ &F \text{ is a convex, lower semi-continuous and proper function on } K. \end{aligned} \tag{4.72}$$

We consider the following minimization problem: find $u \in K$ such that

$$F(u) = \inf_{v \in K} F(v). \tag{4.73}$$

Remark 4.90. Problem (4.73) can be solved throughout the space V, by introducing the extended mapping $\tilde{F} : V \longrightarrow \overline{\mathbb{R}}$ of F by

$$\tilde{F}(u) = F(u) \ \text{ if } \ u \in K \quad \text{and} \quad \tilde{F}(u) = +\infty \ \text{ if } \ u \notin K.$$

Otherwise,

$$\tilde{F}(u) = F(u) + \mathcal{X}_C(u).^3$$

In this way we can relax the constraint $u \in K$ on F to get a proper, convex and lower semi-continuous function $\tilde{F}$ on the whole space V. Therefore, the following problem and its set of solutions:

$$\tilde{F}(u) = \inf_{v \in V} \tilde{F}(v), \tag{4.74}$$

are the same as problem (4.73) and its set of solutions. ◇

[3] $\mathcal{X}_C$ is a convex and lower semi-continuous function since C is a convex and closed set.

Proposition 4.91. *The set of solutions of (4.73) with (4.72) is a closed convex set which is possibly empty.* □

We give now simple conditions of the existence and the uniqueness of a solution of problem (4.73).

Proposition 4.92. *(Existence and uniqueness of an optimal solution) If we assume that the set K is bounded or that the functional J is coercive over K, i.e., satisfies (2.10), then there is at least one $u \in K$ satisfying (4.73). Otherwise $u \in K$ satisfies*

$$F(u) \leq F(v) \quad \text{for all } v \in K. \tag{4.75}$$

Moreover, if we assume that the functional J is strictly convex over K then there is a unique $u \in K$ satisfying (4.73).

Proof. The existence result is given by Theorem 2.101. The uniqueness result is immediate by using the fact that if u_1 and u_2 are two solutions of (4.73) then $(u_1 + u_2)/2$ is also a solution of (4.73) (because Proposition 4.91 holds). □

Let us now give a more analytic condition, for (4.75) to hold, which is said to be the characterization of a solution of (4.73).

Proposition 4.93. *(Characterization of an optimal solution I) If we assume that the functional F is G-differentiable with continuous derivative F', then if $u \in K$, the following conditions are equivalent:*

(i) *u is a solution of (4.73)*

(ii) $\langle F'(u), v - u\rangle_{V',V} \geq 0$ *for all* $v \in K$

(iii) $\langle F'(v), v - u\rangle_{V',V} \geq 0$ *for all* $v \in K$.

Proof. *(i)* $\Leftrightarrow$ *(ii)* is based on the fact that for u a solution of (4.73), $F(u) \leq F((1-t)u + tv)$ for all $v \in K$ and $t \in]0,1[$, and the function F is G-differentiable. The proof of *(ii)* $\Leftrightarrow$ *(iii)* is based on the monotonicity of the function F (because of Proposition 4.46) and on the continuity of F'. The development of the proof is left to the reader as an exercise. □

Proposition 4.94. *(Characterization of an optimal solution II) If we assume that the functional $F = F_1 + F_2$, where F_1 and F_2 are convex and lower semi-continuous functions of K into $\mathbb{R}$, and F_1 is G-differentiable with continuous derivative F_1' (F_2 is not necessarily differentiable), then if $u \in K$, the following conditions are equivalent:*

(i) *u is a solution of (4.73)*

(ii) $\langle F_1'(u), v - u\rangle_{V',V} + F_2(v) - F_2(u) \geq 0$ *for all* $v \in K$

(iii) $\langle F_1'(v), v - u\rangle_{V',V} + F_2(v) - F_2(u) \geq 0$ *for all* $v \in K$.

Proof. The proof is obtained by using the same technique as used in the previous proposition. □

Unilateral Problems or Variational Inequalities

Let $(V, \| \, . \, \|)$ be a reflexive Banach space with norm $\| \, . \, \|$ and let $(V', \| \, . \, \|_*)$ be its dual space with norm $\| \, . \, \|_*$.

Let us consider a non-linear operator A from V into V', a convex, lower semi-continuous and proper function φ of V into $\overline{\mathbb{R}}$ and a given element f of V'.

We consider the following problem: find $u \in V$ such that

$$\langle Au - f, v - u\rangle_{V',V} + \varphi(v) - \varphi(u) \geq 0 \quad \text{for all } v \in V. \tag{4.76}$$

This is the so-called variational inequality.

Theorem 4.95. *We assume that the function φ is a convex, lower semi-continuous and proper function, and the operator A satisfies the following assumptions:*

$$\begin{aligned} &A \text{ is weakly continuous over the subspaces of a finite dimension of } V, \\ &A \text{ is monotone, i.e., } \langle Au - Av, u - v\rangle_{V',V} \geq 0 \ \ \text{for all } u, v \text{ in } V, \end{aligned} \tag{4.77}$$

and there exists an element u_0 of the effective domain of φ domφ *such that*

$$\frac{\langle Av, v - u_0\rangle_{V',V} + \varphi(v)}{\| v \|} \longrightarrow +\infty \quad for \ v \in V, \| v \| \longrightarrow +\infty. \tag{4.78}$$

Then, for f given in V', there is at least one $u \in V$ satisfying the variational inequality (4.76).

Proof. The proof of this theorem and many other existence theorems can be find in Lions [203]. □

Let K be a closed and convex subset of V, we have the following special case as corollary of the previous theorem (by taking φ as the indicator function of K).

Proposition 4.96. *Let K be a closed and convex subset of V and suppose that the operator A satisfies the hypotheses (4.77) and (4.78). Then, for all f in V', there is an element $u \in K$ such that*

$$\langle Au - f, v - u\rangle_{V',V} \geq 0 \quad for \ all \ v \in K. \tag{4.79}$$

If, in addition, $K = V$, then u, the solution of (4.79), satisfies $Au = f$. □

We are now giving some remarks on general boundary value problems.

4.3.6 General Boundary Value Problems

For general boundary value, in addition to the paired function spaces $U(\Omega)$, $U'(\Omega)$ and $X(\Omega)$, $X'(\Omega)$ in duality with respect to certain bilinear forms $\langle .,.\rangle_{U',U}$ and $\langle .,.\rangle_{X',X}$ noted $\langle .,.\rangle_U$ and $\langle .,.\rangle_X$ respectively (to simplify), we introduce some appropriate boundary spaces $U_{\mathrm{B}}(\Gamma), U'_{\mathrm{B}}(\Gamma)$ and $X_{\mathrm{B}}(\Gamma), X'_{\mathrm{B}}(\Gamma)$

in duality with respect to certain bilinear forms $(.,.)_\Gamma$ and $\langle .,.\rangle_\Gamma$ respectively, where Γ is the boundary of the domain Ω. For simplicity, the trace of any element σ of the original spaces on its boundary spaces is also denoted by σ. We can now, for a given linear continuous operator $\Lambda : U \longrightarrow X$, introduce the associated boundary operator $\Lambda_{\rm B} : U_{\rm B} \longrightarrow X_{\rm B}$ and the adjoint operator $\Lambda^* : X' \longrightarrow U'$ by

$$\langle \Lambda u, p^*\rangle_X := \langle u, \Lambda^* p^*\rangle_U + \langle \Lambda_{\rm B} u, p^*\rangle_\Gamma, \tag{4.80}$$

for all u in U and p^* in X'.

We denote by $\Lambda^*_{\rm B} : X'_{\rm B} \longrightarrow U'_{\rm B}$ the adjoint operator of $\Lambda_{\rm B}$ defined by

$$\langle \Lambda_{\rm B} u, p^*\rangle_\Gamma := (u, \Lambda^*_{\rm B} p^*)_\Gamma. \tag{4.81}$$

We introduce now the following operators on the closure of the domain Ω, denoted by $\overline{\Omega}$

$$\begin{aligned} \Lambda_{\rm T} &:= \Lambda \quad \text{on } \Omega, \qquad \Lambda_{\rm T} := -\Lambda_{\rm B} \quad \text{on } \Gamma,\\ \Lambda^*_{\rm T} &:= \Lambda^* \quad \text{on } \Omega, \qquad \Lambda^*_{\rm T} := \Lambda^*_{\rm B} \quad \text{on } \Gamma. \end{aligned} \tag{4.82}$$

In mixed boundary condition value problems, for simplicity, we consider the situation when the boundary B is split into two parts $\Gamma_{\rm u}$ and $\Gamma_{\rm s}$ such that:

$$\Gamma = \Gamma_{\rm u} \cup \Gamma_{\rm s},\;\; \Gamma_{\rm u} \cap \Gamma_{\rm s} = \emptyset \text{ and } \mathrm{mes}(\Gamma_{\rm u}) \neq \emptyset \tag{4.83}$$

and

$$\langle \Lambda_{\rm B} u, p^*\rangle_\Gamma := \langle \Lambda_{\rm B} u, p^*\rangle_{\Gamma_{\rm u}} + (u, \Lambda^*_{\rm B} p^*)_{\Gamma_{\rm s}}. \tag{4.84}$$

We introduce the following bilinear forms:

$$\begin{aligned} \langle \Lambda_{\rm T} u, p^*\rangle_{\overline{X}} &:= \langle \Lambda u, p^*\rangle_X - \langle \Lambda_{\rm B} u, p^*\rangle_{\Gamma_{\rm u}},\\ \langle u, \Lambda^*_{\rm T} p^*\rangle_{\overline{U}} &:= \langle u, \Lambda^* p^*\rangle_U + (u, \Lambda^*_{\rm B} p^*)_{\Gamma_{\rm s}}, \end{aligned} \tag{4.85}$$

where $\overline{X} = X \cap X_{\rm B}$ and $\overline{U} = U \cap U_{\rm B}$. According to (4.81) and (4.84), we obtain the following equality (which is corresponding, for example, to the Green's formula):

$$\langle \Lambda_{\rm T} u, p^*\rangle_{\overline{X}} = \langle u, \Lambda^*_{\rm T} p^*\rangle_{\overline{U}}. \tag{4.86}$$

Let $J_{\rm B}$ be the so-called bilinear concomitant, evaluated on the boundary Γ, (the boundary term generated, for example, by integration by parts) defined by

$$J_{\rm B}(v, q^*) := \langle \Lambda_{\rm B} v, q^*\rangle_\Gamma = \langle \Lambda_{\rm B} v, q^*\rangle_{\Gamma_{\rm u}} + (v, \Lambda^*_{\rm B} q^*)_{\Gamma_{\rm s}}. \tag{4.87}$$

for all v in $U_{\rm B}$ and q^* in $X'_{\rm B}$.

Then, for all u, v in $U_{\rm B}$ and p^*, q^*) in $X'_{\rm B}$, we have

$$\begin{aligned} \frac{\partial J_{\rm B}}{\partial u}(u, p^*)v &= \langle \Lambda_{\rm B} v, p^*\rangle_{\Gamma_{\rm u}} + (v, \Lambda^*_{\rm B} p^*)_{\Gamma_{\rm s}} = J_{\rm B}(v, p^*) = (v, \Lambda^*_{\rm B} p^*)_\Gamma,\\ \frac{\partial J_{\rm B}}{\partial p^*}(u, p^*)q^* &= \langle \Lambda_{\rm B} u, q^*\rangle_{\Gamma_{\rm u}} + (u, \Lambda^*_{\rm B} q^*)_{\Gamma_{\rm s}} = J_{\rm B}(u, q^*) = \langle \Lambda_{\rm B} u, q^*\rangle_\Gamma. \end{aligned}$$

Consequently,

$$\frac{\partial J_{\mathrm{B}}}{\partial u}(u,p^*) = \varLambda_{\mathrm{B}}^* p^* \text{ and } \frac{\partial J_{\mathrm{B}}}{\partial p^*}(u,p^*) = \varLambda_{\mathrm{B}} u. \tag{4.88}$$

In order to finish, we suppose that the boundary conditions $\varLambda_{\mathrm{B}}^* p^* = \phi$ on $\varGamma_{\mathrm{s}}$ and $\varLambda_{\mathrm{B}} u = \xi$ on $\varGamma_{\mathrm{u}}$ are given, then according to (4.88), we can deduce that

$$\frac{\partial J_{\mathrm{B}}}{\partial u}(u,p^*) = \varLambda_{\mathrm{B}}^* p^* = \phi \text{ on } \varGamma_{\mathrm{s}},$$
$$\frac{\partial J_{\mathrm{B}}}{\partial p^*}(u,p^*) = \varLambda_{\mathrm{B}} u = \xi \text{ on } \varGamma_{\mathrm{u}}.$$

Consequently, we have the following abstract equilibrium equation:

$$\begin{aligned} &u^* = \varLambda^* p^* = f \quad \text{on } \varOmega, \qquad \varLambda_{\mathrm{B}}^* p^* = \phi \quad \text{on } \varGamma_{\mathrm{s}}, \\ &p = \varLambda u \quad \text{on } \varOmega, \qquad \varLambda_{\mathrm{B}} u = \xi \quad \text{on } \varGamma_{\mathrm{u}}. \end{aligned} \tag{4.89}$$

We now introduce the following sets: $\mathcal{K}_{\mathrm{ad}} = \{u \in \overline{U} : \ \varLambda_{\mathrm{B}} u = \xi \text{ on } \varGamma_{\mathrm{u}}\}$, which is the admissibility set on u for the mixed boundary-value problem, $\mathcal{A}_{\mathrm{ad}} \subset \overline{X}$, which is the admissibility set on p, and $\mathcal{K}_{\mathrm{kad}} = \{u \in \mathcal{K}_{\mathrm{ad}} : \ \varLambda_{\mathrm{T}} u \in \mathcal{A}_{\mathrm{ad}}\}$, which is the kinematically admissible set.

The external energy $F : \ \mathcal{K}_{\mathrm{ad}} \subset \overline{U} \longrightarrow \mathrm{I\!R}$ is defined by

$$F(u) := \langle f, u \rangle_U + (u, \phi)_{\varGamma_{\mathrm{s}}},$$

and the duality relation between u and u^* is then given by

$$u^* = F'(u) = f \quad \text{on } \varOmega \text{ and } u^* = F'(u) = \phi \quad \text{on } \varGamma_{\mathrm{s}}.$$

Suppose now there exists a G-differentiable and convex operator $W : \ \mathcal{A}_{\mathrm{ad}} \subset \overline{X} \longrightarrow \overline{U}$ such that

$$p^* = W'(p) \quad \text{on } \overline{\varOmega}. \tag{4.90}$$

Thus the total potential energy $J : \ \mathcal{K}_{\mathrm{kad}} \subset \overline{U} \longrightarrow \mathrm{I\!R}$ of the system is defined by

$$J(u) := \mathcal{S}(\varLambda u) + F(u) = \langle W(\varLambda u), 1 \rangle_U + F(u).$$

We can easily prove that, if $\mathcal{K}_{\mathrm{kad}}$ is an open subset of $\mathcal{K}_{\mathrm{ad}}$, u_{c} is a solution of (4.89)-(4.90) if and only if u_{c} is a minimizer of the total potential J, *i.e.*, $J(u_{\mathrm{c}}) = \inf_{u \in \mathcal{K}_{\mathrm{kad}}} J(u)$. Indeed, since the functional J is G-differentiable, then the critical point u_{c} of J, *i.e.*, $J'(u_{\mathrm{c}}) = 0$, satisfies (according to (4.80) and (4.84))

$$\varLambda^* W'(\varLambda u_{\mathrm{c}}) = f \quad \text{on } \varOmega \text{ and } \varLambda_{\mathrm{B}}^* W'(\varLambda_{\mathrm{B}} u_{\mathrm{c}}) = \phi \quad \text{on } \varGamma_{\mathrm{s}}.$$

This shows that the critical points of J are the solution of (4.89) and (4.90) and conversely (by the convexity of J).

We end this chapter by the following remark.

Remark 4.97. The duality plays also a key role in optimal shape design problems. The idea is to find an optimal solution $(\Omega^*, u^*) \in \mathcal{U}_{\text{ad}} \times \mathcal{D}_{\text{ad}}(\Omega^*)$ such that

$$\Phi(u^*, \Lambda(u^*), \Omega^*) = \inf_{\Omega \in \mathcal{U}_{\text{ad}}} \inf_{u \in \mathcal{D}_{\text{ad}}(\Omega)} \Phi(u(\Omega), \Lambda(u(\Omega)), \Omega),$$

where Φ is a given performance (or cost) functional, Λ is some geometrical operator, $\mathcal{U}_{\text{ad}}$ is the set of admissible configurations $\Omega \subset \mathbb{R}^m$ and for any $\Omega \in \mathcal{U}_{\text{ad}}$, and $\mathcal{D}_{\text{ad}}(\Omega)$ denotes the set of admissible deformations u.

For more details, the reader can be referred, for example in the case of the linear geometrical operator, to Haftka and Gurdal [149], Pierre and Henrot [153] and Sokolowski and Zolesio [272]. ♢

5

Lagrange Duality Theory

In this chapter we are interested in the analysis of critical points of Lagrangians with duality. The critical point and duality theories have proven to be one of the most important tools in study of non-linear systems. Among the various critical point methods, minimax principles leading to the existence of saddle points have played an important role in mathematics and science, and together they play a central role for the notion of stability. This chapter describes, first, the Frenchel–Rockafellar duality which is based on a perturbation method and the extended Lagrange duality theory, second, a duality based on the classical minimax theorems, and, finally, the non-convex parametric variational problem for a geometrically non-linear system. For the latter, we have generalized the work of Strang and Gao [127], by introducing a new gap function.

5.1 Frenchel–Rockafellar Duality in Optimization

The Frenchel–Rockafellar duality in convex optimization (see Rockafellar [251]), which is a special case of the generalized Lagrange duality, is based on perturbation techniques. In this section, we associate a so-called dual problem to a given optimization problem, by using conjugate convex function and a family of perturbed problems.

Let U and X be locally convex topological vector spaces with the duals U' and X' respectively. We denote (to simplify the notation) by $\langle .,.\rangle_U$ (respectively $\langle .,.\rangle_X$) the canonical bilinear form $\langle .,.\rangle_{U',U}$ (respectively $\langle .,.\rangle_{X',X}$) with respect to the duality between U' and U (respectively between X' and X); the pairing between $U' \times X'$ and $U \times X$ is written (classically) as

$$\langle (v^*,q^*),(v,q)\rangle_{U\times X} = \langle v^*,v\rangle_U + \langle q^*,q\rangle_X.$$

5.1.1 Primal and Dual Problems

Let us consider a functional J of U into $\overline{\mathbb{R}}$ and consider the following minimization problem $(\mathcal{P})$.

Find the infimum of the functional J, *i.e.*,

$$\inf_{v \in U} J(v). \tag{5.1}$$

Definition 5.1. *The problem* $(\mathcal{P})$ *is called the primal problem, the infimum for problem* $(\mathcal{P})$ *is denoted by* $\inf(\mathcal{P})$ *and every element* u *of* U *such that* $J(u) = \inf(\mathcal{P})$ *is called an optimal solution of* $(\mathcal{P})$*. If there exists an element* u_0 *of* U *such that* $J(u_0) < +\infty$ *(i.e.,* J *is not the functional constant* $+\infty$*), the problem* $(\mathcal{P})$ *is said to be non-trivial.* ■

Before giving the dual problem associated with the primal problem $(\mathcal{P})$, let us consider a functional ϕ of $U \times X$ into $\overline{\mathbb{R}}$ such that

$$\phi(u, 0) = J(u) \quad \forall u \in U. \tag{5.2}$$

Its conjugate function, in the duality between $U' \times X'$ and $U \times X$,

$$\phi^* : U' \times X' \longrightarrow \overline{\mathbb{R}} \;\text{ in } \Gamma(U' \times X')$$

is defined by (for $(u^*, p^*) \in U' \times X'$)

$$\phi^*(u^*, p^*) = \sup_{(v,q) \in U \times X} (\langle (u^*, p^*), (v, q) \rangle_{U \times X} - \phi(v, q)). \tag{5.3}$$

We also introduce, for all $p \in X$, the following sequence of minimization problems $(\mathcal{P}_p)$:

$$\inf_{v \in U} \phi(v, p). \tag{5.4}$$

Classically, the function ϕ is said to be the *perturbation of the function* J, the problem $(\mathcal{P}_p)$ is called the *perturbed problem* and the value p is termed as the *perturbation variable* (or parameter).

Remark 5.2. According to the hypothesis (5.2), it is clear that the problem $(\mathcal{P}_0)$ is the primal problem $(\mathcal{P})$. ◇

Example 5.3. Assume that there exists a continuous linear operator $\Lambda : U \longrightarrow X$, *i.e.*, $\Lambda \in \mathcal{L}(U; X)$, with its adjoint $\Lambda^* \in \mathcal{L}(X'; U')$ such that the functional J can be written in the form

$$J(u) := \Psi(u, \Lambda u),$$

where Ψ is an extended real-valued function on $U \times X$.

Then the primal problem $(\mathcal{P})$ becomes

$$\inf_{u \in U} \Psi(u, \Lambda u).$$

In this case the perturbation function ϕ can be written as

$$\phi(u,p) := \Psi(u, \Lambda u - p).$$

In a particular case, the functional Ψ can be decomposed into $\Psi(u, \Lambda u) := \mathcal{S}(\Lambda u) - F(u)$, and the perturbation function ϕ is then $\phi(u,p) := \mathcal{S}(\Lambda u - p) - F(u)$.

For example, we can take the case of variational problems in the Sobolev spaces $\mathcal{W}^{1,s}(\Omega)$ $(1 \le s \le \infty)$, where Ω is an open bounded subset of $\mathbb{R}^d$ $(d \in \mathbb{N}^*)$ with sufficiently regular boundary $\partial\Omega$ (for example of class $\mathcal{C}^1$). Consider a functional $F: \ \Omega \times \mathbb{R} \times \mathbb{R}^d \longrightarrow \mathbb{R}$, which comes from the boundary value problems and satisfies some growth conditions, and a closed, convex and non-empty subset K of $\mathcal{W}^{1,p}(\Omega)$, for example,

$$K := \{v \in \mathcal{W}^{1,p}(\Omega) : \ v = 0 \text{ on } \partial\Omega\} = \mathcal{W}_0^{1,s}(\Omega).$$

Let $\Lambda = \text{grad}$ and $J(u) = \displaystyle\int_\Omega F(x, u(x), \Lambda u(x))\mathrm{d}x$. The perturbation functional is then

$$\phi(u,p) := \int_\Omega F(x, u(x), \Lambda u(x) - p(x))\mathrm{d}x.$$

A simple example of J is

$$J(u) = \int_\Omega (\frac{1}{2} \mid \Lambda u(x) \mid^2 + \frac{1}{s+1} \mid u(x) \mid^{s+1} - f(x)u(x))\mathrm{d}x$$

where f is given in $L^r(\Omega)$, $r \ge \max(1, 2d/(d+2))$. We can easily prove that the function J is strictly convex, continuously F-differentiable on $U = H_0^1 \cap L^{s+1}(\Omega)$ and satisfies (by using Hölder's and Sobolev's inequalities)

$$\begin{aligned} J(u) &\ge \tfrac{1}{2} \parallel u \parallel^2_{H_0^1(\Omega)} + \tfrac{1}{s+1} \parallel u \parallel^{s+1}_{L^{s+1}(\Omega)} - C_0 \parallel f \parallel_{L^r(\Omega)} \parallel u \parallel_{H_0^1(\Omega)} \\ &\ge \tfrac{1}{4} \parallel u \parallel^2_{H_0^1(\Omega)} + \tfrac{1}{s+1} \parallel u \parallel^{s+1}_{L^{s+1}(\Omega)} - C_1 \parallel f \parallel^2_{L^r(\Omega)} . \end{aligned}$$

Consequently, if $\parallel u \parallel_U \longrightarrow \infty$ then $J(u) \longrightarrow \infty$ and then the infimum of J is attained on U at a unique critical point u_c and we have

$$0 = J'(u_c) = -\Delta u_c + \mid u_c \mid^{s-1} u_c - f \ \text{ on } \Omega \ \text{ and } \ u_c = 0 \text{ on } \partial\Omega.$$

In this case the perturbation functional is given by

$$\phi(u,p) := \int_\Omega (\mid \Lambda u(x) - p(x) \mid^2 + \frac{1}{s+1} \mid u(x) \mid^{s+1} - f(x)u(x))\mathrm{d}x.$$

♣

We are now be able to define a dual or adjoint problem. The dual problem associated with $(\mathcal{P})$ with respect to ϕ is then the following problem $(\mathcal{P}^*)$:

$$\sup_{q \in X'} (-\phi^*(0,q)). \tag{5.5}$$

The supremum for problem $(\mathcal{P}^*)$ is denoted by $\sup(\mathcal{P}^*)$ and every element p^* of X' such that $-\phi^*(0,p^*) = \sup(\mathcal{P}^*)$ is called an optimal solution of $(\mathcal{P}^*)$.

We shall now give some relationships between the primal problem $(\mathcal{P})$ and its dual $(\mathcal{P}^*)$.

Proposition 5.4. *(Weak duality) For problems* $(\mathcal{P})$ *and* $(\mathcal{P}^*)$ *we have:*

(i) $-\infty \leq \sup(\mathcal{P}^*) \leq \inf(\mathcal{P}) \leq +\infty$.

(ii) If $(\mathcal{P})$ *is non-trivial, then* $\sup(\mathcal{P}^*) \leq \inf(\mathcal{P}) < +\infty$.

(iii) If $(\mathcal{P}^*)$ *is non-trivial, then* $-\infty < \sup(\mathcal{P}^*) \leq \inf(\mathcal{P})$.

(iv) If $(\mathcal{P})$ *and* $(\mathcal{P}^*)$ *are non-trivial then* $-\infty < \sup(\mathcal{P}^*) \leq \inf(\mathcal{P}) < +\infty$.

Proof. Let $p^* \in X'$, then we have, by definition, that

$$\phi^*(0,p^*) = \sup_{(v,q)\in U\times X} (\langle p^*, q\rangle_X - \phi(v,q))$$

and, in particular, we have that for all $v \in U$

$$\phi^*(0,p^*) \geq \langle p^*, 0\rangle_X - \phi(v,0) = -\phi(v,0).$$

Consequently, $\sup(\mathcal{P}^*) \leq \inf(\mathcal{P})$ and then (i).

If $(\mathcal{P})$ is non-trivial, there exists u_0 such that $J(u_0) < \infty$ and then $\phi(u_0,0) < +\infty$. Thus with (i) we can deduce the result (ii). In the same way we can prove the result (iii).

The result (iv) follows from (ii) and (iii). □

Definition 5.5. *(Dual gap) The difference between* $\inf(\mathcal{P})$ *and* $\sup(\mathcal{P}^*)$*:*

$$\inf(\mathcal{P}) - \sup(\mathcal{P}^*)$$

is said to be the dual gap of problem $(\mathcal{P})$. ■

Since the technique used to introduce the dual problem corresponding to a minimization problem, is valid for a maximization problem, we can easily introduce the dual of the dual problem $(\mathcal{P}^*)$ and then the bidual problem corresponding to the primal problem $(\mathcal{P})$, $(\mathcal{P}^{**})$, is

$$\inf_{u\in U} (\phi^{**}(u,0)), \tag{5.6}$$

where ϕ^{**} is the conjugate function of ϕ^* and then the biconjugate and the Γ-regularization function of ϕ.

Remark 5.6. (i) Since the dual of ϕ^{**} is none other than ϕ^* (according to Corollary 4.7), then the repetition of the duality operation is limited and the dual problem corresponding to problem $(\mathcal{P}^{**})$ is the problem $(\mathcal{P}^*)$.

(ii) If $\phi^{**} = \phi$ (and then $\phi \in \Gamma(U \times X)$) then problem $(\mathcal{P}^{**})$ is identical to problem $(\mathcal{P})$ and consequently (according to (i)) problem $(\mathcal{P})$ is the dual of $(\mathcal{P}^*)$ (and inversely).

(iii) If $\phi \in \Gamma_0(U \times X)$, *i.e.*, the functional $\phi \in \Gamma(U \times X)$ is not equal to the constant functions $+\infty$ and $-\infty$, then $J \in \Gamma_0(U)$. Consequently, problem $(\mathcal{P})$ is non-trivial and $(\mathcal{P})$ is the dual of $(\mathcal{P}^*)$. ♢

5.1.2 Normal and Stability Problems

Let us consider a functional J of U into $\overline{\mathbb{R}}$, and a functional ϕ of $U \times X$ into $\overline{\mathbb{R}}$ and its conjugate function ϕ^* in the duality between $U' \times X'$ and $U \times X$ defined by (5.3). We make the following assumption for some of our results

$$\phi \in \Gamma_0(U \times X)., \tag{5.7}$$

Let us consider now the infimal value function h of X into $\overline{\mathbb{R}}$ such that: for all element $p \in X$, the value $h(p)$ is defined as the infimum for problem (5.4), *i.e.*,

$$h(p) := \inf_{v \in U} \phi(v, p),$$

and h^* be its conjugate function in the duality between X' and X. It follows that

$$h(0) = \inf(\mathcal{P}) = \inf_{v \in U} J(v).$$

Proposition 5.7. *If the function ϕ is convex then the function h of X into $\overline{\mathbb{R}}$ is also convex.*

Proof. Let p, q be in X and suppose that there exist a, b in $\mathbb{R}$ such that $h(p) < a$ and $h(q) < b$.

Let α be in $(0, 1)$, then, since ϕ is convex on $U \times X$,

$$\begin{aligned} h(\alpha p + (1-\alpha)q) &= \inf_{\delta \in X} \phi(\delta, \alpha p + (1-\alpha)q) \\ &\leq \phi(\alpha w + (1-\alpha)\zeta, \alpha p + (1-\alpha)q) \ \ \forall w, \zeta \in X \\ &\leq \alpha\phi(w, p) + (1-\alpha)\phi(\zeta, q) \ \ \forall w, \zeta \in X. \end{aligned}$$

Consequently,

$$\begin{aligned} h(\alpha p + (1-\alpha)q) &\leq \alpha \inf_{w \in X} \phi(w, p) + (1-\alpha) \inf_{\zeta \in X} \phi(\zeta, q) \\ &= \alpha h(p) + (1-\alpha)h(q) \end{aligned}$$

and then the convexity result.

If $h(p) = +\infty$ or $h(q) = +\infty$ the proof is trivial. □

Remark 5.8. In particular, if ϕ satisfies the assumption (5.7), ϕ is convex and then h is convex, but in general (although $\phi \in \Gamma_0(U \times X)$) the function $h \notin \Gamma_0(X)$. ◊

Now we give the relation between the conjugate function h^* of h and the conjugate function ϕ^* of ϕ.

Lemma 5.9. (*i*) *For all $q^* \in X'$, we have that*

$$h^*(q^*) = \phi^*(0, q^*).$$

(*ii*) *Let h^{**} be the biconjugate function of h, then we have that*

$$\sup(\mathcal{P}^*) = \sup_{q^* \in X'} (-h^*(q^*)) = h^{**}(0).$$

Proof. The proof is left to the reader as an exercise. □

Lemma 5.10. *If h is a subdifferentiable function at point 0, then the set of the solutions to problem $(\mathcal{P}^*)$ is the set $\partial h^{**}(0)$.*

Proof. Let $p^* \in X'$ be an optimal solution to problem $(\mathcal{P}^*)$. It follows that

$$\phi^*(0, p^*) \leq \phi^*(0, q^*) \ \ \forall q^* \in X'.$$

According to Lemma 5.9, we can deduce that

$$h^*(p^*) \leq h^*(q^*) \ \ \forall q^* \in X'$$

and then

$$h^*(p^*) \leq -\sup_{q^* \in X'} (-h^*(q^*)) = -h^{**}(0).$$

Therefore, $h^*(p^*) + h^{**}(0) \leq 0$. Since $h^*(p^*) + h^{**}(0) \geq \langle p^*, 0 \rangle_X$ (Fenchel's inequality), then $h^*(p^*) + h^{**}(0) = 0$ and consequently, $p^* \in \partial h^{**}(0)$ (according to Proposition 4.16).

Conversely, let p^* be in $\partial h^{**}(0)$, then by using the same arguments as to obtain the previous inclusion (but in the reversed direction), it yields that p^* is an optimal solution of $(\mathcal{P}^*)$. □

Definition 5.11. *(Normal problem) The problem $(\mathcal{P})$ defined by (5.1) is said to be normal if h is a finite and lower semi-continuous function at point* 0.■

The following proposition shows some characterization of a normal problem.

Proposition 5.12. *(Characterization of a normal problem) If ϕ satisfies the assumption (5.7), then the following conditions are equivalent:*

(i) problem $(\mathcal{P})$ is normal

(ii) problem $(\mathcal{P}^*)$ is normal

(iii) $-\infty < \inf(\mathcal{P}) = \sup(\mathcal{P}^*) < +\infty$.

Proof. The equivalence *(ii)* ⇔ *(iii)* is immediate and results from the fact that $(\mathcal{P}) = (\mathcal{P}^{**})$, *i.e.*, $(\mathcal{P})$ is the dual problem of $(\mathcal{P}^*)$ (because of Remark 5.6). Prove now that *(i)* ⇔ *(iii)*.

Let $\overline{h}$ be the closure of the convex function h, then (according to Fenchel–Moreau's theorem) we have

$$h^{**} = \overline{h} \leq h.$$

Since $\overline{h}(0) = h(0) \in \mathbb{R}$ (by assumption) then (because of Lemma 5.9)

$$\inf(\mathcal{P}) = h(0) = h^{**}(0) = \sup(\mathcal{P}^*) \in \mathbb{R}, \tag{5.8}$$

and conversely. □

Remark 5.13. In the proof of the equivalence (*i*) $\Leftrightarrow$ (*iii*), we have only used the convexity of h and consequently, only the convexity of ϕ (according to Proposition 5.7). $\diamondsuit$

Let us now introduce the notion of stability.

Definition 5.14. *(Stable problem) The problem* $(\mathcal{P})$ *defined by (5.1) is said to be stable if* h *is a finite and subdifferentiable function at point 0 (i.e.,* $h(0)$ *is finite and* $\partial h(0) \neq \emptyset$*).* ■

The following proposition shows the strong duality.

Proposition 5.15. *(Strong duality I) If* ϕ *satisfies the assumption (5.7), then the following conditions are equivalent:*

(i) problem $(\mathcal{P})$ *is stable*

(ii) problem $(\mathcal{P})$ *is normal and* $(\mathcal{P}^*)$ *has at least one solution.*

Proof. The proof is a simple application of Theorem 4.15 by taking account of Lemma 5.10 and the relation (5.8). □

It is clear that we have the following result (by using Remark 5.6 and Proposition 5.15).

Corollary 5.16. *(Strong duality II) If* ϕ *satisfies the assumption (5.7), then the following conditions are equivalent:*

(i) problems $(\mathcal{P})$ *and* $(\mathcal{P}^*)$ *are normal and have some solutions*

(ii) problems $(\mathcal{P})$ *and* $(\mathcal{P}^*)$ *are stable*

(iii) problem $(\mathcal{P})$ *is stable and has some solutions.* □

To prove the stability by means of the definition is not so easy in general. We give now such a stability criterion.

Proposition 5.17. *(Stability criterion) Let* ϕ *be a convex function and let us assume that* $\inf(\mathcal{P})$ *is finite (i.e.,* $-\infty < \inf_{v\in U} J(v) < \infty$*). Further, let us suppose that*

$$\begin{array}{l} \textit{there exists an element } u_0 \in U \textit{ such that the function} \\ p \in X \longrightarrow \phi(u_0, p) \in \overline{I\!R} \textit{ is finite and continuous at point } 0, \end{array} \tag{5.9}$$

then problem $(\mathcal{P})$ *is stable.*

Proof. According to assumptions we have that $h(0)$ is finite and h is convex (because of Proposition 5.7). Since the function

$$\psi_0 : \; q \in X \longrightarrow \psi_0(q) = \phi(u_0, q) \in \overline{I\!R}$$

is proper convex and continuous at point 0, then there exists a neighborhood of 0 in X on which ψ_0 is bounded above by a finite constant. Consequently (according to Proposition 2.103), the function h that is bounded by the function ψ_0 is then finite and continuous at point 0. Because of Proposition 4.22, we can conclude that h is finite and subdifferentiable at point 0 and then we have the stability of problem $(\mathcal{P})$. □

Now we give some extremality relations or optimality conditions and existence of solutions.

5.1.3 Optimality Conditions and Existence

Proposition 5.18. *(Extremal relations) If problems $(\mathcal{P})$ and $(\mathcal{P}^*)$ have both of the solutions and*

$$-\infty < \inf(\mathcal{P}) = \sup(\mathcal{P}^*) < +\infty, \tag{5.10}$$

then all solutions $u \in U$ of $(\mathcal{P})$ and all solutions $p^* \in X'$ of $(\mathcal{P}^*)$ satisfy the following optimality relations:

$$\phi(u,0) + \phi^*(0,p^*) = 0, \tag{5.11}$$

or (equivalently)

$$(0,p^*) \in \partial\phi(u,0) \text{ or } (u,0) \in \partial\phi^*(0,p^*). \tag{5.12}$$

Conversely, if $u \in U$ and $p^* \in X'$ satisfy the relations (5.11) or (5.12), then u is a solution of problem $(\mathcal{P})$, p^* is a solution of $(\mathcal{P}^*)$ and moreover, we have $-\infty < \inf(\mathcal{P}) = \sup(\mathcal{P}^*) < +\infty$.

Proof. The proof is immediate by using the definition of the conjugate function ϕ^* and the fact that $\inf(\mathcal{P}) = \phi(u,0) = \sup(\mathcal{P}^*) = -\phi^*(0,p^*)$. □

Remark 5.19. Because of Proposition 5.12 and Proposition 5.15, the relation (5.10) is true if problem $(\mathcal{P})$ is normal or stable or if $(\mathcal{P}^*)$ is normal. ◇

We give now a result of existence in the case of the reflexive Banach space.

Proposition 5.20. *Let $(U, \| \,.\, \|)$ be a reflexive Banach space. Under assumptions (5.7) and (5.9), and if*

$$\phi(v,0) \longrightarrow +\infty \text{ for } v \in U, \| v \| \longrightarrow +\infty, \tag{5.13}$$

then problems $(\mathcal{P})$ and $(\mathcal{P}^*)$ have at least one solution,

$$-\infty < \inf(\mathcal{P}) = \sup(\mathcal{P}^*) < +\infty,$$

and the optimality conditions (5.11) (or equivalently (5.12)) are satisfied.

Proof. According to the assumption (5.13), the function $J(.) = \phi(.,0)$ satisfies the condition of the existence result of Proposition 4.92 (here $K := U$) and then we have the existence of a solution of $(\mathcal{P})$. Because of Proposition 5.17, we can deduce the existence of a solution of $(\mathcal{P}^*)$ and then the optimality conditions (according to Proposition 5.18). □

5.1.4 Bidual Problem and Duality in Variational Inequalities

In this section we give some results which concern the bidual problem of a given problem and the duality in variational inequalities.

The Bidual Problem

In addition to the two pairs of topological vector spaces U, U' and X, X' in duality, we take two topological spaces U'' and X'' such that U'' and U' (respectively X'' and X') are in duality and

$$\begin{aligned}&U \subset U'' \text{ is dense in } U'' \text{ (respectively } X \subset X'' \text{ is dense in } X'')\\&\text{where the injection is continuous.}\end{aligned} \tag{5.14}$$

Moreover, we assume that, the function ϕ satisfies (5.7), *i.e.*, $\phi \in \Gamma_0(U \times X)$.

Let ϕ^{**} be the biconjugate function of ϕ in the duality between $U'' \times X''$ and $U' \times X'$. Since $\phi \in \Gamma_0(U \times X)$ then $\phi^{**} \in \Gamma_0(U'' \times X'')$.

We can now give the dual of the dual problem $(\mathcal{P}^*)$ between $U'' \times X''$ and $U' \times X'$ and then the bidual problem corresponding to the primal problem $(\mathcal{P})$, $(\mathcal{P}^{**})$, is

$$\inf_{u \in U''} (\phi^{**}(u, 0)).$$

According to Proposition 5.4, we have that (because of Frenchet–Moreau's theorem)

$$-\infty \leq \sup(\mathcal{P}^*) \leq \inf(\mathcal{P}^{**}) \leq \inf \mathcal{P} \leq +\infty. \tag{5.15}$$

Remark 5.21. (*i*) If $\inf(\mathcal{P}) = \inf(\mathcal{P}^{**})$ then any solution of $(\mathcal{P})$ is a solution of $(\mathcal{P}^{**})$ and any solution in $U \subset U''$ of $(\mathcal{P}^{**})$ is a solution of $(\mathcal{P})$. In this case the problem $(\mathcal{P}^{**})$ is called a *weak formulation* of problem $(\mathcal{P})$ and each solution u in U'' such that $u \notin U$ is called a *weak solution* of problem $(\mathcal{P})$.

(*ii*) If the assumption (5.9) holds then according to Propositions 5.17, 5.15 and 5.12, we have that

$$\inf(\mathcal{P}) = \sup(\mathcal{P}^*)$$

and consequently (according to (5.15)),

$$\sup(\mathcal{P}^*) = \inf(\mathcal{P}^{**}) = \inf(\mathcal{P}). \tag{5.16}$$

(*iii*) If we assume that U (respectively X) is a non-reflexive Banach space, U' (respectively X') its dual and U'' (respectively X'') its bidual, we can observe that the assumption (5.14) is satisfied. ♢

Proposition 5.22. *Let us suppose that U (respectively X) is a non-reflexive Banach space, U' (respectively X') its dual and U'' (respectively X'') its bidual. If the function ϕ satisfies the conditions (5.7) and (5.9) and if* $\mathrm{dom}\phi(.,0)$ *is a non-empty subset of U such that*

$$\phi(v, 0) \longrightarrow +\infty \quad \textit{for } v \in \mathrm{dom}\phi(.,0), \ \| v \|_U \longrightarrow +\infty, \tag{5.17}$$

then problem $(\mathcal{P})$ *admits weak solutions and all accumulation points*[1] *of a minimizing sequence of* $(\mathcal{P})$ *are weak solutions of* $(\mathcal{P})$.

Proof. The proof is left to the reader as an exercise. □

Duality in Variational Inequalities

Let $(U, \| \,.\, \|)$ be a reflexive Banach space with norm $\| \,.\, \|$ and let $(U', \| \,.\, \|_*)$ be its dual space with norm $\| \,.\, \|_*$.

Let us consider an operator A from U into U' satisfying the assumptions (4.77)–(4.78), a function φ that is a convex, lower semi-continuous and proper function of U into $\overline{\mathbb{R}}$ and a given element f of U'.

We consider the following variational inequality problem: find $u \in U$ such that

$$\langle Au - f, v - u \rangle_U + \varphi(v) - \varphi(u) \geq 0 \quad \text{for all } v \in U. \tag{5.18}$$

According to Theorem 4.95, problem (5.18) admits at least one solution u in U. By setting $g = Au - f$ we have that (according to (5.18))

$$\langle g, v \rangle_U + \varphi(v) \geq \langle g, u \rangle_U + \varphi(u) \quad \text{for all } v \in U. \tag{5.19}$$

Problem (5.19) is not an optimization problem because of the dependence of g on u, but once u is known then g (depending on the known u) can be considered as a known element of U' and then problem (5.19) can be considered as an optimization problem $(\mathcal{I})$ (for which u is one solution)

$$\inf_{v \in U} G(v), \tag{5.20}$$

where the functional G of U into $\overline{\mathbb{R}}$ is defined by $G(v) = \langle g, v \rangle_U + \varphi(v)$.

We can use the same technique as in previous section in order to obtain the dual problem corresponding to problem $(\mathcal{I})$.

5.2 Lagrange Duality

The Lagrange duality theory plays an important role in constrained optimization, in general variational problems and in control theory. In this section we examine the extremality relationship between primal and dual functionals, which are linked by an arbitrary extended Lagrangian function, by the comparaison of the infimum with the supremum.

5.2.1 Definitions and Critical Points of Lagrangians

Let us now introduce the notions of saddle point, subcritical point and supercritical point for an extended real-valued function.

[1] Also called cluster points.

Definition 5.23. *(Saddle point) Let U and Y be two topological vector spaces and $L: U \times Y \longrightarrow \overline{I\!R}$ be an extended real-valued function. The pair (u, p^*) of $U \times Y$ is called a saddle point of the function L if*

$$L(u, q^*) \leq L(u, p^*) \leq L(v, p^*), \forall (v, q^*) \in U \times Y. \tag{5.21}$$

■

Definition 5.24. *(Supercritical point) Let U and Y be two topological vector spaces and $L: U \times Y \longrightarrow \overline{I\!R}$ be an extended real-valued function. The pair (u, p^*) of $U \times Y$ is called a supercritical point of the function L if*

$$L(u, q^*) \leq L(u, p^*) \geq L(v, p^*), \forall (v, q^*) \in U \times Y. \tag{5.22}$$

■

Definition 5.25. *(Subcritical point) Let U and Y be two topological vector spaces and $L: U \times Y \longrightarrow \overline{I\!R}$ be an extended real-valued function. The pair (u, p^*) of $U \times Y$ is called a subcritical point of the function L if*

$$L(u, q^*) \geq L(u, p^*) \leq L(v, p^*), \forall (v, q^*) \in U \times Y, \tag{5.23}$$

i.e., (u, p^*) is a supercritical point of the functional $-L$. ■

Definition 5.26. *(Remarquable point) Let U and Y be two topological vector spaces and $L: U \times Y \longrightarrow \overline{I\!R}$ be an extended real-valued function. A point (u, p^*) is said to be a remarquable point of the function L if (u, p^*) is either a saddle point of L or of $-L$, or the supercritical or the subcritical of L.* ■

We shall now give a characterization of remarquable points of an arbitrary given extended real-valued function $L: U \times X' \longrightarrow \overline{I\!R}$ by using its partial subdifferentials.

For a given function L, their partial subdifferentials at point (u, p^*), $\partial_v L(u, p^*)$ and $\partial_{q^*} L(u, p^*)$, are defined by

$$\begin{aligned} &v^* \in \partial_v L(u, p^*) \text{ if } v^* \in U' \text{ and} \\ &\qquad\qquad L(v, p^*) \geq L(u, p^*) + \langle v^*, v - u \rangle_U, \\ &q^{**} \in \partial_{q^*} L(u, p^*) \text{ if } q^{**} \in X'' \text{ and} \\ &\qquad\qquad L(u, q^*) > L(u, p^*) + \langle q^{**}, q^* - p^* \rangle_{X'}, \end{aligned} \tag{5.24}$$

for all v in U and q^* in X' (X'' is the bidual of X).

According to the expressions (5.21), (5.23) and (5.22), we can deduce easily the following characterization.

Lemma 5.27. *Let $L: U \times X' \longrightarrow \overline{I\!R}$ be a given function and (u, p^*) be in $U \times X'$:*

(i) A point (u, p^*) is a saddle point of L if and only if

$$0 \in \partial_v L(u, p^*) \text{ and } 0 \in \partial_{q^*}(-L)(u, p^*). \tag{5.25}$$

(ii) A point (u, p^*) *is a subcritical point of* L *if and only if*

$$0 \in \partial_v L(u, p^*) \ \ \textit{and} \ \ 0 \in \partial_{q^*} L(u, p^*). \tag{5.26}$$

(iii) A point (u, p^*) *is a supercritical point of* L *if and only if*

$$0 \in \partial_v (-L)(u, p^*) \ \ \textit{and} \ \ 0 \in \partial_{q^*} (-L)(u, p^*). \tag{5.27}$$

□

Let us introduce now the Lagrangian forms of problem $(\mathcal{P})$ (given by (5.1)) and give some results concerning the critical points.

Definition 5.28. *(Lagrangian types) Let* $L : \ U \times X' \longrightarrow \overline{I\!R}$ *be a given function:*

(i) L *is said to be a Lagrangian function of type I of problem* $(\mathcal{P})$ *(given by (5.1)), if for each* $v \in U$

$$J(v) = \sup_{q^* \in X'} L(v, q^*). \tag{5.28}$$

(ii) L *is said to be a Lagrangian function of type II of problem* $(\mathcal{P})$ *(given by (5.1)), if for each* $v \in U$

$$J(v) = \inf_{q^* \in X'} L(v, q^*). \tag{5.29}$$

■

If L is a Lagrangian function of type I, the dual problem associated with problem $(\mathcal{P})$ is given by the following problem $(\mathcal{P}_{\mathrm{I}}^*)$:

$$\sup_{q^* \in X'} H(q^*) \tag{5.30}$$

and if L is a Lagrangian function of type II, the dual problem associated with problem $(\mathcal{P})$ is given by the following problem $(\mathcal{P}_{\mathrm{II}}^*)$:

$$\inf_{q^* \in X'} H(q^*), \tag{5.31}$$

where $H : \ X' \longrightarrow \overline{I\!R}$ is defined by

$$H(q^*) := \inf_{v \in U} L(v, q^*). \tag{5.32}$$

Remark 5.29. (*i*) If L is a Lagrangian function of type I, the dual problem $(\mathcal{P}_{\mathrm{I}}^*)$ is a supremum problem and then the problem $(\mathcal{P}_{\mathrm{I}}^*)$ is corresponding to find saddle points of the Lagrangian L. On the other hand, if L is a Lagrangian function of type II, the dual problem $(\mathcal{P}_{\mathrm{II}}^*)$ is an infimum problem and then the problem $(\mathcal{P}_{\mathrm{II}}^*)$ is corresponding to another infimum problem (if u and p^*

are solutions of $(\mathcal{P})$ and $(\mathcal{P}^*_{\mathrm{II}})$ respectively, then (u, p^*) is an infimum of L on $U \times X'$).

(*ii*) If L is a Lagrangian function of type II, then

$$\inf(\mathcal{P}) = \inf(\mathcal{P}^*_{\mathrm{II}}),$$

since $\inf_{v \in U} \left(\inf_{q^* \in X'} L(v, q^*) \right) = \inf_{q^* \in X'} \left(\inf_{v \in U} L(v, q^*) \right)$. ◊

Proposition 5.30. *Let $L : U \times X' \longrightarrow \overline{I\!R}$ be a Lagrangian function of type I of problem $(\mathcal{P})$ and $(\mathcal{P}^*_{\mathrm{I}})$ is the dual problem associated with $(\mathcal{P})$. Then $(u_{\mathrm{s}}, p^*_{\mathrm{s}})$ is a saddle point of L on $U \times X'$ if and only if u_{s} is a solution of $(\mathcal{P})$, p^*_{s} is a solution of $(\mathcal{P}^*_{\mathrm{I}})$ and* $\inf(\mathcal{P}) = \sup(\mathcal{P}^*_{\mathrm{I}})$.

Proof. According to the definition of J and H, it is clear that for all (v, q^*) given in $U \times X'$ we have

$$J(u) \geq H(p^*) \tag{5.33}$$

(since $J(u) - H(p^*) = \sup_{v \in U} \sup_{q^* \in X'} (L(u, q^*) - L(v, p^*)) \geq 0$, because of (5.28) and (5.32)).

Let $(u_{\mathrm{s}}, p^*_{\mathrm{s}})$ be a saddle point of L. Then, according to the definition of J and H, and to the expression (5.21), we have that

$$J(u_{\mathrm{s}}) \leq L(u_{\mathrm{s}}, p^*_{\mathrm{s}}) \leq H(p^*_{\mathrm{s}}).$$

Consequently, because of (5.33), we can deduce that

$$J(u_{\mathrm{s}}) = L(u_{\mathrm{s}}, p^*_{\mathrm{s}}) = H(p^*_{\mathrm{s}})$$

and then u_{s} is a solution of $(\mathcal{P})$, p^*_{s} is a solution of $(\mathcal{P}^*_{\mathrm{I}})$ and $\inf(\mathcal{P}) = \sup(\mathcal{P}^*_{\mathrm{I}})$.

Conversely, let u_{s} be a solution of $(\mathcal{P})$, p^*_{s} be a solution of $(\mathcal{P}^*_{\mathrm{I}})$ and $\inf(\mathcal{P}) = \sup(\mathcal{P}^*_{\mathrm{I}})$. Then, from the definition of J and H

$$\sup_{q^* \in X'} L(u_{\mathrm{s}}, q^*) = J(u_{\mathrm{s}}) = H(p^*_{\mathrm{s}}) = \inf_{v \in U} L(v, p^*_{\mathrm{s}})$$

and we can deduce easily that the expression (5.21) holds, *i.e.*, $(u_{\mathrm{s}}, p^*_{\mathrm{s}})$ is a saddle point of L. □

Similarly we have the following result concerning subcritical points of a Lagrangian function of type II of problem $(\mathcal{P})$.

Proposition 5.31. *Let $L : U \times X' \longrightarrow \overline{I\!R}$ be a Lagrangian function of type II of problem $(\mathcal{P})$ and $(\mathcal{P}^*_{\mathrm{II}})$ is the dual problem associated with $(\mathcal{P})$. Then the following statements hold:*

(*i*) *If $(u_{\mathrm{s}}, p^*_{\mathrm{s}})$ is a subcritical point of L on $U \times X'$ then*

$$J(u_{\mathrm{s}}) = L(u_{\mathrm{s}}, p^*_{\mathrm{s}}) = H(p^*_{\mathrm{s}}).$$

(ii) *A point* (u_s, p_s^*) *minimizes* L *on* $U \times X'$ *if and only if* u_s *is a solution of* $(\mathcal{P})$, p_s^* *is a solution of* $(\mathcal{P}_I^*)$ *and* $\inf(\mathcal{P}) = \inf(\mathcal{P}_{II}^*)$.

Proof. Let (u_s, p_s^*) be a subcritical point of L. According to (5.26) and (5.24), we have that

$$L(v, p_s^*) \geq L(u_s, p_s^*) \text{ and } L(u_s, q^*) \geq L(u_s, p_s^*)$$

for all v in U and q^* in X'.
Consequently,

$$H(p_s^*) = \inf_{v \in U} L(v, p_s^*) = L(u_s, p_s^*) \text{ and } J(u_s) = \inf_{q^* \in X'} L(u_s, q^*) = L(u_s, p_s^*).$$

This proves assertion *(i)*. Assertion *(ii)* can be proved in the same way as the proof of Proposition 5.30. Therefore, the proof is left to the reader as an exercise. □

A direct consequence of Proposition 5.31 is the following result.

Theorem 5.32. *Let* $L: U \times X' \longrightarrow \overline{I\!R}$ *be a Lagrangian function of type II of problem* $(\mathcal{P})$, $(\mathcal{P}_{II}^*)$ *is the dual problem associated with* $(\mathcal{P})$ *and* (u_s, p_s^*) *be a subcritical point of* L. *Then the following conditions are equivalent:*

(i) u_s *is a solution of* $(\mathcal{P})$

(ii) p_s^* *is a solution of* $(\mathcal{P}_{II}^*)$

(iii) $L(u_s, p_s^*) = \inf_{(v,q^*) \in U \times X'} L(v, q^*)$.

Proof. The proof is left to the reader as an exercise. □

We finish this subsection by introducing the notion of critical points of a G-differentiable Lagrangian function.

Definition 5.33. *(Lagrangian-remarquable point) Let* $L: U \times X' \longrightarrow \overline{I\!R}$ *be a Lagrangian function. A point* (u, p^*) *is said to be a Lagrangian-remarquable point of* L *if* (u, p^*) *is a remarquable point of* L. ■

Definition 5.34. *(Critical point of L) Let* $L: U \times X' \longrightarrow \overline{I\!R}$ *be a Lagrangian function. A point* $(u_c, p_c^*) \in U \times X'$ *is said to be a critical point of* L *if* L *is partially G-differentiable at* (u_c, p_c^*) *and*

$$\frac{\partial L}{\partial v}(u_c, p_c^*) = 0 \quad \text{in } U' \quad \text{and} \quad \frac{\partial L}{\partial q^*}(u_c, p_c^*) = 0 \quad \text{in } X'', \tag{5.34}$$

where $\partial L / \partial v$ *and* $\partial L / \partial q^*$ *denote partial derivatives on* U *and* X', *respectively.* ■

We can now give the critical point theorem.

Proposition 5.35. *Let $L : U \times X' \longrightarrow \overline{I\!R}$ be a Lagrangian function of problem $(\mathcal{P})$ (given by (5.1)) and $(u_c, p_c^*) \in U \times X'$ be a Lagrangian-remarquable point of L:*

(*i*) *If L is partially G-differentiable at point (u_c, p_c^*), then (u_c, p_c^*) is a critical point of L.*

(*ii*) *If J (respectively H) is G-differentiable at u_c (respectively p_c^*), then $J'(u_c) = 0$ (respectively $H'(p_c^*) = 0$), where H is given by (5.32).*

Proof. The proof is left to the reader as an exercise[2]. □

In the sequel we will be interested essentially in saddle point results (and consequently, to the Lagrangian function of type I). For details on the supercritical or subcritical point results, the reader can be referred, for instance, to Auchmuty [15].

5.2.2 Lagrangian Duality and Saddle Points

In this section we introduce the so-called Lagrangian duality function and show the relation between the conjugate duality and the well-known Lagrange duality.

Lagrangian Function Associated to a Perturbation

Definition 5.36. *(Lagrangian duality function) The functional $L : U \times X' \longrightarrow \overline{I\!R}$ such that*

$$\begin{aligned} L(u, q^*) &= -\sup_{q \in X} (\langle q^*, q \rangle_X - \phi(u, q)), \forall u \in U, \forall q^* \in X' \\ &= \inf_{q \in X} (\phi(u, q) - \langle q^*, q \rangle_X), \forall u \in U, \forall q^* \in X' \end{aligned} \tag{5.35}$$

is said to be the Lagrangian (duality) function of problem $(\mathcal{P})$ (given by (5.1)) with respect to the associated perturbation $\phi : U \times X \longrightarrow \overline{I\!R}$ (defined by (5.2)).

Moreover, $L(u, q^*)$ can be written as

$$L(u, q^*) = -\psi_u^*(q^*), \forall u \in U, \forall q^* \in X', \tag{5.36}$$

where ψ_u is, for a fixed $u \in U$, the function: $q \longrightarrow \phi(u, q)$, and $\psi_u^* \in \Gamma(X')$ is its conjugate function in the duality between X' and X. ■

Now, we give some properties of the Lagrangian function L.

Lemma 5.37. (*i*) *Let u be given in U. Then the function*

$$R_u : \ q^* \in X' \longrightarrow L(u, q^*) \in \overline{I\!R}$$

is a concave upper and semi-continuous function of X' into $\overline{I\!R}$.

---

[2] By using the fact that if a real-valued function η is G-differentiable at point u_0 and if $\xi \in \partial\eta(u_0)$ then $\xi = \eta'(u_0)$.

(ii) If the function ϕ is convex then, for all $q^* \in X'$, the function

$$S_{q^*} : u \in U \longrightarrow L(u, q^*) \in \overline{\mathbb{R}}$$

is a convex function from U into $\overline{\mathbb{R}}$.

Proof. Since the function $R_u = -\psi_u^* \in \Gamma(X')$, R_u is concave and upper semicontinuous on X' and then the result (i). The result (ii) is a direct consequence of the convexity of ϕ and the expression of L given by (5.35). □

Characterization of Saddle Points

In order to give a characterization of a saddle point of the Lagrangian function L, we first rewrite problem $(\mathcal{P})$ and its dual problem $(\mathcal{P}^*)$ by using the Lagrangian function L with respect to ϕ. We have easily that, for all $(u^*, q^*) \in U' \times X'$ (without supplementary condition on ϕ)

$$\begin{aligned}
\phi^*(u^*, q^*) &= \sup_{(u,p)\in U\times X} (\langle u^*, u\rangle_U + \langle q^*, p\rangle_X - \phi(u,p)) \\
&= \sup_{u\in U}(\langle u^*, u\rangle_U + \sup_{p\in X}(\langle q^*, p\rangle_X - \phi(u,p))) \\
&= \sup_{u\in U}(\langle u^*, u\rangle_U - L(u, q^*))
\end{aligned}$$

and then

$$\phi^*(0, q^*) = \sup_{u\in U}(-L(u, q^*)) = -\inf_{u\in U} L(u, q^*), \forall q^* \in X'. \tag{5.37}$$

According to the expression of problem $(\mathcal{P}^*)$ in (5.5) and the previous relation (5.37), problem $(\mathcal{P}^*)$ can be written as

$$\sup_{q^*\in X'} \inf_{u\in U} L(u, q^*). \tag{5.38}$$

In the same way, if we suppose that $\phi \in \Gamma_0(U \times X)$ then $\forall u \in U$, the function $\psi_u : q \in X \longrightarrow \phi(u, q) \in \overline{\mathbb{R}}$ belongs to $\Gamma(X)$ and consequently, the bidual ψ_u^{**} of ψ_u is exactly ψ_u. Thus (by Fenchel transformation)

$$\begin{aligned}
\phi(u,p) = \psi_u(p) &= \psi_u^{**}(p) \\
&= \sup_{q^*\in X'} (\langle q^*, p\rangle_X - \psi_u^*(q^*))) \\
&= \sup_{q^*\in X'} (\langle q^*, p\rangle_X + L(u, q^*)) \quad \text{(according to (5.36))}
\end{aligned}$$

and then

$$\phi(u, 0) = \sup_{q^*\in X'} L(u, q^*), \forall u \in U. \tag{5.39}$$

According to expression of problem $(\mathcal{P})$ in (5.1) with the conditions (5.2) and (5.39), problem $(\mathcal{P})$ can be written as

$$\inf_{u\in U} \sup_{q^*\in X'} L(u, q^*). \tag{5.40}$$

Remark 5.38. By introducing the Lagrangian function, the weak duality relation $\sup(\mathcal{P}^*) \leq \inf(\mathcal{P})$ (corresponding to problems $(\mathcal{P})$ and $(\mathcal{P}^*)$) is exactly the inequality

$$\sup_{q^* \in X'} \inf_{u \in U} L(u, q^*) < \inf_{u \in U} \sup_{q^* \in X'} L(u, q^*). \tag{5.41}$$

Therefore, the inequality (5.41) is always true. ♢

The following theorem shows a characterization of a saddle point.

Theorem 5.39. *(Saddle point theorem) Let ϕ be a function satisfying $\phi \in \Gamma_0(U \times X)$. Then the following results hold:*

(i) The pair (u, p^*) of $U \times X'$ is a saddle point of the Lagrangian function L if and only if u is a solution of problem $(\mathcal{P})$, p^* is a solution of problem $(\mathcal{P}^*)$ and $\inf(\mathcal{P}) = \sup(\mathcal{P}^*)$ (the strong duality).

(ii) Moreover, if problem $(\mathcal{P})$ is stable then an element $u \in U$ is a solution of $(\mathcal{P})$ if and only if there exists an element $p^* \in X'$ such that (u, p^*) is a saddle point of L (according to (i), p^* is then a solution of $(\mathcal{P}^*)$).

Proof. (*i*) ($\Rightarrow$) From the expressions (5.37), (5.39) and (5.40) we have that

$$L(u, p^*) = \inf_{v \in U} L(v, p^*) = -\phi^*(0, p^*),$$
$$L(u, p^*) = \sup_{q^* \in X'} L(u, q^*) = \phi(u, 0).$$

Therefore, we have the extremality condition

$$\phi(u, 0) + \phi^*(0, p^*) = 0$$

and then (because of Proposition 5.18) the result.

($\Leftarrow$) Conversely, because of Proposition 5.18

$$\phi(u, 0) + \phi^*(0, p^*) = 0.$$

Moreover, according to the relations (5.37) and (5.39) we have that

$$-\phi^*(0, p^*) = \inf_{v \in U} L(v, p^*) \leq L(u, p^*),$$
$$\phi(u, 0) = \sup_{q^* \in X'} L(u, q^*) \geq L(u, p^*).$$

Thus, $\inf_{v \in U} L(v, p^*) = L(u, p^*) = \sup_{q^* \in X'} L(u, q^*)$ and then (u, p^*) is a saddle point of L.

(*ii*) If u is a solution of $(\mathcal{P})$ then (according to Corollary 5.16 and Proposition 5.12), we have that the dual problem $(\mathcal{P}^*)$ has at least one solution p^* and $\inf(\mathcal{P}) = \sup(\mathcal{P}^*)$. Therefore (according to (*i*)), (u, p^*) is a saddle point of L. The converse is immediate. □

Remark 5.40. According to (5.37) and (5.39), the Lagrangian duality function L is a Lagrangian function of type I of problem $(\mathcal{P})$ and then the point (*i*) of Theorem 5.39 is a direct consequence of Proposition 5.30. ♢

5.2.3 Application and Boundary-value Problems

Framework in Linear Geometric Operator

Let U, U' and X, X' be two pairs of real topological vector spaces, in duality with respect to certain bilinear forms $\langle .,. \rangle_U$ and $\langle .,. \rangle_X$, respectively. Let the geometric operator $\Lambda : U \longrightarrow X$ be a linear and continuous operator of U into X, with the adjoint $\Lambda^* : X' \longrightarrow U'$. Let $F : U \longrightarrow \overline{\mathbb{R}}$ and $\mathcal{S} : X \longrightarrow \overline{\mathbb{R}}$ be two extended real-valued functions and finite on convex subsets $\mathcal{K} \subset U$ and $\mathcal{A} \subset X$, respectively.

By using the Fenchel transformation, the conjugate functions corresponding to $\mathcal{S}$ and F are defined respectively by

$$\begin{aligned} \mathcal{S}^*(q^*) &= \sup_{q \in X}(\langle q^*, q \rangle_X - \mathcal{S}(q)) \ \ \forall q^* \in X', \\ F^*(v^*) &= \sup_{v \in U}(\langle v^*, v \rangle_U - F(v)) \ \ \forall v^* \in U'. \end{aligned} \tag{5.42}$$

In this section, we analyze the infimum of the following functional

$$J_\nu(v) = \mathcal{S}(\Lambda v - \nu) + F(v), \tag{5.43}$$

where ν is a given distributed parameter.

The parametric variational problem $(\mathcal{P})$ is then

$$\inf_{v \in \mathcal{K}} J_\nu(v). \tag{5.44}$$

Example 5.41. Let $(U, \| \,.\, \|)$ be a reflexive Banach space with norm $\| \,.\, \|$ and let $(U', \| \,.\, \|_*)$ be its dual space with norm $\| \,.\, \|_*$.

Let us consider an operator A from U into U' satisfying the assumptions (4.77) and (4.78), a function φ that is a convex, lower semi-continuous and proper function of U into $\overline{\mathbb{R}}$ and a given element f of U'. We have proved in Subsection 5.1.4 that the inequalities variational problem (5.19) can be considered as an optimization problem

$$\inf_{v \in U} G(v), \tag{5.45}$$

where the functional G of U into $\overline{\mathbb{R}}$ is defined by

$$G(v) = \langle g, v \rangle_U + \varphi(v),$$

with $g = A\tilde{u} - f$, for some known solution $\tilde{u}$ of (5.19).

If we suppose that $\varphi(v) = \mathcal{S}(\Lambda v - \nu)$, where $\nu \in X$ is a given distributed parameter, $\Lambda \in \mathcal{L}(U; X)$, X is a Banach space and $\mathcal{S} \in \Gamma_0(X)$ then the functional G is similar to J_ν which is given by (5.43), where F is the finite linear function $v \in U \longrightarrow F(v) = \langle g, v \rangle_U$ and problem (5.45) is similar to problem (5.44), with $\mathcal{K} = U$. ♣

In order to study problem (5.44), we consider the perturbation function $\phi : U \times X \longrightarrow \overline{\mathbb{R}}$, which is defined by

$$\phi(v, q) := \mathcal{S}(\Lambda v - \nu - q) + F(v)$$

and its conjugate $\phi^* : U' \times X' \longrightarrow \overline{\mathbb{R}}$, which is defined (using the Fenchel transformation) by

$$\phi^*(v^*, q^*) := \sup_{(v,q)\in U\times X} (\langle (v^*, q^*), (v, q)\rangle_{U\times X} - \phi(v, q))$$

and then

$$\begin{aligned}
\phi^*(0, q^*) &= \sup_{(v,q)\in U\times X} (\langle q^*, q\rangle_X - \phi(v, q)) \\
&= \sup_{v\in U} \sup_{q\in X} (\langle q^*, q\rangle_X - \phi(v, q)) \\
&= \sup_{(v,p)\in U\times X} (\langle q^*, \Lambda v - \nu - p\rangle_X - \mathcal{S}(p) - F(v)) \\
&= \sup_{v\in U} (\langle q^*, \Lambda v - \nu\rangle_X - F(v)) + \sup_{p\in X} (-\langle q^*, p\rangle_X - \mathcal{S}(p)) \\
&= \sup_{v\in U} (\langle \Lambda^* q^*, v\rangle_X - F(v)) + \sup_{p\in X} (\langle -q^*, p\rangle_X - \mathcal{S}(p)) - \langle q^*, \nu\rangle_X \\
&= \mathcal{S}^*(-q^*) + F^*(\Lambda^* q^*) - \langle q^*, \nu\rangle_X .
\end{aligned}$$

The dual problem $(\mathcal{P}^*)$ can be written as

$$\sup_{q^*\in X'} (-\phi^*(0, q^*)) = \sup_{q^*\in X'} (-\mathcal{S}^*(-q^*) - F^*(\Lambda^* q^*) + \langle q^*, \nu\rangle_X). \tag{5.46}$$

Remark 5.42. Let $(u_s, p_s^*) \in U \times X'$ such that $\phi(u_s, 0) + \phi^*(0, p_s^*) = 0$ then

$$\begin{aligned}
&F(u_s) + F^*(\Lambda^* p_s^*) - \langle \Lambda^* p_s^*, u_s\rangle_U = 0, \quad \textit{i.e.,}\ \Lambda^* p_s^* \in \partial F(u_s), \\
&\mathcal{S}(\Lambda u_s - \nu) + \mathcal{S}^*(-p_s^*) - \langle -p_s^*, \Lambda u_s - \nu\rangle_X = 0, \quad \textit{i.e.,}\ -p_s^* \in \partial G(\Lambda u_s - \nu).
\end{aligned}$$

Indeed,

$$\begin{aligned}
0 &= \phi(u_s, 0) + \phi^*(0, p_s^*) \\
&= \mathcal{S}(\Lambda u_s - \nu) + F(u_s) + \mathcal{S}^*(-p_s^*) + F^*(\Lambda^* p_s^*) - \langle p_s^*, \nu\rangle_X \\
&= [F^*(\Lambda^* p_s^*) + F(u_s) - \langle \Lambda^* p_s^*, u_s\rangle_U] \\
&\quad + [\mathcal{S}^*(-p_s^*) + \mathcal{S}(\Lambda u_s - \nu) + \langle p_s^*, \Lambda u_s - \nu\rangle_X].
\end{aligned}$$

According to 5.42, we can deduce that

$$\begin{aligned}
&F^*(\Lambda^* p_s^*) + F(u_s) - \langle \Lambda^* p_s^*, u_s\rangle_U \geq 0, \\
&\mathcal{S}^*(-p_s^*) + \mathcal{S}(\Lambda u_s - \nu) - \langle -p_s^*, \Lambda u_s - \nu\rangle_X \geq 0
\end{aligned}$$

and then the results follow (because of Proposition 4.16). ◊

According to Propositions 5.17, 5.18 and 5.20, and the previous remark, respectively, we can obtain easily the following results.

Theorem 5.43. *Let F and $\mathcal{S}$ be convex functions and let us assume that* $\inf(\mathcal{P})$ *is finite (i.e.,* $-\infty < \inf_{v \in U} J_\nu(v) < \infty$*). Further, let us suppose that*

$$\text{there exists an element } u_0 \in U \text{ such that } F(u_0) \text{ is finite and the function } \mathcal{S} \text{ is finite and continuous at point } \Lambda u_0 - \nu. \tag{5.47}$$

Then problem $(\mathcal{P})$ *is stable.*

Moreover, if $F \in \Gamma_0(U)$ *and* $\mathcal{S} \in \Gamma_0(X)$ *then the dual gap of the problem* $(\mathcal{P})$ *is null (i.e.,* $\inf(\mathcal{P}) = \sup(\mathcal{P}^*)$*) and* $(\mathcal{P}^*)$ *has at least one solution.* □

Theorem 5.44. *If problems* $(\mathcal{P})$ *and* $(\mathcal{P}^*)$ *both have solutions and the dual gap of the problem* $(\mathcal{P})$ *is null then all solutions* $u_s \in \mathcal{K}$ *of* $(\mathcal{P})$ *and all solutions* $p_s^* \in X'$ *of* $(\mathcal{P}^*)$ *satisfy the following optimality conditions*

$$\begin{aligned} &F(u_s) + F^*(\Lambda^* p_s^*) - \langle \Lambda^* p_s^*, u_s \rangle_U = 0, \\ &\mathcal{S}(\Lambda u_s - \nu) + \mathcal{S}^*(-p_s^*) - \langle -p_s^*, \Lambda u_s - \nu \rangle_X = 0, \end{aligned} \tag{5.48}$$

or (equivalently)

$$\Lambda^* p_s^* \in \partial F(u_s) \text{ and } -p_s^* \in \partial \mathcal{S}(\Lambda u_s - \nu). \tag{5.49}$$

Conversely, if $u_s \in U$ *and* $p_s^* \in X'$ *satisfy the relations (5.48) or (5.49), then* u_s *is a solution of problem* $(\mathcal{P})$*,* p_s^* *is a solution of* $(\mathcal{P}^*)$ *and the dual gap of problem* $(\mathcal{P})$ *is null.* □

Theorem 5.45. *Let* $(U, \| \,.\, \|)$ *be a reflexive Banach space. If* $F \in \Gamma_0(U)$ *and* $\mathcal{S} \in \Gamma_0(X)$ *and if*

$$\mathcal{S}(\Lambda v - \nu) + F(v) \longrightarrow +\infty \quad \text{for } v \in U, \| v \| \longrightarrow +\infty, \tag{5.50}$$

then, under assumption (5.47), problems $(\mathcal{P})$ *and* $(\mathcal{P}^*)$ *have at least one solution, the dual gap of problem* $(\mathcal{P})$ *is null and the optimality conditions (5.48) are satisfied.* □

We shall now apply the previous study to an interesting problem of linear elasticity.

Example: Elasticity Problem

Consider a deformable elastic body, occupying in its undeformed state the bounded and open domain $\Omega \subset \mathbb{R}^n$ with boundary $\Gamma = \partial\Omega = \Gamma_u \cup \Gamma_s$ such that $\Gamma_u \cap \Gamma_s = \emptyset$ and $\text{mes}(\Gamma_u) \neq \emptyset$ (see, *e.g.*, Gurtin [147], Landau and Lifschitz [185] for more details). A deformation process of the body, under an external stress, is described by a smooth vector-valued mapping $\mathcal{R}$ from the reference configuration $\overline{\Omega} = \Omega \cup \partial\Omega$ (the closure of Ω) to a configuration $\omega \subset \mathbb{R}^m$ with boundary $\partial\omega$ at time t and is defined by

$$\mathcal{R}(x,t) = (\mathcal{R}_i(x_j,t))_{i=1,\ldots,m;\ j=1,\ldots,n}, \text{ with } x = (x_j)_{j=1,\ldots,n}.$$

The deformation is said to be admissible if $\mathcal{R}(.,t)$ is invertible and then the inverse $\mathcal{R}^{-1}(.,t)$ of $\mathcal{R}(.,t)$ exists. The admissible configuration space, denoted by D_{ad}, is defined by

$$D_{\text{ad}} = \{\mathcal{R}(.,t) \text{ sufficiently regular on } \overline{\Omega} : \operatorname{rank}(\nabla\mathcal{R}(.,t)) = \min(n,m) \text{ on } \overline{\Omega}\}.$$

The deformation gradient tensor is defined on D_{ad} by

$$A(x,t) = \nabla\mathcal{R}(x,t) = (A_i(x_j,t))_{i=1,\ldots,m;\ j=1,\ldots,n}, \text{ with } A_i(x_j,t) = \frac{\partial \mathcal{R}_i}{\partial y}(x_j,t).$$

The metric tensor $\mathbf{C}$ is a symmetric Lagrangian tensor field, with $\operatorname{rank}(\mathbf{C}) = \min(n,m)$, defined by

$$\mathbf{C} = A^T A \in \mathbb{R}^{n\times n}, \text{ with } \mathbf{C}_{kl}(x,t) = \sum_{i=1}^{m} A_i(x_k,t)A_i(x_l,t).$$

Suppose now that $m = n$, then the deformation of the elastic body is usually described in terms of a vector displacement field $u(x,t) \in \mathbb{R}^n$, which is defined by

$$u(x,t) = \mathcal{R}(x,t) - x.$$

In the linearized elasticity theory on the assumption of small strains, the changes in metric induced by the deformation are described by the linearized strain tensor $\epsilon(x,t) \in \mathbb{R}^{n\times n}$ defined by

$$\epsilon[u] = \frac{\nabla u + \nabla u^T}{2}, \quad i.e., \quad \epsilon_{ij} = \frac{1}{2}(\frac{\partial u_j}{\partial x_i} + \frac{\partial u_i}{\partial x_j}) \ \ 1 \le i,j \le n,$$

and by the stress tensor $\sigma(x,t) \in \mathbb{R}^{n\times n}$, which depends linearly on $\epsilon(x,t)$ as

$$\sigma = \mathcal{C} : \epsilon, \quad i.e., \quad \sigma_{ij} = \sum_{1\le k,l\le n} \mathcal{C}_{ijkl}\epsilon_{kl} \ \ 1 \le i,j \le n,$$

where $\mathcal{C}$ is the invertible elasticity tensor, which is supposed only spatially dependent. We assume that $\mathcal{C}$ satisfies the symmetric condition

$$\mathcal{C}_{ijkl} = \mathcal{C}_{klij} = \mathcal{C}_{jikl} = \mathcal{C}_{ijlk} \text{ for } 1 \le i,j,l,k \le n \tag{5.51}$$

(and then $\epsilon_{ij} = \epsilon_{ji}$, $\sigma_{ij} = \sigma_{ji}$ for $1 \le i,j \le n$) and the coercivity condition, *i.e.*, there exists a constant $\alpha > 0$ such that

$$\sum_{1\le i,j,k,l\le n} \mathcal{C}_{ijkl}\xi_{ij}\xi_{kl} \ge \alpha \sum_{1\le i,j\le n} \xi_{ij}\xi_{ij} \ \forall \xi \in E, \tag{5.52}$$

where E is the symmetric second-order tensor space.

In the case of isotropic homogeneous media, the elasticity tensor $\mathcal{C}$ can be defined in terms of the Lamé coefficients, the incompressibility λ and the rigidity μ by

$$\mathcal{C}_{ijkl} = \lambda\delta_{ij}\delta_{kl} + \mu(\delta_{ik}\delta_{jl} + \delta_{jk}\delta_{il} - \frac{2}{3}\delta_{ij}\delta_{kl}) \quad \text{for all} \quad 1 \leq i,j,k,l \leq 3,$$

where δ_{ij} is the kroneker delta, *i.e.*, $\delta_{ij} = 0$ if $i \neq j$ and $\delta_{ij} = 1$ if $i = j$.

We introduce now the kinetic energy and the elastic strain energy, respectively by

$$E_{\mathrm{k}}(u) = \int_\Omega \rho(x) \mid \frac{\partial u}{\partial t} \mid^2 \mathrm{d}x,$$

$$E_{\mathrm{s}}(u) = \int_\Omega \sigma : \epsilon \mathrm{d}x,$$

where ρ is the mass density distribution of the material and

$$\sigma : \epsilon = \sum_{1 \leq i,j \leq n} \sigma_{ij}\epsilon_{ji} = \mathrm{tr}(\sigma\epsilon),$$

where $\mathrm{tr}(\sigma\epsilon)$ is the matrix trace of the product of two tensors.

The fundamental balance equation of the dynamics of deformable bodies is then

$$\rho\frac{\partial^2 u}{\partial t^2} - \mathrm{div}(\sigma) = f,$$

where f is the distribution of body-force.

We then have

$$\rho\frac{\partial^2 u}{\partial t^2} - \mathrm{div}(C : \nabla u) = f \quad \text{on} \quad \mathcal{Q} = \Omega \times (0,T),$$

where $C : \nabla u = \sigma$.

In order to complete the modeling, we give the following initial conditions:

$$(u(.,0), \frac{\partial u}{\partial t}(.,0)) = (u_0, v_0) \quad \text{on} \quad \Omega,$$

and the boundary conditions:

$$u = \xi \quad \text{on } \Gamma_{\mathrm{u}} \times (0,T), \qquad \sigma.\mathbf{n} = \phi \quad \text{on } \Gamma_{\mathrm{s}} \times (0,T), \tag{5.53}$$

where $\sigma.\mathbf{n}$ is a vector function with components

$$(\sigma.\mathbf{n})_i = \sum_{j=1,n} \sigma_{ij}n_j \ i = 1, \ldots, n.$$

In the sequel, in order to simplify the presentation, we suppose that the system is time independent and then the previous system becomes

$$\begin{aligned} -\text{div}(\sigma) &= f, \\ u &= \xi \quad \text{on } \Gamma_{\text{u}}, \\ \sigma.\mathbf{n} &= \phi \quad \text{on } \Gamma_{\text{s}}, \end{aligned} \tag{5.54}$$

with a given elasticity tensor $\mathcal{C}$, a body field f in Ω, the traction ϕ on Γ_{s} and the displacement ξ on Γ_{u}, which satisfies

$$\begin{aligned} &\xi \text{ is the trace of a function } u_\xi \in U_0 \text{ on } \Gamma_{\text{u}}, \\ &\text{where } U_0 = \{v \in H^1(\Omega) : \; v = 0 \text{ on } \Gamma_{\text{s}}\}. \end{aligned} \tag{5.55}$$

Let $U = H^1(\Omega)$ and $X = L^2(\Omega; E) = X'$. Similar to Subsection 3.2.4, we can consider the space

$$H_\sigma(\text{div}; \Omega) = \{v \in L^2(\Omega; E) : \; \text{div}(\sigma) \in L^2(\Omega)\},$$

which is a Hilbert space for the norm

$$\| \sigma \|_{H_\sigma(\text{div};\Omega)} = (\| \sigma \|^2_{L^2(\Omega)} + \| \text{div}(\sigma) \|^2_{L^2(\Omega)})^{1/2},$$

where $\text{div}(\sigma)$ is a vector function with components

$$(\text{div}(\sigma))_i = \sum_{j=1,n} \frac{\partial \sigma_{ij}}{\partial x_j}, \; i = 1, \ldots, n.$$

The following theorem concerns the normal components of boundary values of functions of the space $H_\sigma(\text{div}; \Omega)$.

Theorem 5.46. *Let Ω be a bounded subset of $\mathbb{R}^n$ with Lipschitz boundary Γ, we have the following properties:*

$$\begin{aligned} &\mathcal{D}(\overline{\Omega}; E) \textit{ is dense in } H_\sigma(\text{div}; \Omega) \\ &(\textit{the boundedness of } \Omega \textit{ is not necessary}), \end{aligned} \tag{5.56}$$

$$\begin{aligned} &\textit{there exists a mapping } \gamma_n : \; \sigma \longrightarrow \sigma.\mathbf{n}|_\Gamma \textit{ defined on } \mathcal{D}(\overline{\Omega}; E) \\ &\textit{which can be extended by continuity to a linear and continuous} \\ &\textit{mapping, denoted also by } \gamma_n, \textit{ from } H_\sigma(\text{div}; \Omega) \textit{ into } H^{-1/2}(\Gamma), \end{aligned} \tag{5.57}$$

where $(\sigma.\mathbf{n})_i = \sum_{j=1,n} \sigma_{ij} n_j \; i = 1, \ldots, n.$ □

A consequence of the previous theorem is the following *Green's formula*, for all tensor functions $\sigma \in H_\sigma(\text{div}; \Omega)$ and vector functions $u \in H^1(\Omega)$:

$$\int_\Omega \sigma : \nabla u dx + \int_\Omega u \text{div}(\sigma) dx = \langle \sigma.\mathbf{n}, u \rangle_{H^{-1/2}(\Gamma), H^{1/2}(\Gamma)}. \tag{5.58}$$

Otherwise (for the strain tensor $\epsilon[u] \in L^2(\Omega; E)$ if $u \in H^1(\Omega)$),

$$\int_\Omega \sigma : \epsilon[u] \mathrm{d}x + \int_\Omega u \mathrm{div}(\sigma) \mathrm{d}x = \langle \sigma.\mathbf{n}, u \rangle_{H^{-1/2}(\Gamma), H^{1/2}(\Gamma)}. \tag{5.59}$$

We can also introduce the following space (corresponding to the strain tensor $\epsilon[u]$) by

$$H_\epsilon(\Omega) = \{u \in L^2(\Omega) : \quad \epsilon[u] \in L^2(\Omega; E)\},$$

which is a Hilbert space for the norm

$$\| u \|_{H_\epsilon(\Omega)} = (\| u \|^2_{L^2(\Omega)} + \| \epsilon[u] \|^2_{L^2(\Omega)})^{1/2}.$$

We have the following Korn's inequality result.

Theorem 5.47. *Let Ω be a bounded subset of $I\!R^n$ with Lipschitz boundary Γ, we have the following property:*

$$\mathcal{D}(\overline{\Omega}) \text{ is dense in } H_\epsilon(\Omega) \text{ (the boundedness of } \Omega \text{ is not necessary)}, \tag{5.60}$$

and the classical Korn's inequality:

$$\begin{aligned} &\text{there exists a constant } c_0 > 0 \text{ (depending on } \Omega\text{) such that} \\ &\| u \|^2_{H^1(\Omega)} \leq c_0 (\| u \|^2_{L^2(\Omega)} + \| \epsilon[u] \|^2_{L^2(\Omega)}) = c_0 \| u \|^2_{H_\epsilon(\Omega)} . \end{aligned} \tag{5.61}$$

Proof. For the proof see, *e.g.*, Duvaut and Lions [109]. □

Let now $\mathcal{U}_{\mathrm{ad}}$ and $\mathcal{X}_{\mathrm{ad}}$ be the admissible displacement and stress sets, respectively, defined by

$$\mathcal{U}_{\mathrm{ad}} = \{u \in U : \quad u = \xi \text{ on } \Gamma_{\mathrm{u}}\},$$

$$\mathcal{X}_{\mathrm{ad}} = \{\sigma \in X : \ -\mathrm{div}(\sigma) = f, \quad \sigma.\mathbf{n} = \phi \text{ on } \Gamma_{\mathrm{s}}\},$$

and let $\Lambda : \ U \longrightarrow X$ be the geometrically linear operator, which is defined by

$$\Lambda u = -\frac{\nabla u + \nabla u^T}{2} = -\epsilon[u] \ \text{ on } \ \Omega.$$

The duality between X and X' can be denoted by

$$\langle \sigma, \epsilon \rangle_X = \int_\Omega \sigma : \epsilon \mathrm{d}x.$$

It is well known that the displacement u is a solution of the following problem $(\mathcal{P})$:

$$\inf_{v \in \mathcal{U}_{\mathrm{ad}}} \left(\frac{1}{2} \int_\Omega \Lambda v : C : \Lambda v \mathrm{d}x - \int_\Omega f v \mathrm{d}x - \int_{\Gamma_{\mathrm{s}}} \phi v \mathrm{d}\Gamma \right). \tag{5.62}$$

Otherwise, the displacement u minimizes the potential energy

$$J(v) = \mathcal{S}(\Lambda v) + F(v) \ \text{ for } v \in U,$$

where

$$\begin{aligned}
\mathcal{S}(\epsilon) &= \frac{1}{2}\int_{\Omega} \epsilon : \mathcal{C} : \epsilon \mathrm{d}x, \\
F(v) &= \int_{\Omega} -f v \mathrm{d}x + \int_{\Gamma_{\mathrm{s}}} -\phi v \mathrm{d}\Gamma + \mathcal{X}_{\mathcal{U}_{\mathrm{ad}}}(v) \quad \text{(see Remark 4.90).}
\end{aligned} \tag{5.63}$$

Then problem (5.62) is similar to problem (5.44), with the parameter $\nu = 0$, and to obtain the dual problem it is sufficient to calculate the dual $\mathcal{S}^*$ and F^* of the functionals $\mathcal{S}$ and F at points σ and $\Lambda^*\sigma$, respectively, for $\sigma \in X$.

It is clear that F and $\mathcal{S}$ are G-differentiable functionals on X and $\mathcal{U}_{\mathrm{ad}}$, respectively, and (since $\mathcal{C}$ satisfies $\epsilon_1 : \mathcal{C} : \epsilon_2 = \epsilon_2 : \mathcal{C} : \epsilon_1$)

$$\mathcal{S}'(\epsilon) = C : \epsilon \quad \text{and} \quad (F'(u) = -f \quad \text{on } \Omega; \ F'(u) = -\phi \quad \text{on } \Gamma_{\mathrm{s}}).$$

From Fenchel transformation, the conjugate functions corresponding to $\mathcal{S}$ and F are defined respectively by

$$\begin{aligned}
\mathcal{S}^*(\sigma) &= \sup_{\epsilon \in X}(\langle \sigma, \epsilon \rangle_X - \mathcal{S}(\epsilon)), \\
F^*(\Lambda^*\sigma) &= \sup_{v \in \mathcal{U}_{\mathrm{ad}}}(\langle \Lambda^*\sigma, v \rangle_U - F(v)).
\end{aligned}$$

Calculate first $\mathcal{S}^*$ at point σ. Since the function

$$H : \epsilon \longrightarrow H(\epsilon) = \langle \sigma, \epsilon \rangle_X - \mathcal{S}(\epsilon) = \int_{\Omega} (\sigma : \epsilon - \frac{1}{2}\epsilon : \mathcal{C} : \epsilon) \mathrm{d}x,$$

is strictly concave and upper semi-continuous on X, and $H(\epsilon) \longrightarrow -\infty$ for, $\epsilon \in X$, $\| \epsilon \|_X \longrightarrow \infty$ (by using the coercivity condition (5.52) and Korn's inequality), then, according to Proposition 4.92, H admits a unique supremum ϵ_{s} such that ϵ_{s} is a critical point of H, *i.e.*, $H'(\epsilon_{\mathrm{s}}) = 0$. Consequently, $\epsilon_{\mathrm{s}} = \mathcal{C}^{-1} : \sigma$ and then

$$\mathcal{S}^*(\sigma) = \frac{1}{2}\int_{\Omega} \sigma : \mathcal{C}^{-1} : \sigma \mathrm{d}x. \tag{5.64}$$

We shall now calculate F^* at point $\Lambda^*\sigma$. Since the function

$$v \longrightarrow \int_{\Omega} -f v \mathrm{d}x + \int_{\Gamma_{\mathrm{s}}} -\phi v \mathrm{d}\Gamma,$$

is a linear and continuous mapping from U into $\mathbb{R}$ then there exists $\mathbf{f} \in U'$ such that

$$\langle \mathbf{f}, v \rangle_U = \int_{\Omega} -f v \mathrm{d}x + \int_{\Gamma_{\mathrm{s}}} -\phi v \mathrm{d}\Gamma.$$

Consequently (according to (5.55)),

$$F(v) = \langle \mathbf{f}, v \rangle_U + \mathcal{X}_{\mathcal{U}_{\mathrm{ad}}^0}(v - u_\xi),$$

where $\mathcal{U}_{\mathrm{ad}}^0 = \{v \in U : \ v = 0 \ \text{ on } \Gamma_{\mathrm{u}}\}$.

According to Lemma 4.9, we have that

$$F^*(\Lambda^*\sigma) = \sup_{v\in U}\Big(\langle\Lambda^*\sigma, v\rangle_U - \langle\mathbf{f}, v\rangle_U - \mathcal{X}_{\mathcal{U}^0_{\mathrm{ad}}}(v-u_\xi)\Big)$$
$$= \langle\Lambda^*\sigma - \mathbf{f}, u_\xi\rangle_U + \mathcal{X}_{\mathcal{U}^*_{\mathrm{ad}}}(\Lambda^*\sigma - \mathbf{f}),$$

where $\mathcal{U}^*_{\mathrm{ad}} = \{\mathbf{g}\in U' : \ \langle\mathbf{g}, v\rangle_U = 0 \ \ \forall v\in\mathcal{U}^0_{\mathrm{ad}}\}$. Otherwise,

$$F^*(\Lambda^*\sigma) = \begin{cases} \langle\Lambda^*\sigma - \mathbf{f}, u_\xi\rangle_U \text{ if } \Lambda^*\sigma - \mathbf{f}\in\mathcal{U}^*_{\mathrm{ad}} \\ +\infty \text{ else.} \end{cases} \tag{5.65}$$

We remark, according to Theorem 5.46 and Green's formula (5.59), that $\Lambda^*\sigma - \mathbf{f}\in\mathcal{U}^*_{\mathrm{ad}}$ if $\int_\Omega(\mathrm{div}(\sigma)+f)v\mathrm{d}x + \int_{\Gamma_{\mathrm{s}}}(-\sigma.\mathbf{n}+\phi)v\mathrm{d}\Gamma = 0, \forall v\in\mathcal{U}^0_{\mathrm{ad}}$ and then, if $\sigma\in\mathcal{X}_{\mathrm{ad}}$. Consequently,

$$F^*(\Lambda^*\sigma) = \langle\Lambda^*\sigma - \mathbf{f}, u_\xi\rangle_U \ \ \forall\sigma\in\mathcal{X}_{\mathrm{ad}}.$$

Next calculate the term $\langle\Lambda^*\sigma, u_\xi\rangle_U$. Since

$$\langle\Lambda^*\sigma, u_\xi\rangle_U = \langle\sigma, \Lambda u_\xi\rangle_X = \int_\Omega \mathrm{div}(\sigma)u_\xi\mathrm{d}x - \int_{\Gamma_{\mathrm{u}}}\sigma.\mathbf{n}u_\xi\mathrm{d}\Gamma - \int_{\Gamma_{\mathrm{s}}}\sigma.\mathbf{n}u_\xi\mathrm{d}\Gamma,$$

then (because of $u_\xi = 0$ on Γ_{s} and $u_\xi = \xi$ on Γ_{u}),

$$\langle\Lambda^*\sigma, u_\xi\rangle_U = \int_\Omega \mathrm{div}(\sigma)u_\xi\mathrm{d}x - \int_{\Gamma_{\mathrm{u}}}\sigma.\mathbf{n}\xi\mathrm{d}\Gamma.$$

Since $\sigma\in\mathcal{X}_{\mathrm{ad}}$ and $u_\xi = 0$ on Γ_{s} then $\langle\Lambda^*\sigma - \mathbf{f}, u_\xi\rangle_U = -\int_{\Gamma_{\mathrm{u}}}\sigma.\mathbf{n}\xi\mathrm{d}\Gamma$.

This implies that

$$F^*(\Lambda^*\sigma) = \begin{cases} \int_{\Gamma_{\mathrm{u}}} -\sigma.\mathbf{n}\xi\mathrm{d}\Gamma \ \text{ if } \sigma\in\mathcal{X}_{\mathrm{ad}} \\ +\infty \text{ else.} \end{cases} \tag{5.66}$$

We can now give the dual problem $(\mathcal{P}^*)$ corresponding to the primal problem (5.62). According to (5.46), (5.64) and (5.66), the dual problem $(\mathcal{P}^*)$ can be written as

$$\sup_{\sigma\in\mathcal{X}_{\mathrm{ad}}}\left(-\frac{1}{2}\int_\Omega \sigma : \mathcal{C}^{-1} : \sigma\mathrm{d}x - \int_{\Gamma_{\mathrm{u}}}\sigma.\mathbf{n}\xi\mathrm{d}\Gamma\right). \tag{5.67}$$

It is clear that $F\in\Gamma_0(U)$ and $\mathcal{S}\in\Gamma_0(X)$ and the conditions (5.50) hold (according to the coercivity condition (5.52) and Korn's Inequality) then according to Theorem 5.45 there exists a solution $u_{\mathrm{s}}\in\mathcal{U}_{\mathrm{ad}}$ of (5.62) (which is unique since the functional $\mathcal{S}$ is strictly convex) and a solution $\sigma_s \in \mathcal{X}_{ad}$ (which is also unique) of (5.67) satisfying $-\infty < \inf(\mathcal{P}) = \sup(\mathcal{P}^*) < \infty$ and

$$\begin{aligned} F(u_s) + F^*(\Lambda^*\sigma_s) - \langle \Lambda^*\sigma_s, u_s\rangle_U &= 0, \\ \mathcal{S}(\Lambda u_s) + \mathcal{S}^*(-\sigma_s) + \langle \sigma_s, \Lambda u_s\rangle_X &= 0. \end{aligned} \tag{5.68}$$

Then (we denote by $\epsilon_s := -\Lambda u_s$)

$$\int_\Omega (-\sigma_s : \epsilon_s + \frac{1}{2}\sigma_s : \mathcal{C}^{-1} : \sigma_s + \frac{1}{2}\epsilon_s : \mathcal{C} : \epsilon_s)\mathrm{d}x = 0. \tag{5.69}$$

Let $\epsilon_c := \mathcal{C}^{-1} : \sigma_s$ then (since $\epsilon_s : \mathcal{C} : \epsilon_c = \epsilon_c : \mathcal{C} : \epsilon_s$)

$$(-\sigma_s : \epsilon_s + \frac{1}{2}\sigma_s : \mathcal{C}^{-1} : \sigma_s + \frac{1}{2}\epsilon_s : \mathcal{C} : \epsilon_s) = \frac{1}{2}(\epsilon_c - \epsilon_s) : \mathcal{C} : (\epsilon_c - \epsilon_s) \geq 0.$$

Consequently, $(\epsilon_c - \epsilon_s) : \mathcal{C} : (\epsilon_c - \epsilon_s) = 0$ a.e. in Ω and then $\epsilon_c - \epsilon_s = 0$ a.e. in Ω, *i.e.*,

$$\epsilon_s = \mathcal{C}^{-1} : \sigma_s \quad \text{a.e. in } \Omega,$$

which is called the *Hooke's law.*

Remark 5.48. In addition to the operator Λ, we introduce its boundary operator $\Lambda_B : U_B \longrightarrow X_B$ by $\Lambda_B u = -\mathbf{u}$ on Γ_u, where $(\mathbf{u})_{ij} := u_i n_j$, $1 \leq i, j \leq n$, *i.e.*, $\mathbf{u} = u^t.\mathbf{n}$.

We introduce now the adjoint operators of Λ and Λ_B. By using Green's formula we can deduce that (since $\sigma : \mathbf{u} = \sigma.\mathbf{n}u$)

$$\begin{aligned} \langle \sigma, \Lambda u\rangle_X &= \int_\Omega \sigma : \Lambda u \mathrm{d}x \\ &= \int_\Omega \mathrm{div}(\sigma) u \mathrm{d}x - \int_\Gamma \sigma.\mathbf{n} u \mathrm{d}\Gamma, \\ &= \langle \Lambda^*\sigma, u\rangle_U + \langle \Lambda_B^*\sigma, u\rangle_{\Gamma_s} - \int_{\Gamma_u} \sigma : \mathbf{u}\mathrm{d}\Gamma, \\ &= \langle \Lambda^*\sigma, u\rangle_U + \langle \Lambda_B^*\sigma, u\rangle_{\Gamma_s} + \langle \sigma, \Lambda_B u\rangle_{\Gamma_u}, \end{aligned} \tag{5.70}$$

where the adjoint operator $\Lambda^* : X' = X \longrightarrow U'$ and its boundary operator $\Lambda_B^* : X_B' \longrightarrow U_B'$ are defined by

$$\Lambda^*\sigma = \mathrm{div}(\sigma) \text{ on } \Omega \text{ and } \Lambda_B^*\sigma = -\sigma.\mathbf{n} \text{ on } \Gamma_s.$$

If we denote

$$\begin{aligned} \langle \sigma, \Lambda_T u\rangle_{\overline{X}} &= \langle \sigma, \Lambda u\rangle_X - \langle \sigma, \Lambda_B u\rangle_{\Gamma_u}, \\ \langle \Lambda_T^*\sigma, u\rangle_{\overline{U}} &= \langle \Lambda^*\sigma, u\rangle_U + \langle \Lambda_B^*\sigma, u\rangle_{\Gamma_s}, \end{aligned}$$

then

$$\langle \sigma, \Lambda_T u\rangle_{\overline{X}} = \langle \Lambda_T^*\sigma, u\rangle_{\overline{U}}.$$

The boundary conditions (5.53) can be written as

$$\Lambda_{\mathrm{B}}^{*}\sigma = -\sigma.\mathbf{n} = -\phi \ \text{ on } \Gamma_{\mathrm{s}} \ \text{ and } \ \Lambda_{\mathrm{B}}u = -u^{t}.\mathbf{n} = -\xi^{t}.\mathbf{n} \ \text{ on } \Gamma_{\mathrm{u}}$$

and the total potential energy of the system J is defined by

$$J(u) = \mathcal{S}(\Lambda u) + F(u) = \int_{\Omega} W(\Lambda u)\mathrm{d}x + F(u),$$

where $W(\epsilon) := (1/2)\epsilon : \mathcal{C} : \epsilon$.

According to the result of Section 4.3.6, we have that u_{s} is a critical point of J if and only if (see (4.89) and (4.90))

$$\begin{aligned}
&-\epsilon_{\mathrm{s}} = \Lambda u_{\mathrm{s}} \ \text{ on } \Omega, \ \ u_{\mathrm{s}}^{t}.\mathbf{n} = \xi^{t}.\mathbf{n} \ \text{ on } \Gamma_{\mathrm{u}},\\
&-\sigma_{\mathrm{s}} = W'(-\epsilon_{\mathrm{s}}) \ \text{ on } \overline{\Omega},\\
&\Lambda^{*}\sigma_{\mathrm{s}} = -f \ \text{ on } \Omega, \ \ \sigma_{\mathrm{s}}.\mathbf{n} = \phi \ \text{ on } \Gamma_{\mathrm{s}},
\end{aligned}$$

i.e., if and only if

$$\begin{aligned}
&-\epsilon_{\mathrm{s}} = \Lambda u_{\mathrm{s}} \ \text{ on } \Omega, \ \ u_{\mathrm{s}} = \xi \ \text{ on } \Gamma_{\mathrm{u}},\\
&\sigma_{\mathrm{s}} = \mathcal{C} : \epsilon_{\mathrm{s}} \ \text{ on } \overline{\Omega},\\
&-\mathrm{div}(\sigma_{\mathrm{s}}) = f \ \text{ on } \Omega, \ \ \sigma_{\mathrm{s}}.\mathbf{n} = \phi \ \text{ on } \Gamma_{\mathrm{s}}.
\end{aligned} \tag{5.71}$$

◇

5.3 Minimax Duality

The minimax duality theory plays an important role in constrained optimization, in general variational problems, in numerical analysis, in robust control theory and game theory. In this section, we analyze the connection between an optimization problem where the variable is some function u and some corresponding *inf-sup* problem of a functional type *convex-concave*, where the variables are the function u and some function p. The approach used is based on the classical minimax theorems of Ky-Fan and Sion, which are older than Rockafellar's approach (for more details see Rockafellar [252]).

5.3.1 Motivation

Let U and X be locally convex topological vector spaces and let us consider a function $\Psi : \ U \longrightarrow \overline{\mathrm{IR}}$.

We consider the following minimization problem:

$$\inf_{v\in U} \Psi(v), \tag{5.72}$$

and we assume that we can write $\Psi(v)$ as a supremum on the second variable of some function $L : \ U \times X \longrightarrow \overline{\mathrm{IR}}$ as

$$\Psi(v) = \sup_{q\in X} L(v,q), \qquad \forall v \in U. \tag{5.73}$$

Then problem (5.72) becomes

$$\inf_{v \in U} \sup_{q \in X} L(v, q), \tag{5.74}$$

i.e., seek points u_s in U, which minimize Ψ on U, or points $(u_s, p_s) \in U \times X$ such that u_s minimizes Ψ on U and also

$$\Psi(u_s) = L(u_s, p_s) = \sup_{q \in X} L(u_s, q). \tag{5.75}$$

Definition 5.49. *(Minimax point) A point (u_s, p_s) is said to be a minimax point of L on $U \times X$ if u_s minimizes Ψ on U and (5.75) holds.* ∎

Consequently, if (u_s, p_s) is a minimax point of L on $U \times X$ then

$$L(u_s, q) \leq L(u_s, p_s) = \Psi(u_s) \leq \Psi(v) \tag{5.76}$$

for all $(v, q) \in U \times X$.

Remark 5.50. For L given by (5.73), we have that if (u_s, p_s) is a saddle point of L then (u_s, p_s) is a minimax point of L (because of (5.76) and the fact that $\Psi(v) \leq L(v, p_s)$ for all $v \in U$). But the converse is, in general, false. ◇

Since the convex lower semi-continuous functions are the pointwise supremum of affine continuous functions that they dominates, we can then, in general, rewrite $\Psi(v)$ in the form (5.73). Therefore, a minimization problem becomes an *inf-sup* problem. Moreover, according to the study given in the previous section, it is clear that the study of the connection between

$$\sup_{q \in X} \inf_{v \in U} L(v, q) \tag{5.77}$$

and problem (5.74) is necessary and important. That is the goal of the study which will follow.

First, we will be interested in saddle point results. Second, we will give a situation where a minimax point is not a saddle point.

5.3.2 Saddle Point and Properties

Let $\mathcal{K} \subset U$ and $\mathcal{A} \subset X$ be two arbitrary sets and $L : U \times X \longrightarrow \overline{\mathbb{R}}$ be a real-valued functional, finite on the set $\mathcal{K} \times \mathcal{A} \subset U \times X$.

Game Theory Interpretation

We recall that a point $(u_s, p_s) \in \mathcal{K} \times \mathcal{A}$ is a saddle point of L on $\mathcal{K} \times \mathcal{A}$, if L attains at this point its maximum in $p \in \mathcal{A}$ and its minimum in $u \in \mathcal{K}$, *i.e.*,

$$L(u, p_s) \geq L(u_s, p_s) \geq L(u_s, p). \tag{5.78}$$

Let us consider the so-called two-person zero-sum games, which play a central role in the development of the theory of games. We take the game in which players make payments only to each other. One player's loss is *exactly equal to* the other player's gain, so the total amount of "money" available remains constant.

In order to analyze any game, we make the following assumptions about both players:

(*i*) Each player makes the best possible strategy (or move).
(*ii*) Each player knows that his (or her) opponent is also making the best possible strategy.

This game can be described like this: player P_u chooses a strategy u and player P_p chooses a strategy p ; when the players have decided their strategy, player P_u makes a payment $L(u,p)$ to player P_p. The goal for each player is of course to minimize his (or her) payment and to maximize his (or her) income.

The question is to find the equilibrium of such a game, *i.e.*, to know the strategy (u,p) of the two players such that every one of the players is not interested in varying his (or her) strategy independently whether the strategy of the other player is known. The response is that equilibria are exactly the saddle points of the cost functional L.

Indeed, if $(u_{\mathrm{s}}, p_{\mathrm{s}})$ is such a point, then the first player shall not be interested in taking another strategy u, if the second player keeps his (or her) choice p_{s}; indeed, the first inequality in (5.78) shows that another strategy u cannot decrease the payment of the first player. Similarly, the second player shall not be interested in choosing something different from p_{s}, if the first player keeps his (or her) choice u_{s} (because a different strategy from p_{s} cannot increase the income of the second player).

On the other hand, if a strategy (u,p) is not a saddle point, it is clear that the first player can decrease his payment passing from u to another choice (if the second player keeps his (or her) choice at p); we have similar analysis for the second player; thus, the equilibria are exactly the saddle points.

Let us consider the following two scenarios:

(I) The first player choose first, and the second player makes his (or her) choice already knowing the choice of the first player.
(II) Vice versa, the second player chooses first, and first player makes his (or her) choice already knowing the choice of the second player.

In scenario (I) the reasoning of the first player is: if I choose some strategy u, then the second player will of course choose a strategy p which maximizes my payment $L(u,p)$ which results in a payoff, for the second player, of

$$A(u) = \sup_{p \in \mathcal{A}} L(u,p).$$

Consequently, I should choose u which minimizes the function A, *i.e.*, the one which solves the optimization problem

$$\inf_{u \in \mathcal{K}} A(u).$$

The resulting payoff, from the first player to the second player will then be

$$\inf_{u \in \mathcal{K}} A(u) = \inf_{u \in \mathcal{K}} \sup_{p \in \mathcal{A}} L(u, p). \tag{5.79}$$

In scenario (II), similar reasoning of the second player enforces him (or her) to choose p maximizing the profit function

$$B(p) = \inf_{u \in \mathcal{K}} L(u, p),$$

i.e., the one which solves the optimization problem

$$\sup_{p \in \mathcal{A}} B(p).$$

The resulting payoff, from the first player to the second player, will then be

$$\sup_{p \in \mathcal{A}} B(p) = \sup_{p \in \mathcal{A}} \inf_{u \in \mathcal{K}} L(u, p). \tag{5.80}$$

The difference between the two payoffs can be interpreted as the advantage afforded to the player who makes the second move, with knowledge of the other player's move. The loss of the first player in scenario (II) is less than or equal to the profit of the second player in scenario (I) (since the conditions of game (II) are favorable to the first player and those of game (I) to the second player). Thus, we may guess that, independently of the structure of the function L, there is the following inequality.

Proposition 5.51. *For all real-valued functional L, finite on the set $\mathcal{K} \times \mathcal{A} \subset U \times X$, we have that*

$$\sup_{q \in \mathcal{A}} \inf_{v \in \mathcal{K}} L(v, q) \leq \inf_{v \in \mathcal{K}} \sup_{q \in \mathcal{A}} L(v, q). \tag{5.81}$$

Proof. The proof is trivial and is left to the reader as an exercise. □

If a saddle point exists then games (I) and (II) are similar, *i.e.*, there is no advantage anyway in making the second move (because $L(u_s, p_s)$ is the same value of both payoffs (5.79) and (5.80)).

Existence of Saddle Points

Now we give an existence result of a saddle point of the function L.

Theorem 5.52. *The real-valued functional L, finite on the set $\mathcal{K} \times \mathcal{A} \subset U \times X$, admits a saddle point (u, p) on $\mathcal{K} \times \mathcal{A}$ if and only if*

$$\max_{q \in \mathcal{A}} \inf_{v \in \mathcal{K}} L(v, q) = \min_{v \in \mathcal{K}} \sup_{q \in \mathcal{A}} L(v, q) \tag{5.82}$$

and this quantity is equal to $L(u, p)$.

Here, min *instead of* inf *(respectively* max *instead of* sup*) means that the extremum is taken.*

Proof. Let us assume that there exists a saddle point (u, p) on $\mathcal{K} \times \mathcal{A}$.

It is clear that

$$\inf_{v\in\mathcal{K}} \sup_{q\in\mathcal{A}} L(v,q) \leq \sup_{q\in\mathcal{A}} L(u,q),$$
$$\inf_{v\in\mathcal{K}} L(v,p) \leq \sup_{q\in\mathcal{A}} \inf_{v\in\mathcal{K}} L(v,q).$$

Moreover, by the definition of saddle point, *i.e.*,

$$L(u,q) \leq L(u,p) \leq L(v,p), \qquad \forall (v,q) \in \mathcal{K} \times \mathcal{A},$$

we can deduce that

$$\sup_{q\in\mathcal{A}} L(u,q) = L(u,p) = \inf_{v\in\mathcal{K}} L(v,p). \tag{5.83}$$

Therefore, according to the inequality (5.81), we can deduce that

$$\sup_{q\in\mathcal{A}} \inf_{v\in\mathcal{K}} L(v,q) = L(u,p) = \inf_{v\in\mathcal{K}} \sup_{q\in\mathcal{A}} L(v,q) \tag{5.84}$$

Precisely, according to the relation (5.83), we have

$$\begin{aligned} L(u,p) &= \sup_{q\in\mathcal{A}} L(u,q) = \min_{v\in\mathcal{K}} \sup_{q\in\mathcal{A}} L(v,q) \ (i.e., \text{ the infimum is attained}) \\ &= \inf_{v\in\mathcal{K}} L(v,p) = \max_{q\in\mathcal{A}} \inf_{v\in\mathcal{K}} L(v,q) \ (i.e., \text{ the supremum is attained}). \end{aligned}$$

Conversely, assume that the equality (5.82) holds and let u realize the minimum of $\sup_{q\in\mathcal{A}} L(.,q)$ and p the maximum of $\inf_{v\in\mathcal{K}} L(v,.)$. Therefore,

$$\inf_{v\in\mathcal{K}} L(v,p) \leq L(u,p) \leq \sup_{q\in\mathcal{A}} L(u,q)$$

and then (because of (5.82))

$$\inf_{v\in\mathcal{K}} L(v,p) = L(u,p) = \sup_{q\in\mathcal{A}} L(u,q),$$

meaning that (u,p) is a saddle point of L.

This completes the proof. □

As a direct consequence we have the following proposition.

Proposition 5.53. *Let* $L: U \times X \longrightarrow \overline{\mathbb{R}}$ *be a real-valued functional finite on the set* $\mathcal{K} \times \mathcal{A} \subset U \times X$. *We then have*
if there exists a pair $(u,p) \in \mathcal{K} \times \mathcal{A}$ *and a real value* α *such that*

$$\begin{aligned} L(u,q) &\leq \alpha, \qquad \forall q \in \mathcal{A}, \\ L(v,p) &\geq \alpha, \qquad \forall v \in \mathcal{K}, \end{aligned} \tag{5.85}$$

then the pair (u,p) *is a saddle point of* L *and*

$$\alpha = \sup_{q\in\mathcal{A}} \inf_{v\in\mathcal{K}} L(v,q) = \inf_{v\in\mathcal{K}} \sup_{q\in\mathcal{A}} L(v,q) = L(u,p). \tag{5.86}$$

□

The following proposition gives the form of the set of saddle points of a Lagrangian function.

Proposition 5.54. *(Set of saddle points) Let* $L : U \times X \longrightarrow \overline{I\!R}$ *be a real-valued functional, finite on the set* $\mathcal{K} \times \mathcal{A} \subset U \times X$. *The set of saddle points of* L *is of the form* $\mathcal{K}_0 \times \mathcal{A}_0 \subset \mathcal{K} \times \mathcal{A}$.

Proof. Let (u_i, p_i), for $i = 1, 2$, be two saddle points of the function L on $\mathcal{K} \times \mathcal{A}$ and prove that, for example (u_1, p_2) is also a saddle point of L. From the relation (5.82), we have that

$$\alpha = L(u_1, p_1) = L(u_2, p_2)$$

and then (by the definition of saddle point)

$$L(u_1, q) \leq \alpha, \quad L(v, p_2) \geq \alpha \quad \forall (v, q) \in \mathcal{K} \times \mathcal{A}.$$

Therefore, because of Proposition 5.53, we can deduce that (u_1, p_2) is a saddle point of L. □

5.3.3 Banach Spaces and Saddle Points

In this section we assume that the space U and X are two reflexive Banach spaces, and we make the following assumptions:

$(A1)$ The sets $\mathcal{K}$ and $\mathcal{A}$ verify

$$\begin{aligned} &\mathcal{K} \subset U \ \text{ is convex, closed and non-empty,} \\ &\mathcal{A} \subset X \ \text{ is convex, closed and non-empty.} \end{aligned} \tag{5.87}$$

$(A2)$ The function $L : U \times X \longrightarrow \overline{\mathrm{IR}}$ is finite on $\mathcal{K} \times \mathcal{A}$ and satisfies

$$\begin{aligned} &v \in \mathcal{K} \longrightarrow L(v, q) \text{ is convex and lower semi-continuous } \forall q \in \mathcal{A}, \\ &q \in \mathcal{A} \longrightarrow L(v, q) \text{ is concave and upper semi-continuous } \forall v \in \mathcal{K}. \end{aligned} \tag{5.88}$$

Remark 5.55. According to Asplund [14] we have that

for a reflexive Banach space, we can always find an equivalent norm which is strictly convex. ◇

We can now give some properties of the set of saddle points of the function L.

Proposition 5.56. *Under assumptions* $(A1)$ *and* $(A2)$*, the subset* $\mathcal{K}_0 \times \mathcal{A}_0$ *of saddle points of* L *is convex. Moreover:*

(i) If $\forall q \in \mathcal{A}$*, the function* $v \in \mathcal{K} \longrightarrow L(v, q)$ *is strictly convex then* $\mathcal{K}_0$ *contains at most one point.*
(ii) If $\forall v \in \mathcal{K}$*, the function* $q \in \mathcal{A} \longrightarrow L(v, q)$ *is strictly concave then* $\mathcal{A}_0$ *contains at most one point.*

Proof. If the subset $\mathcal{K}_0 \times \mathcal{A}_0$ is empty the proof is immediate. Suppose now that the subset $\mathcal{K}_0 \times \mathcal{A}_0$ is non-empty then L possesses at least one saddle point. Set α the value

$$\alpha := \sup_{q\in\mathcal{A}} \inf_{v\in\mathcal{K}} L(v,q) = \inf_{v\in\mathcal{K}} \sup_{q\in\mathcal{A}} L(v,q).$$

For $(u_i, p_i) \in \mathcal{K}_0 \times \mathcal{A}_0$, when $i = 1,2$ and $t \in]0,1[$ we have that

$$L(u_i, q) \leq \alpha \leq L(v, p_i) \quad \text{for all } (v,q) \in \mathcal{K} \times \mathcal{A}, \text{ for } i = 1,2$$

and

$(tu_1 + (1-t)u_2, tp_1 + (1-t)p_2) \in \mathcal{K} \times \mathcal{A}$ (because of the assumption (5.87)).

Consequently (because of the assumption (5.88)),

$$L(tu_1 + (1-t)u_2, q) \leq \alpha, \qquad \forall q \in \mathcal{A},$$
$$L(v, tp_1 + (1-t)p_2) \geq \alpha, \qquad \forall v \in \mathcal{K}.$$

With Proposition 5.53 this implies that $(tu_1 + (1-t)u_2, tp_1 + (1-t)p_2)$ is a saddle point of L and then $(tu_1 + (1-t)u_2, tp_1 + (1-t)p_2) \in \mathcal{K}_0 \times \mathcal{A}_0$.

If now L is strictly convex with respect to the first variable, then $\mathcal{K}_0$ is a singleton. Indeed, if $\mathcal{K}_0$ contains two distinct points u_1 and u_2 then for all $t \in]0,1[$ and a saddle point (u_1,p_1), the points (u_2,p_1) and $(tu_1+(1-t)u_2,p_1)$ are saddle points of L (since $\mathcal{K}_0 \times \mathcal{A}_0$ is convex) and we have

$$\alpha = L(tu_1 + (1-t)u_2, p_1) < tL(u_1,p_1) + (1-t)L(u_2,p_1) = \alpha,$$

which is impossible, and we then obtain the result (i).

Similarly, we can obtain the result (ii). □

We shall now give some characterizations of a saddle point of the function L in the case where L is a differentiable function.

Characterization of Saddle Points

Theorem 5.57. *Assume that the function $L = S + R$ such that the function S satisfies*

$$\begin{aligned} &v \in \mathcal{K} \longrightarrow S(v,q) \quad \textit{is convex and G-differentiable}, \quad \forall q \in \mathcal{A}, \\ &q \in \mathcal{A} \longrightarrow S(v,q) \quad \textit{is concave and G-differentiable}, \quad \forall v \in \mathcal{K} \end{aligned} \tag{5.89}$$

and the function R satisfies

$$\begin{aligned} &v \in \mathcal{K} \longrightarrow R(v,q) \quad \textit{is convex}, \quad \forall q \in \mathcal{A}, \\ &q \in \mathcal{A} \longrightarrow R(v,q) \quad \textit{is concave}, \quad \forall v \in \mathcal{K}. \end{aligned} \tag{5.90}$$

Then, the pair $(u,p) \in \mathcal{K} \times \mathcal{A}$ is the saddle point of L if and only if

$$\begin{aligned} &\langle \frac{\partial S}{\partial u}(u,p), v-u\rangle_U + R(v,p) - R(u,p) \geq 0, \quad \forall v \in \mathcal{K}, \\ &\langle \frac{\partial S}{\partial p}(u,p), q-p\rangle_X + R(u,p) - R(u,q) \leq 0, \quad \forall q \in \mathcal{A}, \end{aligned} \tag{5.91}$$

where $\partial S/\partial u$ and $\partial S/\partial p$ denote the partial G-derivatives on U and X, respectively.

Proof. ($\Rightarrow$) Let (u,p) be a saddle point of the function L, then for all $t \in]0,1[$ and for all $v \in \mathcal{K}$ (by (5.88) and (5.90), *i.e.*, the convexity of $R(.,p)$)

$$\begin{aligned} S(u,p) + R(u,p) &= L(u,p) \\ &\leq L(tv + (1-t)u, p) \\ &= S(tv+(1-t)u,p) + R(tv+(1-t)u,p) \\ &\leq S(tv+(1-t)u,p) + tR(v,p) + (1-t)R(u,p). \end{aligned}$$

Thus,

$$R(u,p) - R(v,p) \leq \frac{S(tv+(1-t)u,p) - S(u,p)}{t}$$

and then by passing to the limit in t (because of the G-differentiability of S), we have the first relation of (5.91). In the same way, we can obtain the second relation of (5.91).

($\Leftarrow$) Let (u,p) be an element of $\mathcal{K}\times\mathcal{A}$ such that the inequalities (5.91) hold. Then, because of Proposition 4.45 (by (5.89), *i.e.*, the convexity of $S(.,p)$), we have that

$$-\langle \frac{\partial S}{\partial u}(u,p), v-u\rangle_U + S(v,p) - S(u,p) \geq 0, \quad \forall v \in \mathcal{K}.$$

Thus $R(v,p) - R(u,p) + S(v,p) - S(u,p) \geq 0, \quad \forall v \in \mathcal{K}$, *i.e.*,

$$L(u,p) \leq L(v,p) \quad \forall v \in \mathcal{K}.$$

Similarly, we can prove that

$$L(u,p) \geq L(u,q) \quad \forall q \in \mathcal{A}.$$

This completes the proof. □

As a corollary we have the following theorem.

Theorem 5.58. *In addition to Assumptions (A1) and (A2), we assume that*

$$\begin{aligned} &v \in \mathcal{K} \longrightarrow L(v,q) \quad \text{is G-differentiable}, \quad \forall q \in \mathcal{A}, \\ &q \in \mathcal{A} \longrightarrow L(v,q) \quad \text{is G-differentiable}, \quad \forall v \in \mathcal{K}. \end{aligned} \tag{5.92}$$

Then, the pair $(u,p) \in \mathcal{K} \times \mathcal{A}$ is the saddle point of L if and only if

$$\begin{aligned} &\langle \frac{\partial L}{\partial u}(u,p), v-u\rangle_U \geq 0, \quad \forall v \in \mathcal{K}, \\ &\langle \frac{\partial L}{\partial p}(u,p), q-p\rangle_X \leq 0, \quad \forall q \in \mathcal{A}. \end{aligned} \tag{5.93}$$

□

Existence of Saddle Points

We shall now give some conditions of the existence of a saddle point in a minimax formulation (of Ky Fan–von Neumann).

Theorem 5.59. *In addition to Assumptions (A1) and (A2), we assume that the sets* $\mathcal{K}$ *and* $\mathcal{A}$ *are bounded. Then, the function* L *admits at least one saddle point* $(u,p) \in \mathcal{K} \times \mathcal{A}$ *and*

$$L(u,p) = \max_{q \in \mathcal{A}} \min_{v \in \mathcal{K}} L(v,q) = \min_{v \in \mathcal{K}} \max_{q \in \mathcal{A}} L(v,q). \tag{5.94}$$

Proof. Suppose first that,

$$\forall q \in \mathcal{A}, \text{ the functions } \psi_q : v \in \mathcal{K} \longrightarrow L(v,q) \text{ are strictly convex.} \tag{5.95}$$

Since U and X are Banach reflexive spaces then, since for all $q \in \mathcal{A}$, the functions ψ_q are real-valued, convex and lower semi-continuous, and the set $\mathcal{K}$ is convex, closed, bounded and non-empty, the functional ψ_q is bounded and possesses at least one minimum u_q in $\mathcal{K}$ (according to Proposition 4.92). Moreover, if the functional ψ_q is strictly convex then u_q is unique (according again to Proposition 4.92) and we denote by $r(q)$ the value

$$r(q) := \min_{v \in \mathcal{K}} \psi_q(v) = L(u_q, q). \tag{5.96}$$

According to (5.88), the function $r: q \in \mathcal{A} \longrightarrow r(q)$ is real-valued, concave and upper semi-continuous. Since $\mathcal{A}$ is convex, closed, bounded and non-empty, we can deduce that r is bounded from above and attains its maximum at the point p such that

$$\begin{aligned} &r(p) = \max_{q \in \mathcal{A}} r(q) = \max_{q \in \mathcal{A}} \min_{v \in \mathcal{K}} \psi_q(v) = \max_{q \in \mathcal{A}} \min_{v \in \mathcal{K}} L(v,q), \\ &r(p) \leq L(v,p) \quad \forall v \in \mathcal{K}. \end{aligned} \tag{5.97}$$

According to the second part of the assumption (5.88) we have, for all $(v,q) \in \mathcal{K} \times \mathcal{A}$ and $t \in]0,1[$, that

$$L(v, tq + (1-t)p) \geq tL(v,q) + (1-t)L(v,p).$$

In particular, for $v = v_t := u_{tq+(1-t)p}$, we have (according to (5.97))

$$\begin{aligned} r(p) \geq r(tq + (1-t)p) &= L(v_t, tq + (1-t)p) \\ &\geq tL(v_t, q) + (1-t)L(v_t, p) \\ &\geq tL(v_t, q) + (1-t)r(p) \end{aligned}$$

and then

$$r(p) \geq L(v_t, q) \ \ \forall q \in \mathcal{A}. \tag{5.98}$$

According to Eberlein–Smulian's theorem (Theorem 2.90), for (v_t) in $\mathcal{K}$, we can extract from (v_t) a subsequence, denoted also by (v_t), converging weakly in U (for the topology $\sigma(V, V')$) to a value u in $\mathcal{K}$ and then (because of Proposition 2.65 and the convexity of ψ_q)

$$L(u,q) = \psi_q(u) \leq \liminf_{t \longrightarrow 0} \psi_q(v_t) = \liminf_{t \longrightarrow 0} L(v_t, q) \quad \forall q \in \mathcal{A}. \tag{5.99}$$

The limit u is exactly u_p. Indeed, by the definition of v_t and from the second part of (5.88) (the concavity condition) we have that

$$tL(v_t,q) + (1-t)L(v_t,p) \leq L(v_t, tq + (1-t)p) \leq L(v, tq + (1-t)p) \quad \forall v \in \mathcal{K}.$$

Since $L(v_t, .)$ is uniformly bounded by $r(p)$ then by passing to the limit (according to (5.99)), we can deduce that

$$L(u,p) \leq \liminf_{t \longrightarrow 0} L(v_t, p) \leq \limsup_{t \longrightarrow 0} L(u, tq + (1-t)p)$$

and then $u = u_p$ (according to the strict convexity of ψ_p and Proposition 5.56).

We can now pass to the limit in the inequality (5.98), and we obtain (according to (5.97))

$$L(u,q) \leq r(p) = \max_{s \in \mathcal{A}} \min_{v \in \mathcal{K}} L(v,s) \quad \forall q \in \mathcal{A},$$
$$\max_{q \in \mathcal{A}} \min_{w \in \mathcal{K}} L(w,q) = r(p) \leq L(v,p) \quad \forall v \in \mathcal{K}.$$

Thus (according to Proposition 5.53) the point (u,p) is a saddle point L.

Second, we suppose that the assumption (5.95) does not hold. To show the result, we suppose that U is equipped with a strictly convex norm $\| \, . \, \|$ (see Remark 5.55) and we introduce the following perturbation L_μ, for $\mu > 0$ by

$$L_\mu(v,q) = L(v,q) + \mu \| v \| \quad \forall (v,q) \in \mathcal{K} \times \mathcal{A}.$$

Since ψ_q is convex and $\| \, . \, \|$ is strictly convex, then $\psi_q^\mu = \psi_q + \mu \| \, . \, \|$ is strictly convex and therefore, according to the previous study, we have the existence of a saddle point $(u_\mu, p_\mu) \in \mathcal{K} \times \mathcal{A}$ such that

$$\begin{aligned} L(u_\mu, q) + \mu \| u_\mu \| &\leq L(u_\mu, p_\mu) + \mu \| u_\mu \| \; \forall q \in \mathcal{A}, \\ L(u_\mu, p_\mu) + \mu \| u_\mu \| &\leq L(v, p_\mu) + \mu \| v \| \quad \forall v \in \mathcal{K}. \end{aligned} \tag{5.100}$$

According to Eberlein–Smulian's theorem (Theorem 2.90), for (u_μ, p_μ) in $\mathcal{K} \times \mathcal{A}$, we can extract from (u_μ, p_μ) a subsequence, denoted also by (u_μ, p_μ), converging weakly in $U \times X$ to a value (u,p) in $\mathcal{K} \times \mathcal{A}$ and then (because of Proposition 2.65 and the assumption (5.88))

$$L(u,q) \leq L(u,p) \leq L(v,p) \quad \forall (v,q) \in \mathcal{K} \times \mathcal{A}$$

and then the point (u,p) is a saddle point L.

This completes the proof. □

Remark 5.60. The previous theorem may easily be extended to the case where the sets $\mathcal{K}$ and $\mathcal{A}$ are compact subsets of separated (or Hausdorff) topological vector spaces (by using Weirstrass's theorem), see *e.g.*, Barbu and Precupanu [21]. ◇

Theorem 5.61. *In addition to Assumptions (A1) and (A2), we assume that there exists* $(u_0, p_0) \in \mathcal{K} \times \mathcal{A}$ *such that*

$$\begin{aligned} L(v, p_0) &\longrightarrow +\infty \quad \text{for } v \in \mathcal{K}, \| v \|_U \longrightarrow +\infty, \\ L(u_0, q) &\longrightarrow -\infty \quad \text{for } q \in \mathcal{A}, \| q \|_X \longrightarrow +\infty. \end{aligned} \tag{5.101}$$

Then the function L admits at least one saddle point $(u, p) \in \mathcal{K} \times \mathcal{A}$ *and*

$$L(u, p) = \max_{q \in \mathcal{A}} \inf_{v \in \mathcal{K}} L(v, q) = \min_{v \in \mathcal{K}} \sup_{q \in \mathcal{A}} L(v, q). \tag{5.102}$$

Proof. Assume, for $\mu > 0$ sufficiently large, the following subsets:

$$\begin{aligned} \mathcal{K}_\mu &:= \{v \in \mathcal{K} : \| v \|_U \leq \mu\}, \\ \mathcal{A}_\mu &:= \{q \in \mathcal{A} : \| q \|_X \leq \mu\}. \end{aligned}$$

Since the sets $\mathcal{K}$ and $\mathcal{A}$ are closed, convex and non-empty then the sets $\mathcal{K}_\mu$ and $\mathcal{A}_\mu$ are closed, convex, bounded and non-empty. Consequently, according to Theorem 5.59, the functional L admits a saddle point (u_μ, p_μ) on $\mathcal{K}_\mu \times \mathcal{A}_\mu$ and

$$L(u_\mu, q) \leq L(u_\mu, p_\mu) \leq L(v, p_\mu) \quad \forall (v, q) \in \mathcal{K}_\mu \times \mathcal{A}_\mu. \tag{5.103}$$

Suppose now that μ is sufficiently large so that $(u_0, p_0) \in \mathcal{K}_\mu \times \mathcal{A}_\mu$ then (according to (5.103))

$$L(u_\mu, p_0) \leq L(u_\mu, p_\mu) \leq L(u_0, p_\mu). \tag{5.104}$$

According to assumptions (5.88) and (5.101), we have that the functions $v \in \mathcal{K} \longrightarrow L(v, p_0)$ and $q \in \mathcal{A} \longrightarrow -L(u_0, q)$ are convex, lower semi-continuous and coercive and then (because of Proposition 4.92) there exists a pair of constants $(a, b) \in \mathbb{R}^2$ such that

$$-\infty < a \leq L(v, p_0) \text{ and } L(u_0, q) \leq b < +\infty \quad \forall (v, q) \in \mathcal{K} \times \mathcal{A}.$$

In particular for all $\mu > 0$ ($v := u_\mu$, $q := p_\mu$) we have (according to (5.104))

$$-\infty < a \leq L(u_\mu, p_0) \leq L(u_\mu, p_\mu) \leq L(u_0, p_\mu) \leq b < +\infty$$

and then the sequence $L(u_\mu, p_\mu)$ is uniformly bounded. Moreover, according to (5.101), we can deduce that the sequence (u_μ, p_μ) is also uniformly bounded, and then we can extract a subsequence, denoted also by (u_μ, p_μ) converging weakly to a point (u, p) in $\mathcal{K} \times \mathcal{A}$ and $L(u_\mu, p_\mu)$ converging to a real point λ.

By virtue of (5.104), we can deduce that

$$L(u, q) \leq \lambda \leq L(v, p) \quad \forall (v, q) \in \mathcal{K} \times \mathcal{A}$$

and then (because of Proposition 5.53) (u, p) is a saddle point of L. □

Lemma 5.62. *Let assumptions* $(A1)$ *and* $(A2)$ *hold, then:*

(i) *If* $\mathcal{K}$ *is bounded or else the first part of (5.101) is satisfied then*

$$\sup_{q\in\mathcal{A}}\inf_{v\in\mathcal{K}} L(v,q) = \min_{v\in\mathcal{K}}\sup_{q\in\mathcal{A}} L(v,q).$$

(ii) *If* $\mathcal{A}$ *is bounded or else the second part of (5.101) is satisfied then*

$$\max_{q\in\mathcal{A}}\inf_{v\in\mathcal{K}} L(v,q) = \inf_{v\in\mathcal{K}}\sup_{q\in\mathcal{A}} L(v,q).$$

Proof. The proof is left to the reader as an exercise (by considering, in order to prove for example the result (i), the penalized functional, for $\mu > 0$ sufficiently small, $L_\mu(v,q) = L(v,q) - \mu \parallel q \parallel_X$). □

We can now give the following more general result than Theorems 5.59 and 5.61.

Theorem 5.63. *Let Assumptions* $(A1)$ *and* $(A2)$ *hold, then:*

(i) *If the following condition is true:*

$$\begin{aligned}&\mathcal{K} \text{ is bounded or else the first part of (5.101) is satisfied,}\\ &\mathcal{A} \text{ is bounded or else } \inf_{v\in\mathcal{K}} L(v,q) \longrightarrow -\infty,\ q\in\mathcal{A}, \parallel q \parallel_X \longrightarrow +\infty,\end{aligned} \tag{5.105}$$

then the function L *admits a saddle point on* $\mathcal{K}\times\mathcal{A}$.

(ii) *If the following condition is true:*

$$\begin{aligned}&\mathcal{A} \text{ is bounded or else the second part of (5.101) is satisfied,}\\ &\mathcal{K} \text{ is bounded or else } \sup_{q\in\mathcal{A}} L(v,q) \longrightarrow +\infty,\ v\in\mathcal{K}, \parallel v \parallel_U \longrightarrow +\infty,\end{aligned} \tag{5.106}$$

then the function L *admits a saddle point on* $\mathcal{K}\times\mathcal{A}$.

Proof. Since the functional $\psi : \ q\in\mathcal{A} \longrightarrow \psi(q) := \inf_{v\in\mathcal{K}} L(v,q)$ is real-valued concave, lower semi-continuous and satisfies the same hypotheses as in Proposition 4.92 (according to (5.87) and (5.88), and the second part of (5.105)) we have that ψ attains its maximum on $\mathcal{A}$ and then (by the definition of ψ)

$$\sup_{q\in\mathcal{A}}\inf_{v\in\mathcal{K}} L(v,q) = \sup_{q\in\mathcal{A}}\psi(q) = \max_{q\in\mathcal{A}}\psi(q) = \max_{q\in\mathcal{A}}\inf_{v\in\mathcal{K}} L(v,q).$$

Moreover, from Lemma 5.62 and the first part of (5.105) we have that

$$\sup_{q\in\mathcal{A}}\inf_{v\in\mathcal{K}} L(v,q) = \min_{v\in\mathcal{K}}\sup_{q\in\mathcal{A}} L(v,q).$$

Consequently,

$$\min_{v\in\mathcal{K}}\sup_{q\in\mathcal{A}} L(v,q) = \max_{q\in\mathcal{A}}\inf_{v\in\mathcal{K}} L(v,q)$$

and then the existence of saddle point of L (by Theorem 5.52). So, we have the result (i). In the same way we can prove the result (ii). □

Example 5.64. Let us consider the following boundary problem:

$$\begin{aligned} -\Delta u &= au + f \quad \text{on } \Omega, \\ u &= 0 \quad \text{on } \Gamma, \end{aligned} \tag{5.107}$$

where Ω is an open and bounded subset of $\mathbb{R}^d$, $d \in \mathbb{N}^*$ is sufficiently regular, $\partial\Omega = \Gamma$ is its boundary and $\mathbf{n}$ is the unit outward normal on $\partial\Omega$.

The domain of the operator $-\Delta$ is $\operatorname{dom}(-\Delta) = H_0^1(\Omega) \cap H^2(\Omega)$ which is dense and compact in $L^2(\Omega)$ (see Yosida [306]). Moreover, the operator $-\Delta$ is auto-adjoint on $\operatorname{dom}(-\Delta)$ (*i.e.*, $\int_\Omega -v\Delta u \mathrm{d}x = \int_\Omega -u\Delta v \mathrm{d}x$, by Green's formula) and admits a sequence $(\lambda_i)_{i\geq 1}$ of eigenvalues such that $0 < \lambda_1 \leq \lambda_2 \leq \cdots \leq \lambda_k \leq \cdots$, with the corresponding eigenfunctions $(\varphi_i)_{i\geq 1}$ such that $\| \varphi_i \|_{L^2(\Omega)} = 1$, for all i.

Our problem is to use the min-max results in order to prove the existence and the uniqueness of a solution of problem (5.107) in $\operatorname{dom}(-\Delta)$. To show this result, we suppose that $f \in L^2(\Omega)$ and $a \in L^\infty(\Omega)$ satisfying: there exist $(a_1, a_2) \in \mathbb{R}^+$ and $l_a \in \mathbb{N}$ such that $\lambda_{l_a} < a_1 \leq a \leq a_2 < \lambda_{l_a+1}$ on Ω.

We consider a Hilbert space Y equipped with the norm

$$\| u \|_H := \left(\int_\Omega -u\Delta u \mathrm{d}x\right)^{1/2}$$

and we introduce the following functional:

$$J(u) := \frac{1}{2}\int_\Omega -u\Delta u \mathrm{d}x - \frac{1}{2}\int_\Omega a \mid u \mid^2 \mathrm{d}x - \int_\Omega f u \mathrm{d}x. \tag{5.108}$$

It is clear that the functional J is continuous F-differentiable on Y and, for all $v \in Y$, we have

$$\begin{aligned} \langle J'(u), v\rangle_Y &= \int_\Omega -v\Delta u \mathrm{d}x - \int_\Omega auv \mathrm{d}x - \int_\Omega f v \mathrm{d}x \\ &= \int_\Omega (-\Delta u - au - f)v \mathrm{d}x. \end{aligned} \tag{5.109}$$

If $l_a \geq 1$, the operators J' and $-J'$ are not monotone and then J is not convex and not concave (because of Proposition 4.46). We have then used the minimax duality by decomposing the space Y on the direct sum of two spaces U and X, *i.e.*, $Y := U \oplus^\perp X$ such that

$$\begin{aligned} U &:= \bigoplus_{k\leq l_a} \mathbb{R}\varphi_k = \{u : \ u = \sum_{k\leq l_a} \alpha_k\varphi_k, \text{ where } \alpha_k \in \mathbb{R}\}, \\ X &:= \bigoplus_{k\geq l_a+1} \mathbb{R}\varphi_k = \{v : \ v = \sum_{k\geq l_a+1} \beta_k\varphi_k, \text{ where } \beta_k \in \mathbb{R}\}. \end{aligned} \tag{5.110}$$

Let us consider the following function $L : \ U \times X \longrightarrow \overline{\mathbb{R}}$ such that

$$L(v,q) := J(v+q) \quad \text{for all } (v,q) \in U \times X$$

and prove the existence of a saddle point of L.

The function L is strictly concave on U and L is strictly convex on X. Indeed, since $\lambda_{l_a} < a_1 \leq a \leq a_2 < \lambda_{l_a+1}$ and for all $(w,r) \in U \times X$,

$$\int_\Omega -w\Delta w \mathrm{d}x \leq \lambda_{l_a} \parallel w \parallel^2_{L^2(\Omega)} \quad \text{and} \quad \int_\Omega -r\Delta r \mathrm{d}x \geq \lambda_{l_a+1} \parallel r \parallel^2_{L^2(\Omega)},$$

we can deduce, for all $(u,v) \in U^2$ and $(p,q) \in X^2$, that

$$\begin{aligned}
\langle \frac{\partial L}{\partial u}(u,p) - \frac{\partial L}{\partial u}(v,p), u-v\rangle_U &= \int_\Omega (-\Delta(u-v) - a(u-v))(u-v)\mathrm{d}x \\
&\leq -(a_1 - \lambda_{l_a}) \parallel u-v \parallel^2_{L^2(\Omega)}, \\
\langle \frac{\partial L}{\partial p}(u,p) - \frac{\partial L}{\partial p}(u,q), p-q\rangle_X &= \int_\Omega (-\Delta(p-q) - a(p-q))(p-q)\mathrm{d}x \\
&\geq (\lambda_{l_a+1} - a_2) \parallel p-q \parallel^2_{L^2(\Omega)}
\end{aligned}$$

and then the strict concavity and convexity results on U and X, respectively hold (because of Proposition 4.46).

Moreover, we have, for all $(u_0,p_0) \in U \times X$, that

$$\begin{aligned}
L(u,p_0) &\longrightarrow +\infty \quad \text{for } v \in U, \parallel v \parallel_U \longrightarrow +\infty, \\
L(u_0,q) &\longrightarrow -\infty \quad \text{for } q \in X, \parallel q \parallel_X \longrightarrow +\infty.
\end{aligned}$$

Consequently, according to Theorem 5.61 and Proposition 5.56, the functional L admits a unique saddle point (u,p) on $U \times X$ and we prove easily that $u+p$ is a solution of (5.107) (since $\langle J'(u+p), v\rangle_Y = 0$, because of saddle point characterization theorem 5.92).

To prove the uniqueness of the solution, we can prove that the solution of (5.107) is a saddle point of L and by the uniqueness of the saddle point we can deduce the result. Here, we prove directly the uniqueness of the results.

Let $u \in Y$ and $v \in Y$ be two solutions of (5.107) then $w = u - v \in Y$ is a solution of (5.107) with the second member $f = 0$. Since $Y := U \oplus^\perp X$, then there exists a unique $(w_1, w_2) \in U \times X$ such that $w := w_1 + w_2$ and

$$\int_\Omega -w_1 \Delta w_2 \mathrm{d}x = \int_\Omega -w_2 \Delta w_1 \mathrm{d}x = 0.$$

Consequently,

$$\begin{aligned}
0 = \int_\Omega (-\Delta w - aw)w_1 \mathrm{d}x &= \int_\Omega -w_1\Delta w_1 \mathrm{d}x - \int_\Omega a \mid w_1 \mid^2 \mathrm{d}x - \int_\Omega a w_1 w_2 \mathrm{d}x \\
&\leq -(a_1 - \lambda_{l_a}) \parallel w_1 \parallel^2_{L^2(\Omega)} - \int_\Omega a w_1 w_2 \mathrm{d}x, \\
0 = \int_\Omega (-\Delta w - aw)w_2 \mathrm{d}x &= \int_\Omega -w_2\Delta w_2 \mathrm{d}x - \int_\Omega a \mid w_2 \mid^2 \mathrm{d}x - \int_\Omega a w_1 w_2 \mathrm{d}x \\
&\geq (\lambda_{l_a+1} - a_2) \parallel w_2 \parallel^2_{L^2(\Omega)} - \int_\Omega a w_1 w_2 \mathrm{d}x
\end{aligned}$$

and then $0 \leq (\lambda_{l_a+1} - a_2) \parallel w_2 \parallel^2_{L^2(\Omega)} \leq -(a_1 - \lambda_{l_a}) \parallel w_1 \parallel^2_{L^2(\Omega)} \leq 0$. Therefore, $w_1 = w_2 = 0$ and then $w = 0$.

This completes the proof. ♣

5.3.4 Connection with Duality and Application

We consider a minimization problem $(\mathcal{P})$

$$\inf_{v \in U} \Psi(v), \tag{5.111}$$

which we can write in the form

$$\inf_{v \in \mathcal{K}} \Psi(v), \tag{5.112}$$

where $\mathcal{K} = \mathrm{dom}\Psi$, $\mathcal{K} \subset U$.
Let us assume that we can write $\Psi(v)$ as a supremum on q of a function $L(v, q)$ as

$$\Psi(v) = \sup_{q \in \mathcal{A}} L(v, q), \qquad \forall v \in \mathcal{K}. \tag{5.113}$$

The problem (5.111) then becomes

$$\inf_{v \in \mathcal{K}} \sup_{q \in \mathcal{A}} L(v, q). \tag{5.114}$$

The function Ψ can be written in the form (5.113) by using, for example, the theory of conjugate convex functions. By reducing the problem $(\mathcal{P})$ to the form (5.114), the dual problem $(\mathcal{P}^*)$ associated with problem $(\mathcal{P})$ is then

$$\sup_{q \in \mathcal{A}} \inf_{v \in \mathcal{K}} L(v, q). \tag{5.115}$$

Connection with Duality

In this subsection, we give the analogy between the results obtained in the section 5.3 and the results of Fenchel–Rockfellar given in the section 5.1. Precisely, we give the following interpretations and remarks:

(*i*) Proposition 5.51 means that $-\infty \leq \sup(\mathcal{P}^*) \leq \inf(\mathcal{P}) \leq +\infty$ and it must be compared to Proposition 5.4.

(*ii*) Lemma 5.62 gives criteria which ensure that solutions of problems $(\mathcal{P})$ and $(\mathcal{P}^*)$ exist and the dual gap of $(\mathcal{P})$ is null, *i.e.*, $\sup(\mathcal{P}^*) = \inf(\mathcal{P})$. It must be compared to the stability of problems $(\mathcal{P}^*)$ and $(\mathcal{P})$, respectively, given in Propositions 5.12 and 5.15.

(*iii*) Theorems 5.59 and 5.61 give criteria which determine if problems $(\mathcal{P})$ and $(\mathcal{P}^*)$ are both stable, and it must be compared to Proposition 5.20.

(*iv*) In the case of the equality "inf sup = sup inf" of the Lagrangian L, the existence of a saddle point for L is equivalent to the existence of solutions $u_s \in \mathcal{K}$, $p_s \in \mathcal{A}$, respectively, of problems $(\mathcal{P}^*)$ and $(\mathcal{P})$ with the stability of these problems. Moreover, in this situation, the set of saddle points of L is exactly the set of solutions (u_s, p_s) and we have the following extremality conditions

$$\begin{aligned} L(u_s, p_s) &= \sup_{q \in \mathcal{A}} L(u_s, q), \\ L(u_s, p_s) &= \inf_{v \in \mathcal{K}} L(v, p_s). \end{aligned} \tag{5.116}$$

(*v*) The minimax theorems play a key role in the control theory for problems governed by partial differential equations, in particular in the minimax control (see, *e.g.*, Ahmed and Xiang [6], Arada and Raymond [10], Belmiloudi [44, 47], McMillan and Triggiani [222, 223], Mordukhovich and Zhang [226, 225], Papageorgiou and Yannakakis [234], and the references therein), and in the stability and robust control theory (which will be detailed in Chapter 8, Chapter 9 and Part III).

(*vi*) It is clear that the change from problem (5.111) (or (5.112)) to problem (5.114), can be envisaged, without difficulty, in non-convex optimizations (the duality theorems given in Section 5.3 are valid for non-convex systems) see the applications below.

Applications

In this subsection, we give two interesting applications. More precisely:

1. Suppose that the function Ψ is the sum in U of two functions Φ and ξ, *i.e.*, $\Psi = \Phi + \xi$, where Φ is convex, lower semi-continuous and proper on U. According to the Fenchel–Moreau theorem, we have that $\Phi^{**} = \Phi$ and then

$$\Phi(v) = \sup_{g \in U'} \left(\langle g, v \rangle_U - \Phi^*(g) \right), \quad \forall v \in U,$$

where $\Phi^* \in \Gamma_0(U')$ is the conjugate function of Φ and Φ^{**} is the biconjugate function of Φ. So, for all $u \in U$, we have

$$\Psi(v) = \sup_{g \in U'} \left(\langle g, v \rangle_U + \xi(v) - \Phi^*(g) \right).$$

Put now

$$L(v, g) = \langle g, v \rangle_U + \xi(v) - \Phi^*(g),$$

where $v \in \mathcal{K} = \mathrm{dom}\Phi$ and $g \in \mathcal{A} := U'$.
Then Ψ can now be written in the form (5.113) and we have the formulation (5.114).

2. Let now X be another topological vector space, Λ be an operator of U into X (not necessarily linear) and suppose that $\Psi = \Phi + \xi$, where $\Phi = F \circ \Lambda$ and $F \in \Gamma_0(X)$. In the same way as in the first example, we have that

$$\Psi(v) = F(\Lambda v) + \xi(v) = \sup_{g \in Y}(\langle g, \Lambda v\rangle_{Y,X} + \xi(v) - F^*(g)), \quad \forall v \in U,$$

where X and Y are two locally convex spaces which are placed in duality by the bilinear form $\langle .,.\rangle_{Y,X}$, and $F^* \in \Gamma_0(Y)$ is the conjugate function of F. Consequently, Ψ can be written in the form (5.113), where $\mathcal{A}$ is a subspace of Y and

$$L(v, g) := \langle g, \Lambda v\rangle_{Y,X} + \xi(v) - F^*(g).$$

A more precise and more detailed analysis of this type of non-convex problems will be considered in Section 5.4.

5.3.5 Ky Fan's Minimax Inequality and Non-potential Operators

Let U be a reflexive Banach space with its dual space U' and $\phi : U \longrightarrow \overline{\mathrm{IR}}$ be a proper and weakly lower semi-continuous function on U. The goal of this subsection is to characterize the solutions of the differential inclusion

$$f + \mathcal{F}(u) \in \partial\phi(u) \tag{5.117}$$

by variational problems. The functional $\mathcal{F} : U \longrightarrow U'$ is given, $f \in U'$ and $\partial\phi$ is the subdifferential of ϕ on U.

In order to simplify the presentation, we denote by $\mathcal{G} : U \longrightarrow U'$ the function $f + \mathcal{F}(.)$, *i.e.*, $\mathcal{G}(u) = f + \mathcal{F}(u)$ for all $u \in U$.

Remark 5.65. If the mapping $\mathcal{G}$ is a potential operator, there exists a real-valued and G-differentiable function $\mathcal{R}$ such that $\mathcal{G} = \mathcal{R}'$ and then problem (5.117) is equivalent to seek the minima of the functional $\phi - \mathcal{R}$ on U. ♢

Assume now that the mapping $\mathcal{G}$ is a non-potential operator (*i.e.*, not of "gradient-type"). The variational problem $(\mathcal{P})$ associated with (5.117) is the following minimization problem:

$$\inf_{u \in U} \Psi(u), \tag{5.118}$$

where the functional $\Psi : U \longrightarrow \overline{\mathrm{IR}}$ is defined as a supremum on v of a function $L(u, v)$ as

$$\Psi(u) = \sup_{v \in U} L(u, v) \quad \forall u \in U, \tag{5.119}$$

where $L : U \times U \longrightarrow \overline{\mathrm{IR}}$ is defined by

$$L(u, v) = \phi(u) - \phi(v) + \langle\mathcal{G}(u), v - u\rangle_U. \tag{5.120}$$

The problem (5.118) then becomes

$$\inf_{u \in U} \sup_{v \in U} L(u, v). \tag{5.121}$$

Moreover, according to the definition of the convex conjugate functional, the mapping Ψ can be written as

$$\Psi(u) = \phi(u) - \langle \mathcal{G}(u), u \rangle_U + \phi^*(\mathcal{G}(u)) \geq 0 \tag{5.122}$$

for all u in U.[3] Consequently, because of Proposition 4.16, we can deduce that

$$\mathcal{G}(u_s) \in \partial\phi(u_s) \text{ if and only if } \phi(u_s) + \phi^*(\mathcal{G}(u_s)) - \langle \mathcal{G}(u_s), u_s \rangle_U = 0,$$

i.e.,

$$\mathcal{G}(u_s) \subset \partial\phi(u_s) \text{ if and only if } \Psi(u_s) = 0 \text{ and } u_s \text{ minimizes } \Psi \text{ on } U.$$

Otherwise,

$$\begin{aligned} &u_s \text{ is a solution of (5.117) if and only if} \\ &\Psi(u_s) = 0 \text{ and } u_s \text{ is a solution of (5.118).} \end{aligned} \tag{5.123}$$

As a corollary of the existence result (5.123), we have that if (u_s, v_s) is a minimax point of L on $U \times U$ such that $L(u_s, v_s) = 0$ then u_s is a solution of (5.117).

The following Ky Fan's minimax inequality theorem [118], which is based on the intersection result due to Ky Fan [117], gives some sufficient conditions on $-L$ and then on ϕ and $\mathcal{G}$ in order to obtain directly the existence of minimax points and then the existence of solutions of the differential inclusion problem (5.117). The intersection result of Ky Fan is known in the literature as Ky Fan's lemma.

Lemma 5.66. *(Ky Fan's lemma) Let Y be a Hausdorff topological vector space and K be a subset of Y. For each $u \in K$, consider a closed subset K_u of Y such that:*

(i) there exists $u_0 \in K$ such that the set K_{u_0} is compact

(ii) for any finite set $S_n = \{u_1, \ldots, u_n\}$ of points in K, $\text{co}S_n \subset \bigcup_{i=1}^{n} K_{u_i}$, where $\text{co}S_n$ is the convex hull of S_n.

Then

$$\bigcap_{u \in K} K_u \neq \emptyset.$$

□

Theorem 5.67. *Let K be a non-empty closed, convex subset of a reflexive Banach space U and $L : K \times K \longrightarrow I\!R$ be such that:*

[3] Since, from the definition of ϕ^*, we have that the inequality $\phi(v) + \phi^*(g) \geq \langle g, v \rangle_U$ holds, for all $(v, g) \in U \times U'$ and in particular for all $v \in U$ and $g = \mathcal{G}(v) \in U'$.

(i) $-L(u,u) \geq 0$, *for all* u *in* K

(ii) *for each* $u \in K$, *the mapping* $v \longrightarrow -L(u,v)$ *is quasi-convex on* K

(iii) *for each* $v \in K$, *the mapping* $u \longrightarrow -L(u,v)$ *is weakly upper semi-continuous on* K

(iv) *there exists an element* v_0 *such that* $K_0 = \{u \in K : \ -L(u,v_0) \geq 0\}$ *is a bounded subset of* K.

Then there exists an element u_{s} *in* K *such that*

$$\sup_{v\in K} L(u_{\mathrm{s}},v) \leq 0, \quad i.e., \quad -L(u_{\mathrm{s}},v) \geq 0 \ \ \textit{for all } v \in K, \tag{5.124}$$

i.e., there exists a solution u_{s} *of the so-called equilibrium problem (5.124).*

Proof. For the details of the proof see, *e.g.*, Auchmuty [16], Brezis *et al.* [58] and Ky Fan [118]. □

Remark 5.68. (*i*) Ky Fan's minimax inequality has many applications in various other branches of mathematics: economy, game theory, fixed point theorems and variational inequalities.
(*ii*) If L is defined by (5.120), $L(u,u) = 0$ for all u in K and then the assumption (*i*) of Theorem 5.67 is always true. ◇

As a direct consequence, we have the following result (because of (5.122) and (5.123)).

Corollary 5.69. *Let* K *be a non-empty closed, convex subset of a reflexive Banach space* U *and* L *be defined by (5.120) and satisfying assumptions (ii)–(iv) of Theorem 5.67. Then there exists a minimizer* u_{s} *of* Ψ *on* K *which is a solution of problem (5.117).*

Proof. The proof is left to the reader as an exercise. □

Proposition 5.70. *Let* K *be a non-empty closed, convex subset of a reflexive Banach space* U *and* L *be defined by (5.120). Assume that assumptions (iii)–(iv) of Theorem 5.67 hold and* ϕ *is a weakly lower semi-continuous, quasi-convex and coercive function on* K. *Then there exists a minimax point* $(u_{\mathrm{s}}, v_{\mathrm{s}})$ *of* L *on* $K \times K$ *and* u_{s} *is a solution of problem (5.117).*

Moreover, if we assume that the functional ϕ *is strictly convex over* K, *then there exists a unique* $v_{\mathrm{s}} \in K$ *depending on* u_{s} *(in a unique manner) such that* $\Psi(u_{\mathrm{s}}) = L(u_{\mathrm{s}}, v_{\mathrm{s}})$.

Proof. Since ϕ is a quasi-convex function, then L satisfies the assumption (*ii*) of Theorem 5.67 and from Corollary 5.69, there exists a minimizer u_{s} of Ψ on K and u_{s} is a solution of (5.117).

Consider now the functional $\mathcal{G}_{u_{\mathrm{s}}} : U \longrightarrow \mathbb{R}$ defined by

$$\mathcal{G}_{u_{\mathrm{s}}}(v) := \phi(v) - \langle \mathcal{G}(u_{\mathrm{s}}), v \rangle_U .$$

Since ϕ is a weakly lower semi-continuous, quasi-convex and coercive function on K then $\mathcal{G}_{u_s}$ is a weakly lower semi-continuous and coercive function on K. So, the set K_s of minimizers of $\mathcal{G}_{u_s}$ is a non-empty bounded closed and convex subset of K. Moreover, K_s is also the set of maximizers of the mapping $L(u_s, .)$. Then there exists an element $v_s \in K$ such that $\Psi(u_s) = L(u_s, v_s)$.

This completes the proof of the existence of a minimax point. If ϕ is strictly convex, $\mathcal{G}_{u_s}$ is also strictly convex and K_s is a singleton. This completes the proof. □

Now we present an example of a minimax point problem which is not necessarily a saddle point problem (see Auchmuty [17]).

Example 5.71. Let Ω be an open and bounded subset of $\mathbb{R}^d$, $d \leq 3$ sufficiently regular, $\partial\Omega = \Gamma$ be its boundary, $\mathbf{n}$ be the unit outward normal on $\partial\Omega$. We seek a vector function u representing the velocity of the fluid and a scalar function p representing the pressure of the fluid, which are defined in Ω and satisfy the following stationary non-linear Navier–Stokes system, for the equilibrium of a viscous flow, under a force field $f \in U'$ and subject to no slip boundary conditions on Γ:

$$\begin{aligned} -\nu \Delta u + \nabla p &= f - (u\nabla)u \quad \text{on } \Omega, \\ \operatorname{div}(u) &= 0 \quad \text{on } \Omega, \\ u &= 0 \quad \text{on } \Gamma, \end{aligned} \tag{5.125}$$

where ν is the coefficient of kinematic viscosity (a positive constant).

Let $U := \{u \in H_0^1(\Omega) : \operatorname{div}(u) = 0\}$ be a Hilbert space. U is a reflexive Banach space equipped with the norm $\| u \| = (\int_\Omega | \nabla u |^2 \, dx)^{1/2}$ and satisfies $U \subset L^2(\Omega) \subset U'$.

Let $\phi : U \longrightarrow \mathbb{R}$ be defined by

$$\phi(u) := \frac{\nu}{2} \int_\Omega | \nabla u |^2 \, dx. \tag{5.126}$$

It is clear that

$$\phi \text{ is a continuous, coercive and strictly convex function on } U. \tag{5.127}$$

Moreover, the stationary non-linear Navier–Stokes system (5.125) under the force $f \in U'$ can be written as

$$f - (u\nabla)u \in \partial\phi(u) \quad \text{for all } u \in U. \tag{5.128}$$

By using Green's formula, we have that

$$\langle (u\nabla)v, v \rangle_U = \frac{1}{2} \int_\Omega u.\nabla(| v |^2) dx = 0 \quad \text{for all } u, v \in U. \tag{5.129}$$

Consequently, when we replace v by $v+w$ in (5.129), we have

$$\langle (u\nabla)v, w\rangle_U = -\langle (u\nabla)w, v\rangle_U \quad \text{for all } u, v, w \in U. \tag{5.130}$$

Assume that the force f is in $L^2(\Omega) \subset U'$. The functionals $\mathcal{G} : U \longrightarrow U'$ and $L : U \times U \longrightarrow \mathbb{R}$ corresponding to our problem are defined by (because of (5.129) and (5.130))

$$\begin{aligned} \mathcal{G}(u) &= f - (u\nabla)u, \\ L(u,v) &:= \phi(u) - \phi(v) + \int_\Omega f(v-u)\mathrm{d}x - \int_\Omega (u\nabla)v.u\mathrm{d}x, \end{aligned} \tag{5.131}$$

for all u and v in V.

Remark 5.72. For all $v \in U$ the mapping $u \longrightarrow L(u,v)$ is not, in general, convex. Moreover, from the classical result for Navier–Stokes system, we know (only) that (5.125), or equivalently (5.128), have a finite number of distinct solutions under various conditions on f and ν.[4] Consequently, we can not expect to have a saddle point in $U \times U$, on the other hand we will show the existence of a minimax point in $U \times U$. ◊

Now we will study the existence of the minimax point of L in $U \times U$. Since for all $u \in U$ the mapping $v \longrightarrow -L(u,v) - \phi(v)$ is an affine functional then the mapping $v \longrightarrow -L(u,v)$ is convex on V (since ϕ is convex). Moreover, we have that

$$K_0 = \{u \in U \ : L(u,0) = \phi(u) - \int_\Omega fu\mathrm{d}x \le 0\} \text{ is bounded in } U$$

since, $\forall u \in K_0$,

$$\| u \|^2 = \frac{2}{\nu}\phi(u) \le \frac{2}{\nu} \| f \|_{L^2(\Omega)} \| u \|_{L^2(\Omega)} \le C \| f \|_{L^2(\Omega)} \| u \| .$$

Consequently, L satisfies conditions (i) (by the definition of L), (ii) and (iv) of Ky Fan's theorem 5.67. Prove now condition (iii) of Theorem 5.67, *i.e.*, the weakly upper semi-continuity of the mapping $u \longrightarrow -L(u,v)$ on U, for each fixed $v \in U$. Since the functions ϕ and $u \longrightarrow \int_\Omega fu\mathrm{d}x$ are lower semi-continuous, we need only prove the weak lower semi-continuity of the function

$$u \longrightarrow -\int_\Omega (u\nabla)v.u\mathrm{d}x.$$

Let (u_n) be a sequence of U which converges weakly to u in U. Then, from the compactness of Sobolev embeddings into L^p-spaces, (u_n) converges strongly in $L^p(\Omega)$, for $1 \le p < 6$ if $d = 3$ and $p \in [1, \infty[$ if $d \le 2$. Consequently,

[4] For example, if ν is sufficiently large (or f is sufficiently small) so that $\nu^2 \ge C(\Omega, d) \| f \|_{U'}$ then there exists a unique solution for $u \in U$ of (5.125), see, for instance, Temam [281].

$$-\int_{\Omega}(u_n\nabla)v.u_n\mathrm{d}x \longrightarrow -\int_{\Omega}(u\nabla)v.u\mathrm{d}x$$

and then condition (iii) of Theorem 5.67 follows.

From (5.127) and the fact that L satisfies conditions (i)–(iv) of Ky Fan's Theorem 5.67, we can apply Proposition 5.70 and we obtain the existence of $u_s \in U$ and $v_s \in U$ such that v_s depends in a unique manner on u_s, $0 = \Psi(u_s) = L(u_s, v_s)$ and u_s minimizes Ψ on U. ♣

5.4 Duality and Parametric Variational Problems

In this work we have centered our analysis on the case of the non-convex parametric variational problem, namely the following primal problem (P).

$$\begin{aligned}&\text{Find } u \in \mathcal{U}_{\mathrm{ad}} \text{ the infimum in } \mathcal{U}_{\mathrm{ad}} \text{ of the functional:}\\ &J_\nu(v) = \mathcal{S}(\Lambda v - \nu) + F(v),\end{aligned} \tag{5.132}$$

where ν is a given distributed parameter and the geometric operator Λ is non-linear and of quadratic type. More precisely,

$$\begin{aligned}&\Lambda = \Lambda_0 + \Lambda_{\mathrm{L}}, \text{where } \Lambda_{\mathrm{L}} \text{ is a linear and continuous operator and}\\ &\Lambda_0 \text{ is a non-linear operator as the form } \Lambda_0 v = \mathcal{B}(v,v), \text{ with}\\ &\mathcal{B} \text{ a bilinear and continuous operator.}\end{aligned} \tag{5.133}$$

This type of problems appears in many physical and biological systems such as hysteresis and phase transitions (*e.g.*, supraconductor, solidification, *etc.*), non-convex optimal design and control, non-linear bifurcation and stability analysis, and many others. The operator J_ν is corresponding to the total potential, the operator $\mathcal{S}$ is corresponding to the internal energy and the operator F is corresponding to the external energy. Our approach is a generalization of the work of Strang and Gao [127], by introducing a new gap function.

5.4.1 Abstract Framework

Let U, U' and X, X' be two pairs of reflexive Banach spaces, in duality with respect to certain bilinear forms $\langle .,. \rangle_U$ and $\langle .,. \rangle_X$ respectively. Let the geometric operator $\Lambda : U \longrightarrow X$ be a continuous, G-differential non-linear and quadratic type operator from U into X. Then the so-called geometrical equation can be written as

$$p = \Lambda u. \tag{5.134}$$

Since Λ is G-differentiable then p is G-differentiable and his G-derivative at u is $p'(u) = \Lambda_{\mathrm{t}}(u)$, where $\Lambda_{\mathrm{t}}(u)$ is the G-derivative of Λ at u that is corresponding to the tangent geometric operator. Moreover, since Λ is quadratic such that (5.133) then $p'(u)v = \mathcal{B}(u,v) + \mathcal{B}(v,u) + \Lambda_{\mathrm{L}}v$ because of (4.37), *i.e.*,

$$\Lambda_{\mathrm{t}}(u)v = \mathcal{B}(u,v) + \mathcal{B}(v,u) + \Lambda_{\mathrm{L}}v. \tag{5.135}$$

Remark 5.73. Because of (5.135), we can deduce that the operator

$$u \longrightarrow \Lambda_{\mathrm{t}}(u)u = 2\mathcal{B}(u,u) + \Lambda_{\mathrm{L}}u$$

is non-linear and then the following relation holds.

$$\Lambda_{\mathrm{t}}(u)u = 2\Lambda_0 u + \Lambda_{\mathrm{L}} u. \tag{5.136}$$

◊

Let the operator $\mathcal{S}: \ X \longrightarrow \overline{\mathrm{IR}}$ be an extended real-valued, finite and G-differentiable on a non-empty convex and closed subset $\mathcal{A} \subset X$. In conservative systems, we can usually consider the operator Λ such that the function $\mathcal{S}$ is either convex or concave but the function $\mathcal{S} \circ \Lambda$ may be non-convex and non-concave (by the non-linearity of Λ). The dual relation between the pairing spaces X and X' can be described by the so called constitutive equation

$$p^* = \mathcal{S}'(p - \nu), \tag{5.137}$$

where $\mathcal{S}': \ \mathcal{A} \subset X \longrightarrow X'$ is the G-derivative of $\mathcal{S}$.

In the same way, we suppose that the operator $F: \ U \longrightarrow \overline{\mathrm{IR}}$ is an extended real-valued, finite and G-differentiable functional on a non-empty convex and closed subset $\mathcal{K} \subset U$. The dual relation between the pairing spaces U and U' can be given by the following equation:

$$u^* = F'(u), \tag{5.138}$$

where $F': \ \mathcal{K} \subset U \longrightarrow U'$ is the G-derivative of F.

Definition 5.74. *The parametric variational problem (5.132) is said to be:*

(*i*) *geometrically non-linear if the operator Λ is non-linear*
(*ii*) *physically non-linear if at least one of the duality realations (5.137) and (5.138) is non-linear*
(*iii*) *fully non-linear if it is both geometrically and physically non-linear.* ■

For a given distributed parameter $\nu \in X$, the total potential $J_\nu: \ U \longrightarrow \overline{\mathrm{IR}}$ that is defined by $J_\nu(u) = \mathcal{S}(\Lambda u - \nu) + F(u)$ is finite and G-differentiable at u if and only if the element u is in the admissible space $\mathcal{K}_{\mathrm{ad}}$, where

$$\mathcal{K}_{\mathrm{ad}} := \{v \in \mathcal{K}: \ \Lambda v - \nu \in \mathcal{A}\}.$$

In particular, if $u \in \mathcal{K}_{\mathrm{ad}}$ is a critical point of J_ν, *i.e.*, $J'_\nu(u) = 0$, then we have the so-called equilibrium equation (according to (5.137), (5.138) and (5.132))

$$\Lambda_{\mathrm{t}}^*(u)p^* + u^* = \Lambda_{\mathrm{t}}^*(u)\mathcal{S}'(\Lambda u - \nu) + F'(u) = J'_\nu(u) = 0, \tag{5.139}$$

where the operator $\Lambda_{\mathrm{t}}^*(u): X' \longrightarrow U'$ is the adjoint of $\Lambda_{\mathrm{t}}(u)$ defined by

$$\langle p^*, \Lambda_{\mathrm{t}}(u)v\rangle_X = \langle \Lambda_{\mathrm{t}}^*(u)p^*, v\rangle_U \quad \forall v \in \mathcal{K}_{\mathrm{ad}}.$$

Remark 5.75. (*i*) The set of critical points of J_ν is a subset of $\mathcal{K}_{\mathrm{ad}}$ and is denoted by

$$\mathcal{K}_{\mathrm{c}} := \{u \in \mathcal{K}_{\mathrm{ad}} : J'_\nu(u) = 0\}.$$

(*ii*) The set of the infimum of J_ν is a subset of $\mathcal{K}_{\mathrm{ad}}$ and is denoted by

$$\mathcal{K}_{\mathrm{s}} := \{u \in \mathcal{K}_{\mathrm{ad}} : J_\nu(u) = \inf_{v \in U} J_\nu(v)\}.$$

◇

Remark 5.76. If $\mathcal{K}_{\mathrm{ad}}$ is a non-empty closed and bounded convex subset of a reflexive Banach space U then $\mathcal{K}_{\mathrm{s}}$ is a non-empty convex subset of $\mathcal{K}_{\mathrm{ad}}$. ◇

According to (5.134), (5.136) and (5.139), the relation between the pairing spaces U, U' and X, X' can be written as

$$\begin{aligned}\langle p^*, p\rangle_X + \langle u^*, u\rangle_U &= \langle p^*, \Lambda u\rangle_X - \langle \Lambda_{\mathrm{t}}^*(u)p^*, u\rangle_U \\ &= \langle p^*, \Lambda u\rangle_X - \langle p^*, \Lambda_{\mathrm{t}}(u)u\rangle_X \\ &= -\langle p^*, \Lambda_0 u\rangle_X \\ &= -G(u, p^*),\end{aligned} \tag{5.140}$$

where

$$G(u, p^*) := \langle p^*, \Lambda_0 u\rangle_X. \tag{5.141}$$

G is said to be the *complementary gap function*, which is introduced by Strang and Gao in [127] (see also Gao [128] for different illustrations and physical applications).

Theorem 5.77. *Suppose that the function F is in $\Gamma_0(U)$, finite and G-differentiable on the subset $\mathcal{K}$, and the function $\mathcal{S}$ is in $\Gamma_0(X)$, finite and G-differentiable on the subset $\mathcal{A}$. For u_{c} be a critical point of J_ν on $\mathcal{K}_{\mathrm{ad}}$ and $p_{\mathrm{c}}^* = \mathcal{S}'(\Lambda u_{\mathrm{c}} - \nu)$ we have that*

if $G(v, p_{\mathrm{c}}^*) \geq 0$ for all $v \in U$, then u_{c} is a minimizer of J_ν on $\mathcal{K}_{\mathrm{ad}}$.

Proof. Let u_{c} be a critical point of J_ν then

$$\begin{aligned}p_{\mathrm{c}} &= \Lambda u_{\mathrm{c}}, \\ p_{\mathrm{c}}^* &= \mathcal{S}'(\Lambda u_{\mathrm{c}} - \nu), \\ u_{\mathrm{c}}^* &= -\Lambda_{\mathrm{t}}^*(u_{\mathrm{c}})p_{\mathrm{c}}^* = F'(u_{\mathrm{c}}).\end{aligned} \tag{5.142}$$

Since $\mathcal{S}$ and F are G-differentiable on $\mathcal{A}$ and $\mathcal{K}$, respectively, we have that (because of the convexity of $\mathcal{S}$ and of F)

$$\begin{aligned}&\mathcal{S}(\Lambda v - \nu) - \mathcal{S}(\Lambda u_{\mathrm{c}} - \nu) \geq \langle \mathcal{S}'(\Lambda u_{\mathrm{c}} - \nu), \Lambda v - \Lambda u_{\mathrm{c}}\rangle_X \quad \forall v \in \mathcal{K}_{\mathrm{ad}}, \\ &F(v) - F(u_{\mathrm{c}}) \geq \langle F'(u_{\mathrm{c}}), v - u_{\mathrm{c}}\rangle_U \quad \forall v \in \mathcal{K}.\end{aligned} \tag{5.143}$$

Since Λ is a quadratic operator then for any w and v in U, we can easily obtain the following equality (according to (5.133) and (5.135)):

$$\Lambda(v+w) = \Lambda v + \Lambda_{\rm t}(v)w + \Lambda_0 w. \tag{5.144}$$

By the relations (5.142), (5.143) and (5.144) we can deduce that (for all $v \in \mathcal{K}$)

$$\begin{aligned}
J_\nu\ (u_{\rm c} & +w) - J_\nu(u_{\rm c}) \\
&= (\mathcal{S}(\Lambda(u_{\rm c}+w)-\nu) - \mathcal{S}(\Lambda u_{\rm c}-\nu)) + (F(u_{\rm c}+w) - F(u_{\rm c})) \\
&\geq \langle \mathcal{S}'(\Lambda u_{\rm c}-\nu), \Lambda(u_{\rm c}+w) - \Lambda u_{\rm c}\rangle_X + \langle F'(u_{\rm c}), w\rangle_U \\
&= \langle \mathcal{S}'(\Lambda u_{\rm c}-\nu), \Lambda_{\rm t}(u_{\rm c})w + \Lambda_0 w\rangle_X + \langle F'(u_{\rm c}), w\rangle_U \\
&= \langle p_{\rm c}^*, \Lambda_{\rm t}(u_{\rm c})w + \Lambda_0 w\rangle_X + \langle -\Lambda_{\rm t}^*(u_{\rm c})p_{\rm c}^*, w\rangle_U \\
&= \langle p_{\rm c}^*, \Lambda_0 w\rangle_X + \langle p_{\rm c}^* - p_{\rm c}^*, \Lambda_{\rm t}(u_{\rm c})w\rangle_X \\
&= \langle p_{\rm c}^*, \Lambda_0 w\rangle_X \\
&= G(w, p_{\rm c}^*),
\end{aligned} \tag{5.145}$$

where $w = v - u_{\rm c} \in U$.

Since $G(z, p_{\rm c}^*) \geq 0$ for all $z \in U$ then (according to the previous result) $J_\nu(v) - J_\nu(u_{\rm c}) \geq 0$ for all $v \in \mathcal{K}$. Consequently, $u_{\rm c}$ is a minimizer of J_ν and then the result of the theorem. □

We suppose now that F is a real-valued non-linear convex quadratic function on U (this mathematical formulation represents several realistic situations) in the sense

$$\begin{aligned}
&F = F_0 + F_{\rm L}, \text{ where } F_{\rm L} \text{ is a linear continuous form and} \\
&F_0 \text{ is a non-linear and convex function in the form} \\
&F_0(v) = \mathcal{H}(v,v) \geq 0, \text{ with} \\
&\mathcal{H} \text{ a bilinear positive and continuous form on U.}
\end{aligned} \tag{5.146}$$

Therefore, we have a more precise result, by introducing a *new gap function.*

Theorem 5.78. *Assume that the assumptions of Theorem 5.77 hold and that the function F is a non-linear convex quadratic function such that (5.146). For $u_{\rm c}$ be a critical point of J_ν on $\mathcal{K}_{\rm ad}$ and $p_{\rm c}^* = \mathcal{S}'(\Lambda u_{\rm c} - \nu)$ we have that*

$$\textit{if } G_{\rm q}(v, p_{\rm c}^*) \geq 0 \textit{ for all } v \in U, \textit{ then } u_{\rm c} \textit{ is a minimizer of } J_\nu \textit{ on } \mathcal{K}_{\rm ad},$$

where $G_{\rm q}(v, p_{\rm c}^*) := G(v, p_{\rm c}^*) + F_0(v)$.

This new gap function $G_{\rm q}$ will be called the quadratic complementary gap function.

Proof. Let $u_{\rm c}$ be a critical point of J_ν. Since F is a quadratic form such that (5.146) then for any w and v on U, we can obtain easily the following equality:

$$F(v+w) = F(v) + \langle F'(v), w\rangle_U + F_0(w). \tag{5.147}$$

Applying the G-differentiability of $\mathcal{S}$ and F on $\mathcal{A}$ and $\mathcal{K}$, respectively, we have that (because of the convexity of $\mathcal{S}$ and the expression (5.147))

$$
\begin{aligned}
&\mathcal{S}(\Lambda v-\nu)-\mathcal{S}(\Lambda u_{\rm c}-\nu)\geq\langle\mathcal{S}'(\Lambda u_{\rm c}-\nu),\Lambda v-\Lambda u_{\rm c}\rangle_X\quad\forall v\in\mathcal{K}_{\rm ad},\\
&F(v)-F(u_{\rm c})=F_0(v-u_{\rm c})+\langle F'(u_{\rm c}),v-u_{\rm c}\rangle_U\quad\forall v\in\mathcal{K}.
\end{aligned}
\tag{5.148}
$$

By the relations (5.142), (5.148) and (5.144) we can deduce that (for all $v\in\mathcal{K}$)

$$
\begin{aligned}
J_\nu\ (u_{\rm c}&+w)-J_\nu(u_{\rm c})\\
&=(\mathcal{S}(\Lambda(u_{\rm c}+w)-\nu)-\mathcal{S}(\Lambda u_{\rm c}-\nu))+(F(u_{\rm c}+w)-F(u_{\rm c}))\\
&\geq\langle\mathcal{S}'(\Lambda u_{\rm c}-\nu),\Lambda(u_{\rm c}+w)-\Lambda u_{\rm c}\rangle_X+\langle F'(u_{\rm c}),w\rangle_U+F_0(w)\\
&=\langle\mathcal{S}'(\Lambda u_{\rm c}-\nu),\Lambda_{\rm t}(u_{\rm c})w+\Lambda_0 w\rangle_X+\langle F'(u_{\rm c}),w\rangle_U+F_0(w)\\
&=\langle p_{\rm c}^*,\Lambda_{\rm t}(u_{\rm c})w+\Lambda_0 w\rangle_X+\langle\,-\Lambda_{\rm t}^*(u_{\rm c})p_{\rm c}^*,w\rangle_U+F_0(w)\\
&=\langle p_{\rm c}^*,\Lambda_0 w\rangle_X+\langle p_{\rm c}^*-p_{\rm c}^*,\Lambda_{\rm t}(u_{\rm c})w\rangle_X+F_0(w)\\
&=\langle p_{\rm c}^*,\Lambda_0 w\rangle_X+F_0(w)\\
&=G_{\rm q}(w,p_{\rm c}^*),
\end{aligned}
\tag{5.149}
$$

where $w=v-u_{\rm c}$.

Since $G_{\rm q}(z,p_{\rm c}^*)\geq 0$ for all $z\in U$ then (according to the previous result) $J_\nu(v)-J_\nu(u_{\rm c})\geq 0$ for all $v\in\mathcal{K}$. Consequently, $u_{\rm c}$ is a minimizer of J_ν and the result of the theorem follows. □

5.4.2 Geometrically Non-linear Lagrangian Representation

In order to study the dual problem, we need to find the complementary energy of the non-linear system.

From the Fenchel transformation, we known that the conjugate functionals, of given functionals $\mathcal{S}:\ X\longrightarrow\overline{\rm I\!R}$ and $F:\ U\longrightarrow\overline{\rm I\!R}$, are defined respectively by

$$
\begin{aligned}
\mathcal{S}^*(p^*)&=\sup_{p\in X}(\langle p^*,p\rangle_X-\mathcal{S}(p))\quad\text{for all } p^*\in X',\\
F^*(u^*)&=\sup_{u\in U}(\langle u^*,u\rangle_U-F(u))\quad\text{for all } u^*\in U'.
\end{aligned}
\tag{5.150}
$$

We introduce now the functional $\mathcal{S}_\nu:\ p\in X\longrightarrow\mathcal{S}(p-\nu)\in\overline{\rm I\!R}$. Then the conjugate functional of $\mathcal{S}_\nu$ is given, for all $p^*\in X'$, by

$$
\mathcal{S}_\nu^*(p^*)=\mathcal{S}^*(p^*)+\langle p^*,\nu\rangle_X. \tag{5.151}
$$

Remark 5.79. (*a*) The function $\mathcal{S}_\nu$ is convex (respectively concave) if and only if the function $\mathcal{S}$ is convex (respectively concave).

(*b*) From the theory of convex analysis, the functions $\mathcal{S}_\nu^*$ and F^* are always convex and lower semi-continuous.

(*c*) If the function $\mathcal{S}$ is strictly convex, G-differentiable on $\mathcal{A}$, then the following conditions are equivalent:

(*i*) $p^* = \mathcal{S}'(p)$
(*ii*) $p = (\mathcal{S}^*)'(p^*)$
(*iii*) $\langle p^*, p\rangle_X = \mathcal{S}^*(p^*) + \mathcal{S}(p)$. ◊

Let $\mathcal{K}^* \subset U^{'}$ and $\mathcal{A}^* \subset X^{'}$ be non-empty, convex and closed subsets on which F^* and $\mathcal{S}^*_\nu$ are finite and G-differentiable, respectively. Consider now the geometrical Lagrangian $L_\nu : U \times X' \longrightarrow \overline{\mathrm{IR}}$ defined by

$$L_\nu(v, q^*) := \langle q^*, \Lambda v\rangle_X - \mathcal{S}^*_\nu(q^*) + F(v), \tag{5.152}$$

which is finite and G-differentiable on $\mathcal{K} \times \mathcal{A}^*$. It is easy to obtain that, for all $(u, p^*) \in \mathcal{K} \times \mathcal{A}^*$,

$$\begin{aligned} \frac{\partial L_\nu}{\partial v}(u, p^*) &= \Lambda^*_{\mathrm{t}}(u)p^* + F'(u), \\ \frac{\partial L_\nu}{\partial q^*}(u, p^*) &= \Lambda u - (\mathcal{S}^*_\nu)'(p^*), \end{aligned} \tag{5.153}$$

where $\partial L_\nu/\partial v$ and $\partial L_\nu/\partial q^*$ denote the partial G-derivatives on U and X', respectively.

Remark 5.80. (*i*) Depending on the convexity of the functional $\mathcal{S}_\nu$, there exists an interesting relation between the functional J_ν and the Lagrangian L_ν. More precisely, for any given function F we have that (by using the Fenchel transformation and Corollary 4.6), for all $v \in U$

$$\text{if } \mathcal{S}_\nu \text{ is in } \Gamma(X) \text{ then } J_\nu(v) = \sup_{q^* \in X'} L_\nu(v, q^*), \tag{5.154}$$

i.e., L_ν is a Lagrangian of type I of problem (P) (given by (5.132)).
(*ii*) According to the expressions (5.153) we can deduce that, if $(u_{\mathrm{c}}, p^*_{\mathrm{c}}) \in \mathcal{K} \times \mathcal{A}^*$ is a critical point of L_ν then

$$\begin{aligned} \frac{\partial L_\nu}{\partial v}(u_{\mathrm{c}}, p^*_{\mathrm{c}}) &= \Lambda^*_{\mathrm{t}}(u_{\mathrm{c}})p^*_{\mathrm{c}} + F'(u_{\mathrm{c}}) = 0, \\ \frac{\partial L_\nu}{\partial q^*}(u_{\mathrm{c}}, p^*_{\mathrm{c}}) &= \Lambda(u_{\mathrm{c}}) - (\mathcal{S}^*_\nu)'(p^*_{\mathrm{c}}) = 0. \end{aligned} \tag{5.155}$$

◊

Example 5.81. Let Ω be an open and bounded subset of IR^d, $d \leq 3$ sufficiently regular, $\partial\Omega = \Gamma$ be its boundary, $\mathbf{n}$ be the unit outward normal on $\partial\Omega$. Let $U = H^1_0(\Omega)$ be a Hilbert space. U is a reflexive Banach space equipped with the norm $\| u \| = (\int_\Omega | \nabla u |^2 \,\mathrm{d}x)^{1/2}$ and $U \subset L^2(\Omega) \subset U'$. Moreover, according to Sobolev embeddings into L^p-spaces, we have the embedding $U \subset L^p(\Omega)$, for $1 \leq p < 6$ if $d = 3$ and $p \in [1, \infty[$ if $d \leq 2$ with compactness.

The Landau–Ginzburg energy J_ν for a ferroelectric can be written as

$$J_\nu(u) = \int_\Omega \frac{D(x)}{2} | \nabla u |^2 \,\mathrm{d}x + \int_\Omega \frac{1}{2}(\frac{| u |^2}{2} - \nu(x))^2 \mathrm{d}x - \int_\Omega f u \mathrm{d}x,$$

where u is a phase-field and scalar function whose values describe the phase of the system under consideration, $D > 0$ and $\nu > 0$ are positive and bounded functions, and f is in $L^2(\Omega)$. We introduce now the following quadratic operator $\Lambda : U \longrightarrow X$ by

$$\Lambda v = \frac{1}{2} \mid v \mid^2 = \Lambda_0 v \quad \text{for all } v \in U,$$

where $X = L^2(\Omega) = X'$.

The operator Λ is G-differentiable and its G-derivative at point u, $\Lambda_{\mathrm{t}} : U \longrightarrow X$ is given by

$$\Lambda_{\mathrm{t}}(u) = u = \Lambda_{\mathrm{t}}^*(u) \quad \text{for all } u \in X. \tag{5.156}$$

Then the functional $J_\nu : U \longrightarrow \mathbb{R}$ can be written as

$$J_\nu(u) = \mathcal{S}(\Lambda u - \nu) + F(u),$$

where the functionals $\mathcal{S} : X \longrightarrow \mathbb{R}$ and $F : U \longrightarrow \mathbb{R}$ corresponding to our problem are defined by

$$\begin{aligned} \mathcal{S}(p) &:= \int_\Omega \frac{p^2}{2} \mathrm{d}x \text{ for all } p \in X, \\ F(v) &:= \int_\Omega \frac{D(x)}{2} \mid \nabla v \mid^2 \mathrm{d}x - \int_\Omega f v \mathrm{d}x \text{ for all } v \in U. \end{aligned} \tag{5.157}$$

The functionals $\mathcal{S}$ and F are convex lower semi-continuous finite and G-differentiable, and their G-differentials at points p and u respectively are given by

$$\mathcal{S}'(p) = p \quad \text{and} \quad F'(u) = -\mathrm{div}(D\nabla u) - f. \tag{5.158}$$

Moreover, the gap and the quadratic gap functions are given by (since F is a quadratic form)

$$\begin{aligned} G(v, p^*) &= \frac{1}{2} \int_\Omega p^* \mid v \mid^2 \mathrm{d}x, \\ G_{\mathrm{q}}(v, p^*) &= G(v, p^*) + \int_\Omega \frac{D}{2} \mid \nabla v \mid^2 \mathrm{d}x. \end{aligned} \tag{5.159}$$

Let u_{c} be a critical point of J_ν. Then, according to (5.142), (5.156) and (5.158), we can deduce that u_{c} is a solution of the well-known second-order Ginzburg–Landau equation

$$\begin{aligned} p_{\mathrm{c}} &= \frac{1}{2} \mid u_{\mathrm{c}} \mid^2, \\ p_{\mathrm{c}}^* &= \frac{1}{2} \mid u_{\mathrm{c}} \mid^2 - \nu, \\ -\mathrm{div}(D\nabla u_{\mathrm{c}}) &+ u_{\mathrm{c}}(\frac{1}{2} \mid u_{\mathrm{c}} \mid^2 - \nu) = f \quad \text{on } \Omega \\ u_{\mathrm{c}} &= 0 \quad \text{on } \Gamma. \end{aligned} \tag{5.160}$$

Moreover, according to Theorem 5.78, we have that if $G_{\mathrm{q}}(v, p_{\mathrm{c}}^*) \geq 0$ for all $v \in U$ then u_{c} is a minimizer of J_ν.

We now give some sufficient conditions such that $G_{\mathrm{q}}(v, p_{\mathrm{c}}^*) \geq 0$ for all $v \in U$. Let $\alpha > 0$ be the smallest eigenvalue of the operator $\mathrm{div}(D\nabla .)$, *i.e.*,

$$\alpha = \inf_{v \in U} \frac{\int_\Omega D \mid \nabla v \mid^2 \mathrm{d}x}{\| v \|^2_{L^2(\Omega)}}.$$

We can deduce that if $\alpha + p_{\mathrm{c}}^* \geq 0$, *i.e.*, if $\mid u_{\mathrm{c}} \mid^2 \geq 2(\nu - \alpha)$ then $G_{\mathrm{q}}(v, p_{\mathrm{c}}^*) \geq 0$. Consequently, u_{c} is a minimizer of J_ν. ♣

Next we give the properties of the critical points of L_ν that depend on gap functions.

Theorem 5.82. *Assume that the function F is in $\Gamma_0(U)$, finite and G-differentiable on $\mathcal{K}$, $\Lambda : U \longrightarrow X$ is a quadratic operator and the function $\mathcal{S}_\nu$ is in $\Gamma_0(X)$, finite and G-differentiable on $\mathcal{A}$.*

Let $(u_{\mathrm{c}}, p_{\mathrm{c}}^*)$ be a critical point of L_ν defined by (5.152). Then the following property holds: if $G(v, p_{\mathrm{c}}^*) \geq 0$ for all v in U then $(u_{\mathrm{c}}, p_{\mathrm{c}}^*)$ is a saddle point of L_ν.

Proof. Since F is convex then, for all $v \in U$,

$$F(v) \geq F(u_{\mathrm{c}}) + \langle F'(u_{\mathrm{c}}), v - u_{\mathrm{c}} \rangle_U.$$

According to (5.144) we have that $\Lambda(u_{\mathrm{c}} + w) = \Lambda u_{\mathrm{c}} + \Lambda_{\mathrm{t}}(u_{\mathrm{c}})w + \Lambda_0 w$, for all $w \in U$, and then, according to (5.155) (since $(u_{\mathrm{c}}, p_{\mathrm{c}}^*)$ is a critical point)

$$\begin{aligned} &L_\nu(v, p_{\mathrm{c}}^*) - L_\nu(u_{\mathrm{c}}, p_{\mathrm{c}}^*) \\ &\quad \geq \langle \Lambda_{\mathrm{t}}^*(u_{\mathrm{c}}) p_{\mathrm{c}}^* + F'(u_{\mathrm{c}}), v - u_{\mathrm{c}} \rangle_U + \langle p_{\mathrm{c}}^*, \Lambda_0(v - u_{\mathrm{c}}) \rangle_X \\ &\quad = \langle \frac{\partial L_\nu}{\partial u}(u_{\mathrm{c}}, p_{\mathrm{c}}^*), v - u_{\mathrm{c}} \rangle_U + G(v - u_{\mathrm{c}}, p_{\mathrm{c}}^*) \\ &\quad = G(v - u_{\mathrm{c}}, p_{\mathrm{c}}^*), \end{aligned} \tag{5.161}$$

for all v in $\mathcal{K}$. Consequently (since $G(z, p_{\mathrm{c}}^*) \geq 0$ for all z in U),

$$L_\nu(v, p_{\mathrm{c}}^*) - L_\nu(u_{\mathrm{c}}, p_{\mathrm{c}}^*) \geq G(v - u_{\mathrm{c}}, p_{\mathrm{c}}^*) \geq 0 \quad \text{for all } v \in \mathcal{K}. \tag{5.162}$$

Since the function $\mathcal{S}_\nu^*$ is convex then, for each given $v \in U$, the function $\psi_v : q^* \in X' \longrightarrow \psi_v(q^*) = L_\nu(v, q^*)$ is concave. Moreover, if $(u_{\mathrm{c}}, p_{\mathrm{c}}^*) \in \mathcal{K} \times \mathcal{A}^*$ is a critical point of L_ν, then p_{c}^* is a critical point of $\psi_{u_{\mathrm{c}}}$, *i.e.*, $\psi'_{u_{\mathrm{c}}}(p_{\mathrm{c}}^*) = 0$. Thus, by Proposition 4.45, $\psi_{u_{\mathrm{c}}}(p_{\mathrm{c}}^*) \geq \psi_{u_{\mathrm{c}}}(q^*)$, for all $q^* \in \mathcal{A}^*$.

Consequently,

$$L_\nu(u_{\mathrm{c}}, p_{\mathrm{c}}^*) \geq L_\nu(u_{\mathrm{c}}, q^*) \quad \text{for all } q^* \in \mathcal{A}^*. \tag{5.163}$$

From (5.162) and (5.163) we obtain easily the result of the theorem. □

Nota bene: In the sequel, we will suppose that F is a quadratic form in the sense (5.146).

Theorem 5.83. *Assume that the assumptions of Theorem 5.82 hold and that the function F is a non-linear convex quadratic function such that (5.146).*

Let $(u_{\rm c}, p^*_{\rm c})$ be a critical point of L_ν defined by (5.152). Then the following properties hold:

(i) $(u_{\rm c}, p^*_{\rm c})$ is a saddle point of L_ν if and only if $G_{\rm q}(v, p^*_{\rm c}) \geq 0$ for all v in U

(ii) $(u_{\rm c}, p^*_{\rm c})$ is a supercritical point of L_ν if and only if $G_{\rm q}(v, p^*_{\rm c}) \leq 0$ for all v in U.

Proof. According to (5.144) and (5.147) we have, for all $w \in U$, that

$$\begin{aligned} \Lambda(u_{\rm c} + w) &= \Lambda u_{\rm c} + \Lambda_{\rm t}(u_{\rm c})w + \Lambda_0 w, \\ F(u_{\rm c} + w) &= F(u_{\rm c}) + \langle F'(u_{\rm c}), w\rangle_U + F_0(w). \end{aligned} \tag{5.164}$$

Then for all $v \in \mathcal{K}$, according to (5.155) (since $(u_{\rm c}, p^*_{\rm c})$ is a critical point)

$$\begin{aligned} L_\nu(u_{\rm c} + w, p^*_{\rm c}) - L_\nu(u_{\rm c}, p^*_{\rm c}) \\ &= \langle \Lambda^*_{\rm t}(u_{\rm c})p^*_{\rm c} + F'(u_{\rm c}), w\rangle_U + \langle p^*_{\rm c}, \Lambda_0 w\rangle_X + F_0(w) \\ &= G(w, p^*_{\rm c}) + F_0(w) \\ &= G_{\rm q}(w, p^*_{\rm c}), \end{aligned} \tag{5.165}$$

where $w = v - u_{\rm c} \in U$.

Consequently,

$$L_\nu(v, p^*_{\rm c}) - L_\nu(u_{\rm c}, p^*_{\rm c}) = G_{\rm q}(v - u_{\rm c}, p^*_{\rm c}) \text{ for all } v \text{ in } \mathcal{K}$$

and then

$$\operatorname{sign}(L_\nu(v, p^*_{\rm c}) - L_\nu(u_{\rm c}, p^*_{\rm c})) = \operatorname{sign}(G_{\rm q}(v - u_{\rm c}, p^*_{\rm c})) \ \ \forall v \in \mathcal{K}, \tag{5.166}$$

where $\operatorname{sign}(x) = 1$ if $x > 0$, $\operatorname{sign}(x) = -1$ if $x < 0$ and $\operatorname{sign}(x) = 0$ if $x = 0$.

Since the function $\mathcal{S}^*_\nu$ is convex then, for each given $v \in U$, the following function $\psi_v : q^* \in X' \longrightarrow \psi_v(q^*) = L_\nu(v, q^*)$ is concave. Moreover, if $(u_{\rm c}, p^*_{\rm c})$ is a critical point of L_ν, then $p^*_{\rm c}$ is a critical point of $\psi_{u_{\rm c}}$, *i.e.*, $\psi'_{u_{\rm c}}(p^*_{\rm c}) = 0$. Thus, by Proposition 4.45, $\psi_{u_{\rm c}}(p^*_{\rm c}) \geq \psi_{u_{\rm c}}(q^*)$, for all $q^* \in \mathcal{A}^*$.

Consequently,

$$L_\nu(u_{\rm c}, p^*_{\rm c}) \geq L_\nu(u_{\rm c}, q^*) \text{ for all } q^* \in \mathcal{A}^*. \tag{5.167}$$

From (5.166) and (5.167) we obtain easily results (i) and (ii) of the theorem. □

As a direct consequence we have the following corollary.

Corollary 5.84. *Assume that the assumptions of Theorem 5.83 hold. If the quadratic gap function* G_{q} *is positive on* $\mathcal{E}_{\mathrm{q}} = U \times \mathcal{A}^*_{\mathrm{c}}$*, where*

$$\mathcal{A}^*_{\mathrm{c}} := \{q^* \in \mathcal{A}^* |\ \exists u_q \in \mathcal{K} : (u_q, q^*) \in \mathcal{E}_{\mathrm{ad}}\} \tag{5.168}$$

and

$$\mathcal{E}_{\mathrm{ad}} = \{(v, q^*) \in \mathcal{K} \times \mathcal{A}^* :\ \Lambda^*_{\mathrm{t}}(v)q^* + F'(v) = 0\}, \tag{5.169}$$

then, all critical points of the Lagrangian L_ν *are saddle points of* L_ν*. Similarly if* G_{q} *is negative on* $\mathcal{E}_{\mathrm{q}}$ *then all critical points of the Lagrangian* L_ν *are supercritical points of* L_ν*.* □

The following proposition gives some relations depending on the sign of G_{q} in the case where $\mathcal{E}_{\mathrm{ad}}$ is non-empty.

Proposition 5.85. *Assume that the assumptions of Theorem 5.83 hold. Let* $p^* \in \mathcal{A}^*_{\mathrm{c}}$ *be given. Then there exists* $u_p \in K$ *such that* (u_p, p^*) *is in* $\mathcal{E}_{\mathrm{ad}}$ *and we have the following properties*

$$\begin{aligned} &if\ \ G_{\mathrm{q}}(v, p^*) \geq 0 \ \ \forall v \in U \ \ then \ \ L_\nu(u_p, p^*) = \inf_{v \in K} L_\nu(v, p^*),\\ &if\ \ G_{\mathrm{q}}(v, p^*) \leq 0 \ \ \forall v \in U \ \ then \ \ L_\nu(u_p, p^*) = \sup_{v \in K} L_\nu(v, p^*). \end{aligned} \tag{5.170}$$

Proof. Let $p^* \in \mathcal{A}^*_{\mathrm{c}}$ be given and let us introduce the following function:

$$\psi_{p^*} :\ v \in U \longrightarrow \psi_{p^*}(v) = L_\nu(v, p^*).$$

By hypothesis and the expression (5.153), there exists $u_p \in \mathcal{K}$ such that

$$\psi'_{p^*}(u_p) = \frac{\partial L_\nu}{\partial u}(u_p, p^*) = \Lambda^*_{\mathrm{t}}(u_p)p^* + F'(u_p) = 0.$$

By the relations (5.164), we can deduce that (for all $v \in K$)

$$\begin{aligned} \psi_{p^*}\ & (u_p + w) - \psi_{p^*}(u_p)\\ &= \langle p^*, \Lambda(u_p + w) - \Lambda(u_p)\rangle_X + (F(u_p + w) - F(u_p))\\ &= \langle p^*, \Lambda_{\mathrm{t}}(u_p)w + \Lambda_0 w\rangle_X + \langle F'(u_p), w\rangle_U + F_0(w)\\ &= \langle \Lambda^*_{\mathrm{t}}(u_p)p^* + F'(u_p), w\rangle_U + \langle p^*, \Lambda_0 w\rangle_X + F_0(w)\\ &= G_{\mathrm{q}}(w, p^*) \ \ (\text{since } (u_p, p^*) \in \mathcal{E}_{\mathrm{ad}}), \end{aligned} \tag{5.171}$$

where $w = v - u_p$.

Thus, if $G_{\mathrm{q}}(z, p^*) \geq 0$ (respectively $G_{\mathrm{q}}(z, p^*) \leq 0$) $\forall z \in U$, then $\psi_{p^*}(v) \geq \psi_{p^*}(u_p)$ (respectively $\psi_{p^*}(v) \leq \psi_{p^*}(u_p)$) $\forall v \in K$. Consequently, the value u_p is a minimizer (respectively maximizer) of ψ_{p^*} and $\inf_{v \in K} L_\nu(v, p^*) = L_\nu(u_p, p^*)$ (respectively $\sup_{v \in K} L_\nu(v, p^*) = L_\nu(u_p, p^*)$). □

Let $\mathcal{A}_{\mathrm{q}}^* \subset \mathcal{A}^*$ be the convex hull of the set $\mathcal{A}_{\mathrm{c}}^*$. We introduce the so-called *complementary energy* $J_\nu^* : X' \longrightarrow \overline{\mathbb{R}}$ which is defined by, for $p^* \in \mathcal{A}_{\mathrm{q}}^*$,

$$\begin{aligned} J_\nu^*(p^*) &:= \inf_{v \in K} L_\nu(v, p^*) \text{ if } G_{\mathrm{q}}(z, p^*) \geq 0 \ \forall z \in U, \\ J_\nu^*(p^*) &:= \sup_{v \in K} L_\nu(v, p^*) \text{ if } G_{\mathrm{q}}(z, p^*) \leq 0 \ \forall z \in U. \end{aligned} \tag{5.172}$$

Remark 5.86. If the function F is linear, we prove easily that $\mathcal{A}_{\mathrm{c}}^*$ is convex and then $\mathcal{A}_{\mathrm{q}}^* = \mathcal{A}_{\mathrm{c}}^*$. ♢

According to Proposition 5.85, it is interesting to study the following dual problems associated with problem (P) (given by (5.132)):

$$\begin{aligned} &\text{find } p^* \in X' \text{ such that} \\ &J_\nu^*(p^*) = \sup_{q^* \in X'} J_\nu^*(q^*), \end{aligned} \tag{5.173}$$

and

$$\begin{aligned} &\text{find } p^* \in X' \text{ such that} \\ &J_\nu^*(p^*) = \inf_{q^* \in X'} J_\nu^*(q^*). \end{aligned} \tag{5.174}$$

We can now give the following lemma, which is a corollary of Propositions 5.30, 5.31 and 5.35.

Lemma 5.87. *Assume that the assumptions of Theorem 5.83 hold. Let* $(u_{\mathrm{s}}, p_{\mathrm{s}}^*) \in \mathcal{E}_{\mathrm{ad}}$ *be a saddle or a supercritical point of* L_ν *and assume that* L_ν *is partially G-differentiable at point* $(u_{\mathrm{s}}, p_{\mathrm{s}}^*)$*. Then, if the functions* J_ν^* *and* J_ν *are G-differentiable at* u_{s} *and* p_{s}^* *respectively, then*

$$\begin{aligned} &(J_\nu^*)'(p_{\mathrm{s}}^*) = 0, \quad J_\nu'(u_{\mathrm{s}}) = 0, \\ &J_\nu^*(p_{\mathrm{s}}^*) = J_\nu(u_{\mathrm{s}}) = L_\nu(u_{\mathrm{s}}, p_{\mathrm{s}}^*). \end{aligned}$$

Proof. Let $(u_{\mathrm{s}}, p_{\mathrm{s}}^*) \in \mathcal{E}_{\mathrm{ad}}$. Then, for all $v \in K$ we have

$$L_\nu(v, p_{\mathrm{s}}^*) - L_\nu(u_{\mathrm{s}}, p_{\mathrm{s}}^*) = \langle \Lambda_{\mathrm{t}}^*(u_{\mathrm{s}}) p_{\mathrm{s}}^* + F'(u_{\mathrm{s}}), v - u_{\mathrm{s}} \rangle_U + G_{\mathrm{q}}(v - u_{\mathrm{s}}, p_{\mathrm{s}}^*).$$

Consequently,

$$L_\nu(v, p_{\mathrm{s}}^*) - L_\nu(u_{\mathrm{s}}, p_{\mathrm{s}}^*) = G_{\mathrm{q}}(v - u_{\mathrm{s}}, p_{\mathrm{s}}^*) \ \ \forall v \in K. \tag{5.175}$$

Suppose that $(u_{\mathrm{s}}, p_{\mathrm{s}}^*)$ is a saddle point of L_ν. Then, because of the definition of saddle point and the relation (5.154), we have that $G_{\mathrm{q}}(w, p_{\mathrm{s}}^*) \geq 0 \ \forall w \in U$ (by (5.175) and (5.172)) and

$$\begin{aligned} J_\nu(u_{\mathrm{s}}) &= \sup_{q^* \in \mathcal{A}^*} L_\nu(u_{\mathrm{s}}, q^*) \\ &\leq L_\nu(u_{\mathrm{s}}, p_{\mathrm{s}}^*) \\ &\leq \inf_{v \in \mathcal{K}} L_\nu(v, p_{\mathrm{s}}^*) = J_\nu^*(p_{\mathrm{s}}^*). \end{aligned}$$

Moreover, according to (5.172), we have

$$\begin{aligned} J_\nu(u_s) - J_\nu^*(p_s^*) &= \sup_{q^*\in\mathcal{A}^*} L_\nu(u_s,q^*) - \inf_{v\in\mathcal{K}} L_\nu(v,p_s^*) \\ &= \sup_{q^*\in\mathcal{A}^*}\sup_{v\in\mathcal{K}} (L_\nu(u_s,q^*) - L_\nu(v,p_s^*)) \geq 0. \end{aligned}$$

Consequently,

$$J_\nu(u_s) = L_\nu(u_s,p_s^*) = J_\nu^*(p_s^*).$$

Because of Proposition 5.35, we can deduce that $(J_\nu^*)'(p_s^*) = 0$ and $J_\nu'(u_s) = 0$. Suppose now that (u_s,p_s^*) is a supercritical point of L_ν, then

$$L_\nu(v,p_s^*) \leq L_\nu(u_s,p_s^*) \geq L_\nu(u_s,q^*) \quad \forall (v,q^*) \in K \times \mathcal{A}^*.$$

Thus, because of the relation (5.154), we have that $G_q(w,p_s^*) \leq 0 \ \ \forall w \in U$ (by (5.175) and (5.172)) and

$$J_\nu(u_s) = \sup_{q^*\in\mathcal{A}^*} L_\nu(u_s,q^*) = L_\nu(u_s,p_s^*) = \sup_{v\in\mathcal{K}} L_\nu(v,p_s^*) = J_\nu^*(p_s^*).$$

Consequently,

$$J_\nu(u_s) = L_\nu(u_s,p_s^*) = J_\nu^*(p_s^*).$$

Because of Proposition 5.35, we can deduce that $(J_\nu^*)'(p_s^*) = 0$ and $J_\nu'(u_s) = 0$.
This completes the proof. □

We end this section with the following result.

Theorem 5.88. *(Minimax duality theorem) Assume that the assumptions of Theorem 5.83 hold. Assume also that the set $\mathcal{E}_{ad}$ is non-empty and the quadratic gap function G_q is positive on $U \times \mathcal{A}_q^*$. Then a point $(u_c,p_c^*) \in \mathcal{E}_{ad}$ is a critical point of L_ν if and only if*

$$J_\nu^*(p_c^*) = \sup_{q^*\in\mathcal{A}_q^*} J_\nu^*(q^*) = \inf_{v\in\mathcal{K}} J_\nu(v) = J_\nu(u_c). \tag{5.176}$$

Proof. Assume that the quadratic gap function G_q is positive on the space $U\times\mathcal{A}_q^*$ then, because of Corollary 5.84, we have that all critical points (u_c,p_c^*) of L_ν are saddle points of L_ν. Moreover, since the function

$$\eta_v : \ q^* \in \mathcal{A}_q^* \longrightarrow \eta_v(q^*) := L_\nu(v,q^*) \in \mathbb{R}, \text{ for any } v \in \mathcal{K},$$

is concave, we can deduce that (since $\mathcal{S}_\nu$ is a convex function)

$$J_\nu(v) = \sup_{q^*\in\mathcal{A}_q^*} L_\nu(v,q^*) = \sup_{q^*\in\mathcal{A}_q^*} \eta_v(q^*) = \eta_v(p_c^*) = L_\nu(v,p_c^*) \geq L_\nu(u_c,p_c^*).$$

According to Lemma 5.87, we have that $L_\nu(u_c,p_c^*) = J_\nu(u_c) \leq J_\nu(v)$ for all $v \in \mathcal{K}$ and then

$$J_\nu(u_c) = \inf_{v\in\mathcal{K}} J_\nu(v). \tag{5.177}$$

Since G_q is a positive function on $U \times \mathcal{A}_\mathrm{q}^*$ then, for any $q^* \in \mathcal{A}_\mathrm{q}^*$ (because of Proposition 5.85)

$$J_\nu^*(q^*) = \inf_{v \in \mathcal{K}} L_\nu(v, q^*) = L_\nu(u_\mathrm{c}, q^*) \leq L_\nu(u_\mathrm{c}, p_\mathrm{c}^*).$$

According to Lemma 5.87, we have that $L_\nu(u_\mathrm{c}, p_\mathrm{c}^*) = J_\nu^*(p_\mathrm{c}^*) \leq J_\nu^*(q^*)$ for all $q^* \in \mathcal{A}_\mathrm{q}^*$ and then

$$J_\nu^*(p_\mathrm{c}^*) = \sup_{q^* \in \mathcal{A}_\mathrm{q}^*} J_\nu^*(q^*). \tag{5.178}$$

By the relations (5.177), (5.178) and according again to Lemma 5.87, we can deduce that

$$\inf_{v \in \mathcal{K}} J_\nu(v) = J_\nu(u) = J_\nu^*(p_\mathrm{c}^*) = \sup_{q^* \in \mathcal{A}_\mathrm{q}^*} J_\nu^*(q^*),$$

which gives the relations (5.176).

Conversely, according to the expressions (5.154), (5.170) and since $\mathcal{E}_\mathrm{ad}$ is non-empty, we have that

$$\begin{aligned} J_\nu(u_\mathrm{c}) &= \sup_{s^* \in \mathcal{A}_\mathrm{q}^*} L_\nu(u_\mathrm{c}, s^*) \geq L_\nu(u_\mathrm{c}, q^*) \;\; \forall q^* \in \mathcal{A}_\mathrm{q}^*, \\ J_\nu^*(p_\mathrm{c}^*) &= \inf_{w \in \mathcal{K}} L_\nu(w, p_\mathrm{c}^*) \leq L_\nu(v, p_\mathrm{c}^*) \;\; \forall v \in \mathcal{K}. \end{aligned}$$

Since the relations (5.176) hold then, for all (v, q^*) in $\mathcal{K} \times \mathcal{A}_\mathrm{q}^*$,

$$\alpha = J_\nu(u_\mathrm{c}) = J_\nu^*(p_\mathrm{c}^*) \leq L_\nu(u_\mathrm{c}, q^*) \text{ and } \alpha \geq L_\nu(v, p_\mathrm{c}^*).$$

Consequently, according to Proposition 5.53, we have that $(u_\mathrm{c}, p_\mathrm{c}^*)$ is a saddle point of L_ν and then (because of Corollary 5.84) $(u_\mathrm{c}, p_\mathrm{c}^*)$ is a critical point.

This completes the proof of the theorem. □

Part II

General Results and Concepts on Robust and Optimal Control Theory for Evolutive Systems

6

Studied Systems and General Results

This chapter is devoted to general tools and basic results, on the existence, uniqueness and regularity of the solutions of linear time-dependent systems, which will be used frequently in later chapters.

For the description of various function spaces including Sobolev spaces and different notations, the reader is referred to Chapter 3 for details.

Our objective in this chapter is to recall the general framework and the basic results of existence of solutions for linear evolutive equations of the first order in time of the type

$$\frac{\partial u}{\partial t} + A(t)u = f(t).$$

We consider some operator A (time dependent) and a right-hand side f which is allowed to depend on time. First we give some hypotheses and properties, and second we indicate in what sense such an equation is understood and give the existence, the uniqueness and the regularity of solutions.

6.1 Hypotheses and Properties

Let V and H be two Hilbert spaces on $\mathbb{R}$. We denote by $\| \, . \, \|$ (respectively $| \, . \, |$) the norm on V (respectively on H), and we denote by $((\; , \;))$ (respectively $(\; , \;)$) the inner (or scalar) product for V (respectively for H). We will add indices V or H $((\; , \;)_V$, $\| \, . \, \|_V$, *etc.*) in the event of possible ambiguity. We suppose that

$$\begin{aligned} &V \subset H, \quad \text{with continuous embedding,} \\ &\text{and } V \text{ is dense in } H. \end{aligned} \tag{6.1}$$

We identify the space H to its dual H', *i.e.*, $H \equiv H'$ so that if V' denotes the dual of V we have the following injections:

$$V \subset H \subset V', \text{with continuous and dense embedding.} \tag{6.2}$$

We denote by t the time and suppose that $t \in (0, T)$, where $T < +\infty$ is the final time. Let us now introduce the family of continuous bilinear forms on V by

$$(u, v) \in V \times V \longrightarrow a(t; u, v) \in \mathbb{R}, \text{ for any time } t \in (0, T). \tag{6.3}$$

We suppose that the forms a satisfy the following assumptions:

$$\begin{aligned} &\text{for all } u, v \in V, \text{ the function } t \in (0, T) \longrightarrow a(t; u, v) \text{ is measurable} \\ &\text{and there exists a constant } M > 0 \text{ (independent of } t, u, v) \\ &\text{such that } \quad | a(t; u, v) | \leq M \| u \| \| v \| . \end{aligned} \tag{6.4}$$

It results from these assumptions that the forms $v \in V \longrightarrow a(t; u, v)$ ($t \in (0, T)$ and $u \in V$) are linear and continuous on V and then there exists $A(t) \in \mathcal{L}(V, V')$ (the space of linear and continuous mappings of V into V') such that

$$\begin{aligned} &a(t; u, v) = \langle A(t)u, v \rangle_{V',V} \quad \forall (u, v) \in V \times V, \forall t \in (0, T), \\ &\sup_{t \in (0,T)} \| A(t) \|_{\mathcal{L}(V,V')} \leq M, \end{aligned} \tag{6.5}$$

where the duality pairing $\langle \; , \; \rangle_{V',V}$ is the scalar product between V' and V.

We also suppose the following coercivity assumption:

$$\begin{aligned} &\text{there exist constants } \lambda \geq 0, \alpha > 0, \text{ such that} \\ &a(t; u, u) + \lambda | u |^2 \geq \alpha \| u \|^2 \quad \forall u \in V, \forall t \in (0, T). \end{aligned} \tag{6.6}$$

Remark 6.1. If the operator A is independent of time t, we note its corresponding form $a(t; u, v)$ by $a(u, v)$. ♢

According to (6.5), the linear operator A defined by: if $u \in L^2(0, T; V)$, $A(.)u$ is the function $t \in (0, T) \longrightarrow A(t)u(t) \in V'$, is measurable and $A \in \mathcal{L}(L^2(0, T; V); L^2(0, T; V'))$ and satisfies

$$\| A(t)u(t) \|_{V'} \leq M \| u(t) \|_V .$$

Moreover, for $u \in L^2(0, T; V)$, we can define its time derivative $\partial u / \partial t$. For this, we will need the following simple lemma.

Lemma 6.2. *Let X be a given Banach space with dual X' and let u and g be two functions belonging to $L^1(a, b; X)$. Then the following three conditions are equivalent:*

(*i*) *u is almost everywhere (a.e.) equal to a primitive function of g, i.e., there exists $\psi \in X$ such that*

$$u(t) = \psi + \int_0^t g(s) \mathrm{d}s, \quad \textit{for a.e. } t \in [a, b]. \tag{6.7}$$

(*ii*) *For any test function* $\phi \in \mathcal{D}(]a, b[)$, *we have*

$$\int_a^b u(t)\phi'(t)\mathrm{d}t = -\int_a^b g(t)\phi(t)\mathrm{d}t, \quad (\phi' = \frac{\mathrm{d}\phi}{\mathrm{d}t}). \tag{6.8}$$

(*iii*) *For each* $\eta \in X'$, *we have*

$$\frac{\mathrm{d}}{\mathrm{d}t}\langle u, \eta\rangle = \langle g, \eta\rangle, \quad \textit{in the scalar distribution sense on }]a, b[. \tag{6.9}$$

If one of the conditions (6.7)–(6.9) is satisfied, we say that the function g *is the* (X – *valued) distribution derivative of* u, *and* u *is almost everywhere equal to a continuous function from* $[a, b]$ *into* X.

Proof. For the proof see, *e.g.*, Dautray and Lions [95] and Temam [282]. □

Lemma 6.2 gives a sense to Problem (6.11) if u is in $L^2(0, T; V)$ (which will be the case below). Indeed, A being an isomorphism from V into V', $Au \in L^2(0, T; V')$ by (6.5) and $f \in L^2(0, T; V')$. Thus $\partial u/\partial t = f - Au$ in the distribution sense in V'. In such a case, Lemma 6.2 also implies that u is almost everywhere equal to a continuous function from $[0, T]$ into V' and $u(0) = u_0$ makes sense too.

As a direct consequence of Lemma 6.2, we have the following lemma.

Lemma 6.3. *Let* V, H *be two Hilbert spaces and* V' *be the dual of* V, *such that Assumption (6.2) holds. If a function* w *belongs to* $L^2(0, T; V)$ *and its derivative* $\partial w/\partial t$ *belongs to* $L^2(0, T; V')$, *then* w *is almost everywhere equal to a continuous function from* $[0, T]$ *into* H *and we have the following equality which holds in the scalar distribution sense on* $(0, T)$:

$$\frac{\mathrm{d} \mid w \mid^2}{\mathrm{d}t} = 2\langle \frac{\partial w}{\partial t}, w\rangle_{V',V}.$$

□

Corollary 6.4. *Let* V, H *be two Hilbert spaces and* V' *be the dual of* V, *such that Assumption (6.2) holds. If functions* v *and* w *belong to* $L^2(0, T; V)$ *and their derivatives* $\partial v/\partial t$ *and* $\partial w/\partial t$ *belong to* $L^2(0, T; V')$, *then the following equality holds in the scalar distribution sense on* $(0, T)$:

$$\langle \frac{\partial v}{\partial t}, w\rangle_{V',V} + \langle \frac{\partial w}{\partial t}, v\rangle_{V',V} = \frac{\mathrm{d}}{\mathrm{d}t}(v, w).$$

Proof. According to assumptions, we can deduce that $v - w$ belongs to $L^2(0, T; V)$ and its derivative $\partial(v - w)/\partial t$ belongs to $L^2(0, T; V')$. Then, because of Lemma 6.3, we have the following equalities for v, w and $v - w$:

$$\begin{aligned}\frac{\mathrm{d} \mid v \mid^2}{\mathrm{d}t} &= 2\langle \frac{\partial v}{\partial t}, v\rangle_{V',V},\\ \frac{\mathrm{d} \mid w \mid^2}{\mathrm{d}t} &= 2\langle \frac{\partial w}{\partial t}, w\rangle_{V',V},\\ \frac{\mathrm{d} \mid v-w \mid^2}{\mathrm{d}t} &= 2\langle \frac{\partial (v-w)}{\partial t}, v-w\rangle_{V',V}.\end{aligned} \tag{6.10}$$

By a simple manipulation, we can deduce easily the result of the corollary. □

We now give two other interesting lemmas in order to obtain the continuity result.

Lemma 6.5. *Let X, Y be two Banach spaces such that*

$$X \subset Y \text{ with a continuous embedding.}$$

If a function w belongs to $L^\infty(0,T;X)$ and is weakly continuous with values in Y, then w is weakly continuous with values in X.

Proof. For the proof see, *e.g.*, Strauss [276]. □

The next lemma concerns the compact embedding results for time-dependent functions.

Lemma 6.6. *Let X_0, X and X_1 be reflexive Banach spaces such that*

$$X_0 \subset X \subset X_1,$$

where the first embedding is compact and the second is continuous.
Then, if $T > 0$ is finite, the following compact embeddings hold:

(*i*) $L^{p_0}(0,T;X_0) \cap \{v : \dfrac{\partial v}{\partial t} \in L^{p_1}(0,T;X_1)\} \subset L^{p_0}(0,T;X)$

(*ii*) $L^\infty(0,T;X_0) \cap \{v : \dfrac{\partial v}{\partial t} \in L^{p_2}(0,T;X_1)\} \subset \mathcal{C}([0,T];X)$

(*iii*) $L^{p_0}(0,T;X) \cap \{v : \dfrac{\partial v}{\partial t} \in L^{p_0}(0,T;X)\} \subset \mathcal{C}([0,T];X)$

for any $1 \le p_0$, $p_1 \le \infty$ and $1 < p_2 \le \infty$.

Proof. For the proof, we refer, *e.g.*, to Lions [203] and Simon [271]. □

6.2 Evolution Problems, Existence and Stability Results

We shall now study a linear evolutive problem. For this, let $u_0 \in H$ (for instance), f be a given function from $[0,T]$ in V' satisfying, for example $f \in L^2(0,T;V')$ and let a be a continuous bilinear form satisfying Assumptions

(6.4)-(6.6). We seek a function u from $[0,T]$ in V which is the solution to the initial-value problem

$$\begin{aligned} &\frac{\partial u(t)}{\partial t} + A(t)u(t) = f(t) \quad \text{on } (0,T), \\ &u(0) = u_0. \end{aligned} \tag{6.11}$$

Remark 6.7. If the final time T is finite, the hypothesis (6.6) can always be reduced to the case where $\lambda = 0$. Indeed, if we take $w = \exp(-\lambda t)u$, the problem (6.11) is equivalent to

$$\frac{\partial w(t)}{\partial t} + \tilde{A}(t)w(t) = g(t), \quad w(0) = u_0,$$

where $g(t) = f(t)+\exp(-\lambda t)$ and $\tilde{A}(t) = A(t)+\lambda I$, for which its corresponding bilinear form $\overline{a}$, via (6.5), verifies Assumptions (6.3), (6.4) and (6.6) with $\lambda = 0$. ♢

In the sequel we suppose that the operator $A(t)$ satisfies the hypothesis

$$\begin{aligned} &\text{there exists a constant } \alpha > 0, \text{ such that} \\ &\langle A(t)u, u\rangle_{V',V} = a(t;u,u) \geq \alpha \parallel u \parallel^2 \quad \forall u \in V, \forall t \in (0,T). \end{aligned} \tag{6.12}$$

We shall now give the existence and uniqueness results.

Theorem 6.8. *Suppose that the assumptions given in the previous section hold. Then, for $u_0 \in H$ and $f \in L^2(0,T;V')$, there exists a unique solution u of problem (6.11) such that*

$$\begin{aligned} &u \in L^2(0,T;V) \cap C([0,T];H), \\ &\frac{\partial u}{\partial t} \in L^2(0,T;V'). \end{aligned} \tag{6.13}$$

Moreover, the following estimates hold:

$$\begin{aligned} &\parallel u \parallel_{L^2(0,T;V)\cap L^\infty(0,T;H)} \leq C(\mid u_0 \mid + \parallel f \parallel_{L^2(0,T;V')}), \\ &\parallel \frac{\partial u}{\partial t} \parallel_{L^2(0,T;V')} \leq C(\mid u_0 \mid + \parallel f \parallel_{L^2(0,T;V')}). \end{aligned} \tag{6.14}$$

Proof. This result is proved, *e.g.*, in Lions and Magenes [204]. We give only a sketch of the proof which emphasizes some points needed in the sequel.
Nota bene: Throughout the present proof, we use C to denote a generic constant.

The existence result is proved by using the Faedo–Galerkin method. For simplicity we suppose that the Hilbert space V is separable, then there exists (not in a unique manner) a linearly independent and total sequence $w_1, w_2, \ldots, w_l, \ldots,$ which is complete in V. For each l we denote by V_l the space generated (or spanned) by the elements $w_1, w_2, \ldots, w_l$, *i.e.*, the space $V_l = \text{span}\{w_1, w_2, \ldots, w_m\}$, and we define an approaching problem of (6.11) as follows:

find $u_l(t) \in V_l$ such that (a.e. $t \in (0,T)$)

$$\begin{aligned} &\frac{\mathrm{d}}{\mathrm{d}t}(u_l(t), w_j) + a(t; u_l(t), w_j) = \langle f(t), w_j \rangle_{V',V}, \; j = 1, \ldots, l \\ &u_l(0) = P_l u_0, \end{aligned} \tag{6.15}$$

where P_l is, for example, the H-projector onto V_l (the initial data u_0 satisfies then $\| P_l u_0 \|_H \leq \| u_0 \|_H$ and $P_l u_0 \longrightarrow u_0$ strongly in H as $l \longrightarrow \infty$).

Since the approximate solution u_l is in V_l, we have that

$$u_l(t) = \sum_{i=1,l} g_{il}(t) w_i,$$

where g_{il}, $i = 1, \ldots, l$, are scalar functions on $[0,T]$.

The problem (6.15) is equivalent to an initial-value problem for a linear finite m-dimensional ordinary differential equation for the functions g_{il} in the form:

$$\begin{aligned} &\text{find } G_l = (g_{1l}, \ldots, g_{ll}) \text{ such that} \\ &\mathcal{M}_l \frac{\mathrm{d}G_l}{\mathrm{d}t} + \mathcal{A}_l(t) G_l = F_l, \\ &G_l(0) = \eta_l, \end{aligned} \tag{6.16}$$

where $\mathcal{M}_l$ is the matrix $[(w_i, w_j)]_{1 \leq i,j \leq l}$, $\mathcal{A}_l$ is the matrix $[a(.\,; w_i, w_j)]_{1 \leq i,j \leq l}$, F_l is the vector $[\langle f(.), w_j \rangle_{V',V}]_{1 \leq j \leq l}$ and η_l is the vector $[\xi_{il}]_{1 \leq i \leq l}$ when $P_l u_0 = \sum_{i=1,l} \xi_{il} w_i$.

Clearly $\mathcal{M}_l$ is an invertible matrix (because (w_j) are linearly independent) then, by the theory of ordinary differential equations, Problem (6.16) has a unique solution G_l and consequently the problem (6.15) admits a unique solution $u_l \in \mathcal{C}([0,T]; V_l)$ with $\mathrm{d}u_l/\mathrm{d}t \in L^2(0,T; V_l)$.

We will now show the convergence of the sequence u_l as $l \longrightarrow \infty$, by using some *a priori* estimates and the weak compactness.

Multiplying the system (6.15) by g_{jl} and adding with respect to j, we obtain

$$\langle \frac{\mathrm{d}u_l}{\mathrm{d}t}, u_l \rangle_{V',V} + a(t; u_l, u_l) = \langle f(t), u_l \rangle_{V',V},$$

and then (according to (6.12) and Lemma 6.3)

$$\frac{\mathrm{d} \mid u_l \mid^2}{2\mathrm{d}t} + \alpha \| u_l \|^2 \leq C \| f \|_{V'} \| u_l \| \leq C \| f \|_{V'}^2 + \frac{\alpha}{2} \| u_l \|^2 .$$

So,

$$\frac{\mathrm{d} \mid u_l \mid^2}{\mathrm{d}t} + \alpha \| u_l \|^2 \leq C \| f \|_{V'}^2 . \tag{6.17}$$

After integration with respect to time, we infer from (6.17) an *a priori* estimate (since $\| P_l u_0 \|_H \leq \| u_0 \|_H$):

$$u_l \text{ is uniformly bounded in } L^\infty(0,T;H) \cap L^2(0,T;V). \tag{6.18}$$

According to (6.18) and the relation (6.5), we can deduce that

$$A(.)u_l \text{ is uniformly bounded in } L^2(0,T;V'). \tag{6.19}$$

By weak compactness, we can extract from u_l a subsequence, denoted also by u_l such that

$$\begin{aligned} &u_l \rightharpoonup u \text{ weakly in } L^2(0,T;V), \\ &u_l \rightharpoonup u \text{ weakly star in } L^\infty(0,T;H), \\ &A(.)u_l \rightharpoonup A(.)u \text{ weakly in } L^2(0,T;V'). \end{aligned} \tag{6.20}$$

We are now going to prove that u is a solution of problem (6.11). In order to pass the limit in problem (6.15), let us consider a scalar function φ continuously differentiable on $[0,T]$, such that $\varphi(T) = 0$.

Multiplying the system (6.15) by φ and integrating with respect to time, we obtain (for $i = 1, \ldots, l$)

$$\begin{aligned} -\int_0^T (u_l, w_i \frac{\mathrm{d}\varphi}{\mathrm{d}t})\mathrm{d}t + \int_0^T a(t; u_l, \varphi w_i)\mathrm{d}t \\ = \int_0^T \langle f(t), \varphi w_i \rangle_{V',V} \mathrm{d}t + (P_l u_0, \varphi(0) w_i). \end{aligned} \tag{6.21}$$

According to (6.20) it is easy to pass the limit in (6.21) and we have that (since $P_l u_0 \longrightarrow u_0$ strongly in H as $l \longrightarrow \infty$)

$$\begin{aligned} -\int_0^T (u, w_i \frac{\mathrm{d}\varphi}{\mathrm{d}t})\mathrm{d}t + \int_0^T a(t; u, \varphi w_i)\mathrm{d}t \\ = \int_0^T \langle f(t), \varphi w_i \rangle_{V',V} \mathrm{d}t + (u_0, \varphi(0) w_i). \end{aligned} \tag{6.22}$$

Taking now φ as a $\mathcal{C}^\infty$ function on $(0,T)$ with a compact support in Ω, then (6.22) gives (after adding)

$$\frac{\mathrm{d}}{\mathrm{d}t}(u,v) + a(t;u(t),v) = \langle f(t), v \rangle_{V',V}, \forall l = 1, \ldots, \quad \forall v \in V_l \tag{6.23}$$

in the distribution sense in $(0,T)$.

Since the terms of the system (6.23) depend continuously and linearly on the function $v \in V_l$, for all $l = 1, \ldots$, then (6.23) remains valid for any function $v \in V$ and then

$$\frac{\mathrm{d}}{\mathrm{d}t}(u,v) + a(t;u(t),v) = \langle f(t), v \rangle_{V',V}, \tag{6.24}$$

in the distribution sense in $(0,T)$ and $\forall v \in V$.

According to Lemma 6.2, we can deduce that

$$\frac{\partial u}{\partial t} + A(t)u = f \tag{6.25}$$

and that $\partial u/\partial t \in L^2(0,T;V')$. Moreover, u is almost everywhere equal to a continuous function from $[0,T]$ into V' and the initial condition $u(t=0) = u_0$ follows by passing the limit in the relation $u_l = P_l u_0$.

We will now show that the solution u is unique and that $u \in \mathcal{C}([0,T];H)$. The continuity result is a direct consequence of Lemma 6.3. The proof of uniqueness follows easily also from Lemma 6.3. Indeed, let u and v be two solutions of Problem (6.11) satisfying the regularity (6.13). Then $w = u - v$ satisfies the regularity (6.13) and is a solution of the following system:

$$\begin{aligned} &\frac{\partial w}{\partial t} + A(t)w = 0, \\ &w(0) = 0. \end{aligned} \tag{6.26}$$

Multiplying the first equality of (6.26) by w and using the result of Lemma 6.3, we obtain

$$\frac{\mathrm{d} \mid w \mid^2}{\mathrm{d}t} + 2a(t;w,w) = 0$$

and then (since $a(t;w,w) \geq 0$)

$$\frac{\mathrm{d} \mid w \mid^2}{\mathrm{d}t} \leq 0.$$

By integrating with respect to time, we obtain

$$\mid w(t) \mid^2 \leq \mid w(0) \mid^2 \quad \text{for all } t \in (0,T).$$

According to the null initial condition of (6.26), we can deduce that $w = 0$ and then the uniqueness result.

The estimates given in the theorem are a direct consequence of Proposition 6.9 (see below). □

We shall now give the Lipschitz continuity of the map solution.

Proposition 6.9. *Let (u_0, f) and (v_0, g) be in $H \times L^2(0,T;V')$ and let u and v be the corresponding solutions of problem (6.11), respectively. Then*

$$\begin{aligned} &\| u - v \|_{L^2(0,T;V)\cap L^\infty(0,T;H)} \leq C(\mid u_0 - v_0 \mid + \| f - g \|_{L^2(0,T;V')}), \\ &\| \frac{\partial(u-v)}{\partial t} \|_{L^2(0,T;V')} \leq C(\mid u_0 - v_0 \mid + \| f - g \|_{L^2(0,T;V')}). \end{aligned} \tag{6.27}$$

Proof. Let $w = u - v$, $w_0 = u_0 - v_0$ and $h = f - g$. Then w satisfies the regularity (6.13) and is a solution of the following system:

$$\begin{aligned} &\frac{\partial w}{\partial t} + A(t)w = h, \\ &w(0) = w_0. \end{aligned} \tag{6.28}$$

Multiplying the first equality of (6.28) by w and using the result of Lemma 6.3, we obtain

$$\frac{\mathrm{d} \mid w \mid^2}{\mathrm{d}t} + 2a(t; w, w) = 2\langle h, w\rangle_{V',V}.$$

By using the same method to obtain (6.17), we have

$$\frac{\mathrm{d} \mid w \mid^2}{\mathrm{d}t} + \alpha \parallel w \parallel^2 \leq C \parallel h \parallel^2_{V'} .$$

By integrating with respect to time, we can deduce that (according to the second equality of (6.28))

$$\mid w(t) \mid^2 + \alpha \int_0^T \parallel w \parallel^2 \mathrm{d}s \leq C \int_0^T \parallel h \parallel^2_{V'} \mathrm{d}s + \mid w_0 \mid^2 \quad \text{for } t \in (0, T)$$

and then the first result of (6.27).

Taking now the scalar product of the first equality of (6.28) by $z \in V$, we have

$$\langle \frac{\partial w}{\partial t}, z\rangle_{V',V} = -a(t; w, z) + \langle h, z\rangle_{V',V}.$$

According to Assumption (6.5), we can deduce that

$$\langle \frac{\partial w}{\partial t}, z\rangle_{V',V} \leq C(\parallel w \parallel + \parallel h \parallel_{V'}) \parallel z \parallel$$

and then

$$\parallel \frac{\partial w}{\partial t} \parallel_{V'} \leq C(\parallel w \parallel + \parallel h \parallel_{V'}).$$

Consequently, by integrating with respect to time and by using the first result of (6.27), we can deduce easily the second result of (6.27). □

6.3 Regularity Results

In this section, we shall give two regularity results for the solution u which will be used frequently in non-linear cases. The linear operator $A(t)$ is an isomorphism from V onto V' and from its domain $D(A(t)) = \mathrm{dom}(A(t)) \subset V$ onto H, where the domain

$$\begin{aligned} D(A(t)) = \{u \in V :\ &\text{the function } v \longrightarrow a(t; u, v) \\ &\text{is continuous on V for the topology of } H\}. \end{aligned}$$

Theorem 6.10. *Under the assumptions of Theorem 6.8, we suppose, furthermore, that the bilinear form a satisfies*

$$\begin{aligned}&\textit{for every } u, v \textit{ in } V, \textit{ the function } t \in [0,T] \longrightarrow a(t;u,v) \in I\!R\\ &\quad \textit{is absolutely continuous, its derivative } \frac{\partial}{\partial t}a(t;u,v),\\ &\textit{for a.e. } t \in [0,T], \textit{ is a bilinear continuous form on } V, \textit{ and}\\ &\textit{there exists } M_1 > 0 \textit{ such that for all } u, v \in V \textit{ and a.e. } t \in [0,T]\\ &\qquad \mid \frac{\partial}{\partial t}a(t;u,v) \mid \leq M_1 \parallel u \parallel \parallel v \parallel,\end{aligned} \tag{6.29}$$

and that the right-hand side f and the initial condition u_0 satisfy

$$\begin{aligned}& f, \frac{\partial f}{\partial t} \in L^2(0,T;H),\\ & u_0 \in D(A(0)).\end{aligned} \tag{6.30}$$

Then the solution u of problem (6.11), satisfies the following regularity

$$\begin{aligned}& u(t) \in D(A(t)), \forall t \in [0,T],\\ & t \longrightarrow A(t)u(t) \textit{ is continuous from } [0,T] \textit{ into } H,\\ & \frac{\partial u}{\partial t} \in L^2(0,T;V) \cap \mathcal{C}([0,T];H),\\ & \frac{\partial^2 u}{\partial t^2} \in L^2(0,T;V').\end{aligned} \tag{6.31}$$

Moreover, if the operator $A(t)$ is independent of time t, then the first result in (6.31) reduces to $u \in \mathcal{C}([0,T];D(A))$.

Proof. Introduce the following problem:

$$\begin{aligned}&\langle \frac{\partial w}{\partial t}, v\rangle_{V',V} + a(t;w,v)\\ &\qquad = (\frac{\partial f}{\partial t}, v) - \frac{\partial}{\partial t}a(t;u,v) \;\; \forall v \in V, \;\; \text{a.e. } t \in (0,T),\\ &w(0) = \frac{\partial u}{\partial t}(0) = f(0) - A(0)u_0.\end{aligned} \tag{6.32}$$

Due to the first assumption of (6.30) and Lemma 6.2, the function f is in $\mathcal{C}([0,T];H)$ and $f(0) \in H$ is well defined. Thus, with the second assumption of (6.30), we can deduce that $f(0) - A(0)u_0 \in H$. Moreover, according to (6.29), we can write the right-hand side of (6.32) as $\langle g, v\rangle_{V',V}$, with $g \in L^2(0,T;V')$.

According to Theorem 6.8, problem (6.32) admits a unique solution w such that

$$w \in L^2(0,T;V) \cap \mathcal{C}([0,T];H), \quad \frac{\partial w}{\partial t} \in L^2(0,T;V').$$

Set

$$\omega(t) = \int_0^t w(s)\mathrm{d}s + u_0, \quad \forall t \in [0,T],$$

then $\omega \in \mathcal{C}^1([0,T];H)$ (since $w \in \mathcal{C}([0,T];H)$).

Integrating the first equality of (6.32) with respect to time gives (according to the second equality of (6.32))

$$\langle \frac{\partial \omega}{\partial t}, v\rangle_{V',V} + a(t;\omega,v) = (f,v) \ \ \forall v \in V, \quad \text{a.e. } t \in (0,T),$$
$$\omega(0) = u_0.$$

We have proved that ω is a solution of problem (6.11). Consequently, by the uniqueness result, we have $u = \omega$. On the other hand, $\partial u/\partial t = \partial\omega/\partial t = w$. Therefore, $\partial u/\partial t \in L^2(0,T;V) \cap \mathcal{C}([0,T];H)$ and $\partial^2 u/\partial t^2 \in L^2(0,T;V')$. Then, we have the second and the third results of (6.31).

Thus, with the first equality of (6.11), we can deduce that

$$A(t)u(t) = f - \frac{\partial u}{\partial t} \in \mathcal{C}([0,T];H)$$

and then the first result of (6.31).

If now the operator $A = A(t)$ is independent of t, and since A is an isomorphism from D(A) onto H, then $u \in \mathcal{C}([0,T];D(A))$. □

The last regularity result is the following theorem.

Theorem 6.11. *Under the assumptions of Theorem 6.8, we suppose, furthermore, that the bilinear form a satisfies Assumptions (6.29) and*

$$\begin{aligned} &a \text{ is symmetric, i.e., } a(t;u,v) = a(t;v,u) \quad \forall u,v \in V, \ \forall t \in [0,T], \\ &\text{the injection of } V \text{ in } H \text{ is compact} \end{aligned} \tag{6.33}$$

and that the right-hand side f and the initial condition u_0 satisfy

$$\begin{aligned} &f \in L^2(0,T;H), \\ &u_0 \in V. \end{aligned} \tag{6.34}$$

Then, the solution u of problem (6.11), satisfies the following regularity:

$$\begin{aligned} &A(.)u \in L^2(0,T;H), \\ &u \in L^\infty(0,T;V), \\ &\frac{\partial u}{\partial t} \in L^2(0,T;H). \end{aligned} \tag{6.35}$$

Moreover, if the operator $A(t)$ is independent of time t, then the first and second regularities given in (6.35) become

$$u \in L^2(0,T;D(A)) \cap \mathcal{C}([0,T];V). \tag{6.36}$$

Remark 6.12. According to the first part of (6.33), the operator A is self-adjoint (from V onto V') unbounded operator in H and its inverse A^{-1} is also self-adjoint in H. Moreover, because of the second part of (6.33), the operator A^{-1} can be considered as a self-adjoint compact operator in H and then according to Courant and Hilbert [89], there exist a sequence $(\lambda_i(t))_{i\geq 1}$ of eigenvalues of the operator $A(t)$ such that $0 < \lambda_1 \leq \lambda_2 \leq \cdots \leq \lambda_k \leq \cdots$ and the corresponding, smooth and complete orthonormal basis in H and orthogonal in V, eigenfunctions $(w_i)_{i\geq 1}$, *i.e.*,

$$
\begin{aligned}
&A(t)w_i = \lambda_i w_i \text{ in } V' \text{ and } w_i \in V \quad (\text{for all } i \geq 1),\\
&(w_i, w_j) = \delta_{ij} = \text{ the Kronecker symbol} \quad (\text{for all } i, j \geq 1),\\
&a(t; w_i, w_j) = \langle A(t)w_i, w_j\rangle = \lambda_i \delta_{ij} \quad (\text{for all } i, j \geq 1).
\end{aligned}
\tag{6.37}
$$

◇

Proof of Theorem 6.11. The proof of the theorem is obtained by implementing the Faedo–Galerkin method, with a particular choice of elements w_j. Taking advantage of the properties (6.33), we consider the sequence $(\lambda_i, w_i)_{i\geq 1}$ of eigenvalues–eigenfunctions of the operator $A(t)$ such that (6.37), and we use these eigenfunctions to implement the Faedo–Galerkin method.

We denote by V_l the space generated by $(w_i)_{1\leq i\leq l}$ ($\cup_{l\geq 1} V_l$ is dense in H and also in V), and we introduce the H-orthogonal (and also V-orthogonal) projector P_l on the space V_l.

Remark 6.13. For all $v \in H$ we have $\| P_l v \|_H \leq C \| v \|_H$. Moreover, if $v \in V$ we have $\| P_l v \| \leq C \| v \|$ (the positive constant C is independent of l). ◇

For all $l \geq 1$, we consider $u_l(.,t) = \sum_{i=1,l} g_{il}(t) w_i$ an approximation solution of problem (6.11) with the initial condition u_0 as follows:

$$
\begin{aligned}
&\text{find } u_l(t) \in V_l \text{ such that (a.e. } t \in (0,T))\\
&(\frac{\partial u_l}{\partial t}, w_j) + a(t; u_l(t), w_j) = (f(t), w_j),\ j = 1, \ldots, l,\\
&u_l(0) = P_l u_0.
\end{aligned}
\tag{6.38}
$$

Multiplying the first equality of (6.38) by $\lambda_j g_{jl}$ and adding these equalities for $j = 1, \ldots, l$, we have that (according to (6.37))

$$
(\frac{\partial u_l}{\partial t}, A(t)u_l) + | A(t)u_l |^2 = (f(t), A(t)u_l(t)).
$$

Since $(\mathrm{d}/\mathrm{d}t)(a(t; u_l, u_l)) = (\partial/\partial t) a(t; u_l, u_l) + 2a(t; \partial u_l/\partial t, u_l)$ (since a is symmetric) and $(\partial u_l/\partial t, A(t)u_l) = a(t; \partial u_l/\partial t, u_l)$, then

$$
\frac{\mathrm{d}}{\mathrm{d}t}(a(t; u_l, u_l)) + 2 | A(t)u_l |^2 = 2(f(t), A(t)u_l(t)) + \frac{\partial}{\partial t} a(t; u_l, u_l). \tag{6.39}
$$

According to (6.29) and Young's inequality, we have that

$$
\frac{\mathrm{d}}{\mathrm{d}t}(a(t; u_l, u_l)) + | A(t)u_l |^2 \leq M_1 \| u_l \|^2 + | f |^2 .
$$

This implies (by integration with respect to time)

$$a(t; u_l(t), u_l(t)) + \int_0^t | A(s)u_l(s) |^2 \, \mathrm{d}s \leq C(\| u_l \|^2_{L^2(0,T;V)} + \| f \|^2_{L^2(0,T;H)}) \\ + a(0; P_l u_0, P_l u_0).$$

Since $a(0; P_l u_0, P_l u_0) \leq M \| P_l u_0 \|^2 \leq M \| u_0 \|^2$ and according to (6.18), we can deduce that

$$\begin{aligned} &u_l \ \text{ is uniformly bounded in } L^\infty(0,T;V), \\ &A(.)u_l \ \text{ is uniformly bounded in } L^2(0,T;H). \end{aligned} \tag{6.40}$$

By passing the limit, we can deduce that the limiting solution u belongs to $L^\infty(0,T;V)$ such that $A(.)u$ belongs to $L^2(0,T;H)$.
Multiplying now the first equality of (6.38) by g'_{jl} and adding these equalities for $j = 1, \ldots, l$, we have that (according to (6.37))

$$| \frac{\partial u_l}{\partial t} |^2 = -a(t; u_l, \frac{\partial u_l}{\partial t}) + (f(t), \frac{\partial u_l}{\partial t})$$

and then (since $a(t; u_l, \partial u_l/\partial t) = (A(t)u_l, \partial u_l/\partial t)$)

$$| \frac{\partial u_l}{\partial t} |^2 \leq C(| A(t)u_l |^2 + | f |^2).$$

By integrating with respect to time, according to (6.40) and passing the limit, we can deduce that $\partial u/\partial t \in L^2(0,T;H)$.

If now, the operator A is independent of time, then according to (6.40) and by passing the limit, we have that $u \in L^\infty(0,T;V) \cap L^2(0,T;D(A))$.

Prove now that $u \in \mathcal{C}([0,T];V)$. From Lemma 6.5, we can deduce that

$$\begin{aligned} &t \longrightarrow (u(t), v) \text{ is continuous on } [0,T], \text{ for all } v \in V, \text{ and} \\ &t \longrightarrow a(u(t), v) \text{ is continuous on } [0,T], \text{ for all } v \in V. \end{aligned} \tag{6.41}$$

Moreover, we have for u a similar result to (6.39) (by using the same technique):

$$\frac{\mathrm{d}}{\mathrm{d}t}(a(u,u)) + 2 \, | Au |^2 = 2(f, Au).$$

Consequently, because of (6.41), we can deduce that $t \longrightarrow a(u(t), u(t))$ is a continuous function on $[0,T]$ and then $u \in \mathcal{C}([0,T];V)$.

This completes the proof. □

We shall now give the Lipschitz continuity of the map solution.

Proposition 6.14. *Under the assumptions of Theorem 6.11, we suppose, furthermore, that the bilinear form a is time-independent. Let (u_0, f) and (v_0, g) be in $V \times L^2(0,T;H)$ and let u and v in $L^2(0,T;D(A)) \cap H^1(0,T;H) \cap \mathcal{C}([0,T];V)$ be the corresponding solution of the problem (6.11), respectively.*

Then

$$\| w \|_{L^2(0,T;D(A))\cap L^\infty(0,T;V)} \leq C(\| w_0 \| + \| h \|_{L^2(0,T;H)}),$$
$$\| \frac{\partial w}{\partial t} \|_{L^2(0,T;H)} \leq C(\| w_0 \| + \| h \|_{L^2(0,T;H)}), \tag{6.42}$$

where $w = u - v$, $w_0 = u_0 - v_0$ *and* $h = f - g$.

Proof. Let $w = u - v$, $w_0 = u_0 - v_0$ and $h = f - g$. Then w satisfies the regularity $w \in L^2(0,T;D(A)) \cap H^1(0,T;H) \cap \mathcal{C}([0,T];V)$ and is a solution of the following system:

$$\frac{\partial w}{\partial t} + Aw = h,$$
$$w(0) = w_0. \tag{6.43}$$

Multiplying the first equality of (6.43) by Aw and using the same method to obtain (6.39), we obtain

$$\frac{\mathrm{d}}{\mathrm{d}t}(a(w,w)) + 2 \mid Aw \mid^2 = 2(h, Aw).$$

This implies

$$\frac{\mathrm{d}}{\mathrm{d}t}(a(w,w)) + \mid Aw \mid^2 \leq \mid h \mid^2 .$$

After integration with respect to time we obtain that, for all $t \in (0,T)$,

$$a(w(t),w(t)) + \int_0^t \mid Aw(s) \mid^2 \mathrm{d}s \leq \int_0^t \mid h(s) \mid^2 \mathrm{d}s + a(w_0, w_0).$$

By using the continuity and the coercivity of the bilinear form a, we can deduce that

$$\alpha \| w(t) \|^2 + \int_0^t \mid Aw(s) \mid^2 \mathrm{d}s \leq \int_0^t \mid h(s) \mid^2 \mathrm{d}s + C \| w_0 \|^2 \quad \text{for all } t \in (0,T).$$

Consequently,

$$\| w \|^2_{L^2(0,T;D(A))\cap L^\infty(0,T;V)} \leq C(\| h \|^2_{L^2(0,T;H)} + \| w_0 \|^2). \tag{6.44}$$

Then the first result of (6.42) follows.

Since w is a solution of (6.43) then

$$\frac{\partial w}{\partial t} = -Aw + h.$$

Consequently (since Aw and h are in $L^2(0,T;H)$),

$$\mid \frac{\partial w}{\partial t} \mid^2 \leq C(\mid Aw(t) \mid^2 + \mid h(t) \mid^2)$$

and then (according to (6.44))

$$\| \frac{\partial w}{\partial t} \|^2_{L^2(0,T;H)} \leq C(\| h \|^2_{L^2(0,T;H)} + \| w_0 \|^2). \tag{6.45}$$

Then, we have the second result of (6.42).

This completes the proof. □

6.4 Examples of Operators and Spaces

Let Ω be an open and bounded domain in $\mathbb{R}^n$, $n \geq 1$, with a smooth boundary $\Gamma = \partial\Omega$ with the regularity (3.1) or (3.2), and $T > 0$ be a fixed constant (a given final time). We take for V a closed subspace of $H^1(\Omega)$ such that $H_0^1(\Omega) \subset V \subset H^1(\Omega)$ (with continuous injection) and $H = L^2(\Omega)$.

We denote by $\mathcal{Q}$ the cylinder $\Omega \times (0,T)$ and by Σ the lateral (or side) boundary of $\mathcal{Q}$, *i.e.*, $\Gamma \times (0,T)$.

The generic vector point of $\mathbb{R}^n$ is denoted by $x = (x_1, x_2, \ldots, x_n)$ (or $y = (y_1, y_2, \ldots, y_n)$, *etc.*) and the Lebesgue measure on $\mathbb{R}^n$ is denoted by $\mathrm{d}x = \mathrm{d}x_1 \mathrm{d}x_2 \cdots \mathrm{d}x_n$.

Let $a_{ij} : \mathcal{Q} \longrightarrow \mathbb{R}$, for $i, j = 1, \ldots, n$ and $a_0 : \mathcal{Q} \longrightarrow \mathbb{R}$ be functions in $L^\infty(\mathcal{Q})$ such that the functions a_{ij} satisfy the ellipticity conditions: there exists $\mu > 0$ such that

$$\sum_{i,j=1,n} a_{ij}(x,t)\xi_i\xi_j \geq \mu \sum_{k=1,n} \xi_k^2 \quad \forall \xi = (\xi)_{i=1,n} \in \mathbb{R}^n \quad \text{a.e. in } \mathcal{Q} \tag{6.46}$$

and a_0 satisfies the pointwise constraint: there exist $\lambda_i \in \mathbb{R}$, for $i = 1, 2$ such that

$$\lambda_1 \leq a_0 \leq \lambda_2 \quad \text{a.e. in } \mathcal{Q}. \tag{6.47}$$

We introduce the bilinear form a_1 by

$$a_1(t; u, v) = \sum_{i,j=1,n} \int_\Omega a_{ij}(x,t) \frac{\partial u}{\partial x_i} \frac{\partial v}{\partial x_j} \mathrm{d}x + \int_\Omega a_0(x,t) uv \mathrm{d}x, \tag{6.48}$$

for all u and v in $H^1(\Omega)$.

According to the boundedness of a_{ij} and a_0, there exists $M > 0$ such that, a.e. $t \in (0,T)$,

$$| a_1(t; u, v) | \leq M \| u \| \| v \| \qquad \text{for all } u, v \text{ in } V$$

and by the condition (6.46) we have, for $\lambda \geq 0$ sufficiently large and a.e. $t \in (0,T)$, that

$$a_1(t; u, u) + \lambda \mid u \mid^2 > \nu \parallel u \parallel^2 \qquad \text{for all } u \text{ in } H^1(\Omega), \text{ with } \nu > 0.$$

We will now present three simple examples, which will enable us to specify the functional spaces in which we work.

6.4.1 Dirichlet Boundary Condition

We take $V = H_0^1(\Omega)$,[1] $H = L^2(\Omega)$, $V' = H^{-1}(\Omega)$, and we denote

[1] The norm in $H_0^1(\Omega)$ is equivalent to the seminorm.

$$\mathcal{W} = \{w \in L^2(0,T;V) : \frac{\partial w}{\partial t} \in L^2(0,T;V')\}.$$

Assume that the function a_0 is positive,[2] then the bilinear form a_1 satisfies the assumptions of Theorem 6.8 (*i.e.*, (6.3), (6.4) and (6.6)) and so there exists a unique solution $u \in \mathcal{W} \cap \mathcal{C}([0,T];H)$ of the problem

$$\begin{aligned}
&\frac{\partial u}{\partial t} + A_1(.)u = f \quad \text{on } \mathcal{Q}, \\
&\text{subject to the homogeneous Dirichlet boundary condition} \\
&u = 0 \quad \text{on } \Sigma, \\
&\text{with the initial condition} \\
&u(0) = u_0 \quad \text{on } \Omega,
\end{aligned} \tag{6.49}$$

where

$$\begin{aligned}
A(t)u &= -\sum_{i,j=1,n} \frac{\partial}{\partial x_i}(a_{ij}(x,t)\frac{\partial u}{\partial x_j}), \\
A_1(t)u &= A(t)u + a_0(x,t)u
\end{aligned} \tag{6.50}$$

and $f \in L^2(0,T;V')$ and $u_0 \in H$.

Remark 6.15. (*i*) The homogeneous Dirichlet boundary condition given in (6.49) is included in the space of solution $\mathcal{W}$, precisely in space $L^2(0,T;V)$. (*ii*) for a given $t \in (0,T)$, $v \in D(A(t))$ is equivalent to $v \in V = H_0^1(\Omega)$ and $A(t)v \in H = L^2(\Omega)$. ♢

6.4.2 Neumann Boundary Condition

We take $V = H^1(\Omega)$, $H = L^2(\Omega)$. So the space $V' = (H^1(\Omega))'$ is not a distribution space on Ω and then the space $L^2(0,T;V')$ is not also a distribution space on $\mathcal{Q}$. We denote

$$\mathcal{W} = \{w \in L^2(0,T;V) : \frac{\partial w}{\partial t} \in L^2(0,T;V')\}.$$

Then the bilinear form a_1 satisfies the assumptions of Theorem 6.8 (*i.e.*, (6.3), (6.4) and (6.6)) and so, "formally," Theorem 6.8 gives the solution of the following problem

$$\begin{aligned}
&\frac{\partial u}{\partial t} + A_1(.)u = f \quad \text{on } \mathcal{Q}, \\
&\text{subject to the homogeneous Neumann boundary condition} \\
&\frac{\partial u}{\partial \eta_A} = 0 \quad \text{on } \Sigma, \\
&\text{with the initial condition} \\
&u(0) = u_0 \quad \text{on } \Omega,
\end{aligned} \tag{6.51}$$

[2] If a_0 is negative, we put the term $a_0 u$ in the right-hand side of (6.49) and we can obtain similar results.

where

$$\begin{aligned} A(t)u &= -\sum_{i,j=1,n} \frac{\partial}{\partial x_i}(a_{ij}(x,t)\frac{\partial u}{\partial x_j}), \\ A_1(t)u &= A(t)u + a_0(x,t)u, \end{aligned} \tag{6.52}$$

and the normal derivative $\partial u/\partial \eta_A$ at Γ, directed towards the exterior of Ω, is given by (3.22), $f \in L^2(0,T;V')$ and $u_0 \in H$.

More precisely, Theorem 6.8 proves the existence and the uniqueness of the solution $u \in \mathcal{W} \cap \mathcal{C}([0,T];H)$ of the problem (the weak formulation of (6.51))

$$\begin{aligned} &\langle \frac{\partial u}{\partial t}, v\rangle_{V',V} + a_1(.;u,v) = \langle f, v\rangle_{V',V} \quad \text{a.e. in } (0,T),\ \forall v \in V, \\ &\text{with the initial condition} \\ &u(0) = u_0 \ \text{ on } \Omega. \end{aligned} \tag{6.53}$$

Remark 6.16. (*i*) The homogeneous Neumann boundary condition given in (6.51) is included (by using the Green's formula) in the formulation (6.53). (*ii*) for a given $t \in (0,T)$, $u \in D(A_1(t))$ is equivalent to $u \in V = H^1(\Omega)$, $A_1(t)u \in H = L^2(\Omega)$ and $\int_\Omega A_1(t)uv\mathrm{d}x = a_1(t;u,v)$ for all $v \in V$. ♢

6.4.3 Robin Boundary Condition

We take $V = H^1(\Omega)$, $H = L^2(\Omega)$ and $V' = (H^1(\Omega))'$. We denote by

$$\mathcal{W} = \{w \in L^2(0,T;V) : \frac{\partial w}{\partial t} \in L^2(0,T;V')\}.$$

Then the bilinear form a satisfies the assumptions of Theorem 6.8 (*i.e.*, (6.3), (6.4) and (6.6)) and so, "formally," Theorem 6.8 gives the solution of the following problem

$$\begin{aligned} &\frac{\partial u}{\partial t} + A_1(.)u = f \ \text{ on } \mathcal{Q}, \\ &\text{subject to the homogeneous Robin boundary condition} \\ &\frac{\partial u}{\partial \eta_A} + \alpha u = 0 \ \text{ on } \Sigma, \\ &\text{with the initial condition} \\ &u(0) = u_0 \ \text{ on } \Omega, \end{aligned} \tag{6.54}$$

where

$$\begin{aligned} A(t)u &= -\sum_{i,j=1,n} \frac{\partial}{\partial x_i}(a_{ij}(x,t)\frac{\partial u}{\partial x_j}), \\ A_1(t)u &= A(t)u + a_0(x,t)u, \end{aligned} \tag{6.55}$$

α is a positive constant, $f \in L^2(0,T;V')$ and $u_0 \in H$.

More precisely, Theorem 6.8 proves the existence and the uniqueness of the solution $u \in \mathcal{W} \cap \mathcal{C}([0,T];H)$ of the problem (the weak formulation of (6.54))

$$\begin{aligned} &\langle \frac{\partial u}{\partial t}, v \rangle_{V',V} + \tilde{a}(.;u,v) = \langle f, v \rangle_{V',V} \quad \text{a.e. in } (0,T),\ \forall v \in V, \\ &\text{with the initial condition} \\ &u(0) = u_0 \ \text{ on } \Omega, \end{aligned} \tag{6.56}$$

where $\tilde{a}(t;u,v) = a_1(t;u,v) + \alpha \int_\Gamma uv \mathrm{d}\Gamma$.

Remark 6.17. (*i*) The homogeneous Robin boundary condition given in (6.54) is included (by using the Green's formula) in the formulation (6.56).
(*ii*) for a given $t \in (0,T)$, $u \in D(A(t))$ is equivalent to $u \in V = H^1(\Omega)$, $A_1(t)u \in H = L^2(\Omega)$ and $\int_\Omega A_1(t)uv\mathrm{d}x = \tilde{a}(t;u,v)$ for all $v \in V$. ♢

6.4.4 Non-homogeneous Neumann and Dirichlet Boundary Conditions

Suppose now that the domain Ω is bounded, $\mathcal{C}^\infty$-manifold of dimension $(n-1)$ and locally totally one side of the boundary Γ and that the coefficients a_{ij} for $i,j = 1,n$ are sufficiently regular functions (for example are $\mathcal{C}^\infty$ functions on $\overline{\mathcal{Q}}$, the closure of $\mathcal{Q}$) and a_0 is in $L^\infty(\mathcal{Q})$.

Let us consider the following problem

$$\begin{aligned} &\frac{\partial u}{\partial t} + A(.)u + a_0 u = f \ \text{ on } \mathcal{Q}, \\ &\text{with the initial condition} \\ &u(0) = u_0 \ \text{ on } \Omega, \end{aligned} \tag{6.57}$$

subject to one of the following boundary conditions:

$$\frac{\partial u}{\partial \eta_A} = g \ \text{ on } \Sigma, \tag{6.58}$$

or

$$u = u_B \ \text{ on } \Sigma, \tag{6.59}$$

where

$$A(t)u = -\sum_{i,j=1,n} \frac{\partial}{\partial x_i}(a_{ij}(x,t)\frac{\partial u}{\partial x_j}) \tag{6.60}$$

and the normal derivative $\partial u / \partial \eta_A$ at Γ, directed towards the exterior of Ω, is given by (3.22).

We shall formulate sufficient conditions for the existence of a unique solution of the mixed initial-boundary value problem (6.57) with boundary conditions (6.58) or (6.59). For this purpose, for any positive real numbers r and s, we introduce the Sobolev space $H^{r,s}(\mathcal{Q})$ by

$$H^{r,s}(\mathcal{Q}) = L^2(0,T;H^r(\Omega)) \cap H^s(0,T;L^2(\Omega)), \tag{6.61}$$

which is a Hilbert space equipped with the norm

$$\| u \|_{H^{r,s}(\mathcal{Q})} := \left(\int_0^T \| u(t) \|^2_{H^r(\Omega)} \, dt + \| u \|^2_{H^s(0,T;L^2(\Omega))} \right)^{1/2} \tag{6.62}$$

where $H^s(0,T;L^2(\Omega)) = [H^m(0,T;L^2(\Omega)), L^2(0,T;L^2(\Omega))]_\theta$ (interpolation between $H^m(0,T;L^2(\Omega))$ and $L^2(0,T;L^2(\Omega))$), $0 < \theta < 1$, m is an integer such that $s < m$ and $(1-\theta)m = s$, and

$$H^m(0,T;L^2(\Omega)) := \{u \in L^2(\mathcal{Q}) : \frac{\partial^k u}{\partial t^k} \in L^2(\mathcal{Q}), \text{ for all } k = 1, \ldots, m\}.$$

In the spaces $H^{r,s}(\mathcal{Q})$, we have the following trace theorem (for the proof see, for instance, Lions and Magenes [204])

Theorem 6.18. *For $u \in H^{r,s}(\mathcal{Q})$ with $r > 1/2$ and $s \geq 0$, we may define*

$$\begin{aligned} &\frac{\partial^j u}{\partial \eta_A^j} \quad \text{on } \Sigma \quad \text{for } j < r - 1/2 \text{ and } j \in \mathbb{N}, \\ &\frac{\partial^j u}{\partial \eta_A^j} \in H^{r_j, s_j}(\Sigma), \end{aligned} \tag{6.63}$$

where $r_j/r = s_j/s = (r - j - 1/2)/r$ if $s > 0$ and $s_j = 0$ if $s = 0$. Then $u \longrightarrow \partial^j u/\partial \eta_A^j$ is a continuous linear mapping from $H^{r,s}(\mathcal{Q})$ into $H^{r_j,s_j}(\Sigma)$. □

We can now give the results of existence and uniqueness of the solution to Problem (6.57).

First, we give the following theorem concerning the existence of a unique solution in the case of non-homogeneous Neumann boundary conditions (6.58) (which can be found in Lions and Magenes [204]).

Theorem 6.19. *Let u_0, g and f be given:*

(i) If $u_0 \in H^1(\Omega)$, $g \in H^{1/2,1/4}(\Sigma)$ and $f \in L^2(\mathcal{Q})$ then there exists a unique solution $u \in H^{2,1}(\mathcal{Q})$ for problems (6.57) and (6.58).

(ii) If $u_0 \in H^{1/2}(\Omega)$, $g \in L^2(\Sigma)$ and $f \in H^{-1/2,-1/4}(\mathcal{Q})$ then there exists a unique solution $u \in H^{3/2,3/4}(\mathcal{Q})$ for problems (6.57) and (6.58). □

We may also formulate the theorem about the existence of a unique solution in the case of non-homogeneous Dirichlet boundary conditions (which can be found also in Lions and Magenes [204]).

Theorem 6.20. *Let u_0, u_B and f be given:*

(i) If $u_0 \in H^1(\Omega)$, $u_B \in H^{3/2,3/4}(\Sigma)$ and $f \in L^2(Q)$ and the compatibility condition $u(t = 0) = u_B(t = 0)$ on Γ holds, then there exists a unique solution $u \in H^{2,1}(Q)$ for problems (6.57) and (6.59).

(ii) If $u_0 \in H^{-1/2}(\Omega)$, $u_B \in L^2(\Sigma)$ and $f \in H^{-3/2,-3/4}(Q)$ then there exists a unique solution $u \in H^{1/2,1/4}(Q)$ for problems (6.57) and (6.59). □

Remark 6.21. By similar technique as used in the proof of Propositions 6.9 and 6.14, we can obtain the uniform Lipschitz continuity of the map solution in the different situations enunciated in Theorems 6.19 and 6.20. ♢

7

Optimal Control Problems

The objective of this chapter is to describe optimal control theory in different situations. First we use a very basic problem in order to explain the theory as simply as possible, then we present the theory for general linear evolutive problems and for some classes of non-linear evolutive problems. In each section, we study the existence, the uniqueness and the optimality conditions for the optimal solution. We develop, our study for different realistic cases of observations and controls. The numerical aspect will be presented in a later chapter.

7.1 Introduction

The theory of the optimal control of partial differential equations (PDEs) is based on the minimization (or maximization) of a calculus function, depending on the solution of the PDE, called the cost (or objective) functional. The control problem can be formulated, for example, like the minimization of the variation between the experimental observations and the corresponding quantities calculated by resolution of the system of equations. The control variables are the parameters or the functions to be identified (or estimate). They can intervene in the initial conditions, or in the boundary conditions or in the equation itself.

The essential data used in optimal control problems are the following:

1. A "control" φ in a set U_{ad} (known as set of "admissible controls").
2. The state $u(\varphi)$ of the system to be controlled, which is given, for a chosen control φ, by the resolution of the problem:

$$\mathcal{F}(t)(u(\varphi)) = \text{"given function of } \varphi\text{"}, \tag{7.1}$$

   where $\mathcal{F}(.)$ is an (known) operator which represents the system to be controlled ($\mathcal{F}(.)$ is the model of the studied system). In this context problem (7.1) is called *primal* or *direct* problem.

3. An "observation" u_{obs} which is supposed to be known exactly (for example given by measurements).
4. A "cost" functional $J(\varphi)$ which is defined from a real-valued and positive function $G(\varphi, \psi)$ by $J(\varphi) = G(\varphi, u(\varphi))$.

We want to find the optimal control, *i.e.*, the infimum of the functional J:

$$\inf_{\varphi \in U_{\text{ad}}} J(\varphi).$$

This infimum is said to be the *optimal control* or *optimal solution.*

We lay stress upon the fact that there is no general method to analyze the problems of control (it is necessary to adapt it to each situation). On the other hand, we can define a process to be followed for each problem:

(*i*) solve the direct problem (analysis of PDE, existence of solutions, stability according to the data, regularity, differentiability of the operators, *etc.*)

(*ii*) define the function or the parameter to be controlled

(*iii*) define the cost functional

(*iv*) obtain and analyse the necessary (and if possible the sufficient) conditions of optimality

(*v*) characterize the optimal solutions

(*vi*) define an algorithm allowing to solve numerically the control problem.

We shall now present a basic framework for an optimal control problem.

7.2 Basic Framework

In this section, we take an abstract boundary value problem. We give the space of controls, the admissibility set of controls, the control variable himself and the observation operator. The optimal control we consider is to maintain target state variables. We define the cost functional and the adjoint problem, corresponding to the primal problem. The main result of this study is the characterization of optimal solutions.

Let $\mathcal{F}$ be a continuous linear partial differential operator from X into E, where X and E are, for example, Hilbert spaces. The space X contains in its definition some appropriate boundary conditions. We assume that the corresponding boundary value problem is *well-posed* (or correctly-set) in a Hadamard sense, *i.e.*, $\mathcal{F}$ is an isomorphism from X onto E.

Let U be the space of controls, which is assumed to be, for example, a Hilbert space. Let also U_{ad} (the admissibility set of controls) be a closed convex non-empty subset of U and $\mathcal{B}$ be a linear and continuous operator (*i.e.*, $\mathcal{B} \in \mathcal{L}(U; E)$).

For every $\varphi \in U$ and $f \in E$, we consider the abstract boundary value problem by

$$\mathcal{F}(u) = f + \mathcal{B}\varphi. \tag{7.2}$$

The problem (7.2) admits a unique solution $u \in X$ such that

$$u = u(\varphi) = \mathcal{F}^{-1}(f + \mathcal{B}\varphi). \tag{7.3}$$

Let now $\mathcal{C} \in \mathcal{L}(X; M)$ (where M is, for example, a Hilbert space), be the *observation* operator and we take φ as the control. The optimal control we consider, for example, is to maintain the target state variables while the desired power level and adjustment costs are taken into consideration. We will study the following optimal control problem.

find $(u, \varphi) \in X \times U$ such that the following cost (or objective) functional, in the reduced form

$$J(\varphi) = \frac{1}{2} \parallel \mathcal{C}u(\varphi) - u_{\text{obs}} \parallel_M^2 + \frac{1}{2}(\mathcal{N}\varphi, \varphi)_U. \tag{7.4}$$

is minimized subject to the problem (7.2), with $u_{\text{obs}} \in M$ is the target (the observation given, for example, by experiment measurements) and $\mathcal{N} \in \mathcal{L}(U; U)$ is a symmetric positive definite operator, such that there exists a constant $\beta > 0$,

$$(\mathcal{N}\psi, \psi)_U \geq \beta \parallel \psi \parallel_U^2, \ \forall \psi \in U.$$

Remark 7.1. The cost functional J depends on the control φ and the state function u. In order to simplify the presentation we have used, in the expression (7.4), the reduced form of the functional, *i.e.*, $J(\varphi)$ in place of the classical form $J(\varphi, u(\varphi))$. ◇

Since $v \longrightarrow \mathcal{C}v - u_{\text{obs}}$ is a continuous affine function from X into M, the norm is a continuous function and $(\varphi, \psi) \longrightarrow (\mathcal{N}\varphi, \psi)_U$ is a bilinear form and continuous on U, then the functional J is continuous. Moreover, according to the nature of J, the functional J is convex and even *strictly convex*, and we have $J(\varphi) \longrightarrow +\infty$ if $\varphi \in U_{\text{ad}}, \ \parallel \varphi \parallel_U \longrightarrow \infty$. Consequently, there exists a unique optimal solution $(u, \varphi) \in X \times U_{\text{ad}}$ such that φ is a solution of (see Proposition 4.92)

$$J(\varphi) = \inf_{\psi \in U_{\text{ad}}} J(\psi) \tag{7.5}$$

and $u = u(\varphi)$ is the solution of (7.2), corresponding to φ.

It follows directly that the functional J is differentiable, and then the optimal control φ satisfies the following inequality (see Proposition 4.93)

$$J'(\varphi).(\psi - \varphi) \geq 0 \quad \forall \psi \in U_{\text{ad}}. \tag{7.6}$$

Since $u(\varphi)$ satisfies (7.3) then, by differentiation, we have that

$$u'(\varphi).\theta = \mathcal{F}^{-1}(\mathcal{B}\theta) \ \ \forall \theta \in U,$$

and, in particular

$$u'(\varphi).(\psi - \varphi) = \mathcal{F}^{-1}(\mathcal{B}(\psi - \varphi)) = u(\psi) - u(\varphi) \;\; \forall \psi \in U_{\text{ad}}. \tag{7.7}$$

According to (7.7), the condition (7.6) can be written as

$$(\mathcal{C}u(\varphi) - u_{\text{obs}}, \mathcal{C}(u(\psi) - u(\varphi)))_M + (\mathcal{N}\varphi, \psi - \varphi)_U \geq 0 \;\; \forall \psi \in U_{\text{ad}}. \tag{7.8}$$

In order to simplify the relation (7.8), we introduce the adjoint problem

$$\mathcal{F}^*(\tilde{u}) = \mathcal{C}^* \Lambda (\mathcal{C}u - u_{\text{obs}}), \tag{7.9}$$

where $\tilde{u}$ is the adjoint state corresponding to the primal state u, $\mathcal{C}^* \in \mathcal{L}(M', X')$ is the adjoint of the operator $\mathcal{C}$, $\mathcal{F}^*$ is the adjoint of the operator solution $\mathcal{F}$ and Λ is the canonical isomorphism from M onto M' (such that $\langle p, q^* \rangle_{M,M'} = (\Lambda p, q^*)_{M'} = (p, \Lambda^{-1} q^*)_M$, $\forall (p, q^*) \in M \times M'$), where M' (respectively X') is the dual of M (respectively of X).

Then,

$$\begin{aligned}
(\mathcal{C}u(\varphi) - u_{\text{obs}}, \mathcal{C}(u(\psi) - u(\varphi)))_M &= \langle \mathcal{C}^* \Lambda (\mathcal{C}u(\varphi) - u_{\text{obs}}), u(\psi) - u(\varphi) \rangle_{X',X} \\
&= \langle \mathcal{F}^*(\tilde{u}(\varphi)), u(\psi) - u(\varphi) \rangle_{X',X} \\
&= \langle \tilde{u}(\varphi), \mathcal{F}(u(\psi) - u(\varphi)) \rangle_{E',E} \\
&= \langle \tilde{u}(\varphi), \mathcal{B}(\psi - \varphi) \rangle_{E',E} \\
&= (\Lambda_U^{-1} \mathcal{B}^* \tilde{u}(\varphi), \psi - \varphi)_U, \forall \psi \in U_{\text{ad}},
\end{aligned}$$

where $\mathcal{B}^* \in \mathcal{L}(E', U')$ is the adjoint of $\mathcal{B}$, U' (respectively E') is the dual of U (respectively of E), Λ_U is the canonical isomorphism from U onto U' and $\tilde{u}(\varphi)$ is the solution of the adjoint problem (7.9) corresponding to the primal solution $u(\varphi)$.

Thus, the condition (7.8) can be written as

$$(\Lambda_U^{-1} \mathcal{B}^* \tilde{u}(\varphi) + \mathcal{N}\varphi, \psi - \varphi)_U \geq 0, \forall \psi \in U_{\text{ad}}.$$

Remark 7.2. Since the operator $\mathcal{F}^*$ is an isomorphism then Problem (7.9) admits a unique solution. ♢

We have proved that the optimal control $\varphi \in U_{\text{ad}}$ (the solution of the problem (7.5)) is characterized by the following optimality system which include the direct (or primal) problem and the adjoint problem, linked by an inequality

$$\begin{aligned}
&\mathcal{F}(u(\varphi)) = f + \mathcal{B}\varphi, \\
&\mathcal{F}^*(\tilde{u}(\varphi)) = \mathcal{C}^* \Lambda (\mathcal{C}u(\varphi) - u_{\text{obs}}), \\
&(\Lambda_U^{-1} \mathcal{B}^* \tilde{u}(\varphi) + \mathcal{N}\varphi, \psi - \varphi)_U \geq 0, \forall \psi \in U_{\text{ad}}.
\end{aligned} \tag{7.10}$$

It is clear that the analysis of control problems depends naturally on the nature of the operator $\mathcal{F}$. In the next study we consider problems described by partial differential operators. More precisely, we study systems whose state $u(\varphi)$ is given by the resolution of linear evolutive partial differential equations to which we add the boundary conditions and the initial conditions.

7.3 Linear Control Problems

In this section, we consider linear control problems subject to parabolic type systems in a more general framework. The notations, the assumptions and the spaces are the same as in Chapter 6.

7.3.1 Position of the Problem, Existence and Uniqueness of the Optimal Solution

Let U be the space of controls, which is assumed to be a Hilbert space, and let U_{ad} (the admissibility set of controls) be a closed convex non-empty subset of U. Let also $\mathcal{B}$ be a linear continuous operator such that $\mathcal{B} \in \mathcal{L}(U; L^2(0,T;V'))$.

For f and u_0 be given such that $f \in L^2(0,T;V')$ and $u_0 \in H$, and φ given in U, we consider the following linear evolutive equation (a.e. $t \in (0,T)$):

$$\begin{aligned} &\frac{\partial u(t)}{\partial t} + A(t)u(t) = f + \mathcal{B}\varphi \quad (\in L^2(0,T;V')), \\ &u(0) = u_0 \quad (\in H). \end{aligned} \tag{7.11}$$

According to Theorem 6.8, the problem (7.11) admits a unique solution $u \in L^2(0,T;V) \cap \mathcal{C}([0,T];H)$ such that $\partial u/\partial t \in L^2(0,T;V')$, for every $f \in L^2(0,T;V')$, $u_0 \in H$ and $\varphi \in U$.

Let us now introduce the space

$$\mathcal{W} = \{u \in L^2(0,T;V) : \frac{\partial u}{\partial t} \in L^2(0,T;V')\},$$

and consider the *observation* operator $\mathcal{C} \in \mathcal{L}(\mathcal{W}; M)$, where M is a Hilbert space. The optimal control that we consider is to find $(u,\varphi) \in \mathcal{W} \times U$ such that the following cost (or objective) functional, in the reduced form

$$J(\varphi) = \frac{1}{2} \parallel \mathcal{C}u(\varphi) - u_{\mathrm{obs}} \parallel_M^2 + \frac{1}{2}(\mathcal{N}\varphi, \varphi)_U, \tag{7.12}$$

is minimized subject to problem (7.11), where $u_{\mathrm{obs}} \in M$ is the target and $\mathcal{N} \in \mathcal{L}(U;U)$ is a symmetric positive definite operator such that there exists a constant $\beta > 0$ such that $(\mathcal{N}\psi, \psi)_U \geq \beta \parallel \psi \parallel_U^2$, $\forall \psi \in U$.

Since $v \longrightarrow \mathcal{C}v - u_{\mathrm{obs}}$ is a continuous affine function from X into M, the norm is a continuous function and $(\varphi, \psi) \longrightarrow (\mathcal{N}\varphi, \psi)_U$ is a bilinear form and continuous on U, then the functional J is continuous. Moreover, according to the nature of J, the functional J is *strictly convex*, and $J(\varphi) \longrightarrow +\infty$ if $\varphi \in U_{\mathrm{ad}}$, $\parallel \varphi \parallel_U \longrightarrow \infty$. Consequently, there exists a unique optimal solution $(u, \varphi) \in \mathcal{W} \times U_{\mathrm{ad}}$ such that φ is a solution of

$$J(\varphi) = \inf_{\psi \in U_{\mathrm{ad}}} J(\psi) \tag{7.13}$$

and $u = u(\varphi)$ is the solution of (7.11), corresponding to φ.

7.3.2 Optimality Conditions and Identification of the Gradients

As in the general framework (see Section 7.2), the functional J is differentiable and the control $\varphi \in U_{\rm ad}$ is optimal if and only if

$$J'(\varphi).(\psi - \varphi) \geq 0 \quad \forall \psi \in U_{\rm ad},$$

$$i.e., \quad (\mathcal{C}u(\varphi) - u_{\rm obs}, \mathcal{C}(u(\psi) - u(\varphi)))_M + (\mathcal{N}\varphi, \psi - \varphi)_U \geq 0 \quad \forall \psi \in U_{\rm ad}.$$

Otherwise for all $\psi \in U_{\rm ad}$, we have

$$(\mathcal{C}^* \Lambda(\mathcal{C}u(\varphi) - u_{\rm obs}), u(\psi) - u(\varphi))_{\mathcal{W}', \mathcal{W}} + (\mathcal{N}\varphi, \psi - \varphi)_U \geq 0, \qquad (7.14)$$

where Λ is the canonical isomorphism from M onto M' and $\mathcal{C}^* \in \mathcal{L}(M', \mathcal{W}')$ is the adjoint of $\mathcal{C}$. In order to simplify the relation (7.14), we introduce the adjoint problem associated with the primal problem (7.11). For this, we consider the following two situations:

(i) the case of $\mathcal{C} \in \mathcal{L}(L^2(0,T;V);M) \subset \mathcal{L}(\mathcal{W};M)$

(ii) the case of the final observation: $\mathcal{C}u = \mathcal{D}u(t = T)$, where $\mathcal{D} \in \mathcal{L}(H;H)$ and $M = H$.

Remark 7.3. It is clear that the operator $\mathcal{C}$ can be the linear combination of operators which correspond to (i) and (ii). ♢

The Case (i) and the Optimality System

In this case, the adjoint of $\mathcal{C}$, $\mathcal{C}^*$ is in $\mathcal{L}(M', L^2(0,T;V'))$ and the inequality (7.14) becomes, for all $\psi \in U_{\rm ad}$

$$\int_0^T \langle \mathcal{C}^* \Lambda(\mathcal{C}u(\varphi) - u_{\rm obs}), u(\psi) - u(\varphi)\rangle_{V',V} {\rm d}t + (\mathcal{N}\varphi, \psi - \varphi)_U \geq 0. \qquad (7.15)$$

Let us introduce the following adjoint problem:

$$\begin{aligned} -\frac{\partial \tilde{u}}{\partial t} + A^*(t)\tilde{u} &= \mathcal{C}^* \Lambda(\mathcal{C}u - u_{\rm obs}) \in L^2(0,T;V'), \\ \tilde{u}(t = T) &= 0 \in H, \end{aligned} \qquad (7.16)$$

which admits a unique solution $\tilde{u} \in \mathcal{W}$ with A^* the adjoint of A. To prove this result, we change the variables of System (7.16) by reversing the sense of time, *i.e.*, $t := T - t$ and we apply the same way to obtain the result of Theorem 6.8. Indeed, by setting $w(.,t) := u(.,T-t)$, $w_{\rm obs}(.,t) := u_{\rm obs}(.,T-t)$, $\tilde{w}(.,t) := \tilde{u}(.,T-t)$ and $\tilde{A}^*(t) := A^*(T-t)$, problem (7.16) can be written as

$$\begin{aligned} \frac{\partial \tilde{w}}{\partial t} + \tilde{A}^*(t)\tilde{w} &= \mathcal{C}^* \Lambda(\mathcal{C}w - w_{\rm obs}) \in L^2(0,T;V'), \\ \tilde{w}(0) &= 0. \end{aligned} \qquad (7.17)$$

Since the operator $\tilde{A}^*$ satisfies the same assumptions as the operator A then problem (7.17) is similar to problem (7.11). Consequently, by Theorem 6.8, we have the existence and the uniqueness of the solution $\tilde{w}$ in $\mathcal{W}$ and then the existence and the uniqueness of the solution $\tilde{u}$ in $\mathcal{W}$.

Next we give the optimality conditions. Multiplying the first equation of (7.16) by $u(\varphi) - u(\psi)$ and integrating with respect to time we have that

$$\begin{aligned} -\int_0^T \langle \frac{\partial \tilde{u}(\varphi)}{\partial t}, u(\varphi) - u(\psi)\rangle_{V',V} \mathrm{d}t \\ + \int_0^T \langle A^*(t)\tilde{u}(\varphi), u(\varphi) - u(\psi)\rangle_{V',V} \mathrm{d}t \\ = \int_0^T \langle \mathcal{C}^* \Lambda(\mathcal{C}u(\varphi) - u_{\mathrm{obs}}), u(\varphi) - u(\psi)\rangle_{V',V} \mathrm{d}t, \end{aligned} \tag{7.18}$$

with $\tilde{u}(\varphi)$ solution of the adjoint problem (7.16), corresponding to the primal solution $u(\varphi)$.

By using the result of Corollary 6.4, we can deduce that

$$\begin{aligned} -\int_0^T \langle \frac{\partial \tilde{u}(\varphi)}{\partial t}, u(\varphi) - u(\psi)\rangle_{V',V} \mathrm{d}t = \int_0^T \langle \tilde{u}(\varphi), \frac{\partial u(\varphi)}{\partial t} - \frac{\partial u(\psi)}{\partial t}\rangle_{V,V'} \mathrm{d}t \\ -(\tilde{u}(\varphi)(T), (u(\varphi) - u(\psi))(T)) + (\tilde{u}(\varphi)(0), (u(\varphi) - u(\psi))(0)). \end{aligned}$$

Since $\tilde{u}(\varphi)(t = T) = 0$ and $(u(\varphi) - u(\psi))(0) = u_0 - u_0 = 0$ then

$$-\int_0^T \langle \frac{\partial \tilde{u}(\varphi)}{\partial t}, u(\varphi) - u(\psi)\rangle_{V',V} \mathrm{d}t = \int_0^T \langle \frac{\partial u(\varphi)}{\partial t} - \frac{\partial u(\psi)}{\partial t}, \tilde{u}(\varphi)\rangle_{V',V} \mathrm{d}t.$$

Consequently, the relation (7.18) becomes

$$\begin{aligned} \int_0^T \langle \frac{\partial (u(\varphi) - u(\psi))}{\partial t}, \tilde{u}(\varphi)\rangle_{V',V} \mathrm{d}t \\ + \int_0^T \langle A(t)(u(\varphi) - u(\psi)), \tilde{u}(\varphi)\rangle_{V',V} \mathrm{d}t \\ = \int_0^T \langle \mathcal{C}^* \Lambda(\mathcal{C}u(\varphi) - u_{\mathrm{obs}}), u(\varphi) - u(\psi)\rangle_{V',V} \mathrm{d}t. \end{aligned} \tag{7.19}$$

Since $u(\varphi) - u(\psi)$ is the solution of

$$\begin{aligned} &\frac{\partial (u(\varphi) - u(\psi))}{\partial t} + A(t)(u(\varphi) - u(\psi)) = \mathcal{B}(\varphi - \psi), \\ &(u(\varphi) - u(\psi))(0) = 0, \end{aligned}$$

then, (7.19) gives

$$\begin{aligned} \int_0^T \langle \mathcal{C}^* \Lambda(\mathcal{C}u(\varphi) - u_{\mathrm{obs}}), u(\varphi) - u(\psi)\rangle_{V',V} \mathrm{d}t &= \int_0^T \langle \mathcal{B}(\varphi - \psi), \tilde{u}(\varphi)\rangle_{V',V} \mathrm{d}t \\ &= \langle \mathcal{B}^* \tilde{u}(\varphi), \varphi - \psi\rangle_{U',U} \\ &= (\Lambda_U^{-1} \mathcal{B}^* \tilde{u}(\varphi), \varphi - \psi)_U, \end{aligned}$$

where $\mathcal{B}^* \in \mathcal{L}(L^2(0,T;V);U')$ is the adjoint of $\mathcal{B}$ and Λ_U is the canonical isomorphism from U onto U'.

So, the inequality (7.15) can be written as

$$(\Lambda_U^{-1}\mathcal{B}^*\tilde{u}(\varphi) + \mathcal{N}\varphi, \psi - \varphi)_U \geq 0 \quad \forall \psi \in U_{\text{ad}} \tag{7.20}$$

and then we have proved the following theorem.

Theorem 7.4. *Under the assumptions of Theorem 6.8, we suppose furthermore that the operator $\mathcal{C} \in \mathcal{L}(L^2(0,T;V);M)$ and the symmetric positive operator $\mathcal{N} \in \mathcal{L}(U;U)$ satisfies $(\mathcal{N}\psi,\psi)_U \geq \beta \parallel \psi \parallel_U^2$, $\forall \psi \in U$, for a constant $\beta > 0$. Then there exists a unique optimal solution $(\varphi^*, u^*) \in U \times \mathcal{W}$ such that*

$$\frac{\partial u^*}{\partial t} + A(t)u^* = f + \mathcal{B}\varphi^*,$$
$$u^*(0) = u_0$$

and

$$(\Lambda_U^{-1}\mathcal{B}^*\tilde{u}^* + \mathcal{N}\varphi^*, \psi - \varphi^*)_U \geq 0 \quad \forall \psi \in U_{\text{ad}},$$

where Λ_U is the canonical isomorphism from U onto U' and $\tilde{u}^*$ is the unique solution, which is in $\mathcal{W}$, of the adjoint problem

$$-\frac{\partial \tilde{u}^*}{\partial t} + A^*(t)\tilde{u}^* = \mathcal{C}^*\Lambda(\mathcal{C}u^* - u_{\text{obs}}),$$
$$\tilde{u}^*(t = T) = 0,$$

where A^* is the adjoint operator of A and Λ is the canonical isomorphism from M onto M'. □

The Case (ii) and the Optimality System

In this case ($\mathcal{C}u = \mathcal{D}u(t = T)$ where $\mathcal{D} \in \mathcal{L}(H;H)$ and $M = H = M'$ since $H' = H$) the final observation u_{obs} is in H, $\mathcal{D}^* \in \mathcal{L}(H;H)$, the cost functional is defined by

$$J(\varphi) = \frac{1}{2} \parallel \mathcal{D}u(\varphi)(T) - u_{\text{obs}} \parallel_H^2 + \frac{1}{2}(\mathcal{N}\varphi, \varphi)_U \tag{7.21}$$

and the inequality (7.14) is equivalent, for all $\psi \in U_{\text{ad}}$, to

$$(\mathcal{D}^*(\mathcal{D}u(\varphi)(T) - u_{\text{obs}}), (u(\psi) - u(\varphi))(T))_H + (\mathcal{N}\varphi, \psi - \varphi)_U \geq 0, \tag{7.22}$$

where $u(\varphi)$ and $u(\psi)$ are solutions of the primal problem (7.11), corresponding respectively with φ and ψ.

Let us introduce the following adjoint problem:

$$\begin{aligned} -\frac{\partial \tilde{u}}{\partial t} + A^*(t)\tilde{u} &= 0, \\ \tilde{u}(t = T) &= \mathcal{D}^*(\mathcal{D}u(\varphi)(t = T) - u_{\text{obs}}) \quad (\in H), \end{aligned} \tag{7.23}$$

which admits a unique solution $\tilde{u} \in \mathcal{W}$. To prove this result, we change the variables of System (7.23) by reversing the sense of time, *i.e.*, $t := T - t$ and we apply the same way to obtain the result of Theorem 6.8.

Multiplying the first equation of (7.23) by $u(\varphi) - u(\psi)$ and integrating with respect to time we have that

$$-\int_0^T \langle \frac{\partial \tilde{u}(\varphi)}{\partial t}, u(\varphi) - u(\psi)\rangle_{V',V} \mathrm{d}t + \int_0^T \langle A^*(t)\tilde{u}(\varphi), u(\varphi) - u(\psi)\rangle_{V',V} \mathrm{d}t = 0.$$

By using the result of Corollary 6.4, we can deduce that

$$\begin{aligned}-\int_0^T \langle \frac{\partial \tilde{u}(\varphi)}{\partial t}, u(\varphi) - u(\psi)\rangle_{V',V} \mathrm{d}t &= \int_0^T \langle \tilde{u}(\varphi), \frac{\partial u(\varphi)}{\partial t} - \frac{\partial u(\psi)}{\partial t}\rangle_{V,V'} \mathrm{d}t \\ &\quad -(\tilde{u}(\varphi)(T), (u(\varphi) - u(\psi))(T))_H + (\tilde{u}(\varphi)(0), (u(\varphi) - u(\psi))(0))_H.\end{aligned}$$

Since $\tilde{u}(\varphi)(T) = \mathcal{D}^*(\mathcal{D}u(\varphi)(T) - u_{\text{obs}})$ and $(u(\varphi) - u(\psi))(0) = u_0 - u_0 = 0$ then

$$\begin{aligned}-\int_0^T \langle \frac{\partial \tilde{u}(\varphi)}{\partial t}, u(\varphi) - u(\psi)\rangle_{V',V} \mathrm{d}t &= \int_0^T \langle \tilde{u}(\varphi), \frac{\partial u(\varphi)}{\partial t} - \frac{\partial u(\psi)}{\partial t}\rangle_{V,V'} \mathrm{d}t \\ &\quad -(\mathcal{D}^*(\mathcal{D}u(\varphi)(T) - u_{\text{obs}}), (u(\varphi) - u(\psi))(T))_H\end{aligned}$$

and consequently

$$\begin{aligned}&\int_0^T \langle \tilde{u}(\varphi), \frac{\partial u(\varphi)}{\partial t} - \frac{\partial u(\psi)}{\partial t}\rangle_{V,V'} \mathrm{d}t \\ &+ \int_0^T \langle \tilde{u}(\varphi), A(t)(u(\varphi) - u(\psi))\rangle_{V,V'} \mathrm{d}t \\ &= (\mathcal{D}^*(\mathcal{D}u(\varphi)(T) - u_{\text{obs}}), (u(\varphi) - u(\psi))(T))_H.\end{aligned} \tag{7.24}$$

Since $u(\varphi) - u(\psi)$ is the solution of

$$\begin{aligned}&\frac{\partial(u(\varphi) - u(\psi))}{\partial t} + A(t)(u(\varphi) - u(\psi)) = \mathcal{B}(\varphi - \psi), \\ &(u(\varphi) - u(\psi))(0) = 0,\end{aligned}$$

then, (7.24) gives

$$\begin{aligned}(\mathcal{D}^*(\mathcal{D}u(\varphi)(T) - u_{\text{obs}}), (u(\varphi) - u(\psi))(T))_H &= \int_0^T \langle \tilde{u}(\varphi), \mathcal{B}(\varphi - \psi)\rangle_{V,V'} \mathrm{d}t \\ &= \langle \mathcal{B}^*\tilde{u}(\varphi), \varphi - \psi\rangle_{U',U} \\ &= (\Lambda_U^{-1}\mathcal{B}^*\tilde{u}(\varphi), \varphi - \psi)_U,\end{aligned}$$

where $\mathcal{B}^* \in \mathcal{L}(L^2(0,T;V); U')$ the adjoint of $\mathcal{B}$ and Λ_U is the canonical isomorphism from U onto U'.

So, the inequality (7.22) can be written as

$$(\Lambda_U^{-1}\mathcal{B}^*\tilde{u}(\varphi) + \mathcal{N}\varphi, \psi - \varphi)_U \geq 0 \quad \forall \psi \in U_{\text{ad}} \tag{7.25}$$

and then we have proved the following theorem.

Theorem 7.5. *Under the assumptions of Theorem 6.8, we suppose furthermore that the cost J is given by (7.21) with the operator $\mathcal{D} \in \mathcal{L}(H;H)$ and the symmetric positive operator $\mathcal{N} \in \mathcal{L}(U;U)$ satisfies $(\mathcal{N}\psi,\psi)_U \geq \beta \parallel \psi \parallel_U^2$, $\forall \psi \in U$, for a constant $\beta > 0$. Then there exists a unique optimal solution $(\varphi^*, u^*) \in U \times \mathcal{W}$ such that*

$$\frac{\partial u^*}{\partial t} + A(t)u^* = f + \mathcal{B}\varphi^*,$$
$$u^*(0) = u_0$$

and

$$(\Lambda_U^{-1}\mathcal{B}^*\tilde{u}^* + \mathcal{N}\varphi^*, \psi - \varphi^*)_U \geq 0 \quad \forall \psi \in U_{\mathrm{ad}},$$

where Λ_U is the canonical isomorphism from U onto U' and $\tilde{u}^*$ is the unique solution, which is in $\mathcal{W}$, of the adjoint problem

$$-\frac{\partial \tilde{u}^*}{\partial t} + A^*(t)\tilde{u}^* = 0,$$
$$\tilde{u}^*(t = T) = \mathcal{D}^*(\mathcal{D}u^*(t = T) - u_{\mathrm{obs}}).$$

□

More generally, if we suppose that we have two observations: the final observation v_{obs}, which is in H, and the state observation u_{obs}, which is in M, the cost functional can be defined by

$$J(\varphi) = \frac{\theta_1}{2} \parallel \mathcal{C}u - u_{\mathrm{obs}} \parallel_M^2 + \frac{\theta_2}{2} \parallel \mathcal{D}u(T) - v_{\mathrm{obs}} \parallel_H^2 + \frac{1}{2}(\mathcal{N}\varphi, \varphi)_U, \qquad (7.26)$$

where $\theta_i \geq 0, \ \ i = 1, 2$, $\theta_1 + \theta_2 > 0$, $\mathcal{C} \in \mathcal{L}(L^2(0,T;V);M)$ and $\mathcal{D} \in \mathcal{L}(H;H)$. Then we have the following theorem.

Theorem 7.6. *Under the assumptions of Theorem 6.8, we suppose furthermore that the cost functional is given by (7.26) with the operators $\mathcal{D} \in \mathcal{L}(H;H)$ and $\mathcal{C} \in \mathcal{L}(L^2(0,T;V);M)$, and that the symmetric positive operator $\mathcal{N} \in \mathcal{L}(U;U)$ satisfies $(\mathcal{N}\psi,\psi)_U \geq \beta \parallel \psi \parallel_U^2$, $\forall \psi \in U$, for a constant $\beta > 0$. Then there exists a unique optimal solution $(\varphi^*, u^*) \in U \times \mathcal{W}$ such that*

$$\frac{\partial u^*}{\partial t} + A(t)u^* = f + \mathcal{B}\varphi^*,$$
$$u^*(0) = u_0$$

and

$$(\Lambda_U^{-1}\mathcal{B}^*\tilde{u}^* + \mathcal{N}\varphi^*, \psi - \varphi^*)_U \geq 0 \quad \forall \psi \in U_{\mathrm{ad}},$$

where Λ_U is the canonical isomorphism from U onto U' and $\tilde{u}^*$ is the unique solution, which is in $\mathcal{W}$, of the adjoint problem

$$-\frac{\partial \tilde{u}^*}{\partial t} + A^*(t)\tilde{u}^* = \mathcal{C}^*\Lambda(\mathcal{C}u^* - u_{\mathrm{obs}}),$$
$$\tilde{u}^*(t = T) = \mathcal{D}^*(\mathcal{D}u^*(t = T) - v_{\mathrm{obs}}),$$

where Λ is the canonical isomorphism from M onto M'. □

In the next section, we give some applications where we specify more precisely the control and the observation.

7.4 Examples of Controls and Observations

In this section we present optimal control problems with different types of controls and observations, namely, problems with distributed control, boundary control, boundary observation, pointwise control, pointwise observation and/or data assimilation.

For this, we consider the same operators and assumptions as in Section 6.4 and we take the following problem:

$$\frac{\partial u}{\partial t} + A(t)u = f + \phi - a_0 u \quad \text{on } \mathcal{Q} = \Omega \times (0,T),$$

subject to the linear Robin boundary condition

$$\alpha \frac{\partial u}{\partial \eta_A} + \beta u = \alpha \psi \quad \text{on } \Sigma = \Gamma \times (0,T), \tag{7.27}$$

with the initial condition

$$u(0) = u_0 \quad \text{on } \Omega,$$

under the pointwise constraint

$$r_1 \leq a_0 \leq r_2 \quad \text{a.e. in } \mathcal{Q}, \tag{7.28}$$

where the normal derivative $\partial u/\partial \eta_A$ at Γ, directed towards the exterior of Ω, is given by (3.22), the functions f, ϕ, a_0 are in $L^2(\mathcal{Q})$, the function ψ is in $L^2(\Sigma)$, the initial condition u_0 is in $L^2(\Omega)$, the real constants α, β are such that $\alpha\beta \geq 0$ and $\alpha + \beta \neq 0$, the real constants r_1, r_2 are such that $r_1 < r_2$ and

$$A(t)u = -\sum_{i,j=1,n} \frac{\partial}{\partial x_i}(a_{ij}(x,t)\frac{\partial u}{\partial x_j}). \tag{7.29}$$

Next, let us denote by V the space $H_0^1(\Omega)$ in the case of homogeneous Dirichlet boundary condition (*i.e.*, in the case of $\alpha = 0$) and the space $H^1(\Omega)$ in the case of linear Robin or Neumann boundary condition (*i.e.*, if $\alpha \neq 0$). We denote its dual by V' and by $\langle .,. \rangle_{V',V}$ the duality product between V and V'. Then the embedding $V \subset L^2(\Omega) \subset V'$ are continuous. Finally, we denote by $H = L^2(\Omega)$ and by $\mathcal{W} = \{w \in L^2(0,T;V) : \partial w/\partial t \in L^2(0,T;V')\}$.

We introduce now the following form by:

$$\tilde{a}(t;u,v) = a(t;u,v) + \int_\Omega a_0 u.v\mathrm{d}x + \int_\Gamma \theta u.v\mathrm{d}\Gamma,$$

where a is the bilinear form corresponding to the operator A (see section 6.4), $\theta = \beta/\alpha \geq 0$ if $\alpha \neq 0$ and $\theta = 0$ else.

Then the bilinear form $\tilde{a}$ satisfies the assumptions of Theorem 6.8 (*i.e.*, (6.3), (6.4) and (6.6), since $\| v \|_{L^2(\Gamma)} \leq \| v \|_V$, $\forall v \in V$, $\theta \geq 0$ and a_0 satisfies (7.28)) and so, Theorem 6.8 gives the solution of problem (7.27). More precisely, Theorem 6.8 and Lemma 6.6 prove the existence and the uniqueness of the solution $u \in \mathcal{W} \cap \mathcal{C}([0,T];H)$ of the problem, a.e. $t \in (0,T)$ (the weak formulation of (7.27))

$$\langle \frac{\partial u}{\partial t}, v\rangle_{V',V} + \tilde{a}(t;u,v) = (f+\phi, v) + \alpha \int_{\Gamma} \psi v \mathrm{d}\Gamma \qquad \forall v \in V,$$

with the initial condition (7.30)

$$u(0) = u_0 \quad \text{on } \Omega$$

such that the following estimate holds.

$$\begin{aligned} \| u \|^2_{\mathcal{W}\cap\mathcal{C}([0,T];H)} \leq C_1(\| f \|^2_{L^2(\mathcal{Q})} + \| \phi \|^2_{L^2(\mathcal{Q})}) \\ + C_2(\| u_0 \|^2_{L^2(\Omega)} + \| \psi \|^2_{L^2(\Sigma)}). \end{aligned} \tag{7.31}$$

7.4.1 Boundary Control

In this subsection, we consider the optimal control problem where the control is both in the state equation and in the boundary condition.

We suppose that the observation operator $\mathcal{C}$ is the injection of $L^2(0,T;V)$ onto $L^2(\mathcal{Q})$ (*i.e.*, $M = L^2(\mathcal{Q}) = M'$, so the canonical isomorphism from M onto M' is the identity operator) and consider the following two cases for the control:

(*i*) The control is the function ϕ, *i.e.*, the control is distributed in $\mathcal{Q}$.
(*ii*) The control is the function ψ, *i.e.*, the control is on the boundary.

Let $\mathcal{K}_1$ and $\mathcal{K}_2$ be given non-empty, closed and convex subsets of $L^2(\mathcal{Q})$ and $L^2(\Sigma)$ respectively, and the observation $u_{\text{obs}} \in L^2(\mathcal{Q})$. Our problem is then, to find $\Phi \in \mathcal{U}_{\text{ad}}$ such that the cost functional

$$J(\Phi) = \frac{1}{2} \| \mathcal{C}u(\Phi) - u_{\text{obs}} \|^2_{L^2(\mathcal{Q})} + \frac{\delta_1}{2} \| \phi \|^2_{L^2(\mathcal{Q})} + \frac{\delta_2}{2} \| \psi \|^2_{L^2(\Sigma)} \tag{7.32}$$

is minimized with respect to Φ subject to the problem (7.27), where $\delta_i \geq 0,\ i = 1,2$ are fixed such that $\delta_1 + \delta_2 > 0$. The control Φ plays the role of:

(*i*) the function ϕ (*i.e.*, $\Phi = \phi$) if $\delta_2 = 0$ and then $\mathcal{U}_{\text{ad}} = \mathcal{K}_1$
(*ii*) the function ψ (*i.e.*, $\Phi = \psi$) if $\delta_1 = 0$ and then $\mathcal{U}_{\text{ad}} = \mathcal{K}_2$
(*iii*) the function (ϕ, ψ) (*i.e.*, $\Phi = (\phi, \psi)$) if $\delta_i > 0, i = 1,2$ and then $\mathcal{U}_{\text{ad}} = \mathcal{K}_1 \times \mathcal{K}_2$.

The adjoint problem corresponding to the problem (7.27) is given by the following system:

$$-\frac{\partial \tilde{u}(\Phi)}{\partial t} + A^*(t)\tilde{u}(\Phi) = -a_0 u(\Phi) + \mathcal{C}^*(\mathcal{C}u(\Phi) - u_{\text{obs}}) \ \text{ on } \mathcal{Q},$$

subject to the linear Robin boundary condition

$$\alpha \frac{\partial \tilde{u}(\Phi)}{\partial \eta_{A^*}} + \beta \tilde{u}(\Phi) = 0 \ \text{ on } \Sigma, \tag{7.33}$$

with the final condition

$$\tilde{u}(\Phi)(t = T) = 0 \ \text{ on } \Omega,$$

where A^* is the adjoint of the operator A.

Moreover, the following optimality conditions apply:

(*i*) If $\delta_2 = 0$ (*i.e.*, the control is distributed on $\mathcal{Q}$), are given by

$$\iint_{\mathcal{Q}} (\tilde{u}(\phi^*) + \delta_1 \phi^*)(\phi - \phi^*)\mathrm{d}x\mathrm{d}t \geq 0 \quad \forall \phi \in \mathcal{K}_1. \tag{7.34}$$

(*ii*) If $\delta_1 = 0$ (*i.e.*, the control is in the boundary), are given by

$$\iint_{\Sigma} (\tilde{u}(\psi^*) + \delta_2 \psi^*)(\psi - \psi^*)\mathrm{d}\Gamma\mathrm{d}t \geq 0 \quad \forall \psi \in \mathcal{K}_2. \tag{7.35}$$

(*iii*) If $\delta_i > 0, i = 1, 2$, are given by, $\forall(\phi, \psi) \in \mathcal{K}_1 \times \mathcal{K}_2$

$$\begin{aligned} &\iint_{\mathcal{Q}} (\tilde{u}(\Phi^*) + \delta_1 \phi^*)(\phi - \phi^*)\mathrm{d}x\mathrm{d}t \geq 0, \\ &\iint_{\Sigma} (\tilde{u}(\Phi^*) + \delta_2 \psi^*)(\psi - \psi^*)\mathrm{d}\Gamma\mathrm{d}t \geq 0. \end{aligned} \tag{7.36}$$

Where $(\Phi^* = (\phi^*, \psi^*), u(\Phi^*))$ is the optimal solution.

Remark 7.7. In case (*iii*), for example, and without constraints, *i.e.*, $\mathcal{U}_{\mathrm{ad}} = L^2(\mathcal{Q}) \times L^2(\Sigma)$, the optimality conditions (7.36) become

$$\frac{\tilde{u}(\Phi^*)}{\delta_1} = -\phi^*, \frac{\tilde{u}(\Phi^*)}{\delta_2} = -\psi^*$$

and then, we can obtain the optimal control by the resolution of the following coupled system:

$$\begin{aligned} &\frac{\partial u^*}{\partial t} + A(t)u^* + a_0 u^* = f - \frac{\tilde{u}^*}{\delta_1} \quad \text{on } \mathcal{Q}, \\ &-\frac{\partial \tilde{u}^*}{\partial t} + A^*(t)\tilde{u}^* + a_0 \tilde{u}^* = \mathcal{C}^*(\mathcal{C}u^* - u_{\mathrm{obs}}) \quad \text{on } \mathcal{Q}, \\ &\alpha \frac{\partial u^*}{\partial \eta_A} + \beta u^* = -\frac{\alpha \tilde{u}^*}{\delta_2} \quad \text{on } \Sigma, \\ &\alpha \frac{\partial \tilde{u}^*}{\partial \eta_{A^*}} + \beta \tilde{u}^* = 0 \quad \text{on } \Sigma, \\ &u^*(0) = u_0, \quad \tilde{u}^*(T) = 0 \quad \text{on } \Omega. \end{aligned} \tag{7.37}$$

We can use the same method in order to study the case of the final observation. ◊

7.4.2 Pointwise Observations

In this subsection we consider pointwise observations (*i.e.*, concentrated on internal points in the domain Ω) and a distributed control.

Let $(x_i)_{i=1,\ldots,d}$ be given d points of the domain Ω. We suppose that the observation is as the form $(u(x_i,t))_{i=1,\ldots,d}$ (if $u(x_i,t)$ has a sense).

Suppose now that the operator A is independent of time, the space $\Omega \subset \mathbb{R}^n$ with $n \leq 3$, the boundary Γ of Ω and the coefficients a_{ij} of the operator A are sufficiently regular, and the function $\psi = 0$. Moreover, if $u_0 \in V$, then the solution u of (7.27) is in $H^{2,1}(\mathcal{Q}) = L^2(0,T;H^2(\Omega)) \cap H^1(0,T;L^2(\Omega))$ and satisfies

$$\| u \|_{H^{2,1}(\mathcal{Q})} \leq c(\| f \|_{L^2(\mathcal{Q})} + \| \phi \|_{L^2(\mathcal{Q})} + \| u_0 \|_V). \tag{7.38}$$

Moreover, since $\Omega \subset \mathbb{R}^n$, $n \leq 3$, we have that $H^2(\Omega) \subset \mathcal{C}(\overline{\Omega})$ (the space of continuous functions on $\overline{\Omega}$) and then $u(x_i,t) \in L^2(0,T)$, for $i = 1,\ldots,d$, and

$$\int_0^T | u(x_i,t) |^2 \,\mathrm{d}t \leq c(\| f \|^2_{L^2(\mathcal{Q})} + \| \phi \|^2_{L^2(\mathcal{Q})} + \| u_0 \|^2_V). \tag{7.39}$$

Let $\mathcal{K}$ be a given non-empty, closed and convex subset of $L^2(\mathcal{Q})$ and the observation $u_{\mathrm{obs}} = (u_{i,\mathrm{obs}})_{i=1,\ldots,d} \in (L^2(0,T))^d$. Our problem is then, to find $\phi \in \mathcal{K}$ such that the cost functional

$$J(\phi) = \frac{1}{2}\sum_{i=1}^{d} \| u(x_i,.) - u_{i,\mathrm{obs}} \|^2_{L^2(0,T)} + \frac{\gamma}{2} \| \phi \|^2_{L^2(\mathcal{Q})} \tag{7.40}$$

is minimized with respect to ϕ subject to the problem (7.27) (with $\psi = 0$), where $\gamma > 0$ is a fixed constant.

We have the existence and the uniqueness of the optimal control ϕ^* which is characterized by the following optimality conditions, given by

$$\iint_{\mathcal{Q}} (\tilde{u}(\phi^*) + \gamma\phi^*)(\phi - \phi^*)\mathrm{d}x\mathrm{d}t \geq 0 \quad \forall \phi \in \mathcal{K}, \tag{7.41}$$

where $\tilde{u}(\phi^*)$ is the solution of the following adjoint problem corresponding to problem (7.27) (with $\psi = 0$).

$$-\frac{\partial \tilde{u}}{\partial t} + A^*\tilde{u} + a_0\tilde{u} = \sum_{i=1}^{d}(u(\phi)(x_i,.) - u_{i,\mathrm{obs}})\delta_{x_i} \text{ on } \mathcal{Q},$$

subject to the linear Robin boundary condition

$$\alpha\frac{\partial \tilde{u}}{\partial \eta_{A^*}} + \beta\tilde{u} = 0 \text{ on } \Sigma, \tag{7.42}$$

with the final condition

$$\tilde{u}(t = T) = 0 \text{ on } \Omega,$$

where A^* is the adjoint of the operator A, δ_{x_i} is the usual Dirac function at point x_i and $h(t)\delta_{x_i}$ is the distribution

$$\mathcal{S} \in \mathcal{D}(\mathcal{Q}) \longrightarrow \int_0^T h(t)\mathcal{S}(x_i,t)\mathrm{d}t.$$

The problem (7.42) admits a unique solution $\tilde{u}$ on $L^2(\mathcal{Q})$, given by transposition (see, *e.g.*, Lions and Magenes [204]). For this let $g \in L^2(\Omega)$ be given and let $w \in H^{2,1}(\mathcal{Q})$ be the unique solution of

$$\begin{aligned} &\frac{\partial w}{\partial t} + Aw + a_0 w = g \quad \text{on } \mathcal{Q}, \\ &\alpha \frac{\partial w}{\partial \eta_A} + \beta w = 0 \quad \text{on } \Sigma, \\ &w(0) = 0 \quad \text{on } \Omega. \end{aligned} \tag{7.43}$$

Moreover, we have that

$$\| w \|^2_{H^{2,1}(\mathcal{Q})} \leq c \| g \|^2_{L^2(\mathcal{Q})} \tag{7.44}$$

and then (since $H^2(\Omega) \subset \mathcal{C}(\overline{\Omega})$ and therefore $w(x_i, .) \in L^2(0,T)$)

$$\int_0^T | w(x_i, t) |^2 \,\mathrm{d}t \leq c \| g \|^2_{L^2(\mathcal{Q})} . \tag{7.45}$$

Multiplying now (7.42) by w, integrating over $\mathcal{Q}$, integrating by parts in time, using Green's formula and taking into account the boundary, initial and final conditions, we obtain

$$\int_0^T \int_\Omega \tilde{u} g \mathrm{d}x\mathrm{d}t = \sum_{i=1}^d \int_0^T (u(\phi)(x_i, t) - u_{i,\text{obs}}(t)) w(x_i, t) \mathrm{d}t. \tag{7.46}$$

Because of the relations (7.45), (7.39) and Hölder's inequality, we have that

$$\begin{aligned} &| \sum_{i=1}^d \int_0^T (u(\phi)(x_i, t) - u_{i,\text{obs}}(t)) w(x_i, t) \mathrm{d}t | \\ &\leq c_1 \| g \|_{L^2(\mathcal{Q})} \sum_{i=1}^d (\| u(\phi)(x_i, \cdot) \|_{L^2(0,T)} + \| u_{i,\text{obs}} \|_{L^2(0,T)}) \\ &\leq c_2 \| g \|_{L^2(\mathcal{Q})} \end{aligned} \tag{7.47}$$

and then the right-hand side of (7.46) is a linear and continuous form on $H^{2,1}(\mathcal{Q})$. Consequently, Problem (7.46) admits a unique solution $\tilde{u}$ on $L^2(\mathcal{Q})$.

Remark 7.8. In the case without constraints, *i.e.*, $\mathcal{K} = L^2(\mathcal{Q})$, the optimality conditions (7.41) become

$$\frac{\tilde{u}(\phi^*)}{\gamma} = -\phi^*$$

and then, we can obtain the optimal control by the resolution of the following coupled system:

$$
\begin{aligned}
&\frac{\partial u^*}{\partial t} + Au^* + a_0 u^* = f - \frac{\tilde{u}^*}{\gamma} \quad \text{on } \mathcal{Q},\\
&-\frac{\partial \tilde{u}^*}{\partial t} + A^*\tilde{u}^* + a_0\tilde{u}^* = \sum_{i=1}^{d} (u^*(x_i, .) - u_{i,\text{obs}})\delta_{x_i} \quad \text{on } \mathcal{Q},\\
&\alpha \frac{\partial u^*}{\partial \eta_A} + \beta u^* = 0 \quad \text{on } \Sigma,\\
&\alpha \frac{\partial \tilde{u}^*}{\partial \eta_{A^*}} + \beta \tilde{u}^* = 0 \quad \text{on } \Sigma,\\
&u^*(0) = u_0, \quad \tilde{u}^*(T) = 0 \quad \text{on } \Omega.
\end{aligned}
\tag{7.48}
$$

◇

7.4.3 Pointwise Controls

In this subsection we consider pointwise controls (*i.e.*, concentrated on internal points in the domain Ω) and a distributed observation.

Let $(x_i)_{i=1,\dots,d}$ be given d points of the domain Ω, and we suppose that the control function ϕ is as the form

$$
\phi = \sum_{i=1}^{d} \varphi_i \delta_{x_i}
$$

where $\varphi_i \in L^2(0,T)$, for $i = 1, \dots, d$ and δ_{x_i} is the usual Dirac function at point x_i.

Similarly as in the previous subsection, we suppose that the operator A is independent of time, the space $\Omega \subset \mathbb{R}^n$ with $n \leq 3$, the boundary Γ of Ω and the coefficients a_{ij} of the operator A are sufficiently regular, and the function $\psi = 0$.

By using the same technique (the transposition technique) as to prove the unique solution of the adjoint problem (7.42), we obtain the existence and the uniqueness of the solution u in $L^2(\mathcal{Q})$.

Let $\mathcal{K}$ be a given non-empty, closed and convex subset of $(L^2(0,T))^d$ and the observation $u_{\text{obs}} \in L^2(\mathcal{Q})$. Our problem is then, to find $(\varphi_i)_{i=1,\dots,d} \in \mathcal{K}$ such that the cost functional

$$
J(\phi) = \frac{1}{2} \parallel u(\phi) - u_{\text{obs}} \parallel^2_{L^2(\mathcal{Q})} + \frac{1}{2} \sum_{i=1}^{d} \gamma_i \parallel \varphi_i \parallel^2_{L^2(0,T)} \tag{7.49}
$$

is minimized with respect to ϕ subject to the problem (7.27) (with $\psi = 0$), where $\gamma_i > 0$, for $i = 1, \dots, d$ are fixed constants.

We have the existence and the uniqueness of the optimal control $\phi^* = \sum_{i=1}^{d} \varphi_i^* \delta_{x_i}$ which is characterized by the following optimality conditions:

$$
\sum_{i=1}^{d} \int_0^T (\tilde{u}(\phi^*)(x_i,t) + \gamma_i \varphi_i^*(t))(\varphi_i(t) - \varphi_i^*(t))\text{d}t \geq 0 \quad \forall (\varphi_i)_{i=1,d} \in \mathcal{K}, \tag{7.50}
$$

where $\tilde{u} = \tilde{u}(\phi^*)$ is the solution of the following adjoint problem corresponding to problem (7.27) (with $\psi = 0$)

$$\begin{aligned}
&-\frac{\partial \tilde{u}}{\partial t} + A^*\tilde{u} + a_0\tilde{u} = u(\phi^*) - u_{\text{obs}} \quad \text{on } \mathcal{Q}, \\
&\text{subject to the linear Robin boundary condition} \\
&\alpha \frac{\partial \tilde{u}}{\partial \eta_{A^*}} + \beta \tilde{u} = 0 \quad \text{on } \Sigma, \\
&\text{with the final condition} \\
&\tilde{u}(t = T) = 0 \quad \text{on } \Omega,
\end{aligned} \tag{7.51}$$

where A^* is the adjoint of the operator A.

Since $u(\phi) - u_{\text{obs}} \in L^2(\mathcal{Q})$ and the final condition (*i.e.*, the null function) is in V, then Problem (7.51) admits a unique solution $\tilde{u}$ in $H^{2,1}(\mathcal{Q})$. Moreover, since $H^2(\Omega) \subset C(\overline{\Omega})$, we have $\tilde{u}(x_i, t) \in L^2(0, T)$, for $i = 1, \ldots, d$.

Remark 7.9. In the case without constraints, *i.e.*, $\mathcal{K} = (L^2(0, T))^d$, the optimality conditions (7.50) become

$$\frac{\tilde{u}(\phi^*)(x_i, .)}{\gamma_i} = -\varphi_i^*$$

and then we can obtain the optimal control by the resolution of the following coupled system:

$$\begin{aligned}
&\frac{\partial u^*}{\partial t} + Au^* + a_0 u^* = f - \sum_{i=1}^{d} \frac{\tilde{u}^*(x_i, .)\delta_{x_i}}{\gamma_i} \quad \text{on } \mathcal{Q}, \\
&-\frac{\partial \tilde{u}^*}{\partial t} + A^*\tilde{u}^* + a_0\tilde{u}^* = u^* - u_{\text{obs}} \quad \text{on } \mathcal{Q}, \\
&\alpha \frac{\partial u^*}{\partial \eta_A} + \beta u^* = 0 \quad \text{on } \Sigma, \\
&\alpha \frac{\partial \tilde{u}^*}{\partial \eta_{A^*}} + \beta \tilde{u}^* = 0 \quad \text{on } \Sigma, \\
&u^*(0) = u_0, \quad \tilde{u}^*(T) = 0 \quad \text{on } \Omega.
\end{aligned} \tag{7.52}$$

◇

7.4.4 Boundary Controls and Boundary Observations

In this subsection we consider a problem when controls and observations act both on the boundary during a time T. We denote, in the system (7.27), by g the sum of f and ϕ, *i.e.*, $g := f + \phi$, and we suppose that the parameter $\alpha \neq 0$. The control is the function ψ defined on Σ and the observation is given by

$$\mathcal{C}u = \mathcal{D}(u|_\Sigma), \tag{7.53}$$

where $u|_\Sigma$ is the trace of u on Σ and $\mathcal{D} \in \mathcal{L}(L^2(\Sigma); L^2(\Sigma))$ (then Λ is the identity operator).

Example 7.10. If $\mathcal{D}$ is the identity, then we observe u on Σ. If $\mathcal{D}$ is the multiplication by the characteristic function of the subdomain Σ_1 of Σ, then we observe u on Σ_1. ♣

Let $\mathcal{K}$ be a given non-empty, closed and convex subset of $L^2(\Sigma)$ and the observation $u_{\text{obs}} \in L^2(\Sigma)$. Our problem is then, to find $\psi \in \mathcal{K}$ such that the cost functional

$$J(\psi) = \frac{1}{2} \parallel \mathcal{D}(u|_\Sigma) - u_{\text{obs}} \parallel^2_{L^2(\Sigma)} + \frac{\gamma}{2} \parallel \psi \parallel^2_{L^2(\Sigma)} \tag{7.54}$$

is minimized with respect to ψ subject to the problem (7.27), where $\gamma > 0$ is a fixed constant.

We have the existence and the uniqueness of the optimal control $\psi^* \in \mathcal{K}$, which is characterized by the optimality condition

$$\iint_\Sigma (\tilde{u}(\psi^*) + \gamma\psi^*)(\psi - \psi^*) \mathrm{d}\Gamma \mathrm{d}t \geq 0 \quad \forall \psi \in \mathcal{K}, \tag{7.55}$$

where $\tilde{u}(\psi^*)$ is the solution of the following adjoint problem corresponding to the primal problem (7.27):

$$\begin{aligned}
&-\frac{\partial \tilde{u}}{\partial t} + A^*(t)\tilde{u} + a_0\tilde{u} = 0 \quad \text{on } \mathcal{Q}, \\
&\text{subject to the linear Robin boundary condition} \\
&\alpha \frac{\partial \tilde{u}}{\partial \eta_{A^*}} + \beta\tilde{u} = \alpha \mathcal{D}^*(\mathcal{D}(u(\psi)|_\Sigma) - u_{\text{obs}}) \quad \text{on } \Sigma, \\
&\text{with the final condition} \\
&\tilde{u}(t = T) = 0 \quad \text{on } \Omega,
\end{aligned} \tag{7.56}$$

where A^* is the adjoint of the operator A and $\mathcal{D}^*$ is the adjoint of $\mathcal{D}$.

Since the function $\mathcal{D}^*(\mathcal{D}(u(\psi)|_\Sigma) - u_{\text{obs}})$ is in $L^2(\Sigma)$ then the problem (7.56) admits a unique solution in $\mathcal{W} \cap \mathcal{C}([0, T]; H)$ (in the weak formulation sense).

Remark 7.11. In the case without constraints, *i.e.*, $\mathcal{K} = L^2(\Sigma)$, the optimality conditions (7.50) become

$$\frac{\tilde{u}(\psi^*)|_\Sigma}{\gamma} = -\psi^*$$

and then, we can obtain the optimal control by the resolution of the coupled system:

$$\begin{aligned}
&\frac{\partial u^*}{\partial t} + Au^* + a_0u^* = g \quad \text{on } \mathcal{Q}, \\
&-\frac{\partial \tilde{u}^*}{\partial t} + A^*\tilde{u}^* + a_0\tilde{u}^* = 0 \quad \text{on } \mathcal{Q}, \\
&\alpha \frac{\partial u^*}{\partial \eta_A} + \beta u^* = -\frac{\alpha \tilde{u}(\psi^*)|_\Sigma}{\gamma} \quad \text{on } \Sigma, \\
&\alpha \frac{\partial \tilde{u}^*}{\partial \eta_{A^*}} + \beta\tilde{u}^* = \alpha \mathcal{D}^*(\mathcal{D}(u(\psi^*)|_\Sigma) - u_{\text{obs}}) \quad \text{on } \Sigma, \\
&u^*(0) = u_0, \quad \tilde{u}^*(T) = 0 \quad \text{on } \Omega.
\end{aligned} \tag{7.57}$$

◇

7.4.5 Data Assimilation Problem and Initial Condition Control

In this subsection we consider an application when the control is in the initial condition and the observation acts on the boundary. We denote, in the system (7.27), by g the sum of f and ϕ, *i.e.*, $g := f + \phi$, and we suppose that the parameter $\alpha \neq 0$ and that the initial condition u_0 is the sum of v_0 (assumed to be known) and θ (assumed to be not well known), *i.e.*, $u_0 = v_0 + \theta$. The control is the function θ defined on Ω and the observation is given by

$$\mathcal{C}u = \mathcal{D}(u|_\Sigma), \tag{7.58}$$

where $u|_\Sigma$ is the trace of u on Σ and $\mathcal{D} \in \mathcal{L}(L^2(\Sigma); L^2(\Sigma))$ (then Λ is the identity operator).

Remark 7.12. For the types of boundary observation operator $\mathcal{D}$, see, *e.g.*, Example 7.10. ◇

Let $\mathcal{K}$ be a given non-empty, closed and convex subset of $L^2(\Omega)$ and the observation $u_{\text{obs}} \in L^2(\Sigma)$. Our problem is then, to find $\psi \in \mathcal{K}$ such that the cost functional

$$J(\theta) = \frac{1}{2} \parallel \mathcal{D}(u|_\Sigma) - u_{\text{obs}} \parallel^2_{L^2(\Sigma)} + \frac{\gamma}{2} \parallel \theta \parallel^2_{L^2(\Omega)} \tag{7.59}$$

is minimized with respect to θ subject to the problem (7.27), where $\gamma > 0$ is a fixed constant.

We have the existence and the uniqueness of the optimal control $\theta^* \in \mathcal{K}$, which is characterized by the optimality condition

$$\int_\Omega (\tilde{u}(\theta^*) + \gamma\theta^*)(\theta - \theta^*)\mathrm{d}x \geq 0 \quad \forall \theta \in \mathcal{K}, \tag{7.60}$$

where $\tilde{u}(\theta^*)$ is the solution of the following adjoint problem corresponding to the primal problem (7.27):

$$\begin{aligned}
&-\frac{\partial \tilde{u}}{\partial t} + A^*(t)\tilde{u} + a_0\tilde{u} = 0 \quad \text{on } Q, \\
&\text{subject to the linear Robin boundary condition} \\
&\alpha\frac{\partial \tilde{u}}{\partial \eta_{A^*}} + \beta\tilde{u} = \alpha\mathcal{D}^*(\mathcal{D}(u(\theta)|_\Sigma) - u_{\text{obs}}) \quad \text{on } \Sigma, \\
&\text{with the final condition} \\
&\tilde{u}(t = T) = 0 \quad \text{on } \Omega,
\end{aligned} \tag{7.61}$$

where A^* is the adjoint of the operator A and $\mathcal{D}^*$ is the adjoint of the operator $\mathcal{D}$.

Since the function $\mathcal{D}^*(\mathcal{D}(u(\psi)|_\Sigma) - u_{\text{obs}})$ is in $L^2(\Sigma)$ then the problem (7.61) admits a unique solution in $\mathcal{W} \cap \mathcal{C}([0, T]; H)$ (in the weak formulation sense).

Remark 7.13. In the case without constraints, *i.e.*, $\mathcal{K} = L^2(\Omega)$, the optimality conditions (7.60) become

$$\frac{\tilde{u}(\theta^*)}{\gamma} = -\theta^*$$

and then, we can obtain the optimal control by the resolution of the following coupled system:

$$\begin{aligned}
&\frac{\partial u^*}{\partial t} + Au^* + a_0 u^* = g \quad \text{on } \mathcal{Q},\\
&-\frac{\partial \tilde{u}^*}{\partial t} + A^* \tilde{u}^* + a_0 \tilde{u}^* = 0 \quad \text{on } \mathcal{Q},\\
&\alpha \frac{\partial u^*}{\partial \eta_A} + \beta u^* = \alpha \psi \quad \text{on } \Sigma,\\
&\alpha \frac{\partial \tilde{u}^*}{\partial \eta_{A^*}} + \beta \tilde{u}^* = \alpha \mathcal{D}^* (\mathcal{D}(u(\psi^*)|_\Sigma) - u_{\text{obs}}) \quad \text{on } \Sigma,\\
&u^*(0) = v_0 - \frac{\tilde{u}(\theta^*)}{\gamma}, \quad \tilde{u}^*(T) = 0 \quad \text{on } \Omega.
\end{aligned} \tag{7.62}$$

◇

Remark 7.14. We can consider other control and observation cases, the technique remains the same (see, *e.g.*, Lions [202] for other linear problems). ◇

In the next two sections we study two classes of non-linear control problems. More precisely, we analyze a class of bilinear problems (the primal problem is linear on the state variable when the control is fixed, and conversely) and a class of non-linear evolutive problems (which arise from the modeling, for example, of pollutants in liquid or atmospheric systems).

7.5 Parameter Estimations and Bilinear Control Problems

In this section we present a first problem of non-linear control, namely an estimate parameter problem. The problem presented here is a simple but non-trivial application of the more general problems of estimate parameter models (see below for different biological and physical models). The problem is treated as an optimal control form with boundary observations. The control problems arising in this context are bilinear (this adds difficulties from a mathematical viewpoint).[1]

7.5.1 State Problem

For this, we consider the linear primal problem used in Section 7.4, precisely the problem (7.27). In System (7.27) we use g to denote the sum of f and

[1] For references on bilinear problems, see Section 8.5.

ϕ, *i.e.*, $g := f + \phi$, and we suppose that the parameter $\alpha = 1$ and that the function $\theta := a_0$ is supposed to be not well known. The control is then the function θ defined on $\mathcal{Q}$ and the observation is given by

$$\mathcal{C}u = \mathcal{D}(u|_\Sigma), \tag{7.63}$$

where $u|_\Sigma$ is the trace of u on Σ and $\mathcal{D} \in \mathcal{L}(L^2(\Sigma); L^2(\Sigma))$.

Then the problem (7.27) becomes

$$\frac{\partial u}{\partial t} + A(t)u = g - \theta(x,t)u \quad \text{on } \mathcal{Q},$$

subject to the linear Robin boundary condition

$$\frac{\partial u}{\partial \eta_A} + \beta u = \psi \quad \text{on } \Sigma, \tag{7.64}$$

with the initial condition

$$u(0) = u_0 \quad \text{on } \Omega,$$

under the pointwise constraint

$$r_1 \leq \theta \leq r_2 \quad \text{a.e. in } \mathcal{Q}, \tag{7.65}$$

where the interval $[r_1, r_2]$ contains 0.

The system (7.64) is linear on the state variable when the control is fixed, and linear on the control when the state variable is fixed, but the solution of (7.64) is depending non-linearily on the control θ. Consequently, the control problem is non-linear. Let $U_{\text{ad}} := \{\theta \in L^2(\mathcal{Q}) : \ r_1 \leq \theta \leq r_2 \quad \text{a.e. in } \mathcal{Q}\}$ and $\mathcal{F} : \ \theta \in U_{\text{ad}} \longrightarrow \mathcal{W} \cap \mathcal{C}([0,T]; H)$ such that $u = \mathcal{F}(\theta)$ is the unique solution of (7.64), corresponding to θ.

Remark 7.15. (i) It is clear that U_{ad} is a subset of $L^\infty(\mathcal{Q})$.
(ii) According to Theorem 6.8 and Lemma 6.6 we have, for any given $\theta \in U_{\text{ad}}$, the following estimate:

$$\| u \|^2_{\mathcal{W}\cap\mathcal{C}([0,T];H)} \leq C(\| g \|^2_{L^2(\mathcal{Q})} + \| u_0 \|^2_{L^2(\Omega)} + \| \psi \|^2_{L^2(\Sigma)}). \tag{7.66}$$

(iii) For types of the boundary observation operator $\mathcal{D}$, see, *e.g.*, Example 7.10. ♢

Our problem is then, to find $\theta \in U_{\text{ad}}$ such that the cost functional

$$J(\theta) = \frac{1}{2} \| \mathcal{D}(u|_\Sigma) - u_{\text{obs}} \|^2_{L^2(\Sigma)} + \frac{\gamma}{2} \| \theta \|^2_{L^2(\mathcal{Q})} \tag{7.67}$$

is minimized with respect to θ subject to the problem (7.64), where $\gamma > 0$ is a fixed constant and the function $u_{\text{obs}} \in L^2(\Sigma)$ is a given observation.

7.5.2 Existence of Optimal Solutions

Now, let us study the following existence of an optimal solution.

Theorem 7.16. *There exists $\theta^* \in U_{\rm ad}$ and $u^* \in \mathcal{W}$ such that θ^* is defined by (7.67) and $u^* = \mathcal{F}(\theta^*)$ is a solution of (7.64).*

Proof. Let $\theta_k \in U_{\rm ad}$ be a minimizing sequence of J, *i.e.*,

$$\liminf_{k \longrightarrow \infty} J(\theta_k) = \inf_{\theta \in L^2(\mathcal{Q})} J(\theta).$$

Then, according to the nature of the cost function J, we can deduce that θ_k is uniformly bounded in $U_{\rm ad}$ and we can extract from θ_k a subsequence also denoted by θ_k such that $\theta_k \rightharpoonup \theta$ weakly in $U_{\rm ad}$. Therefore, $u_k = \mathcal{F}(\theta_k)$ is uniformly bounded in $\mathcal{W}$. Indeed, by using the weak form of (7.64), we can deduce easily that

$$\frac{\mathrm{d} \parallel u_k \parallel^2_{L^2(\Omega)}}{2\mathrm{d}t} + a(t; u_k, u_k) + \int_\Omega \theta_k \mid u_k \mid^2 \mathrm{d}x + \beta \int_\Gamma \mid u_k \mid^2 \mathrm{d}\Gamma = \int_\Omega g u_k \mathrm{d}x.$$

Then, by integrating with respect to time and because of the uniform boundedness of θ_k in $L^\infty(\mathcal{Q})$, we have that

$$\parallel u_k(t) \parallel^2_{L^2(\Omega)} + \int_0^t \parallel u_k \parallel^2_V \mathrm{d}s + \int_0^t \parallel u_k \parallel^2_{L^2(\Gamma)} \mathrm{d}s \le c_1 \int_0^t \parallel u_k \parallel^2_{L^2(\Omega)} \mathrm{d}s + c_2.$$

By using Gronwall's lemma, we can deduce first that u_k is uniformly bounded in $L^\infty(0, T; L^2(\Omega))$ and finally that u_k is uniformly bounded in $L^2(0, T; V)$. Using the previous result and Equation (7.64) we obtain easily that $\partial u_k / \partial t$ is uniformly bounded in $L^2(0, T; V')$ and then u_k is uniformly bounded in $\mathcal{W}$. According to Lemma 6.6, the injection of $\mathcal{W}$ into $L^2(\mathcal{Q})$ is compact. Therefore, this result makes it possible to extract from u_k a subsequence also denoted by u_k such that

$$\begin{aligned} u_k &\rightharpoonup \tilde{u} \text{ weakly in } L^2(0, T; V), \\ u_k &\longrightarrow \tilde{u} \text{ strongly in } L^2(\mathcal{Q}), \\ \theta_k &\rightharpoonup \theta \text{ weakly in } L^2(\mathcal{Q}) \text{ and } \theta \in U_{\rm ad}. \end{aligned} \tag{7.68}$$

Prove now that $u_k \theta_k \longrightarrow u\theta$ weakly in $L^2(\mathcal{Q})$. Since $u_k\theta_k - u\theta = (u_k - u)\theta_k + u(\theta_k - \theta)$, then, according to the first and second parts of (7.68), we obtain the result. We can prove easily that $\tilde{u}$ is a solution of (7.64) with a parameter θ and according to the uniqueness of the solution of (7.64), we have then $\tilde{u} = u = \mathcal{F}(\theta)$.

Finally, since the norm is lower semi-continuous we have that the map $J : \theta \longrightarrow J(\theta)$ is lower semi-continuous and then the function θ is an optimal solution, *i.e.*,

$$\inf_{\theta \in U_{\rm ad}} J(\theta) = \liminf_k J(\theta_k) = J(\theta^*).$$

□

7.5.3 First-order Optimality Conditions

Before giving the optimality conditions for an optimal solution, we study the following G-differentiability results.

Proposition 7.17. *The function $\mathcal{F} : \theta \longrightarrow u = \mathcal{F}(\theta)$ solution of (7.64) is G-differentiable, with respect to θ from U_{ad} to $\mathcal{W} \cap \mathcal{C}([0,T];H)$ where the G-derivative $\mathcal{F}'(\theta) : h \longrightarrow w = \mathcal{F}'(\theta)h$ is the unique solution in $\mathcal{W} \cap \mathcal{C}([0,T];H)$ of the following linear parabolic problem:*

$$\begin{aligned} &\frac{\partial w}{\partial t} + A(t)w = -\theta w - hu \quad \text{on } Q \\ &\frac{\partial w}{\partial \eta_A} + \beta w = 0 \quad \text{on } \Sigma, \\ &w(0) = 0 \quad \text{on } \Omega. \end{aligned} \tag{7.69}$$

Proof. Problem (7.69) is similar to Problem (7.64) (where the function $-hu$ is in $L^2(Q)$, since $u \in L^2(Q)$ and $h \in L^\infty(Q)$, plays the role of g in (7.64)). Then Problem (7.69) admits a unique solution $w \in \mathcal{W} \cap \mathcal{C}([0,T];H)$ such that (because of the estimate (7.66))

$$\| w \|_{\mathcal{W}\cap\mathcal{C}([0,T];H)} \leq C \| u \|_{L^2(Q)} .$$

Let $(\theta, h) \in U_{\text{ad}} \times L^\infty(Q)$ and $\epsilon > 0$ such that $\epsilon h + \theta \in U_{\text{ad}}$. Let $u = \mathcal{F}(\theta)$ and $u_\epsilon = \mathcal{F}(\theta + \epsilon h)$.

Step 1: Prove that

$$u_\epsilon \longrightarrow u \text{ strongly in } \mathcal{W} \cap \mathcal{C}([0,T];H) \text{ as } \epsilon \longrightarrow 0.$$

Let $v_\epsilon := u_\epsilon - u$, obviously, v_ϵ satisfies

$$\begin{aligned} &\frac{\partial v_\epsilon}{\partial t} + A(t)v_\epsilon = -(\theta + \epsilon h)v_\epsilon - \epsilon hu \quad \text{on } Q, \\ &\frac{\partial v_\epsilon}{\partial \eta_A} + \beta v_\epsilon = 0 \quad \text{on } \Sigma, \\ &v_\epsilon(0) = 0 \quad \text{on } \Omega. \end{aligned} \tag{7.70}$$

Since $h \in L^\infty(Q)$ and $u \in L^2(Q)$ then $g_\epsilon := -\epsilon hu$ is in $L^2(Q)$. Consequently, since $(\theta + \epsilon h) \in L^\infty(Q)$, we have that, there exists a constant $C > 0$ (independent of ϵ) such that

$$\| v_\epsilon \|_{\mathcal{W}\cap\mathcal{C}([0,T];H)} \leq C \| g_\epsilon \|_{L^2(Q)}$$

and then

$$\| v_\epsilon \|_{\mathcal{W}\cap\mathcal{C}([0,T];H)} \leq \epsilon C_1 \| u \|_{L^2(Q)} .$$

Consequently, $v_\epsilon \longrightarrow 0$ strongly in $\mathcal{W} \cap \mathcal{C}([0,T];H)$ as $\epsilon \longrightarrow 0$.

Step 2: Prove now that

$$w_\epsilon := \frac{v_\epsilon}{\epsilon} = \frac{(u_\epsilon - u)}{\epsilon} \longrightarrow w \text{ strongly in } \mathcal{W} \cap \mathcal{C}([0,T];H) \text{ as } \epsilon \longrightarrow 0.$$

Obviously, $\tilde{w}_\epsilon := w_\epsilon - w$ satisfies

$$\begin{aligned} &\frac{\partial \tilde{w}_\epsilon}{\partial t} + A(t)\tilde{w}_\epsilon = -\theta\tilde{w}_\epsilon - hv_\epsilon \text{ on } \mathcal{Q}, \\ &\frac{\partial \tilde{w}_\epsilon}{\partial \eta_A} + \beta\tilde{w}_\epsilon = 0 \text{ on } \Sigma, \\ &\tilde{w}_\epsilon(0) = 0 \text{ on } \Omega. \end{aligned} \tag{7.71}$$

The problem (7.71) is similar to problem (7.70), so

$$\| \tilde{w}_\epsilon \|_{\mathcal{W}\cap\mathcal{C}([0,T];H)} \leq C \| v_\epsilon \|_{L^2(\mathcal{Q})}$$

and then $\| \tilde{w}_\epsilon \|_{\mathcal{W}\cap\mathcal{C}([0,T];H)} \longrightarrow 0$ as $\epsilon \longrightarrow 0$ (since $\| v_\epsilon \|_{L^2(\mathcal{Q})} \longrightarrow 0$ as $\epsilon \longrightarrow 0$). This completes the proof. □

With the aid of this proposition, we can easily show the first-order necessary conditions (optimality conditions).

Theorem 7.18. *Let $\theta^* \in U_{\rm ad}$ be an optimal control defined by (7.67) and $u^* \in \mathcal{W}\cap\mathcal{C}([0,T];H)$ be the optimal state such that $u^* = \mathcal{F}(\theta^*)$ is the solution of (7.64), with the parameter θ^*. Then there exists a unique solution $\tilde{u}^* \in \mathcal{W}\cap\mathcal{C}([0,T];H)$ for the following adjoint problem corresponding to the primal problem (7.64):*

$$\begin{aligned} &-\frac{\partial \tilde{u}}{\partial t} + A^*(t)\tilde{u} + \theta^*\tilde{u} = 0 \quad \text{on } \mathcal{Q}, \\ &\text{subject to the linear Robin boundary condition} \\ &\frac{\partial \tilde{u}}{\partial \eta_{A^*}} + \beta\tilde{u} = \mathcal{D}^*(\mathcal{D}(u^*|_\Sigma) - u_{\rm obs}) \quad \text{on } \Sigma, \\ &\text{with the final condition} \\ &\tilde{u}(t=T) = 0 \quad \text{on } \Omega, \end{aligned} \tag{7.72}$$

where A^* is the adjoint of the operator A and $\mathcal{D}^*$ is the adjoint of $\mathcal{D}$. Moreover,

$$\begin{aligned} &\theta^* = \max\left(r_1, \min\left(\frac{u^*\tilde{u}^*}{\gamma}, r_2\right)\right) \\ &\text{or in the variational inequality formulation} \\ &\int_0^T\int_\Omega (-u^*\tilde{u}^* + \gamma\theta^*)(\theta - \theta^*)\mathrm{d}x\mathrm{d}t \geq 0 \quad \forall\theta \in U_{\rm ad}, \end{aligned} \tag{7.73}$$

Proof. Since the function $\mathcal{D}^*(\mathcal{D}(u^*|_\Sigma) - u_{\rm obs})$ is in $L^2(\Sigma)$ then the problem (7.72) admits a unique solution in $\mathcal{W}\cap\mathcal{C}([0,T];H)$ (in the weak formulation sense).

The cost functional J is a composition of G-differentiable mappings then J is G-differentiable and in particular the G-derivative of J at the optimal point $\theta^* \in U_{\rm ad}$ is given by (for all $h \in L^\infty(\mathcal{Q})$ such that $(\theta^* + \epsilon h) \in U_{\rm ad}$ for ϵ small)

$$0 \leq J'(\theta^*) = \lim_{\epsilon \longrightarrow 0} \frac{J(\theta^* + \epsilon h) - J(\theta^*)}{\epsilon}$$
$$= \int_0^T \int_\Gamma (\mathcal{D}^*(\mathcal{D}(u|_\Sigma) - u_{\text{obs}}).w\text{d}\Gamma\text{d}s + \gamma \int_0^T \int_\Omega \theta^* h\text{d}x\text{d}t, \tag{7.74}$$

where $w = \mathcal{F}'(\theta^*)\theta$.

Multiplying (7.69) by $\tilde{u}^*$ and integrating over $\mathcal{Q}$, this gives (using Green's formula and integrating by parts in time)

$$\int_0^T \int_\Omega (-\frac{\partial \tilde{u}^*}{\partial t} + A^*(t)\tilde{u}^* + \theta^*\tilde{u}^*)w\text{d}x\text{d}t + \int_0^T \int_\Gamma (\frac{\partial \tilde{u}^*}{\partial \eta_{A^*}} + \beta\tilde{u}^*)w\text{d}\Gamma\text{d}t$$
$$= - \int_0^T \int_\Omega hu^*\tilde{u}^*\text{d}x\text{d}t + \int_\Omega w(0)\tilde{u}^*(0)\text{d}x - \int_\Omega w(T)\tilde{u}^*(T)\text{d}x.$$

Since $\tilde{u}^*$ is the solution of (7.72) and $w(0) = 0$, therefore we can deduce that

$$\int_0^T \int_\Gamma \mathcal{D}^*(\mathcal{D}(u^*|_\Sigma) - u_{\text{obs}}).w\text{d}\Gamma\text{d}s = - \int_0^T \int_\Omega hu^*\tilde{u}^*\text{d}x\text{d}t \tag{7.75}$$

and then (according to (7.74))

$$0 \leq \int_0^T \int_\Omega (\gamma\theta^* - u^*\tilde{u}^*)h\text{d}x\text{d}t. \tag{7.76}$$

By using a standard control argument concerning the sign of the variation h (depending on the size of θ^*), we obtain that

$$\theta^* = \max\left(r_1, \min\left(\frac{u^*\tilde{u}^*}{\gamma}, r_2\right)\right).$$

Indeed, by taking $h = \max(r_1, \min(u^*\tilde{u}^*/\gamma, r_2)) - \theta^*$, we prove easily that $h(\gamma\theta^* - u^*\tilde{u}^*)$ is negative, and then

$$\left(\max\left(r_1, \min\left(\frac{u^*\tilde{u}^*}{\gamma}, r_2\right)\right) - \theta^*\right)\left(\theta^* - \frac{u^*\tilde{u}^*}{\gamma}\right) = 0.$$

So,

(*i*) if $r_2 \leq u^*\tilde{u}^*/\gamma$, we have $(r_2 - \theta^*)(\theta^* - u^*\tilde{u}^*/\gamma) = 0$ and then (since $\theta^* \leq r_2$),
$$\theta^* = r_2$$

(*ii*) if $r_2 \geq u^*\tilde{u}^*/\gamma$ and $r_1 \geq u^*\tilde{u}^*/\gamma$, we have $(r_1 - \theta^*)(\theta^* - u^*\tilde{u}^*/\gamma) = 0$ and then (since $\theta^* \geq r_1$)
$$\theta^* = r_1$$

(*iii*) if $r_2 \geq u^*\tilde{u}^*/\gamma$ and $r_1 \leq u^*\tilde{u}^*/\gamma$, we have $(u^*\tilde{u}^*/\gamma - \theta^*)(\theta^* - u^*\tilde{u}^*/\gamma) = 0$ and then
$$\theta^* = \frac{u^*\tilde{u}^*}{\gamma}.$$

We can conclude that $\theta^* = \max(r_1, \min(u^*\tilde{u}^*/\gamma, r_2))$.

This completes the proof. □

7.6 Non-linear Control for Non-linear Evolutive PDE Problems

In this section we analyze the full non-linear control problems. For this we consider, in this study, some non-linear evolutive problems where we can prove existence and uniqueness (under extra assumptions) theorems for the optimal control problems, and we give the optimality conditions characterizing the optimal solutions. In contrast to linear systems, the dynamic of evolutive non-linear systems obeys complicated laws that, in general, cannot be found by intuitive and direct calculations.

The main result of the section includes the existence, the uniqueness and the first-order necessary conditions of optimality for the optimal controllers. The plan of this section is as follows. First, we give the existence and the uniqueness of the state equation and obtain some *a priori* estimates. Second, we formulate the optimal control problem, prove the existence and the uniqueness of the optimal solution, and give the appropriate optimality system. Next, we consider a data assimilation problem because in many situations the initial condition is not well known. We reformulate the optimal control problem. As in previous sections, the existence, the uniqueness and the optimality conditions are described. Finally, we present an example of convection–diffusion in the case of pollutants in liquid or atmospheric systems.

7.6.1 State Problem and Assumptions

In this subsection we will be consider the non-linear parabolic partial differential equations of the form

$$\begin{aligned} &\frac{\partial u}{\partial t} + Au + F(x,t,u) + K(\varpi, u) = f \quad \text{on } \mathcal{Q} = (0,T) \times \Omega, \\ &u(0) = u_0 \quad \text{on } \Omega, \end{aligned} \tag{7.77}$$

where Ω is an open and bounded subset of $\mathbb{R}^m$ $(m \geq 1)$ sufficiently regular, $T > 0$ is a fixed constant (a given final time), A is an elliptic, selfadjoint operator, $K(\varpi, .)$ is a linear operator with ϖ a given and sufficiently regular vector field, and $F : \ \mathcal{Q} \times \mathbb{R} \longrightarrow \mathbb{R}$ is a Nemytsky operator on $L^2(\mathcal{Q})$ for which we state the following hypothesis (see, *e.g.*, Seidman and Zhou [265]):

(*i*) $F(.,0) = 0$

(*ii*) F satisfies Carathéodory conditions and a one-sided Lipschitz condition:

$$-2(F(.,u) - F(.,v))(u-v) \leq \gamma_0 \mid u - v \mid^2 \ \ \forall (u,v) \in \mathbb{R}^2 \tag{7.78}$$

(*iii*) F is differentiable with $G(.,u) = F'(.,u)$ Lipschitz continuous:

$$\mid G(.,u) - G(.,v) \mid \leq \lambda \mid u - v \mid \ \ \forall (u,v) \in \mathbb{R}^2.$$

Remark 7.19. According to (7.78) we can deduce that the operator F is bounded and continuous, and that

$$-2G(.,u) \leq \gamma_0 \ \forall u \in \mathbb{R}.$$

◇

We assume that we can introduce a reflexive Banach space $\mathcal{D}$ of functions on Ω satisfying the boundary conditions such that $\mathcal{D} \subset L^2(\Omega)$, with $\mathcal{D}$ dense in $L^2(\Omega)$ and the embedding of $\mathcal{D}$ in $L^2(\Omega)$ being compact. We can identify $L^2(\Omega)$ with its dual $L^2(\Omega)$ and then $\mathcal{D} \subset L^2(\Omega) \subset \mathcal{D}'$ where the injections are continuous and each space is dense in the following one. We assume also that the linear and continuous operator A: $\mathcal{D} \longrightarrow \mathcal{D}'$ with $\langle Av, v\rangle \geq \nu \parallel v \parallel_{\mathcal{D}}^2 \ \forall v \in \mathcal{D}$ and $\langle Au, v\rangle \leq M \parallel u \parallel_{\mathcal{D}} \parallel v \parallel_{\mathcal{D}} \ \forall u, v \in \mathcal{D}$ ($\langle .,. \rangle$ is the duality pairing on $\mathcal{D}'$ and $\mathcal{D}$, $\parallel . \parallel_{\mathcal{D}}$ is the norm on $\mathcal{D}$, $\parallel . \parallel_*$ is the dual norm on $\mathcal{D}'$ and $\nu, M > 0$ are constants). Moreover, we impose the condition that $\mathcal{D}$ is embedded in $L^4(\Omega)$, *i.e.*, $\mathcal{D} \subset L^4(\Omega)$ and

$$\exists \ c_e > 0 \text{ such that } \parallel v \parallel_{L^4(\Omega)} \leq c_e \parallel v \parallel_{\mathcal{D}} \ \ \forall v \in \mathcal{D}. \tag{7.79}$$

We introduce the following spaces: $\mathcal{H} = L^\infty(0,T;L^2(\Omega)), \mathcal{V} = L^2(0,T;\mathcal{D})$, $\mathcal{H}_1 = H^1(0,T;\mathcal{D}') = \{v : \partial v/\partial t \in L^2(0,T;\mathcal{D}')\}$, $\mathcal{W} = \mathcal{H} \cap \mathcal{V}$ and we denote $\parallel . \parallel_{\mathcal{W}} = \max(\parallel . \parallel_{\mathcal{H}}, \nu \parallel . \parallel_{\mathcal{V}})$ and by c_I the constant of the embedding map: $\mathcal{D} \longrightarrow L^2(\Omega)$, *i.e.*,

$$\parallel v \parallel_{L^2(\Omega)} \leq c_I \parallel v \parallel_{\mathcal{D}} \ \ \forall v \in \mathcal{D}. \tag{7.80}$$

Moreover, we assume that there exists a constant $\gamma_\infty \geq 0$ such that:

$$\parallel K(\varpi, v) \parallel_{L^2(\Omega)} \leq \sqrt{\nu\gamma_\infty} \parallel v \parallel_{\mathcal{D}} \ \ \forall v \in \mathcal{D}. \tag{7.81}$$

Remark 7.20. (*i*) The parameter γ_∞, introduced in (7.81), is depending on the norm of the given field ϖ.
(*ii*) If one has $\langle Av, v\rangle + \theta \parallel v \parallel_{L^2(\Omega)}^2 \geq \nu \parallel v \parallel_{\mathcal{D}}^2 \ \forall v \in \mathcal{D}$, then we can shift the term θv from the expression of the operator F to the expression of the Av. Consequently, the value of γ_0 in assumption (*ii*) of (7.78) is modified; however, the other assumptions of (7.78) remain unchanged. ◇

Remark 7.21. We denote by γ the value of the sum: $\gamma_0 + \gamma_\infty$, *i.e.*,

$$\gamma = \gamma_0 + \gamma_\infty \tag{7.82}$$

and by K^* the adjoint of K, *i.e.*,

$$\langle K^*(\varpi, u), v\rangle = \langle K(\varpi, v), u\rangle \ \ \forall (u,v) \in \mathcal{D}^2. \tag{7.83}$$

◇

In this work, the cost functional J describing the control problem depends on the control ϕ and the solution $u(\phi)$ in the domain Ω over the time interval under consideration $[0,T]$. The optimal control corresponds to obtain the minimum point of a function J which measures the distance between the pronostic variable u and the observation $(u_{\text{obs}}, v_{\text{obs}})$. Precisely we will study the following optimal control problem: find (u, ϕ) such that the cost functional

$$J(\phi) = \frac{1}{2}\int_0^T \| \mathcal{C}(u - u_{\text{obs}}) \|^2_{L^2(\Omega)} + \frac{\mu}{2} \| u(T) - v_{\text{obs}} \|^2_{L^2(\Omega)} + \frac{\alpha}{2} \| \phi \|^2_{Rs}$$

is minimized with respect to ϕ subject to the solution of the problem (7.77), where the space Rs is a subset of some Banach space and $\mathcal{C}$ is an unbounded operator on $L^2(\Omega)$.

7.6.2 Existence and Uniqueness of the Solution

Proposition 7.22. *Assume that $u_0 \in L^2(\Omega)$ and $f \in L^2(\mathcal{Q})$. Then the problem (7.77) admits a unique solution u such that $u \in \mathcal{W} \cap \mathcal{H}_1$ and*

$$\| u \|^2_{\mathcal{W}\cap\mathcal{H}_1} \leq C(\| u_0 \|^2_{L^2} + \| f \|^2_{L^2(\mathcal{Q})}).$$

Proof. We will just sketch the proof based on suitable *a priori* estimates. Multiplying (7.77) by u ($u \in \mathcal{D}$) and integrating over Ω, this gives

$$\frac{1}{2}\frac{\mathrm{d} \mid u \mid^2}{\mathrm{d}t} + \langle Au, u\rangle + \langle F(.,t,u), u\rangle + \langle K(\varpi, u), u\rangle = \langle f, u\rangle.$$

According to assumption (i) of (7.78), we obtain

$$\frac{\mathrm{d} \mid u \mid^2}{\mathrm{d}t} + 2\langle Au, u\rangle = -2\langle F(.,t,u) - F(.,t,0), u - 0\rangle - 2\langle K(\varpi, u), u\rangle + 2\langle f, u\rangle.$$

By using the definition of the norm in $\mathcal{D}$ and according to assumption (ii) of (7.78) and assumption (7.81) we have

$$\frac{\mathrm{d} \mid u \mid^2}{\mathrm{d}t} + \nu \| u \|^2_{\mathcal{D}} \leq \gamma \mid u \mid^2 + 2 \mid f \mid\mid u \mid,$$

with γ given by the notation (7.82).

Integrating with respect to time, over $(0, t)$ for all $t \in (0, T)$, gives (by using Hölder's inequality)

$$\mid u(t) \mid^2 + \nu \int_0^t \| u \|^2_{\mathcal{D}}\, \mathrm{d}s \leq (1 + \gamma) \int_0^t \mid u \mid^2 \mathrm{d}s + \| f \|^2_{L^2(\mathcal{Q})} + \mid u_0 \mid^2$$

and then (according to Gronwall's lemma)

$$\| u \|^2_{\mathcal{W}} \leq C(\| u_0 \|^2_{L^2} + \| f \|^2_{L^2(\mathcal{Q})}). \tag{7.84}$$

Using the result (7.84) and (7.77), we prove easily that u satisfies the following estimate

$$\| u \|^2_{\mathcal{H}_1} \leq C(\| u_0 \|^2_{L^2} + \| f \|^2_{L^2(\mathcal{Q})}). \tag{7.85}$$

The proof of the existence result can be completed by implementing the Galerkin method, by taking advantage of the above estimate and by using the continuity and the uniform boundedness of the operator F. Uniqueness of the solution of (7.77) is a corollary of Proposition 7.23 (see below). □

Proposition 7.23. *Let u_{01}, u_{02} be two functions in $L^2(\Omega)$ and let f_1, f_2 be two functions in $L^2(\mathcal{Q})$. If u_1 (respectively u_2) is a solution of (7.77) with data (f_1, u_{01}) (respectively with data (f_2, u_{02})) then*

$$\| u_1 - u_2 \|_{\mathcal{W}\cap\mathcal{H}_1}^2 \leq C(\| u_{01} - u_{02} \|_{L^2}^2 + \| f_1 - f_2 \|_{L^2(\mathcal{Q})}^2). \tag{7.86}$$

Proof. Let $(u_i)_{i=1,2}$ be two solutions of (7.77), with the given data $(f_i, u_{0i})_{i=1,2}$, respectively. We denote by $u = u_1 - u_2$, $u_0 = u_{01} - u_{02}$ and $f = f_1 - f_2$. Then u is the solution of the following problem

$$\begin{aligned} &\frac{\partial u}{\partial t} + Au + F(x,t,u_1) - F(x,t,u_2) + K(\varpi, u) = f \quad \text{on } \mathcal{Q}, \\ &u(0) = u_0 \quad \text{on } \Omega. \end{aligned} \tag{7.87}$$

By using the same way to obtain the estimations (7.84) and (7.85) and the assumption (*ii*) of (7.78), we obtain the result of the proposition. □

7.6.3 The Control Framework

In the control framework, the value f is decomposed into a known function $g \in L^2(\mathcal{Q})$ and the control $\phi \in L^2(\mathcal{Q})$, *i.e.*, $g = f + \mathcal{B}\phi$, where $\mathcal{B}$ is a given linear continuous operator on $H = L^2(\Omega)$:

$$\mathcal{B} \in \mathcal{L}(H; H) \text{ such that } \forall h \in L^2(\Omega), \| Bh \|_{L^2} \leq b \| h \|_{L^2}, \text{ for } b > 0. \tag{7.88}$$

Then Problem (7.77) becomes

$$\begin{aligned} &\frac{\partial u}{\partial t} + Au + F(x,t,u) + K(\varpi, u) = g + \mathcal{B}\phi \quad \text{on } \mathcal{Q}, \\ &u(0) = u_0 \text{ (given)} \quad \text{on } \Omega. \end{aligned} \tag{7.89}$$

We suppose the following hypothesis: $u_0 \in L^2(\Omega)$, $(g, \phi) \in L^2(\mathcal{Q})^2$.

Let $\mathbf{U} : \phi \longrightarrow u = \mathbf{U}(\phi)$ be the map: $L^2(\mathcal{Q}) \longrightarrow \mathcal{W}$ defined by (7.89). We introduce the cost functional defined by

$$J(\phi) = \frac{1}{2} \| \mathcal{C}(u - u_{\text{obs}}) \|_{L^2(\mathcal{Q})}^2 + \frac{\mu}{2} \| u(T) - v_{\text{obs}} \|_{L^2}^2 + \frac{\alpha}{2} \| \phi \|_{L^2(\mathcal{Q})}^2, \tag{7.90}$$

where $\mu \geq 0$ and $\alpha > 0$ are fixed parameters, $(u_{\text{obs}}, v_{\text{obs}}) \in \mathcal{V} \times L^2(\Omega)$ is the (given) observation, and $\mathcal{C}$ is an unbounded linear operator on $L^2(\Omega)$ satisfying the condition

$$\| \mathcal{C}v \|_{L^2}^2 \leq \delta_1 \| v \|_{L^2}^2 + \delta_2 \| v \|_{\mathcal{D}}^2 \quad \forall v \in \mathcal{D}. \tag{7.91}$$

Then the optimal control problem is to minimize the functional J with respect to ϕ, *i.e.*, to find $\phi^* \in \mathcal{U}_{\text{ad}}$ such that

$$J(\phi^*) = \inf_{\phi \in \mathcal{U}_{\text{ad}}} J(\phi), \tag{7.92}$$

with $\mathcal{U}_{\text{ad}}$ is a (given) non-empty, closed, convex subset of $L^2(\mathcal{Q})$.

Proposition 7.24. *Let F be an operator that satisfies assumptions (7.78). Then the function $\mathbf{U} : \phi \longrightarrow u = \mathbf{U}(\phi)$ solution of (7.89) is continuously F-differentiable from $L^2(\mathcal{Q})$ to $\mathcal{W}$ and its derivative $w = \mathbf{U}'(\phi)h$ at point $\phi \in L^2(\mathcal{Q})$ in the direction $h \in L^2(\mathcal{Q})$ is given by the linear parabolic problem*

$$\begin{aligned} &\frac{\partial w}{\partial t} + Aw + G(x,t,\mathbf{U}(\phi))w + K(\varpi, w) = Bh \quad \text{on } \mathcal{Q} \\ &w(0) = 0 \quad \text{on } \Omega. \end{aligned} \tag{7.93}$$

Moreover, for all $\phi \in L^2(\mathcal{Q})$ we have the following estimate:

$$\| \mathbf{U}'(\phi) \|_{\mathcal{L}(L^2(\mathcal{Q});\mathcal{W})} \leq C.$$

Proof. Let $(\phi, h) \in (L^2(\mathcal{Q}))^2$, let $u = \mathbf{U}(\phi)$ and $u_h = \mathbf{U}(\phi + h) = u + w_h$. Let v_h be the solution of

$$\begin{aligned} &\frac{\partial v_h}{\partial t} + Av_h + G(x,t,u)v_h + K(\varpi, v_h) = Bh \quad \text{on } \mathcal{Q}, \\ &v_h(0) = 0 \quad \text{on } \Omega. \end{aligned} \tag{7.94}$$

Multiplying (7.94) by v_h and integrating over $(0,t) \times \Omega$, for all $t \in (0,T)$, we obtain (according to $-2G(.,u) \leq \gamma_0$)

$$\begin{aligned} &\| v_h(t) \|_{L^2}^2 + \nu \int_0^t \| v_h \|_{\mathcal{D}}^2 \, \mathrm{d}s \\ &\quad \leq \gamma \int_0^t \| v_h \|_{L^2}^2 \, \mathrm{d}s + 2 \| Bh \|_{L^2(\mathcal{Q})} \Big(\int_0^t \| v_h \|_{L^2}^2 \, \mathrm{d}s \Big)^{1/2}. \end{aligned} \tag{7.95}$$

By using Gronwall's formula we then have (according to (7.88))

$$\| v_h \|_{\mathcal{W}}^2 \leq c_1 \| h \|_{L^2(\mathcal{Q})}^2 . \tag{7.96}$$

This shows that the map: $h \longrightarrow v_h$ solution of (7.94) is continuous.

Moreover, by using Proposition 7.23 and (7.88) we deduce easily that:

$$\| w_h \|_{\mathcal{W}}^2 \leq c_2 \| h \|_{L^2(\mathcal{Q})}^2 . \tag{7.97}$$

Let $v = w_h - v_h$ and according to the equations satisfied by u, u_h and v_h we can deduce that v satisfies the following system:

$$\begin{aligned} &\frac{\partial v}{\partial t} + Av + G(x,t,u)v + K(\varpi, v) = g \quad \text{on } \mathcal{Q}, \\ &v(0) = 0 \quad \text{on } \Omega, \end{aligned} \tag{7.98}$$

where $g = -(F(.,u + w_h) - F(.,u)) + G(.,u)w_h$.

Multiplying (7.98) by v and integrating over $(0,t) \times \Omega$, for all $t \in (0,T)$, we obtain (according to $-2G(.,u) \leq \gamma_0$)

$$\| v(t) \|_{L^2}^2 + \nu \int_0^t \| v \|_{\mathcal{D}}^2 \, \mathrm{d}s \leq \gamma \int_0^t \| v \|_{L^2}^2 \, \mathrm{d}s + 2 \int_0^t \langle g, v \rangle(s) \mathrm{d}s. \tag{7.99}$$

We shall now estimate the term $2\int_0^t \langle g, v\rangle(s)\mathrm{d}s$ for all $t \in (0,T)$. By using a simple manipulation we obtain

$$g = \int_0^1 (G(.,u) - G(.,u + sw_h))w_h \mathrm{d}s.$$

Applying assumption (iii) of (7.78) we can deduce that $2 \mid g \mid \leq \lambda \mid w_h \mid^2$. So,

$$2\int_0^t \langle g, v\rangle(s)\mathrm{d}s \leq 2\int_0^T \langle g, v\rangle(s)\mathrm{d}s \leq \lambda \int_0^T \int_\Omega \mid w_h \mid^2 \mid v \mid \mathrm{d}x\mathrm{d}s$$

and then

$$2\int_0^t \langle g, v\rangle(s)\mathrm{d}s \leq \lambda \parallel w_h \parallel^2_{L^2(0,T;L^4(\Omega))} \parallel v \parallel_{L^\infty(0,T;L^2(\Omega))} .$$

According to assumption (7.79) we obtain

$$2\int_0^t \langle g, v\rangle(s)\mathrm{d}s \leq c_3 \parallel w_h \parallel^2_{\mathcal{W}} \parallel v \parallel_{\mathcal{W}} . \tag{7.100}$$

From (7.99) and (7.100), we deduce that

$$\parallel v(t) \parallel^2_{L^2} + \nu \int_0^t \parallel v \parallel^2_{\mathcal{D}} \leq \gamma \int_0^t \parallel v \parallel^2_{L^2} + c_3 \parallel w_h \parallel^2_{\mathcal{W}} \parallel v \parallel_{\mathcal{W}} .$$

Using Gronwall's formula we then have

$$\parallel v \parallel^2_{\mathcal{W}} \leq c_4 \parallel w_h \parallel^2_{\mathcal{W}} \parallel v \parallel_{\mathcal{W}} .$$

According to (7.97) we can deduce that $\parallel v \parallel_{\mathcal{W}} \leq c_5 \parallel h \parallel^2_{L^2(Q)}$ and then

$$\parallel v \parallel_{\mathcal{W}} = o(\parallel h \parallel_{L^2(Q)}).$$

Therefore, $\mathbf{U}'(\phi)$ defined by (7.93) is the F-derivative of $\mathbf{U}$ at point ϕ and verifies $\parallel \mathbf{U}'(\phi) \parallel_{\mathcal{L}(L^2(Q);\mathcal{W})} \leq C$. □

Proposition 7.25. *Let F be an operator that satisfies assumptions (7.78). Then for each $t \in [0,T]$, the function $\mathbf{V}_t : \phi \longrightarrow u(t) = \mathbf{V}_t(\phi)$ solution of (7.89) is continuously F-differentiable from $L^2(Q)$ to $L^2(\Omega)$ and its derivative $w(t) = \mathbf{V}'_t(\phi).h$ at point $\phi \in L^2(Q)$ in the direction $h \in L^2(Q)$ is given by the linear parabolic problem*

$$\begin{aligned} &\frac{\partial w}{\partial t} + Aw + G(x,t,\mathbf{V}_t(\phi))w + K(\varpi, w) = Bh \quad \text{on } Q, \\ &w(0) = 0 \quad \text{on } \Omega. \end{aligned} \tag{7.101}$$

Moreover, for all $\phi \in L^2(Q)$ the following estimate holds:

$$\parallel \mathbf{V}'_t(\phi) \parallel_{\mathcal{L}(L^2(Q);L^2(\Omega))} \leq C.$$

Proof. The functional $\phi \longrightarrow u(t)$ is continuous from $L^2(\mathcal{Q})$ to $L^2(\Omega)$ (the corollary of Proposition 7.23). The rest of the proposition is a corollary of Proposition 7.24. □

Before proving the existence of the optimal solution, we study the following result.

Proposition 7.26. *Let F be an operator that satisfies assumptions (7.78). Then the maps $\mathbf{U} : \phi \longrightarrow u = \mathbf{U}(\phi)$ and $\mathbf{V}_t : \phi \longrightarrow u(t) = \mathbf{V}_t(\phi)$, for $t \in (0,T)$ solutions of (7.89) are continuous from the weak topology of $L^2(\mathcal{Q})$ to the strong topology of $L^2(\mathcal{Q})$ and the weak topology of $L^2(\Omega)$, respectively.*

Proof. Let ϕ be given in $L^2(\mathcal{Q})$ and let the sequence ϕ_k such that ϕ_k is weakly convergent in $L^2(\mathcal{Q})$ to ϕ.

Set $u = \mathbf{U}(\phi)$, $u_k = \mathbf{U}(\phi_k)$ and $v_k = u - u_k$. Since $\phi_k \rightharpoonup \phi$ weakly in $L^2(\mathcal{Q})$ then the sequence ϕ_k is uniformly bounded in $L^2(\mathcal{Q})$ and therefore (according to Proposition 7.23) u_k is uniformly bounded in $\mathcal{W}$. By using assumption (ii) of (7.78) we deduce that $F(.,u_k)$ is uniformly bounded in $L^2(\mathcal{Q})$.

Using this result and Equation (7.89) we obtain easily that $\partial u_k/\partial t$ is uniformly bounded in $L^1(0,T;\mathcal{D}')$. Let us introduce the space

$$\mathcal{Y} = \{v \in L^2(0,T;\mathcal{D}), \frac{\partial v}{\partial t} \in L^1(0,T;\mathcal{D}')\}.$$

According to Lemma 6.6, the injection of $\mathcal{Y}$ into $L^2(\mathcal{Q})$ is compact. Therefore u_k is uniformly bounded in $\mathcal{Y}$. This result makes it possible to extract from $(u_k, \phi_k, F(.,u_k))$ a subsequence also denoted by $(u_k, \phi_k, F(.,u_k))$ such that[2]

$$\begin{aligned}
&\phi_k \rightharpoonup \phi \text{ weakly in } L^2(\mathcal{Q})\\
&u_k \rightharpoonup \tilde{u} \text{ weakly in } L^2(0,T;\mathcal{D})\\
&u_k \longrightarrow \tilde{u} \text{ strongly in } L^2(\mathcal{Q})\\
&F(.,u_k) \longrightarrow F(.,\tilde{u}) \text{ strongly in } L^2(\mathcal{Q}).
\end{aligned} \tag{7.102}$$

It is easy for us to prove that $\tilde{u}$ is a solution of (7.89) with a forcing ϕ and according to the uniqueness of the solution of (7.89), we have then $\tilde{u} = u = \mathbf{U}(\phi)$.

In the same way we prove that for each $t \in [0,T]$, $\mathbf{V}_t(\phi_k) \rightharpoonup \mathbf{V}_t(\phi)$ weakly in $L^2(\Omega)$. □

Now, let us study the following existence of an optimal solution.

Theorem 7.27. *Let F be an operator that satisfies assumptions (7.78). Then there exists $\phi^* \in L^2(\mathcal{Q})$ and $u^* \in \mathcal{W}$ such that ϕ^* is defined by (7.92) and $u^* = \mathbf{U}(\phi^*)$ is a solution of (7.89).*

---

[2] The operator F is continuous then $F(.,u_k) \longrightarrow F(.,\tilde{u})$ strongly in $L^2(\mathcal{Q})$ (according to assumption (iii) of (7.102)).

Proof. Let $\phi_k \in \mathcal{U}_{\rm ad}$ be a minimizing sequence of J, *i.e.*,

$$\liminf_{k\longrightarrow\infty} J(\phi_k) = \inf_{\phi\in L^2(\mathcal{Q})} J(\phi).$$

Then ϕ_k is uniformly bounded in $\mathcal{U}_{\rm ad}$ and we can extract from ϕ_k a subsequence also denoted by ϕ_k such that $\phi_k \rightharpoonup \phi$ weakly in $\mathcal{U}_{\rm ad}$. By using Proposition 7.26 we then have

$$\begin{aligned}
&\mathbf{U}(\phi_k) \rightharpoonup u = \mathbf{U}(\phi) \text{ weakly in } L^2(0,T;\mathcal{D}),\\
&\mathbf{U}(\phi_k) \longrightarrow u \text{ strongly in } L^2(\mathcal{Q}),\\
&\mathbf{V}_t(\phi_k) \rightharpoonup u(t) = \mathbf{V}_t(\phi) \text{ weakly in } L^2(\Omega), \forall t \in [0,T]
\end{aligned}$$

and u satisfies the system (7.89).

Since the norm is lower semi-continuous therefore the map $J : \phi \longrightarrow J(\phi)$ is lower semi-continuous and then the function ϕ is an optimal solution, *i.e.*,

$$\inf_{\psi\in\mathcal{U}_{\rm ad}} J(\psi) = \liminf_{k} J(\phi_k) = J(\phi).$$

□

In order to characterize the solution of the optimal control problem, we introduce the "adjoint" problem corresponding to the primal problem (7.89) (we denote by $u = \mathbf{U}(\phi)$ the solution of Problem (7.89) with the forcing ϕ):

$$\begin{aligned}
&-\frac{\partial \tilde{u}}{\partial t} + A\tilde{u} + (G(x,t,u))^*\tilde{u} + K^*(\varpi,\tilde{u}) = \mathcal{C}^*\mathcal{C}(u-u_{\rm obs}) \text{ on } \mathcal{Q},\\
&\text{subject to the final condition}\\
&\tilde{u}(t=T) = \mu(u(t=T) - v_{\rm obs}) \text{ on } \Omega,
\end{aligned} \tag{7.103}$$

where $\mathcal{C}^*$ (respectively $(G(.,u))^*$ and K^*) is the adjoint of the operator $\mathcal{C}$ (respectively $G(.,u)$ and K).

Proposition 7.28. *Let F be an operator that satisfies assumptions (7.78) and $u \in \mathcal{W}$ then the solution of (7.103) is in $\mathcal{W}$ and satisfies the following estimate:*

$$\| \tilde{u} \|_{\mathcal{H}}^2 + \| \tilde{u} \|_{\mathcal{V}}^2 \leq C_d^2(\mu^2 \| u(T) - v_{\rm obs} \|_{L^2}^2 + \| \mathcal{C}(u-u_{\rm obs}) \|_{L^2(\mathcal{Q})}^2).$$

Proof. Multiplying (7.103) by $\tilde{u}$ and integrating over $(t,T)\times\Omega$, for all $t \in (0,T)$, we obtain for all $r > 0$ (according to $-2G(.,u) \leq \gamma_0$ and to (7.81))

$$\begin{aligned}
\| \tilde{u}(t) \|_{L^2}^2 + \nu \int_t^T \| \tilde{u} \|_{\mathcal{D}}^2 \, ds \leq \mu^2 \| u(T) - v_{\rm obs} \|_{L^2}^2 + \gamma \int_t^T \| \tilde{u} \|_{L^2}^2 \, ds\\
+ \frac{1}{r^2} \int_t^T \| \mathcal{C}(u-u_{\rm obs}) \|_{L^2}^2 \, ds + r^2 \int_t^T \| \mathcal{C}\tilde{u} \|_{L^2}^2 \, ds.
\end{aligned}$$

According to assumption (7.91) we obtain

$$\begin{aligned} \| \tilde{u}(t) \|_{L^2}^2 + \nu \int_t^T \| \tilde{u} \|_{\mathcal{D}}^2 \,\mathrm{d}s \\ \leq \mu^2 \| u(T) - v_{\mathrm{obs}} \|_{L^2}^2 + (\delta_1 r^2 + \gamma) \int_t^T \| \tilde{u} \|_{L^2}^2 \,\mathrm{d}s \\ + \delta_2 r^2 \int_t^T \| \tilde{u} \|_{\mathcal{D}}^2 \,\mathrm{d}s + \frac{1}{r^2} \| \mathcal{C}(u - u_{\mathrm{obs}}) \|_{L^2(\mathcal{Q})}^2 . \end{aligned} \tag{7.104}$$

Choosing $r^2 := \nu/2\delta_2$ we can deduce that

$$\begin{aligned} \| \tilde{u} \|_{L^2}^2 + \frac{\nu}{2} \int_t^T \| \tilde{u} \|_{\mathcal{D}}^2 \,\mathrm{d}s \leq \mu^2 \| u(T) - v_{\mathrm{obs}} \|_{L^2}^2 + \frac{2\delta_2}{\nu} \| \mathcal{C}(u - u_{\mathrm{obs}}) \|_{L^2(\mathcal{Q})}^2 \\ + (\frac{\nu\delta_1}{2\delta_2} + \gamma) \int_t^T \| \tilde{u} \|_{L^2}^2 \,\mathrm{d}s. \end{aligned}$$

By using Gronwall's formula we can conclude that

$$\| \tilde{u} \|_{\mathcal{H}}^2 + \| \tilde{u} \|_{\mathcal{V}}^2 \leq C_d^2 (\mu^2 \| u(T) - v_{\mathrm{obs}} \|_{L^2}^2 + \| \mathcal{C}(u - u_{\mathrm{obs}}) \|_{L^2(\mathcal{Q})}^2).$$

This completes the proof. □

In the sequel, we will denote by $\tilde{\mathbf{U}} : \phi \longrightarrow \tilde{u} = \tilde{\mathbf{U}}(\phi)$ the map defined by the solution of the adjoint problem (7.103).

We can now give the first-order optimality conditions for the optimal control problem (7.92).

Theorem 7.29. *Let F be an operator that satisfies assumptions (7.78) and $(\phi^*, u^*) \in L^2(\mathcal{Q}) \times \mathcal{W}$ such that ϕ^* is defined by (7.92) and $u^* = \mathbf{U}(\phi^*)$ is a solution of (7.89). Then*

$$\int_0^T \int_\Omega (\alpha\phi^* + B^*\tilde{u})(\phi - \phi^*) \mathrm{d}x\mathrm{d}t \geq 0, \forall \phi \in \mathcal{U}_{\mathrm{ad}},$$

where $\tilde{u} = \tilde{\mathbf{U}}(\phi^*)$ is a solution of the adjoint problem (7.103), corresponding to the primal solution u^*. Moreover, the gradient of the cost functional J at ϕ^* is given by

$$J'(\phi^*) = \alpha\phi^* + B^*\tilde{u}.$$

Proof. From Proposition 7.25 we know that $\mathbf{U}$ is differentiable. Therefore, the cost functional J is a composition of F-differentiable mappings and, then, J is F-differentiable and, for all $h \in L^2(\mathcal{Q})$, we have

$$\begin{aligned} J'(\phi)h = \int_0^T \langle C(u - u_{\mathrm{obs}}), C\mathbf{U}'(\phi)h \rangle \mathrm{d}s + \langle \mu(u(T) - v_{\mathrm{obs}}), \mathbf{V}_T'(\phi)h \rangle \\ + \int_0^T \langle \alpha\phi, h \rangle \mathrm{d}s, \end{aligned}$$

which implies

$$J'(\phi)h = \int_0^T \langle \mathcal{C}^*\mathcal{C}(u - u_{\text{obs}}), \mathbf{U}'(\phi)h\rangle \mathrm{d}s + \langle \mu(u(T) - v_{\text{obs}}), \mathbf{V}'_T(\phi)h\rangle + \int_0^T \langle \alpha\phi, h\rangle \mathrm{d}s.$$

Multiplying (7.93) by $\tilde{u}$, integrating over $\mathcal{Q}$, we obtain that, for all $h \in L^2(\mathcal{Q})$,

$$\int_0^T \langle B^*\tilde{u}, h\rangle \mathrm{d}s = \langle \mathbf{V}'_T(\phi^*)h, \tilde{u}(T)\rangle + \int_0^T \langle -\frac{\partial \tilde{u}}{\partial t} + A\tilde{u} + (G(., u^*))^*\tilde{u} + K^*(\varpi, \tilde{u}), \mathbf{U}'(\phi^*)h\rangle \mathrm{d}s.$$

Since $\tilde{u}$ is a solution of the adjoint problem (7.103) we can deduce that, for all $h \in L^2(\mathcal{Q})$,

$$\int_0^T \langle B^*\tilde{u}, h\rangle \mathrm{d}s = \langle \mathbf{V}'_T(\phi^*)h, \mu(u^*(T) - v_{\text{obs}})\rangle + \int_0^T \langle \mathcal{C}^*\mathcal{C}(u^* - u_{\text{obs}}), \mathbf{U}'(\phi^*)h\rangle \mathrm{d}s.$$

Using the previous equalities and the expression of J', we see that, for all $h \in L^2(\mathcal{Q})$,

$$J'(\phi).h = \int_0^T \langle \alpha\phi + B^*\tilde{u}, h\rangle \mathrm{d}s. \qquad (7.105)$$

Since ϕ^* is a solution of (7.92), we have $J'(\phi^*).(\phi - \phi^*) \geq 0 \ \forall \phi \in \mathcal{U}_{\text{ad}}$, and therefore, according to (7.105), we can deduce that

$$\int_0^T \int_\Omega (\alpha\phi^* + B^*\tilde{u})(\phi - \phi^*) \mathrm{d}x\mathrm{d}t \geq 0, \ \forall \phi \in \mathcal{U}_{\text{ad}}.$$

So the proof is complete. □

In the sequel, we will assume that there exists a pair (ϕ^*, u^*) such that ϕ^* is defined by (7.92), $u^* = \mathbf{U}(\phi^*)$ is a solution of (7.89) and

$$\int_0^T \int_\Omega (\alpha\phi^* + B^*\tilde{u})(\phi - \phi^*) \mathrm{d}x\mathrm{d}t \geq 0$$

for all $\phi \in \mathcal{U}_{\text{ad}}$, where $\tilde{u} = \tilde{\mathbf{U}}(\phi^*)$ is a solution of (7.103)). Now we give some conditions to obtain the uniqueness of the optimal solution (ϕ^*, u^*).

Theorem 7.30. *Suppose that F satisfies assumptions (7.78) and $\mu < 1$ holds. If the following assumptions hold:*

(*i*) $\theta = (\nu - \delta_2 - c_I^2(\gamma + \delta_1) - 2b^2 c_I^2/\alpha) > 0$

(*ii*) $\lambda c_e^2 e^{(\delta_1+\gamma+1)T} \left(\| u^*(T) - v_{\text{obs}} \|_{L^2}^2 + \| \mathcal{C}(u^* - u_{\text{obs}}) \|_{L^2(\mathcal{Q})}^2 \right)^{1/2} < \theta,$

then the optimal solution (ϕ^*, u^*)*, of Problem (7.92) subject to Problem (7.89), is unique.*

Proof. Assume that (ϕ_1^*, u_1^*) is another solution. Then, ϕ_1^* satisfies (7.92), $u_1^* = \mathbf{U}(\phi_1^*)$ is a solution of (7.89) and

$$\int_0^T \int_\Omega (\alpha \phi_1^* + B^* \tilde{u}_1)(\phi - \phi_1^*) \mathrm{d}x \mathrm{d}t \geq 0$$

for all $\phi \in \mathcal{U}_{\text{ad}}$, where $\tilde{u}_1 = \tilde{\mathbf{U}}(\phi_1^*)$ is a solution of (7.103)).

We set $\phi = \phi^* - \phi_1^*$, $v = u^* - u_1^*$ and $\tilde{v} = \tilde{u} - \tilde{u}_1$. Then we have

$$\begin{aligned} &\frac{\partial v}{\partial t} + Av + (F(.,u^*) - F(.,u_1^*)) + K(\varpi, v) = B\phi \quad \text{on } \mathcal{Q}, \\ &v(0) = 0 \quad \text{on } \Omega, \end{aligned} \tag{7.106}$$

$$\begin{aligned} &-\frac{\partial \tilde{v}}{\partial t} + A\tilde{v} + (G(.,u_1^*))^* \tilde{v} + K^*(\varpi, \tilde{v}) \\ &\qquad = \mathcal{C}^* \mathcal{C} v - (G(.,u^*) - G(.,u_1^*))^* \tilde{u} \quad \text{on } \mathcal{Q}, \\ &\tilde{v}(T) = \mu v(T) \quad \text{on } \Omega \end{aligned} \tag{7.107}$$

and

$$\alpha \| \phi \|_{L^2(\mathcal{Q})}^2 + \int_0^T \int_\Omega B^* \tilde{v} \phi \mathrm{d}x \mathrm{d}t \leq 0. \tag{7.108}$$

According to assumptions (7.78) and (7.80) we have

$$\begin{aligned} &-2\langle F(.,u^*) - F(.,u_1^*), v \rangle \leq \gamma_0 c_I^2 \| v \|_{\mathcal{D}}^2, \\ &-2\langle (G(.,u_1^*))^* \tilde{v}, \tilde{v} \rangle \leq \gamma_0 c_I^2 \| \tilde{v} \|_{\mathcal{D}}^2, \\ &| \langle (G(.,u^*) - G(.,u_1^*))^* \tilde{u}, \tilde{v} \rangle | \leq \lambda \int_\Omega | v || \tilde{v} || \tilde{u} | \, \mathrm{d}x. \end{aligned} \tag{7.109}$$

Multiplying (7.106) by v, (7.107) by $\tilde{v}$ and integrating over $\mathcal{Q}$, we obtain (according to (7.108), (7.109), and assumption (7.91))

$$\begin{aligned} &\int_0^T \frac{\mathrm{d}}{\mathrm{d}t} \| v \|_{L^2}^2 \, \mathrm{d}s + \nu \int_0^T \| v \|_{\mathcal{D}}^2 \, \mathrm{d}s \leq \gamma c_I^2 \int_0^T \| v \|_{\mathcal{D}}^2 \, \mathrm{d}s \\ &\qquad\qquad + \frac{2}{\alpha} \| B^* \tilde{v} \|_{L^2(\mathcal{Q})} \| B^* v \|_{L^2(\mathcal{Q})}, \\ &-\int_0^T \frac{\mathrm{d}}{\mathrm{d}t} \| \tilde{v} \|_{L^2}^2 \, \mathrm{d}s + \nu \int_0^T \| \tilde{v} \|_{\mathcal{D}}^2 \, \mathrm{d}s \leq \gamma c_I^2 \int_0^T \| \tilde{v} \|_{\mathcal{D}}^2 \, \mathrm{d}s \\ &\qquad + (\delta_1 c_I^2 + \delta_2) \int_0^T (\| v \|_{\mathcal{D}}^2 + \| \tilde{v} \|_{\mathcal{D}}^2) \mathrm{d}s + 2\lambda \int_0^T \int_\Omega | \tilde{u} | \, | \tilde{v} | \, | v | \, \mathrm{d}x \mathrm{d}s, \\ &\tilde{v}(T) = \mu v(T) \text{ and } v(0) = 0 \quad \text{on } \Omega. \end{aligned}$$

By using Hölder's inequality and the relationship (7.79) we obtain

$$\begin{aligned}
&\int_0^T \frac{\mathrm{d}}{\mathrm{d}t} \| v \|_{L^2}^2 \,\mathrm{d}s + (\nu - \gamma c_I^2) \int_0^T \| v \|_{\mathcal{D}}^2 \,\mathrm{d}s \\
&\qquad \leq \frac{2}{\alpha} b^2 c_I^2 \int_0^T (\| v \|_{\mathcal{D}}^2 + \| \tilde{v} \|_{\mathcal{D}}^2) \mathrm{d}s, \\
&-\int_0^T \frac{\mathrm{d}}{\mathrm{d}t} \| \tilde{v} \|_{L^2}^2 \,\mathrm{d}s + (\nu - \gamma c_I^2) \int_0^T \| \tilde{v} \|_{\mathcal{D}}^2 \,\mathrm{d}s \\
&\qquad \leq (\delta_1 c_I^2 + \delta_2 + \lambda c_e^2 \| \tilde{u} \|_{\mathcal{H}}) \int_0^T (\| v \|_{\mathcal{D}}^2 + \| \tilde{v} \|_{\mathcal{D}}^2) \mathrm{d}s, \\
&\tilde{v}(T) = \mu v(T) \text{ and } v(0) = 0 \text{ on } \Omega.
\end{aligned} \tag{7.110}$$

Adding first and second inequalities of (7.110) we obtain

$$\begin{aligned}
&\int_0^T \frac{\mathrm{d}}{\mathrm{d}t} (\| v \|_{L^2}^2 - \| \tilde{v} \|_{L^2}^2) \mathrm{d}s + \theta \int_0^T (\| v \|_{\mathcal{D}}^2 + \| \tilde{v} \|_{\mathcal{D}}^2) \mathrm{d}s \\
&\qquad \leq \lambda c_e^2 \| \tilde{u} \|_{\mathcal{H}} \int_0^T (\| v \|_{\mathcal{D}}^2 + \| \tilde{v} \|_{\mathcal{D}}^2) \mathrm{d}s,
\end{aligned} \tag{7.111}$$

where $\theta = \nu - \delta_2 - c_I^2(\gamma + (2b^2/\alpha) + \delta_1) > 0$ (according to assumption (i)). According to the third equalities of (7.110) we have

$$\begin{aligned}
(1 - \mu^2) \| v(T) \|_{L^2}^2 + \| \tilde{v}(0) \|_{L^2}^2 &+ \theta(\| v \|_{\mathcal{V}}^2 + \| \tilde{v} \|_{\mathcal{V}}^2) \\
&\leq \lambda c_e^2 \| \tilde{u} \|_{\mathcal{H}} (\| v \|_{\mathcal{V}}^2 + \| \tilde{v} \|_{\mathcal{V}}^2).
\end{aligned}$$

Since $1 - \mu^2 > 0$ then $\theta(\| v \|_{\mathcal{V}}^2 + \| \tilde{v} \|_{\mathcal{V}}^2) \leq \lambda c_e^2 \| \tilde{u} \|_{\mathcal{H}} (\| v \|_{\mathcal{V}}^2 + \| \tilde{v} \|_{\mathcal{V}}^2)$.

By choosing $r = 1$ in (7.104) (since $\nu > \delta_2$, according to assumption (i)) and by using Gronwall's formula, we can deduce that

$$\| \tilde{u} \|_{\mathcal{H}}^2 \leq \mathrm{e}^{(\delta_1 + \gamma + 1)T} (\mu^2 \| u^*(T) - v_{\mathrm{obs}} \|_{L^2}^2 + \| \mathcal{C}(u^* - u_{\mathrm{obs}}) \|_{L^2(\mathcal{Q})}^2), \tag{7.112}$$

and then $\theta^*(\| v \|_{\mathcal{V}}^2 + \| \tilde{v} \|_{\mathcal{V}}^2) \leq 0$, where (since $\mu < 1$)

$$\theta^* = \theta - \lambda c_e^2 \mathrm{e}^{(\delta_1 + \gamma + 1)T/2} \left(\| u^*(T) - v_{\mathrm{obs}} \|_{L^2}^2 + \| \mathcal{C}(u^* - u_{\mathrm{obs}}) \|_{L^2(\mathcal{Q})}^2 \right)^{1/2}.$$

Since $\theta^* > 0$ (according to assumption (ii)), we have $\| v \|_{\mathcal{V}}^2 + \| \tilde{v} \|_{\mathcal{V}}^2 = 0$ and then $v = 0$ and $\tilde{v} = 0$.

The proof of the uniqueness is complete. □

7.6.4 Initial Condition Control

In this subsection the objective of the optimal control problem is to find the best estimate of the initial state u_0.

We suppose now that the value u_0 is decomposed into a known function $g \in L^2(\Omega)$ and the control $\phi \in L^2(\Omega)$, *i.e.*, $u_0 = g + B\phi$, where B is a given bounded operator on $L^2(\Omega)$.

So the function u is assumed to be related to the control ϕ through Problem (7.77):

$$\begin{aligned} &\frac{\partial u}{\partial t} + Au + F(x,t,u) + K(\varpi, u) = f \text{ (given) } \text{ on } \mathcal{Q}, \\ &u(0) = g + B\phi \text{ on } \Omega. \end{aligned} \tag{7.113}$$

To obtain the regularity of Proposition 7.23, we suppose the following hypothesis: $f \in L^2(\mathcal{Q})$, $g \in L^2(\Omega)$ and $\phi \in L^2(\Omega)$. Let $\mathbf{U} : \phi \longrightarrow u = \mathbf{U}(\phi)$ be the map: $L^2(\Omega) \longrightarrow \mathcal{W}$ defined by (7.113) and let us introduce the cost function defined by

$$J(\phi) = \frac{1}{2} \parallel \mathcal{C}(u - u_{\text{obs}}) \parallel^2_{L^2(\mathcal{Q})} + \frac{\mu}{2} \parallel u(T) - v_{\text{obs}} \parallel^2_{L^2} + \frac{\alpha}{2} \parallel \phi \parallel^2_{L^2}, \tag{7.114}$$

where $\mu \geq 0$ and $\alpha > 0$ are fixed parameters, $(u_{\text{obs}}, v_{\text{obs}}) \in \mathcal{V} \times L^2(\Omega)$ is the observation and $\mathcal{C}$ is an unbounded linear operator on $L^2(\Omega)$ satisfying assumption (7.91).

We want to minimize the functional J with respect to ϕ subject to Problem (7.113), *i.e.*, to find $\phi^* \in \mathcal{U}_{\text{ad}}$ such that

$$J(\phi^*) \leq J(\phi), \forall \phi \in \mathcal{U}_{\text{ad}} \tag{7.115}$$

with $\mathcal{U}_{\text{ad}}$ is a given non-empty, closed and convex subset of $L^2(\Omega)$.

By using the same technique as used in Propositions 7.23 and 7.24 and Theorem 7.27, we have the following results (with no further estimates required).

Proposition 7.31. *Let F be an operator that satisfies assumptions (7.78). Then, the function $\mathbf{U} : \phi \longrightarrow u = \mathbf{U}(\phi)$ solution of (7.113) is continuously F-differentiable from $L^2(\Omega)$ to $\mathcal{W}$ and its derivative $w = \mathbf{U}'(\phi)h$ at point $\phi \in L^2(\Omega)$ in the direction $h \in L^2(\Omega)$ is given by the linear parabolic problem*

$$\begin{aligned} &\frac{\partial w}{\partial t} + Aw + G(x,t,\mathbf{U}(\phi))w + K(\varpi, w) = 0 \quad \text{on } \mathcal{Q}, \\ &w(0) = Bh \quad \text{on } \Omega. \end{aligned} \tag{7.116}$$

Moreover, for all $\phi \in L^2(\Omega)$ we have the following estimate:

$$\parallel \mathbf{U}'(\phi) \parallel_{\mathcal{L}(L^2(\Omega);\mathcal{W})} \leq C.$$

□

Proposition 7.32. *Let F be an operator that satisfies assumptions (7.78). Then, for each $t \in [0,T]$, the function $\mathbf{V}_t : \phi \longrightarrow u(t) = \mathbf{V}_t(\phi)$ solution of (7.113) is continuously F-differentiable from $L^2(\Omega)$ to $L^2(\Omega)$ and its derivative $w(t) = \mathbf{V}_t'(\phi).h$ at point $\phi \in L^2(\Omega)$ in the direction $h \in L^2(\Omega)$ is given by* *the linear parabolic problem*

$$\begin{aligned} &\frac{\partial w}{\partial t} + Aw + G(x,t,\mathbf{V}_t(\phi))w + K(\varpi, w) = 0 \quad \textit{on } \mathcal{Q}, \\ &w(0) = Bh \quad \textit{on } \Omega. \end{aligned} \tag{7.117}$$

Moreover, for all $\phi \in L^2(\mathcal{Q})$ the following estimate holds.

$$\| \mathbf{V}_t'(\phi) \|_{\mathcal{L}(L^2(\Omega);L^2(\Omega))} \leq C.$$

□

Theorem 7.33. *Let F be an operator that satisfies assumptions (7.78). Then, there exist $\phi^* \in L^2(\Omega)$ and $u^* \in \mathcal{W}$ such that ϕ^* is defined by (7.115) and $u^* = \mathbf{U}(\phi^*)$ is a solution of (7.113).* □

In order to characterize the solution of the optimal control problem, we use the "adjoint" problem corresponding to the primal problem (7.113) (we denote by $u = \mathbf{U}(\phi)$ the solution of problem (7.113) where the initial condition is ϕ):

$$\begin{aligned} &-\frac{\partial \tilde{u}}{\partial t} + A\tilde{u} + (G(x,t,u))^*\tilde{u} + K^*(\varpi, \tilde{u}) = \mathcal{C}^*\mathcal{C}(u - u_{\text{obs}}) \quad \text{on } \mathcal{Q}, \\ &\text{subject to the final condition} \\ &\tilde{u}(T) = \mu(u(T) - v_{\text{obs}}) \quad \text{on } \Omega, \end{aligned} \tag{7.118}$$

where $\mathcal{C}^*$ (resp. $(G(.,u))^*$) is the adjoint of the operator $\mathcal{C}$ (resp. $G(.,u)$).

Remark 7.34. The adjoint problem (7.118) is the same as the problem (7.103), then the result of Proposition 7.28 remains valid. ◇

We can now give the optimality system for the optimal control problem (7.115).

Theorem 7.35. *Let F be an operator that satisfies assumptions (7.78), $\nu > \delta_2$ and $(\phi^*, u^*) \in \mathcal{U}_{\text{ad}} \times \mathcal{W}$ such that ϕ^* is defined by (7.115) and $u^* = \mathbf{U}(\phi^*)$ is a solution of (7.113). Then*

$$\int_\Omega (\alpha\phi^* + B^*\tilde{u}(0))(\phi - \phi^*)\mathrm{d}x \geq 0 \quad \forall \phi \in \mathcal{U}_{\text{ad}},$$

where $\tilde{u} = \tilde{\mathbf{U}}(\phi^*)$ is a solution of the adjoint problem (7.118), corresponding to the primal solution u^*. Moreover, the gradient of the cost functional J at ϕ^* is given by

$$J'(\phi^*) = \alpha\phi^* + B^*\tilde{u}(0).$$

Proof. The cost functional J is a composition of F-differentiable mappings then J is F-differentiable and for all $h \in L^2(\Omega)$ we have

$$J'(\phi).h = \int_0^T \langle C(u - u_{\text{obs}}), C\mathbf{U}'(\phi)h\rangle \mathrm{d}s + \langle \mu(u(T) - v_{\text{obs}}), \mathbf{V}'_T(\phi)h\rangle + \langle \alpha\phi, h\rangle$$

and then

$$J'(\phi).h = \int_0^T \langle \mathcal{C}^*\mathcal{C}(u - u_{\text{obs}}), \mathbf{U}'(\phi)h\rangle \mathrm{d}s + \langle \mu(u(T) - v_{\text{obs}}), \mathbf{V}'_T(\phi)h\rangle + \langle \alpha\phi, h\rangle.$$

Multiplying (7.116) by $\tilde{u}$ and integrating over $\mathcal{Q}$, we obtain that for all $h \in L^2(\mathcal{Q})$

$$\begin{aligned}\langle g + Bh, \tilde{u}(0)\rangle &= \langle \mathbf{V}'_T(\phi^*)\mathbf{h}, \tilde{u}(T)\rangle \\ &\quad + \int_0^T \langle -\frac{\partial \tilde{u}}{\partial t} + A\tilde{u} + (G(.,u^*))^*\tilde{u} + K^*(\varpi, \tilde{u}), \mathbf{U}'(\phi^*)\mathbf{h}\rangle \mathrm{d}s.\end{aligned}$$

Since $\tilde{u}$ is a solution of the adjoint problem we can deduce that for all $h \in L^2(\Omega)$

$$\begin{aligned}\langle g + Bh, \tilde{u}(0)\rangle &= \langle \mathbf{V}'_T(\phi^*)\mathbf{h}, \mu(u^*(T) - v_{\text{obs}})\rangle \\ &\quad + \int_0^T \langle \mathcal{C}^*\mathcal{C}(u^* - u_{\text{obs}}), \mathbf{U}'(\phi^*)h\rangle \mathrm{d}s.\end{aligned}$$

Then we infer from the expression of J' that for all $h \in L^2(\Omega)$

$$J'(\phi).h = \langle \alpha\phi + B^*\tilde{u}(0), h\rangle. \tag{7.119}$$

Since ϕ^* is a solution of (7.115) then $J'(\phi^*).(\phi - \phi^*) \geq 0$, $\forall \phi \in \mathcal{U}_{\text{ad}}$ and therefore, from (7.119), we can deduce the result of this theorem.

This completes the proof. □

In the sequel, we will assume that there exists a pair (ϕ^*, u^*) such that ϕ^* is defined by (7.115), $u^* = \mathbf{U}(\phi^*)$ is a solution of (7.113) and

$$\int_\Omega (\alpha\phi^* + B^*\tilde{u}(0))(\phi - \phi^*)\mathrm{d}x \geq 0$$

for all $\phi \in \mathcal{U}_{\text{ad}}$, where $\tilde{u} = \tilde{\mathbf{U}}(\phi^*)$ is a solution of (7.118).

We can now give some conditions to obtain the uniqueness of the optimal solution (ϕ^*, u^*).

Theorem 7.36. *Suppose that F satisfies assumptions (7.78), $\alpha \geq b^2$ and $\mu < 1$ holds. If the following assumptions hold:*

(i) $\theta = (\nu - \delta_2 - c_I^2(\gamma + \delta_1)) > 0$

(ii) $\lambda c_e^2 e^{(\delta_1+\gamma+1)T/2} \left(\| u^*(T) - v_{\text{obs}} \|_{L^2}^2 + \| \mathcal{C}(u^* - u_{\text{obs}}) \|_{L^2(\mathcal{Q})}^2 \right)^{1/2} < \theta,$

then the optimal solution (ϕ^*, u^*), of the problem (7.115) subject to the problem (7.113), is unique.

Proof. Assume that (ϕ_1^*, u_1^*) is another optimal solution, then ϕ_1^* satisfies (7.115), $u_1^* = \mathbf{U}(\phi_1^*)$ is a solution of (7.113) and

$$\int_\Omega (\alpha\phi_1^* + B^*\tilde{u}_1(0))(\phi - \phi_1^*)\mathrm{d}x \geq 0$$

for all $\phi \in \mathcal{U}_{\mathrm{ad}}$, where $\tilde{u}_1 = \tilde{\mathbf{U}}(\phi_1^*)$ is a solution of (7.118).

We set $\phi = \phi^* - \phi_1^*$, $v = u^* - u_1^*$ and $\tilde{v} = \tilde{u} - \tilde{u}_1$, and can deduce that v satisfies

$$\begin{aligned} &\frac{\partial v}{\partial t} + Av + (F(x,t,u^*) - F(x,t,u_1^*)) + K(\varpi, v) = 0 \text{ on } \mathcal{Q}, \\ &v(0) = B\phi \text{ on } \Omega, \end{aligned} \tag{7.120}$$

$$\begin{aligned} &-\frac{\partial \tilde{v}}{\partial t} + A\tilde{v} + (G(.,u_1^*))^*\tilde{v} + K^*(\varpi, \tilde{v}) \\ &\qquad\qquad = \mathcal{C}^*\mathcal{C}v - (G(.,u^*) - G(.,u_1^*))^*\tilde{u} \text{ on } \mathcal{Q}, \\ &\tilde{v}(t=T) = \mu v(t=T) \text{ on } \Omega \end{aligned} \tag{7.121}$$

and

$$\alpha \parallel \phi \parallel^2_{L^2(\Omega)} + \int_\Omega B^*\tilde{v}(0)\phi \mathrm{d}x \leq 0, \tag{7.122}$$

Multiplying (7.120) by v, (7.121) by $\tilde{v}$ and integrating over Ω we obtain that (according to (7.122), (7.91) and (7.109))

$$\begin{aligned} &\frac{\mathrm{d}}{\mathrm{d}t} \parallel v \parallel^2_{L^2} + \nu \parallel v \parallel^2_{\mathcal{D}} \leq \gamma c_I^2 \parallel v \parallel^2_{\mathcal{D}}, \\ &-\frac{\mathrm{d}}{\mathrm{d}t} \parallel \tilde{v} \parallel^2_{L^2} + \nu \parallel \tilde{v} \parallel^2_{\mathcal{D}} \leq \gamma c_I^2 \parallel \tilde{v} \parallel^2_{\mathcal{D}} \\ &\qquad\qquad + (\delta_1 c_I^2 + \delta_2)(\parallel v \parallel^2_{\mathcal{D}} + \parallel \tilde{v} \parallel^2_{\mathcal{D}}) + 2\lambda \int_\Omega |\, \tilde{u} \,|\,|\, \tilde{v} \,|\,|\, v \,| \,\mathrm{d}x, \\ &\tilde{v}(T) = \mu v(T) \text{ and } v(0) = B\phi \text{ on } \Omega. \end{aligned}$$

Using Hölder's inequality and the relationship (7.79) we have that

$$\begin{aligned} &\frac{\mathrm{d}}{\mathrm{d}t} \parallel v \parallel^2_{L^2} + (\nu - \gamma c_I^2) \parallel v \parallel^2_{\mathcal{D}} \leq 0, \\ &-\frac{\mathrm{d}}{\mathrm{d}t} \parallel \tilde{v} \parallel^2_{L^2} + (\nu - \gamma c_I^2) \parallel \tilde{v} \parallel^2_{\mathcal{D}} \\ &\qquad\qquad \leq (\delta_1 c_I^2 + \delta_2 + \lambda c_e^2 \parallel \tilde{u} \parallel_{\mathcal{H}})(\parallel v \parallel^2_{\mathcal{D}} + \parallel \tilde{v} \parallel^2_{\mathcal{D}}), \\ &\tilde{v}(T) = \mu v(T) \text{ and } v(0) = B\phi \text{ on } \Omega. \end{aligned} \tag{7.123}$$

Summing the first and the second inequality of (7.123) we obtain that (according to $\nu > \delta_2$)

$$\begin{aligned} &\frac{\mathrm{d}}{\mathrm{d}t}(\parallel v \parallel^2_{L^2} - \parallel \tilde{v} \parallel^2_{L^2}) + \theta(\parallel v \parallel^2_{\mathcal{D}} + \parallel \tilde{v} \parallel^2_{\mathcal{D}}) \\ &\qquad\qquad \leq \lambda c_e^2 \parallel \tilde{u} \parallel_{\mathcal{H}} (\parallel v \parallel^2_{\mathcal{D}} + \parallel \tilde{v} \parallel^2_{\mathcal{D}}), \\ &\tilde{v}(T) = \mu v(T) \text{ and } v(0) = B\phi, \end{aligned} \tag{7.124}$$

where $\theta = \nu - \delta_2 - c_I^2(2\gamma + \delta_1) > 0$ (according to the assumption (i)).

Integrating over $[0, T]$ and according to the third part of (7.123) we have that

$$(1-\mu^2) \parallel v(T) \parallel_{L^2}^2 + \parallel \tilde{v}(0) \parallel_{L^2}^2 + \theta(\parallel v \parallel_{\mathcal{V}}^2 + \parallel \tilde{v} \parallel_{\mathcal{V}}^2)$$
$$\leq \lambda c_e^2 \parallel \tilde{u} \parallel_{\mathcal{H}} (\parallel v \parallel_{\mathcal{V}}^2 + \parallel \tilde{v} \parallel_{\mathcal{V}}^2) + \frac{b^4}{\alpha^2} \parallel \tilde{v}(0) \parallel_{L^2}^2 .$$

Since $1 - \mu^2 > 0$ and $1 - b^4/\alpha^2 \geq 0$ then

$$\theta(\parallel v \parallel_{\mathcal{V}}^2 + \parallel \tilde{v} \parallel_{\mathcal{V}}^2) \leq \lambda c_e^2 \parallel \tilde{u} \parallel_{\mathcal{H}} (\parallel v \parallel_{\mathcal{V}}^2 + \parallel \tilde{v} \parallel_{\mathcal{V}}^2).$$

According to the relation (7.112) (see Remark 7.34) and that $\mu < 1$ we can deduce that

$$\theta^*(\parallel v \parallel_{\mathcal{V}}^2 + \parallel \tilde{v} \parallel_{\mathcal{V}}^2) \leq 0,$$

where

$$\theta^* = \theta - \lambda c_e^2 e^{(\delta_1+\gamma+1)T/2} \left(\parallel u^*(T) - v_{\text{obs}} \parallel_{L^2}^2 + \parallel \mathcal{C}(u^* - u_{\text{obs}}) \parallel_{L^2(\mathcal{Q})}^2 \right)^{1/2} .$$

Since $\theta^* > 0$ (according to assumption (ii)), we have $\parallel v \parallel_{\mathcal{V}}^2 + \parallel \tilde{v} \parallel_{\mathcal{V}}^2 = 0$ and then $v = 0$ and $\tilde{v} = 0$.

The uniqueness result is proved. □

7.6.5 Example

We present here a more practical example dealing with a reaction–diffusion–transport system. The system is governed by:

$$\begin{aligned} &\frac{\partial u}{\partial t} - \text{div}(\beta \nabla u) + \varpi . \nabla u + F(x,t,u) = f(x,t), \quad (x,t) \in \mathcal{Q} \\ &u(.,t) = 0 \quad \text{on } \Sigma = (0,T) \times \partial\Omega \\ &u(0) = u_0 \quad \text{on } \Omega. \end{aligned} \tag{7.125}$$

Here, Ω is an open bounded set in $\mathbb{R}^m$, $m \geq 1$, with the boundary $\partial\Omega$ sufficiently regular, for example the oceanic (or atmospheric) domain: U represents the concentration, for example, of some biochemical pollutants in the studied domain Ω, β denotes the coefficient of eddy diffusivity, and ϖ is the three-dimensional velocity field of the fluid or the air. The second term in the equation accounts for pollutant movement by diffusion, the third term represents the transport of the pollutant by the flow field and the fourth term F (non-linear function) represents the reaction term which models the interaction between the pollutant and others biological elements (*e.g.*, phytoplankton, zooplankton, nutrients, *etc.*). The function F is depended, in addition to the concentration U, on the concentrations of the other biological elements, represented by a vector function $\mathbf{C}$, and on some parameters which describe the interaction between the pollutant and these other biological elements, represented by a vector function $\mathbf{P}$, i.e., $F(x,t,U) := \mathcal{F}(\mathbf{C}(x,t), \mathbf{P}(x,t), U)$. The right-hand side of the equation, f, may consist of agents (for example, biological agents capable of producing biodegradation of the pollutant), or chemical or physical extraction.

Remark 7.37. The oceanic and atmospheric circulations are strongly sensitive to the parametrization of the vertical turbulent diffusion. The turbulent flux are usually modeled by the dissipative term (which corresponds to Reynolds stresses with a coefficient of eddy viscosity β) and linked to large-scale oceans (or atmospheres) by using the mixing coefficients β_h and β_v (which have very different behaviors): β_h (resp. β_v) denotes the coefficient of horizontal (resp. vertical) eddy viscosity. ◇

For the mathematical setting, we take $\mathcal{D} = H_0^1(\Omega)$ and $\mathcal{D}' = H^{-1}(\Omega)$. According to Sobolev embedding theorem, we have that $\mathcal{D}$ is embedded in $L^4(\Omega)$ provided $m \leq 4$. Physically this condition on the dimension is not too restrictive, since in this case $m = 3$. The operator A is $-\mathrm{div}(\beta\nabla.)$ with Dirichlet boundary condition and K is the operator $\varpi.\nabla$. For the non-linear operators we suppose that they satisfy the hypothesis (7.78).

If we assume that the coefficient β is positive and bounded function, then the operator A is continuous and coercive and we denote the coercivity constant by $\nu = \min_{\Omega}(\beta)$.

Assume now that the velocity field ϖ is known and satisfies $\varpi \in L^\infty(Q)$ and $\mathrm{div}(\varpi) = 0$, so we have easily the estimate (7.81), where $\gamma_\infty = \| \varpi \|^2_{L^\infty(Q)}/\nu$ and that $K^* = -K$.

Remark 7.38. If we suppose that the initial condition is in $H_0^1(\Omega)$ (compatibility condition) we can obtain for the non-linear problem (7.125), by removing the terms $F(.,u) + K(\varpi, u)$ on the right-hand side of (7.125) and by using Theorem 6.20, the regularity result $u \in H^{2,1}(\mathcal{Q})$. ◇

In the cost functional the operator $\mathcal{C}$ represents the regional and temporal variation in the cost of pollutant extraction. The observation denotes the acceptable pollution level in the studied region. Since all the assumptions of our abstract results are satisfied by the example in this particular case, so our study applies.

Remark 7.39. The vector function $\mathbf{P}$, in the real application, is often badly known. So, we can use the parameter estimation technique in order to identify this vector by using, for example, satellite data or the ocean colour satellites, which measure the colour of the ocean's surface (this instrument is often used to measure the concentration of chlorophyll in surface waters, from which the concentration of the desired biological or biochemical element can be inferred). ◇

8

Stabilization and Robust Control Problem

This chapter contains the essential and fundamental developments of the robust control theory of distributed parameter systems. This area concerns the investigation of the control, stability and adjoint control optimization of infinite-dimensional dissipative dynamical systems. The considered systems are derived from spatially and time-dependent partial differential equations associated with boundary-value problems. We recall that in our approach it is not assumed that the system is stabilizable or detectable. Moreover, we are interested in the robust regulation of the deviation of the systems from the desired target, by analyzing *full non-linear* systems, which models large perturbations to the desired target.

The objective of this chapter is to describe robust control theory in different situations. First, we use a very basic problem to explain the theory as simply as possible. Second, we present the theory for general linear evolutive problems. Then, we present an estimate parameter problem in the case of bilinear systems. Finally, we analyze full non-linear robust control problems, with particular attention to time-delay problems. In each section, we study the existence, uniqueness and optimality conditions for the optimal solution (by using adjoint problems). We develop our study for different realistic cases of observations and controls. The numerical aspect (based on the adjoint problem) is presented in the next chapter.

8.1 Motivation and Objectives

The goal of the robust control theory of partial differential equations (PDEs) is to take into account uncertainty (such as discrepancy or errors between reality and mathematical models used for controller design, and unmeasured noises and disturbances that act on the physical, biological or economical plants, fluctuations, *etc.*) and instability (because uncertainty modifies the system behavior).

In our approach, we transform robust stability and performance problems into constrained game-type minimax optimization ones, and in turn transform these into unconstrained game-type problems. The objective of robust control is then to compensate for the undesirable effects of system disturbances through control actions such that a cost function achieves its minimum for the worst disturbances, *i.e.*, to find the best control which takes into account the worst-case disturbance. The problem is then to find a saddle point of a functional calculus (called *cost* or *objective* or *performance* functional) depending on the control and the disturbance (intervening either in the initial conditions, or in the boundary conditions or in the equation itself) and on the solution of the perturbed PDE.

For more convenience, we recall here the data using in robust control problems (see Chapter 1):

1. A "control" variable φ in a set $U_{\rm ad}$ (set of "admissible controls") and a "disturbance" variable ψ in a set $V_{\rm ad}$ (set of "admissible disturbances").
2. The state $u(\varphi, \psi)$ of the system to be controlled, which is given, for a chosen control-disturbance (φ, ψ), by the resolution of a perturbed equation
$$\tilde{\mathcal{F}}u(\varphi, \psi) = \text{"given function of } (\varphi, \psi)\text{"}$$
where $\tilde{\mathcal{F}}$ is an operator (supposed to be known) which represents the system to be controlled ($\tilde{\mathcal{F}}$ is the perturbation of the model $\mathcal{F}$ of the studied system).
3. An "observation" $u_{\rm obs}$ which is supposed to be known exactly[1] (for example the desired tolerance for the perturbation or the offset given by measurements).
4. A "cost" functional $J(\varphi, \psi)$ which is defined from a numerical and positive function $G(X, Y)$ by
$$J(\varphi, \psi) = G((\varphi, \psi), u(\varphi, \psi)).$$
We want to find a saddle point of J, *i.e.*, a solution $(\varphi^*, \psi^*) \in U_{\rm ad} \times V_{\rm ad}$ of
$$J(\varphi^*, \psi) \leq J(\varphi^*, \psi^*) \leq J(\varphi, \psi^*) \quad \forall \varphi \in U_{\rm ad}, \psi \in V_{\rm ad}.$$

In a similar way to optimal control problems, it should be noted that there is no general method to analyze the problems of robust control but only a process to be followed for each situation:

(*i*) solve the initial problem (analysis of PDEs, existence of solutions, stability according to the data, regularity, differentiability of the operators, *etc.*)

(*ii*) define the function or the parameter to be identified and the type of disturbance to be controlled

[1] In order to simplify, the reader can assume that this function is null.

(*iii*) introduce and solve the perturbed problem which plays the role of the primal problem (analysis of PDEs, existence of solutions, stability according to the data, regularity, differentiability of the operators, *etc.*)

(*iv*) define the cost functional, which is dependent on control and disturbance functions

(*v*) obtain the existence of an optimal solution (as a saddle point of the cost functional) and analyze the necessary (and if possible the sufficient) conditions of optimality (which require to obtain before a very fine regularity on the state functions)

(*vi*) characterize the optimal solutions

(*vii*) define an algorithm that solves numerically the robust control problem.

We shall now present a basic framework for a robust control problem.

8.2 Basic Framework

In this section, we take an abstract boundary value problem. We give the space of controls and disturbances, the admissibility set of controls and of disturbances, the control and the disturbance variables, and the observation operator. The robust control we consider is to maintain the target state variables by taking into account the influence of data noise, while the desired power level and adjustment costs are taken into consideration. We define the cost functional and the adjoint problem, corresponding to the primal problem. The main result of this study is the characterization of optimal solutions.

Let $\mathcal{F}$ be a continuous linear partial differential operator from X into E, where X and E are, for example, Hilbert spaces. The space X contains in its definition some appropriate boundary conditions. We assume that the corresponding boundary value problem is *well-posed* (or correctly-set) in Hadamard sense, *i.e.*, $\mathcal{F}$ is an isomorphism from X onto E.

Let us consider the abstract boundary-value problem as the form

$$\mathcal{F}(w) = h \in E. \tag{8.1}$$

Problem (8.1) admits a unique solution $w \in X$.

In the following, the solution w of problem (8.1) will be treated as the target function. We are then interested in the robust regulation of the deviation of the problem from the desired target w. We study the system, which models large perturbation u to the target w. Hence we consider the following perturbed system:

$$\mathcal{F}(w + u) = h + g, \tag{8.2}$$

where g is in E.

Since the operator $\mathcal{F}$ is linear and w is a solution of (8.1) then the perturbation u satisfies

$$\mathcal{F}(u) = g, \tag{8.3}$$

where g is in E.

Let U_1 and U_2 be two spaces of controls and disturbances, respectively, which are assumed to be, for example, Hilbert spaces. Assume now that the value g is decomposed into the disturbance $\psi \in U_2$, the control $\varphi \in U_1$ and a known function $f \in E$, *i.e.*,

$$g := f + \mathcal{B}_1\varphi + \mathcal{B}_2\psi,$$

where $\mathcal{B}_i$, for $i = 1, 2$, are bounded operators and such that $\mathcal{B}_i \in \mathcal{L}(U_i; E)$, for $i = 1, 2$.

So, the function u is assumed to be related to the disturbance ψ and the control φ through the problem (8.3):

$$\mathcal{F}(u) = f + \mathcal{B}_1\varphi + \mathcal{B}_2\psi. \tag{8.4}$$

For every $\varphi \in U_1$, $\psi \in U_2$ and $f \in E$, the problem (8.4) admits a unique solution $u \in X$ such that

$$u = \mathcal{F}^{-1}(f + \mathcal{B}_1\varphi + \mathcal{B}_2\psi). \tag{8.5}$$

Let U_{ad} be a closed convex non-empty subset of U_1 (the admissibility set of controls), V_{ad} be a closed convex non-empty subset of U_2 (the admissibility set of disturbances) and $\mathcal{C} \in \mathcal{L}(X; M)$ (where M, for example, is a Hilbert space), be the *observation* operator. The robust control we consider, for example, is to maintain the target state variables while the desired power level and adjustment costs are taken into consideration. In particular, we will study the following robust control problem: find $(u, \varphi, \psi) \in X \times U_{\text{ad}} \times V_{\text{ad}}$ such that the following cost functional, in the reduced form

$$J(\varphi, \psi) = \frac{1}{2} \parallel \mathcal{C}u(\varphi, \psi) - u_{\text{obs}} \parallel_X^2 + \frac{\beta}{2} \parallel \varphi \parallel_{U_1}^2 - \frac{\gamma}{2} \parallel \psi \parallel_{U_2}^2 \tag{8.6}$$

is minimized with respect to φ and maximized with respect to ψ subject to the problem (8.4), where $u_{\text{obs}} \in M$ is the target (the observation given, for example, by experiment measurements) and the constant $\beta > 0$ and $\gamma > 0$.

Remark 8.1. (*i*) The coefficient β can be interpreted as the measure of the price of the control (that the engineer can afford) and the coefficient γ can be interpreted as the measure of the price of the disturbance (that the nature or environnement can afford).
(*ii*) The cost functional J depends on the control φ, the disturbance ψ and the state function $u(\varphi, \psi)$ (solution of problem (8.4) corresponding to the data (φ, ψ)). In order to simplify the presentation we have used, in the expression (8.6), the reduced form of the functional J, *i.e.*, $J(\varphi, \psi)$ in place of the form $J(\varphi, \psi, u(\varphi, \psi))$. ♢

Since $v \longrightarrow \mathcal{C}v - u_{\text{obs}}$ is a continuous affine function from X into M and the norm is a continuous function then the functional J is continuous. Moreover, according to the expression of J, the functional $J_\psi : \varphi \in U_1 \longrightarrow J(\varphi, \psi)$ is strictly convex, the functional $J_\varphi : \psi \in U_2 \longrightarrow J(\varphi, \psi)$ is strictly concave, and $J_\psi(\varphi) \longrightarrow +\infty$ if $\varphi \in U_{\text{ad}}$, $\| \varphi \|_{U_1} \longrightarrow \infty$, $J_\varphi(\psi) \longrightarrow -\infty$ if $\psi \in V_{\text{ad}}$, $\| \psi \|_{U_2} \longrightarrow \infty$. Consequently, there exists a unique optimal solution (see Section 5.3) $(\varphi^*, \psi^*, u^*) \in U_{\text{ad}} \times V_{\text{ad}} \times X$ such that (φ^*, ψ^*) is the saddle point of J, *i.e.*, the solution of

$$J(\varphi^*, \psi) \leq J(\varphi^*, \psi^*) \leq J(\varphi, \psi^*) \quad \forall \varphi \in U_{\text{ad}}, \psi \in V_{\text{ad}} \tag{8.7}$$

and $u^* = u(\varphi^*, \psi^*)$ is the solution of (8.4), corresponding to (φ^*, ψ^*).

It follows directly that the functional J is G-differentiable, and then the saddle point solution (φ^*, ψ^*) satisfies the following inequalities (see also Section 5.3):

$$\begin{aligned} &\frac{\partial J}{\partial \varphi}(\varphi^*, \psi^*).(\varphi - \varphi^*) \geq 0 \quad \forall \varphi \in U_{\text{ad}}, \\ &\frac{\partial J}{\partial \psi}(\varphi^*, \psi^*).(\psi - \psi^*) \leq 0 \quad \forall \psi \in V_{\text{ad}}. \end{aligned} \tag{8.8}$$

Since $u(\varphi, \psi)$ satisfies (8.5) then, by differentiation, we have that

$$\begin{aligned} &\frac{\partial u}{\partial \varphi}(\varphi^*, \psi^*).\theta = \mathcal{F}^{-1}(\mathcal{B}_1 \theta), \quad \forall \theta \in U_1, \\ &\frac{\partial u}{\partial \psi}(\varphi^*, \psi^*).\zeta = \mathcal{F}^{-1}(\mathcal{B}_2 \zeta), \quad \forall \zeta \in U_2 \end{aligned}$$

and, in particular, for all $\varphi \in U_{\text{ad}}$ and $\psi \in V_{\text{ad}}$, we have

$$\begin{aligned} &\frac{\partial u}{\partial \varphi}(\varphi^*, \psi^*).(\varphi - \varphi^*) = \mathcal{F}^{-1}(\mathcal{B}_1(\varphi - \varphi^*)) = u(\varphi, \psi^*) - u(\varphi^*, \psi^*), \\ &\frac{\partial u}{\partial \psi}(\varphi^*, \psi^*).(\psi - \psi^*) = \mathcal{F}^{-1}(\mathcal{B}_2(\psi - \psi^*)) = u(\varphi^*, \psi) - u(\varphi^*, \psi^*). \end{aligned} \tag{8.9}$$

According to (8.9), the condition (8.8) can be written as (for all $\varphi \in U_{\text{ad}}$ and $\psi \in V_{\text{ad}}$)

$$\begin{aligned} &(\mathcal{C}u(\varphi^*, \psi^*) - u_{\text{obs}}, \mathcal{C}(u(\varphi, \psi^*) - u(\varphi^*, \psi^*)))_X + \beta(\varphi^*, \varphi - \varphi^*)_{U_1} \geq 0, \\ &(\mathcal{C}u(\varphi^*, \psi^*) - u_{\text{obs}}, \mathcal{C}(u(\varphi^*, \psi) - u(\varphi^*, \psi^*)))_X - \gamma(\psi^*, \psi - \psi^*)_{U_2} \leq 0. \end{aligned} \tag{8.10}$$

In order to simplify the relation (8.10), we introduce the adjoint problem

$$\mathcal{F}^*(\tilde{u}(\varphi, \psi)) = \mathcal{C}^* \Lambda(\mathcal{C}u(\varphi, \psi) - u_{\text{obs}}), \tag{8.11}$$

where $u(\varphi, \psi)$ is the solution of the primal problem (8.4), corresponding to data $(\varphi, \psi) \in U_1 \times U_2$, $\mathcal{C}^* \in \mathcal{L}(M', X')$ is the adjoint of the operator $\mathcal{C}$, $\mathcal{F}^*$ is the adjoint of the operator solution $\mathcal{F}$ and Λ is the canonical isomorphism from M onto M' (such that $\langle p, q^* \rangle_{M,M'} = (\Lambda p, q^*)_{M'} = (p, \Lambda^{-1} q^*)_M$, $\forall (p, q^*) \in M \times M'$), where M' (respectively X') is the dual of M (respectively of X).

Then

$$\begin{aligned}
(\mathcal{C}u(\varphi^*,\psi^*) - u_{\text{obs}}, \mathcal{C}(u(\varphi,\psi^*) - u(\varphi^*,\psi^*)))_X \\
&= \langle \mathcal{C}^*\Lambda(\mathcal{C}u(\varphi^*,\psi^*) - u_{\text{obs}}), u(\varphi,\psi^*) - u(\varphi^*,\psi^*)\rangle_{X',X} \\
&= \langle \mathcal{F}^*(\tilde{u}^*), u(\varphi,\psi^*) - u(\varphi^*,\psi^*)\rangle_{X',X} \\
&= (\tilde{u}^*, \mathcal{F}(u(\varphi,\psi^*) - u(\varphi^*,\psi^*)))_{E',E} \\
&= (\tilde{u}^*, \mathcal{B}_1(\varphi - \varphi^*))_{E',E} \\
&= (\Lambda_{U_1}^{-1}\mathcal{B}_1^*\tilde{u}^*, \varphi - \varphi^*)_{U_1}, \forall \varphi \in U_{\text{ad}}
\end{aligned}$$

and (in the same way)

$$\begin{aligned}
(\mathcal{C}u(\varphi^*,\psi^*) - u_{\text{obs}}, \mathcal{C}(u(\varphi^*,\psi) - u(\varphi^*,\psi^*)))_X \\
&= (\Lambda_{U_2}^{-1}\mathcal{B}_2^*\tilde{u}^*, \psi - \psi^*)_{U_2}, \forall \psi \in V_{\text{ad}},
\end{aligned}$$

where $\mathcal{B}_i^* \in \mathcal{L}(E', U_i')$ is the adjoint of $\mathcal{B}_i$, U_i' (respectively E') is the dual of U_i (respectively of E) and Λ_{U_i} is the canonical isomorphism from U_i onto U_i', for $i = 1, 2$, and $\tilde{u}^* = \tilde{u}(\varphi^*, \psi^*)$ is the solution of the adjoint problem (8.11), corresponding to the primal solution $u(\varphi^*, \psi^*)$.

Thus, the condition (8.10) can be written as

$$\begin{aligned}
&(\Lambda_{U_1}^{-1}\mathcal{B}_1^*\tilde{u}^* + \beta\varphi^*, \varphi - \varphi^*)_{U_1} \geq 0, \forall \varphi \in U_{\text{ad}}, \\
&(\Lambda_{U_2}^{-1}\mathcal{B}_2^*\tilde{u}^* - \gamma\psi^*, \psi - \psi^*)_{U_2} \leq 0, \forall \psi \in V_{\text{ad}}.
\end{aligned}$$

We have proved that the optimal solution $(\varphi^*, \psi^*, u^*) \in U_{\text{ad}} \times V_{\text{ad}} \times X$, solution of the problem (8.7) subject to the problem (8.4), is characterized by the following optimality system which includes the direct (or primal) problem and the adjoint problem, linked by inequalities:

$$\begin{aligned}
&\mathcal{F}(u) = f + \mathcal{B}_1\varphi + \mathcal{B}_2\psi, \\
&\mathcal{F}^*(\tilde{u}^*) = \mathcal{C}^*\Lambda(\mathcal{C}u^* - u_{\text{obs}}), \\
&(\Lambda_{U_1}^{-1}\mathcal{B}_1^*\tilde{u}^* + \beta\varphi^*, \varphi - \varphi^*)_{U_1} \geq 0, \forall \varphi \in U_{\text{ad}}, \\
&(\Lambda_{U_2}^{-1}\mathcal{B}_2^*\tilde{u}^* - \gamma\psi^*, \psi - \psi^*)_{U_2} \leq 0, \forall \psi \in V_{\text{ad}}.
\end{aligned} \tag{8.12}$$

8.3 Linear Robust Control Problems

In this section, we consider problems of linear parabolic type in a more general framework (the same problems as in the previous chapter). The notations, the assumptions and the spaces are the same as in Chapter 6.

8.3.1 Position of the Problem, and the Existence and Uniqueness of the Optimal Solution

For f and U_0 given such that $f \in L^2(0, T; V')$ and $U_0 \in H$, and f_0 given in the space $L^2(0,T;V')$, we consider the following linear evolutive equation:

$$\begin{aligned} &\frac{\partial U}{\partial t} + A(t)U = f_0 + f \in L^2(0,T;V'), \\ &U(0) = U_0 \in H. \end{aligned} \tag{8.13}$$

According to Theorem 6.8 and Proposition 6.9, the problem (8.13) admits a unique solution $U \in \mathcal{W} \cap \mathcal{C}([0,T];H)$ (since $f \in L^2(0,T;V')$, $U_0 \in H$ and $g \in L^2(0,T;V')$), where $\mathcal{W} = \{u \in L^2(0,T;V) : \partial u/\partial t \in L^2(0,T;V')\}$. Moreover, if $v_1 \in \mathcal{W}$ (respectively $v_2 \in \mathcal{W}$) is the solution of (8.13), with data $(f_0, f_1, v_{01}) \in L^2(0,T;V') \times L^2(0,T;V') \times H$ (respectively with data $(f_0, f_2, v_{02}) \in L^2(0,T;V') \times L^2(0,T;V') \times H$), then

$$\| v_1 - v_2 \|_{\mathcal{W}}^2 \leq C(\| f_1 - f_2 \|_{L^2(0,T;V')}^2 + \| v_{01} - v_{02} \|_H^2).$$

The solution U with the corresponding forcing f will be referred as the target function for the control problem. We are interested in the robust regulation of the deviation of the problem from the desired target (U, f). We analyze the full equation which models large perturbations (u, g) to the target (U, f) with known initial conditions. Hence we consider the equation of the perturbation (since the problem (8.13) is linear)

$$\begin{aligned} &\frac{\partial u}{\partial t} + A(t)u = g \quad \text{on } \mathcal{Q}, \\ &u(0) = u_0 \quad \text{on } \Omega, \end{aligned} \tag{8.14}$$

where g and u_0 are perturbations of f and U_0 respectively.

The problem (8.14) is similar to (8.13) so we have the existence and the uniqueness of the solution u in $\mathcal{W}$, and the following Lipschitz continuity result:

$$\| u_1 - u_2 \|_{\mathcal{W}}^2 \leq C(\| g_1 - g_2 \|_{L^2(0,T;V')}^2 + \| u_{01} - u_{02} \|_H^2), \tag{8.15}$$

where $u_1 \in \mathcal{W}$ (respectively $u_2 \in \mathcal{W}$) is the solution of (8.14) with data $(g_1, u_{01}) \in L^2(0,T;V') \times H$ (respectively with $(g_2, u_{02}) \in L^2(0,T;V') \times H$).

In our control framework, the value g is decomposed into the disturbance $\psi \in U_2$ and the control $\phi \in U_1$. Thus, we write

$$g = \mathcal{B}_1 \phi + \mathcal{B}_2 \psi,$$

where $\mathcal{B}_i$ are given in $\mathcal{L}(U_i, L^2(0,T;V'))$ for $i = 1, 2$, U_1 is the space of controls and U_2 is the space of disturbances which are assumed to be Hilbert spaces.

The objective in the robust control problem is to find the best control ϕ in the presence of the disturbance (or noise) ψ which maximally spoils the control objective. The function u is assumed to be related to the disturbance ψ and control ϕ through the problem (8.14)

$$
\begin{aligned}
&\frac{\partial u}{\partial t} + A(t)u = \mathcal{B}_1\phi + \mathcal{B}_2\psi \quad \text{on } \mathcal{Q}, \\
&u(0) = u_0 \quad \text{on } \Omega.
\end{aligned}
\tag{8.16}
$$

For u_0 given in H, we consider the mapping solution

$$\mathbf{U} : (\phi, \psi) \longrightarrow u = \mathbf{U}(\phi, \psi) \text{ from } U_1 \times U_2 \text{ into } \mathcal{W} \text{ defined by (8.16).}$$

In this section, the cost functional is given, in the reduced form, by

$$J(\phi, \psi) = \frac{1}{2} \parallel \mathcal{C}u(\phi, \psi) - u_{\mathrm{obs}} \parallel_M^2 + \frac{\beta}{2} \parallel \phi \parallel_{U_1}^2 - \frac{\gamma}{2} \parallel \psi \parallel_{U_2}^2, \tag{8.17}$$

where the scalar control parameters $\alpha, \beta > 0$ are fixed and $u_{\mathrm{obs}} \in M$ is a known observation. The *observation* operator $\mathcal{C}$ is such that $\mathcal{C} \in \mathcal{L}(\mathcal{W}; M)$, with M a Hilbert space.

The robust control problem corresponds to seek saddle points on $U_{\mathrm{ad}} \times V_{\mathrm{ad}}$ of the functional J, *i.e.*, to find $(\phi^*, \psi^*) \in U_{\mathrm{ad}} \times V_{\mathrm{ad}}$ such that

$$J(\phi^*, \psi) \leq J(\phi^*, \psi^*) \leq J(\phi, \psi^*) \;\; \forall (\phi, \psi) \in U_{\mathrm{ad}} \times V_{\mathrm{ad}}, \tag{8.18}$$

with the admissibility set of controls U_{ad} (respectively the admissibility set of disturbances V_{ad}) is (given) non-empty, closed, convex, bounded subsets of U_1 (respectively of U_2).

Since $v \longrightarrow \mathcal{C}v - u_{\mathrm{obs}}$ is a continuous affine function from $L^2(0, T; V)$ into M, the norm is a lower semi-continuous function we have that the map $\mathcal{P}_\psi : \phi \longrightarrow J(\phi, \psi)$ is lower semi-continuous for all $\psi \in U_2$ and the map $\mathcal{Q}_\phi : \psi \longrightarrow J(\phi, \psi)$ is upper semi-continuous for all $\phi \in U_1$. Moreover, according to the nature of J, we have that the maps $\mathcal{P}_\psi$ and $\mathcal{Q}_\phi$ are strictly convex and strictly concave respectively. Consequently, (by using the minimax theorems in infinite dimensions presented in Chapter 5) there exists a unique optimal solution $(u^*, \phi^*, \psi^*) \in \mathcal{W} \times U_{\mathrm{ad}} \times V_{\mathrm{ad}}$ such that (ϕ^*, ψ^*) is the solution of (8.18) and $u^* = \mathbf{U}(\phi^*, \psi^*)$ is the solution of (8.16), corresponding to (ϕ^*, ψ^*).

Remark 8.2. In order to obtain the existence and the uniqueness of the optimal solution, it is not necessary to impose the boundedness condition on the spaces U_{ad} and V_{ad}. Indeed, $\mathcal{P}_\psi(\phi) \longrightarrow +\infty$ if $\phi \in U_{\mathrm{ad}}, \;\; \parallel \phi \parallel_{U_1} \longrightarrow \infty$, and $\mathcal{Q}_\phi(\psi) \longrightarrow -\infty$ if $\psi \in V_{\mathrm{ad}}, \;\; \parallel \psi \parallel_{U_2} \longrightarrow \infty$. ♢

8.3.2 Optimality Conditions and Identification of the Gradients

As in the basic framework (Section 8.2), the functional J is differentiable and $(\phi^*, \psi^*, u^*) \in U_{\mathrm{ad}} \times V_{\mathrm{ad}} \times \mathcal{W}$ is an optimal solution if and only if

$$
\begin{aligned}
&\frac{\partial J}{\partial \phi}(\phi^*, \psi^*).(\phi - \phi^*) \geq 0 \quad \forall \phi \in U_{\mathrm{ad}}, \\
&\frac{\partial J}{\partial \psi}(\phi^*, \psi^*).(\psi - \psi^*) \leq 0 \quad \forall \psi \in V_{\mathrm{ad}}
\end{aligned}
$$

and then

$$(\mathcal{C}u^* - u_{\mathrm{obs}}, \mathcal{C}(\frac{\partial u}{\partial \phi}(\phi^*,\psi^*).(\phi-\phi^*))_M + \beta(\phi^*,\phi-\phi^*)_{U_1} \geq 0 \quad \forall \phi \in U_{\mathrm{ad}},$$
$$(\mathcal{C}u^* - u_{\mathrm{obs}}, \mathcal{C}(\frac{\partial u}{\partial \psi}(\phi^*,\psi^*).(\psi-\psi^*))_M - \gamma(\psi^*,\psi-\psi^*)_{U_2} \leq 0 \quad \forall \psi \in V_{\mathrm{ad}},$$

where $u^* = \mathbf{U}(\phi^*,\psi^*)$ is the solution of (8.16), corresponding to the solution (ϕ^*,ψ^*).

Since, for all $(\phi,\psi) \in U_1 \times U_2$, the function $\mathbf{U}(\phi,\psi)$ is a solution of (8.16) then the derivative of the operator solution $\mathbf{U}$ at point (ϕ^*,ψ^*) is defined by $\mathbf{U}'(\phi^*,\psi^*) : (h_1,h_2) \in U_1 \times U_2 \longrightarrow w \in \mathcal{W}$ such that

$$w = \frac{\partial u}{\partial \phi}(\phi^*,\psi^*).h_1 + \frac{\partial u}{\partial \psi}(\phi^*,\psi^*).h_2$$

is the unique solution of the following problem:

$$\begin{aligned} &\frac{\partial w}{\partial t} + A(t)w = \mathcal{B}_1 h_1 + \mathcal{B}_2 h_2 \quad \text{on } \mathcal{Q}, \\ &w(0) = 0 \quad \text{on } \Omega. \end{aligned} \tag{8.19}$$

The problem (8.19) is the same as the problem (8.13), with the null initial condition. Then we can deduce that

$$\frac{\partial u}{\partial \phi}(\phi^*,\psi^*).(\phi-\phi^*) = \mathbf{U}(\phi,\psi^*) - \mathbf{U}(\phi^*,\psi^*) = \mathbf{U}(\phi,\psi^*) - u^*,$$
$$\frac{\partial u}{\partial \psi}(\phi^*,\psi^*).(\psi-\psi^*) = \mathbf{U}(\phi^*,\psi) - \mathbf{U}(\phi^*,\psi^*) = \mathbf{U}(\phi^*,\psi) - u^*$$

and then

$$(\mathcal{C}u^* - u_{\mathrm{obs}}, \mathcal{C}(\mathbf{U}(\phi,\psi^*) - u^*))_M + \beta(\phi^*,\phi-\phi^*)_{U_1} \geq 0 \quad \forall \phi \in U_{\mathrm{ad}},$$
$$(\mathcal{C}u^* - u_{\mathrm{obs}}, \mathcal{C}(\mathbf{U}(\phi^*,\psi) - u^*))_M - \gamma(\psi^*,\psi-\psi^*)_{U_2} \leq 0 \quad \forall \psi \in V_{\mathrm{ad}}.$$

Otherwise, for all $\phi \in U_{\mathrm{ad}}$ and $\psi \in V_{\mathrm{ad}}$,

$$\begin{aligned} &\langle \mathcal{C}^* \Lambda(\mathcal{C}u^* - u_{\mathrm{obs}}), \mathbf{U}(\phi,\psi^*) - u^* \rangle_{\mathcal{W}',\mathcal{W}} + \beta(\phi^*,\phi-\phi^*)_{U_1} \geq 0, \\ &\langle \mathcal{C}^* \Lambda(\mathcal{C}u^* - u_{\mathrm{obs}}), \mathbf{U}(\phi^*,\psi) - u^* \rangle_{\mathcal{W}',\mathcal{W}} - \gamma(\psi^*,\psi-\psi^*)_{U_2} \leq 0, \end{aligned} \tag{8.20}$$

where Λ is the canonical isomorphism from M onto M' and $\mathcal{C}^* \in \mathcal{L}(M',\mathcal{W}')$ the adjoint of $\mathcal{C}$.

In order to simplify the relation (8.20), we introduce the adjoint problem associated with the primal problem (8.16). For this, we consider the two following situations (as in the previous chapter):

(*i*) the case of $\mathcal{C} \in \mathcal{L}(L^2(0,T;V);M) \subset \mathcal{L}(\mathcal{W};M)$

(*ii*) the case of the final observation: $\mathcal{C}u = \mathcal{D}u(T)$, where $\mathcal{D} \in \mathcal{L}(H;H)$ and $M = H = M'$, since $H' = H$.

Case (i) and the Optimality System

In the case of the operator $\mathcal{C} \in \mathcal{L}(L^2(0,T;V);M)$, the adjoint operator $\mathcal{C}^*$ of $\mathcal{C}$ is in $\mathcal{L}(M', L^2(0,T;V'))$ and the inequality (8.20), for all $\phi \in U_{\text{ad}}$ and $\psi \in V_{\text{ad}}$), becomes

$$\begin{aligned}&\int_0^T \langle \mathcal{C}^* \Lambda(\mathcal{C}u^* - u_{\text{obs}}), \mathbf{U}(\phi, \psi^*) - u^* \rangle_{V',V} \mathrm{d}t + \beta(\phi^*, \phi - \phi^*)_{U_1} \geq 0,\\ &\int_0^T \langle \mathcal{C}^* \Lambda(\mathcal{C}u^* - u_{\text{obs}}), \mathbf{U}(\phi^*, \psi) - u^* \rangle_{V',V} \mathrm{d}t - \gamma(\psi^*, \psi - \psi^*)_{U_2} \leq 0.\end{aligned} \tag{8.21}$$

Let us introduce the following adjoint problem:

$$\begin{aligned}&-\frac{\partial \tilde{u}}{\partial t} + A^*(t)\tilde{u} = \mathcal{C}^* \Lambda(\mathcal{C}u - u_{\text{obs}}) \in L^2(0,T;V'),\\ &\tilde{u}(t = T) = 0 \in H,\end{aligned} \tag{8.22}$$

which admits a unique solution $\tilde{u} \in \mathcal{W}$, where $u = \mathbf{U}(\phi, \psi) \in \mathcal{W}$ and Λ is the canonical isomorphism from M onto M'. To prove this result, we change the variables of System (8.22) by reversing the sense of time, *i.e.*, $t := T - t$ and we apply the same way to obtain the result of Theorem 6.8.

Multiplying the first equation of (8.22) by $u^* - \mathbf{U}^*(\psi)$, where $\mathbf{U}^*(\psi) = \mathbf{U}(\phi^*, \psi)$, and integrating with respect to time we have that

$$\begin{aligned}&-\int_0^T \langle \frac{\partial \tilde{u}^*}{\partial t}, u^* - \mathbf{U}^*(\psi) \rangle_{V',V} \mathrm{d}t + \int_0^T \langle A^*(t)\tilde{u}^*, u^* - \mathbf{U}^*(\psi) \rangle_{V',V} \mathrm{d}t\\ &= \int_0^T \langle \mathcal{C}^* \Lambda(\mathcal{C}u^* - u_{\text{obs}}), u^* - \mathbf{U}^*(\psi) \rangle_{V',V} \mathrm{d}t,\end{aligned}$$

where $\tilde{u}^*$ is the solution of (8.22) which corresponds to the primal solution u^*.

By using the result of Corollary 6.4, we can deduce that

$$\begin{aligned}-\int_0^T \langle \frac{\partial \tilde{u}^*}{\partial t}, u^* - \mathbf{U}^*(\psi) \rangle_{V',V} \mathrm{d}t &= \int_0^T \langle \tilde{u}^*, \frac{\partial u^*}{\partial t} - \frac{\partial \mathbf{U}^*(\psi)}{\partial t} \rangle_{V,V'} \mathrm{d}t\\ &\quad -(\tilde{u}^*(T), (u^* - \mathbf{U}^*(\psi))(T))_H + (\tilde{u}^*(0), (u^* - \mathbf{U}^*(\psi))(0))_H.\end{aligned}$$

Since $\tilde{u}^*(T) = 0$ and $(u^* - \mathbf{U}^*(\psi))(0) = u_0 - u_0 = 0$ then

$$-\int_0^T \langle \frac{\partial \tilde{u}^*}{\partial t}, u^* - \mathbf{U}^*(\psi) \rangle_{V',V} \mathrm{d}t = \int_0^T \langle \tilde{u}^*, \frac{\partial u^*}{\partial t} - \frac{\partial \mathbf{U}^*(\psi)}{\partial t} \rangle_{V,V'} \mathrm{d}t$$

and consequently

$$\begin{aligned}&\int_0^T \langle \tilde{u}^*, \frac{\partial u^*}{\partial t} - \frac{\partial \mathbf{U}^*(\psi)}{\partial t} \rangle_{V,V'} \mathrm{d}t + \int_0^T \langle \tilde{u}^*, A(t)(u^* - \mathbf{U}^*(\psi)) \rangle_{V,V'} \mathrm{d}t\\ &= \int_0^T \langle \mathcal{C}^* \Lambda(\mathcal{C}u^* - u_{\text{obs}}), u^* - \mathbf{U}^*(\psi) \rangle_{V',V} \mathrm{d}t.\end{aligned} \tag{8.23}$$

Since $u^* - \mathbf{U}^*(\psi)$ is the solution of

$$\frac{\partial(u^* - \mathbf{U}^*(\psi))}{\partial t} + A(t)(u^* - \mathbf{U}^*(\psi)) = \mathcal{B}_2(\psi^* - \psi),$$
$$(u^* - \mathbf{U}^*(\psi))(0) = 0,$$

then (8.23) gives

$$\int_0^T \langle \mathcal{C}^* \Lambda(\mathcal{C}u^* - u_{\text{obs}}), u^* - \mathbf{U}^*(\psi)\rangle_{V',V} \mathrm{d}t = \int_0^T \langle \tilde{u}^*, \mathcal{B}_2(\psi^* - \psi)\rangle_{V,V'} \mathrm{d}t$$
$$= \langle \mathcal{B}_2^* \tilde{u}^*, \psi^* - \psi\rangle_{U_2',U_2}$$
$$= (\Lambda_{U_2}^{-1} \mathcal{B}_2^* \tilde{u}^*, \psi^* - \psi)_{U_2},$$

where the operator $\mathcal{B}_i^* \in \mathcal{L}(L^2(0,T;V); U_i')$ is the adjoint of $\mathcal{B}_i$ and Λ_{U_i} is the canonical isomorphism from U_i onto U_i', for $i = 1, 2$.

Therefore, the second inequality of (8.21) can be written as

$$(\Lambda_{U_2}^{-1} \mathcal{B}_2^* \tilde{u}^* - \gamma\psi^*, \psi - \psi^*)_{U_2} \leq 0 \quad \forall \psi \in V_{\text{ad}}. \tag{8.24}$$

In the same way we can prove also that the first inequality of (8.21) can be written as

$$(\Lambda_{U_1}^{-1} \mathcal{B}_1^* \tilde{u}^* + \beta\phi^*, \phi - \phi^*)_{U_1} \geq 0 \quad \forall \phi \in U_{\text{ad}} \tag{8.25}$$

and then we have proved the following theorem, which gives the existence, the uniqueness and the first-order optimality conditions of the optimal solution.

Theorem 8.3. *Under the assumptions of Theorem 6.8, we suppose furthermore that the operator* $\mathcal{C}$ *is in* $\mathcal{L}(L^2(0,T;V); M)$ *and that the operators* $\mathcal{B}_i$ *are in* $\mathcal{L}(U_i; L^2(0,T;V'))$, *for* $i = 1, 2$. *Then there exists a unique optimal solution* $(\phi^*, \psi^*, u^*) \in U_1 \times U_2 \times \mathcal{W}$ *such that*

$$\frac{\partial u^*}{\partial t} + A(t)u^* = \mathcal{B}_1\phi^* + \mathcal{B}_2\psi^*,$$
$$u^*(0) = u_0$$

and

$$(\Lambda_{U_1}^{-1} \mathcal{B}_1^* \tilde{u}^* + \beta\phi^*, \phi - \phi^*)_{U_1} \geq 0 \quad \forall \phi \in U_{\text{ad}},$$
$$(\Lambda_{U_2}^{-1} \mathcal{B}_2^* \tilde{u}^* - \gamma\psi^*, \psi - \psi^*)_{U_2} \leq 0 \quad \forall \psi \in V_{\text{ad}},$$

where Λ_{U_i} *is the canonical isomorphism from* U_i *onto* U_i' *and* $\tilde{u}^*$ *is the unique solution, which is in* $\mathcal{W}$, *of the adjoint problem*

$$-\frac{\partial \tilde{u}^*}{\partial t} + A^*(t)\tilde{u}^* = \mathcal{C}^* \Lambda(\mathcal{C}u^* - u_{\text{obs}}),$$
$$\tilde{u}^*(t = T) = 0,$$

where Λ *is the canonical isomorphism from* M *onto* M'. □

Case (*ii*) and the Optimality System

In the case of $\mathcal{C}u = \mathcal{D}u(T)$, where $\mathcal{D} \in \mathcal{L}(H;H)$ and $M = H$, the observation u_{obs} is in H, $\mathcal{D}^* \in \mathcal{L}(H;H)$ and the cost functional is defined by

$$J(\phi,\psi) = \frac{1}{2} \parallel \mathcal{D}u(T) - u_{\text{obs}} \parallel_H^2 + \frac{\beta}{2} \parallel \phi \parallel_{U_1}^2 - \frac{\gamma}{2} \parallel \psi \parallel_{U_2}^2 \tag{8.26}$$

and for all $(\phi,\psi) \in U_{\text{ad}} \times V_{\text{ad}}$ the inequality (8.20) is equivalent to

$$\begin{aligned} &(\mathcal{D}^*(\mathcal{D}u^*(T) - u_{\text{obs}}), (\mathbf{U}(\phi,\psi^*) - u^*)(T))_H + \beta(\phi^*, \phi - \phi^*)_{U_1} \geq 0, \\ &(\mathcal{D}^*(\mathcal{D}u^*(T) - u_{\text{obs}}), (\mathbf{U}(\phi^*,\psi) - u^*)(T))_H - \gamma(\psi^*, \psi - \psi^*)_{U_2} \leq 0 \end{aligned} \tag{8.27}$$

where $u^* = \mathbf{U}(\phi^*,\psi^*)$ is the primal solution which corresponds to the optimal solution (ϕ^*,ψ^*).

Let us introduce the following adjoint problem:

$$\begin{aligned} &-\frac{\partial \tilde{u}}{\partial t} + A^*(t)\tilde{u} = 0, \\ &\tilde{u}(t=T) = \mathcal{D}^*(\mathcal{D}u(t=T) - u_{\text{obs}}) \in H, \end{aligned} \tag{8.28}$$

which admits a unique solution $\tilde{u} \in \mathcal{W}$, where $u = \mathbf{U}(\phi,\psi) \in \mathcal{W}$. To prove this result, we change the variables of System (8.28) by reversing the sense of time, *i.e.*, $t := T - t$ and we apply the same way to obtain the result of Theorem 6.8.

Multiplying the first equation of (8.28) by $u^* - \mathbf{U}^*(\psi)$, where $\mathbf{U}^*(\psi) = \mathbf{U}(\phi^*,\psi)$, and integrating with respect to time we have that

$$-\int_0^T \langle \frac{\partial \tilde{u}^*}{\partial t}, u^* - \mathbf{U}^*(\psi)\rangle_{V',V} \mathrm{d}t + \int_0^T \langle A^*(t)\tilde{u}^*, u^* - \mathbf{U}^*(\psi)\rangle_{V',V} \mathrm{d}t = 0,$$

where $\tilde{u}^*$ is the solution of (8.28) which corresponds to the primal solution $u^* = \mathbf{U}(\phi^*,\psi^*)$.

By using the result of Corollary 6.4, we can deduce that

$$\begin{aligned} -\int_0^T \langle \frac{\partial \tilde{u}^*}{\partial t}, u^* - \mathbf{U}^*(\psi)\rangle_{V',V} \mathrm{d}t &= \int_0^T \langle \tilde{u}^*, \frac{\partial u^*}{\partial t} - \frac{\partial \mathbf{U}^*(\psi)}{\partial t}\rangle_{V,V'} \mathrm{d}t \\ &\quad -(\tilde{u}^*(T), (u^* - \mathbf{U}^*(\psi))(T))_H + (\tilde{u}^*(0), (u^* - \mathbf{U}^*(\psi))(0))_H, \end{aligned}$$

Since $\tilde{u}^*(T) = \mathcal{D}^*(\mathcal{D}u^*(T) - u_{\text{obs}})$ and $(u^* - \mathbf{U}^*(\psi))(0) = u_0 - u_0 = 0$ then

$$\begin{aligned} -\int_0^T \langle \frac{\partial \tilde{u}^*}{\partial t}, u^* - \mathbf{U}^*(\psi)\rangle_{V',V} \mathrm{d}t &= \int_0^T \langle \tilde{u}^*, \frac{\partial u^*}{\partial t} - \frac{\partial \mathbf{U}^*(\psi)}{\partial t}\rangle_{V,V'} \mathrm{d}t \\ &\quad -(\mathcal{D}^*(\mathcal{D}u^*(T) - u_{\text{obs}}), (u^* - \mathbf{U}^*(\psi))(T))_H. \end{aligned}$$

Consequently,

$$\int_0^T \langle \tilde{u}^*, \frac{\partial(u^* - \mathbf{U}^*(\psi))}{\partial t} \rangle_{V,V'} \mathrm{d}t + \int_0^T \langle \tilde{u}^*, A(t)(u^* - \mathbf{U}^*(\psi)) \rangle_{V,V'} \mathrm{d}t \\ = (\mathcal{D}^*(\mathcal{D}u^*(T) - u_{\mathrm{obs}}), (u^* - \mathbf{U}^*(\psi))(T))_H. \tag{8.29}$$

Since $u^* - \mathbf{U}^*(\psi)$ is a solution of

$$\frac{\partial(u^* - \mathbf{U}^*(\psi))}{\partial t} + A(t)(u^* - \mathbf{U}^*(\psi)) = \mathcal{B}_2(\psi^* - \psi),$$
$$(u^* - \mathbf{U}^*(\psi))(0) = 0,$$

then (8.29) gives

$$\begin{aligned}(\mathcal{D}^*(\mathcal{D}u^*(T) - u_{\mathrm{obs}}, (u^* - \mathbf{U}^*(\psi))(T))_H &= \int_0^T \langle \tilde{u}^*, \mathcal{B}_2(\psi^* - \psi) \rangle_{V,V'} \mathrm{d}t \\ &= \langle \mathcal{B}_2^* \tilde{u}^*, \psi^* - \psi \rangle_{U_2',U_2} \\ &= (\Lambda_{U_2}^{-1} \mathcal{B}_2^* \tilde{u}^*, \psi^* - \psi)_{U_2},\end{aligned}$$

where the operator $\mathcal{B}_i^* \in \mathcal{L}(L^2(0,T;V); U_i')$ is the adjoint of $\mathcal{B}_i$ and Λ_{U_i} is the canonical isomorphism from U_i onto U_i', for $i = 1, 2$.

Therefore, the second inequality of (8.27) can be written as

$$(\Lambda_{U_2}^{-1} \mathcal{B}_2^* \tilde{u}^* - \gamma\psi^*, \psi - \psi^*)_{U_2} \leq 0 \quad \forall \psi \in V_{\mathrm{ad}}. \tag{8.30}$$

In the same way we can prove that the first inequality of (8.27) can also be written as

$$(\Lambda_{U_1}^{-1} \mathcal{B}_1^* \tilde{u}^* + \beta\phi^*, \phi - \phi^*)_{U_1} \geq 0 \quad \forall \phi \in U_{\mathrm{ad}} \tag{8.31}$$

and then we have proved the following characterization of the robust control theorem.

Theorem 8.4. *Under the assumptions of Theorem 6.8, we suppose furthermore that the operators* $\mathcal{B}_i \in \mathcal{L}(U_i; L^2(0,T;V'))$, *for* $i = 1, 2$. *Then there exists a unique optimal solution* $(\phi^*, \psi^*, u^*) \in U_1 \times U_2 \times \mathcal{W}$ *such that*

$$\frac{\partial u^*}{\partial t} + A(t)u^* = \mathcal{B}_1\phi^* + \mathcal{B}_2\psi^*,$$
$$u^*(0) = u_0$$

and

$$(\Lambda_{U_1}^{-1} \mathcal{B}_1^* \tilde{u}^* + \beta\phi^*, \phi - \phi^*)_{U_1} \geq 0 \quad \forall \phi \in U_{\mathrm{ad}},$$
$$(\Lambda_{U_2}^{-1} \mathcal{B}_2^* \tilde{u}^* - \gamma\psi^*, \psi - \psi^*)_{U_2} \leq 0 \quad \forall \psi \in V_{\mathrm{ad}},$$

where Λ_{U_i} *is the canonical isomorphism from* U_i *onto* U_i' *and* $\tilde{u}^*$ *is the unique solution, which is in* $\mathcal{W}$, *of the adjoint problem*

$$-\frac{\partial \tilde{u}^*}{\partial t} + A^*(t)\tilde{u}^* = 0,$$
$$\tilde{u}^*(t = T) = \mathcal{D}^*(\mathcal{D}u(t = T) - u_{\mathrm{obs}}).$$

□

We conclude this section by the following results. Assume that we have two observations: the final observation v_{obs}, which is in H, and the state observation u_{obs}, which is in M, and therefore the cost functional can be defined by

$$\begin{aligned} J(\phi,\psi) = & \frac{\theta_1}{2} \parallel \mathcal{C}u - u_{\text{obs}} \parallel_M^2 + \frac{\theta_2}{2} \parallel \mathcal{D}u(T) - v_{\text{obs}} \parallel_H^2 \\ & + \frac{\beta}{2} \parallel \phi \parallel_{U_1}^2 - \frac{\gamma}{2} \parallel \psi \parallel_{U_2}^2, \end{aligned} \tag{8.32}$$

where $\theta_i \geq 0, \quad i = 1, 2, \theta_1 + \theta_2 > 0, \mathcal{C} \in \mathcal{L}(L^2(0,T;V);M)$ and $\mathcal{D} \in \mathcal{L}(H;H)$. Then we have the following theorem.

Theorem 8.5. *Under the assumptions of Theorem 6.8, we suppose furthermore that the cost J is given by (8.32) with the operators $\mathcal{D} \in \mathcal{L}(H;H)$ and $\mathcal{C} \in \mathcal{L}(L^2(0,T;V);M)$, and that the operators $\mathcal{B}_i \in \mathcal{L}(U_i;L^2(0,T;V'))$, for $i = 1,2$. Then there exists a unique optimal solution $(\phi^*,\psi^*,u^*) \in U_1 \times U_2 \times \mathcal{W}$ such that*

$$\begin{aligned} & \frac{\partial u^*}{\partial t} + A(t)u^* = \mathcal{B}_1\phi^* + \mathcal{B}_2\psi^*, \\ & u^*(0) = u_0 \end{aligned}$$

and

$$\begin{aligned} & (\Lambda_{U_1}^{-1}\mathcal{B}_1^*\tilde{u}^* + \beta\phi^*, \phi - \phi^*)_{U_1} \geq 0 \quad \forall \phi \in U_{\text{ad}}, \\ & (\Lambda_{U_2}^{-1}\mathcal{B}_2^*\tilde{u}^* - \gamma\psi^*, \psi - \psi^*)_{U_2} \leq 0 \quad \forall \psi \in V_{\text{ad}}, \end{aligned}$$

where Λ_{U_i} is the canonical isomorphism from U_i onto U_i' and $\tilde{u}^*$ is the unique solution, which is in $\mathcal{W}$, of the adjoint problem

$$\begin{aligned} & -\frac{\partial \tilde{u}^*}{\partial t} + A^*(t)\tilde{u}^* = \mathcal{C}^*\Lambda(\mathcal{C}u^* - u_{\text{obs}}), \\ & \tilde{u}^*(t=T) = \mathcal{D}^*(\mathcal{D}u^*(t=T) - v_{\text{obs}}), \end{aligned}$$

where Λ is the canonical isomorphism from M onto M'. □

8.4 Examples of Controls, Disturbances and Observations

In this section we present some robust optimal control problems with different types of controls, disturbances and observations. For this, we consider the same operators and assumptions as in Section 6.4 and take the following problem:

$$\begin{aligned} & \frac{\partial U}{\partial t} + A(t)U = f + \phi - a_0(x,t)U \quad \text{on } \mathcal{Q}, \\ & \text{subject to the linear Robin boundary condition} \\ & \alpha\frac{\partial U}{\partial \eta_A} + \beta U = \alpha\Psi \quad \text{on } \Sigma, \\ & \text{with the initial condition} \\ & U(0) = U_0 \quad \text{on } \Omega, \end{aligned} \tag{8.33}$$

under the pointwise constraint

$$r_1 \leq a_0 \leq r_2 \quad \text{a.e. in } \mathcal{Q}, \tag{8.34}$$

where f, ϕ, $a_0 \in L^2(\mathcal{Q})$, $\Psi \in L^2(\Sigma)$, $U_0 \in L^2(\Omega)$, the real constants α, β are such that $\alpha\beta > 0$, the real constants r_1, r_2 are such that $r_1 < r_2$, and

$$A(t)U = -\sum_{i,j=1,n} \frac{\partial}{\partial x_i}(a_{ij}(x,t)\frac{\partial U}{\partial x_j}). \tag{8.35}$$

Next, let us denote by V the space $H_0^1(\Omega)$ in the case of homogeneous Dirichlet boundary condition (*i.e.*, in the case of $\alpha = 0$) and the space $H^1(\Omega)$ in the case of linear Robin or Neumann boundary condition (*i.e.*, if $\alpha \neq 0$). We denote its dual by V' and by $\langle .,. \rangle_{V',V}$ the duality product between V and V'. Then the embedding $V \subset L^2(\Omega) \subset V'$ are continuous. Finally, we denote by $H = L^2(\Omega)$ and by $\mathcal{W} = \{w \in L^2(0,T;V) : \partial w/\partial t \in L^2(0,T;V')\}$.

We introduce now the following form by

$$\tilde{a}(t;u,v) = a(t;u,v) + \int_\Omega a_0 u.v \mathrm{d}x + \int_\Gamma \theta u.v \mathrm{d}\Gamma,$$

where a is the bilinear form corresponding to the operator A (see Section 6.4), $\theta = \beta/\alpha \geq 0$ if $\alpha \neq 0$ and $\theta = 0$ otherwise.

As in Section 7.4, problem (8.33) admits a unique solution $U \in \mathcal{W} \cap \mathcal{C}([0,T];H)$ of the problem, a.e. $t \in (0,T)$ (the weak formulation of (8.33))

$$\begin{aligned} &\langle \frac{\partial U}{\partial t}, v \rangle_{V',V} + \tilde{a}(t;U,v) = (f+\phi,v) + \alpha\int_\Gamma \Psi v \mathrm{d}\Gamma \qquad \forall v \in V, \\ &\text{with the initial condition} \\ &U(0) = U_0 \quad \text{on } \Omega \end{aligned} \tag{8.36}$$

and satisfies the following estimate:

$$\begin{aligned} \| U \|^2_{\mathcal{W}\cap\mathcal{C}([0,T];H)} \leq\; & C_1(\| f \|^2_{L^2(\mathcal{Q})} + \| \phi \|^2_{L^2(\mathcal{Q})}) \\ & + C_2(\| U_0 \|^2_{L^2(\Omega)} + \| \Psi \|^2_{L^2(\Sigma)}). \end{aligned} \tag{8.37}$$

Moreover, for a_0 and f be given, $v_1 \in \mathcal{W}\cap\mathcal{C}([0,T];H)$ (respectively $v_2 \in \mathcal{W}\cap\mathcal{C}([0,T];H)$) is the solution of (8.33), with data $(\phi_1, \Psi_1, v_{01}) \in L^2(\mathcal{Q})\times L^2(\Sigma)\times L^2(\Omega)$ (respectively with data $(\phi_2, \Psi_2, v_{02}) \in L^2(\mathcal{Q}) \times L^2(\Sigma) \times L^2(\Omega)$), then

$$\| v_1 - v_2 \|^2_{\mathcal{W}\cap\mathcal{C}([0,T];H)} \leq C(\| \phi \|^2_{L^2(\mathcal{Q})} + \| \Psi \|^2_{L^2(\Sigma)} + \| v_0 \|^2_{L^2(\Omega)}),$$

where $\phi = \phi_1 - \phi_2$, $\Psi = \Psi_1 - \Psi_2$ and $v_0 = v_{01} - v_{02}$.

In the following, the solution U of (8.33) will be treated as the target function. We are then interested in the robust regulation of the deviation of the problem from the desired target (U, ϕ, Ψ, U_0). We analyze the full equation which models large perturbations (u, φ, ψ, u_0) to the target (U, ϕ, Ψ, U_0). We consider then the following equation (since the problem (8.33) is linear):

$$\begin{aligned} &\frac{\partial u}{\partial t} + A(t)u = \varphi - a_0(x,t)u \text{ on } \mathcal{Q}, \\ &\text{subject to the linear Robin boundary condition} \\ &\alpha \frac{\partial u}{\partial \eta_A} + \beta u = \alpha \psi \text{ on } \Sigma, \\ &\text{with the initial condition} \\ &u(0) = u_0 \text{ on } \Omega, \end{aligned} \tag{8.38}$$

under the pointwise constraint (8.34).

The problem (8.38) is similar to (8.33), so we have the existence and uniqueness of the solution $u \in \mathcal{W} \cap \mathcal{C}([0,T];H)$, and the following Lipschitz continuity result holds:

$$\| u_1 - u_2 \|^2_{\mathcal{W}\cap\mathcal{C}([0,T];H)} \leq C(\| \varphi \|^2_{L^2(\mathcal{Q})} + \| \psi \|^2_{L^2(\Sigma)} + \| u_0 \|^2_{L^2(\Omega)}), \tag{8.39}$$

where $u_i \in \mathcal{W} \cap \mathcal{C}([0,T];H)$ is the solution of (8.38), with data $(\varphi_i, \psi_i, u_{0i}) \in L^2(\mathcal{Q}) \times L^2(\Sigma) \times L^2(\Omega)$, for $i = 1,2$ and $\varphi = \varphi_1 - \varphi_2$, $\psi = \psi_1 - \psi_2$ and $u_0 = u_{01} - u_{02}$.

8.4.1 Boundary Disturbance

Suppose that the observation operator $\mathcal{C}$ is the injection of $L^2(0,T;V)$ onto $L^2(\mathcal{Q})$ (*i.e.*, $M = L^2(\mathcal{Q}) = M'$ and then the canonical isomorphism Λ from M onto M' is the identity operator) and consider the case where the control is the function φ *i.e.*, the control is distributed in $\mathcal{Q}$, and the disturbance is the function ψ, *i.e.*, the disturbance is on the boundary condition.

Let $\mathcal{K}_1$ and $\mathcal{K}_2$ be given non-empty, closed and convex subsets of $L^2(\mathcal{Q})$ and $L^2(\Sigma)$, respectively, and the observation $u_{\text{obs}} \in L^2(\mathcal{Q})$. Our problem is then, to find $(\varphi, \psi) \in \mathcal{K}_1 \times \mathcal{K}_2$ such that the following cost functional

$$J(\varphi, \psi) = \frac{1}{2} \| \mathcal{C}u - u_{\text{obs}} \|^2_{L^2(\mathcal{Q})} + \frac{\delta_1}{2} \| \varphi \|^2_{L^2(\mathcal{Q})} - \frac{\delta_2}{2} \| \psi \|^2_{L^2(\Sigma)} \tag{8.40}$$

is minimized with respect to φ and maximized with respect to ψ subject to the problem (8.38), where $\delta_i > 0, i = 1,2$ are fixed parameters, *i.e.*, to find a saddle point of J in $\mathcal{K}_1 \times \mathcal{K}_2$.

The adjoint problem corresponding to problem (8.38) is given by the following system:

$$\begin{aligned} &-\frac{\partial \tilde{u}(\varphi,\psi)}{\partial t} + A^*(t)\tilde{u}(\varphi,\psi) \\ &\qquad = -a_0\tilde{u}(\varphi,\psi) + \mathcal{C}u(\varphi,\psi) - u_{\text{obs}} \text{ on } \mathcal{Q}, \\ &\text{subject to the linear Robin boundary condition} \\ &\alpha \frac{\partial \tilde{u}(\varphi,\psi)}{\partial \eta_{A^*}} + \beta\tilde{u}(\varphi,\psi) = 0 \text{ on } \Sigma, \\ &\text{with the final condition} \\ &\tilde{u}(\varphi,\psi)(t = T) = 0 \text{ on } \Omega, \end{aligned} \tag{8.41}$$

where A^* is the adjoint of the operator A.

Moreover, the optimality conditions are given by, $\forall(\varphi,\psi) \in \mathcal{K}_1 \times \mathcal{K}_2$

$$\begin{aligned} &\iint_Q (\tilde{u}^* + \delta_1\varphi^*)(\varphi - \varphi^*)\mathrm{d}x\mathrm{d}t \geq 0, \\ &\iint_\Sigma (\tilde{u}^* - \delta_2\psi^*)(\psi - \psi^*)\mathrm{d}\Gamma\mathrm{d}t \leq 0, \end{aligned} \tag{8.42}$$

where (φ^*, ψ^*) is a saddle point of J, *i.e.*, a solution of (8.40), $u^* = u(\varphi^*, \psi^*)$ is the solution of (8.38) (corresponding to (φ^*, ψ^*)) and $\tilde{u}^* = \tilde{u}(\varphi^*, \psi^*)$ is the solution of the adjoint problem (8.41) (corresponding to the optimal solution (φ^*, ψ^*, u^*)).

Remark 8.6. If without constraints, *i.e.*, $\mathcal{K}_1 \times \mathcal{K}_2 = L^2(\mathcal{Q}) \times L^2(\Sigma)$, the optimality conditions (8.42) become

$$\frac{\tilde{u}^*}{\delta_1} = -\varphi^*, \frac{\tilde{u}^*}{\delta_2} = \psi^*,$$

and then we can obtain the optimal control by the resolution of the following coupled system

$$\begin{aligned} &\frac{\partial u^*}{\partial t} + A(t)u^* + a_0 u^* = -\frac{\tilde{u}^*}{\delta_1} \quad \text{on } \mathcal{Q}, \\ &-\frac{\partial \tilde{u}^*}{\partial t} + A^*(t)\tilde{u}^* + a_0\tilde{u}^* = u^* - u_{\mathrm{obs}} \quad \text{on } \mathcal{Q}, \\ &\alpha\frac{\partial u^*}{\partial \eta_A} + \beta u^* = \frac{\alpha\tilde{u}^*}{\delta_2} \quad \text{on } \Sigma, \\ &\alpha\frac{\partial \tilde{u}^*}{\partial \eta_{A^*}} + \beta\tilde{u}^* = 0 \quad \text{on } \Sigma, \\ &u^*(0) = u_0, \quad \tilde{u}^*(T) = 0 \quad \text{on } \Omega. \end{aligned} \tag{8.43}$$

We can use the same method in order to study the case of the final observation, or the case where the control and the disturbance are in the boundary condition, for example, for φ known, we suppose that ψ is decomposed into the disturbance $\psi_{\mathrm{d}} \in V_{\mathrm{ad}} \subset L^2(\Sigma)$ and the control $\psi_{\mathrm{c}} \in U_{\mathrm{ad}} \subset L^2(\Sigma)$ (*i.e.*, $\psi := \psi_{\mathrm{c}} + \psi_{\mathrm{d}}$). ◇

8.4.2 Pointwise Observations

In this subsection we consider the pointwise observation and distributed control and disturbance.

Let $(x_i)_{i=1,\ldots,d}$ be given d points of the domain Ω, and we suppose that the observation is as the form $(u(x_i,t))_{i=1,\ldots,d}$ (if $u(x_i,t)$ has a sense).

Suppose now that the operator A is independent of time, the space $\Omega \subset \mathbb{R}^n$ with $n \leq 3$, the boundary Γ of Ω and the coefficients a_{ij} of the operator A are sufficiently regular, and the function $\psi = 0$. Moreover, we suppose that the function φ is decomposed into the control $g \in L^2(\mathcal{Q})$ and the disturbance $\xi \in L^2(\mathcal{Q})$, *i.e.*, $\varphi := g + \xi$.

The function u is assumed to be related to disturbance ξ and control g through the problem (8.38):

$$\frac{\partial u}{\partial t} + A(t)u = \xi + g - a_0(x,t)u \quad \text{on } \mathcal{Q},$$

subject to the linear Robin boundary condition

$$\alpha \frac{\partial u}{\partial \eta_A} + \beta u = 0 \quad \text{on } \Sigma, \tag{8.44}$$

with the initial condition

$$u(0) = u_0 \quad \text{on } \Omega.$$

Moreover, if $u_0 \in V$, then the unique solution of the problem (8.44) is in $H^{2,1}(\mathcal{Q}) = L^2(0,T; H^2(\Omega)) \cap H^1(0,T; L^2(\Omega))$ and we have the following Lipschitz continuity result:

$$\| u \|^2_{H^{2,1}(\mathcal{Q})} \leq c(\| g \|^2_{L^2(\mathcal{Q})} + \| \xi \|^2_{L^2(\mathcal{Q})} + \| u_0 \|^2_V). \tag{8.45}$$

where $u_i \in H^{2,1}(\mathcal{Q})$ is the solution of (8.44), with data $(g_j, \xi_j, u_{0j}) \in L^2(\mathcal{Q}) \times L^2(\mathcal{Q}) \times L^2(\Omega)$, for j=1,2 and $u = u_1 - u_2$, $g = g_1 - g_2$, $\xi = \xi_1 - \xi_2$ and $u_0 = u_{01} - u_{02}$.

Moreover, since $\Omega \subset \mathbb{R}^n$, $n \leq 3$, we have that $H^2(\Omega) \subset \mathcal{C}(\overline{\Omega})$ (the space of continuous functions on $\overline{\Omega}$) and then $u_j(x_i,t) \in L^2(0,T)$, for $i = 1, \ldots, d$, $j = 1, 2$, and

$$\int_0^T | u(x_i,t) |^2 \, \mathrm{d}t \leq c(\| g \|^2_{L^2(\mathcal{Q})} + \| \xi \|^2_{L^2(\mathcal{Q})} + \| u_0 \|^2_V), \tag{8.46}$$

where $u = u_1 - u_2$.

Let $\mathcal{K}_j, j = 1,2$ be given non-empty, closed and convex subsets of $L^2(\mathcal{Q})$ and the observation $u_{\text{obs}} = (u_{i,\text{obs}})_{i=1,\ldots,d} \in (L^2(0,T))^d$. Our problem, then, is to find $(g,\xi) \in \mathcal{K}_1 \times \mathcal{K}_2$ such that the cost functional:

$$\begin{aligned} J(g,\xi) = \frac{1}{2} \sum_{i=1}^{d} \| u(x_i,.) - u_{i,\text{obs}} \|^2_{L^2(0,T)} \\ + \frac{\gamma_1}{2} \| g \|^2_{L^2(\mathcal{Q})} - \frac{\gamma_2}{2} \| \xi \|^2_{L^2(\mathcal{Q})} . \end{aligned} \tag{8.47}$$

is minimized with respect to g and maximized with respect to ξ subject to the problem (8.44), where $\gamma_j > 0$ for $j = 1,2$, are fixed constants, to find a saddle point of J in $\mathcal{K}_1 \times \mathcal{K}_2$.

We have the existence and the uniqueness of the optimal solution (g^*, ξ^*) which is characterized by the following optimality conditions, given by

$$\begin{aligned} \iint_{\mathcal{Q}} (\tilde{u}^* + \gamma_1 g^*)(g - g^*) \mathrm{d}x\mathrm{d}t \geq 0 \quad \forall g \in \mathcal{K}_1, \\ \iint_{\mathcal{Q}} (\tilde{u}^* - \gamma_2 \xi^*)(\xi - \xi^*) \mathrm{d}x\mathrm{d}t \leq 0 \quad \forall \xi \in \mathcal{K}_2, \end{aligned} \tag{8.48}$$

where $\tilde{u}^* = \tilde{u}(g^*, \xi^*)$ is the unique solution of the following adjoint problem corresponding to problem (8.44):

$$
\begin{aligned}
&-\frac{\partial \tilde{u}}{\partial t} + A^*\tilde{u} + a_0\tilde{u} = \sum_{i=1}^{d}(u^*(x_i, .) - u_{i,\mathrm{obs}})\delta_{x_i} \quad \text{on } \mathcal{Q},\\
&\text{subject to the linear Robin boundary condition}\\
&\alpha \frac{\partial \tilde{u}}{\partial \eta_{A^*}} + \beta\tilde{u} = 0 \quad \text{on } \Sigma,\\
&\text{with the final condition}\\
&\tilde{u}(t = T) = 0 \quad \text{on } \Omega,
\end{aligned}
\tag{8.49}
$$

where $u^* = u(g^*, \xi^*)$ is the solution of (8.44) (corresponding to (g^*, ξ^*)), A^* is the adjoint of the operator A, δ_{x_i} is the usual Dirac function at point x_i and $h(t)\delta_{x_i}$ is the distribution:

$$
\mathcal{S} \in \mathcal{D}(\mathcal{Q}) \longrightarrow \int_0^T h(t)\mathcal{S}(x_i, t)\mathrm{d}t.
$$

As the problem (8.49) is similar to (7.42) then it admits a unique solution $\tilde{u}^*$ on $L^2(\mathcal{Q})$, given by the transposition technique.

Remark 8.7. (*i*) In the case without constraints, *i.e.*, $\mathcal{K}_1 \times \mathcal{K}_2 = (L^2(\mathcal{Q}))^2$, the optimality conditions (8.48) become

$$
\frac{\tilde{u}^*}{\gamma_1} = -g^* \quad \text{and} \quad \frac{\tilde{u}^*}{\gamma_2} = \xi^*,
$$

so, we can obtain the optimal control by the resolution of the following coupled system:

$$
\begin{aligned}
&\frac{\partial u^*}{\partial t} + Au^* + a_0u^* = -\frac{\tilde{u}^*}{\gamma_1} + \frac{\tilde{u}^*}{\gamma_2} \quad \text{on } \mathcal{Q},\\
&-\frac{\partial \tilde{u}^*}{\partial t} + A^*\tilde{u}^* + a_0\tilde{u}^* = \sum_{i=1}^{d}(u^*(x_i, .) - u_{i,\mathrm{obs}})\delta_{x_i} \quad \text{on } \mathcal{Q},\\
&\alpha \frac{\partial u^*}{\partial \eta_A} + \beta u^* = 0 \quad \text{on } \Sigma,\\
&\alpha \frac{\partial \tilde{u}^*}{\partial \eta_{A^*}} + \beta\tilde{u}^* = 0 \quad \text{on } \Sigma,\\
&u^*(0) = u_0, \quad \tilde{u}^*(T) = 0 \quad \text{on } \Omega.
\end{aligned}
\tag{8.50}
$$

(*ii*) If we suppose that the boundary function ψ is sufficiently regular, such that the solutions of (8.33) are in $H^{2,1}(\mathcal{Q})$, we can use the same technique in order to study problems with boundary controls and/or boundary disturbances. ◇

8.4.3 Pointwise Controls and Pointwise Disturbances

In this section we consider the pointwise controls and pointwise disturbances and a distributed observation, *i.e.*, we consider the case where the value φ is decomposed into a pointwise disturbance ξ and a pointwise control g. More precisely, let $(x_i)_{i=1,\dots,d_c}$ (respectively $(y_i)_{i=1,\dots,d_d}$) be given d_c points (respectively d_d points) of the domain Ω, and suppose that the control function g and the disturbance function ξ are in the form

$$g := \sum_{i=1}^{d_c} g_i \delta_{x_i}, \ \xi := \sum_{j=1}^{d_d} \xi_j \delta_{y_j} \ \text{ and } \ \varphi := g + \xi,$$

where $g_i, \xi_j \in L^2(0,T)$, for $i = 1, \dots, d_c$ and $j = 1, \dots, d_d$ and δ_{x_i} (respectively δ_{y_j}) is the usual Dirac function at point x_i (respectively point y_j).

Similarly, as in the previous subsection, we suppose that the operator A is independent of time, the space $\Omega \subset \mathbb{R}^n$ with $n \leq 3$, the boundary Γ of Ω and the coefficients a_{ij} of the operator A are sufficiently regular, and the function $\psi = 0$.

Then the function u is related to disturbance ξ and control g through the problem (8.44). By using the same technique (the transposition technique) as used to prove the unique solution of the adjoint problem (7.42), we obtain the existence and the uniqueness of the solution u in $L^2(\mathcal{Q})$.

Let $\mathcal{K}_1$ (respectively $\mathcal{K}_2$) be a given non-empty, closed and convex subset of $(L^2(0,T))^{d_c}$ (respectively $(L^2(0,T))^{d_d}$) and the observation $u_{\text{obs}} \in L^2(\mathcal{Q})$. Our problem, then, is to find $(g_i)_{i=1,\dots,d_c} \in \mathcal{K}_1$ and $(\xi_j)_{j=1,\dots,d_d} \in \mathcal{K}_2$ such that the cost functional:

$$\begin{aligned} J(g,\xi) = &\frac{1}{2} \parallel u - u_{\text{obs}} \parallel^2_{L^2(\mathcal{Q})} \\ &+\frac{1}{2}\sum_{i=1}^{d_c} \alpha_i \parallel g_i \parallel^2_{L^2(0,T)} - \frac{1}{2}\sum_{j=1}^{d_d} \beta_j \parallel \xi_j \parallel^2_{L^2(0,T)}, \end{aligned} \tag{8.51}$$

is minimized with respect to g and maximized with respect to ξ subject to the problem (8.44), where $\alpha_i, \beta_j > 0$, for $i = 1, \dots, d_c$ and $j = 1, \dots, d_d$ are fixed constants.

We have the existence and the uniqueness of the optimal solution $g^* = (g_i^*)_{i=1,\dots,d_c}$ and $\xi^* = (\xi_j^*)_{j=1,\dots,d_d}$ which is characterized by the following optimality conditions:

$$\begin{aligned} &\sum_{i=1}^{d_c} \int_0^T (\tilde{u}^*(x_i,t) + \alpha_i g_i^*(t))(g_i(t) - g_i^*(t))\mathrm{d}t \geq 0 \ \forall g = (g_i)_{i=1,d_c} \in \mathcal{K}_1, \\ &\sum_{j=1}^{d_d} \int_0^T (\tilde{u}^*(y_j,t) - \beta_j \xi_j^*(t))(\xi_j(t) - \xi_j^*(t))\mathrm{d}t \leq 0 \ \forall \xi = (\xi_j)_{j=1,d_d} \in \mathcal{K}_2, \end{aligned} \tag{8.52}$$

where $\tilde{u}^* = \tilde{u}(g^*, \xi^*)$ is the solution of the following adjoint problem corresponding to problem (8.44):

$$-\frac{\partial \tilde{u}}{\partial t} + A^*\tilde{u} + a_0\tilde{u} = u^* - u_{\text{obs}} \quad \text{on } \mathcal{Q},$$

subject to the linear Robin boundary condition

$$\alpha \frac{\partial \tilde{u}}{\partial \eta_{A^*}} + \beta \tilde{u} = 0 \quad \text{on } \Sigma, \tag{8.53}$$

with the final condition

$$\tilde{u}(t = T) = 0 \quad \text{on } \Omega,$$

where $u^* = u(g^*, \xi^*)$ is the optimal state and A^* is the adjoint of the operator A.

Since $u^* - u_{\text{obs}} \in L^2(\mathcal{Q})$ and the final condition (*i.e.*, the null function) is in V, then the problem (8.53) admits a unique solution $\tilde{u}$ in $H^{2,1}(\mathcal{Q})$. Moreover, since $H^2(\Omega) \subset \mathcal{C}(\overline{\Omega})$, we have $\tilde{u}(x_i, t)$ (respectively $\tilde{u}(y_j, t)$) is in $L^2(0, T)$, for $i = 1, \ldots, d_{\text{c}}$ (respectively $j = 1, \ldots, d_{\text{d}}$).

Remark 8.8. In the case without constraints, *i.e.*, $\mathcal{K}_1 = (L^2(0,T))^{d_{\text{c}}}$ and $\mathcal{K}_2 = (L^2(0,T))^{d_{\text{d}}}$, the optimality conditions (8.52) become

$$\frac{\tilde{u}^*(x_i, .)}{\alpha_i} = -g_i^* \quad \text{and} \quad \frac{\tilde{u}(y_j, .)}{\beta_j} = \xi_j^* \quad \text{for} \quad i = 1, \ldots, d_{\text{c}}; \; j = 1, \ldots, d_{\text{d}},$$

so, we can obtain the optimal control by the resolution of the coupled system:

$$\begin{aligned}
&\frac{\partial u^*}{\partial t} + Au^* + a_0 u^* = -\sum_{i=1}^{d_{\text{c}}} \frac{\tilde{u}^*(x_i, .)\delta_{x_i}}{\alpha_i} + \sum_{j=1}^{d_{\text{d}}} \frac{\tilde{u}^*(y_j, .)\delta_{y_j}}{\beta_j} \quad \text{on } \mathcal{Q},\\
&-\frac{\partial \tilde{u}^*}{\partial t} + A^*\tilde{u}^* + a_0\tilde{u}^* = u^* - u_{\text{obs}} \quad \text{on } \mathcal{Q},\\
&\alpha \frac{\partial u^*}{\partial \eta_A} + \beta u^* = 0 \quad \text{on } \Sigma,\\
&\alpha \frac{\partial \tilde{u}^*}{\partial \eta_{A^*}} + \beta \tilde{u}^* = 0 \quad \text{on } \Sigma,\\
&u^*(0) = u_0, \quad \tilde{u}^*(T) = 0 \quad \text{on } \Omega.
\end{aligned} \tag{8.54}$$

◇

8.4.4 Boundary Controls and Boundary Observations

In this subsection we consider a problem where controls and observations act on the boundary, and we suppose that the parameter $\alpha \neq 0$. The disturbance can act on the boundary or on the state equation and the observation is given by

$$\mathcal{C}u = \mathcal{D}(u|_{\Sigma}), \tag{8.55}$$

where $u|_{\Sigma}$ is the trace of u on Σ and $\mathcal{D} \in \mathcal{L}(L^2(\Sigma); L^2(\Sigma))$. For the types of boundary observation operator $\mathcal{D}$, see, *e.g.*, Example 7.10.

Boundary Disturbances

We assume that the function ψ is decomposed on the disturbance ξ and the control g defined on Σ, *i.e.*, $\psi := g + \xi$. The function u is assumed to be related to disturbance ξ and control g through the problem (8.38):

$$\begin{aligned} &\frac{\partial u}{\partial t} + A(t)u = \varphi - a_0(x,t)u \quad \text{on } Q, \\ &\text{subject to the linear Robin boundary condition} \\ &\alpha \frac{\partial u}{\partial \eta_A} + \beta u = \alpha(g + \xi) \quad \text{on } \Sigma, \\ &\text{with the initial condition} \\ &u(0) = u_0 \quad \text{on } \Omega. \end{aligned} \tag{8.56}$$

Let $\mathcal{K}_j$, for $j = 1, 2$, be given non-empty, closed and convex subsets of $L^2(\Sigma)$ and the observation $u_{\text{obs}} \in L^2(\Sigma)$. Our problem, then, is to find $(g, \xi) \in \mathcal{K}_1 \times \mathcal{K}_2$ such that the following cost functional:

$$J(g,\xi) = \frac{1}{2} \parallel \mathcal{D}(u|_\Sigma) - u_{\text{obs}} \parallel^2_{L^2(Q)} + \frac{\gamma_1}{2} \parallel g \parallel^2_{L^2(\Sigma)} - \frac{\gamma_2}{2} \parallel \xi \parallel^2_{L^2(\Sigma)}, \tag{8.57}$$

is minimized with respect to g and maximized with respect to ξ subject to the problem (8.56), where $\gamma_j > 0$, for $j = 1, 2$ are fixed constants.

We have the existence and the uniqueness of the optimal control $(g^*, \xi^*) \in \mathcal{K}_1 \times \mathcal{K}_2$, which is characterized by the following optimality condition:

$$\begin{aligned} &\int\!\!\int_\Sigma (\tilde{u}^* + \gamma_1 g^*)(g - g^*) \mathrm{d}\Gamma \mathrm{d}t \geq 0 \quad \forall g \in \mathcal{K}_1, \\ &\int\!\!\int_\Sigma (\tilde{u}^* - \gamma_2 \xi^*)(\xi - \xi^*) \mathrm{d}\Gamma \mathrm{d}t \leq 0 \quad \forall \xi \in \mathcal{K}_2, \end{aligned} \tag{8.58}$$

where $\tilde{u}^* = \tilde{u}(g^*, \xi^*)$ is the solution of the following adjoint problem:

$$\begin{aligned} &-\frac{\partial \tilde{u}}{\partial t} + A^*(t)\tilde{u} + a_0 \tilde{u} = 0 \quad \text{on } Q, \\ &\text{subject to the linear Robin boundary condition} \\ &\alpha \frac{\partial \tilde{u}}{\partial \eta_{A^*}} + \beta \tilde{u} = \mathcal{D}^*(\mathcal{D}(u^*|_\Sigma) - u_{\text{obs}}) \quad \text{on } \Sigma, \\ &\text{with the final condition} \\ &\tilde{u}(t = T) = 0 \quad \text{on } \Omega, \end{aligned} \tag{8.59}$$

where $u^* = u(g^*, \xi^*)$ is the solution (8.38) (corresponding to (g^*, ξ^*)), A^* is the adjoint of the operator A and $\mathcal{D}^*$ is the adjoint of the operator $\mathcal{D}$.

Since the function $\mathcal{D}^*(\mathcal{D}(u^*|_\Sigma) - u_{\text{obs}})$ is in $L^2(\Sigma)$ then the problem (8.59) admits a unique solution in $\mathcal{W} \cap \mathcal{C}([0, T]; H)$ (in the weak formulation sense).

Remark 8.9. In the case without constraints, *i.e.*, $\mathcal{K}_j = L^2(\Sigma)$, for $j = 1, 2$, the optimality conditions (8.58) become

$$\frac{\tilde{u}^*|_\Sigma}{\gamma_1} = -g^*, \quad \frac{\tilde{u}^*|_\Sigma}{\gamma_2} = \xi^*,$$

so, we can obtain the optimal control by the resolution of the coupled system:

$$\begin{aligned}
&\frac{\partial u^*}{\partial t} + A(t)u^* + a_0 u^* = \varphi \ \text{ on } \mathcal{Q}, \\
&-\frac{\partial \tilde{u}^*}{\partial t} + A^*(t)\tilde{u}^* + a_0 \tilde{u}^* = 0 \ \text{ on } \mathcal{Q}, \\
&\alpha \frac{\partial u^*}{\partial \eta_A} + \beta u^* = -\frac{\alpha \tilde{u}^*|_\Sigma}{\gamma_1} + \frac{\alpha \tilde{u}^*|_\Sigma}{\gamma_2} \ \text{ on } \Sigma, \\
&\alpha \frac{\partial \tilde{u}^*}{\partial \eta_{A^*}} + \beta \tilde{u}^* = \mathcal{D}^*(\mathcal{D}(u^*|_\Sigma) - u_{\text{obs}}) \ \text{ on } \Sigma, \\
&u^*(0) = u_0, \quad \tilde{u}^*(T) = 0 \ \text{ on } \Omega.
\end{aligned} \tag{8.60}$$

◇

Distributed Disturbances

We assume that the disturbance ξ, defined on $\mathcal{Q}$, is in the function φ and the control g, defined on Σ, is in the boundary function ψ, *i.e.*, $\psi := g + \psi_0$ and $\varphi := \xi + \varphi_0$, where ψ_0 and φ_0 are supposed known. The function u is assumed to be related to disturbance ξ and control g through the problem (8.38):

$$\begin{aligned}
&\frac{\partial u}{\partial t} + A(t)u = \xi + \varphi_0 - a_0(x,t)u \ \text{ on } \mathcal{Q}, \\
&\text{subject to the linear Robin boundary condition} \\
&\alpha \frac{\partial u}{\partial \eta_A} + \beta u = \alpha(g + \psi_0) \ \text{ on } \Sigma, \\
&\text{with the initial condition} \\
&u(0) = u_0 \ \text{ on } \Omega.
\end{aligned} \tag{8.61}$$

Let $\mathcal{K}_1$ (respectively $\mathcal{K}_2$) be a given non-empty, closed and convex subset of $L^2(\Sigma)$ (respectively $L^2(\mathcal{Q})$) and the observation $u_{\text{obs}} \in L^2(\Sigma)$. Our problem, then, is to find $(g, \xi) \in \mathcal{K}_1 \times \mathcal{K}_2$ such that the cost functional:

$$J(g,\xi) = \frac{1}{2} \parallel \mathcal{D}(u|_\Sigma) - u_{\text{obs}} \parallel^2_{L^2(\mathcal{Q})} + \frac{\gamma_1}{2} \parallel g \parallel^2_{L^2(\Sigma)} - \frac{\gamma_2}{2} \parallel \xi \parallel^2_{L^2(\mathcal{Q})}, \tag{8.62}$$

is minimized with respect to g and maximized with respect to ξ subject to the problem (8.61), where $\gamma_j > 0$, for $j = 1, 2$ are fixed constants.

We have the existence and the uniqueness of the optimal control $(g^*, \xi^*) \in \mathcal{K}_1 \times \mathcal{K}_2$, which is characterized by the following optimality conditions:

$$\begin{aligned}
&\iint_{\Sigma} (\tilde{u}^* + \gamma_1 g^*)(g - g^*)\mathrm{d}\Gamma\mathrm{d}t \geq 0 \quad \forall g \in \mathcal{K}_1, \\
&\iint_{\mathcal{Q}} (\tilde{u}^* - \gamma_2 \xi^*)(\xi - \xi^*)\mathrm{d}x\mathrm{d}t \leq 0 \quad \forall \xi \in \mathcal{K}_2,
\end{aligned} \tag{8.63}$$

where $\tilde{u}^* = \tilde{u}(g^*, \xi^*)$ is the solution of the adjoint problem (8.59) corresponding to the primal problem (8.61).

Moreover, in the case without constraints, *i.e.*, $\mathcal{K}_1 = L^2(\Sigma)$ and $\mathcal{K}_2 = L^2(\mathcal{Q})$ the optimality conditions (8.63) become

$$\frac{\tilde{u}^*|_\Sigma}{\gamma_1} = -g^*, \quad \frac{\tilde{u}^*}{\gamma_2} = \xi^*,$$

so, we can obtain the optimal control by the resolution of the coupled system:

$$\begin{aligned}
&\frac{\partial u^*}{\partial t} + Au^* + a_0 u^* = \varphi_0 + \frac{\tilde{u}^*}{\gamma_2} \quad \text{on } \mathcal{Q}, \\
&-\frac{\partial \tilde{u}^*}{\partial t} + A^*\tilde{u}^* + a_0 \tilde{u}^* = 0 \quad \text{on } \mathcal{Q}, \\
&\alpha \frac{\partial u^*}{\partial \eta_A} + \beta u^* = \alpha(\psi_0 - \frac{\tilde{u}^*|_\Sigma}{\gamma_1}) \quad \text{on } \Sigma, \\
&\alpha \frac{\partial \tilde{u}^*}{\partial \eta_{A^*}} + \beta \tilde{u}^* = \mathcal{D}^*(\mathcal{D}(u^*|_\Sigma) - u_{\text{obs}}) \quad \text{on } \Sigma, \\
&u^*(0) = u_0, \quad \tilde{u}^*(T) = 0 \quad \text{on } \Omega.
\end{aligned} \tag{8.64}$$

Remark 8.10. We can replace in (8.56) and (8.61), for example, the control g by Bq, where $B : U \longrightarrow L^2(\Sigma)$ denotes a bounded linear control operator which maps abstract controls of a Hilbert space U to feasible boundary controls. Typical control operators and control spaces are given by

- $U = L^2(\Sigma)$ and $B = \text{Identity}$
- $U = (L^2(0,T))^d$ and $Bq(t) = \sum_{i=1}^{d} q_i(t)\alpha_i$, where $\alpha_i \in L^2(\Gamma)$. ◊

8.4.5 Data Assimilation Problem and Initial Condition Control

In this subsection, we suppose that the parameter $\alpha \neq 0$ and we consider an application when the control is in the initial condition, the disturbance is in φ and the observation acts on the boundary. We assume that φ is the sum of ξ (assumed to be not well known) and φ_0 (assumed to be known), *i.e.*, $\varphi := \xi + \varphi_0$ and that the initial condition u_0 is the sum of v_0 (assumed to be known) and θ (assumed to be not well known), *i.e.*, $u_0 = v_0 + \theta$. The control is the function θ defined on Ω, the disturbance is the function ξ defined on $\mathcal{Q}$ and the observation is given by

$$\mathcal{C}u = \mathcal{D}(u|_\Sigma), \tag{8.65}$$

where $u|_\Sigma$ is the trace of u on Σ and $\mathcal{D} \in \mathcal{L}(L^2(\Sigma); L^2(\Sigma))$. For the types of boundary observation operator $\mathcal{D}$, see, e.g., Example 7.10.

The function u is assumed to be related to disturbance ξ and control θ through the problem (8.38):

$$\frac{\partial u}{\partial t} + A(t)u = \xi + \varphi_0 - a_0(x,t)u \quad \text{on } \mathcal{Q},$$

subject to the linear Robin boundary condition

$$\alpha \frac{\partial u}{\partial \eta_A} + \beta u = \alpha \psi \quad \text{on } \Sigma, \tag{8.66}$$

with the initial condition

$$u(0) = v_0 + \theta \quad \text{on } \Omega.$$

Let $\mathcal{K}_1$ (respectively $\mathcal{K}_2$) be a given non-empty, closed and convex subset of $L^2(\Omega)$ (respectively $L^2(\mathcal{Q})$) and the observation $u_{\text{obs}} \in L^2(\Sigma)$. Our problem, then, is to find $(\theta, \xi) \in \mathcal{K}_1 \times \mathcal{K}_2$ such that the following cost functional:

$$J(\theta,\xi) = \frac{1}{2} \parallel \mathcal{D}(u|_\Sigma) - u_{\text{obs}} \parallel^2_{L^2(\Sigma)} + \frac{\gamma_1}{2} \parallel \theta \parallel^2_{L^2(\Omega)} - \frac{\gamma_2}{2} \parallel \xi \parallel^2_{L^2(\mathcal{Q})}, \tag{8.67}$$

is minimized with respect to θ and maximized with respect to ξ subject to the problem (8.66), where $\gamma_j > 0$, for $j = 1, 2$ are fixed constants.

We have the existence and the uniqueness of the optimal control $(\theta^*, \xi^*) \in \mathcal{K}_1 \times \mathcal{K}_2$, which is characterized by the following optimality conditions:

$$\begin{aligned} &\int_\Omega (\tilde{u}^* + \gamma_1 \theta^*)(\theta - \theta^*)\mathrm{d}x \geq 0 \quad \forall \theta \in \mathcal{K}_1, \\ &\iint_{\mathcal{Q}} (\tilde{u}^* - \gamma_2 \xi^*)(\xi - \xi^*)\mathrm{d}x\mathrm{d}t \leq 0 \quad \forall \xi \in \mathcal{K}_2, \end{aligned} \tag{8.68}$$

where $\tilde{u}^*$ is the solution of the following adjoint problem:

$$-\frac{\partial \tilde{u}}{\partial t} + A^*(t)\tilde{u} + a_0 \tilde{u} = 0 \quad \text{on } \mathcal{Q},$$

subject to the linear Robin boundary condition

$$\alpha \frac{\partial \tilde{u}}{\partial \eta_{A^*}} + \beta \tilde{u} = \mathcal{D}^*(\mathcal{D}(u^*|_\Sigma) - u_{\text{obs}}) \quad \text{on } \Sigma, \tag{8.69}$$

with the final condition

$$\tilde{u}(t = T) = 0 \quad \text{on } \Omega,$$

where $u^* = u(\theta^*, \xi^*)$, A^* is the adjoint of the operator A and $\mathcal{D}^*$ is the adjoint of the operator $\mathcal{D}$.

Since the function $\mathcal{D}^*(\mathcal{D}(u(\psi)|_\Sigma) - u_{\text{obs}})$ is in $L^2(\Sigma)$ then the problem (8.69) admits a unique solution in $\mathcal{W} \cap \mathcal{C}([0,T]; H)$ (in the weak formulation sense).

Remark 8.11. (*i*) In the case without constraints, *i.e.*, $\mathcal{K} = L^2(\Omega)$, the optimality conditions (8.68) become

$$\frac{\tilde{u}^*}{\gamma_1} = -\theta^* \text{ and } \frac{\tilde{u}^*}{\gamma_2} = \xi^*,$$

so, we can obtain the optimal control by the resolution of the coupled system:

$$\begin{aligned}
&\frac{\partial u^*}{\partial t} + A(t)u^* + a_0 u^* = \varphi_0 + \frac{\tilde{u}^*}{\gamma_2} \quad \text{on } \mathcal{Q},\\
&-\frac{\partial \tilde{u}^*}{\partial t} + A^*(t)\tilde{u}^* + a_0\tilde{u}^* = 0 \quad \text{on } \mathcal{Q},\\
&\alpha\frac{\partial u^*}{\partial \eta_A} + \beta u^* = \alpha\psi \quad \text{on } \Sigma,\\
&\alpha\frac{\partial \tilde{u}^*}{\partial \eta_{A^*}} + \beta\tilde{u}^* = \mathcal{D}^*(\mathcal{D}(u^*|_\Sigma) - u_{\text{obs}}) \quad \text{on } \Sigma,\\
&u^*(0) = v_0 - \frac{\tilde{u}^*}{\gamma_1}, \quad \tilde{u}^*(T) = 0 \quad \text{on } \Omega.
\end{aligned} \tag{8.70}$$

(*ii*) In the case where we consider the boundary disturbance, *i.e.*, for example, we suppose that the function φ is completely known and that the boundary function ψ is the sum of the disturbance ξ and the known function $\psi_0 \in L^2(\Sigma)$: $\psi := \xi + \psi_0$, we obtain the same results. Indeed in this case we can give the cost functional J, on $\mathcal{K}_1 \times \mathcal{K}_2$, by

$$J(\theta, \xi) = \frac{1}{2} \parallel \mathcal{D}(u|_\Sigma) - u_{\text{obs}} \parallel^2_{L^2(\Sigma)} + \frac{\gamma_1}{2} \parallel \theta \parallel^2_{L^2(\Omega)} - \frac{\gamma_2}{2} \parallel \xi \parallel^2_{L^2(\Sigma)},$$

where $\gamma_j > 0$, for $j = 1, 2$ are fixed constants, $\mathcal{K}_1$ (respectively $\mathcal{K}_2$) is a given non-empty, closed and convex subset of $L^2(\Omega)$ (respectively $L^2(\Sigma)$), and the observation $u_{\text{obs}} \in L^2(\Sigma)$.

We have the following optimality conditions:

$$\begin{aligned}
&\int_\Omega (\tilde{u}^* + \gamma_1\theta^*)(\theta - \theta^*)\mathrm{d}x \geq 0 \quad \forall\theta \in \mathcal{K}_1,\\
&\iint_\Sigma (\tilde{u}^* - \gamma_2\xi^*)(\xi - \xi^*)\mathrm{d}\Gamma\mathrm{d}t \leq 0 \quad \forall\xi \in \mathcal{K}_2,
\end{aligned} \tag{8.71}$$

where $\tilde{u}^*$ is the solution of the adjoint problem (8.69) corresponding to the primal problem:

$$\frac{\partial u}{\partial t} + A(t)u = \varphi - a_0 u \quad \text{on } \mathcal{Q},$$

subject to the linear Robin boundary condition

$$\alpha\frac{\partial u}{\partial \eta_A} + \beta u = \alpha(\xi^* + \psi_0) \quad \text{on } \Sigma,$$

with the initial condition

$$u(0) = v_0 + \theta^* \quad \text{on } \Omega.$$

Moreover, in the case without constraints, *i.e.*, $\mathcal{K}_1 = L^2(\Omega)$ and $\mathcal{K}_2 = L^2(\Sigma)$ the optimality conditions (8.71) become

$$\frac{\tilde{u}^*}{\gamma_1} = -\theta^*, \quad \frac{\tilde{u}^*|_\Sigma}{\gamma_2} = \xi^*,$$

so, we can obtain the optimal control by the resolution of the coupled system:

$$\begin{aligned}
&\frac{\partial u^*}{\partial t} + A(t)u^* + a_0 u^* = \varphi \quad \text{on } \mathcal{Q},\\
&-\frac{\partial \tilde{u}^*}{\partial t} + A^*(t)\tilde{u}^* + a_0\tilde{u}^* = 0 \quad \text{on } \mathcal{Q},\\
&\alpha\frac{\partial u^*}{\partial \eta_A} + \beta u^* = \frac{\alpha\tilde{u}^*|_\Sigma}{\gamma_2} + \alpha\psi_0 \quad \text{on } \Sigma,\\
&\alpha\frac{\partial \tilde{u}^*}{\partial \eta_{A^*}} + \beta\tilde{u}^* = \mathcal{D}^*(\mathcal{D}(u^*|_\Sigma) - u_{\text{obs}}) \quad \text{on } \Sigma,\\
&u^*(0) = -\frac{\tilde{u}^*}{\gamma_1} + v_0, \quad \tilde{u}^*(T) = 0 \quad \text{on } \Omega.
\end{aligned} \tag{8.72}$$

◇

In the two next sections we study two classes of non-linear robust control problems, more precisely a class of bilinear problems (the primal problem is linear on the state variable when the control is fixed, and conversely) and a class of non-linear evolutive problems (which come from the modeling, for example, of pollutants in liquid or atmospheric systems).

8.5 Bilinear-type Robust Control Problems

In this section we present a first type of non-linear robust control problem. The robust control problems arising in this context are bilinear. This adds fundamental difficulties from a mathematical viewpoint and makes these problems extremely challenging. A solution of bilinear control problems was proposed at first for the investigation of the dynamic processes of nuclear reactors, the kinetics of neutrons and heat transfer. Further investigations show that many processes in engineering, biology, ecology, medicine, chemical reactions, human population growth and many other areas can be described by *bilinear systems* (the literature on such models and methods such as controllability, observability and stabilization is vast, see, *e.g.*, Baciotti [18], Christensen *et al.* [84], Isidori [165], Khapalov [173] and the references therein). We can mention also Ball *et al.* [19], in which the authors study controllability for wave equations; Lenhart *et al.* [57, 195] in which the authors treat bilinear control for the Kirchhoff plate equation and for a wave equation with viscous damping; Sachkov [255] for the controllability of bilinear systems governed by ordinary differential equations; Zuazua [313] for the controllability of the Schrödinger equation; Leung and Chen [198] and Sadek and Vedantham [256], in which the authors consider the optimal control problem of nuclear fission reactors; and finally Belmiloudi [47], in which the author considers the minimax control problem of nuclear fission reactors.

8.5.1 State Problem

For this, we consider the linear primal problem used in Section 8.4, precisely the problem (8.33). We suppose that the operator A is independent of time, the space $\Omega \subset \mathbb{R}^n$ with $n \leq 3$, the boundary Γ of Ω and the coefficients a_{ij} of the operator A are sufficiently regular. Moreover, we assume that the function $\Psi = 0$ and $\alpha = 1$, and we denote by g the sum of f and ϕ, *i.e.*, $g := f + \phi$. More precisely,

$$\begin{aligned}
&\frac{\partial U}{\partial t} + AU + a_0 U = g \quad \text{on } Q,\\
&\text{subject to the linear Robin boundary condition}\\
&\frac{\partial U}{\partial \eta_A} + \beta U = 0 \quad \text{on } \Sigma, \\
&\text{with the initial condition}\\
&U(0) = U_0 \quad \text{on } \Omega.
\end{aligned} \tag{8.73}$$

According to the regularity results of Theorem 6.19, we have that the solution U of problem (8.73) is in $H^{2,1}(Q) \times L^\infty(0, T; V)$ for any initial condition $U_0 \in V$, the right-hand side $g \in L^2(Q)$ and $a_0 \in L^\infty(Q)$.

In our study, the solution U of (8.73) will be treated as the target function. We are then interested in the robust regulation of the deviation of the problem from the desired target (U, g, U_0, a_0). We analyze the full equation which models large perturbations $(u, \varphi, u_0, \theta)$ to the target (U, g, U_0, a_0), *i.e.*, we assume $U + u$ satisfies problem (8.73) with the data $(g+\varphi, U_0+u_0, a_0+\theta)$. Here we consider the following system:

$$\begin{aligned}
&\frac{\partial u}{\partial t} + Au + a_0 u = \varphi - \theta(u + U) \quad \text{on } Q,\\
&\text{subject to the linear Robin boundary condition}\\
&\frac{\partial u}{\partial \eta_A} + \beta u = 0 \quad \text{on } \Sigma, \\
&\text{with the initial condition}\\
&u(0) = u_0 \quad \text{on } \Omega,
\end{aligned} \tag{8.74}$$

In the following, we assume that the control is the parameter θ and the disturbance is in the initial condition u_0, *i.e.*, $u_0 = B\xi$. The functions $\varphi \in L^2(Q)$ and $a_0 \in L^\infty(Q)$ are given data, and B denotes the linear and bounded operator which maps $L^2(\Omega)$ into V. The control is then the parameter θ defined on Q, the disturbance is the function ξ defined on Ω.

Then the problem (8.74) becomes

$$\begin{aligned}
&\frac{\partial u}{\partial t} + Au + a_0 u = \varphi - \theta(u+U) \quad \text{on } \mathcal{Q}, \\
&\text{subject to the linear Robin boundary condition} \\
&\frac{\partial u}{\partial \eta_A} + \beta u = 0 \quad \text{on } \Sigma, \\
&\text{with the initial condition} \\
&u(0) = B\xi \quad \text{on } \Omega,
\end{aligned} \tag{8.75}$$

under the pointwise constraint

$$\tau_1 \leq \theta \leq \tau_2 \quad \text{a.e. in } \mathcal{Q}, \tag{8.76}$$

where the interval $[\tau_1, \tau_2]$ contains 0.

The systems (8.75) is linear on the state variable when the control θ is fixed, and linear on the control when the state variable u is fixed, but the solution of (8.75) is depending non-linearily on the control θ. Consequently, the robust control problem is non-linear.

Let $U_{\text{ad}} = \{\theta \in L^2(\mathcal{Q}) : \tau_1 \leq \theta \leq \tau_2 \quad \text{a.e. in } \mathcal{Q}\} \subset L^\infty(\mathcal{Q})$, $V_{\text{ad}} := L^2(\Omega)$ and $\mathcal{F} : (\theta, \xi) \in U_{\text{ad}} \times V_{\text{ad}} \longrightarrow \mathcal{W}$ such that $u = \mathcal{F}(\theta, \xi)$ is the unique solution of (8.75), corresponding to (θ, ξ). More precisely, since $B\xi \in V$, $\varphi \in L^2(\mathcal{Q})$, $a_0 \in L^\infty(\mathcal{Q})$ and $\theta \in L^\infty(\mathcal{Q})$, the solution u of (8.75) is in $H^{2,1}(\mathcal{Q}) \cap L^\infty(0,T;V)$.

Our problem is then:

find $(\theta, \xi) \in U_{\text{ad}} \times V_{\text{ad}}$ such that the cost functional

$$J(\theta,\xi) = \frac{1}{2} \parallel \mathcal{C}(u - u_{\text{obs}}) \parallel^2_{L^2(\mathcal{Q})} + \frac{\gamma_1}{2} \parallel \theta \parallel^2_{L^2(\mathcal{Q})} - \frac{\gamma_2}{2} \parallel \xi \parallel^2_{L^2(\Omega)}, \tag{8.77}$$

is minimized with respect to θ and maximized with respect to ξ subject to the problem (8.75),

where $\gamma_j > 0$, for $j = 1, 2$ are fixed constants and the function $u_{\text{obs}} \in L^2(\mathcal{Q})$ is a given observation.

Before studying the saddle point problem (8.77), we give the Lipschitz continuity result and study the G-differentiability of the mapping $\mathcal{F}$.

Proposition 8.12. *Let $\varphi \in L^2(\mathcal{Q})$ and $a_0 \in L^\infty(\mathcal{Q})$ be fixed functions. Let $\xi_i \in V_{\text{ad}}$, $\theta_i \in U_{\text{ad}}$ and $u_i = \mathcal{F}(\theta_i, \xi_i)$ be the solution of* (8.75)*, corresponding to (θ_i, ξ_i), for $i = 1, 2$. We have the following estimates:*

$$\parallel u \parallel^2_{\mathcal{W} \cap \mathcal{C}([0,T];H)} \leq C(\parallel \theta \parallel^2_{L^2(\mathcal{Q})} + \parallel \xi \parallel^2_{L^2(\Omega)}). \tag{8.78}$$

where $u := u_1 - u_2$, $\theta := \theta_1 - \theta_2$ and $\xi := \xi_1 - \xi_2$.

Proof. Let $\xi_i \in V_{\text{ad}}$, $\theta_i \in U_{\text{ad}}$ and $u_i = \mathcal{F}(\theta_i, \xi_i)$, for $i = 1, 2$ then $u = u_1 - u_2$ is a solution of the problem:

$$\frac{\partial u}{\partial t} + Au + a_0 u = -\theta_2 u - \theta(u_1 + U) \quad \text{on } Q,$$

subject to the linear Robin boundary condition

$$\frac{\partial u}{\partial \eta_A} + \beta u = 0 \quad \text{on } \Sigma, \tag{8.79}$$

with the initial condition

$$u(0) = B\xi \quad \text{on } \Omega.$$

Multiplying now (8.79) by u and integrating over Ω (taking into account the boundary conditions), we obtain (by using Hölder's inequalities)

$$\begin{aligned} &\frac{\mathrm{d}}{\mathrm{d}t} \| u \|^2_{L^2(\Omega)} + 2 \| u \|^2_V + \| u \|^2_{L^2(\Sigma)} \\ &\qquad \leq c_1 \| u \|^2_{L^2(\Omega)} + c_2 \| \theta \|_{L^2(\Omega)} \| u_1 + U \|_{L^4(\Omega)} \| u \|_{L^4(\Omega)} . \end{aligned} \tag{8.80}$$

Since $V \subset H^1(\Omega) \subset L^4(\Omega)$ (because $n \leq 3$) and according to the regularity of $u_1 + U$ (*i.e.*, $u_1 + U \in L^\infty(0,T;V)$), we can deduce that

$$\begin{aligned} &\frac{\mathrm{d}}{\mathrm{d}t} \| u \|^2_{L^2(\Omega)} + \| u \|^2_V + \| u \|^2_{L^2(\Sigma)} \\ &\qquad \leq c_1 \| u \|^2_{L^2(\Omega)} + c_3 \| \theta \|^2_{L^2(\Omega)} . \end{aligned}$$

Integrating now with respect to time over $(0,t)$, $\forall t \in (0,T)$, we can deduce that

$$\begin{aligned} &\| u \|^2_{L^2(\Omega)} + \int_0^t \| u \|^2_V \, \mathrm{d}s + \int_0^t \| u \|^2_{L^2(\Sigma)} \, \mathrm{d}s \\ &\qquad \leq c_1 \int_0^t \| u \|^2_{L^2(\Omega)} \, \mathrm{d}s + c_4(\| \theta \|^2_{L^2(Q)} + \| \xi \|^2_{L^2(\Omega)}). \end{aligned}$$

By using Gronwall's formula we can deduce that

$$\| u \|^2_{L^\infty(0,T;L^2(\Omega)) \cap L^2(0,T;V)} \leq C(\| \theta \|^2_{L^2(Q)} + \| \xi \|^2_{L^2(\Omega)}). \tag{8.81}$$

Multiplying now (8.79) by $v \in V$ and integrating over Ω (taking into account the boundary conditions), we obtain (by using Hölder's inequalities)

$$\langle \frac{\partial u}{\partial t}, v \rangle_{V',V} \leq c_1 \| u \|_V \| v \|_V + c_2 \| \theta \|_{L^2(\Omega)} \| u_1 + U \|_{L^4(\Omega)} \| v \|_{L^4(\Omega)} .$$

Since $V \subset H^1(\Omega) \subset L^4(\Omega)$ (because $n \leq 3$) and according to the regularity of $u_1 + U$ (*i.e.*, $u_1 + U \in L^\infty(0,T;V)$), we can deduce that

$$\langle \frac{\partial u}{\partial t}, v \rangle_{V',V} \leq c_1 \| u \|_V \| v \|_V + c_3 \| \theta \|_{L^2(\Omega)} \| v \|_V .$$

Integrating now with respect to time over $(0,t)$, for $t \in (0,T)$, we can deduce that

$$\| \frac{\partial u}{\partial t} \|^2_{L^2(0,T;V')} \leq C(\int_0^t \| u \|^2_V \, \mathrm{d}s + \| \theta \|^2_{L^2(Q)}).$$

According to (8.81), we can conclude that

$$\| u \|^2_{\mathcal{W} \cap \mathcal{C}([0,T];H)} \leq C(\| \theta \|^2_{L^2(Q)} + \| \xi \|^2_{L^2(\Omega)}).$$

□

8.5.2 Differentiability of the Mapping Solution

We will now state the G-differentiability of the mapping $\mathcal{F}$ which maps the source term (θ, ξ) of (8.75) to the corresponding solution u, and we will obtain some estimates for the derivative of the map $\mathcal{F}$. Namely we will prove the following proposition.

Proposition 8.13. *The function $\mathcal{F} : (\theta, \xi) \longrightarrow u = \mathcal{F}(\theta, \xi)$ solution of (8.75) is G-differentiable, with respect to (θ, ξ) from $U_{\text{ad}} \times V_{\text{ad}}$ to $\mathcal{W} \cap \mathcal{C}([0,T]; H)$ where the G-derivative $\mathcal{F}'(\theta, \xi)$: $\mathbf{h} = (h_1, h_2) \in L^\infty(\mathcal{Q}) \times V_{\text{ad}} \longrightarrow w = \mathcal{F}'(\theta, \xi)\mathbf{h}$ is the unique solution in $\mathcal{W} \cap \mathcal{C}([0,T]; H)$ of the following linear parabolic problem:*

$$\begin{aligned} &\frac{\partial w}{\partial t} + Aw + a_0 w = -\theta w - h_1(u+U) \quad on\ \mathcal{Q} \\ &\frac{\partial w}{\partial \eta_A} + \beta w = 0 \quad on\ \Sigma, \\ &w(0) = Bh_2 \quad on\ \Omega \end{aligned} \tag{8.82}$$

such that

$$\| w \|^2_{\mathcal{W} \cap \mathcal{C}([0,T];H)} \leq C(\| h_1 \|^2_{L^2(\mathcal{Q})} + \| h_2 \|^2_{L^2(\Omega)}). \tag{8.83}$$

Moreover, for all $(\theta_i, \xi_i) \in U_{\text{ad}} \times V_{\text{ad}}$, for $i = 1, 2$, we have the following estimate (for all $\mathbf{h} \in L^\infty(\mathcal{Q}) \times V_{\text{ad}}$):

$$\begin{aligned} &\| \mathcal{F}'(\theta_1, \xi_1)\mathbf{h} - \mathcal{F}'(\theta_2, \xi_2)\mathbf{h} \|^2_{\mathcal{W} \cap \mathcal{C}([0,T];H)} \\ &\leq C_1 \left(\| \theta \|^2_{L^2(\mathcal{Q})} + \| \xi \|^2_{L^2(\Omega)} \right) \left(\| h_1 \|^2_{L^2(\mathcal{Q})} + \| h_2 \|^2_{L^2(\Omega)} \right)^{1/2} \\ &+C_2 \left(\| h_1 \|^2_{L^2(\mathcal{Q})} + \| h_2 \|^2_{L^2(\Omega)} \right) \left(\| \theta \|^2_{L^2(\mathcal{Q})} + \| \xi \|^2_{L^2(\Omega)} \right)^{1/2}. \end{aligned} \tag{8.84}$$

Proof. The problem (8.82) is similar to the problem ((8.75)) (with $-h_1(u + U) + \theta U \in L^2(\mathcal{Q})$, since $(u, U) \in L^2(\mathcal{Q})^2$ and $(h_1, \theta) \in (L^\infty(\mathcal{Q}))^2$, plays the role of φ, and $h_2 \in L^2(\Omega)$ plays the role of ξ in (8.75)). Then (8.82) admits a unique solution $w \in \mathcal{W} \cap \mathcal{C}([0,T]; H)$ such that

$$\| w \|_{\mathcal{W} \cap \mathcal{C}([0,T];H)} \leq C_1 \| u \|_{L^2(\mathcal{Q})} + C_2.$$

Let $(\theta, h_1) \in U_{\text{ad}} \times L^\infty(\mathcal{Q})$, $(\xi, h_2) \in (V_{\text{ad}})^2$ and $\epsilon > 0$ such that $\epsilon h_1 + \theta \in U_{\text{ad}}$. Let $u = \mathcal{F}(\theta, \xi)$ and $u_\epsilon = \mathcal{F}(\theta + \epsilon h_1, \xi + \epsilon h_2)$.

Step 1: Prove that $u_\epsilon \longrightarrow u$ strongly in $\mathcal{C}([0,T]; H)$ as $\epsilon \longrightarrow 0$. Let $v_\epsilon := u_\epsilon - u$, obviously, v_ϵ satisfies

$$\begin{aligned} &\frac{\partial v_\epsilon}{\partial t} + Av_\epsilon + a_0 v_\epsilon = -(\theta + \epsilon h_1)v_\epsilon - \epsilon h_1(u+U) \quad \text{on } \mathcal{Q}, \\ &\frac{\partial v_\epsilon}{\partial \eta_A} + \beta v_\epsilon = 0 \quad \text{on } \Sigma, \\ &v_\epsilon(0) = \epsilon B h_2 \quad \text{on } \Omega. \end{aligned} \tag{8.85}$$

Since $h_1 \in L^\infty(\mathcal{Q})$, $h_2 \in L^2(\Omega)$ and $(u+U) \in L^2(\mathcal{Q})$ then $g_\epsilon := -\epsilon h_1(u+U)$ is in $L^2(\mathcal{Q})$ and $v_{0\epsilon} := \epsilon h_2$ is in $L^2(\Omega)$. Consequently, since $(\theta+\epsilon h_1) \in L^\infty(\mathcal{Q})$ and the problem (8.85) is similar to problem (8.79), we have that, there exists a constant $C > 0$ (independent of ϵ) such that

$$\| v_\epsilon \|_{\mathcal{W}\cap\mathcal{C}([0,T];H)} \leq C(\| g_\epsilon \|_{L^2(\mathcal{Q})} + \| v_{0\epsilon} \|_{L^2(\Omega)})$$

and then

$$\| v_\epsilon \|_{\mathcal{W}\cap\mathcal{C}([0,T];H)} \leq \epsilon C_1(\| u+U \|_{L^2(\mathcal{Q})} + \| h_2 \|_{L^2(\Omega)}).$$

Consequently, $v_\epsilon \longrightarrow 0$ strongly in $\mathcal{W}\cap\mathcal{C}([0,T];H)$ as $\epsilon \longrightarrow 0$.

Step 2: Prove now that $w_\epsilon := (u_\epsilon - u)/\epsilon \longrightarrow w$ strongly in $\mathcal{W}\cap\mathcal{C}([0,T];H)$ as $\epsilon \longrightarrow 0$. Let $\tilde{w}_\epsilon := w_\epsilon - w$, obviously, $\tilde{w}_\epsilon$ satisfies

$$\begin{aligned} &\frac{\partial \tilde{w}_\epsilon}{\partial t} + A\tilde{w}_\epsilon + a_0\tilde{w}_\epsilon = -\theta\tilde{w}_\epsilon - h_1 v_\epsilon \ \text{ on } \mathcal{Q},\\ &\frac{\partial \tilde{w}_\epsilon}{\partial \eta_A} + \beta\tilde{w}_\epsilon = 0 \ \text{ on } \Sigma,\\ &\tilde{w}_\epsilon(0) = 0 \ \text{ on } \Omega. \end{aligned} \tag{8.86}$$

The problem (8.86) is similar to problem (8.85), so

$$\| \tilde{w}_\epsilon \|_{\mathcal{W}\cap\mathcal{C}([0,T];H)} \leq C \| v_\epsilon \|_{L^2(\mathcal{Q})} .$$

Since $\| v_\epsilon \|_{L^2(\mathcal{Q})} \longrightarrow 0$ as $\epsilon \longrightarrow 0$, we can deduce the convergence result.

The problem (8.82) is similar to problem (8.79), so we can apply the estimate (8.78) and we obtain the result

$$\| w \|^2_{\mathcal{W}\cap\mathcal{C}([0,T];H)} \leq C(\| h_1 \|^2_{L^2(\mathcal{Q})} + \| h_2 \|^2_{L^2(\Omega)}).$$

We shall now prove the estimate given in the second part of Proposition 8.13.

For $i = 1, 2$, let $(\theta_i, \xi_i) \in U_{\mathrm{ad}} \times V_{\mathrm{ad}}$, $w_i = \mathcal{F}'(\theta_i, \xi_i)\mathbf{h}$ (solution of (8.82)) and $u_i = \mathcal{F}(\theta_i, \xi_i)\mathbf{h}$ (solution of (8.75)). Then $w = w_1 - w_2$ is the solution of

$$\begin{aligned} &\frac{\partial w}{\partial t} + Aw + (a_0+\theta_2)w = -\theta w_1 - h_1 u \ \text{ on } \mathcal{Q}\\ &\frac{\partial w}{\partial \eta_A} + \beta w = 0 \ \text{ on } \Sigma,\\ &w(0) = 0 \ \text{ on } \Omega. \end{aligned} \tag{8.87}$$

where $u = u_1 - u_2$, $\theta = \theta_1 - \theta_2$.

Multiplying now (8.87) by w and integrating over Ω (taking into account the boundary conditions), we obtain (by using Hölder's inequalities)

$$\begin{aligned} &\frac{\mathrm{d}}{\mathrm{d}t} \| w \|^2_{L^2(\Omega)} + 2 \| w \|^2_V + \| w \|^2_{L^2(\Sigma)}\\ &\quad \leq c_1 \| w \|^2_{L^2(\Omega)} + c_2 \| \theta \|_{L^2(\Omega)} \| w_1 \|_{L^3(\Omega)} \| w \|_{L^6(\Omega)}\\ &\quad\quad + c_3 \| h_1 \|_{L^2(\Omega)} \| u \|_{L^3(\Omega)} \| w \|_{L^6(\Omega)} . \end{aligned} \tag{8.88}$$

Since $V \subset H^1(\Omega) \subset L^6(\Omega)$ (because $n \leq 3$), we can deduce that

$$\begin{aligned} \frac{\mathrm{d}}{\mathrm{d}t} \| w \|^2_{L^2(\Omega)} + \| w \|^2_V + \| w \|^2_{L^2(\Sigma)} \\ \leq c_1 \| w \|^2_{L^2(\Omega)} + c_4 \| \theta \|^2_{L^2(\Omega)} \| w_1 \|^2_{L^3(\Omega)} \\ + c_5 \| h_1 \|^2_{L^2(\Omega)} \| u \|^2_{L^3(\Omega)} . \end{aligned}$$

Because of the Gagliardo–Nirenberg inequalities we have that, for all $v \in H^1(\Omega)$,

$$\| v \|^2_{L^3(\Omega)} \leq C \| v \|^{\frac{n}{3}}_V \| v \|^{\frac{6-n}{3}}_{L^2(\Omega)}$$

and then (since w_1 and u are in $L^\infty(0,T;H) = L^\infty(0,T;L^2(\Omega))$)

$$\begin{aligned} \frac{\mathrm{d}}{\mathrm{d}t} \| w \|^2_{L^2(\Omega)} + \| w \|^2_V + \| w \|^2_{L^2(\Sigma)} \\ \leq c_1 \| w \|^2_{L^2(\Omega)} + c_4 \| \theta \|^2_{L^2(\Omega)} \| w_1 \|^{\frac{n}{3}}_V \| w_1 \|^{\frac{6-n}{3}}_{L^\infty(0,T;H)} \\ + c_5 \| h_1 \|^2_{L^2(\Omega)} \| u \|^{\frac{n}{3}}_V \| u \|^{\frac{6-n}{3}}_{L^\infty(0,T;H)} . \end{aligned}$$

Integrating now with respect to time over $(0,t)$, $\forall t \in (0,T)$ we can deduce, in particular, that (by using Hölder's inequalities)

$$\begin{aligned} \| w(t) \|^2_{L^2(\Omega)} + \int_0^t \| w \|^2_V \,\mathrm{d}s \\ \leq c_1 \int_0^t \| w \|^2_{L^2(\Omega)} \,\mathrm{d}s \\ + c_4 \left(\int_0^T \| \theta \|^{\frac{12}{6-n}}_{L^2(\Omega)} \,\mathrm{d}s \right)^{\frac{6-n}{6}} \| w_1 \|^{\frac{n}{6}}_{L^2(0,T;V)} \| w_1 \|^{\frac{6-n}{3}}_{L^\infty(0,T;H)} \\ + c_5 \left(\int_0^T \| h_1 \|^{\frac{12}{6-n}}_{L^2(\Omega)} \,\mathrm{d}s \right)^{\frac{6-n}{6}} \| u \|^{\frac{n}{6}}_{L^2(0,T;V)} \| u \|^{\frac{6-n}{3}}_{L^\infty(0,T;H)} . \end{aligned}$$

Since (because $L^2(0,T;V)$ and $L^\infty(0,T;H)$ are subsets of $\mathcal{W} \cap \mathcal{C}([0,T];H)$),

$$\begin{aligned} & \| w_1 \|^{\frac{n}{6}}_{L^2(0,T;V)} \| w_1 \|^{\frac{6-n}{3}}_{L^\infty(0,T;H)} \leq C \| w_1 \|^2_{\mathcal{W} \cap \mathcal{C}([0,T];H)}, \\ & \| u \|^{\frac{n}{6}}_{L^2(0,T;V)} \| u \|^{\frac{6-n}{3}}_{L^\infty(0,T;H)} \leq C \| u \|^2_{\mathcal{W} \cap \mathcal{C}([0,T];H)}, \\ & \left(\int_0^T \| \theta \|^{\frac{12}{6-n}}_{L^2(\Omega)} \,\mathrm{d}s \right)^{\frac{6-n}{6}} \leq C \left(\int_0^T \| \theta \|^4_{L^2(\Omega)} \,\mathrm{d}s \right)^{\frac{1}{2}}, \\ & \left(\int_0^T \| h_1 \|^{\frac{12}{6-n}}_{L^2(\Omega)} \,\mathrm{d}s \right)^{\frac{6-n}{6}} \leq C \left(\int_0^T \| h_1 \|^4_{L^2(\Omega)} \,\mathrm{d}s \right)^{\frac{1}{2}}, \end{aligned}$$

we can deduce that (since θ and h_1 are in $L^\infty(\mathcal{Q})$)

$$\| w \|^2_{L^2(\Omega)} + \int_0^t \| w \|^2_V \,\mathrm{d}s \leq c_1 \int_0^t \| w \|^2_{L^2(\Omega)} \,\mathrm{d}s$$
$$+c_6 \| w_1 \|^2_{\mathcal{W}\cap\mathcal{C}([0,T];H)} \| \theta \|_{L^2(Q)}$$
$$+c_7 \| u \|^2_{\mathcal{W}\cap\mathcal{C}([0,T];H)} \| h_1 \|_{L^2(Q)}$$

By using Gronwall's formula and the estimates (8.78) and (8.83), we can deduce that

$$\| w \|^2_{L^\infty(0,T;L^2(\Omega))\cap L^2(0,T;V)}$$
$$\leq C_1(\| \theta \|^2_{L^2(Q)} + \| \xi \|^2_{L^2(\Omega)})(\| h_1 \|^2_{L^2(Q)} + \| h_2 \|^2_{L^2(\Omega)})^{1/2} \tag{8.89}$$
$$+C_2(\| h_1 \|^2_{L^2(Q)} + \| h_2 \|^2_{L^2(\Omega)})(\| \theta \|^2_{L^2(Q)} + \| \xi \|^2_{L^2(\Omega)})^{1/2}.$$

Because of Equation (8.88), and the previous estimate (8.89), we deduce easily the estimate

$$\| w \|^2_{\mathcal{W}\cap\mathcal{C}([0,T];H)}$$
$$\leq C_1(\| \theta \|^2_{L^2(Q)} + \| \xi \|^2_{L^2(\Omega)})(\| h_1 \|^2_{L^2(Q)} + \| h_2 \|^2_{L^2(\Omega)})^{1/2}$$
$$+C_2(\| h_1 \|^2_{L^2(Q)} + \| h_2 \|^2_{L^2(\Omega)})(\| \theta \|^2_{L^2(Q)} + \| \xi \|^2_{L^2(\Omega)})^{1/2}.$$

This completes the proof. □

We can now study the following existence of an optimal solution.

8.5.3 Existence of an Optimal Solution

Let K_d be a convex, closed, non-empty and bounded subset of V_ad. We have the following existence theorem.

Theorem 8.14. *For $\gamma_i, i = 1, 2$ sufficiently large, there exists $(\theta^*, \xi^*) \in U_\mathrm{ad} \times K_\mathrm{d}$ and $u^* \in \mathcal{W}\cap\mathcal{C}([0,T];H)$ such that (θ^*, ξ^*) is the optimal solution of (8.77) and $u^* = \mathcal{F}(\theta^*, \xi^*)$ is the solution of (8.75).*

Proof. Let P_ξ be the mapping: $\theta \longrightarrow J(\theta, \xi)$ and Q_θ be the mapping: $\xi \longrightarrow J(\theta, \xi)$. To obtain the existence of the robust control problem, we prove first that P_ξ is convex and lower semi-continuous for all $\xi \in K_\mathrm{d}$, second that Q_θ is concave and upper semi-continuous for all $\theta \in U_\mathrm{ad}$ and finally we use the minimax theorems in infinite dimensions presented in Chapter 5.

In order to prove the convexity, it is sufficient to show that for $(\theta_1, \theta_2) \in U_\mathrm{ad} \times U_\mathrm{ad}$, we have $(P'_\xi(\theta_1) - P'_\xi(\theta_2)).\theta \geq 0$, where $\theta := \theta_1 - \theta_2$.

From the expression of G-differentiable cost functional J (a composition of G-differentiable mappings), it follows that P_ξ is G-differentiable and for $i = 1, 2$,

$$P'_\xi(\theta_i).\theta = \lim_{\epsilon \longrightarrow 0} \frac{J(\theta_i + \epsilon\theta, \xi) - J(\theta_i, \xi)}{\epsilon}$$
$$= \int\!\!\int_Q (u_i - u_\mathrm{obs}) w_i \mathrm{d}x\mathrm{d}t + \gamma_1 \int\!\!\int_Q \theta_i \theta \mathrm{d}x\mathrm{d}t,$$

where $u_i = \mathcal{F}(\theta_i, \xi)$ and $w_i = \mathcal{F}'(\theta_i, \xi).(\theta, 0)$.

Consequently,

$$\begin{aligned}(P'_\xi(\theta_1) - P'_\xi(\theta_2)).\theta = \gamma_1 \parallel \theta \parallel^2_{L^2(\mathcal{Q})} + \iint_{\mathcal{Q}} (u_1 - u_2) w_1 \mathrm{d}x\mathrm{d}t \\ + \iint_{\mathcal{Q}} (u_2 - u_{\mathrm{obs}})(w_1 - w_2)\mathrm{d}x\mathrm{d}t.\end{aligned} \tag{8.90}$$

The estimates (8.84), (8.83) and (8.78) imply that

$$\begin{aligned}\iint_{\mathcal{Q}} (u_1 - u_2) w_1 \mathrm{d}x\mathrm{d}t &\leq \parallel u_1 - u_2 \parallel_{L^2(\mathcal{Q})} \parallel w_1 \parallel_{L^2(\mathcal{Q})} \\ &\leq C_0 \parallel \theta \parallel^2_{L^2(\mathcal{Q})}, \\ \iint_{\mathcal{Q}} (u_2 - u_{\mathrm{obs}})(w_1 - w_2)\mathrm{d}x\mathrm{d}t &\leq \parallel u_2 - u_{\mathrm{obs}} \parallel_{L^2(\mathcal{Q})} \parallel w_1 - w_2 \parallel_{L^2(\mathcal{Q})} \\ &\leq C_1 \parallel \theta \parallel^{3/2}_{L^2(\mathcal{Q})} .\end{aligned} \tag{8.91}$$

From (8.90) and the previous results (8.91), we can deduce that there exists a constant $\gamma_{1l} > 0$ such that for $\gamma_1 \geq \gamma_{1l}$ we have

$$(P'_\xi(\theta_1) - P'_\xi(\theta_2)).\theta \geq (\gamma_1 - C_0) \parallel \theta \parallel^2_{L^2(\mathcal{Q})} - C_1 \parallel \theta \parallel^{3/2}_{L^2(\mathcal{Q})} \geq 0$$

and then the convexity of P_ξ is established.

In the same way, we can find $\gamma_{2l} > 0$ such that for $\gamma_2 \geq \gamma_{2l}$, Q_θ is concave. We prove now that P_ξ (respectively Q_θ) is lower (respectively upper) semi-continuous for all $\xi \in K_{\mathrm{d}}$ (respectively $\theta \in U_{\mathrm{ad}}$). Let $\theta_k \in U_{\mathrm{ad}}$ be a minimizing sequence of P_ξ, *i.e.*,

$$\liminf_{k \longrightarrow \infty} J(\theta_k, \xi) = \inf_{\theta \in L^2(\mathcal{Q})} J(\theta, \xi).$$

Then, according to the nature of the cost function J, we can deduce that θ_k is uniformly bounded in U_{ad} and we can extract from θ_k a subsequence also denoted by θ_k such that $\theta_k \rightharpoonup \theta_\xi$ weakly in U_{ad}. Therefore, by using the same technique as used in the proof of the estimate (8.78), $u_k = \mathcal{F}(\theta_k, \xi)$ is uniformly bounded in $\mathcal{W} \cap \mathcal{C}([0, T]; H)$. Moreover, according now to Lemma 6.6, the injection of $\mathcal{W}$ into $L^2(\mathcal{Q})$ is compact. Consequently, these results make it possible to extract from u_k a subsequence also denoted by u_k such that

$$\begin{aligned}&u_k \rightharpoonup u_\xi \text{ weakly in } L^2(0, T; V), \\ &u_k \longrightarrow u_\xi \text{ strongly in } L^2(\mathcal{Q}), \\ &\theta_k \rightharpoonup \theta_\xi \text{ weakly in } L^2(\mathcal{Q}) \text{ and } \theta_\xi \in U_{\mathrm{ad}}.\end{aligned} \tag{8.92}$$

Prove now that $u_k \theta_k \longrightarrow u_\xi \theta_\xi$ weakly in $L^2(\mathcal{Q})$. Since $u_k \theta_k - u_\xi \theta_\xi = (u_k - u_\xi)\theta_k + u_\xi(\theta_k - \theta_\xi)$, then according to the first and second parts of (8.92), we then obtain the result. Is is easy to prove that u_ξ is a solution of (8.75) with a parameter (θ_ξ, ξ) and according to the uniqueness of the solution of (8.75), we have then $u_\xi = \mathcal{F}(\theta_\xi, \xi)$.

Since the norm is lower semi-continuous, therefore we have that the map P_ξ is lower semi-continuous for all $\xi \in K_{\mathrm{d}}$. By applying similar argument as in the proof of the previous result we obtain that Q_θ is upper semi-continuous for all $\theta \in U_{\mathrm{ad}}$.

This completes the proof. □

We next wish to show the appropriate first-order necessary conditions (optimality conditions) of the saddle point problem (8.77).

8.5.4 First-order Necessary Conditions

Theorem 8.15. *Let $\gamma_i, i = 1, 2$ be sufficiently large, $(\theta^*, \xi^*) \in U_{\mathrm{ad}} \times K_{\mathrm{d}}$ be an optimal control defined by (8.77) and $u^* \in \mathcal{W} \cap \mathcal{C}([0,T];H)$ be the optimal state such that $u^* = \mathcal{F}(\theta^*, \xi^*)$ is the solution of (8.75), with the data (θ^*, ξ^*). Then there exists a unique solution $\tilde{u}^* \in \mathcal{W} \cap \mathcal{C}([0,T];H)$ to the following adjoint problem corresponding to the primal problem (8.75):*

$$
\begin{aligned}
&-\frac{\partial \tilde{u}}{\partial t} + A^*\tilde{u} + a_0\tilde{u} = -\theta^*\tilde{u} + \mathcal{C}^*\mathcal{C}(u^* - u_{\mathrm{obs}}) \quad on\ \mathcal{Q},\\
&\textit{subject to the linear Robin boundary condition}\\
&\frac{\partial \tilde{u}}{\partial \eta_{A^*}} + \beta\tilde{u} = 0 \quad on\ \Sigma,\\
&\textit{with the final condition}\\
&\tilde{u}(t = T) = 0 \quad on\ \Omega,
\end{aligned}
\tag{8.93}
$$

where A^* is the adjoint of A. Moreover, for all $\theta \in U_{\mathrm{ad}}$ and $\xi \in K_{\mathrm{d}}$,

$$
\begin{aligned}
&\theta^* = \max\left(\tau_1, \min\left((u^* + U)\tilde{u}^*/\gamma_1, \tau_2\right)\right)\\
&\left(\ or \iint_{\mathcal{Q}} (\gamma_1\theta^* - (u^* + U)\tilde{u}^*)(\theta - \theta^*)\mathrm{d}x\mathrm{d}t \geq 0\right),\\
&\int_{\Omega} (-\gamma_2\xi^* + B^*\tilde{u}^*(0))(\xi - \xi^*)\mathrm{d}x \leq 0,
\end{aligned}
$$

where B^* is the adjoint of the operator B.

Proof. Since the function $\mathcal{C}^*\mathcal{C}(u^* - u_{\mathrm{obs}})$ is in $L^2(0,T;V')$ then the problem (8.101) admits a unique solution in $\mathcal{W} \cap \mathcal{C}([0,T];H)$ (in the weak formulation sense).

The cost functional J is a composition of G-differentiable mappings then J is G-differentiable and in particular the G-derivative of J at the optimal point $(\theta^*, \xi^*) \in U_{\mathrm{ad}} \times K_{\mathrm{d}}$ is given by (for all $\mathbf{h} = (h_1, h_2) \in L^\infty(\mathcal{Q}) \times K_{\mathrm{d}}$ such that $(\theta^* + \epsilon h_1) \in U_{\mathrm{ad}}$ for ϵ small)

$$
\begin{aligned}
J'(\theta^*, \xi^*)\mathbf{h} &= \lim_{\epsilon \longrightarrow 0} \frac{J(\theta^* + \epsilon h_1, \xi^* + \epsilon h_2) - J(\theta^*, \xi^*)}{\epsilon}\\
&= \iint_{\mathcal{Q}} \mathcal{C}(u - u_{\mathrm{obs}}).\mathcal{C}w\mathrm{d}x\mathrm{d}t\\
&\quad + \gamma_1 \iint_{\mathcal{Q}} \theta^* h_1 \mathrm{d}x\mathrm{d}t - \gamma_2 \int_{\Omega} \xi^* h_2 \mathrm{d}x,
\end{aligned}
\tag{8.94}
$$

where $w = \mathcal{F}'(\theta^*, \xi^*)\mathbf{h}$ is the solution of (8.82).

Multiplying the first equation of (8.82) by $\tilde{u}^*$ and integrating over $\mathcal{Q}$, this gives (using Green's formula and integrating by parts in time)

$$\int_0^T \langle -\frac{\partial \tilde{u}^*}{\partial t} + A^*\tilde{u}^* + (a_0 + \theta^*)\tilde{u}^*, w\rangle \mathrm{d}t + \int_0^T \int_\Gamma (\frac{\partial \tilde{u}^*}{\partial \eta_{A^*}} + \beta\tilde{u}^*) w \mathrm{d}\Gamma \mathrm{d}t$$
$$= -\int_0^T \int_\Omega h_1(u^* + U)\tilde{u}^* \mathrm{d}x\mathrm{d}t + \int_\Omega w(0)\tilde{u}^*(0)\mathrm{d}x - \int_\Omega w(T)\tilde{u}^*(T)\mathrm{d}x.$$

Since $\tilde{u}^*$ is the solution of (8.101) and $w(0) = Bh_2$, therefore, we can deduce that

$$\begin{aligned}&\iint_{\mathcal{Q}} \mathcal{C}(u^* - u_{\mathrm{obs}}).\mathcal{C}w \mathrm{d}x\mathrm{d}t \\ &\qquad = -\iint_{\mathcal{Q}} h_1(u^* + U)\tilde{u}^* \mathrm{d}x\mathrm{d}t + \int_\Omega (Bh_2)\tilde{u}^*(0)\mathrm{d}x\end{aligned} \tag{8.95}$$

and then (according to (8.94))

$$\begin{aligned}J'(\theta^*, \xi^*)\mathbf{h} = &\iint_{\mathcal{Q}} (\gamma_1\theta^* - (u^* + U)\tilde{u}^*) h_1 \mathrm{d}x\mathrm{d}t \\ &+ \int_\Omega (-\gamma_2\xi^* + B^*\tilde{u}^*(0)) h_2 \mathrm{d}x,\end{aligned} \tag{8.96}$$

where B^* the adjoint operator of the linear and bounded operator B.

Since (θ^*, ξ^*) is an optimal solution, then

$$\begin{aligned}&\iint_{\mathcal{Q}} (\gamma_1\theta^* - (u^* + U)\tilde{u}^*)(\theta - \theta^*)\mathrm{d}x\mathrm{d}t \geq 0 \quad \forall \theta \in U_{\mathrm{ad}}, \\ &\int_\Omega (-\gamma_2\xi^* + B^*\tilde{u}^*(0))(\xi - \xi^*)\mathrm{d}x \leq 0 \quad \forall \xi \in K_{\mathrm{d}}.\end{aligned} \tag{8.97}$$

Moreover, by using a standard control argument concerning the sign of the variation h (depending on the size of θ^*), we obtain that

$$\theta^* = \max\left(\tau_1, \min\left(\frac{(u^* + U)\tilde{u}^*}{\gamma_1}, \tau_2\right)\right).$$

This completes the proof. □

8.5.5 Other Situations and Applications

We assume now that the parameter θ is decomposed into the disturbance ξ and the control ϑ, *i.e.*, $\theta = B_1\vartheta + B_2\xi$, where $B_1 : K_1 \longrightarrow U_{\mathrm{ad}}$ (respectively $B_2 : K_2 \longrightarrow V_{\mathrm{ad}}$) is the bounded and linear operator which maps abstract control (respectively disturbance) of the Hilbert space K_1 (respectively K_2) to feasible parameter control (respectively disturbance) space U_{ad} (respectively V_{ad}), with

$$U_{\rm ad} = \{\vartheta \in K_1 : \ \tau_{11} \le B_1\vartheta \le \tau_{12} \quad \text{a.e. in } \mathcal{Q}\},$$
$$V_{\rm ad} = \{\xi \in K_2 : \ \tau_{21} \le B_2\xi \le \tau_{22} \quad \text{a.e. in } \mathcal{Q}\}, \tag{8.98}$$

such that the intervals $[\tau_{11}, \tau_{12}]$ and $[\tau_{21}, \tau_{22}]$ contain 0.

The function u is assumed to be related to the disturbance ξ and control ϑ through the problem (8.74):

$$\begin{aligned}
&\frac{\partial u}{\partial t} + Au + a_0 u = \varphi - (B_1\vartheta + B_2\xi)(u + U) \ \text{ on } \mathcal{Q},\\
&\text{subject to the linear Robin boundary condition}\\
&\frac{\partial u}{\partial \eta_A} + \beta u = 0 \ \text{ on } \Sigma,\\
&\text{with the initial condition}\\
&u(0) = u_0 \ \text{ on } \Omega.
\end{aligned} \tag{8.99}$$

Our robust control problem is then:

$$\begin{aligned}
&\text{find } (\vartheta, \xi) \in K_{1\rm ad} \times K_{2\rm ad} \text{ such that the cost functional}\\
&J(\vartheta, \xi) = \frac{1}{2} \parallel \mathcal{C}(u - u_{\rm obs}) \parallel^2_{L^2(\mathcal{Q})} + \frac{\gamma_1}{2} \parallel \vartheta \parallel^2_{K_1} - \frac{\gamma_2}{2} \parallel \xi \parallel^2_{K_2},\\
&\text{is minimized with respect to } \vartheta \text{ and maximized with respect to } \xi\\
&\text{subject to the problem (8.99),}
\end{aligned} \tag{8.100}$$

where $\gamma_j > 0$, for $j = 1, 2$ are fixed constants, $K_{1\rm ad}$ (respectively $K_{2\rm ad}$) is a non-empty, closed, convex, bounded subset of K_1 (respectively of K_2) and the function $u_{\rm obs} \in L^2(\mathcal{Q})$ is a given observation.

The arguments of the previous section extend directly to the present case without further estimates. We have then the following results

Theorem 8.16. *Let $\gamma_i, i = 1, 2$ be sufficiently large, $(\vartheta^*, \xi^*) \in K_{1\rm ad} \times K_{2\rm ad}$ be an optimal control defined by (8.100) and $u^* \in \mathcal{W} \cap \mathcal{C}([0, T]; H)$ be the optimal state such that u^* is the solution of (8.99), with the data (ϑ^*, ξ^*). Then there exists a unique solution $\tilde{u}^* \in \mathcal{W} \cap \mathcal{C}([0, T]; H)$ to the following adjoint problem corresponding to the primal problem (8.99):*

$$\begin{aligned}
&-\frac{\partial \tilde{u}}{\partial t} + A^*\tilde{u} + a_0\tilde{u} = -(B_1\vartheta^* + B_2\xi^*)\tilde{u} + \mathcal{C}^*\mathcal{C}(u^* - u_{\rm obs}) \ \textit{ on } \mathcal{Q},\\
&\textit{subject to the linear Robin boundary condition}\\
&\frac{\partial \tilde{u}}{\partial \eta_{A^*}} + \beta\tilde{u} = 0 \ \textit{ on } \Sigma,\\
&\textit{with the final condition}\\
&\tilde{u}(t = T) = 0 \ \textit{ on } \Omega,
\end{aligned} \tag{8.101}$$

where A^* is the adjoint of the operator A. Moreover,

$$(\gamma_1\vartheta^* - B_1^*((u^* + U)\tilde{u}^*), \vartheta - \vartheta^*)_{K_1} \ge 0 \ \forall \vartheta \in K_{1\rm ad},$$
$$(-\gamma_2\xi^* - B_2^*((u^* + U)\tilde{u}^*), \xi - \xi^*)_{K_2} \le 0 \ \forall \xi \in K_{2\rm ad},$$

where B_i^* is the adjoint of the operator B_i, for $i = 1, 2$. □

Application: One-Neutron Diffusion Equation

Here, we present an example illustrative of the abstract result of this section. We consider a one-neutron diffusion equation, describing the neutron reaction inside a nuclear fission reactor, when pointwise controllers are applied at locations $x = (x_i)_{i=1,l} \in \Omega$, and pointwise disturbances are applied at locations $y = (y_j)_{j=1,m} \in \Omega$.

We suppose that the reactor core is a slab geometry with length L. The perturbed system that we consider is then

$$\frac{\partial u}{\partial t} - \nu \frac{\partial^2 u}{\partial x^2} = \mu u - \sum_{i=1}^{l} f_i(t)\phi_0(x)\delta_{x_i}(u+U) - \sum_{j=1}^{m} g_j(t)\phi_1(x)\delta_{y_j}(u+U), \text{ in } \mathcal{Q},$$

subject to the boundary conditions

$$-\frac{\partial u}{\partial x}(0,t) + \beta u(0,t) = 0, \quad \frac{\partial u}{\partial x}(L,t) + \beta u(L,t) = 0, \quad t \in (0,T) \tag{8.102}$$

and the initial condition

$$u(x,0) = u_0 \quad (x \in (0,L)),$$

where δ_x is the usual Dirac function, $\mathcal{Q} = (0,L) \times (0,T)$, $(\phi_k, k = 0,1)$ are given positive functions bounded above and below by two positive constants, $\nu = VD$, $\mu = V(r\sigma_f - \sigma_a)$, D is the neutron diffusion coefficient, and σ_f and σ_a denote the fission and absorption macroscopic cross-sections, respectively. The parameter V is the average neutron velocity and r denotes the number of neutrons generated per nuclear fission (for more details see, *e.g.*, Christensen *et al.* [84]). The control (respectively disturbance) absorption cross-section at location x_i (respectively y_j) is represented by $f_i(t)$ (respectively $g_j(t)$) such that

$$a_i \leq f_i(t) \leq b_i, \text{ for } i = 1,l, \quad \text{a.e. } t \in (0,T),$$

$$c_j \leq g_j(t) \leq d_j, \text{ for } j = 1,m, \quad \text{a.e. } t \in (0,T).$$

We propose the following objective functional:

$$J(\vartheta,\xi) = \frac{k_0}{2}\int_0^T\int_0^L |\, u - u_{\text{obs}} \,|^2 \, \mathrm{d}x\mathrm{d}t + \frac{\gamma_1}{2}\sum_{i=1,l}\int_0^T |\, f_i \,|^2 \, \mathrm{d}t - \frac{\gamma_2}{2}\sum_{j=1,m}\int_0^T |\, g_j \,|^2 \, \mathrm{d}t,$$

where $\vartheta = (f_i)_{i=1,l} \in K_1$ and $\xi = (g_j)_{j=1,m} \in K_2$, with

$$K_1 = \{\vartheta = (f_i)_{i=1,l} \in [L^2(0,T)]^l : a_i \leq f_i(t) \leq b_i \quad \text{a.e. in } (0,T)\},$$

$$K_2 = \{\xi = (g_j)_{i=1,m} \in [L^2(0,T)]^m : c_j \leq g_j(t) \leq d_j \quad \text{a.e. in } (0,T)\}.$$

Since all the assumptions of the previous abstract results are satisfied by the model in this particular case, therefore we can apply the previous result of Theorem 8.16 and conclude the existence and the optimality conditions. Moreover, the gradients of J are

$$\frac{\partial J}{\partial \vartheta}(\vartheta,\xi)$$
$$= [\gamma_1 f_1(t) - \phi_0(x_1)u(x_1,t)\tilde{u}(x_1,t), \ldots, \gamma_1 f_l(t) - \phi_0(x_l)u(x_l,t)\tilde{u}(x_l,t)]^T ,$$
$$\frac{\partial J}{\partial \xi}(\vartheta,\xi)$$
$$= [-\gamma_2 g_1(t) - \phi_1(y_1)u(y_1,t)\tilde{u}(y_1,t), \ldots, -\gamma_2 g_m(t) - \phi_1(y_m)u(y_m,t)\tilde{u}(y_m,t)]^T ,$$

where $\tilde{u}$ is the solution of following adjoint problem:

$$-\frac{\partial \tilde{u}}{\partial t} - \nu\frac{\partial^2 \tilde{u}}{\partial x^2} = \mu\tilde{u} - \sum_{i=1,l} f_i(t)\phi_0(x)\delta_{x_i}\tilde{u}$$
$$- \sum_{j=1,m} g_j(t)\phi_1(x)\delta_{y_j}\tilde{u} + k_0(u - u_{\text{obs}}), \text{ in } Q ,$$

subject to the boundary conditions

$$-\frac{\partial \tilde{u}}{\partial x}(0,t) + \beta\tilde{u}(0,t) = 0, \; \frac{\partial \tilde{u}}{\partial x}(L,t) + \beta\tilde{u}(L,t) = 0, \quad t \in (0,T)$$

and the final condition

$$\tilde{u}(x,T) = 0 \quad (x \in (0,L)).$$

Remark 8.17. (*i*) Another application is the cancer chemotherapy model (one-compartment) in which the function u describes the number of cancer cells and the function θ represents the drug dosage administered in the system (8.74).

(*ii*) The same technique is valid in the case where the state variable u is a vector function (coupled system), see Belmiloudi [47]. We can then consider applications like the multi-neutron diffusion system (a problem in nuclear fission reactors) and the multi-compartment model for chemotherapy. ◇

8.6 Non-linear Robust Control for Non-linear Evolutive Problems

In this section we analyze full non-linear robust control problems. For this we consider, the non-linear evolutive problems where we can prove the existence and stability (under extra assumptions) theorems. In contrast to linear systems, the dynamics of evolutive non-linear systems obey complicated laws that, in general, cannot be arrived at by intuitive and direct calculations.

The main result of this study includes the existence, uniqueness and first-order necessary conditions of optimality for the worst disturbance and optimal controllers. The plan of this study is as follows. First, we study the existence and the uniqueness of the perturbation of the problem (8.103) and obtain some *a priori* estimates. Second, we formulate the robust control problem in the case of the forcing f decomposed into a disturbance ψ and a control ϕ. We prove the existence and uniqueness and give the appropriate optimality system. Next, we consider an initial perturbation problem, *i.e.*, the adjustment of initial conditions in order to obtain a system that agrees with a desired target system (because in many situations the initial condition is not well known). We reformulate the robust control problem for two cases: where the control is the initial condition and the disturbance is the forcing f, and where the initial condition is decomposed into a disturbance ψ and a control ϕ. As in previous sections we study the existence and uniqueness and give the optimality conditions. Finally, we present an example of convection–diffusion in the case of pollutants in liquid or atmospheric systems.

The notations, assumptions and spaces are the same as in Section 7.6.1.

8.6.1 State Equations

We will be consider the non-linear parabolic partial differential equations of the form (as in Section 7.6.1)

$$\begin{aligned} &\frac{\partial U}{\partial t} + AU + F(x,t,U) + K(\varpi,U) = f \quad \text{on } \mathcal{Q} = (0,T)\times\Omega \\ &U(0) = U_0 \quad \text{on } \Omega, \end{aligned} \tag{8.103}$$

where Ω is a boundary subset of $\mathbb{R}^m$, $m \geq 1$, with boundary Γ is sufficiently regular, A: $\mathcal{D} \longrightarrow \mathcal{D}'$ is an elliptic, selfadjoint operator with $\langle Av, v\rangle \geq \nu \parallel v \parallel_{\mathcal{D}}^2$, $\langle Au, v\rangle \leq M \parallel u \parallel_{\mathcal{D}} \parallel v \parallel_{\mathcal{D}}$ $\forall u, v \in \mathcal{D}$ ($\langle .,. \rangle$ is the duality pairing on $\mathcal{D}'$ and $\mathcal{D}$, $\parallel . \parallel_{\mathcal{D}}$ is the norm on $\mathcal{D}$, $\parallel . \parallel_*$ is the dual norm on $\mathcal{D}'$ and $\nu, M > 0$ are constants), $K(\varpi, .)$: $\mathcal{D} \longrightarrow L^2(\Omega)$ is a linear operator satisfying (7.81), ϖ is a given sufficiently regular function and F: $\mathcal{Q}\times\mathbb{R} \longrightarrow \mathbb{R}$ is a Nemytsky operator on $L^2(\mathcal{Q})$ satisfying Assumptions (7.78). The Banach space $\mathcal{D}$ of functions on Ω (satisfying the boundary conditions) such that $\mathcal{D} \subset L^2(\Omega) \subset \mathcal{D}'$ satisfies Assumption (7.79).

According to Theorems 7.22 and 7.23, the problem (8.103) admits a unique solution $U \in \mathcal{W} \cap \mathcal{H}_1$ for every $U_0 \in L^2(\Omega)$ and $f \in L^2(\mathcal{Q})$, where $\mathcal{H}_1 = H^1(0,T;\mathcal{D}') = \{v : \partial v/\partial t \in L^2(0,T;\mathcal{D}')\}$ and $\mathcal{W} = \mathcal{H} \cap \mathcal{V}$ such that $\mathcal{H} = L^\infty(0,T;L^2(\Omega))$ and $\mathcal{V} = L^2(0,T;\mathcal{D})$. Moreover, if U_1 (respectively U_2) is a solution of (8.103) where the data in $L^2(\mathcal{Q}) \times L^2(\Omega)$ is (f_1, U_0) (respectively (f_2, V_0)) then

$$\parallel U_1 - U_2 \parallel_{\mathcal{W}\cap\mathcal{H}_1} \leq C \left(\parallel U_0 - V_0 \parallel_{L^2}^2 + \parallel f_1 - f_2 \parallel_{L^2(\mathcal{Q})}^2 \right)^{1/2}.$$

In the following section, we formulate the perturbation problem and present the existence, uniqueness and regularity results of the perturbation solution.

8.6.2 The Perturbation Problem

In the following, the solution $U \in \mathcal{W}$ of problem (8.103) will be treated as the target function. We are then interested in the robust regulation of the deviation of the problem from the desired target (U, f, U_0). We analyze the full non-linear equation which models large perturbations (u, g, u_0) to the target (U, f, U_0). Hence we consider the equation:

$$\begin{aligned} &\frac{\partial u}{\partial t} + Au + F(x,t,u+U) - F(x,t,U) + K(\varpi, u) = g \text{ on } \mathcal{Q}, \\ &u(0) = u_0 \text{ on } \Omega. \end{aligned} \tag{8.104}$$

If we set $\tilde{F}(.,y) = F(.,y+U) - F(.,U)$ then (8.104) reduces to

$$\begin{aligned} &\frac{\partial u}{\partial t} + Au + \tilde{F}(x,t,u) + K(\varpi, u) = g \text{ on } \mathcal{Q}, \\ &u(0) = u_0 \text{ on } \Omega. \end{aligned} \tag{8.105}$$

Remark 8.18. (*i*) We easily verify that $\tilde{F}$ satisfies the same hypothesis that F, *i.e.*, (7.78).
(*ii*) For simplicity of future reference, we omit the "$\tilde{}$" on $\tilde{F}$ for (8.105). ◇

The problem (8.105) is of the same type as (8.103) so we have then the following proposition.

Proposition 8.19. (*i*) *Assume that $g \in L^2(\mathcal{Q})$ and $u_0 \in L^2(\Omega)$. Then the problem (8.105) admits a unique solution u such that $u \in \mathcal{W}$ and*

$$\| u \|^2_{\mathcal{W}\cap\mathcal{H}_1} \leq C(\| u_0 \|^2_{L^2} + \| g \|^2_{L^2(\mathcal{Q})}).$$

(*ii*) *Let u_0, v_0 be two functions in $L^2(\Omega)$ and let g_1, g_2 be two functions of $L^2(\mathcal{Q})$. If u_1 (resp. u_2) is a solution of (8.105) with data (g_1, u_0) (respectively with data (g_2, v_0)) then*

$$\| u_1 - u_2 \|^2_{\mathcal{W}\cap\mathcal{H}_1} \leq C(\| u_0 - v_0 \|^2_{L^2} + \| g_1 - g_2 \|^2_{L^2(\mathcal{Q})}).$$

□

Remark 8.20. According to Lemma 6.6 we can deduce that $\mathcal{W} \cap \mathcal{H}_1 \subset \mathcal{C}([0,T], H)$ and then (because of Proposition 8.19) for u (resp. v) the solution of (8.105) with data (g, u_0) (resp. (g_1, v_0)) we have the following estimates

$$\begin{aligned} &\| u \|^2_{\mathcal{W}\cap\mathcal{H}_1\cap\mathcal{C}([0,T];H)} \leq C(\| u_0 \|^2_{L^2} + \| g \|^2_{L^2(\mathcal{Q})}). \\ &\| u - v \|^2_{\mathcal{W}\cap\mathcal{H}_1\cap\mathcal{C}([0,T];H)} \leq C(\| u_0 - v_0 \|^2_{L^2} + \| g - g_1 \|^2_{L^2(\mathcal{Q})}). \end{aligned} \tag{8.106}$$

◇

8.6.3 The Control Framework

In the control framework, the value g is decomposed into the disturbance $\psi \in L^2(\mathcal{Q})$ and the control $\phi \in L^2(\mathcal{Q})$, *i.e.*, $g = B_1\phi + B_2\psi$, where B_i, for $i = 1, 2$, are given (linear) bounded operators on $L^2(\Omega)$ such that (for $i = 1, 2$):

$$\text{there exists } b_i > 0 \text{ such that } \forall h_i \in L^2(\Omega), \| B_i h_i \|^2_{L^2} \leq b_i^2 \| h_i \|^2_{L^2}. \tag{8.107}$$

Remark 8.21. If we put the linear operator $\mathcal{B} = (B_1, B_2)$ defined by $\mathcal{B}\mathbf{h} = B_1 h_1 + B_2 h_2$ where $\mathbf{h} = (h_1, h_2)^t$, and $b = \sqrt{2}\max(b_1, b_2)$ then

$$\| \mathcal{B}\mathbf{h} \|_{L^2}^2 \leq (b_1 \| h_1 \|_{L^2} + b_2 \| h_2 \|_{L^2})^2 \leq b^2 \| \mathbf{h} \|_{L^2}^2, \tag{8.108}$$

where $\| \mathbf{h} \|_{L^2}^2 = \| h_1 \|_{L^2}^2 + \| h_2 \|_{L^2}^2$. ◇

The objective in the robust control problem is to find the best control ϕ in the presence of the disturbance ψ which maximally spoils the control objective. The function u is assumed to be related to the disturbance ψ and control ϕ through the problem (8.105):

$$\begin{aligned} &\frac{\partial u}{\partial t} + Au + F(x,t,u) + K(\varpi, u) = B_1\phi + B_2\psi \quad \text{on } \mathcal{Q}, \\ &u(0) = u_0 \text{ (given)} \quad \text{on } \Omega. \end{aligned} \tag{8.109}$$

To obtain the regularity of Proposition 8.19, we suppose the following hypothesis: $u_0 \in L^2(\Omega)$, $(\phi, \psi) \in L^2(\mathcal{Q})^2$. Let $\mathbf{U} : (\phi, \psi) \longrightarrow u = \mathbf{U}(\phi, \psi)$ be the mapping: $(L^2(\mathcal{Q}))^2 \longrightarrow \mathcal{W}$ defined by (8.109). We introduce the cost functional defined by

$$\begin{aligned} J(\phi, \psi) = \frac{1}{2} \| \mathcal{C}(u - u_{\text{obs}}) \|_{L^2(\mathcal{Q})}^2 + \frac{\mu}{2} \| u(T) - v_{\text{obs}} \|_{L^2}^2 \\ + \frac{\alpha}{2} \| \phi \|_{L^2(\mathcal{Q})}^2 - \frac{\beta}{2} \| \psi \|_{L^2(\mathcal{Q})}^2, \end{aligned} \tag{8.110}$$

where $\mu, \alpha, \beta > 0$ are fixed, $(u_{\text{obs}}, v_{\text{obs}}) \in \mathcal{V} \times L^2(\Omega)$ is the observation (given) and $\mathcal{C}$ is an unbounded, linear operator on $L^2(\Omega)$ satisfying the condition (7.91)

$$\| \mathcal{C}v \|_{L^2}^2 \leq \delta_1 \| v \|_{L^2}^2 + \delta_2 \| v \|_{\mathcal{D}}^2 \quad \forall v \in \mathcal{D}.$$

The robust control problem, then, is to minimize the functional J with respect to ϕ and maximize J with respect to ψ, *i.e.*, to find $(\phi^*, \psi^*) \in \mathcal{U}_{\text{ad}} \times \mathcal{V}_{\text{ad}}$ such that

$$J(\phi^*, \psi) \leq J(\phi^*, \psi^*) \leq J(\phi, \psi^*), \forall (\phi, \psi) \in \mathcal{U}_{\text{ad}} \times \mathcal{V}_{\text{ad}}, \tag{8.111}$$

where $\mathcal{U}_{\text{ad}}$ and $\mathcal{V}_{\text{ad}}$ are non-empty, closed, convex, bounded subsets of $L^2(\mathcal{Q})$.

We are now going to show the differentiability result of the operator solution of (8.109).

Proposition 8.22. *Let F be an operator that satisfies Assumptions (7.78). Then the function $\mathbf{U} : (\phi, \psi) \longrightarrow u = \mathbf{U}(\phi, \psi)$ solution of (8.109) is continuously F-differentiable from $(L^2(\mathcal{Q}))^2$ to $\mathcal{W}$ with the derivative $\mathbf{U}'(\phi, \psi) : (h_1, h_2) \longrightarrow w$ given by the linear parabolic problem*

$$\begin{aligned} &\frac{\partial w}{\partial t} + Aw + G(., \mathbf{U}(\phi, \psi))w + K(\varpi, w) = B_1 h_1 + B_2 h_2 \quad \text{on } \mathcal{Q}, \\ &w(0) = 0 \quad \text{on } \Omega, \end{aligned} \tag{8.112}$$

with $G(.,v) = F'(.,v)$.

Moreover, for all $\Phi_i = (\phi_i, \psi_i) \in (L^2(\mathcal{Q}))^2, i = 1, 2$, *the following estimates hold:*

(*i*) $\| \mathbf{U}'(\phi_1, \psi_1) \|_{\mathcal{L}((L^2(\mathcal{Q}))^2;\mathcal{W})} \leq C$

(*ii*) $\| \mathbf{U}'(\phi_1, \psi_1) - \mathbf{U}'(\phi_2, \psi_2) \|_{\mathcal{L}((L^2(\mathcal{Q}))^2;\mathcal{W})} \leq C \| \Phi_1 - \Phi_2 \|_{L^2(\mathcal{Q})}$.

Proof. Let $(\phi, \psi, h_1, h_2) \in (L^2(\mathcal{Q}))^4$, $u = \mathbf{U}(\phi, \psi)$ and $u_h = u + w_h = \mathbf{U}(\phi + h_1, \psi + h_2)$. Let $v = w_h - v_h$, where v_h is the solution of

$$\begin{aligned} &\frac{\partial v_h}{\partial t} + Av_h + G(.,u)v_h + K(\varpi, v_h) = B_1 h_1 + B_2 h_2 \quad \text{on } \mathcal{Q}, \\ &v_h(0) = 0 \quad \text{on } \Omega. \end{aligned} \tag{8.113}$$

Then (according to the equations satisfied by u, u_h and v_h)

$$\begin{aligned} &\frac{\partial v}{\partial t} + Av + G(.,u)v + K(\varpi, v) = g \quad \text{on } \mathcal{Q} \\ &v(0) = 0 \quad \text{on } \Omega, \end{aligned} \tag{8.114}$$

where $g = -(F(.,u + w_h) - F(.,u)) + G(.,u)w_h$.

By using similar argument as in the proof of Proposition 7.24, Chapter 7, and the estimate (8.108) we have the following estimates (where $\mathbf{h} = (h_1, h_2)$):

$$\| v_h \|_{\mathcal{W}} \leq C \| \mathbf{h} \|_{L^2(\mathcal{Q})} \tag{8.115}$$

and

$$\| v \|_{\mathcal{W}} \leq C \| \mathbf{h} \|^2_{L^2(\mathcal{Q})} . \tag{8.116}$$

Consequently,

$$\| v \|_{\mathcal{W}} = o(\| \mathbf{h} \|_{L^2(\mathcal{Q})}).$$

Therefore, $\mathbf{U}'(\phi, \psi)$ defined by (8.112), is the F-derivative of $\mathbf{U}$ at point (ϕ, ψ) and verifies $\| \mathbf{U}'(\phi, \psi) \|_{\mathcal{L}((L^2(\mathcal{Q}))^2,\mathcal{W})} \leq C$.

We shall now prove the second part of the proposition. Let $\Phi_i = (\phi_i, \psi_i) \in (L^2(\mathcal{Q}))^2$, for $i = 1, 2$, be given and $w_i = \mathbf{U}'(\phi_i, \psi_i)\mathbf{h}$, for $i = 1, 2$, be the solution of the problem (8.112) (we denote by $u_i = \mathbf{U}(\phi_i, \psi_i)$, for $i = 1, 2$, and by $\mathbf{h} = (h_1, h_2)$).

We set $w = w_1 - w_2$. Then, according to the equations satisfied by w_1 and w_2, we can deduce that w satisfies

$$\begin{aligned} &\frac{\partial w}{\partial t} + Aw + G(.,u_1)w + K(\varpi, w) = (G(.,u_2) - G(.,u_1))w_2 \quad \text{on } \mathcal{Q}, \\ &w(0) = 0 \quad \text{on } \Omega. \end{aligned} \tag{8.117}$$

Multiplying (8.117) by w and integrating over $(0,t) \times \Omega$, for all $t \in (0,T)$, we obtain (according to $-2G(.,u) \leq \gamma_0$ and Assumptions (7.78) and (7.81))

$$\| w(t) \|_{L^2}^2 + \nu \int_0^t \| w \|_{\mathcal{D}}^2 \, \mathrm{d}s \leq \gamma \int_0^t \| w \|_{L^2}^2 \, \mathrm{d}s + 2\lambda \int_0^t \int_\Omega | u_2 - u_1 || w_2 || w | \, \mathrm{d}x\mathrm{d}s,$$

with γ given by (7.82).

By using Hölder's inequality and the relationship (7.79), we have

$$\| w(t) \|_{L^2}^2 + \nu \int_0^t \| w \|_{\mathcal{D}}^2 \, \mathrm{d}s \leq \gamma \int_0^t \| w \|_{L^2}^2 \, \mathrm{d}s + 2\lambda c_e^2 \| u_2 - u_1 \|_{\mathcal{W}} \| w_2 \|_{\mathcal{W}} \| w \|_{\mathcal{W}},$$

with γ given by (7.82).

Using Gronwall's formula we have then

$$\| w \|_{\mathcal{W}} \leq C \| u_2 - u_1 \|_{\mathcal{W}} \| w_2 \|_{\mathcal{W}} .$$

According to Proposition 8.19 and (8.115) we can deduce that

$$\| w \|_{\mathcal{W}} \leq C \| \Phi_1 - \Phi_2 \|_{(L^2(\mathcal{Q}))^2} \| \mathbf{h} \|_{(L^2(\mathcal{Q}))^2} .$$

Therefore,

$$\| \mathbf{U}'(\phi_1, \psi_1) - \mathbf{U}'(\phi_2, \psi_2) \|_{\mathcal{L}((L^2(\mathcal{Q}))^2, \mathcal{W})} \leq C \| \Phi_1 - \Phi_2 \|_{(L^2(\mathcal{Q}))^2}$$

and then result *(ii)* of Proposition 8.22 follows. □

Proposition 8.23. *Let F be an operator that satisfies Assumptions (7.78). Then for each $t \in [0, T]$, the function $\mathbf{V}_t : (\phi, \psi) \longrightarrow u(t) = \mathbf{V}_t(\phi, \psi)$ solution of (8.109) is continuously F-differentiable from $(L^2(\mathcal{Q}))^2$ to $L^2(\Omega)$ with the derivative $\mathbf{V}_t'(\phi, \psi) : (h_1, h_2) \longrightarrow w(t)$ given by the linear parabolic problem*

$$\begin{aligned} &\frac{\partial w}{\partial t} + Aw + G(., \mathbf{V}_t(\phi, \psi))w + K(\varpi, w) = B_1 h_1 + B_2 h_2 \quad \text{on } \mathcal{Q} \\ &w(0) = 0 \quad \text{on } \Omega \end{aligned} \tag{8.118}$$

which satisfies, for all $\Phi_i = (\phi_i, \psi_i) \in (L^2(\mathcal{Q}))^2, i = 1, 2$, the estimates:

(i) $\| \mathbf{V}_t'(\phi_1, \psi_1) \|_{\mathcal{L}((L^2(\mathcal{Q}))^2; L^2(\Omega))} \leq C$

(ii) $\| \mathbf{V}_t'(\phi_1, \psi_1) - \mathbf{V}_t'(\phi_2, \psi_2) \|_{\mathcal{L}((L^2(\mathcal{Q}))^2; L^2(\Omega))} \leq C \| \Phi_1 - \Phi_2 \|_{L^2(\mathcal{Q})}$.

Proof. The functional $(\phi, \psi) \longrightarrow u(t)$ is continuous from $L^2(0, T; L^2(\Omega))$ to $L^2(\Omega)$ (corollary of Proposition 8.19). The rest of this proposition is a corollary of Proposition 8.22. □

Proposition 8.24. *Let F be an operator that satisfies Assumptions (7.78). Then the maps $\mathbf{U}$ and $\mathbf{V}_t$ defined by (8.109) are continuous from the weak topology of $(L^2(\mathcal{Q}))^2$ to the strong topology of $L^2(\mathcal{Q})$ and the weak topology of $L^2(\Omega)$, respectively.*

Proof. Let $\Phi = (\phi, \psi)$ be given in $(L^2(\mathcal{Q}))^2$ and let the sequence $\Phi_k = (\phi_k, \psi_k)$ such that Φ_k is weakly convergent in $(L^2(\mathcal{Q}))^2$ to Φ.

Set $u = \mathbf{U}(\phi, \psi)$, $u_k = \mathbf{U}(\phi_k, \psi_k)$ and $v_k = u - u_k$. Since $\Phi_k \rightharpoonup \Phi$ weakly in $(L^2(\mathcal{Q}))^2$ then the sequence Φ_k is uniformly bounded in $(L^2(\mathcal{Q}))^2$ and therefore (according to Proposition 8.19) u_k is uniformly bounded in $\mathcal{W}$. By using Assumption 2 of (7.78) we deduce that $F(., u_k)$ is uniformly bounded in $L^2(\mathcal{Q})$. Using this result and Equation (8.109) we obtain easily that $\partial u_k/\partial t$ is uniformly bounded in $L^1(0,T;\mathcal{D}')$. Let us introduce the space $\mathcal{Y} = \{v \in L^2(0,T;\mathcal{D}), \partial v/\partial t \in L^1(0,T;\mathcal{D}')\}$. According to Lemma 6.6, the injection of $\mathcal{Y}$ into $L^2(0,T;L^2(\Omega))$ is compact. Therefore, u_k is uniformly bounded in $\mathcal{Y}$. This result makes it possible to extract from $(u_k, \phi_k, \psi_k, F(., u_k))$ a subsequence also denoted by $(u_k, \phi_k, \psi_k, F(., u_k))$ and such that:[2]

$$\begin{aligned} &(\phi_k, \psi_k) \rightharpoonup (\phi, \psi) \text{ weakly in } (L^2(\mathcal{Q}))^2, \\ &u_k \rightharpoonup \tilde{u} \text{ weakly in } L^2(0,T;\mathcal{D}), \\ &u_k \longrightarrow \tilde{u} \text{ strongly in } L^2(\mathcal{Q}), \\ &F(., u_k) \longrightarrow F(., \tilde{u}) \text{ strongly in } L^2(\mathcal{Q}). \end{aligned} \tag{8.119}$$

We easily prove that $\tilde{u}$ is a solution of (8.109) with a forcing (ϕ, ψ) and according to the uniqueness of the solution of (8.109), we have then $\tilde{u} = u = \mathbf{U}(\phi, \psi)$.

In the same way we prove that for each $t \in [0,T]$, $\mathbf{V}_t(\phi_k, \psi_k) \rightharpoonup \mathbf{V}_t(\phi, \psi)$ weakly in $L^2(\Omega)$. □

Theorem 8.25. *Let F be an operator that satisfies Assumptions (7.78). Then, for α and β sufficiently large, there exists $(\phi^*, \psi^*) \in (L^2(\mathcal{Q}))^2$ and $u^* \in \mathcal{W}$ such that (ϕ^*, ψ^*) is defined by (8.111) and $u^* = \mathbf{U}(\phi^*, \psi^*)$ is a solution of (8.109).*

Proof. Let P_ψ be the map: $\phi \longrightarrow J(\phi, \psi)$ and Q_ϕ be the map: $\psi \longrightarrow J(\phi, \psi)$. To obtain the existence of the robust control problem we prove that P_ψ is convex and lower semi-continuous for all $\psi \in \mathcal{V}_{\text{ad}}$, and P_ϕ is concave and upper semi-continuous for all $\phi \in \mathcal{U}_{\text{ad}}$ and we use the minimax theorems in infinite dimensions presented in Chapter 5.

First we prove that for α and β sufficiently large we have the convexity of the map P_ψ and the concavity of the map Q_ϕ.

In order to prove the convexity, it is sufficient to show that for all $(\phi_1, \phi_2) \in \mathcal{U}_{\text{ad}}$ we have: $(P'_\psi(\phi_1) - P'_\psi(\phi_2)).\phi \geq 0$, where $\phi = \phi_1 - \phi_2$. According to the definition of J, we have that

$$\begin{aligned} (P'_\psi(\phi_1) - P'_\psi(\phi_2)).\phi = \alpha \parallel \phi \parallel^2_{L^2(\mathcal{Q})} &+ \int_0^T \langle \mathcal{C}(u_1 - u_2), \mathcal{C}w_2 \rangle \mathrm{d}t \\ &+ \int_0^T \langle \mathcal{C}(u_1 - u_{\text{obs}}), \mathcal{C}(w_1 - w_2) \rangle \mathrm{d}t \\ &+ \mu \langle u_1(T) - u_2(T), w_{2T} \rangle \\ &+ \mu \langle u_1(T) - v_{\text{obs}}, w_{1T} - w_{2T} \rangle, \end{aligned} \tag{8.120}$$

[2] The operator F is continuous then $F(., u_k) \longrightarrow F(., \tilde{u})$ strongly in $L^2(\mathcal{Q})$ (according to the thirth result of (8.119)).

where $u_i = U(\phi_i, \psi)$, $w_i = U'(\phi_i, \psi).(\phi, 0)$ and $w_{iT} = V_T'(\phi_i, \psi).(\phi, 0)$, for $i = 1, 2$.

According to (7.80), (7.91) and the result of Propositions 8.19, 8.22 and 8.23, we have

$$\begin{aligned}\int_0^T & \langle \mathcal{C}(u_1 - u_2), \mathcal{C}w_2 \rangle dt + \mu \langle u_1(T) - u_2(T), w_{2T} \rangle \\ &\leq \| \mathcal{C}(u_1 - u_2) \|_{L^2(\mathcal{Q})} \| \mathcal{C}w_2 \|_{L^2(\mathcal{Q})} \\ &\quad + \mu \| u_1(T) - u_2(T) \|_{L^2} \| w_{2T} \|_{L^2} \\ &\leq c_1 \| u_1 - u_2 \|_{\mathcal{V}} \| w_2 \|_{\mathcal{V}} \\ &\quad + \mu \| u_1(T) - u_2(T) \|_{L^2} \| w_{2T} \|_{L^2} \\ &\leq C_0 \| \phi \|^2_{L^2(\mathcal{Q})}\end{aligned} \tag{8.121}$$

and

$$\begin{aligned}\int_0^T & \langle \mathcal{C}(u_1 - u_{\text{obs}}), \mathcal{C}(w_1 - w_2) \rangle dt \\ &\quad + \mu \langle u_1(T) - v_{\text{obs}}, w_{1T} - w_{2T} \rangle \\ &\leq \| \mathcal{C}(u_1 - u_{\text{obs}}) \|_{L^2(\mathcal{Q})} \| \mathcal{C}(w_1 - w_2) \|_{L^2(\mathcal{Q})} \\ &\quad + \mu \| u_1(T) - v_{\text{obs}} \|_{L^2} \| w_{1T} - w_{2T} \|_{L^2} \\ &\leq c_1 \| u_1 - u_{\text{obs}} \|_{\mathcal{V}} \| w_1 - w_2 \|_{\mathcal{V}} \\ &\quad + \mu \| u_1(T) - v_{\text{obs}} \|_{L^2} \| w_{1T} - w_{2T} \|_{L^2} \\ &\leq c_2 (\| u_1 - u_{\text{obs}} \|_{\mathcal{V}} + \| u_1(T) - v_{\text{obs}} \|_{L^2}) \| \phi \|^2_{L^2(\mathcal{Q})} \\ &\leq C_1 C_2 \| \phi \|^2_{L^2(\mathcal{Q})}.\end{aligned} \tag{8.122}$$

Remark 8.26. The generic constants C_0 and C_1 depend on the parameters $\mu, \delta_1, c_I, c_e, \lambda, \delta_2, \gamma, b$ and the final time T. The generic constant C_2 depend on the observation $(u_{\text{obs}}, v_{\text{obs}})$ and on the initial data u_0. ◇

From (8.120)–(8.122) we deduce that for $\alpha \geq \alpha_l = C_0 + C_1 C_2$ we have $(P_\psi'(\phi_1) - P_\psi'(\phi_2)).\phi \geq 0$ and then the convexity of P_ψ. In the same way, we can find β_l such that for $\beta \geq \beta_l$ we have the concavity of Q_ϕ.

We prove now that P_ψ is lower semi-continuous for all $\psi \in \mathcal{V}_{\text{ad}}$, and P_ϕ is upper semi-continuous for all $\phi \in \mathcal{U}_{\text{ad}}$. Let ϕ_k be a minimizing sequence of J, *i.e.*,

$$\liminf_k J(\phi_k, \psi) = \min_{\phi \in L^2(\mathcal{Q})} J(\phi, \psi) \; (\forall \psi \in \mathcal{V}_{\text{ad}}).$$

Then ϕ_k is uniformly bounded in $\mathcal{U}_{\text{ad}}$ and we can extract from ϕ_k a subsequence also denoted by ϕ_k such that $\phi_k \rightharpoonup \phi_\psi$ weakly in $\mathcal{U}_{\text{ad}}$. By using Proposition 8.24 we then have

$$\begin{aligned}&\mathbf{U}(\phi_k, \psi) \rightharpoonup u_\psi = \mathbf{U}(\phi_\psi, \psi) \text{ weakly in } L^2(0, T; \mathcal{D}), \\ &\mathbf{U}(\phi_k, \psi) \longrightarrow u_\psi \text{ strongly in } L^2(\mathcal{Q}), \\ &\mathbf{V}_t(\phi_k, \psi) \rightharpoonup u_\psi(t) = \mathbf{V}_t(\phi_\psi, \psi) \text{ weakly in } L^2(\Omega), \forall t \in [0, T].\end{aligned}$$

Since the norm is lower semi-continuous, therefore the map $P_\psi : \phi \longrightarrow J(\phi,\psi)$ is lower semi-continuous for all $\psi \in \mathcal{V}_{\text{ad}}$. By using the same technique we then obtain that P_ϕ is upper semi-continuous for all $\phi \in \mathcal{U}_{\text{ad}}$. □

In order to characterize the solution of the robust control problem, we introduce the adjoint problem corresponding to the primal problem (8.109) (we denote by $u = \mathbf{U}(\phi,\psi)$ the solution of problem (8.109) where the forcing is (ϕ,ψ)):

$$\begin{aligned} &-\frac{\partial\tilde{u}}{\partial t} + A\tilde{u} + (G(.,u))^*\tilde{u} + K^*(\varpi,\tilde{u}) = \mathcal{C}^*\mathcal{C}(u - u_{\text{obs}}) \;\text{ on } \mathcal{Q}, \\ &\tilde{u}(t=T) = \mu(u(t=T) - v_{\text{obs}}) \;\text{ on } \Omega, \end{aligned} \tag{8.123}$$

where $\mathcal{C}^*$ (resp. $(G(.,u))^*$) is the adjoint of the operator $\mathcal{C}$ (resp. $G(.,u)$) (the adjoint A^* of A is itself, *i.e.*, $A^* = A$ since A is a self-adjoint operator).

The adjoint problem (8.123) is the same as the adjoint problem (7.103). Then, according to Proposition 7.28, the problem (8.123) admits a unique solution $\tilde{u} \in \mathcal{W}$, if F satisfies Assumptions (7.78), $u \in \mathcal{W}$, such that

$$\| \tilde{u} \|_{\mathcal{H}}^2 + \| \tilde{u} \|_{\mathcal{V}}^2 \leq C_d^2(\mu^2 \| u(T) - v_{\text{obs}} \|_{L^2}^2 + \| \mathcal{C}(u - u_{\text{obs}}) \|_{L^2(\mathcal{Q})}^2). \tag{8.124}$$

In the sequel, we will denote by $\tilde{\mathbf{U}} : (\phi,\psi) \longrightarrow \tilde{u} = \tilde{\mathbf{U}}(\phi,\psi)$ the map defined by (8.123). We can now give the optimality system for the robust control problem (8.111).

Theorem 8.27. *Let F be an operator that satisfies Assumptions (7.78) and the optimal solution $(\phi^*,\psi^*,u^*) \in (L^2(\mathcal{Q}))^2 \times \mathcal{W}$ such that (ϕ^*,ψ^*) is defined by (8.111) and $u^* = \mathbf{U}(\phi^*,\psi^*)$ is the solution of (8.109). Then, for α and β sufficiently large, we have for all $(\phi,\psi) \in \mathcal{U}_{\text{ad}} \times \mathcal{V}_{\text{ad}}$*

$$\begin{aligned} &\int_0^T \int_\Omega (\alpha\phi^* + B_1^*\tilde{u})(\phi - \phi^*)\mathrm{d}x\mathrm{d}t \geq 0, \\ &\int_0^T \int_\Omega (-\beta\psi^* + B_2^*\tilde{u})(\psi - \psi^*)\mathrm{d}x\mathrm{d}t \leq 0, \end{aligned}$$

where $\tilde{u} = \tilde{\mathbf{U}}(\phi^*,\psi^*)$ is a solution of adjoint problem (8.123). Moreover, the gradient of the functional J at (ϕ^*,ψ^*) in any direction $\mathbf{h} = (h_1,h_2) \in (L^2(\mathcal{Q}))^2$ is given by

$$J'(\phi^*,\psi^*).\mathbf{h} = \int_0^T \int_\Omega (\alpha\phi^* + B_1^*\tilde{u})h_1\mathrm{d}x\mathrm{d}t + \int_0^T \int_\Omega (-\beta\psi^* + B_2^*\tilde{u})h_2\mathrm{d}x\mathrm{d}t.$$

Otherwise (in the weak sense),

$$\frac{\partial J}{\partial\phi}(\phi^*,\psi^*) = \alpha\phi^* + B_1^*\tilde{u} \quad \textit{and} \quad \frac{\partial J}{\partial\psi}(\phi^*,\psi^*) = -\beta\psi^* + B_2^*\tilde{u}.$$

Proof. Since the cost functional J is a composition of F-differentiable maps then J is F-differentiable and we have, for all $\mathbf{h} = (h_1, h_2) \in (L^2(\mathcal{Q}))^2)$,

$$J'(\phi,\psi).\mathbf{h} = \int_0^T \langle C(u-u_{\text{obs}}), C\mathbf{U}'(\phi,\psi)\mathbf{h}\rangle + \langle \mu(u(T)-v_{\text{obs}}), \mathbf{V}'_T(\phi,\psi)\mathbf{h}\rangle$$
$$+ \int_0^T (\langle \alpha\phi, h_1\rangle - \langle \beta\psi, h_2\rangle)$$

and then

$$J'(\phi,\psi).\mathbf{h} = \int_0^T \langle \mathcal{C}^*\mathcal{C}(u-u_{\text{obs}}), \mathbf{U}'(\phi,\psi)\mathbf{h}\rangle + \langle \mu(u(T)-v_{\text{obs}}), \mathbf{V}'_T(\phi,\psi)\mathbf{h}\rangle$$
$$+ \int_0^T (\langle \alpha\phi, h_1\rangle - \langle \beta\psi, h_2\rangle).$$

Multiplying (8.112) by $\tilde{u}$, integrating over $\mathcal{Q}$ and (integrating) by parts in time t, we then obtain, for all $\mathbf{h} = (h_1, h_2) \in (L^2(\mathcal{Q}))^2$,

$$\int_0^T (\langle B_1^*\tilde{u}, h_1\rangle + \langle B_2^*\tilde{u}, h_2\rangle) = \langle \mathbf{V}'_T(\phi^*,\psi^*)\mathbf{h}, \tilde{u}(T)\rangle$$
$$+ \int_0^T \langle -\frac{\partial \tilde{u}}{\partial t} + A\tilde{u} + (G(.,u))^*\tilde{u} + K^*(\varpi,\tilde{u}), \mathbf{U}'(\phi^*,\psi^*)\mathbf{h}\rangle.$$

Since $\tilde{u}$ is a solution of the adjoint problem (8.123) we then obtain, for all $\mathbf{h} = (h_1, h_2) \in (L^2(\mathcal{Q}))^2$,

$$\begin{aligned} \int_0^T (\langle B_1^*\tilde{u}, h_1\rangle + \langle B_2^*\tilde{u}, h_2\rangle)\mathrm{d}t &= \langle \mathbf{V}'_T(\phi^*,\psi^*)\mathbf{h}, \mu(u^*(T)-v_{\text{obs}})\rangle \\ &+ \int_0^T \langle \mathcal{C}^*\mathcal{C}(u^*-u_{\text{obs}}), \mathbf{U}'(\phi^*,\psi^*)\mathbf{h}\rangle. \end{aligned} \tag{8.125}$$

As (ϕ^*, ψ^*) is a solution of (8.111), we have

$$J'(\phi^*,\psi^*)(\phi-\phi^*,0) \geq 0, \quad J'(\phi^*,\psi^*)(0,\psi-\psi^*) \leq 0,$$

for all $(\phi,\psi) \in \mathcal{U}_{\text{ad}} \times \mathcal{V}_{\text{ad}}$ and, therefore (according to (8.125) and to the expression of J'), we deduce that

$$\int_0^T\int_\Omega (\alpha\phi^* + B_1^*\tilde{u})(\phi-\phi^*)\mathrm{d}x\mathrm{d}t \geq 0, \quad \int_0^T\int_\Omega (-\beta\psi^* + B_2^*\tilde{u})(\psi-\psi^*)\mathrm{d}x\mathrm{d}t \leq 0$$

for all $(\phi,\psi) \in \mathcal{U}_{\text{ad}} \times \mathcal{V}_{\text{ad}}$.

The proof is complete. □

In the sequel, we will assume that there exists an optimal solution (ϕ^*, ψ^*, u^*) such that (ϕ^*, ψ^*) is defined by (8.111), $u^* = \mathbf{U}(\phi^*, \psi^*)$ is a solution of (8.109) and

$$\int_0^T\int_\Omega (\alpha\phi^* + B_1^*\tilde{u})(\phi-\phi^*)\mathrm{d}x\mathrm{d}t \geq 0, \quad \int_0^T\int_\Omega (-\beta\psi^* + B_2^*\tilde{u})(\psi-\psi^*)\mathrm{d}x\mathrm{d}t \leq 0$$

for all $(\phi, \psi) \in \mathcal{U}_{\mathrm{ad}} \times \mathcal{V}_{\mathrm{ad}}$, where $\tilde{u} = \tilde{\mathbf{U}}(\phi^*, \psi^*)$ is a solution of (8.123).

Now we give some conditions to obtain the uniqueness of the solution (ϕ^*, ψ^*).

Theorem 8.28. *Suppose that F satisfies the assumptions (7.78) and $\mu < 1$ holds. Then, the solution (ϕ^*, ψ^*, u^*) is unique if the following conditions hold:*

(i) $\theta = (\nu - \delta_2 - c_I^2(\gamma + \delta_1)) - 2b^2 c_I^2 (1/\alpha + 1/\beta) > 0$

(ii) $\sigma \left(\| u^*(T) - v_{\mathrm{obs}} \|_{L^2}^2 + \| \mathcal{C}(u^* - u_{\mathrm{obs}}) \|_{L^2(\mathcal{Q})}^2 \right)^{1/2} < \theta,$
where $\sigma = \lambda c_e^2 \exp((\delta_1 + \gamma + 1)T/2)$.

Proof. Assume that $(\phi_1^*, \psi_1^*, u_1^*)$ is another solution. Then (ϕ_1^*, ψ_1^*) satisfies (8.111), $u_1^* = \mathbf{U}(\phi_1^*, \psi_1^*)$ is a solution of (8.109), and

$$\int_0^T \int_\Omega (\alpha \phi_1^* + B_1^* \tilde{u}_1)(\phi - \phi_1^*) \mathrm{d}x \mathrm{d}t \geq 0,$$
$$\int_0^T \int_\Omega (-\beta \psi_1^* + B_2^* \tilde{u}_1)(\psi - \psi_1^*) \mathrm{d}x \mathrm{d}t \leq 0,$$

for all $(\phi, \psi) \in \mathcal{U}_{\mathrm{ad}} \times \mathcal{V}_{\mathrm{ad}}$, where $\tilde{u}_1 = \tilde{\mathbf{U}}(\phi_1^*, \psi_1^*)$ is a solution of (8.123).

We set $\phi = \phi^* - \phi_1^*$, $\psi = \psi^* - \psi_1^*$, $v = u^* - u_1^*$ and $\tilde{v} = \tilde{u} - \tilde{u}_1$. Then we have

$$\begin{aligned} &\frac{\partial v}{\partial t} + Av + (F(.,u^*) - F(.,u_1^*)) + K(\varpi, v) = B_1 \phi + B_2 \psi \quad \text{on } \mathcal{Q}, \\ &v(0) = 0 \quad \text{on } \Omega, \end{aligned} \tag{8.126}$$

$$\begin{aligned} &-\frac{\partial \tilde{v}}{\partial t} + A\tilde{v} + (G(.,u^*))^* \tilde{v} + K^*(\varpi, \tilde{v}) \\ &\qquad = \mathcal{C}^* \mathcal{C} v - (G(.,u^*) - G(.,u_1^*))^* \tilde{u} \quad \text{on } \mathcal{Q}, \\ &\tilde{v}(t = T) = \mu v(t = T) \quad \text{on } \Omega \end{aligned} \tag{8.127}$$

and

$$\begin{aligned} \alpha \| \phi \|_{L^2(\mathcal{Q})}^2 + \int_0^T \int_\Omega B_1^* \tilde{v} \phi \mathrm{d}x \mathrm{d}t \leq 0, \\ \beta \| \psi \|_{L^2(\mathcal{Q})}^2 - \int_0^T \int_\Omega B_2^* \tilde{v} \psi \mathrm{d}x \mathrm{d}t \leq 0. \end{aligned} \tag{8.128}$$

According to Assumptions (7.78) and (7.80), we have

$$\begin{aligned} &-2\langle F(.,u^*) - F(.,u_1^*), v \rangle \leq \gamma_0 c_I^2 \| v \|_{\mathcal{D}}^2, \\ &-2\langle (G(.,u^*))^* \tilde{v}, \tilde{v} \rangle \leq \gamma_0 c_I^2 \| \tilde{v} \|_{\mathcal{D}}^2, \\ &| \langle (G(.,u^*) - G(.,u_1^*))^* \tilde{u}, \tilde{v} \rangle | \leq \lambda \int_\Omega | v || \tilde{v} || \tilde{u} | . \end{aligned} \tag{8.129}$$

Multiplying (8.126) by v and (8.127) by $\tilde{v}$ and integrating over $\mathcal{Q}$ we obtain (according to (8.128), (8.129), (7.81) and (7.91))

$$\begin{aligned}
&\int_0^T \frac{\mathrm{d}}{\mathrm{d}t} \| v \|_{L^2}^2 \,\mathrm{d}t + \nu \int_0^T \| v \|_{\mathcal{D}}^2 \,\mathrm{d}t \\
&\qquad \le \gamma c_I^2 \int_0^T \| v \|_{\mathcal{D}}^2 \,\mathrm{d}t + \frac{2}{\alpha} \| B_1^* \tilde{v} \|_{L^2(\mathcal{Q})} \| B_1^* v \|_{L^2(\mathcal{Q})} \\
&\qquad\qquad + \frac{2}{\beta} \| B_2^* \tilde{v} \|_{L^2(\mathcal{Q})} \| B_2^* v \|_{L^2(\mathcal{Q})}, \\
&-\int_0^T \frac{\mathrm{d}}{\mathrm{d}t} \| \tilde{v} \|_{L^2}^2 \,\mathrm{d}t + \nu \int_0^T \| \tilde{v} \|_{\mathcal{D}}^2 \,\mathrm{d}t \\
&\qquad \le \gamma c_I^2 \int_0^T \| \tilde{v} \|_{\mathcal{D}}^2 \,\mathrm{d}t + (\delta_1 c_I^2 + \delta_2) \int_0^T (\| v \|_{\mathcal{D}}^2 + \| \tilde{v} \|_{\mathcal{D}}^2) \mathrm{d}t \\
&\qquad\qquad + 2\lambda \int_0^T \int_\Omega | \tilde{u} | \, | \tilde{v} | \, | v | \,\mathrm{d}x\mathrm{d}t, \\
&\tilde{v}(T) = \mu v(T) \quad \text{and} \quad v(0) = 0.
\end{aligned}$$

By using Hölder's inequality, the relationship (7.79), the assumption (8.107) and the estimate (8.108) we obtain

$$\begin{aligned}
&\int_0^T \frac{\mathrm{d}}{\mathrm{d}t} \| v \|_{L^2}^2 \,\mathrm{d}t + (\nu - \gamma c_I^2) \int_0^T \| v \|_{\mathcal{D}}^2 \,\mathrm{d}t \\
&\qquad \le 2\left(\frac{1}{\alpha} + \frac{1}{\beta}\right) b^2 c_I^2 \int_0^T (\| v \|_{\mathcal{D}}^2 + \| \tilde{v} \|_{\mathcal{D}}^2) \mathrm{d}t, \\
&-\int_0^T \frac{\mathrm{d}}{\mathrm{d}t} \| \tilde{v} \|_{L^2}^2 \,\mathrm{d}t + (\nu - \gamma c_I^2) \int_0^T \| \tilde{v} \|_{\mathcal{D}}^2 \,\mathrm{d}t \\
&\qquad \le (\delta_1 c_I^2 + \delta_2 + \lambda c_e^2 \| \tilde{u} \|_{\mathcal{H}}) \int_0^T (\| v \|_{\mathcal{D}}^2 + \| \tilde{v} \|_{\mathcal{D}}^2) \mathrm{d}t, \\
&\tilde{v}(T) = \mu v(T) \quad \text{and} \quad v(0) = 0.
\end{aligned} \tag{8.130}$$

Summing the first and the second part of (8.130) we obtain

$$\begin{aligned}
&\int_0^T \frac{\mathrm{d}}{\mathrm{d}t} (\| v \|_{L^2}^2 - \| \tilde{v} \|_{L^2}^2) \mathrm{d}t + \theta \int_0^T (\| v \|_{\mathcal{D}}^2 + \| \tilde{v} \|_{\mathcal{D}}^2) \mathrm{d}t \\
&\qquad \le \lambda c_e^2 \| \tilde{u} \|_{\mathcal{H}} \int_0^T (\| v \|_{\mathcal{D}}^2 + \| \tilde{v} \|_{\mathcal{D}}^2) \mathrm{d}t,
\end{aligned} \tag{8.131}$$

where (according to assumption (i))

$$\theta = \nu - \delta_2 - c_I^2 \left(\gamma + \frac{2b^2}{\alpha} + \frac{2b^2}{\beta} + \delta_1\right) > 0.$$

According to the third part of (8.130), we have

$$(1-\mu^2) \parallel v(T) \parallel_{L^2}^2 + \parallel \tilde{v}(0) \parallel_{L^2}^2 + \theta(\parallel v \parallel_{\mathcal{V}}^2 + \parallel \tilde{v} \parallel_{\mathcal{V}}^2)$$
$$\leq \lambda c_e^2 \parallel \tilde{u} \parallel_{\mathcal{H}} (\parallel v \parallel_{\mathcal{V}}^2 + \parallel \tilde{v} \parallel_{\mathcal{V}}^2).$$

Since $1-\mu^2 > 0$ then $\theta(\parallel v \parallel_{\mathcal{V}}^2 + \parallel \tilde{v} \parallel_{\mathcal{V}}^2) \leq \lambda c_e^2 \parallel \tilde{u} \parallel_{\mathcal{H}} (\parallel v \parallel_{\mathcal{V}}^2 + \parallel \tilde{v} \parallel_{\mathcal{V}}^2)$. By applying the estimate (7.112) and $\mu < 1$ we can deduce that

$$\theta^*(\parallel v \parallel_{\mathcal{V}}^2 + \parallel \tilde{v} \parallel_{\mathcal{V}}^2) \leq 0,$$

where

$$\theta^* = \theta - \lambda c_e^2 e^{(\delta_1+\gamma+1)T/2} \left(\parallel u^*(T) - v_{\text{obs}} \parallel_{L^2}^2 + \parallel \mathcal{C}(u^* - u_{\text{obs}}) \parallel_{L^2(\mathcal{Q})}^2 \right)^{1/2}.$$

Since $\theta^* > 0$ (according to assumption (*ii*)), we have $\parallel v \parallel_{\mathcal{V}}^2 + \parallel \tilde{v} \parallel_{\mathcal{V}}^2 = 0$ and then $v = 0$ and $\tilde{v} = 0$. We obtain then the uniqueness result. □

8.6.4 Initial Condition Control

In this section the objective of the robust control problem is to find the best estimate of the initial state u_0 in the presence of the disturbance which maximally spoils the control objective. For this, we study two problems: first, the case where the worst disturbance is in the initial condition, and, second, the case where the worst disturbance is in the forcing term.

Distributed Disturbance in the Initial Condition

We suppose now that the value u_0 is decomposed into the disturbance $\psi \in L^2(\Omega)$ and the control $\phi \in L^2(\Omega)$, *i.e.*, $u_0 = B_1\phi + B_2\psi$, where $B_i, i = 1, 2$ are given bounded operators on $L^2(\Omega)$ satisfying the assumption (8.107) and the estimate (8.108).
Therefore, the function u is assumed to be related to the disturbance ψ and control ϕ through the problem (8.105):

$$\begin{aligned} &\frac{\partial u}{\partial t} + Au + F(x,t,u) + K(\varpi, u) = g \text{ (given) on } \mathcal{Q}, \\ &u(0) = B_1\phi + B_2\psi \text{ on } \Omega. \end{aligned} \tag{8.132}$$

To obtain the regularity of Proposition 8.19, we suppose the following hypothesis: $g \in L^2(\mathcal{Q})$, $(\phi, \psi) \in L^2(\Omega)^2$. Let $\mathbf{U} : (\phi, \psi) \longrightarrow u = \mathbf{U}(\phi, \psi)$ be the map: $(L^2(\Omega))^2 \longrightarrow \mathcal{W}$ defined by (8.132) and introducing the cost functional defined by

$$\begin{aligned} J(\phi, \psi) = &\frac{1}{2} \parallel \mathcal{C}(u - u_{\text{obs}}) \parallel_{L^2(\mathcal{Q})}^2 + \frac{\mu}{2} \parallel u(T) - v_{\text{obs}} \parallel_{L^2(\Omega)}^2 \\ &+ \frac{\alpha}{2} \parallel \phi \parallel_{L^2(\Omega)}^2 - \frac{\beta}{2} \parallel \psi \parallel_{L^2(\Omega)}^2, \end{aligned} \tag{8.133}$$

where $\mu, \alpha, \beta > 0$ are fixed, $(u_{obs}, v_{obs}) \in \mathcal{V} \times L^2(\Omega)$ is the observation and $\mathcal{C}$ is an unbounded, linear operator on $L^2(\Omega)$ satisfying the hypothesis (7.91).

We want to minimize the functional J with respect to ϕ and maximize J with respect to ψ, *i.e.*, to find $(\phi^*, \psi^*) \in \mathcal{U}_{ad} \times \mathcal{V}_{ad}$ such that

$$J(\phi^*, \psi) \leq J(\phi^*, \psi^*) \leq J(\phi, \psi^*), \forall (\phi, \psi) \in \mathcal{U}_{ad} \times \mathcal{V}_{ad}, \tag{8.134}$$

where $\mathcal{U}_{ad}$ and $\mathcal{V}_{ad}$ are non-empty, closed, convex, bounded subsets of $L^2(\Omega)$.

By using the same technique as used in the proof of Propositions 8.19 and 8.22 and Theorem 8.25, we have the following results (with no further estimates required).

Proposition 8.29. *Let F be an operator that satisfies the assumptions (7.78). Then the function $\mathbf{U} : (\phi, \psi) \longrightarrow u = \mathbf{U}(\phi, \psi)$ solution of (8.132) is continuously F-differentiable from $(L^2(\Omega))^2$ to $\mathcal{W}$ with the derivative $\mathbf{U}'(\phi, \psi) : (h_1, h_2) \longrightarrow w$ given by the linear parabolic problem*

$$\begin{aligned} &\frac{\partial w}{\partial t} + Aw + G(., \mathbf{U}(\phi, \psi))w + K(\varpi, w) = 0 \quad \text{on } \mathcal{Q} \\ &w(0) = B_1 h_1 + B_2 h_2 \quad \text{on } \Omega, \end{aligned} \tag{8.135}$$

which satisfies, for all $\Phi_i = (\phi_i, \psi_i) \in (L^2(\Omega)^2$ for $i = 1, 2$, the estimates:

(i) $\| \mathbf{U}'(\phi_1, \psi_1) \|_{\mathcal{L}((L^2(\Omega))^2; \mathcal{W})} \leq C$

(ii) $\| \mathbf{U}'(\phi_1, \psi_1) - \mathbf{U}'(\phi_2, \psi_2) \|_{\mathcal{L}((L^2(\Omega))^2; \mathcal{W})} \leq C \| \Phi_1 - \Phi_2 \|_{L^2(\mathcal{Q})}$. □

Proposition 8.30. *Let F be an operator that satisfies the assumptions (7.78). Then, for each $t \in [0, T]$, the function $\mathbf{V}_t : (\phi, \psi) \longrightarrow u(t) = \mathbf{V}_t(\phi, \psi)$ solution of (8.132) is continuously F-differentiable from $(L^2(\Omega))^2$ to $L^2(\Omega)$ with the derivative $\mathbf{V}'_t(\phi, \psi) : (h_1, h_2) \longrightarrow w(t)$ given by the linear parabolic problem*

$$\begin{aligned} &\frac{\partial w}{\partial t} + Aw + G(., \mathbf{V}_t(\phi, \psi))w + K(\varpi, w) = 0 \quad \text{on } \mathcal{Q} \\ &w(0) = B_1 h_1 + B_2 h_2 \quad \text{on } \Omega, \end{aligned} \tag{8.136}$$

which satisfies, for all $\Phi_i = (\phi_i, \psi_i) \in (L^2(\mathcal{Q}))^2$ for $i = 1, 2$, the estimates:

(i) $\| \mathbf{V}'_t(\phi_1, \psi_1) \|_{\mathcal{L}((L^2(\Omega))^2; L^2(\Omega))} \leq C$

(ii) $\| \mathbf{V}'_t(\phi_1, \psi_1) - \mathbf{V}'_t(\phi_2, \psi_2) \|_{\mathcal{L}((L^2(\Omega))^2; L^2(\Omega))} \leq C \| \Phi_1 - \Phi_2 \|_{L^2(\mathcal{Q})}$. □

Theorem 8.31. *Let F be an operator that satisfies the assumptions (7.78). Then, for α and β sufficiently large, there exists $(\phi^*, \psi^*) \in (L^2(\Omega))^2$ and $u^* \in \mathcal{W}$ such that (ϕ^*, ψ^*) is defined by (8.134) and $u^* = \mathbf{U}(\phi^*, \psi^*)$ is a solution of (8.132).* □

Remark 8.32. The adjoint problem associated with the primal problem (8.132) is exactly the adjoint problem (8.123). ◊

Now we give the optimality system for the robust control problem (8.134).

Theorem 8.33. *Let F be an operator that satisfies the assumptions (7.78), $\nu > \delta_2$, and the optimal solution $(\phi^*, \psi^*, u^*) \in (L^2(\mathcal{Q}))^2 \times \mathcal{W}$ such that (ϕ^*, ψ^*) is defined by (8.134) and $u^* = \mathbf{U}(\phi^*, \psi^*)$ is the solution of (8.132). Then, for α and β sufficiently large, we have $(\forall(\phi, \psi) \in \mathcal{U}_{\mathrm{ad}} \times \mathcal{V}_{\mathrm{ad}})$*

$$\int_{\Omega} (\alpha\phi^* + B_1^*\tilde{u}(0))(\phi - \phi^*)\mathrm{d}x \geq 0,$$
$$\int_{\Omega} (-\beta\psi^* + B_2^*\tilde{u}(0))(\psi - \psi^*)\mathrm{d}x \leq 0,$$

where $\tilde{u} = \tilde{\mathbf{U}}(\phi^*, \psi^*)$ is a solution of the adjoint problem (8.123). Moreover, the gradient of the functional J at (ϕ^*, ψ^*) in any direction $\mathbf{h} = (h_1, h_2) \in (L^2(\mathcal{Q}))^2$ is given by

$$J'(\phi^*, \psi^*).\mathbf{h} = \int_{\Omega} (\alpha\phi^* + B_1^*\tilde{u}(0))h_1\mathrm{d}x + \int_{\Omega} (-\beta\psi^* + B_2^*\tilde{u}(0))h_2\mathrm{d}x.$$

Otherwise (in the weak sense),

$$\frac{\partial J}{\partial \phi}(\phi^*, \psi^*) = \alpha\phi^* + B_1^*\tilde{u}(0) \quad \textit{and} \quad \frac{\partial J}{\partial \psi}(\phi^*, \psi^*) = -\beta\psi^* + B_2^*\tilde{u}(0).$$

Proof. The cost functional J is a composition of F-differentiable mappings then J is F-differentiable and we have $(\forall \mathbf{h} = (h_1, h_2) \in (L^2(\Omega))^2)$:

$$J'(\phi, \psi)\mathbf{h} = \int_0^T \langle C(u - u_{\mathrm{obs}}), C\mathbf{U}'(\phi, \psi)\mathbf{h}\rangle + \langle \mu(u(T) - v_{\mathrm{obs}}), \mathbf{V}_T'(\phi, \psi)\mathbf{h}\rangle + \langle \alpha\phi, h_1\rangle - \langle \beta\psi, h_2\rangle$$

and then

$$J'(\phi, \psi)\mathbf{h} = \int_0^T \langle \mathcal{C}^*\mathcal{C}(u - u_{\mathrm{obs}}), \mathbf{U}'(\phi, \psi)\mathbf{h}\rangle + \langle \mu(u(T) - v_{\mathrm{obs}}), \mathbf{V}_T'(\phi, \psi)\mathbf{h}\rangle + \langle \alpha\phi, h_1\rangle - \langle \beta\psi, h_2\rangle.$$

Multiplying (8.135) by $\tilde{u}$, integrating over $\mathcal{Q}$ and (integrating) by parts in time t, we obtain $(\forall \mathbf{h} = (h_1, h_2) \in (L^2(\mathcal{Q}))^2)$:

$$\langle B_1h_1 + B_2h_2, \tilde{u}(0)\rangle = \langle \mathbf{V}_T'(\phi^*, \psi^*)\mathbf{h}, \tilde{u}(T)\rangle + \int_0^T \langle -\frac{\partial \tilde{u}}{\partial t} + A\tilde{u} + (G(.,u))^*\tilde{u} + K^*(\varpi, \tilde{u}), \mathbf{U}'(\phi^*, \psi^*)\mathbf{h}\rangle.$$

Since $\tilde{u}$ is a solution of adjoint problem we then obtain $(\forall \mathbf{h} = (h_1, h_2) \in (L^2(\Omega))^2)$:

$$\langle B_1h_1 + B_2h_2, \tilde{u}(0)\rangle = \langle \mathbf{V}_T'(\phi^*, \psi^*)\mathbf{h}, \mu(u^*(T) - v_{\mathrm{obs}})\rangle + \int_0^T \langle \mathcal{C}^*\mathcal{C}(u^* - u_{\mathrm{obs}}), \mathbf{U}'(\phi^*, \psi^*)\mathbf{h}\rangle. \tag{8.137}$$

As (ϕ^*, ψ^*) is a solution of (8.134) then

$$J'(\phi^*, \psi^*)(\phi - \phi^*, 0) \geq 0, \quad J'(\phi^*, \psi^*)(0, \psi - \psi^*) \leq 0,$$

for all $(\phi, \psi) \in \mathcal{U}_{\text{ad}} \times \mathcal{V}_{\text{ad}}$, and we can then deduce the result of this theorem (according to the expression of J'). The proof is complete. □

In the following, we will assume that there exists an optimal solution (ϕ^*, ψ^*, u^*) such that (ϕ^*, ψ^*) is defined by (8.134), $u^* = \mathbf{U}(\phi^*, \psi^*)$ is a solution of (8.132) and

$$\int_\Omega (\alpha\phi^* + B_1^*\tilde{u}(0))(\phi - \phi^*)\mathrm{d}x \geq 0, \quad \int_\Omega (-\beta\psi^* + B_2^*\tilde{u}(0))(\psi - \psi^*)\mathrm{d}x \leq 0$$

for all $(\phi, \psi) \in \mathcal{U}_{\text{ad}} \times \mathcal{V}_{\text{ad}}$, where $\tilde{u} = \tilde{\mathbf{U}}(\phi^*, \psi^*)$ is a solution of (8.123).

Now we give some conditions to obtain the uniqueness of the optimal solution (ϕ^*, ψ^*).

Theorem 8.34. *Suppose that F satisfies the assumptions (7.78), $\nu > \gamma_2$ and $\mu < 1$ holds. The optimal solution (ϕ^*, ψ^*, u^*) is unique if the following conditions hold:*

(i) $\theta = (\nu - \delta_2 - c_I^2(\gamma + \delta_1)) > 0$ *and* $1 - (b^4/\alpha^2) - (b^4/\beta^2) \geq 0$

(ii) $\lambda c_e^2 c_T \left(\| u^*(T) - v_{\text{obs}} \|_{L^2}^2 + \| \mathcal{C}(u^* - u_{\text{obs}}) \|_{L^2(\mathcal{Q})}^2 \right)^{1/2} < \theta,$

where $c_T = \exp((\delta_1 + \gamma + 1)T/2)$.

Proof. Suppose $(\phi_1^*, \psi_1^*, u_1^*)$ is another solution, then (ϕ_1^*, ψ_1^*) satisfies (8.134), $u_1^* = \mathbf{U}(\phi_1^*, \psi_1^*)$ is a solution of (8.132) and

$$\int_\Omega (\alpha\phi_1^* + B_1^*\tilde{u}_1(0))(\phi - \phi_1^*)\mathrm{d}x \geq 0, \quad \int_\Omega (-\beta\psi_1^* + B_2^*\tilde{u}_1(0))(\psi - \psi_1^*)\mathrm{d}x \leq 0$$

for all $(\phi, \psi) \in \mathcal{U}_{\text{ad}} \times \mathcal{V}_{\text{ad}}$, where $\tilde{u}_1 = \tilde{\mathbf{U}}(\phi_1^*, \psi_1^*)$ is a solution of (8.123).

We set $\phi = \phi^* - \phi_1^*$, $\psi = \psi^* - \psi_1^*$, $v = u^* - u_1^*$ and $\tilde{v} = \tilde{u} - \tilde{u}_1$. Then we have

$$\begin{aligned} &\frac{\partial v}{\partial t} + Av + (F(.,u^*) - F(.,u_1^*)) + K(\varpi, v) = 0 \quad \text{on } \mathcal{Q} \\ &v(0) = B_1\phi + B_2\psi \quad \text{on } \Omega, \end{aligned} \tag{8.138}$$

$$\begin{aligned} &-\frac{\partial \tilde{v}}{\partial t} + A\tilde{v} + (G(.,u^*))^*\tilde{v} + K^*(\varpi, \tilde{v}) \\ &\qquad = \mathcal{C}^*\mathcal{C}v - (G(.,u^*) - G(.,u_1^*))^*\tilde{u} \quad \text{on } \mathcal{Q}, \\ &\tilde{v}(t = T) = \mu v(t = T) \quad \text{on } \Omega \end{aligned} \tag{8.139}$$

and

$$\begin{aligned} &\alpha \| \phi \|_{L^2(\Omega)}^2 + \int_\Omega B_1^*\tilde{v}(0)\phi\mathrm{d}x \leq 0, \\ &\beta \| \psi \|_{L^2(\Omega)}^2 - \int_\Omega B_2^*\tilde{v}(0)\psi\mathrm{d}x \leq 0. \end{aligned} \tag{8.140}$$

Multiplying (8.138) by v, (8.139) by $\tilde{v}$ and integrating over Ω, this gives (according to (7.80), (7.81), (7.91) and (8.129))

$$\frac{\mathrm{d}}{\mathrm{d}t} \| v \|_{L^2}^2 + \nu \| v \|_{\mathcal{D}}^2 \leq \gamma c_I^2 \| v \|_{\mathcal{D}}^2,$$

$$-\frac{\mathrm{d}}{\mathrm{d}t} \| \tilde{v} \|_{L^2}^2 + \nu \| \tilde{v} \|_{\mathcal{D}}^2 \leq \gamma c_I^2 \| \tilde{v} \|_{\mathcal{D}}^2 + (\delta_1 c_I^2 + \delta_2)(\| v \|_{\mathcal{D}}^2 + \| \tilde{v} \|_{\mathcal{D}}^2) + 2\lambda \int_\Omega | \tilde{u} | \, | \tilde{v} | \, | v |,$$

$$\tilde{v}(T) = \mu v(T) \text{ and } v(0) = B_1\phi + B_2\psi.$$

By using Hölder's inequality and the relationship (7.79), we obtain

$$\begin{aligned} &\frac{\mathrm{d}}{\mathrm{d}t} \| v \|_{L^2}^2 + (\nu - \gamma c_I^2) \| v \|_{\mathcal{D}}^2 \leq 0, \\ &-\frac{\mathrm{d}}{\mathrm{d}t} \| \tilde{v} \|_{L^2}^2 + (\nu - \gamma c_I^2) \| \tilde{v} \|_{\mathcal{D}}^2 \\ &\qquad \leq (\delta_1 c_I^2 + \delta_2 + \lambda c_e^2 \| \tilde{u} \|_{\mathcal{H}})(\| v \|_{\mathcal{D}}^2 + \| \tilde{v} \|_{\mathcal{D}}^2), \\ &\tilde{v}(T) = \mu v(T) \text{ and } v(0) = B_1\phi + B_2\psi. \end{aligned} \tag{8.141}$$

Summing the first and second parts of (8.141) we obtain

$$\begin{aligned} &\frac{\mathrm{d}}{\mathrm{d}t}(\| v \|_{L^2}^2 - \| \tilde{v} \|_{L^2}^2) + \theta(\| v \|_{\mathcal{D}}^2 + \| \tilde{v} \|_{\mathcal{D}}^2) \\ &\qquad \leq \lambda c_e^2 \| \tilde{u} \|_{\mathcal{H}} (\| v \|_{\mathcal{D}}^2 + \| \tilde{v} \|_{\mathcal{D}}^2), \\ &\tilde{v}(T) = \mu v(T) \text{ and } v(0) = B_1\phi + B_2\psi \end{aligned} \tag{8.142}$$

where $\theta = \nu - \delta_2 - c_I^2(\gamma + \delta_1) > 0$ (according to assumption (i)).

By integrating over $[0, T]$ the first part of (8.142) and according to the second and third parts of (8.142), to the assumption (8.107), and the relations (8.108) and (8.140) we can deduce that

$$\begin{aligned} &(1 - \mu^2) \| v(T) \|_{L^2}^2 + \| \tilde{v}(0) \|_{L^2}^2 + \theta(\| v \|_{\mathcal{V}}^2 + \| \tilde{v} \|_{\mathcal{V}}^2) \\ &\qquad \leq \lambda c_e^2 \| \tilde{u} \|_{\mathcal{H}} (\| v \|_{\mathcal{V}}^2 + \| \tilde{v} \|_{\mathcal{V}}^2) + (\frac{b^4}{\alpha^2} + \frac{b^4}{\beta^2}) \| \tilde{v}(0) \|_{L^2}^2 . \end{aligned}$$

Since $1 - \mu^2 > 0$ and $1 - (b^4/\alpha^2) - (b^4/\beta^2) \geq 0$ then

$$\theta(\| v \|_{\mathcal{V}}^2 + \| \tilde{v} \|_{\mathcal{V}}^2) \leq \lambda c_e^2 \| \tilde{u} \|_{\mathcal{H}} (\| v \|_{\mathcal{V}}^2 + \| \tilde{v} \|_{\mathcal{V}}^2).$$

According to the estimate (7.112) and $\mu < 1$ we have

$$\theta^*(\| v \|_{\mathcal{V}}^2 + \| \tilde{v} \|_{\mathcal{V}}^2) \leq 0,$$

where

$$\theta^* = \theta - \lambda c_e^2 \exp((\delta_1 + \gamma + 1)T/2)(\| u^*(T) - v_{\text{obs}} \|_{L^2}^2 + \| \mathcal{C}(u^* - u_{\text{obs}}) \|_{L^2(Q)}^2)^{1/2}.$$

Since $\theta^* > 0$ (according to assumption (ii)), we have $\| v \|_{\mathcal{V}}^2 + \| \tilde{v} \|_{\mathcal{V}}^2 = 0$ and then $v = 0$ and $\tilde{v} = 0$. Therefore, the uniqueness result follows. □

Distributed Disturbance in the Forcing

In this section, the value g is the disturbance $\psi \in L^2(\Omega)$ and the initial condition u_0 is the control $\phi \in L^2(\Omega)$, *i.e.*, $g = B_2\psi, u_0 = B_1\phi$, where $B_i, i = 1, 2$ are given bounded operators on $L^2(\Omega)$ such that Assumption (8.107) holds. The function u is assumed to be related to the disturbance ψ and control ϕ through the problem (8.105):

$$\begin{aligned} &\frac{\partial u}{\partial t} + Au + F(., u) + K(\varpi, u) = B_2\psi \ \text{ on } \mathcal{Q}, \\ &u(0) = B_1\phi \ \text{ on } \Omega. \end{aligned} \tag{8.143}$$

To obtain the regularity of Proposition 8.19, we suppose the following hypothesis: $(\phi, \psi) \in L^2(\Omega) \times L^2(\mathcal{Q})$. Let $\mathbf{U} : (\phi, \psi) \longrightarrow u = \mathbf{U}(\phi, \psi)$ be the map: $L^2(\Omega) \times L^2(\mathcal{Q}) \longrightarrow \mathcal{W}$ defined by (8.143) and introducing the cost functional defined by

$$\begin{aligned} J(\phi, \psi) = \frac{1}{2} \parallel \mathcal{C}(u - u_{\text{obs}}) \parallel^2_{L^2(\mathcal{Q})} + \frac{\mu}{2} \parallel u(T) - v_{\text{obs}} \parallel^2_{L^2} \\ + \frac{\alpha}{2} \parallel \phi \parallel^2_{L^2(\Omega)} - \frac{\beta}{2} \parallel \psi \parallel^2_{L^2(\mathcal{Q})}, \end{aligned} \tag{8.144}$$

where $\mu, \alpha, \beta > 0$ are fixed, $(u_{\text{obs}}, v_{\text{obs}}) \in \mathcal{V} \times L^2(\Omega)$ is the observation (given) and $\mathcal{C}$ is unbounded, linear operator on $L^2(\Omega)$ satisfying the hypothesis (7.91).

We want to minimize the functional J with respect to ϕ and maximize J with respect to ψ, *i.e.*, to find $(\phi^*, \psi^*) \in \mathcal{U}_{\text{ad}} \times \mathcal{V}_{\text{ad}}$ such that

$$J(\phi^*, \psi) \leq J(\phi^*, \psi^*) \leq J(\phi, \psi^*), \forall (\phi, \psi) \in \mathcal{U}_{\text{ad}} \times \mathcal{V}_{\text{ad}}, \tag{8.145}$$

with $\mathcal{U}_{\text{ad}}$ and $\mathcal{V}_{\text{ad}}$ are (given) non-empty, closed, convex, bounded subsets of $L^2(\Omega)$ and $L^2(\mathcal{Q})$ respectively.

The arguments of the previous sections can be extended directly to the present case without further estimates. We have then the following results.

Proposition 8.35. *Let F be an operator that satisfies the assumptions (7.78). Then the function $\mathbf{U} : (\phi, \psi) \longrightarrow u = \mathbf{U}(\phi, \psi)$ which is a solution of (8.143) is continuously F-differentiable from $L^2(\Omega) \times L^2(\mathcal{Q})$ to $\mathcal{W}$ with the derivative $\mathbf{U}'(\phi, \psi) : (h_1, h_2) \longrightarrow w$ given by the linear parabolic problem*

$$\begin{aligned} &\frac{\partial w}{\partial t} + Aw + G(., \mathbf{U}(\phi, \psi))w + K(\varpi, w) = B_2 h_2 \ \ on \ \mathcal{Q}, \\ &w(0) = B_1 h_1 \ \ on \ \Omega, \end{aligned} \tag{8.146}$$

which satisfies, for all $\Phi_i = (\phi_i, \psi_i) \in L^2(\Omega) \times L^2(\mathcal{Q})$, for $i = 1, 2$, the estimates:

(*i*) $\parallel \mathbf{U}'(\phi_1, \psi_1) \parallel_{\mathcal{L}(L^2(\Omega) \times L^2(\mathcal{Q}); \mathcal{W})} \leq C$

(*ii*) $\parallel \mathbf{U}'(\phi_1, \psi_1) - \mathbf{U}'(\phi_2, \psi_2) \parallel_{\mathcal{L}(L^2(\Omega) \times L^2(\mathcal{Q}); \mathcal{W})} \leq C \parallel \Phi_1 - \Phi_2 \parallel_{L^2(\Omega) \times L^2(\mathcal{Q})}$.

□

Proposition 8.36. *Let F be an operator that satisfies Assumptions (7.78). Then for each $t \in [0,T]$, the function $\mathbf{V}_t : (\phi,\psi) \longrightarrow u(t) = \mathbf{V}_t(\phi,\psi)$ which is a solution of (8.143) is continuously F-differentiable from $(L^2(\Omega))^2$ to $L^2(\Omega)$ with the derivative $\mathbf{V}'_t(\phi,\psi) : (h_1,h_2) \longrightarrow w(t)$ given by the linear parabolic problem*

$$\begin{aligned} &\frac{\partial w}{\partial t} + Aw + G(.,\mathbf{V}_t(\phi,\psi))w + K(\varpi,w) = B_2h_2 \quad \text{on } Q \\ &w(0) = B_1h_1 \quad \text{on } \Omega \end{aligned} \tag{8.147}$$

which satisfies, for all $\Phi_i = (\phi_i,\psi_i) \in L^2(\Omega)\times L^2(Q)$, for $i = 1,2$, the estimates:

(i) $\| \mathbf{V}'_t(\phi_1,\psi_1) \|_{\mathcal{L}(L^2(\Omega)\times L^2(Q);L^2(\Omega))} \leq C$

(ii) $\| \mathbf{V}'_t(\phi_1,\psi_1) - \mathbf{V}'_t(\phi_2,\psi_2) \|_{\mathcal{L}(L^2(\Omega)\times L^2(Q);L^2(\Omega))}$

$$\leq C \| \Phi_1 - \Phi_2 \|_{L^2(\Omega)\times L^2(Q)} .$$ □

Theorem 8.37. *Let F be an operator that satisfies the assumptions (7.78). Then, for α and β sufficiently large, there exists $(\phi^*,\psi^*) \in \mathcal{U}_{\rm ad}\times\mathcal{V}_{\rm ad}$ and $u^* \in \mathcal{W}$ such that (ϕ^*,ψ^*) is defined by (8.145) and $u^* = \mathbf{U}(\phi^*,\psi^*)$ is a solution of (8.143).* □

Remark 8.38. The adjoint problem associated with the primal problem (8.143) is exactly the adjoint problem (8.123). ♢

Now we give the optimality system for the robust control problem (8.145).

Theorem 8.39. *Let F be an operator that satisfies Assumptions (7.78), $\nu > \delta_2$, and an optimal solution $(\phi^*,\psi^*,u^*) \in \mathcal{U}_{\rm ad}\times\mathcal{V}_{\rm ad}\times\mathcal{W}$ such that (ϕ^*,ψ^*) is defined by (8.134) and $u^* = \mathbf{U}(\phi^*,\psi^*)$ is the solution of (8.143). Then, for α and β sufficiently large, we have $(\forall(\phi,\psi) \in \mathcal{U}_{\rm ad}\times\mathcal{V}_{\rm ad})$*

$$\int_\Omega (\alpha\phi^* + B_1^*\tilde{u}(0))(\phi-\phi^*)\mathrm{d}x \geq 0,$$

$$\int_0^T\int_\Omega (-\beta\psi^* + B_2^*\tilde{u})(\psi-\psi^*)\mathrm{d}x\mathrm{d}t \leq 0,$$

where $\tilde{u} = \tilde{\mathbf{U}}(\phi^*,\psi^*)$ is a solution of the adjoint problem (8.123). Moreover, the gradient of the functional J at (ϕ^*,ψ^*) in any direction $\mathbf{h} = (h_1,h_2) \in L^2(\Omega)\times L^2(Q)$ is given by

$$J'(\phi^*,\psi^*).\mathbf{h} = \int_\Omega (\alpha\phi^* + B_1^*\tilde{u}(0))h_1\mathrm{d}x + \int_0^T\int_\Omega (-\beta\psi^* + B_2^*\tilde{u})h_2\mathrm{d}x.$$

Otherwise (in the weak sense),

$$\frac{\partial J}{\partial\phi}(\phi^*,\psi^*) = \alpha\phi^* + B_1^*\tilde{u}(0) \quad \text{and} \quad \frac{\partial J}{\partial\psi}(\phi^*,\psi^*) = -\beta\psi^* + B_2^*\tilde{u}.$$

Proof. We use the same technique as used in the proof of Theorem 8.33. So, we skip the details. ∎

In the following, we will assume that there exists an optimal solution (ϕ^*, ψ^*, u^*) such that (ϕ^*, ψ^*) is defined by (8.145), $u^* = \mathbf{U}(\phi^*, \psi^*)$ is a solution of (8.143) and

$$\int_\Omega (\alpha\phi^* + B_1^*\tilde{u}(0))(\phi - \phi^*)\mathrm{d}x \geq 0, \quad \int_0^T \int_\Omega (-\beta\psi^* + B_2^*\tilde{u})(\psi - \psi^*)\mathrm{d}x\mathrm{d}t \leq 0$$

for all $(\phi, \psi) \in \mathcal{U}_{\mathrm{ad}} \times \mathcal{V}_{\mathrm{ad}}$, where $\tilde{u} = \tilde{\mathbf{U}}(\phi^*, \psi^*)$ is a solution of (8.123).

Now we give some conditions to obtain the uniqueness of the solution (ϕ^*, ψ^*).

Theorem 8.40. *Suppose that F satisfies Assumptions (7.78), $\mu < 1$ and $\alpha \geq b^2$ holds. The optimal solution (ϕ^*, ψ^*, u^*) is unique if the following conditions hold:*

(i) $\theta = (\nu - \delta_2 - c_I^2(\gamma + \delta_1) - b^2 c_I^2/\beta) > 0$

(ii) $\lambda c_e^2 c_T \left(\| u^*(T) - v_{\mathrm{obs}} \|_{L^2}^2 + \| \mathcal{C}(u^* - u_{\mathrm{obs}}) \|_{L^2(\mathcal{Q})}^2 \right)^{1/2} < \theta,$
where $c_T = \exp((\delta_1 + \gamma + 1)T/2)$.

Proof. Suppose $(\phi_1^*, \psi_1^*, u_1^*)$ is another solution, then (ϕ_1^*, ψ_1^*) satisfies (8.134), $u_1^* = \mathbf{U}(\phi_1^*, \psi_1^*)$ is a solution of (8.143) and

$$\int_\Omega (\alpha\phi^* + B_1^*\tilde{u}_1(0))(\phi - \phi_1^*)\mathrm{d}x \geq 0,$$
$$\int_0^T \int_\Omega (-\beta\psi_1^* + B_2^*\tilde{u}_1)(\psi - \psi_1^*)\mathrm{d}x\mathrm{d}t \leq 0,$$

for all $(\phi, \psi) \in \mathcal{U}_{\mathrm{ad}} \times \mathcal{V}_{\mathrm{ad}}$, where $\tilde{u}_1 = \tilde{\mathbf{U}}(\phi_1^*, \psi_1^*)$ is a solution of (8.123).

We set $\phi = \phi^* - \phi_1^*$, $\psi = \psi^* - \psi_1^*$, $v = u^* - u_1^*$ and $\tilde{v} = \tilde{u} - \tilde{u}_1$. We then have

$$\begin{aligned} &\frac{\partial v}{\partial t} + Av + (F(.,u^*) - F(.,u_1^*)) = B_2\psi \quad \text{on } \mathcal{Q} \\ &v(0) = B_1\phi \quad \text{on } \Omega, \end{aligned} \tag{8.148}$$

$$\begin{aligned} &-\frac{\partial \tilde{v}}{\partial t} + A\tilde{v} + (G(.,u^*))^*\tilde{v} \\ &\qquad = \mathcal{C}^*\mathcal{C}v - (G(.,u^*) - G(.,u_1^*))^*\tilde{u} \quad \text{on } \mathcal{Q}, \\ &\tilde{v}(T) = \mu v(T) \quad \text{on } \Omega \end{aligned} \tag{8.149}$$

and

$$\begin{aligned} &\alpha \| \phi \|_{L^2(\Omega)}^2 + \int_\Omega B_1^*\tilde{v}(0)\phi\mathrm{d}x \leq 0, \\ &\beta \| \psi \|_{L^2(\mathcal{Q})}^2 - \int_0^T \int_\Omega B_2^*\tilde{v}\psi\mathrm{d}x\mathrm{d}t \leq 0. \end{aligned} \tag{8.150}$$

Multiplying (8.148) by v, (8.149) by $\tilde{v}$ and integrating over $\mathcal{Q}$, this gives (according to (8.150), (8.129), (7.80), (7.81) and (7.91))

$$\begin{aligned}
&\int_0^T \frac{\mathrm{d}}{\mathrm{d}t} \| v \|_{L^2}^2 \,\mathrm{d}s + \nu \int_0^T \| v \|_{\mathcal{D}}^2 \,\mathrm{d}s \\
&\qquad \leq \gamma c_I^2 \int_0^T \| v \|_{\mathcal{D}}^2 \,\mathrm{d}s + \frac{2}{\beta} \| B_2^* \tilde{v} \|_{L^2(\mathcal{Q})} \| B_2^* v \|_{L^2(\mathcal{Q})}, \\
&-\int_0^T \frac{\mathrm{d}}{\mathrm{d}t} \| \tilde{v} \|_{L^2}^2 \,\mathrm{d}s + \nu \int_0^T \| \tilde{v} \|_{\mathcal{D}}^2 \,\mathrm{d}s \\
&\qquad \leq \gamma c_I^2 \int_0^T \| \tilde{v} \|_{\mathcal{D}}^2 \,\mathrm{d}s + (\delta_1 c_I^2 + \delta_2) \int_0^T (\| v \|_{\mathcal{D}}^2 \,\mathrm{d}s + \| \tilde{v} \|_{\mathcal{D}}^2) \\
&\qquad\qquad + 2\lambda \int_0^T \int_\Omega | \tilde{u} | \, | \tilde{v} | \, | v | \,\mathrm{d}s, \\
&\tilde{v}(T) = \mu v(T) \text{ and } v(0) = B_1 \phi.
\end{aligned}$$

By using Hölder's inequality, the relationship (7.79), the assumption (8.107) and the estimate (8.108) we obtain

$$\begin{aligned}
&\int_0^T \frac{\mathrm{d}}{\mathrm{d}t} \| v \|_{L^2}^2 \,\mathrm{d}s + (\nu - \gamma c_I^2) \int_0^T \| v \|_{\mathcal{D}}^2 \,\mathrm{d}s \\
&\qquad \leq \frac{1}{\beta} b^2 c_I^2 \int_0^T (\| v \|_{\mathcal{D}}^2 \,\mathrm{d}s + \| \tilde{v} \|_{\mathcal{D}}^2), \\
&-\int_0^T \frac{\mathrm{d}}{\mathrm{d}t} \| \tilde{v} \|_{L^2}^2 \,\mathrm{d}s + (\nu - \gamma c_I^2) \int_0^T \| \tilde{v} \|_{\mathcal{D}}^2 \,\mathrm{d}s \\
&\qquad \leq (\delta_1 c_I^2 + \delta_2 + \lambda c_e^2 \| \tilde{u} \|_{\mathcal{H}}) \int_0^T (\| v \|_{\mathcal{D}}^2 \,\mathrm{d}s + \| \tilde{v} \|_{\mathcal{D}}^2), \\
&\tilde{v}(T) = \mu v(T) \text{ and } v(0) = B_1 \phi.
\end{aligned} \tag{8.151}$$

Summing the first and second parts of (8.151) we obtain

$$\begin{aligned}
&\int_0^T \frac{\mathrm{d}}{\mathrm{d}t} (\| v \|_{L^2}^2 - \| \tilde{v} \|_{L^2}^2) \mathrm{d}s + \theta \int_0^T (\| v \|_{\mathcal{D}}^2 + \| \tilde{v} \|_{\mathcal{D}}^2) \mathrm{d}s \\
&\qquad \leq \lambda c_e^2 \| \tilde{u} \|_{\mathcal{H}} \int_0^T (\| v \|_{\mathcal{D}}^2 + \| \tilde{v} \|_{\mathcal{D}}^2) \mathrm{d}s,
\end{aligned} \tag{8.152}$$

where $\theta = \nu - \delta_2 - c_I^2(\gamma + \delta_1 + b^2/\beta) > 0$ (according to assumption (i)). According to third part of (8.151), the first part of (8.150) and the estimate (8.108), we have

$$\begin{aligned}
&(1 - \mu^2) \| v(T) \|_{L^2}^2 + \| \tilde{v}(0) \|_{L^2}^2 + \theta (\| v \|_{\mathcal{V}}^2 + \| \tilde{v} \|_{\mathcal{V}}^2) \\
&\qquad \leq \lambda c_e^2 \| \tilde{u} \|_{\mathcal{H}} (\| v \|_{\mathcal{V}}^2 + \| \tilde{v} \|_{\mathcal{V}}^2) + \frac{b^4}{\alpha^2} \| \tilde{v}(0) \|_{L^2}^2.
\end{aligned}$$

Since $1 - \mu^2 > 0$ and $1 - b^4/\alpha^2 \geq 0$ then

$$\theta(\| v \|_{\mathcal{V}}^2 + \| \tilde{v} \|_{\mathcal{V}}^2) \leq \lambda c_e^2 \| \tilde{u} \|_{\mathcal{H}} (\| v \|_{\mathcal{V}}^2 + \| \tilde{v} \|_{\mathcal{V}}^2).$$

Applying the estimate (7.112) and $\mu < 1$ we have

$$\theta^* (\| v \|_{\mathcal{V}}^2 + \| \tilde{v} \|_{\mathcal{V}}^2) \leq 0,$$

where

$$\theta^* = \theta - \lambda c_e^2 e^{(\delta_1+\gamma+1)T/2} \left(\| u^*(T) - v_{\text{obs}} \|_{L^2}^2 + \| \mathcal{C}(u^* - u_{\text{obs}}) \|_{L^2(\mathcal{Q})}^2 \right)^{1/2}.$$

Since $\theta^* > 0$ (according to assumption (*ii*)), we have $\| v \|_{\mathcal{V}}^2 + \| \tilde{v} \|_{\mathcal{V}}^2 = 0$ and then $v = 0$ and $\tilde{v} = 0$. The uniqueness result is proved. □

8.6.5 A Remark on the Robust Boundary Control Problem

We conclude with an indication as to how the previous arguments can be made to apply to the robust boundary control problem. Let us consider the following problem:

$$\begin{aligned} &\frac{\partial u}{\partial t} - \operatorname{div}(\eta \nabla u) + F(x,t,u) + K(\varpi, u) = g \quad \text{on } \mathcal{Q}, \\ &\eta \frac{\partial u}{\partial \mathbf{n}} + au = f \quad \text{on } \Sigma = (0,T) \times \partial\Omega, \\ &u(0) = u_0 \quad \text{on } \Omega, \end{aligned} \tag{8.153}$$

where Ω is an open bounded in IR^m, $m \geq 1$, with the boundary $\partial\Omega$ sufficiently regular, g is in $L^2(\mathcal{Q})$, u_0 is in $L^2(\Omega)$ and F satisfies Assumptions (7.78) as before. We assume that f is in $L^2(\Sigma)$, a is in $L^\infty(\Sigma)$ and non-negative, $\partial u / \partial \mathbf{n}$ denotes the exterior normal derivative of u at the boundary $\partial\Omega$, the function η is positive and bounded function above and below by two non-negative constants in Ω, and we denote $\nu = \min_{\Omega}(\eta)$.

Here, we assume that $m \leq 4$,[3] $\mathcal{D} = H^1(\Omega) \subset L^4(\Omega)$ and $\mathcal{D}' = (H^1(\Omega))'$. Multiplying now (8.153) by u, integrating over Ω, using Green's formula and the first assumption of (7.78), we have that

$$\begin{aligned} \frac{\mathrm{d} \mid u \mid^2}{\mathrm{d}t} + 2 \int_\Omega \eta \mid \nabla u \mid^2 \mathrm{d}x + 2 \int_\Gamma a \mid u \mid^2 \mathrm{d}\Gamma - -2 \int_\Omega K(\varpi, u) u \mathrm{d}x \\ -2 \int_\Omega (F(.,t,u) - F(.,t,0))(u - 0) \mathrm{d}x + 2 \int_\Omega g u \mathrm{d}x + 2 \int_\Gamma f u \mathrm{d}\Gamma. \end{aligned}$$

According to the second assumption of (7.78), the assumption (7.81) and the trace theory we can deduce that

$$\frac{\mathrm{d} \mid u \mid^2}{\mathrm{d}t} + \nu \| u \|_{\mathcal{D}}^2 \leq c_1 \mid u \mid^2 + c_2 (\mid g \mid^2 + \| f \|_{L^2(\Gamma)}^2).$$

[3] $\mathcal{D} \subset L^4(\Omega)$ provided $m \leq 4$ according to the Sobolev embedding theorem.

Integrating over $(0,t)$, for all $t \in (0,T)$, gives

$$| u(t) |^2 + \nu \int_0^t \| u \|_{\mathcal{D}}^2 \, \mathrm{d}s \leq c_1 \int_0^t | u |^2 \, \mathrm{d}s + c_3(\| g \|_{L^2(\mathcal{Q})}^2 + \| f \|_{L^2(\Sigma)}^2 + | u_0 |^2)$$

and then (according to Gronwall's lemma)

$$\| u \|_{\mathcal{W}}^2 \leq C(\| u_0 \|_{L^2}^2 + \| g \|_{L^2(\mathcal{Q})}^2 + \| f \|_{L^2(\Sigma)}^2). \tag{8.154}$$

Using the results (8.154) and (8.153), we prove easily that u satisfies also the following estimate:

$$\| u \|_{\mathcal{H}_1}^2 \leq C(\| u_0 \|_{L^2}^2 + \| g \|_{L^2(\mathcal{Q})}^2 + \| f \|_{L^2(\Sigma)}^2). \tag{8.155}$$

Consequently, $u \in \mathcal{W} \cap \mathcal{H}_1$. One similarly gets a uniform Lipschitz continuous solution for the mapping solution in $\mathcal{W} \cap \mathcal{H}_1$.

Remark 8.41. If we suppose that the initial condition is in $H^{1/2}(\Omega)$ we can obtain for the solution u of the non-linear problem (8.153), by removing the terms $F(.,u)+K(\varpi,u)$ on the right-hand side of (8.153) and by using Theorem 6.19, the regularity result $u \in H^{3/2,3/4}(\mathcal{Q})$. ♢

We suppose now that the value g is the disturbance $\psi \in L^2(\mathcal{Q})$ and the boundary condition f is the control $\phi \in L^2(\Sigma)$, *i.e.*, $f = B_1\phi$ and $g = B_2\psi$, where B_1 and B_2 are given bounded operators on $L^2(\Sigma)$ and on $L^2(\mathcal{Q})$ respectively. So the function u is assumed to be related to the disturbance ψ and control ϕ through the problem (8.153):

$$\begin{aligned} &\frac{\partial u}{\partial t} - \operatorname{div}(\eta \nabla u) + F(x,t,u) + K(\varpi,u) = B_2\psi \quad \text{on } \mathcal{Q}, \\ &\eta \frac{\partial u}{\partial \mathbf{n}} + au = B_1\phi \quad \text{on } \Sigma = (0,T) \times \partial\Omega, \\ &u(0) = u_0 \quad \text{on } \Omega. \end{aligned} \tag{8.156}$$

Let $\mathbf{U} : (\phi,\psi) \longrightarrow u = \mathbf{U}(\phi,\psi)$ be the map: $L^2(\Sigma) \times L^2(\mathcal{Q}) \longrightarrow \mathcal{W}$ defined by (8.156) and introducing the cost functional defined by

$$\begin{aligned} J(\phi,\psi) = \frac{1}{2} \| \mathcal{C}(u - u_{\text{obs}}) \|_{L^2(\mathcal{Q})}^2 + \frac{\mu}{2} \| u(T) - v_{\text{obs}} \|_{L^2(\Omega)}^2 \\ + \frac{\alpha}{2} \| \phi \|_{L^2(\Sigma)}^2 - \frac{\beta}{2} \| \psi \|_{L^2(\mathcal{Q})}^2, \end{aligned} \tag{8.157}$$

where $\mu, \alpha, \beta > 0$ are fixed, $(u_{\text{obs}}, v_{\text{obs}}) \in \mathcal{V} \times L^2(\Omega)$ is the observation and $\mathcal{C}$ is an unbounded, linear operator on $L^2(\Omega)$ satisfying the hypothesis (7.91).

We want to minimize the functional J with respect to ϕ and maximize J with respect to ψ, *i.e.*, to find $(\phi^*, \psi^*) \in \mathcal{U}_{\text{ad}} \times \mathcal{V}_{\text{ad}}$ such that

$$J(\phi^*,\psi) \leq J(\phi^*,\psi^*) \leq J(\phi,\psi^*), \forall (\phi,\psi) \in \mathcal{U}_{\text{ad}} \times \mathcal{V}_{\text{ad}} \tag{8.158}$$

with $\mathcal{U}_{\rm ad}$ and $\mathcal{V}_{\rm ad}$ are (given) non-empty, closed, convex, bounded subsets of $L^2(\Sigma)$ and $L^2(Q)$, respectively.

By using the same technique as used in the proof of Propositions 8.19 and 8.22 and Theorems 8.25 and 8.27, we have the following results.

Proposition 8.42. *Let F be an operator that satisfies the assumptions (7.78). Then the function $\mathbf{U} : (\phi,\psi) \longrightarrow u = \mathbf{U}(\phi,\psi)$ solution of (8.156) is continuously F-differentiable from $L^2(\Sigma) \times L^2(Q)$ to $\mathcal{W}$ with the derivative $\mathbf{U}'(\phi,\psi) : (h_1,h_2) \longrightarrow w$ given by the linear parabolic problem*

$$\begin{aligned}&\frac{\partial w}{\partial t} - \operatorname{div}(\eta\nabla w) + G(.,\mathbf{U}(\phi,\psi))w + K(\varpi,w) = B_2h_2 \quad on\ Q\\ &\eta\frac{\partial w}{\partial \mathbf{n}} + au = B_1h_1 \quad on\ \Sigma,\\ &w(0) = 0 \quad on\ \Omega,\end{aligned} \tag{8.159}$$

which satisfies, for all $\Phi_i = (\phi_i,\psi_i) \in L^2(\Sigma) \times L^2(Q)$, for $i = 1,2$, the following estimates:

(i) $\| \mathbf{U}'(\phi_1,\psi_1) \|_{\mathcal{L}(L^2(\Sigma)\times L^2(Q);\mathcal{W})} \leq C$

(ii) $\| \mathbf{U}'(\phi_1,\psi_1) - \mathbf{U}'(\phi_2,\psi_2) \|_{\mathcal{L}(L^2(\Sigma)\times L^2(Q);\mathcal{W})} \leq C \| \Phi_1 - \Phi_2 \|_{L^2(\Sigma)\times L^2(Q)}$. □

Proposition 8.43. *Let F be an operator that satisfies the assumptions (7.78). Then for each $t \in [0,T]$, the function $\mathbf{V}_t : (\phi,\psi) \longrightarrow u(t) = \mathbf{V}_t(\phi,\psi)$ solution of (8.156) is continuously F-differentiable from $(L^2(\Omega))^2$ to $L^2(\Omega)$ with the derivative $\mathbf{V}_t'(\phi,\psi) : (h_1,h_2) \longrightarrow w(t)$ given by the linear parabolic problem*

$$\begin{aligned}&\frac{\partial w}{\partial t} - \operatorname{div}(\eta\nabla w) + G(.,\mathbf{U}(\phi,\psi))w + K(\varpi,w) = B_2h_2 \quad on\ Q\\ &\eta\frac{\partial w}{\partial \mathbf{n}} + au = B_1h_1 \quad on\ \Sigma,\\ &w(0) = 0 \quad on\ \Omega,\end{aligned} \tag{8.160}$$

which satisfies, for all $\Phi_i = (\phi_i,\psi_i) \in L^2(\Sigma) \times L^2(Q)$, for $i = 1,2$, the following estimates:

(i) $\| \mathbf{V}_t'(\phi_1,\psi_1) \|_{\mathcal{L}(L^2(\Sigma)\times L^2(Q);L^2(\Omega))} \leq C$

(ii) $\| \mathbf{V}_t'(\phi_1,\psi_1) - \mathbf{V}_t'(\phi_2,\psi_2) \|_{\mathcal{L}(L^2(\Sigma)\times L^2(Q);L^2(\Omega))}$

$$\leq C \| \Phi_1 - \Phi_2 \|_{L^2(\Sigma)\times L^2(Q)}.$$ □

Theorem 8.44. *Let F be an operator that satisfies the assumptions (7.78). Then, for α and β sufficiently large, there exists $(\phi^*,\psi^*) \in L^2(\Sigma) \times L^2(Q)$ and $u^* \in \mathcal{W}$ such that (ϕ^*,ψ^*) is defined by (8.158) and $u^* = \mathbf{U}(\phi^*,\psi^*)$ is a solution of (8.156).* □

Now we give the optimality system for the robust control problem (8.158).

Theorem 8.45. *Let F be an operator that satisfies the assumptions (7.78) and an optimal solution $(\phi^*, \psi^*, u^*) \in L^2(\Sigma) \times L^2(\mathcal{Q}) \times \mathcal{W}$ such that (ϕ^*, ψ^*) is defined by (8.158) and $u^* = \mathbf{U}(\phi^*, \psi^*)$ is the solution of (8.156). Then, for α and β sufficiently large, we have, for all $(\phi, \psi) \in \mathcal{U}_{\text{ad}} \times \mathcal{V}_{\text{ad}}$,*

$$\begin{aligned}
&\int_0^T \int_\Gamma (\alpha\phi^* + B_1^* \tilde{u}|_\Sigma)(\phi - \phi^*) \mathrm{d}\Gamma \mathrm{d}t \geq 0, \\
&\int_0^T \int_\Omega (-\beta\psi^* + B_2^* \tilde{u})(\psi - \psi^*) \mathrm{d}x \mathrm{d}t \leq 0,
\end{aligned} \tag{8.161}$$

where $\tilde{u} = \tilde{\mathbf{U}}(\phi^*, \psi^*)$ is a solution of the following adjoint problem

$$\begin{aligned}
&-\frac{\partial \tilde{u}}{\partial t} - \mathrm{div}(\eta \nabla \tilde{u}) + (G(.,u))^* \tilde{u} + K^*(\varpi, \tilde{u}) \\
&\qquad\qquad = \mathcal{C}^* \mathcal{C}(u^* - u_{\text{obs}}) \quad \text{on } \mathcal{Q}, \\
&\eta \frac{\partial \tilde{u}}{\partial \mathbf{n}} + a\tilde{u} = 0 \quad \text{on } \Sigma, \\
&\tilde{u}(T) = \mu(u^*(T) - v_{\text{obs}}) \quad \text{on } \Omega,
\end{aligned} \tag{8.162}$$

where $\mathcal{C}^*$ (resp. $(G(.,u))^*$) is the adjoint of the operator $\mathcal{C}$ (resp. $G(.,u)$).

Moreover, the gradient of the functional J at (ϕ^*, ψ^*) in any direction $\mathbf{h} = (h_1, h_2) \in L^2(\Sigma) \times L^2(\mathcal{Q})$ is given by

$$J'(\phi^*, \psi^*).\mathbf{h} = \int_0^T \int_\Gamma (\alpha\phi^* + B_1^* \tilde{u}|_\Sigma) h_1 \mathrm{d}\Gamma \mathrm{d}t + \int_0^T \int_\Omega (-\beta\psi^* + B_2^* \tilde{u}) h_2 \mathrm{d}x \mathrm{d}t.$$

Otherwise (in the weak sense),

$$\frac{\partial J}{\partial \phi}(\phi^*, \psi^*) = \alpha\phi^* + B_1^* \tilde{u}|_\Sigma \quad \text{and} \quad \frac{\partial J}{\partial \psi}(\phi^*, \psi^*) = -\beta\psi^* + B_2^* \tilde{u}.$$

□

Remark 8.46. It is clear that we can consider other controls and disturbances (which can appear in the boundary condition or in the state system) and we can obtain the same results by using the same technique as in the previous work of this section. ◇

8.6.6 Contraction Mapping and Fixed-point Formulation

In this section we rewrite the robust control problem as a fixed-point problem (in the case of the equality-type constraints).

Let U_i, $i = 1, 2$ be Hilbert spaces who will play the role of $L^2(\mathcal{Q})$ or $L^2(\Omega)$ depending on different situations control-disturbance presented previously. For $(\phi, \psi) \in \mathcal{U} = U_1 \times U_2$, we set

$$\mathcal{G}_\rho(\phi, \psi) := (\phi - \rho(\alpha\phi + B_1^* \tilde{u}), \psi - \rho(\beta\psi - B_2^* \tilde{u})), \tag{8.163}$$

where, $\rho > 0$ is a positive constant and $\tilde{u} = \tilde{\mathbf{U}}(\phi^*, \psi^*)$ is a solution of the adjoint problem (8.123).

Our goal is to prove that $\mathcal{G}_\rho$ is a contraction mapping on some subset $\mathcal{Y}$ of $\mathcal{U}$. Let $\epsilon > 0$ and $\mathcal{Y}_\epsilon$ be a ball of radius ϵ, *i.e.*,

$$\mathcal{Y}_\epsilon = \{(\phi, \psi) \in \mathcal{U} : \| (\phi, \psi) \|_{\mathcal{U}} \leq \epsilon\}.$$

Proposition 8.47. *Assume that α and β are sufficiently large and that ϵ is sufficiently small. Then there exists $\rho_0 > 0$ (depending on $\epsilon, u_{\text{obs}}, v_{\text{obs}}, U$ and given fixed parameters) such that $\mathcal{G}_\rho$ is a mapping from $\mathcal{Y}_\epsilon$ into $\mathcal{Y}_\epsilon$, for all $\rho \leq \rho_0$.*

Proof. Let $(\phi, \psi) \in \mathcal{Y}_\epsilon$. We have easily that

$$\begin{aligned}\| \phi - \rho(\alpha\phi + B_1^*\tilde{u}) \|_{U_1}^2 &= \rho^2(\alpha^2 \| \phi \|_{U_1}^2 + \| B_1^*\tilde{u} \|_{U_1}^2 + 2\alpha(\phi, B_1^*\tilde{u})_{U_1}) \\ &\quad -2\rho(\alpha \| \phi \|_{U_1}^2 + (\phi, B_1^*\tilde{u})_{U_1}) + \| \phi \|_{U_1}^2 \\ \| \psi - \rho(\beta\psi - B_2^*\tilde{u}) \|_{U_2}^2 &= \rho^2(\beta^2 \| \psi \|_{U_2}^2 + \| B_2^*\tilde{u} \|_{U_2}^2 - 2\beta(\psi, B_2^*\tilde{u})_{U_2}) \\ &\quad -2\rho(\beta \| \psi \|_{U_2}^2 - (\psi, B_2^*\tilde{u})_{U_2}) + \| \psi \|_{U_2}^2\end{aligned}$$

and then

$$\begin{aligned}\| \mathcal{G}_\rho(\phi, \psi) \|_{\mathcal{U}}^2 &= \| \phi - \rho(\alpha\phi + B_1^*\tilde{u}) \|_{U_1}^2 + \| \psi - \rho(\beta\psi - B_2^*\tilde{u}) \|_{U_2}^2 \\ &\leq (1 + 2\rho^2\alpha^2 + (1 - 2\alpha)\rho) \| \phi \|_{U_1}^2 + \rho(2\rho + 1) \| B_1^*\tilde{u} \|_{U_1}^2 \\ &\quad + (1 + 2\rho^2\beta^2 + (1 - 2\beta)\rho) \| \psi \|_{U_2}^2 + \rho(2\rho + 1) \| B_2^*\tilde{u} \|_{U_2}^2 .\end{aligned}$$

According to the continuity of the operators B_i^*, $i = 1, 2$, we can deduce that (because of (8.124) which is valid for all the control-disturbance cases)

$$\| B_i^*\tilde{u} \|_{U_i}^2 \leq C_0 \| \tilde{u} \|_{U_i}^2 \leq C_1(\| u(T) - v_{\text{obs}} \|_{L^2}^2 + \| \mathcal{C}(u - u_{\text{obs}}) \|_{L^2(\mathcal{Q})}^2).$$

According to (8.106) and (8.107) and Assumption (7.91), we can deduce that

$$C_1(\| u(T) - v_{\text{obs}} \|_{L^2}^2 + \| \mathcal{C}(u - u_{\text{obs}}) \|_{L^2(\mathcal{Q})}^2) \leq C_2(\| \phi \|_{U_1}^2 + \| \psi \|_{U_2}^2) + C_3,$$

where the generic constant C_3 depends on the observation $(u_{\text{obs}}, v_{\text{obs}})$ and data (initial condition or forcing) and the generic constant C_2 depends on the function U. Consequently,

$$\begin{aligned}\| \mathcal{G}_\rho(\phi, \psi) \|_{\mathcal{U}}^2 \leq\ & 2\rho(2\rho + 1)C_3 + (1 + 2\rho^2\alpha^2 + (1 - 2\alpha)\rho + \rho(2\rho + 1)C_2) \| \phi \|_{U_1}^2 \\ & + (1 + 2\rho^2\beta^2 + (1 - 2\beta)\rho + \rho(2\rho + 1)C_2) \| \psi \|_{U_2}^2 .\end{aligned}$$

Since $(\phi, \psi) \in \mathcal{Y}_\epsilon$, we can deduce that $\| \mathcal{G}_\rho(\phi, \psi) \|_{\mathcal{U}} \leq \epsilon$ if

$$2\rho\alpha^2 + (1 - 2\alpha) + (2\rho + 1)C_2 + \frac{2(2\rho + 1)C_3}{\epsilon^2} \leq 0,$$

$$2\rho\beta^2 + (1 - 2\beta) + (2\rho + 1)C_2 + \frac{2(2\rho + 1)C_3}{\epsilon^2} \leq 0.$$

So,

$$(2\rho+1)\alpha^2 + 2 - (1+\alpha)^2 + C_2(2\rho+1) + \frac{2(2\rho+1)C_3}{\epsilon^2} \le 0,$$
$$(2\rho+1)\beta^2 + 2 - (1+\beta)^2 + C_2(2\rho+1) + \frac{2(2\rho+1)C_3}{\epsilon^2} \le 0.$$

Consequently,

$$(2\rho+1) \le \frac{-2+(1+\alpha)^2}{\alpha^2 + C_2 + 2C_3/\epsilon^2},$$
$$(2\rho+1) \le \frac{-2+(1+\beta)^2}{\beta^2 + C_2 + 2C_3/\epsilon^2}$$

and then

$$\begin{aligned} 0 \le 2\rho \le \frac{-2+(1+\alpha)^2}{\alpha^2 + C_2 + 2C_3/\epsilon^2} - 1 = \frac{-1+2\alpha - C_2 - 2C_3/\epsilon^2}{\alpha^2 + C_2 + 2C_3/\epsilon^2}, \\ 0 \le 2\rho \le \frac{-2+(1+\beta)^2}{\beta^2 + C_2 + 2C_3/\epsilon^2} - 1 = \frac{-1+2\beta - C_2 - 2C_3/\epsilon^2}{\beta^2 + C_2 + 2C_3/\epsilon^2}. \end{aligned} \tag{8.164}$$

Consequently, if $\min(\alpha, \beta)$ is sufficiently large such that

$$\min(\alpha,\beta) \ge \frac{1+C_2}{2} + \frac{C_3}{\epsilon^2},$$

then for

$$\rho_0 = \frac{1}{2}\min\left(\frac{-1+2\alpha - C_2 - 2C_3/\epsilon^2}{\alpha^2 + C_2 + 2C_3/\epsilon^2}, \frac{-1+2\beta - C_2 - 2C_3/\epsilon^2}{\beta^2 + C_2 + 2C_3/\epsilon^2}\right)$$

we can deduce that

$$\mathcal{G}_\rho \text{ maps } \mathcal{Y}_\epsilon \text{ into } \mathcal{Y}_\epsilon, \quad \text{for all } \rho \le \rho_0. \tag{8.165}$$

This completes the proof. □

Remark 8.48. Since the result (8.165) is valid for $\min(\alpha, \beta)$ sufficiently large. In order to avoid and then to compensate for this large value of $\min(\alpha, \beta)$ (which makes expensive the optimization problem in practice), we can take the given data, the observations and ϵ sufficiently small (which is not very restrictive because our problem corresponds to a small perturbation). For this, we suppose that $2C_3 \le \epsilon^{2+r}$ and $r \in (0,1)$. Then (8.164) is valid if

$$0 \le 2\rho \le \frac{-1+2\alpha - C_2 - \epsilon^r}{\alpha^2 + C_2 + \epsilon^r},$$
$$0 \le 2\rho \le \frac{-1+2\beta - C_2 - \epsilon^r}{\beta^2 + C_2 + \epsilon^r}.$$

Consequently, if

$$\min(\alpha,\beta) \geq \frac{1+C_2+\epsilon^r}{2},$$
$$\rho_0 = \frac{1}{2}\min\left(\frac{-1+2\alpha-C_2-\epsilon^r}{\alpha^2+C_2+\epsilon^r}, \frac{-1+2\beta-C_2-\epsilon^r}{\beta^2+C_2+\epsilon^r}\right)$$

then $\mathcal{G}_\rho$ maps $\mathcal{Y}_\epsilon$ into $\mathcal{Y}_\epsilon$, for all $\rho \leq \rho_0$. ♢

Proposition 8.49. *Assume that α and β are sufficiently large. Then there exists $\rho_2 > 0$ such that $\mathcal{G}_\rho$ is a K-contraction mapping from $\mathcal{Y}_\epsilon$ into $\mathcal{Y}_\epsilon$, i.e., there exists $0 < K < 1$ such that for all $(\phi_i, \psi_i) \in \mathcal{Y}_\epsilon$, $i = 1, 2$,*

$$\| \mathcal{G}_\rho(\phi_1,\psi_1) - \mathcal{G}_\rho(\phi_2,\psi_2) \|_{\mathcal{U}} \leq K \| (\phi_1-\phi_2, \psi_2-\psi_2) \|_{\mathcal{U}},$$

for all $0 < \rho < \rho_2$.

Proof. Let $(\phi_i, \psi_i) \in \mathcal{Y}_\epsilon$, $u_i = \mathbf{U}(\phi_i,\psi_i)$ the solution of the primal problem and $\tilde{u}_i = \tilde{\mathbf{U}}(\phi_i,\psi_i)$ the adjoint solution corresponding to the primal solution u_i, for $i = 1, 2$.

We set, $(\phi,\psi) = (\phi_1,\psi_1) - (\phi_2,\psi_2)$, $u = u_1 - u_2$ and $\tilde{u} = \tilde{u}_1 - \tilde{u}_2$. We can prove easily that $\mathcal{G}_\rho(\phi_1,\psi_1) - \mathcal{G}_\rho(\phi_2,\psi_2) = \mathcal{G}_\rho(\phi,\psi)$ and that

$$\begin{aligned}
\| \phi - \rho(\alpha\phi + B_1^*\tilde{u}) \|_{U_1}^2 &= \rho^2(\alpha^2 \| \phi \|_{U_1}^2 + \| B_1^*\tilde{u} \|_{U_1}^2 + 2\alpha(\phi, B_1^*\tilde{u})_{U_1}) \\
&\quad -2\rho(\alpha \| \phi \|_{U_1}^2 + (\phi, B_1^*\tilde{u})_{U_1}) + \| \phi \|_{U_1}^2 \\
\| \psi - \rho(\beta\psi - B_2^*\tilde{u}) \|_{U_2}^2 &= \rho^2(\beta^2 \| \psi \|_{U_2}^2 + \| B_2^*\tilde{u} \|_{U_2}^2 - 2\beta(\psi, B_2^*\tilde{u})_{U_2}) \\
&\quad -2\rho(\beta \| \psi \|_{U_2}^2 - (\psi, B_2^*\tilde{u})_{U_2}) + \| \psi \|_{U_2}^2 .
\end{aligned}$$

Then

$$\begin{aligned}
\| \mathcal{G}_\rho(\phi,\psi) \|_{\mathcal{U}}^2 &= \| \phi - \rho(\alpha\phi + B_1^*\tilde{u}) \|_{U_1}^2 + \| \psi - \rho(\beta\psi - B_2^*\tilde{u}) \|_{U_2}^2 \\
&\leq (1 + 2\rho^2\alpha^2 + (1-2\alpha)\rho) \| \phi \|_{U_1}^2 + \rho(2\rho+1) \| B_1^*\tilde{u} \|_{U_1}^2 \\
&\quad + (1 + 2\rho^2\beta^2 + (1-2\beta)\rho) \| \psi \|_{U_2}^2 + \rho(2\rho+1) \| B_2^*\tilde{u} \|_{U_2}^2 .
\end{aligned}$$

According to the continuity of the operators B_i^*, $i = 1, 2$, we can deduce that (because of (8.124))

$$\| B_i^*\tilde{u} \|_{U_i}^2 \leq C_0 \| \tilde{u} \|_{U_i}^2 \leq C_1(\| u(T) \|_{L^2}^2 + \| Cu \|_{L^2(\mathcal{Q})}^2).$$

According to (8.106) and (8.107) and Assumption (7.91), we can deduce that

$$C_1(\| u(T) \|_{L^2}^2 + \| Cu \|_{L^2(\mathcal{Q})}^2) \leq C_2(\| \phi \|_{U_1}^2 + \| \psi \|_{U_2}^2),$$

where the generic constant C_2 depends on the U.

Consequently,

$$\begin{aligned}
\| \mathcal{G}_\rho(\phi,\psi) \|_{\mathcal{U}}^2 &\leq (1 + 2\rho^2\alpha^2 + (1-2\alpha)\rho + C_2\rho(2\rho+1)) \| \phi \|_{U_1}^2 \\
&\quad + (1 + 2\rho^2\beta^2 + (1-2\beta)\rho + C_2\rho(2\rho+1)) \| \psi \|_{U_2}^2
\end{aligned}$$

and then we have $\| \mathcal{G}_\rho(\phi, \psi) \|_{\mathcal{U}} \leq K\epsilon$ if

$$\rho(2\rho\alpha^2 + (1-2\alpha) + C_2(2\rho+1)) + \tilde{K} \leq 0,$$
$$\rho(2\rho\beta^2 + (1-2\beta) + C_2(2\rho+1)) + \tilde{K} \leq 0,$$

where $\tilde{K} = 1 - K^2 > 0$.

So,

$$2\rho^2(\alpha^2 + C_2) + (1 - 2\alpha + C_2)\rho + \tilde{K} \leq 0,$$
$$2\rho^2(\beta^2 + C_2) + (1 - 2\beta + C_2)\rho + \tilde{K} \leq 0.$$

Consequently, if $\min(\alpha, \beta)$ is sufficiently large such that

$$\min(\alpha, \beta) > \frac{1 + C_2}{2},$$
$$\tilde{K} = \delta^2 \min\left(\frac{(1 - 2\alpha + C_2)^2}{8(\alpha^2 + C_2)}, \frac{(1 - 2\beta + C_2)^2}{8(\beta^2 + C_2)}\right) < 1, \text{ where } 0 < \delta < 1,$$

then for

$$\rho_2 = \frac{1}{4}\min\left(\frac{-(1 - 2\alpha + C_2) + \sqrt{D_\alpha}}{\alpha^2 + C_2}, \frac{-(1 - 2\beta + C_2) + \sqrt{D_\beta}}{\beta^2 + C_2}\right),$$
$$\rho_1 = \frac{1}{4}\max\left(\frac{-(1 - 2\alpha + C_2) - \sqrt{D_\alpha}}{\alpha^2 + C_2}, \frac{-(1 - 2\beta + C_2) - \sqrt{D_\beta}}{\beta^2 + C_2}\right),$$

where

$$D_\alpha = (1 - 2\alpha + C_2)^2 - 8\tilde{K}(\alpha^2 + C_2),$$
$$D_\beta = (1 - 2\beta + C_2)^2 - 8\tilde{K}(\beta^2 + C_2),$$

we have that

$$\mathcal{G}_\rho \text{ is a contraction from } \mathcal{Y}_\epsilon \text{ into } \mathcal{Y}_\epsilon, \quad \text{for all } \rho_1 < \rho < \rho_2. \tag{8.166}$$

Since we can choose δ^2 as small as we want, then ρ_1 can be taken as small as we want, this completes the proof. □

Remark 8.50. (*i*) Since the function $\mathcal{G}_\rho$ is a K-contraction from $\mathcal{Y}_\epsilon$ into $\mathcal{Y}_\epsilon$, for all $0 < \rho < \rho_2$, then there exists a unique *fixed-point* (ϕ^*, ψ^*) of $\mathcal{G}_\rho$ on $\mathcal{Y}_\epsilon$, *i.e.*,

$$(\phi^*, \psi^*) = \mathcal{G}_\rho(\phi^*, \psi^*) = (\phi^* - \rho(\alpha\phi^* + B_1^*\tilde{u}^*), \psi^* - \rho(\beta\psi^* - B_2^*\tilde{u}^*)),$$

where, $\tilde{u}^* = \tilde{\mathbf{U}}(\phi^*, \psi^*)$ is a solution of the adjoint problem (8.123), corresponding to the primal solution $u^* = \mathbf{U}(\phi^*, \psi^*)$. Consequently, $(u^*, \tilde{u}^*, \phi^*, \psi^*)$ is a unique solution of the robust control problem.

(*ii*) In order to approximate the solution (ϕ^*, ψ^*), we can use the well known fixed-point iterative algorithm: for $k = 1, \ldots,$ (iterative index), we denote by (ϕ_k, ψ_k) the numerical approximation of the optimal solution at the kth iteration of the algorithm.

$$\begin{aligned} &\text{Initialization}: \ (\phi_0, \psi_0) \in \mathcal{Y}_\epsilon \text{ (given)} \\ &(\phi_{k+1}, \psi_{k+1}) := \mathcal{G}_\rho(\phi_k, \psi_k). \end{aligned} \tag{8.167}$$

Since the function $\mathcal{G}_\rho$ is a K-contraction, we can obtain easily the classical error estimate:[4]

$$\| (\phi_k - \phi^*, \psi_k - \psi^*) \|_{\mathcal{U}} \leq \frac{K^k}{1-K} \| (\phi_0 - \phi^*, \psi_0 - \psi^*) \|_{\mathcal{U}}$$

and then the convergence result (because $K < 1$).

(*iii*) By taking into account the expression of the mapping $\mathcal{G}_\rho$, the fixed-point method is performed as follows:

1. Initialization: $k = 0$ and $(\phi_0, \psi_0) \in \mathcal{Y}_\epsilon$ (given).
2. Resolution of the direct problem with the source term (ϕ_k, ψ_k), gives
$$u_k = \mathbf{U}(\phi_k, \psi_k).$$
3. Resolution of the adjoint problem (based on (ϕ_k, ψ_k, u_k)), gives
$$\tilde{u}_k = \tilde{\mathbf{U}}(\phi_k, \psi_k).$$
4. Gradient of J at point (ϕ_k, ψ_k):
$$(\mathbf{GJ}) \begin{cases} c_k \overset{def}{=} \dfrac{\partial J}{\partial \phi}(\phi_k, \psi_k) = \alpha\phi_k + B_1^* \tilde{u}_k, \\ d_k \overset{def}{=} \dfrac{\partial J}{\partial \psi}(\phi_k, \psi_k) = -\beta\psi_k + B_2^* \tilde{u}_k. \end{cases}$$
5. Determine (ϕ_{k+1}, ψ_{k+1}):
$$\begin{cases} \phi_{k+1} := \phi_k - \rho c_k, \\ \psi_{k+1} := \psi_k + \rho d_k. \end{cases}$$
6. Set $k := k+1$ and goto 2. until convergence. The approximation of the optimal solution (ϕ^*, ψ^*, u^*) is (ϕ_k, ψ_k, u_k).

(*iv*) We remark that the fixed-point iterative method is a particular case of the general iterative methods. Consequently, a more precise and more detailed analysis of some iterative methods to solve robust control problem will be considered in Chapter 9. ♢

[4] The proof is left to the reader as an exercise.

8.7 Non-linear Time-varying Delay Systems

In this section, we study a robust control problem for a class of systems governed by non-linear parabolic equations with multiple time-varying delays appearing in the state equation and in the boundary conditions. The introduction of retarded arguments is to reflect an after-effect. The delays occur naturally in biological and biochemical systems (*e.g.*, the chemostat in which the delay is due to the cell cycle and to the fact that the organism stores the nutrient), in population modeling (*e.g.*, the gestation period), in the area of plasma control (*e.g.*, the confinement plasma in a bounded domain, where an electric potential barrier or magnetic mirror surrounds this domain, and where the particle reflection at this domain boundary does not act instantaneously), or in technical devices (*e.g.*, a control circuit in which the delay is a measurable physical quantity).

First, we introduce the studied model and analyze some well-posedness properties. Second, we introduce the perturbation problem and we formulate a distributed robust control problem when delays appear only in the state equation. We prove the existence and the uniqueness of the solution and we give necessary optimality conditions with the quadratic performance functionals. Finally, we consider a boundary control problem where delays appear in the state equation and in the boundary conditions. We reformulate a robust control problem, and derive the existence and the conditions of the uniqueness of the optimal solution, obtaining first-order necessary conditions of optimality.

8.7.1 Mathematical Setting

Motivation

The delay systems or systems with after-effects (delays in the state and/or in the input) represent a class of infinite-dimensional dynamical systems where the solution depends on an evolutive initial function (which includes informations on the past history) rather than an initial value. These equations constitute a universal mathematical model for many diffusion processes in which time-delayed feedback signals are introduced and are used to describe propagation phenomena, biological and chemical systems, population dynamics or mechanical engineering.

The delay effects on the stability and control of dynamical systems are problems of recurring interest since the delay presence may induce complex behaviors, *e.g.*, instability and poor performance, for the dynamical systems.[5] Therefore, these behaviors and aspects motivate the study of time delay effects on properties of dynamical systems.

[5] It turns out that time delay has been one of the major sources of instability in control.

Our concern here is to present methods and techniques for the analysis and control of non-linear dynamical systems in the presence of multiple time-varying delays in the state systems and/or boundary conditions of systems.

Various problems associated with theory of delay equations have been studied over the last few years see, *e.g.*, the following references and the references therein: for the general theory see, *e.g.*, Kolmanovskii and Myshkis [177], Hale and Verduyn-Lunel [150]; for problems of population dynamics and neutral equations see, *e.g.*, Cushing [91, 92], Diekmann *et al.* [101], Kuang [183] and Banks [20]; for questions of stability, oscillations and control see, *e.g.*, Gopalsamy [135], Gu *et al.* [143], Kolmanovskii and Shaikhet [178], MacDonald [212] and Niculescu [230]; for water quality management and pollution problems see, *e.g.*, Lee and Leitmann [191], *etc.* Here we consider robust control problems for a class of non-linear parabolic systems with disturbances and controls in which multiple time-varying delays appear in the state systems and/or boundary conditions. For the optimal control problem of this type of system, for the linear case see Ichikawa [164], Kowalewski [179, 180] and Wang [292], and for the robust control problem in the non-linear case, see Belmiloudi [38, 39, 42, 43].

Studied Models with Delays

In this section we consider non-linear parabolic partial differential delay equations of the form

$$\begin{aligned} &\frac{\partial U}{\partial t} + AU + F(x,t,U) + K(\varpi,U) \\ &\qquad + \sum_{i=1,n} d_i(x,t)U(x,t-e_i(t)) = f(x,t) \quad (x,t) \in \mathcal{Q}, \\ &U(x,t') = R_0(x,t') \quad (x,t') \in \mathcal{Q}_0 = \Omega \times [-\delta(0),0), \\ &U(x,0) = S_0(x) \quad x \in \Omega, \end{aligned} \tag{8.168}$$

where $\mathcal{Q}$ is the cylinder $\Omega \times (0,T)$ with Ω a boundary subset of $\mathbb{R}^m$, $m \geq 1$, Γ its sufficiently regular boundary, and Ω is totally on one side of Γ. The operator A: $\mathcal{D} \longrightarrow \mathcal{D}'$ is elliptic and selfadjoint with $\langle Av, v\rangle \geq \nu \parallel v \parallel_{\mathcal{D}}^2$, $\langle Au, v\rangle \leq M \parallel u \parallel_{\mathcal{D}} \parallel v \parallel_{\mathcal{D}}$ $\forall u, v \in \mathcal{D}$ ($\langle .,. \rangle$ is the duality pairing on $\mathcal{D}'$ and $\mathcal{D}$, $\parallel . \parallel_{\mathcal{D}}$ is the norm on $\mathcal{D}$, $\parallel . \parallel_*$ is the dual norm on $\mathcal{D}'$ and $\nu, M > 0$ are constants). The operator $K(\varpi, .)$: $\mathcal{D} \longrightarrow L^2(\Omega)$ is linear and satisfies (7.81), ϖ is a given sufficiently regular function. The operator F: $\mathcal{Q} \times \mathbb{R} \longrightarrow \mathbb{R}$ is a Nemytsky operator on $L^2(\mathcal{Q})$ satisfying Assumptions (7.78). The Banach space $\mathcal{D}$ of functions on Ω (satisfying the boundary conditions) such that $\mathcal{D} \subset L^2(\Omega) \subset \mathcal{D}'$ satisfies Assumption (7.79). The functions $(d_i, i = 1, n)$: $\overline{\mathcal{Q}} \longrightarrow \mathbb{R}$ are given $\mathcal{C}^\infty$ on $\overline{\mathcal{Q}}$, the functions $(e_i, i = 1, n)$ are sufficiently regular and represent multiple time-varying delays, and $\delta(0) = \max_{i=1,n}(e_i(0))$.

We introduce the following spaces: $\mathcal{H} = L^\infty(0,T,L^2(\Omega))$, $\mathcal{V} = L^2(0,T,\mathcal{D})$, $\mathcal{W} = \mathcal{H} \cap \mathcal{V}$ and $\mathcal{H}_1 = H^1(0,T;\mathcal{D}') = \{v : \partial v/\partial t \in L^2(0,T;\mathcal{D}')\}$.

We state the following assumptions for the delay functions e_i and for the functions r_i defined by $r_i : t \in [0,T) \longrightarrow r_i(t) = t - e_i(t)$, for $i = 1, n$:

$$\begin{aligned} &(r_i, i = 1, n) \quad \text{are strictly increasing functions and} \\ &(e_i, i = 1, n) \quad \text{are } \mathcal{C}^1 \text{ non-negative functions on } [0,T). \end{aligned} \tag{8.169}$$

Remark 8.51. According to Assumptions (8.169), we have the existence of the inverse functions $(f_i, i = 1, n)$ of the functions $(r_i, i = 1, n)$. ♢

Thus we define the following subdivision:

$$\begin{aligned} s_{-1} &= -\delta(0), s_0 = 0, \\ s_j &= \min_{i=1,n} (f_i(s_{j-1})) \quad \text{for all } j \in \mathbb{N} - \{0\}. \end{aligned} \tag{8.170}$$

Finally, we introduce the following notations: $I_j =]s_{-1}, s_j[$ and $\mathcal{Q}_j = \Omega \times I_j$ for $j \in \mathbb{N}$, and $T_j = s_j - s_{j-1}$, for $j \in \mathbb{N} - \{0\}$.

Remark 8.52. According to Assumptions (8.169), we can prove easily the following results:

(*i*) the sequence $(s_j)_{j\in\mathbb{N}}$ is strictly increasing and $s_j \leq T$, for all $j \in \mathbb{N}$

(*ii*) for any integer $j \geq 2$, if $t \in (s_{j-1}, s_j)$ then for all $i = 1, \dots, n$,

$$r_i(t) \leq s_{j-1}$$

(*iii*) if $t \in (s_0, s_1)$ then for all $i = 1, \dots, n$, $r_i(t) \in (s_{-1}, s_0)$. ♢

Since the functions $(d_i, i = 1, n)$ are in $\mathcal{C}(\overline{\mathcal{Q}})$ and $(f_i', i = 1, n)$ are in $\mathcal{C}([0,T])$, then we can introduce the following constants $(D_i, i = 1, n)$, C_∞ and E_∞ such that

$$\begin{aligned} D_i &= \| d_i \|_\infty (\| f_i' \|_\infty)^{1/2} \text{ for } i = 1, \dots, n, \\ C_\infty &= \sum_{i=1}^{n} D_i, \\ E_\infty &= (1 + C_\infty)^2. \end{aligned} \tag{8.171}$$

8.7.2 Existence and Uniqueness of the Solution

Theorem 8.53. *Let F be an operator that satisfies the assumptions (7.78). Let S_0, R_0 and f be given such that $S_0 \in L^2(\Omega)$, $R_0 \in L^2(\mathcal{Q}_0)$ and $f \in L^2(\mathcal{Q})$. Then the problem (8.168) admits a unique solution U such that $U \in \mathcal{W} \cap \mathcal{H}_1 \subset \mathcal{C}([0,T]; L^2(\Omega))$ and there exists a constant $C > 0$ such that*

$$\| U \|^2_{\mathcal{W}\cap\mathcal{H}_1} \leq C \exp((\gamma + E_\infty)T)(\| S_0 \|^2_{L^2} + \| f \|^2_{L^2(\mathcal{Q})} + \| R_0 \|^2_{L^2(\mathcal{Q}_0)}),$$

where E_∞ is given by (8.171).

Proof. The proof of this theorem can be found in Belmiloudi [38]. Let us recall the proof for the reader's convenience.

To prove the existence of a unique solution on $\mathcal{Q}$, we first establish the existence of a unique solution on $\mathcal{Q}_j, j \geq 1$ and obtain some estimates. We solve the problem on $\mathcal{Q}_1$ and obtain the existence of a unique solution on $\mathcal{Q}_1$. Then, the existence of a unique solution on $\mathcal{Q}_2$ is proved by using the solution on $\mathcal{Q}_1$ to generate the initial data at s_1. This advancing process is repeated for $\mathcal{Q}_3, \mathcal{Q}_4, \dots, \mathcal{Q}_j, \mathcal{Q}_{j+1}, \dots$ until the final set is reached (in this way the solution in the step j determines the solution in the step $j+1$). Hereafter, the solution on $\mathcal{Q}_j$ will be denoted by U_j for $j = 1, \dots$.

We take $U_0(., s_0) = S_0$ and

$$U_0(., r_i(t)) = R_0(., r_i(t)), \text{ for all } i = 1, n \text{ and } t \in (s_0, s_1). \tag{8.172}$$

Now we introduce the following problems $(\mathcal{P}_j)$ for $j \in \mathbb{N} - \{0\}$

$$\begin{aligned}&\frac{\partial V_j}{\partial t} + AV_j + F(., V_j) + K(\varpi, V_j) = g_j \quad \text{on } \Omega \times (s_{j-1}, s_j),\\ &V_j(x, s_{j-1}) = U_{j-1}(x, s_{j-1}) \in L^2(\Omega),\end{aligned}$$

where $g_j(x,t) = f(x,t) - \sum_{i=1,n} d_i(x,t) U_{j-1}(x, r_i(t))$ and $U_{j-1} \in L^2(\mathcal{Q}_{j-1})$. Since $f \in L^2(\mathcal{Q})$, $U_{j-1} \in L^2(\mathcal{Q}_{j-1})$ and $(d_i, i = 1, n) \in \mathcal{C}^\infty(\overline{\mathcal{Q}})$, and according to the remark (8.52) we have that $g_j \in L^2(s_{j-1}, s_j, L^2(\Omega))$.

According to Propositions 7.22 and 7.23 and Lemma 6.6, the problem $(\mathcal{P}_j)$ admits a unique solution $V_j \in \mathcal{W}^{(j)} \cap \mathcal{H}_1^{(j)} \subset \mathcal{C}([s_{j-1}, s_j]; L^2(\Omega))$, where

$$\begin{aligned}&\mathcal{H}^{(j)} = L^\infty(s_{j-1}, s_j, L^2), \; \mathcal{V}^{(j)} = L^2(s_{j-1}, s_j; \mathcal{D}), \; \mathcal{W}^{(j)} = \mathcal{H}^{(j)} \cap \mathcal{V}^{(j)},\\ &\mathcal{H}_1^{(j)} = H^1(s_{j-1}, s_j; \mathcal{D}') = \{v : \frac{\partial v}{\partial t} \in L^2(s_{j-1}, s_j; \mathcal{D}')\}.\end{aligned}$$

We can then extend the result to the cylinder set $\mathcal{Q}_{j+1}$ by taking $U_j = U_{j-1}$ on $\mathcal{Q}_{j-1}$ and $U_j = V_j$ on $\Omega \times (s_{j-1}, s_j)$.

We shall now prove some estimates of V_j in $\mathcal{W}_j$. Multiplying the first equation of $(\mathcal{P}_j)$ by $2V_j$ and integrating over $\Omega \times (s_{j-1}, t)$, for $t \in (s_{j-1}, s_j)$, this gives, according to the relations (7.81) and (7.82),

$$\begin{aligned}\| V_j(., t) \|_{L^2}^2 + \nu \int_{s_{j-1}}^t \| V_j \|_{\mathcal{D}}^2 \mathrm{d}s \leq \gamma \int_{s_{j-1}}^t \| V_j \|_{L^2}^2 \mathrm{d}s&\\ + \| U_{j-1}(., s_{j-1}) \|_{L^2}^2 + 2 \int_{s_{j-1}}^t \langle g_j, V_j \rangle \mathrm{d}s.&\end{aligned} \tag{8.173}$$

According to the regularity of $d_i, i = 1, n$ and Remark 8.52 we obtain

$$\begin{aligned}&\int_{s_{j-1}}^t \langle d_i(., s) U_{j-1}(., r_i(s)), V_j \rangle \mathrm{d}s\\ &\qquad \leq \| d_i \|_\infty \left(\int_{s_{j-1}}^t \| U_{j-1}(., r_i) \|_{L^2}^2 \mathrm{d}s \right)^{1/2} \left(\int_{s_{j-1}}^t \| V_j \|_{L^2}^2 \mathrm{d}s \right)^{1/2}.\end{aligned}$$

Put $a = r_i(s)$, we have $s = f_i(a)$ and then $\mathrm{d}s = f_i'(a)\mathrm{d}a$. So, according to Remarks 8.52 and the relations (8.171),

$$\int_{s_{j-1}}^{t} \langle d_i(.,s)U_{j-1}(.,r_i(s)), V_j\rangle \mathrm{d}s \\ \leq D_i \parallel U_{j-1} \parallel_{L^2(\mathcal{Q}_{j-1})} (\int_{s_{j-1}}^{t} \parallel V_j \parallel_{L^2}^2 \mathrm{d}s)^{1/2}. \tag{8.174}$$

Since

$$g_j(x,t) = f(x,t) - \sum_{i=1,n} d_i(x,t)U_{j-1}(x, r_i(t)),$$

we obtain then, according to (8.174) and (8.171)

$$\int_{s_{j-1}}^{t} \langle g_j, V_j\rangle \mathrm{d}s \leq \parallel f \parallel_{L^2(\mathcal{Q}_j)} (\int_{s_{j-1}}^{t} \parallel V_j \parallel_{L^2}^2 \mathrm{d}s)^{1/2} \\ +C_\infty \parallel U_{j-1} \parallel_{L^2(\mathcal{Q}_{j-1})} (\int_{s_{j-1}}^{t} \parallel V_j \parallel_{L^2}^2 \mathrm{d}s)^{1/2}.$$

Consequently, the relation (8.173) becomes

$$\parallel V_j \parallel_{L^2}^2 + \nu \int_{s_{j-1}}^{t} \parallel V_j \parallel_{\mathcal{D}}^2 \mathrm{d}s \\ \leq \gamma \int_{s_{j-1}}^{t} \parallel V_j \parallel_{L^2}^2 \mathrm{d}s + \parallel U_{j-1}(., s_{j-1}) \parallel_{L^2}^2 \\ +2 \parallel f \parallel_{L^2(\mathcal{Q}_j)} (\int_{s_{j-1}}^{t} \parallel V_j \parallel_{L^2}^2 \mathrm{d}s)^{1/2} \\ +2C_\infty \parallel U_{j-1} \parallel_{L^2(\mathcal{Q}_{j-1})} (\int_{s_{j-1}}^{t} \parallel V_j \parallel_{L^2}^2 \mathrm{d}s)^{1/2}. \tag{8.175}$$

Applying Gronwall's lemma now gives

$$\parallel V_j \parallel_{\mathcal{W}_j}^2 \leq D_\infty^{(j)} (\parallel U_{j-1}(., s_{j-1}) \parallel_{L^2}^2 + \parallel f \parallel_{L^2(\mathcal{Q}_j)}^2 + \parallel U_{j-1} \parallel_{L^2(\mathcal{Q}_{j-1})}^2),$$

where $D_\infty^{(j)} = \max(1, C_\infty^2)\exp((\gamma+2)T_j)$.

We shall now prove the existence and uniqueness result of the problem (8.168). We observe that for $j = 1$, we have

$$g_1(x,t) = f(x,t) - \sum_{i=1,n} d_i(x,t)U_0(x, r_i(t))$$

and according to (8.172) we can deduce that

$$g_1(x,t) = f(x,t) - \sum_{i=1,n} d_i(x,t)R_0(x, r_i(t)).$$

Consequently, $g_1 \in L^2(\mathcal{Q}_0)$. By using the previous result, the second and the third relations of (8.168) we can deduce that the problem $(\mathcal{P}_1)$ admits a unique solution V_1 such that $V_1 \in \mathcal{W}_1$ and $V_1 \in \mathcal{C}([s_0, s_1], L^2(\Omega))$. We obtain then the solution U_1. We inject now U_1 in the problem $(\mathcal{P}_2)$ and by using the same approach, we obtain the existence and the uniqueness of $V_2 \in \mathcal{W}_2$ and $V_2 \in \mathcal{C}([s_1, s_2], L^2(\Omega))$ (solution of $(\mathcal{P}_2)$).

We can now iterate the process for any domain $\mathcal{Q}_j$, for $j \geq 1$ (until the procedure covers the whole cylinder $\mathcal{Q}$) and we obtain the existence and the uniqueness of $V_j \in \mathcal{W}_j$ and $V_j \in \mathcal{C}([s_{j-1}, s_j], L^2(\Omega))$, solution of $(\mathcal{P}_j)$. We deduce then the existence and the uniqueness of the solution $U \in \mathcal{W}$ of (8.168) such that $U|_{\mathcal{Q}_j} = U_j, j \geq 1$.

We are now going to prove the estimate given in the proposition. Multiplying the first part of (8.168) by $2U$ and integrating over $\Omega \times (0, t)$, for $t \in (0, T)$, this gives

$$\begin{aligned} \| U \|_{L^2}^2 + \nu \int_0^t \| U \|_{\mathcal{D}}^2 \, \mathrm{d}s \\ \leq \gamma \int_0^t \| U \|_{L^2}^2 \, \mathrm{d}s + 2 \int_0^t \langle g, U \rangle \mathrm{d}s + \| S_0 \|_{L^2}^2, \end{aligned} \tag{8.176}$$

where

$$g(x,t) = f(x,t) - \sum_{i=1,n} d_i(x,t) U(x, r_i(t)). \tag{8.177}$$

According to the expression (8.177) of the function g, the second relation of (8.168) and the estimate (8.180) given in Lemma 8.54 (see below), we can deduce that

$$\begin{aligned} \int_0^t \langle g, U \rangle \mathrm{d}s \leq C_\infty \int_0^t \| U \|_{L^2}^2 \, \mathrm{d}s + \| f \|_{L^2(\mathcal{Q})} \left(\int_0^t \| U \|_{L^2}^2 \, \mathrm{d}s \right)^{1/2} \\ + C_\infty \| R_0 \|_{L^2(\mathcal{Q}_0)} \left(\int_0^t \| U \|_{L^2}^2 \, \mathrm{d}s \right)^{1/2}. \end{aligned}$$

Consequently, the relation (8.176) becomes

$$\begin{aligned} \| U \|_{L^2}^2 + \nu \int_0^t \| U \|_{\mathcal{D}}^2 \, \mathrm{d}s \leq \| S_0 \|_{L^2}^2 + (\gamma + 2C_\infty) \int_0^t \| U \|_{L^2}^2 \, \mathrm{d}s \\ + 2 \| f \|_{L^2(\mathcal{Q})} \left(\int_0^t \| U \|_{L^2}^2 \, \mathrm{d}s \right)^{1/2} \\ + 2C_\infty \| R_0 \|_{L^2(\mathcal{Q}_0)} \left(\int_0^t \| U \|_{L^2}^2 \, \mathrm{d}s \right)^{1/2}. \end{aligned}$$

By using Gronwall's lemma we can deduce

$$\| U \|_{\mathcal{W}}^2 \leq \exp((\gamma + E_\infty)T)(\| S_0 \|_{L^2}^2 + \| f \|_{L^2(\mathcal{Q})}^2 + \| R_0 \|_{L^2(\mathcal{Q}_0)}^2), \tag{8.178}$$

where E_∞ is given by (8.171).

By using the system (8.168) and the relation (8.178), we can obtain easily that the solution U satisfies the following estimate:

$$\| U \|^2_{\mathcal{H}_1} \leq C \exp((\gamma+E_\infty)T)(\| S_0 \|^2_{L^2} + \| f \|^2_{L^2(\mathcal{Q})} + \| R_0 \|^2_{L^2(\mathcal{Q}_0)}). \quad (8.179)$$

This completes the proof. □

We shall now prove the lemma previously used.

Lemma 8.54. *Let $v \in L^2(\mathcal{Q}_0 \cup \mathcal{Q})$ such that $v = k_0$ on $\mathcal{Q}_0$, and $w \in L^2(\mathcal{Q})$ we have for all $t \in (0, T)$*

$$\begin{aligned} &\int_0^t \langle d_i(., s)v(., r_i(s)), w\rangle \mathrm{d}s \\ &\quad \leq D_i \left(\| k_0 \|_{L^2(\mathcal{Q}_0)} + (\int_0^t \| v \|^2_{L^2} \mathrm{d}s)^{1/2} \right) (\int_0^t \| w \|^2_{L^2} \mathrm{d}s)^{1/2}, \end{aligned} \quad (8.180)$$

for all $i = 1, \ldots, n$.

Proof. According to the regularity of $d_i, i = 1, n$ and to Remark 8.52 we obtain

$$\begin{aligned} &\int_0^t \langle d_i(., s)v(., r_i(s)), w\rangle \mathrm{d}s \\ &\qquad \leq \| d_i \|_\infty (\int_0^t \| v(., r_i) \|^2_{L^2} \mathrm{d}s)^{1/2} (\int_0^t \| w \|^2_{L^2} \mathrm{d}s)^{1/2}. \end{aligned}$$

Set $a = r_i(s)$, according to Remark 8.51 we can deduce that $s = f_i(a)$ and then $\mathrm{d}s = f_i'(a)\mathrm{d}a$. So, according to (8.171)

$$\int_0^t \langle d_i(., s)v(., r_i(s)), w\rangle \mathrm{d}s \leq D_i (\int_{-e_i(0)}^{t-e_i(t)} \| v \|^2_{L^2} \mathrm{d}a)^{1/2} (\int_0^t \| w \|^2_{L^2} \mathrm{d}s)^{1/2}.$$

Since $-\delta(0) \leq -e_i(0)$, for all $i = 1, \ldots, n$ and according to the assumptions (8.169) we can deduce

$$\begin{aligned} &\int_0^t \langle d_i(., s)v(., r_i(s)), w\rangle \mathrm{d}s \\ &\qquad \leq D_i \left(\| k_0 \|_{L^2(\mathcal{Q}_0)} + (\int_0^t \| v \|^2_{L^2} \mathrm{d}a)^{1/2} \right) (\int_0^t \| w \|^2_{L^2} \mathrm{d}s)^{1/2}. \end{aligned}$$

This completes the proof. □

The following proposition shows the Lipschiz continuous result.

Proposition 8.55. *Let F be an operator that satisfies the assumptions (7.78). Let S_{01}, S_{02} be two functions in $L^2(\Omega)$, R_{01}, R_{02} be two functions in $L^2(\mathcal{Q}_0)$ and let f_1, f_2 be two functions of $L^2(\mathcal{Q})$. If U_1 (respectively U_2) is the solution of (8.168) with data (f_1, R_{01}, S_{01}) (respectively with data (f_2, R_{02}, S_{02})) then*

$$\| U \|^2_{\mathcal{W}\cap\mathcal{H}_1} \leq C \exp((\gamma + E_\infty)T)(\| f \|^2_{L^2(\mathcal{Q})} + \| S_0 \|^2_{L^2} + \| R_0 \|^2_{L^2(\mathcal{Q}_0)}),$$

where $U = U_1 - U_2$, $S_0 = S_{01} - S_{02}$, $f = f_1 - f_2$ and $R_0 = R_{01} - R_{02}$.

Proof. Using the same way to obtain the estimates (8.178) and (8.179) and the second assumption of (7.78), we obtain the result of the proposition. So we omit the details. □

In the following, the solution U will be treated as the target function. We are then interested in the robust regulation of the deviation of the problem from the desired target U. We analyze the full non-linear equation which models large perturbations u to the target U. We assume that U satisfies the problem (8.168) with the data (f, R_0, S_0) and U satisfies the problem (8.168) with the data $(f+g, R_0+r_0, S_0+u_0)$. Hence, we consider the system with multiple time-varying delays

$$\begin{aligned} &\frac{\partial u}{\partial t} + Au + F(.,u+U) - F(.,U) + K(\varpi, u) \\ &\qquad + \sum_{i=1,n} d_i u(.,r_i) = g \ \text{ on } \mathcal{Q}, \\ &u(x,t') = r_0(x,t'), \ (x,t') \in \mathcal{Q}_0, \\ &u(x,0) = u_0(x), \quad x \in \Omega. \end{aligned} \tag{8.181}$$

If we set $\tilde{F}(.,y) = F(.,y+U) - F(.,U)$ then (8.181) reduce to

$$\begin{aligned} &\frac{\partial u}{\partial t} + Au + \tilde{F}(.,u) + K(\varpi,u) + \sum_{i=1,n} d_i u(.,r_i) = g \ \text{ on } \mathcal{Q}, \\ &u(x,t') = r_0(x,t'), \ (x,t') \in \mathcal{Q}_0, \\ &u(x,0) = u_0(x), \quad x \in \Omega. \end{aligned} \tag{8.182}$$

In the following we assume that the solution U satisfies the regularity $U \in \mathcal{W} \cap \mathcal{C}([0,T]; L^2(\Omega))$.

Similarly as in previous section, we verify easily that the operator $\tilde{F}$ satisfies the same hypotheses that the operator F, *i.e.*, (7.78). For simplicity of future reference, we omit the "$\tilde{\ }$" on $\tilde{F}$ for (8.182).

Problem (8.182) is similar to (8.168), and consequently the following proposition holds.

Proposition 8.56. *Let F be an operator that satisfies the assumptions (7.78). (i) Assume that $g \in L^2(\mathcal{Q})$, $r_0 \in L^2(\mathcal{Q}_0)$ and $u_0 \in L^2(\Omega)$. Then the problem (8.182) admits a unique solution u such that $u \in \mathcal{W} \cap \mathcal{H}_1$ with the following estimate:*

$$\| u \|^2_{\mathcal{W}\cap\mathcal{H}_1} \leq C \exp((\gamma + E_\infty)T)(\| g \|^2_{L^2(\mathcal{Q})} + \| u_0 \|^2_{L^2} + \| r_0 \|^2_{L^2(\mathcal{Q}_0)}).$$

(ii) Let u_{01}, u_{02} be two functions in $L^2(\Omega)$, r_{01}, r_{02} be two functions in $L^2(\mathcal{Q}_0)$ and let g_1, g_2 be two functions of $L^2(\mathcal{Q})$. If u_1 (respectively u_2) is the solution of (8.182) with data (g_1, r_{01}, u_{01}) (respectively with data (g_2, r_{02}, u_{02})) then

$$\| u \|^2_{\mathcal{W}\cap\mathcal{H}_1} \leq C \exp((\gamma + E_\infty)T)(\| g \|^2_{L^2(\mathcal{Q})} + \| u_0 \|^2_{L^2} + \| r_0 \|^2_{L^2(\mathcal{Q}_0)}),$$

where $u = u_1 - u_2$, $u_0 = u_{01} - u_{02}$, $g = g_1 - g_2$ and $r_0 = r_{01} - r_{02}$.

8.7.3 The Control Framework

In order to illustrate our robust control problem, we assume here that the disturbance $\psi \in L^2(\Omega)$ is in the initial condition u_0 and the control $\phi \in L^2(\mathcal{Q})$ is in the forcing g, *i.e.*, $u_0 = B_2\psi$ and $g = B_1\phi$, where $B_i, i = 1,2$ are given bounded operators on $L^2(\mathcal{Q})$ and $L^2(\Omega)$, respectively, satisfying the assumption (8.107) and the estimate (8.108).

Remark 8.57. For other types of controls and disturbances, the reader should refer to Belmiloudi [38]. ◇

The function u is then assumed to be related to the disturbance ψ and control ϕ through the problem (8.182)

$$
\begin{aligned}
&\frac{\partial u}{\partial t} + Au + F(.,u) + K(\varpi, u) \\
&\qquad\qquad + \sum_{i=1,n} d_i u(., r_i) = B_1\phi \ \text{ on } \mathcal{Q}, \\
&u(x,t') = r_0(x,t'), \quad (x,t') \in \mathcal{Q}_0, \\
&u(.,0) = B_2\psi \ \text{ on } \Omega.
\end{aligned}
\tag{8.183}
$$

To obtain the regularity of Proposition 8.56, we suppose that the given function r_0 is in $L^2(\mathcal{Q}_0)$. Let $\mathbf{U} : (\phi,\psi) \longrightarrow u = \mathbf{U}(\phi,\psi)$ be the map: $L^2(\mathcal{Q}) \times L^2(\Omega) \longrightarrow \mathcal{W}$ defined by (8.183) and introducing the cost functional J by

$$
\begin{aligned}
J(\phi,\psi) = \frac{1}{2} \parallel \mathcal{C}(u - u_{\text{obs}}) \parallel^2_{L^2(\mathcal{Q})} + \frac{\mu}{2} \parallel u(T) - v_{\text{obs}} \parallel^2_{L^2(\Omega)} \\
+ \frac{\alpha}{2} \parallel \phi \parallel^2_{L^2(\mathcal{Q})} - \frac{\beta}{2} \parallel \psi \parallel^2_{L^2(\Omega)},
\end{aligned}
\tag{8.184}
$$

where $\mu, \alpha, \beta > 0$ are fixed, $(u_{\text{obs}}, v_{\text{obs}}) \in \mathcal{V} \times L^2(\Omega)$ is the observation (given) and $\mathcal{C}$ is an unbounded, linear operator on $L^2(\Omega)$ satisfying Assumption (7.91).

We want to minimize the functional J with respect to ϕ and maximize J with respect to ψ, *i.e.*, to find $(\phi^*, \psi^*) \in \mathcal{U}_{\text{ad}} \times \mathcal{V}_{\text{ad}}$ such that

$$
J(\phi^*, \psi) \le J(\phi^*, \psi^*) \le J(\phi, \psi^*), \forall (\phi,\psi) \in \mathcal{U}_{\text{ad}} \times \mathcal{V}_{\text{ad}}, \tag{8.185}
$$

where $\mathcal{U}_{\text{ad}}$ and $\mathcal{V}_{\text{ad}}$ are (given) non-empty, closed, convex, bounded subsets of $L^2(\mathcal{Q})$ and $L^2(\Omega)$, respectively.

Proposition 8.58. *Let F be an operator that satisfies the assumptions (7.78). Then the function $\mathbf{U} : (\phi,\psi) \longrightarrow u = \mathbf{U}(\phi,\psi)$ solution of (8.183) is continuously F-differentiable from $L^2(\mathcal{Q}) \times L^2(\Omega)$ to $\mathcal{W}$ with the derivative $\mathbf{U}'(\phi,\psi) : (h_1,h_2) \longrightarrow w$ given by the linear parabolic problem with multiple time-varying delays:*

$$\begin{aligned}
&\frac{\partial w}{\partial t} + Aw + G(.,\mathbf{U}(\phi,\psi))w + K(\varpi,w) \\
&\qquad\qquad + \sum_{i=1,n} d_i(x,t)w(x,r_i(t)) = B_1h_1 \quad on\ \mathcal{Q}, \\
&w(x,t') = 0, \quad (x,t') \in \mathcal{Q}_0, \\
&w(.,0) = B_2h_2 \quad on\ \Omega,
\end{aligned} \tag{8.186}$$

where the operator G is the differential of the operator F.

Moreover, for all $\Phi_i = (\phi_i,\psi_i) \in L^2(\mathcal{Q}) \times L^2(\Omega)$, for $i = 1,2$, the following estimates hold:

(i) $\| \mathbf{U}'(\phi_1,\psi_1) \|_{\mathcal{L}(L^2(\mathcal{Q})\times L^2(\Omega);\mathcal{W})} \leq C\sqrt{M_\infty}b$

(ii) $\| \mathbf{U}'(\phi_1,\psi_1) - \mathbf{U}'(\phi_2,\psi_2) \|_{\mathcal{L}(L^2(\mathcal{Q})\times L^2(\Omega);\mathcal{W})}$

$$\leq CM_\infty^2 \lambda c_e^2 b^2 \| \Phi_1 - \Phi_2 \|_{L^2(\mathcal{Q})\times L^2(\Omega)},$$

where E_∞ is given by (8.171), $M_\infty = \exp((\gamma + E_\infty)T)$ and the constants λ, c_e and b are given by the relations (7.78),(7.79) and (8.108).

Proof. The proof is obtained by using similar technique as to obtain the results of Proposition 8.22 and by taking into account the estimate (8.180) given in Lemma 8.54. So we omit the details. □

Proposition 8.59. *Let F be an operator that satisfies the assumptions (7.78). Then for each $t \in [0,T]$, the function $\mathbf{V}_t : (\phi,\psi) \longrightarrow u(t) = \mathbf{V}_t(\phi,\psi)$ solution of (8.183) is uniformly Lipschitz continuous and continuously F-differentiable from $L^2(\mathcal{Q}) \times L^2(\Omega)$ to $L^2(\Omega)$ with the derivative $\mathbf{V}_t'(\phi,\psi) : (h_1,h_2) \longrightarrow w(t)$, where the function w is given by the parabolic system with time-varying delays*

$$\begin{aligned}
&\frac{\partial w}{\partial t} + Aw + G(.,\mathbf{V}_t(\phi,\psi))w + K(\varpi,w) \\
&\qquad\qquad + \sum_{i=1,n} d_i w(.,r_i) = B_1h_1 \quad on\ \mathcal{Q}, \\
&w(x,t') = 0 \quad (x,t') \in \mathcal{Q}_0, \\
&w(x,0) = B_2h_2 \quad x \in \Omega.
\end{aligned} \tag{8.187}$$

Moreover, for all $\Phi_i = (\phi_i,\psi_i) \in L^2(\mathcal{Q}) \times L^2(\Omega)$, for $i = 1,2$, the following estimates hold:

(i) $\| \mathbf{V}_t'(\phi_1,\psi_1) \|_{\mathcal{L}(L^2(\mathcal{Q})\times L^2(\Omega),;L^2(\Omega))} \leq Cb\sqrt{M_\infty}$

(ii) $\| \mathbf{V}_t'(\phi_1,\psi_1) - \mathbf{V}_t'(\phi_2,\psi_2) \|_{\mathcal{L}(L^2(\mathcal{Q})\times L^2(\Omega);L^2(\Omega))}$

$$\leq CM_\infty^2 \lambda c_e^2 b^2 \| \Phi_1 - \Phi_2 \|_{L^2(\mathcal{Q})\times L^2(\Omega)},$$

where E_∞ is given by (8.171), $M_\infty = \exp((\gamma + E_\infty)T)$ and the constants λ, c_e and b are given by the relations (7.78),(7.79) and (8.108).

Proof. The proof of this proposition is a direct consequence of Propositions 8.56 and 8.58. □

Remark 8.60. Problem (8.187) is exactly the problem (8.186). □

Proposition 8.61. *Let F be an operator that satisfies the assumptions (7.78). Then the maps $\mathbf{U}$ and $\mathbf{V}_t$ defined by (8.183) are continuous from the weak topology of $L^2(\mathcal{Q}) \times L^2(\Omega)$ to the strong topology of $L^2(\mathcal{Q})$ and the weak topology of $L^2(\Omega)$, respectively.*

Proof. The proof of this proposition is similar to Proposition 8.24, so we omit the details. □

We can now obtain the existence of an optimal solution of the robust control problem.

Theorem 8.62. *Let F be an operator that satisfies the assumptions (7.78). Then, for α and β sufficiently large, there exists $(\phi^*, \psi^*) \in L^2(\mathcal{Q}) \times L^2(\Omega)$ and $u^* \in \mathcal{W}$ such that (ϕ^*, ψ^*) is defined by (8.185) and $u^* = \mathbf{U}(\phi^*, \psi^*)$ is a solution of (8.183).*

Proof. The proof of this theorem can be obtained by using similar existence result showed in Theorem 8.25, by taking advantage of the results given in Propositions 8.56–8.61. □

Now we give the characterization of the solution of the robust control problem. *For simplicity we suppose that $e_i(T) = \delta(T)$ for all $i = 1, n$, where $\delta(T)$ is a given non-negative constant.* In order to characterize the robust control,[6] we introduce the following adjoint problem corresponding to the primal problem (8.183) (we denote by $u = \mathbf{U}(\phi, \psi)$ the solution of problem (8.183) with the forcing (ϕ, ψ)):

$$\begin{aligned}
&-\frac{\partial \tilde{u}}{\partial t} + A\tilde{u} + (G(.,u))^*\tilde{u} + K^*(\varpi, \tilde{u}) \\
&\qquad + \sum_{i=1,n} d_i(., f_i(t))\tilde{u}(., f_i(t)) f_i'(t) = \mathcal{C}^*\mathcal{C}(u - u_{\text{obs}}) \;\text{ on } \tilde{\mathcal{Q}}, \\
&-\frac{\partial \tilde{u}}{\partial t} + A\tilde{u} + (G(.,u))^*\tilde{u} + K^*(\varpi, \tilde{u}) = \mathcal{C}^*\mathcal{C}(u - u_{\text{obs}}) \;\text{ on } \mathcal{Q}_T, \\
&\tilde{u}(T) = \mu(u(T) - v_{\text{obs}}) \;\text{ on } \Omega,
\end{aligned} \tag{8.188}$$

where $\mathcal{C}^*$ (respectively $(G(.,u))^*$) is the adjoint of the operator $\mathcal{C}$ (respectively $G(.,u)$), $\tilde{\mathcal{Q}} = \Omega \times (0, T - \delta(T))$ and $\mathcal{Q}_T = \Omega \times (T - \delta(T), T)$.

Proposition 8.63. *Let F be an operator that satisfies Assumptions (7.78) and $u \in \mathcal{W}$. Then the solution of (8.188) is in $\mathcal{W}$ and satisfies the estimate:*

$$\| \tilde{u} \|_{\mathcal{H}}^2 + \| \tilde{u} \|_{\mathcal{V}}^2 \leq K_\infty(\| u(T) - v_{\text{obs}} \|_{L^2}^2 + \| \mathcal{C}(u - u_{\text{obs}}) \|_{L^2(\mathcal{Q})}^2),$$

where K_∞ is a constant depending on the parameters $\nu, \mu, \delta_1, \delta_2, \gamma$ and C_∞.

[6] More precisely, in order to simplify the gradient of the cost functional J which depends on the derivative of the operator solution $\mathbf{U}$.

Proof. To prove the existence of a unique solution $\tilde{u}$ of (8.188), we change the variables of problem (8.188) by reversing the sense of time, *i.e.*, $t := T - t$, and we follow the same procedure used to obtain the result of Theorem 8.53. Setting $U(.,t) = u(.,T-t)$, $U_{\text{obs}}(.,t) = u_{\text{obs}}(.,T-t)$, $W(.,t) = \varpi(.,T-t)$, $\tilde{U}(.,t) = \tilde{u}(.,T-t)$, $E_i(t) = f_i(T-t) - (T-t)$, $R_i(t) = t - E_i(t)$ and $F_i(t) = T - r_i(T-t)$, problem (8.188) can be written as

$$\begin{aligned}
&\frac{\partial \tilde{U}}{\partial t} + A\tilde{U} + (G(.,U))^*\tilde{U} + K^*(W,\tilde{U}) \\
&\qquad + \sum_{i=1,n} d_i(.,f_i(T-t))f_i'(T-t)\tilde{U}(.,R_i(t)) \\
&\qquad = \mathcal{C}^*\mathcal{C}(U - U_{\text{obs}}) \ \text{ on } \Omega \times (\delta(T),T), \qquad (8.189)\\
&\frac{\partial \tilde{U}}{\partial t} + A\tilde{U} + (G(.,U))^*\tilde{U} + K^*(W,\tilde{U}) \\
&\qquad = \mathcal{C}^*\mathcal{C}(U - U_{\text{obs}}) \ \text{ on } \Omega \times (0,\delta(T)), \\
&\tilde{U}(0) = \mu(U(0) - v_{\text{obs}}) \ \text{ on } \Omega.
\end{aligned}$$

According to the properties of the functions $(r_i, i = 1,n)$ and $(f_i, i = 1,n)$ we can prove easily that $(E_i, i = 1,n)$ are $\mathcal{C}^1$ non-negative functions, $(R_i, i = 1,n)$ are strictly increasing functions and $(F_i, i = 1,n)$ are the inverse functions of $(R_i, i = 1,n)$. Thus we define the following subdivision: $S_{-1} = 0$, $S_0 = \delta(T)$ and $\forall j \geq 1, S_j = \min_{i=1,n}(F_i(S_{j-1}))$, and we denote by $\mathcal{I}_j =]S_{-1}, S_j[$ and by $\mathcal{Q}_j = \Omega \times \mathcal{I}_j$ for $j \geq 0$.

We can now follow the same approach used to obtain the existence and uniqueness of the problem (8.168). Indeed, since $\mathcal{C}^*\mathcal{C}(U - U_{\text{obs}}) \in L^2(\mathcal{Q})$ and $(U(0) - v_{\text{obs}}) \in L^2(\Omega)$ then the existence and uniqueness of the solution $\tilde{U} \in \mathcal{W}$ of the problem (8.189) hold and, therefore, the existence and uniqueness of $\tilde{u} \in \mathcal{W}$ solution of (8.188) hold also.

Now we prove the estimate given in the proposition. Multiplying the first equation of (8.188) by $\tilde{u}$ and integrating over $(t,T) \times \Omega$, we obtain, according to $-2G(.,u) \leq \gamma_0$ and to the second and third relations of (8.188), that (using Young's inequality and Hölder's inequality)

$$\begin{aligned}
&\| \tilde{u} \|_{L^2}^2 + \nu \int_t^T \| \tilde{u} \|_{\mathcal{D}}^2 \,\mathrm{d}s \leq \mu^2 \| u(T) - v_{\text{obs}} \|_{L^2}^2 + \gamma \int_t^T \| \tilde{u} \|_{L^2}^2 \,\mathrm{d}s \\
&\quad + \frac{1}{\epsilon^2} \int_t^T \| \mathcal{C}(u - u_{\text{obs}}) \|_{L^2}^2 \,\mathrm{d}s + \epsilon^2 \int_t^T \| \mathcal{C}\tilde{u} \|_{L^2}^2 \,\mathrm{d}s \qquad (8.190)\\
&\quad + 2\sum_{i=1}^n \int_{\min(t,T-\delta(T))}^{T-\delta(T)} \langle d_i(.,f_i(s))\tilde{u}(.,f_i(s))f_i'(s), \tilde{u}(.,s)\rangle \mathrm{d}s.
\end{aligned}$$

We can now estimate the term

$$\int_{\min(t,T-\delta(T))}^{T-\delta(T)} \langle d_i(.,f_i(s))\tilde{u}(.,f_i(s))f_i'(s), \tilde{u}(.,s)\rangle \mathrm{d}s.$$

According to the regularity of d_i and f_i we then obtain

$$\int_{\min(t,T-\delta(T))}^{T-\delta(T)} \langle d_i(.,f_i(s))\tilde{u}(.,f_i(s))f_i'(s),\tilde{u}(.,s)\rangle \mathrm{d}s$$
$$\leq D_i \left(\int_{\min(t,T-\delta(T))}^{T-\delta(T)} \| \tilde{u}(.,f_i(s)) \|_{L^2} f_i'(s)\mathrm{d}s \right)^{1/2} \left(\int_t^T \| \tilde{u} \|_{L^2}^2 \mathrm{d}s\right)^{1/2}.$$

Set $a = f_i(s)$, we have $s = r_i(a)$, $\mathrm{d}a = f_i'(s)\mathrm{d}s$ and then, since $r_i(T) = T - \delta(T)$ and $r_i(t) \leq t$,

$$\begin{aligned} &\int_{\min(t,T-\delta(T))}^{T-\delta(T)} \langle d_i(.,f_i(s))\tilde{u}(.,f_i(s))f_i'(s),\tilde{u}(.,s)\rangle \mathrm{d}s \\ &\qquad \leq D_i \int_t^T \| \tilde{u} \|_{L^2}^2 \, da. \end{aligned} \tag{8.191}$$

According to (7.91) and (8.191) the relation (8.190) becomes

$$\begin{aligned} \| \tilde{u} \|_{L^2}^2 + \nu \int_t^T \| \tilde{u} \|_{\mathcal{D}}^2 \,\mathrm{d}s &\leq \mu^2 \| u(T) - v_{\text{obs}} \|_{L^2}^2 \\ &+ (\gamma + \epsilon^2\delta_1 + 2C_\infty) \int_t^T \| \tilde{u} \|_{L^2}^2 \,\mathrm{d}s \\ &+ \frac{1}{\epsilon^2} \int_t^T \| \mathcal{C}(u - u_{\text{obs}}) \|_{L^2}^2 \,\mathrm{d}s + \epsilon^2\delta_2 \int_t^T \| \tilde{u} \|_{\mathcal{D}}^2 \,\mathrm{d}s. \end{aligned} \tag{8.192}$$

By choosing $\epsilon^2 = \nu/(2\delta_2)$,

$$\begin{aligned} \| \tilde{u} \|_{L^2}^2 + \frac{\nu}{2} \int_t^T \| \tilde{u} \|_{\mathcal{D}}^2 \,\mathrm{d}s &\leq \mu^2 \| u(T) - v_{\text{obs}} \|_{L^2}^2 \\ &+ \left(\frac{\nu\delta_1}{2\delta_2} + \gamma + 2C_\infty \right) \int_t^T \| \tilde{u} \|_{L^2}^2 \,\mathrm{d}s + \frac{2\delta_2}{\nu} \| \mathcal{C}(u - u_{\text{obs}}) \|_{L^2(\mathcal{Q})}^2 . \end{aligned}$$

By using Gronwall's formula we then have

$$\| \tilde{u} \|_{\mathcal{H}}^2 + \| \tilde{u} \|_{\mathcal{V}}^2 \leq K_\infty(\| u(T) - v_{\text{obs}} \|_{L^2}^2 + \| \mathcal{C}(u - u_{\text{obs}}) \|_{L^2(\mathcal{Q})}^2),$$

where K_∞ is a constant depending on the parameters ν, μ, δ_1, δ_2, γ and C_∞.
This completes the proof. □

We will denote in the sequel by $\tilde{\mathbf{U}} : (\phi,\psi) \longrightarrow \tilde{u} = \tilde{\mathbf{U}}(\phi,\psi)$ the map defined by (8.188). We can now give the optimality system for the robust control problem (8.185).

Theorem 8.64. *Let F be an operator that satisfies the assumptions (7.78), α and β sufficiently large, and an optimal solution $(\phi^*,\psi^*,u^*) \in \mathcal{U}_{\text{ad}} \times \mathcal{V}_{\text{ad}} \times \mathcal{W}$*

such that (ϕ^*, ψ^*) *is defined by (8.185) and* $u^* = \mathbf{U}(\phi^*, \psi^*)$ *is a solution of (8.183). Then*

$$\begin{aligned} &\int_0^T \int_\Omega (\alpha\phi^* + B_1^* \tilde{u})(\phi - \phi^*) \mathrm{d}x\mathrm{d}t \geq 0, \\ &\int_\Omega (-\beta\psi^* + B_2^* \tilde{u}(0))(\psi - \psi^*) \mathrm{d}x \leq 0, \end{aligned} \tag{8.193}$$

for all $(\phi, \psi) \in \mathcal{U}_{\mathrm{ad}} \times \mathcal{V}_{\mathrm{ad}}$, *where* $\tilde{u} = \tilde{\mathbf{U}}(\phi^*, \psi^*)$ *is the solution of the adjoint problem (8.188).*

Proof. The cost functional J is a composition of differentiable maps then J is differentiable and we have, for all $\mathbf{h} = (h_1, h_2) \in \mathcal{U}_{\mathrm{ad}} \times \mathcal{V}_{\mathrm{ad}}$

$$\begin{aligned} J'(\phi, \psi)\mathbf{h} = &\int_0^T \langle C(u - u_{\mathrm{obs}}), C\mathbf{U}'(\phi, \psi)\mathbf{h} \rangle \mathrm{d}t + \langle \mu(u(T) - v_{\mathrm{obs}}), \mathbf{V}'_T(\phi, \psi)\mathbf{h} \rangle \\ &+ \int_0^T \langle \alpha\phi, h_1 \rangle \mathrm{d}t - \langle \beta\psi, h_2 \rangle, \end{aligned}$$

where $\mathbf{U}'(\phi, \psi)\mathbf{h} = w$ and $\mathbf{V}'_T(\phi, \psi)\mathbf{h} = w(T)$, where w is the solution of (8.187). Then

$$\begin{aligned} J'(\phi, \psi)\mathbf{h} = &\int_0^T \langle \mathcal{C}^*\mathcal{C}(u - u_{\mathrm{obs}}), \mathbf{U}'(\phi, \psi)\mathbf{h} \rangle \mathrm{d}t + \langle \mu(u(T) - v_{\mathrm{obs}}), \mathbf{V}'_T(\phi, \psi)\mathbf{h} \rangle \\ &+ \int_0^T \langle \alpha\phi, h_1 \rangle \mathrm{d}t - \langle \beta\psi, h_2 \rangle. \end{aligned}$$

In order to simplify the expression of $J'(\phi, \psi)$, multiplying (8.186) by $\tilde{u}$, integrating over $\mathcal{Q}$ and (integrating) by parts in time t, we obtain, for all $\mathbf{h} = (h_1, h_2) \in \mathcal{U}_{\mathrm{ad}} \times \mathcal{V}_{\mathrm{ad}}$

$$\begin{aligned} &\int_0^T \langle B_1^* \tilde{u}, h_1 \rangle \mathrm{d}t + \langle B_2^* \tilde{u}(0), h_2 \rangle \\ &\quad = \langle \mathbf{V}'_T(\phi^*, \psi^*)\mathbf{h}, \tilde{u}(T) \rangle \\ &\qquad + \int_0^T \langle -\frac{\partial \tilde{u}}{\partial t} + A\tilde{u} + (G(., u^*))^* \tilde{u} + K^*(\varpi, \tilde{u}), \mathbf{U}'(\phi^*, \psi^*)\mathbf{h} \rangle \mathrm{d}t \\ &\qquad + \sum_{i=1,n} \int_0^T \langle d_i(., t)\tilde{u}(., t), \mathbf{U}'(\phi^*, \psi^*)\mathbf{h}(., r_i(t)) \rangle \mathrm{d}t. \end{aligned} \tag{8.194}$$

We can now calculate the term

$$\int_0^T \langle d_i(., t)\tilde{u}(., t), \mathbf{U}'(\phi^*, \psi^*)\mathbf{h}(., r_i(t)) \rangle \mathrm{d}t.$$

Let $s = r_i(t)$, then $t = f_i(s)$ and $\mathrm{d}t = f_i'(s)\mathrm{d}s$. So

$$\int_0^T \langle d_i(.,t)\tilde{u}(.,t), \mathbf{U}'(\phi^*,\psi^*)\mathbf{h}(.,r_i(t))\rangle \mathrm{d}t$$
$$= \int_{-e_i(0)}^{T-e_i(T)} \langle d_i(.,f_i(s))\tilde{u}(.,f_i(s))f_i'(s), \mathbf{U}'(\phi^*,\psi^*)\mathbf{h}\rangle \mathrm{d}s.$$

Since $e_i(T) = \delta(T), \forall i = 1, n$ and according to the second term of (8.186) we have

$$\begin{aligned}\int_0^T \langle d_i(.,t)\tilde{u}(.,t), \mathbf{U}'(\phi^*,\psi^*)\mathbf{h}(.,r_i(t))\rangle \mathrm{d}t \\ = \int_0^{T-\delta(T)} \langle d_i(.,f_i(s))\tilde{u}(.,f_i(s))f_i'(s), \mathbf{U}'(\phi^*,\psi^*)\mathbf{h}\rangle \mathrm{d}s.\end{aligned} \tag{8.195}$$

Since $\tilde{u}$ is a solution of the adjoint problem (8.188) and according to (8.195), the equality (8.194) becomes

$$\begin{aligned}\int_0^T \langle B_1^*\tilde{u}, h_1\rangle \mathrm{d}t + \langle B_2^*\tilde{u}(0), h_2\rangle = \langle \mathbf{V}_T'(\phi^*,\psi^*)\mathbf{h}, \mu(u^*(T) - v_{\mathrm{obs}})\rangle \\ + \int_0^T \langle \mathcal{C}^*\mathcal{C}(u^* - u_{\mathrm{obs}}), \mathbf{U}'(\phi^*,\psi^*)\mathbf{h}\rangle \mathrm{d}t.\end{aligned} \tag{8.196}$$

Applying (8.196) and according to the expression of J' one has, for all $\mathbf{h} = (h_1, h_2) \in \mathcal{U}_{\mathrm{ad}} \times \mathcal{V}_{\mathrm{ad}}$,

$$J'(\phi^*,\psi^*).\mathbf{h} = \int_0^T\int_\Omega (\alpha\phi^* + B_1^*\tilde{u})h_1 \mathrm{d}x\mathrm{d}t + \int_\Omega (-\beta\psi^* + B_2^*\tilde{u}(0))h_2 \mathrm{d}x. \tag{8.197}$$

As (ϕ^*,ψ^*) is a solution of the saddle point problem (8.185) then

$$J'(\phi^*,\psi^*)(\phi - \phi^*, 0) \geq 0, \quad J'(\phi^*,\psi^*)(0, \psi - \psi^*) \leq 0,$$

for all $(\phi,\psi) \in \mathcal{U}_{\mathrm{ad}} \times \mathcal{V}_{\mathrm{ad}}$ and therefore, according to the expression (8.197), we can deduce that

$$\iint_Q (\alpha\phi^* + B_1^*\tilde{u})(\phi - \phi^*)\mathrm{d}x\mathrm{d}t \geq 0,$$
$$\int_\Omega (-\beta\psi^* + B_2^*\tilde{u}(0))(\psi - \psi^*)\mathrm{d}x \leq 0,$$

for all $(\phi,\psi) \in \mathcal{U}_{\mathrm{ad}} \times \mathcal{V}_{\mathrm{ad}}$.

This completes the proof. □

Now we give some conditions to obtain the uniqueness of the saddle point (ϕ^*,ψ^*). More precisely, we show that, for small data or large coefficient ν, α and β, one has uniqueness.

Theorem 8.65. *Assume that F satisfies the assumptions (7.78), $\nu > \delta_2$, $\mu < 1$ and $b^2/\beta \leq 1$ hold. Subject to the following conditions:*

(*i*) $\theta = (\nu - \delta_2 - c_I^2(\gamma + \delta_1 + 2C_\infty)) - 2b^2c_I^2/\alpha > 0$

(*ii*) $N_\infty \left(\| u^*(T) - v_{\text{obs}} \|_{L^2}^2 + \| \mathcal{C}(u^* - u_{\text{obs}}) \|_{L^2(\mathcal{Q})}^2\right)^{1/2} < \theta,$
where C_∞ *is given by (8.171),* $N_\infty = \lambda c_e^2 \exp((\delta_1 + \gamma + 2C_\infty)T/2)$ *and the constants* λ, c_e, c_I, δ_1, δ_2 *and* b *are given by the relations (7.78), (7.79), (7.80), (7.82) and (8.108),*

the optimal solution (ϕ^*, ψ^*, u^*) *is unique.*

Proof. Suppose that there exist (ϕ^*, ψ^*, u^*) and $(\phi_1^*, \psi_1^*, u_1^*)$ two optimal solutions where (ϕ^*, ψ^*) and (ϕ_1^*, ψ_1^*) satisfy (8.185), $u_1^* = \mathbf{U}(\phi_1^*, \psi_1^*)$ and $u^* = \mathbf{U}(\phi^*, \psi^*)$ are solutions of (8.183) such that the optimality conditions (8.193) hold, *i.e.*, for all $(\phi, \psi) \in \mathcal{U}_{\text{ad}} \times \mathcal{V}_{\text{ad}}$,

$$\begin{aligned}
&\iint_{\mathcal{Q}} (\alpha\phi_1^* + B_1^*\tilde{u}_1)(\phi - \phi_1^*)\mathrm{d}x\mathrm{d}t \geq 0,\\
&\int_{\Omega} (-\beta\psi_1^* + B_2^*\tilde{u}_1(0))(\psi - \psi_1^*)\mathrm{d}x \leq 0,\\
&\iint_{\mathcal{Q}} (\alpha\phi^* + B_1^*\tilde{u})(\phi - \phi^*)\mathrm{d}x\mathrm{d}t \geq 0,\\
&\int_{\Omega} (-\beta\psi^* + B_2^*\tilde{u}(0))(\psi - \psi^*)\mathrm{d}x \leq 0,
\end{aligned} \tag{8.198}$$

where $\tilde{u}_1 = \tilde{\mathbf{U}}({\phi^*}_1, {\psi^*}_1)$ (respectively $\tilde{u} = \tilde{\mathbf{U}}(\phi^*, \psi^*)$) is a solution of the adjoint problem (8.188) corresponding to the primal solution u_1^* (respectively u^*).

Set $\phi = \phi^* - \phi_1^*$, $\psi = \psi^* - \psi_1^*$, $v = u^* - u_1^*$ and $\tilde{v} = \tilde{u} - \tilde{u}_1$. Then v satisfies the following system:

$$\begin{aligned}
&\frac{\partial v}{\partial t} + Av + (F(.,u^*) - F(.,u_1^*)) + K(\varpi, v)\\
&\qquad\qquad + \sum_{i=1,n} d_i v(.,r_i) = B_1\phi \quad \text{on } \mathcal{Q},\\
&v(.,.) = 0 \quad \text{on } \mathcal{Q}_0,\\
&v(.,0) = B_2\psi \quad \text{on } \Omega,
\end{aligned} \tag{8.199}$$

$\tilde{v}$ satisfies the following system:

$$\begin{aligned}
&-\frac{\partial \tilde{v}}{\partial t} + A\tilde{v} + (G(.,u^*))^*\tilde{v} + K^*(\varpi, \tilde{v}) + \sum_{i=1,n} d_i(.,f_i)\tilde{v}(.,f_i)f_i'\\
&\qquad = \mathcal{C}^*\mathcal{C}v - (G(.,u^*) - G(.,u_1^*))^*\tilde{u} \quad \text{on } \tilde{\mathcal{Q}},\\
&-\frac{\partial \tilde{v}}{\partial t} + A\tilde{v} + (G(.,u^*))^*\tilde{v} + K^*(\varpi, \tilde{v})\\
&\qquad = \mathcal{C}^*\mathcal{C}v - (G(.,u^*) - G(.,u_1^*))^*\tilde{u} \quad \text{on } \mathcal{Q}_T,\\
&\tilde{v}(.,T) = \mu v(.,T) \quad \text{on } \Omega
\end{aligned} \tag{8.200}$$

and (ϕ, ψ) satisfies the following inequalities:

$$\begin{aligned} &\alpha \parallel \phi \parallel^2_{L^2(\mathcal{Q})} + \iint_{\mathcal{Q}} B_1^* \tilde{v} \phi \mathrm{d}x\mathrm{d}t \leq 0, \\ &\beta \parallel \psi \parallel^2_{L^2(\Omega)} - \int_{\Omega} B_2^* \tilde{v}(0) \psi \mathrm{d}x \leq 0. \end{aligned} \tag{8.201}$$

By multiplying (8.199) by v, (8.200) by $\tilde{v}$ and integrating over $\mathcal{Q}$, this gives, according to (8.201), the estimates (8.129) and (7.91) and the relations (7.81) and (7.82)

$$\begin{aligned} &\int_0^T \frac{\partial}{\partial t} \parallel v \parallel^2_{L^2} \mathrm{d}t + \nu \int_0^T \parallel v \parallel^2_{\mathcal{D}} \mathrm{d}t \\ &\qquad \leq \gamma c_I^2 \int_0^T \parallel v \parallel^2_{\mathcal{D}} \mathrm{d}t + \frac{2}{\alpha} \parallel B_1^* \tilde{v} \parallel_{L^2(\mathcal{Q})} \parallel B_1^* v \parallel_{L^2(\mathcal{Q})} \\ &\qquad +2 \mid \sum_{i=1,n} \int_0^T \langle d_i v(.,r_i), v \rangle \mathrm{d}t \mid, \\ &-\int_0^T \frac{\partial}{\partial t} \parallel \tilde{v} \parallel^2_{L^2} \mathrm{d}t + \nu \int_0^T \parallel \tilde{v} \parallel^2_{\mathcal{D}} \mathrm{d}t \\ &\qquad \leq \gamma c_I^2 \int_0^T \parallel \tilde{v} \parallel^2_{\mathcal{D}} \mathrm{d}t + (\delta_1 c_I^2 + \delta_2) \int_0^T (\parallel v \parallel^2_{\mathcal{D}} + \parallel \tilde{v} \parallel^2_{\mathcal{D}}) \mathrm{d}t \\ &\qquad +2\lambda \int_0^T \int_{\Omega} \mid \tilde{u} \mid \mid \tilde{v} \mid \mid v \mid \mathrm{d}t \\ &\qquad +2 \mid \sum_{i=1,n} \int_0^{T-\delta(T)} \langle d_i(.,f_i) \tilde{v}(.,f_i) f_i', \tilde{v} \rangle \mathrm{d}t \mid, \\ &\tilde{v}(.,T) = \mu v(T) \text{ and } v(.,0) = B_2 \psi. \end{aligned} \tag{8.202}$$

We can now estimate the terms

$$\mid \sum_{i=1,n} \int_0^T \langle d_i v(.,r_i), v \rangle \mathrm{d}t \mid, \ \mid \sum_{i=1,n} \int_0^{T-\delta(T)} \langle d_i(.,f_i) \tilde{v}(.,f_i) f_i', \tilde{v} \rangle \mathrm{d}t \mid .$$

According to Lemma 8.54 and since $v = 0$ on $\mathcal{Q}_0$, we can deduce that

$$\mid \sum_{i=1,n} \int_0^T \langle d_i v(.,r_i), v \rangle \mathrm{d}t \mid \leq C_\infty \parallel v \parallel^2_{L^2(\mathcal{Q})}, \tag{8.203}$$

where C_∞ is given in (8.171). Moreover, by using a similar technique to that given for Lemma 8.54, we can deduce that

$$\mid \sum_{i=1,n} \int_0^{T-\delta(T)} \langle d_i(.,f_i) \tilde{v}(.,f_i) f_i', \tilde{v} \rangle \mathrm{d}t \mid \leq C_\infty \parallel \tilde{v} \parallel^2_{L^2(\mathcal{Q})} . \tag{8.204}$$

Applying Hölder's inequality and the relations (7.79), (8.108), (8.107), (8.203) and (8.204), the estimates (8.202) become

$$\begin{aligned}
&\int_0^T \frac{\partial}{\partial t} \| v \|_{L^2}^2 \,\mathrm{d}t + (\nu - (\gamma + 2C_\infty)c_I^2) \int_0^T \| v \|_{\mathcal{D}}^2 \,\mathrm{d}t \\
&\qquad \le \frac{2b^2c_I^2}{\alpha} \int_0^T (\| v \|_{\mathcal{D}}^2 \,\mathrm{d}t + \| \tilde{v} \|_{\mathcal{D}}^2)\mathrm{d}t, \\
&- \int_0^T \frac{\partial}{\partial t} \| \tilde{v} \|_{L^2}^2 \,\mathrm{d}t + (\nu - (\gamma + 2C_\infty)c_I^2) \int_0^T \| \tilde{v} \|_{\mathcal{D}}^2 \,\mathrm{d}t \\
&\qquad \le (\delta_1 c_I^2 + \delta_2 + \lambda c_e^2 \| \tilde{u} \|_{\mathcal{H}}) \int_0^T (\| v \|_{\mathcal{D}}^2 + \| \tilde{v} \|_{\mathcal{D}}^2)\mathrm{d}t, \\
&\tilde{v}(.,T) = \mu v(.,T) \text{ and } v(.,0) = B_2\psi.
\end{aligned} \tag{8.205}$$

Adding the first and the second estimate of (8.205) we obtain

$$\begin{aligned}
&\int_0^T \frac{\partial}{\partial t}(\| v \|_{L^2}^2 - \| \tilde{v} \|_{L^2}^2)\mathrm{d}t + \theta \int_0^T (\| v \|_{\mathcal{D}}^2 + \| \tilde{v} \|_{\mathcal{D}}^2)\mathrm{d}t \\
&\qquad \le \lambda c_e^2 \| \tilde{u} \|_{\mathcal{H}} \int_0^T (\| v \|_{\mathcal{D}}^2 + \| \tilde{v} \|_{\mathcal{D}}^2)\mathrm{d}t,
\end{aligned} \tag{8.206}$$

where $\theta = \nu - \delta_2 - c_I^2(\gamma + 2C_\infty + (2b^2/\alpha) + \delta_1) > 0$, because of assumption (i).

By using the third relation of (8.205) we can deduce that

$$\begin{aligned}
&(1-\mu^2) \| v(T) \|_{L^2}^2 + \| \tilde{v}(0) \|_{L^2}^2 + \theta(\| v \|_{\mathcal{V}}^2 + \| \tilde{v} \|_{\mathcal{V}}^2) \\
&\qquad \le \lambda c_e^2 \| \tilde{u} \|_{\mathcal{H}} (\| v \|_{\mathcal{V}}^2 + \| \tilde{v} \|_{\mathcal{V}}^2) + \| B_2\psi \|_{L^2}^2 .
\end{aligned}$$

According to the relation (8.107) and the second estimate of (8.201), we can deduce that

$$\| B_2\psi \|_{L^2}^2 \le \frac{b^4}{\beta^2} \| \tilde{v}(0) \|_{L^2}^2$$

and then

$$\begin{aligned}
&(1-\mu^2) \| v(T) \|_{L^2}^2 + (1 - \frac{b^4}{\beta^2}) \| \tilde{v}(0) \|_{L^2}^2 + \theta(\| v \|_{\mathcal{V}}^2 + \| \tilde{v} \|_{\mathcal{V}}^2) \\
&\qquad \le \lambda c_e^2 \| \tilde{u} \|_{\mathcal{H}} (\| v \|_{\mathcal{V}}^2 + \| \tilde{v} \|_{\mathcal{V}}^2).
\end{aligned}$$

Since $1-\mu^2 > 0$ and $(1 - b^4/\beta^2) > 0$ then

$$\theta(\| v \|_{\mathcal{V}}^2 + \| \tilde{v} \|_{\mathcal{V}}^2) \le \lambda c_e^2 \| \tilde{u} \|_{\mathcal{H}} (\| v \|_{\mathcal{V}}^2 + \| \tilde{v} \|_{\mathcal{V}}^2). \tag{8.207}$$

By choosing $\epsilon^2 = 1$ in the inequality (8.192) and by using Gronwall's formula we can deduce, since $\mu < 1$ and $\delta_2 < \nu$, that

$$\| \tilde{u} \|_{\mathcal{H}}^2 \le \mathrm{e}(\delta_1 + \gamma + 2C_\infty)T \left(\| u^*(T) - v_{\mathrm{obs}} \|_{L^2}^2 + \| \mathcal{C}(u^* - u_{\mathrm{obs}}) \|_{L^2(\mathcal{Q})}^2 \right).$$

Therefore, according to (8.207), $\theta^*(\| v \|_{\mathcal{V}}^2 + \| \tilde{v} \|_{\mathcal{V}}^2) \leq 0$, where

$$\theta^* = \theta - N_\infty \left(\| u^*(T) - v_{\text{obs}} \|_{L^2}^2 + \| \mathcal{C}(u^* - u_{\text{obs}}) \|_{L^2(\mathcal{Q})}^2 \right)^{1/2},$$

$$N_\infty = \lambda c_e^2 \exp\left(\frac{(\delta_1 + \gamma + 2C_\infty)T}{2} \right).$$

Since $\theta^* > 0$ (according to assumption (*ii*)), we have $\| v \|_{\mathcal{V}}^2 + \| \tilde{v} \|_{\mathcal{V}}^2 = 0$ and then $v = 0$ and $\tilde{v} = 0$. We obtain then the uniqueness result. □

Remark 8.66. We can consider other time-varying delays, for example in the convolution (or integral) form or for some combination of the additional and the convolution forms. For more details in the case of the convolution form, for similar non-linear PDEs treated in this book, the reader is referred to Belmiloudi [39]. ♢

Example 8.67. We can consider the problem of biochemical pollutants presented in Section 7.6.5, by taking into account the time delays of pollution. In this case the concentration of pollutants can be governed by the following time delay reaction–diffusion–transport system

$$\begin{aligned}
&\frac{\partial u}{\partial t} - \operatorname{div}(\beta \nabla u) + \varpi.\nabla u + F(x,t,u) \\
&\qquad + \sum_{i=1,n} d_i(x,t) u(x, r_i(t)) = f(x,t) \quad (x,t) \in \mathcal{Q} \\
&u = 0 \quad \text{on } \Sigma = \partial\Omega \times (0,T) \\
&u(x,t') = r_0(x,t') \quad (x,t') \in \mathcal{Q}_0, \\
&u(0) = u_0 \quad \text{on } \Omega.
\end{aligned} \tag{8.208}$$

In the cost functional J given by (8.184) the operator $\mathcal{C}$ represents the regional and temporal variation in the cost of pollutant extraction.

Since all the assumptions of our abstract results are satisfied by the example in this particular case, our study applies. ♣

8.7.4 Remarks on Time-varying Delays and Control in the Boundary Conditions

We conclude with an indication as to how the previous arguments can be made to apply to robust boundary control problem with boundary multiple time-varying delays. Let us consider the following perturbed problem (by using a similar process as used to introduce the perturbation system in Section 8.7.2):

$$\begin{aligned}
&\frac{\partial u}{\partial t} - \operatorname{div}(\eta \nabla u) + F(x,t,u) + K(\varpi, u) \\
&\qquad + \sum_{i=1}^{n} d_i(x,t) u(x, t - e_i(t)) = g(x,t) \quad (x,t) \in \mathcal{Q}, \\
&u(x,t') = r_0(x,t') \quad (x,t') \in \mathcal{Q}_0 = (-\delta(0), 0) \times \Omega, \\
&u(x,0) = u_0(x) \quad x \in \Omega,
\end{aligned} \tag{8.209}$$

with the boundary conditions:

$$\begin{aligned} &\eta\frac{\partial u}{\partial \mathbf{n}} = \sum_{j=1}^{l} a_i(x,t)u(x,t-k_i(t)) + f(x,t) \quad (x,t)\in\Sigma,\\ &u(x,t') = \pi_0(x,t') \quad (x,t')\in\Sigma_0 = (-\delta(0),0)\times\partial\Omega, \end{aligned} \tag{8.210}$$

where $\Sigma = \partial\Omega\times(0,T)$, Ω is open bounded in $\mathbb{R}^m$, $m\geq 1$, with the boundary $\partial\Omega$ sufficiently regular, g is in $L^2(\mathcal{Q})$, u_0 is in $L^2(\Omega)$ and F satisfies Assumptions (7.78) as before. Assume that f is in $L^2(\Sigma)$, $\partial u/\partial\mathbf{n}$ denotes the exterior normal derivative of u at the boundary $\partial\Omega$, the function η is positive and bounded function above and below by two non-negative constants in Ω, and we denote by $\nu = \min_{\Omega}(\eta)$. The terms $(d_i, i=1,n) : \overline{\mathcal{Q}} \longrightarrow \mathbb{R}$ are given $\mathcal{C}^\infty$ functions on $\overline{\mathcal{Q}}$, $(a_j, j=1,l) : \Sigma \longrightarrow \mathbb{R}$ are given $\mathcal{C}^\infty$ functions on Σ, $(e_i, i=1,n)$ and $(k_j, j=1,l)$ are sufficiently regular functions representing multiple time-varying delays and $\delta(0) = \max(\max_{i=1,n}(e_i(0)), \max_{j=1,l}(k_j(0)))$.

Remark 8.68. Such a problem arises for example in control of a collision-dominated plasma (for the linear case see, *e.g.*, Wang [292]). The goal is to confine plasma in a bounded region Ω by introducing, for example, a finite electric potential barrier surrounding the domain Ω. In this case, the *particle density* is governed, in general, by parabolic equation. Because of the finiteness of electric potential the particle reflection is not instantaneous. Indeed the particle flux at the boundary of the considered domain at any time depends on particle flux which escaped earlies with various velocities and were reflected back into the domain Ω at later time. So in order to take into account this phenomenon, we introduce time delays in the boundary conditions. ♢

We state the following assumptions for the delay functions e_i, r_i which is defined by $r_i : t\in[0,T) \longrightarrow r_i(t) = t - e_i(t)$, k_j and q_j which is defined by $q_j : t\in[0,T) \longrightarrow q_j(t) = t - k_j(t)$, for $i = 1,\ldots,n$ and $j=1,\ldots,l$:

$$\begin{aligned} &(r_i)_{i=1,n},\ (q_j)_{j=1,l} \quad \text{are strictly increasing functions,}\\ &(e_i)_{i=1,n},\ (k_j)_{j=1,l} \quad \text{are } \mathcal{C}^1 \text{ non-negative functions on } [0,T). \end{aligned} \tag{8.211}$$

Remark 8.69. According to Assumptions (8.211), we have the existence of the inverse functions $(f_i, i=1,n)$ and $(g_j, j=1,l)$ of the functions $(r_i, i=1,n)$ and $(q_j, j=1,l)$, respectively. ♢

Thus we define the following subdivision:

$$\begin{aligned} &s_{-1} = -\delta(0), s_0 = 0,\\ &s_j = \min(\min_{i=1,n}(f_i(s_{j-1})), \min_{k=1,n}(g_k(s_{j-1}))) \quad \text{for all } j\in\mathbb{N}-\{0\}. \end{aligned} \tag{8.212}$$

Finally, we introduce the following notations: $I_j =]s_{-1}, s_j[$ and $\mathcal{Q}_j = \Omega\times I_j$ for $j\in\mathbb{N}$, and $T_j = s_j - s_{j-1}$, for $j\in\mathbb{N}-\{0\}$. We assume that $m\leq 4$,[7] $\mathcal{D} = H^1(\Omega)\subset L^4(\Omega)$ and $\mathcal{D}' = (H^1(\Omega))'$.

[7] $\mathcal{D}\subset L^4(\Omega)$ provided $m\leq 4$ according to the Sobolev embedding theorem.

By using similar technique as used to prove Theorem 8.53 and Remark 8.41, we obtain the following proposition.

Proposition 8.70. *Assume that F satisfies the assumptions (7.78). Let u_0, r_0, π_0, f and g be given such that $u_0 \in H^{1/2}(\Omega)$, $r_0 \in H^{3/2,3/4}(\mathcal{Q}_0)$, $\pi_0 \in L^2(\Sigma_0)$,$f \in L^2(\Sigma)$ and $g \in L^2(\mathcal{Q})$. Then, there exists a unique solution $u \in H^{3/2,3/4}(\mathcal{Q})$ of problem (8.209) with the boundary conditions (8.210).* □

We suppose now that the disturbance $\psi \in L^2(\mathcal{Q})$ is in the value g and the control $\phi \in L^2(\Sigma)$ is in the boundary condition f, *i.e.*, $f = B_1\phi$ and $g = B_2\psi$, where B_1 and B_2 are given bounded operators on $L^2(\Sigma)$ and on $L^2(\mathcal{Q})$ respectively. So the function u is assumed to be related to the disturbance ψ and control ϕ through the problem (8.209) with the boundary conditions (8.210):

$$
\begin{aligned}
&\frac{\partial u}{\partial t} - \mathrm{div}(\eta\nabla u) + F(.,u) + K(\varpi, u) \\
&\qquad\qquad + \sum_{i=1,n} d_i u(., r_i) = B_2\psi \quad \text{on } \mathcal{Q}, \\
&u(x,t') = r_0(x,t') \quad (x,t') \in \mathcal{Q}_0, \\
&u(.,0) = u_0 \quad \text{on } \Omega, \\
&\eta\frac{\partial u}{\partial \mathbf{n}} = \sum_{j=1,l} a_i u(., q_i) + B_1\phi \quad \text{on } \Sigma, \\
&u(x,t') = \pi_0(x,t') \quad (x,t') \in \Sigma_0.
\end{aligned}
\tag{8.213}
$$

Let $\mathbf{U} : (\phi, \psi) \longrightarrow u = \mathbf{U}(\phi, \psi)$ be the map: $L^2(\Sigma) \times L^2(\mathcal{Q}) \longrightarrow \mathcal{W}$ defined by (8.213) and introducing the cost functional defined by

$$
\begin{aligned}
J(\phi,\psi) = \frac{1}{2} \parallel \mathcal{C}(u - u_{\mathrm{obs}}) \parallel^2_{L^2(\mathcal{Q})} &+ \frac{\mu}{2} \parallel u(T) - v_{\mathrm{obs}} \parallel^2_{L^2(\Omega)} \\
&+ \frac{\alpha}{2} \parallel \phi \parallel^2_{L^2(\Sigma)} - \frac{\beta}{2} \parallel \psi \parallel^2_{L^2(\mathcal{Q})},
\end{aligned}
\tag{8.214}
$$

where $\mu, \alpha, \beta > 0$ are fixed, $(u_{\mathrm{obs}}, v_{\mathrm{obs}}) \in \mathcal{V} \times L^2(\Omega)$ is the observation and $\mathcal{C}$ is an unbounded, linear operator on $L^2(\Omega)$ satisfying the hypothesis (7.91).

We want to minimize the functional J with respect to ϕ and maximize J with respect to ψ, *i.e.*, to find $(\phi^*, \psi^*) \in \mathcal{U}_{\mathrm{ad}} \times \mathcal{V}_{\mathrm{ad}}$ such that

$$
J(\phi^*, \psi) \leq J(\phi^*, \psi^*) \leq J(\phi, \psi^*), \forall (\phi, \psi) \in \mathcal{U}_{\mathrm{ad}} \times \mathcal{V}_{\mathrm{ad}} \tag{8.215}
$$

with $\mathcal{U}_{\mathrm{ad}}$ and $\mathcal{V}_{\mathrm{ad}}$ are (given) non-empty, closed, convex, bounded subsets of $L^2(\Sigma)$ and $L^2(\mathcal{Q})$, respectively.

By using the same technique as used in the proof of Theorems 8.62 and 8.64, we have the following results.

Theorem 8.71. *Let F be an operator that satisfies Assumptions (7.78). Then, for the coefficients α and β sufficiently large, there exists an optimal* *solution* $(\phi^*, \psi^*, u^*) \in L^2(\Sigma) \times L^2(\mathcal{Q}) \times \mathcal{W}$ *such that* (ϕ^*, ψ^*) *is defined by (8.215) and* $u^* = \mathbf{U}(\phi^*, \psi^*)$ *is the solution of (8.213), which satisfies the following optimality conditions, for all* $(\phi, \psi) \in \mathcal{U}_{\mathrm{ad}} \times \mathcal{V}_{\mathrm{ad}}$,

$$\begin{aligned} &\iint_\Sigma (\alpha\phi^* + B_1^*\tilde{u}|_\Sigma)(\phi - \phi^*)\mathrm{d}\Gamma\mathrm{d}t \geq 0, \\ &\iint_Q (-\beta\psi^* + B_2^*\tilde{u})(\psi - \psi^*)\mathrm{d}x\mathrm{d}t \leq 0, \end{aligned} \tag{8.216}$$

where $\tilde{u} = \tilde{\mathbf{U}}(\phi^*, \psi^*)$ *is the unique solution of the following adjoint problem:*

$$\begin{aligned} &-\frac{\partial \tilde{u}}{\partial t} - \mathrm{div}(\eta \nabla \tilde{u}) + (G(.,u^*))^*\tilde{u} + K^*(\varpi, \tilde{u}) \\ &\qquad + \sum_{i=1,n} d_i(., f_i)\tilde{u}(., f_i) f_i' = \mathcal{C}^*\mathcal{C}(u - u_{\mathrm{obs}}) \quad on\ \tilde{\mathcal{Q}}, \\ &-\frac{\partial \tilde{u}}{\partial t} - \mathrm{div}(\eta \nabla \tilde{u}) + (G(.,u^*))^*\tilde{u} + K^*(\varpi, \tilde{u}) \\ &\qquad = \mathcal{C}^*\mathcal{C}(u - u_{\mathrm{obs}}) \quad on\ \mathcal{Q}_T, \\ &\tilde{u}(T) = \mu(u(T) - v_{\mathrm{obs}}) \quad on\ \Omega, \\ &\eta\frac{\partial \tilde{u}}{\partial \mathbf{n}} = \sum_{j=1,l} a_j(., g_j)\tilde{u}(., g_j) g_j' \quad on\ \tilde{\Sigma} = \partial\Omega \times (0, T - \delta(T)), \\ &\eta\frac{\partial \tilde{u}}{\partial \mathbf{n}} = 0 \quad on\ \Sigma_T = \partial\Omega \times (T - \delta(T), T). \end{aligned} \tag{8.217}$$

Moreover, the gradient of the functional J *at* (ϕ^*, ψ^*) *in any direction* $\mathbf{h} = (h_1, h_2) \in L^2(\Sigma) \times L^2(\mathcal{Q})$ *is given by*

$$J'(\phi^*, \psi^*).\mathbf{h} = \iint_\Sigma (\alpha\phi^* + B_1^*\tilde{u}|_\Sigma)h_1\mathrm{d}\Gamma\mathrm{d}t + \iint_Q (-\beta\psi^* + B_2^*\tilde{u})h_2\mathrm{d}x\mathrm{d}t.$$

Otherwise (in the weak sense),

$$\frac{\partial J}{\partial \phi}(\phi^*, \psi^*) = \alpha\phi^* + B_1^*\tilde{u}|_\Sigma \quad and \quad \frac{\partial J}{\partial \psi}(\phi^*, \psi^*) = -\beta\psi^* + B_2^*\tilde{u}.$$

□

Remark 8.72. It is clear that we can consider other observations, controls and/or disturbances (which can appear in the boundary condition or in the state system) and we obtain similar results by using the same technique as used in the previous work of this section. ♢

9

Remarks on Numerical Techniques

We can now give some numerical strategies in order to solve *robust* control problems, by using the adjoint variables. In this chapter, we do not pretend to give a general theory for numerical resolution, but only assist the reader with possible strategies based on adjoint control optimization.

9.1 Introduction and Studied Problem

We present algorithms where the descent direction is calculated using the adjoint variables, particularly by choosing an admissible step size. The primal and adjoint problems are discretized by a combination of Galerkin and finite element methods for space discretization and by using the classical first-order Euler method for time discretization.

In order to illustrate the numerical methods, we consider the non-linear problem treated in Section 8.6, more precisely the non-linear robust control problem defined by the following systems:

1. *The primal problem is given by*

$$\begin{aligned} &\frac{\partial u}{\partial t} + Au + F(.,u) + K(\varpi, u) = B_1\phi + B_2\psi \ \text{ on } \mathcal{Q}, \\ &u(0) = u_0 \text{ (given) } \ \text{ on } \Omega. \end{aligned} \tag{9.1}$$

   The operator solution $\mathbf{U} : (\phi, \psi) \in (L^2(\mathcal{Q}))^2 \longrightarrow \mathbf{U}(\phi, \psi)$ is defined by $\mathbf{U}(\phi, \psi) = u$, which is a solution of (9.1).

2. *The objective functional is given by*

$$\begin{aligned} J(\phi, \psi) = \frac{1}{2} \parallel \mathcal{C}(u - u_{\text{obs}}) \parallel^2_{L^2(\mathcal{Q})} + \frac{\mu}{2} \parallel u(T) - v_{\text{obs}} \parallel^2_{L^2} \\ + \frac{\alpha}{2} \parallel \phi \parallel^2_{L^2(\mathcal{Q})} - \frac{\beta}{2} \parallel \psi \parallel^2_{L^2(\mathcal{Q})}, \end{aligned} \tag{9.2}$$

where $\mu, \alpha, \beta > 0$ are fixed, $(u_{\text{obs}}, v_{\text{obs}}) \in \mathcal{V} \times L^2(\Omega)$ is the observation (given) and $\mathcal{C}$ is an unbounded, linear operator on $L^2(\Omega)$ satisfying (7.91). The robust control problem is to find a saddle point (ϕ^*, ψ^*) of J in a space $\mathcal{U}_{\text{ad}} \times \mathcal{V}_{\text{ad}}$ subject to the perturbation problem (9.1), where $\mathcal{U}_{\text{ad}}$ and $\mathcal{V}_{\text{ad}}$ are (given) non-empty, closed, convex, bounded subsets of $L^2(\mathcal{Q})$, *i.e.*, to find $(\phi^*, \psi^*) \in \mathcal{U}_{\text{ad}} \times \mathcal{V}_{\text{ad}}$ such that

$$J(\phi^*, \psi) \leq J(\phi^*, \psi^*) \leq J(\phi, \psi^*), \forall (\phi, \psi) \in \mathcal{U}_{\text{ad}} \times \mathcal{V}_{\text{ad}}. \tag{9.3}$$

3. *The derivative of the operator solution is given by*

$$\begin{aligned} &\frac{\partial w}{\partial t} + Aw + G(., \mathbf{U}(\phi, \psi))w + K(\varpi, w) = B_1 h_1 + B_2 h_2 \quad \text{on } \mathcal{Q}, \\ &w(0) = 0 \quad \text{on } \Omega, \end{aligned} \tag{9.4}$$

where

$$w = \frac{\partial \mathbf{U}}{\partial \phi}(\phi, \psi).h_1 + \frac{\partial \mathbf{U}}{\partial \psi}(\phi, \psi).h_2.$$

4. *The adjoint operator* $\tilde{\mathbf{U}}$ *corresponding to the operator solution* $\mathbf{U}$ *is given by* $\tilde{\mathbf{U}} : (\phi, \psi) \longrightarrow \tilde{\mathbf{U}}(\phi, \psi) = \tilde{u}$ such that $\tilde{u}$ satisfies

$$\begin{aligned} &-\frac{\partial \tilde{u}}{\partial t} + A\tilde{u} + (G(., u))^* \tilde{u} + K^*(V, \tilde{u}) = \mathcal{C}^* \mathcal{C}(u - u_{\text{obs}}) \quad \text{on } \mathcal{Q}, \\ &\tilde{u}(T) = \mu(u(T) - v_{\text{obs}}) \quad \text{on } \Omega, \end{aligned} \tag{9.5}$$

where $u = \mathbf{U}(\phi, \psi)$ is the solution of problem (9.1) with the forcing (ϕ, ψ), $\mathcal{C}^*$ (resp. $(G(., u))^*$) is the adjoint of the operator $\mathcal{C}$ (resp. $G(., u)$) (the adjoint A^* of A is itself, *i.e.*, $A^* = A$ since A is a self-adjoint operator).

5. *The optimality conditions are given by*

$$\begin{aligned} &\int_0^T \int_\Omega (\alpha \phi^* + B_1^* \tilde{u}^*)(\phi - \phi^*) \mathrm{d}x\mathrm{d}t \geq 0 \\ &\int_0^T \int_\Omega (-\beta \psi^* + B_2^* \tilde{u}^*)(\psi - \psi^*) \mathrm{d}x\mathrm{d}t \leq 0, \end{aligned} \tag{9.6}$$

for all $(\phi, \psi) \in \mathcal{U}_{\text{ad}} \times \mathcal{V}_{\text{ad}}$, where $\tilde{u}^* = \tilde{\mathbf{U}}(\phi^*, \psi^*)$ is the solution of the adjoint problem (9.5) and $(\phi^*, \psi^*) \in \mathcal{U}_{\text{ad}} \times \mathcal{V}_{\text{ad}}$ is an optimal solution (a saddle point of the objective functional J).

6. *The gradient of the objective functional* J *is given by*

$$\frac{\partial J}{\partial \phi}(\phi, \psi) = \alpha \phi + B_1^* \tilde{u} \quad \text{and} \quad \frac{\partial J}{\partial \psi}(\phi, \psi) = -\beta \psi + B_2^* \tilde{u}, \tag{9.7}$$

where $\tilde{u} = \tilde{\mathbf{U}}(\phi, \psi)$ is the solution of the adjoint problem (9.5), corresponding to the solution $u = \mathbf{U}(\phi, \psi)$ of the primal problem (9.1).

9.2 Continuous Case

In this section the descent method is formulated in terms of continuous variables that are independent of a specific discretization.

We propose three classical algorithms: gradient algorithm, conjugate gradient algorithm and Lagrange–Newton algorithm.

9.2.1 Gradient Algorithm

The gradient algorithm for the resolution of the optimization problem (9.3) is given by: for $k = 1, \ldots$ (the iteration index) we denote by (ϕ_k, ψ_k) the numerical approximation of the control-disturbance at the kth iteration of the algorithm.

1. Initialization: $k = 0$ and (ϕ_0, ψ_0) (given for example by $(\phi_0, \psi_0) = (0, 0)$ on $t \in [0, T]$).
2. Resolution of the direct problem (9.1) with the source term (ϕ_k, ψ_k), gives $u_k = \mathbf{U}(\phi_k, \psi_k)$.
3. Resolution of the adjoint problem (9.5) (based on (ϕ_k, ψ_k, u_k)), gives $\tilde{u}_k$.
4. Local expression of the gradient of J at point (ϕ_k, ψ_k):

$$(GJ) \begin{cases} c_k \stackrel{def}{=} \dfrac{\partial J}{\partial \phi}(\phi_k, \psi_k) = \alpha \phi_k + B_1^* \tilde{u}_k, \\ d_k \stackrel{def}{=} \dfrac{\partial J}{\partial \psi}(\phi_k, \psi_k) = -\beta \psi_k + B_2^* \tilde{u}_k. \end{cases}$$

5. Determine (ϕ_{k+1}, ψ_{k+1}):

$$\begin{cases} \psi_{k+1} := \psi_k + \delta_k d_k, \\ \phi_{k+1} := \phi_k - \lambda_k c_k, \end{cases}$$

   where $0 < m \leq \lambda_k, \delta_k \leq M$ are the sequences of step lengths.
6. IF the gradient is sufficiently small (convergence) THEN end; ELSE set $k := k + 1$ and REPEAT from 2 UNTIL convergence. The approximation of the optimal solution (ϕ^*, ψ^*, u^*) is then given by (ϕ_k, ψ_k, u_k).

In order to obtain an algorithm that is numerically efficient, the best choice of λ_k and δ_k will be the result of a line minimization and maximization algorithm, respectively. Otherwise, at each iteration step k of the previous algorithm, we solve one-dimensional optimization problems of the parameters (λ_k, δ_k):

$$\min_{\lambda > 0} J(\phi_k - \lambda c_k, \psi_k) \tag{9.8}$$

and

$$\max_{\delta > 0} J(\phi_k, \psi_k + \delta d_k). \tag{9.9}$$

From the numerical computational viewpoint, it is most efficient to compute (λ_k, δ_k) only approximately, in order to reduce the computational cost. To derive an approximation for (λ_k, δ_k) we can use, for example, the linearization of $\mathbf{U}(\phi_k - \lambda c_k, \psi_k)$ and $\mathbf{U}(\phi_k, \psi_k + \delta d_k)$ at (ϕ_k, ψ_k) by

$$\begin{aligned} \mathbf{U}(\phi_k - \lambda c_k, \psi_k) &\approx \mathbf{U}(\phi_k, \psi_k) - \frac{\partial \mathbf{U}}{\partial \phi}(\phi_k, \psi_k).(\lambda c_k), \\ \mathbf{U}(\phi_k, \psi_k + \delta d_k) &\approx \mathbf{U}(\phi_k, \psi_k) + \frac{\partial \mathbf{U}}{\partial \psi}(\phi_k, \psi_k).(\delta d_k), \end{aligned} \tag{9.10}$$

otherwise

$$\begin{aligned} \mathbf{U}(\phi_k - \lambda c_k, \psi_k) &\approx u_k - \lambda w_\phi^k, \\ \mathbf{U}(\phi_k, \psi_k - \delta d_k) &\approx u_k + \delta w_\psi^k, \end{aligned} \tag{9.11}$$

where $w_\phi^k = (\partial \mathbf{U}/\partial \phi)(\phi_k, \psi_k).c_k$ and $w_\psi^k = (\partial \mathbf{U}/\partial \psi)(\phi_k, \psi_k).d_k$ are the solution of (9.4). According to the previous approximation (9.11), we can approximate the problems (9.8) and (9.9) by

$$\min_{\lambda > 0} H(\lambda), \tag{9.12}$$

and

$$\max_{\delta > 0} R(\delta), \tag{9.13}$$

where

$$\begin{aligned} H(\lambda) = \frac{1}{2} \| \mathcal{C}(u_k - u_{\text{obs}}) - \lambda \mathcal{C} w_\phi^k \|_{L^2(Q)}^2 &+ \frac{\mu}{2} \| (u_k(T) - v_{\text{obs}}) - \lambda w_\phi^k(T) \|_{L^2}^2 \\ &+ \frac{\alpha}{2} \| \phi_k - \lambda c_k \|_{L^2(Q)}^2 - \frac{\beta}{2} \| \psi_k \|_{L^2(Q)}^2, \\ R(\delta) = \frac{1}{2} \| \mathcal{C}(u_k - u_{\text{obs}}) + \delta \mathcal{C} w_\psi^k \|_{L^2(Q)}^2 &+ \frac{\mu}{2} \| (u_k(T) - v_{\text{obs}}) + \delta w_\psi^k(T) \|_{L^2}^2 \\ &+ \frac{\alpha}{2} \| \phi_k \|_{L^2(Q)}^2 - \frac{\beta}{2} \| \psi_k + \delta d_k \|_{L^2(Q)}^2 . \end{aligned}$$

As the functions H and R are polynomial functions of degree 2, the problems (9.12) and (9.13) can be solved exactly. Consequently, we obtain explicitly the value of the parameters λ_k and δ_k.

9.2.2 Conjugate Gradient Algorithm

Another strategy to numerically solve the optimization problem (8.111) is to use a conjugate gradient type algorithm (CG-algorithm) combined with the Wolfe–Powell line search procedure for computing admissible step sizes along the descent direction. The advantage of this method, compared to the gradient method, is that it performs a soft reset whenever the GC search direction yields no significant progress. In general, the method has the following form:

$$D_k = Dz_k = \begin{cases} -G_k & \text{for } k = 0, \\ -G_k + \xi_{k-1} D_{k-1} & \text{for } k \geq 1, \end{cases}$$

$$z_{k+1} = z_k + \gamma_k D_k,$$

where G_k denotes the gradient of the functional to be minimized at point z_k, γ_k is a step length obtained by a line search, D_k is the search direction and ξ_k is a constant. Several varieties of this method differ in the way of selecting ξ_k. Some well-known formula for ξ_k are given by Fletcher–Reeves, Polak–Ribière, Hestenes–Stiefel and, recently, by Dai and Yuan [94].

The GC-algorithm for the resolution of the saddle point problem (8.111) is given below.

For $k = 1, \ldots$ (the iteration index) we denote by $X_k = (\phi_k, \psi_k)$ the numerical approximation of the control-disturbance at the kth iteration of the algorithm.

1. Initialization: (ϕ_0, ψ_0) (given), $\xi_{-1} = 0$, $\eta_{-1} = 0$ and $C_0 = 0$, $D_{-1} = 0$.
   (a) Resolution of the direct problem (9.1), where the source term is (ϕ_0, ψ_0), gives u_0.
   (b) Resolution of the adjoint problem (9.5) (based on (ϕ_0, ψ_0, u_0)), gives $\tilde{u}_0$.
   (c) Gradient of J at (ϕ_0, ψ_0), the vector (c_0, d_0) is given by the system (GJ).
   (d) Determine the direction: $C_0 = -c_0$, $D_0 = -d_0$.
   (e) Determine (ϕ_1, ψ_1): $\phi_1 = \phi_0 + \lambda_0 C_0$ and $\psi_1 = \psi_0 - \delta_0 D_0$.
2. Resolution of the problem (9.1) with the source term (ϕ_k, ψ_k), gives u_k.
3. Resolution of the problem (9.5) (based on (ϕ_k, ψ_k, u_k)), gives $\tilde{u}_k$.
4. Gradient of J at X_k, the vector (c_k, d_k) is given by the system (GJ).
5. Determine ξ_{k-1}, η_{k-1} by one of the following expressions:

$$\xi_{k-1} = \frac{\| c_k \|^2_{U_{\mathrm{ad}}}}{\| c_{k-1} \|^2_{U_{\mathrm{ad}}}}, \quad \eta_{k-1} = \frac{\| d_k \|^2_{V_{\mathrm{ad}}}}{\| d_{k-1} \|^2_{V_{\mathrm{ad}}}} \quad \text{(Fletcher–Reeves)},$$

$$\xi_{k-1} = \frac{(c_k - c_{k-1}, c_k)_{U_{\mathrm{ad}}}}{\| c_{k-1} \|^2_{U_{\mathrm{ad}}}}, \quad \eta_{k-1} = \frac{(d_k - d_{k-1}, d_k)_{V_{\mathrm{ad}}}}{\| d_{k-1} \|^2_{V_{\mathrm{ad}}}} \quad \text{(Polak–Ribière)},$$

$$\xi_{k-1} = \frac{(c_k, c_k - c_{k-1})_{U_{\mathrm{ad}}}}{(C_{k-1}, c_k - c_{k-1})_{U_{\mathrm{ad}}}},$$

$$\eta_{k-1} = \frac{(d_k, d_k - d_{k-1})_{V_{\mathrm{ad}}}}{(D_{k-1}, d_k - d_{k-1})_{V_{\mathrm{ad}}}} \quad \text{(Hestenes–Stiefel)},$$

$$\xi_{k-1} = \frac{\| c_k \|^2_{U_{\mathrm{ad}}}}{(C_{k-1}, c_k - c_{k-1})_{U_{\mathrm{ad}}}}, \quad \eta_{k-1} = \frac{\| d_k \|^2_{V_{\mathrm{ad}}}}{(D_{k-1}, d_k - d_{k-1})_{V_{\mathrm{ad}}}} \quad \text{(Dai–Yuan)}.$$

6. Determine the direction: $C_k = -c_k + \xi_{k-1} C_{k-1}$, $D_k = -d_k + \xi_{k-1} D_{k-1}$.

7. Determine (ϕ_{k+1}, ψ_k): $\phi_{k+1} = \phi_k + \lambda_k C_k$, $\psi_{k+1} = \psi_k - \delta_k D_k$ where $0 < m \leq \lambda_k \leq M$ are the sequences of step lengths.
8. IF the gradient is sufficiently small (convergence) THEN end; ELSE set $k := k+1$ and GOTO 2. The approximation of the optimal solution (ϕ^*, ψ^*, u^*) is (ϕ_k, ψ_k, u_k).

These methods are convergent if the following line search conditions for λ_k are satisfied:

$$J(\phi_k + \lambda_k C_k, \psi_k) \leq J(\phi_k, \psi_k) + \sigma_1 \lambda_k (c_k, C_k)_{U_{\rm ad}} + \tau_{1k},$$

$$J(\phi_k, \psi_k - \delta_k D_k) \leq J(\phi_k, \psi_k) - \sigma_2 \delta_k (d_k, D_k)_{V_{\rm ad}} + \tau_{2k},$$

$$\rho_{11} < c_k, C_k >_{U_{\rm ad}} \leq (c_{k+1}, C_k)_{U_{\rm ad}} \leq -\rho_{12}(c_k, C_k)_{U_{\rm ad}} + \tau_{3k},$$

$$\rho_{21} < d_k, D_k >_{V_{\rm ad}} \leq (d_{k+1}, D_k)_{V_{\rm ad}} \leq -\rho_{22}(d_k, D_k)_{V_{\rm ad}} + \tau_{4k},$$

where $(\tau_{ik})_{i=1,4}$ are the tolerances, $0 < \sigma_i < \rho_{i1} < 1$ and $0 < \rho_{i2} < 1$.

In practice the tolerances $(\tau_{ik})_{i=1,4}$ are of the order of the roundoff error. For $\sigma_i, i = 1, 2$ and $(\rho_{ij})_{i=1,2;\ j=1,2}$, the choices $\sigma_i = 10^{-p}$ for $p = 4, 5$ or 6 and $\rho_{ij} = 10^{-q}$ for $q = 1$ or 2 are effective when double precision arithmetic is used.

Remark 9.1. We can adapt other strategies in order to solve the robust control problem by using descent algorithms combined with the line search procedure for computing admissible step sizes along the descent direction (in every iteration) (see, *e.g.*, Gill *et al.* [133] and Hestenes [154]).

9.2.3 Lagrange–Newton Method

The robust control problem (9.1), (9.5) and (9.7) with the unknown $X = (\phi, \psi, u, \tilde{u})$ with $u = \mathbf{U}(\phi, \psi)$ and $\tilde{u} = \tilde{\mathbf{U}}(\phi, \psi)$, can be reformulated as an inclusion, or the so-called generalized equation of the form

$$0 \in \mathcal{F}(X) + \mathcal{T}(X), \tag{9.14}$$

where $\mathcal{F} : \mathcal{M} \longrightarrow \mathcal{E}$ is F-differentiable and $\mathcal{T} : \mathcal{M} \longrightarrow 2^{\mathcal{E}}$ (the power set of $\mathcal{E}$) is a set-valued map with closed graph. More precisely,

$$\begin{aligned}
\mathcal{F}_1(X) &:= \frac{\partial u}{\partial t} + Au + F(.,u) + K(\varpi, u) - B_1\phi - B_2\psi,\\
\mathcal{F}_2(X) &:= u(t=0) - u_0,\\
\mathcal{F}_3(X) &= -\frac{\partial \tilde{u}}{\partial t} + A\tilde{u} + (G(.,u))^*\tilde{u} + K^*(V, \tilde{u}) - \mathcal{C}^*\mathcal{C}(u - u_{\rm obs}),\\
\mathcal{F}_4(X) &:= \tilde{u}(T) - \mu(u(T) - v_{\rm obs}),\\
\mathcal{F}_5(X) &:= \alpha\phi + B_1^*\tilde{u},\\
\mathcal{F}_6(X) &:= -\beta\psi + B_2^*\tilde{u},
\end{aligned} \tag{9.15}$$

and

$$\mathcal{T}(X) := (\{0\}, \{0\}, \{0\}, \{0\}, \mathcal{N}_1(\phi), \mathcal{N}_2(\psi)) \tag{9.16}$$

where the multi-valued maps $\mathcal{N}_1$ and $\mathcal{N}_2$, called normal cone operators, are defined by

$$\begin{aligned}
&\mathcal{N}_1(\phi) = \{z \in L^2(\mathcal{Q}) : \int_0^T \int_\Omega z(\phi_1 - \phi)\mathrm{d}x\mathrm{d}t \leq 0 \;\; \forall \phi_1 \in \mathcal{U}_{\mathrm{ad}}\}, \\
&\qquad\qquad\qquad\qquad\qquad\qquad\qquad\qquad \text{if } \phi \in \mathcal{U}_{\mathrm{ad}}, \\
&\mathcal{N}_1(\phi) = \emptyset, \;\; \text{if } \phi \notin \mathcal{U}_{\mathrm{ad}}, \\
&\mathcal{N}_2(\psi) = \{z \in L^2(\mathcal{Q}) : \int_0^T \int_\Omega z(\psi_1 - \psi)\mathrm{d}x\mathrm{d}t \geq 0 \;\; \forall \psi_1 \in \mathcal{V}_{\mathrm{ad}}\}, \\
&\qquad\qquad\qquad\qquad\qquad\qquad\qquad\qquad \text{if } \psi \in \mathcal{V}_{\mathrm{ad}}, \\
&\mathcal{N}_2(\psi) = \emptyset, \;\; \text{if } \psi \notin \mathcal{V}_{\mathrm{ad}},
\end{aligned} \tag{9.17}$$

Remark 9.2. The cone-constrained problem (9.14) is an abstract model for various problems:

- when $\mathcal{T} = 0$, the problem (9.14) corresponds to finding a zero of a mapping $\mathcal{F}$ defined in a Banach space
- when $\mathcal{T}$ is the positive orthant in $\mathbb{R}^p, p \geq 1$, the problem (9.14) is a system of inequalities
- when $\mathcal{T}$ is the normal cone to a convex and closed set in a space $\mathcal{M}$, the problem (9.14) may represent variational inequalities.

For other situations the reader should refer, for instance, to Dontchev [104].

◇

We assume that F is a twice-differentiable operator. The generalized Newton method for solving (9.14) is similar to the Newton method for non-linear equations in Banach spaces. More precisely, the Newton procedure for (9.14) is defined by constructing the sequence (X_k) as follows: assume that we have already computed $X_i,\ i = 1, k$, the value X_{k+1} is given by

$$0 \in \mathcal{F}(X_k) - \mathcal{F}'(X_k)(X_{k+1} - X_k) + \mathcal{T}(X_{k+1}), \tag{9.18}$$

where $X_k = (\phi_k, \psi_k, u_k, \tilde{u}_k)$ is the approximation of X at the kth iteration of the algorithm in $\mathcal{M} = (L^2(\mathcal{Q}))^2 \times (\mathcal{W} \cap \mathcal{H}_1)^2$ and X_0 is given in $\mathcal{M}$.

In this work, we are not interested in the general cone constraints but we give the algorithm to resolve the problem (9.18) in the case of the equality-type constraints, *i.e.*, when $\mathcal{T} = 0$. In this case the generalized Newton method can be formulated as (we assume that F satisfies sufficient conditions such that the F-derivative of $\mathcal{F}$ is invertible)

$$X_{k+1} = X_k - (\mathcal{F}'(X_k))^{-1}\mathcal{F}(X_k), \tag{9.19}$$

The Newton procedure (9.19) can be decomposed into the following two steps:

1. Find the solution $Z_k := (p_k, q_k, v_k, \tilde{v}_k)$ of the problem

$$\mathcal{F}'(X_k)Z_k = \mathcal{F}(X_k). \tag{9.20}$$

2. Calculate X_{k+1} by

$$X_{k+1} = X_k - Z_k. \tag{9.21}$$

By using (9.15) and (9.4), the mapping Z_k satisfies the following system:

$$\begin{aligned}
&\frac{\partial v_k}{\partial t} + Av_k + G(.,u_k)v_k + K(\varpi, v_k) - B_1 p_k - B_2 q_k \\
&\qquad = \frac{\partial u_k}{\partial t} + Au_k + F(.,u_k) + K(\varpi, u_k) - B_1\phi_k - B_2\psi_k, \\
&v_k(0) = u_k(0) - u_0, \\
&-\frac{\partial \tilde{v}_k}{\partial t} + A\tilde{v}_k + (G(.,u_k))^*\tilde{v}_k + K^*(V,\tilde{v}_k) \\
&\qquad +(G'(.,u_k))^*(\tilde{u}_k)(v_k) - \mathcal{C}^*\mathcal{C}v_k = -\frac{\partial \tilde{u}_k}{\partial t} + A\tilde{u}_k \\
&\qquad\qquad +(G(.,u_k))^*\tilde{u}_k + K^*(V,\tilde{u}_k) - \mathcal{C}^*\mathcal{C}(u_k - u_{\text{obs}}), \\
&\tilde{v}_k(T) - \mu v_k(T) = \tilde{u}_k(T) - \mu(u_k(T) - v_{\text{obs}}), \\
&\alpha p_k + B_1^*\tilde{v}_k = \alpha\phi_k + B_1^*\tilde{u}_k, \\
&-\beta q_k + B_2^*\tilde{v}_k = -\beta\psi_k + B_2^*\tilde{u}_k.
\end{aligned} \tag{9.22}$$

According to (9.22) and (9.21) the mapping $X_{k+1} = (\phi_{k+1}, \psi_{k+1}, u_{k+1}, \tilde{u}_{k+1})$, with $\phi_{k+1} = \phi_k - p_k$, $\psi_{k+1} = \psi_k - q_k$, $u_{k+1} = u_k - v_k$ and $\tilde{u}_{k+1} = \tilde{u}_k - \tilde{v}_k$, satisfies the following system:

$$\begin{aligned}
&\frac{\partial u_{k+1}}{\partial t} + Au_{k+1} + G(.,u_k)u_{k+1} + K(\varpi, u_{k+1}) \\
&\qquad = (G(.,u_k)u_k - F(.,u_k)) + B_1\phi_{k+1} + B_2\psi_{k+1}, \\
&u_{k+1}(0) = u_0, \\
&-\frac{\partial \tilde{u}_{k+1}}{\partial t} + A\tilde{u}_{k+1} + (G(.,u_k))^*\tilde{u}_{k+1} + (G'(.,u_k))^*(\tilde{u}_k)(u_{k+1}) \\
&\qquad -(G'(.,u_k))^*(\tilde{u}_k)(u_k) + K^*(V,\tilde{u}_{k+1}) = \mathcal{C}^*\mathcal{C}(u_{k+1} - u_{\text{obs}}), \\
&\tilde{u}_{k+1}(T) = \mu(u_{k+1}(T) - v_{\text{obs}}), \\
&\alpha\phi_{k+1} + B_1^*\tilde{u}_{k+1} = 0, \\
&-\beta\psi_{k+1} + B_2^*\tilde{u}_{k+1} = 0.
\end{aligned} \tag{9.23}$$

Consequently, at each iteration of the Newton method, we have to solve the following system:

$$
\begin{aligned}
&\frac{\partial w}{\partial t} + Aw + G(.,u_k)w + K(\varpi, w) = f_k + B_1\Phi + B_2\Psi, \\
&w(0) = u_0, \\
&-\frac{\partial \tilde{w}}{\partial t} + A\tilde{w} + (G(.,u_k))^*\tilde{w} + (G'(.,u_k))^*(\tilde{u}_k)(w) + K^*(V,\tilde{w}) \\
&\qquad\qquad\qquad = g_k + \mathcal{C}^*\mathcal{C}(w - u_{\mathrm{obs}}), \\
&\tilde{w}(T) = \mu(w(T) - v_{\mathrm{obs}}), \\
&\alpha\Phi + B_1^*\tilde{w} = 0, \\
&-\beta\Psi + B_2^*\tilde{w} = 0.
\end{aligned}
\tag{9.24}
$$

where $f_k = G(.,u_k)u_k - F(.,u_k)$ and $g_k = (G'(.,u_k))^*(\tilde{u}_k)(u_k)$.

Remark 9.3. (*i*) It was shown by Robinson [250], that Newton's method applied to the general equation is locally quadratically convergent to the exact solution, provided that a propriety called strong regularity is satisfied. In our situation and in the case of the equality-type contraints we can prove the following result:

Let X_0 be given in $\mathcal{M}$ and $\mathcal{M}_\epsilon(X_0) := \{X \in \mathcal{M} : \| X - X_0 \|_{\mathcal{M}} < \epsilon\}$ be the open ball in $\mathcal{M}$ of radius ϵ centred on X_0. Under adequate assumptions, there exists $h < 1$, $1 < \beta \leq 2$ and $c > 0$ such the Newton sequence X_k is well defined and the following rate of convergence holds.

$$
\| X^* - X_k \|_{\mathcal{M}} \leq c h^{\beta^k - 1} \quad \text{for } k \geq 2,
$$

where X^* is the exact optimal solution.

(*ii*) In general, the advantage of the Newton method is that the convergence is almost quadratic but its disadvantage resides in the fact that it is expensive (because it requires a lot of computing time) and if the structure is far from the optimal solution the method can become unstable. Therefore, Newton's method is reserved mainly for cases of rapid convergence (*i.e.*, the number of iterations is small) with a well-defined optimal solution.
(*iii*) We can adapt other popular iterative techniques of Newton's type to resolve the robust control problems, for example, SQP (sequential quadratic programming) (see, *e.g.*, Alt [8, 9], Bergounioux and Kunisch [50], Ito and Kunisch [166, 167], Malanowski [214], Troltzsch [284], and the references therein), the Gauss–Newton method (see, *e.g.*, Dennis and Schnabel [98], Le Dimet and Shutyaev [188]) and Landweber's method (when the idea is to replace, in (9.19), the mapping $(\mathcal{F}'(X_k))^{-1}$ by the adjoint $(\mathcal{F}'(X_k))^*$ of $\mathcal{F}'(X_k)$, which is calculated from the adjoint problem, and then for Landweber one need not to compute also the inverse of $\mathcal{F}'(X_k)$). For more discussion on this last method see, *e.g.*, Engl and Scherezer [114] and Scherezer [260]. ♢

In the following, we will be only interested in the discretization of the robust control problem in the case of the gradient procedure (in order to numerically solve the optimization problem).

9.3 Discrete Problem

9.3.1 Approximation of Robust Control Problems

Next we propose a finite element method combined with the first-order Euler scheme for solving the continuous optimization problem. Since the primal and dual systems are (linear or non-linear) PDEs, the approximation analysis of these problems is a non-negligible part of total numerical analysis.

For space discretization, we give a space discretization parameter h (that is converging to 0). To describe the space discretization scheme, we introduce the finite dimensional subspace V_h of $\mathcal{D}$ associated with a partition $\mathcal{T}_h$ of the domain Ω and piecewise polynomial functions of some fixed degree l.

Example 9.4. In the case of $\mathcal{D} = H_0^1(\Omega)$, we introduce the finite dimensional subspace V_h of $\mathcal{D}$ associated with a partition $\mathcal{T}_h$ of the domain Ω and piecewise polynomial functions of some fixed degree l with all functions vanishing on the boundary $\partial\Omega$.

Before expressing the space–time discretization, we recall the following hypotheses (see, *e.g.*, Ciarlet [85]):

$(H_{1\mathrm{E}})$ There exists $C > 0$, such that for all $1 \leq r \leq l$ and $v \in H^{r+1}(\Omega)$

$$\inf_{v_h \in V_h} \| v - v_k \|_{H^1} \leq C h^r \| v \|_{H^{r+1}} .$$

$(H_{2\mathrm{E}})$ There exists $C > 0$, such that for all $(m,k) \in \mathbb{N}^2$ with $0 \leq k \leq m$ and $v_h \in V_h$,

$$\| v_h \|_{H^m} \leq C h^{k-m} \| v_h \|_{H^k} .$$

$(H_{3\mathrm{E}})$ The approximation u_{0h} in V_h of u_0 satisfies: if $u_0 \in H^{s+1}(\Omega)$ then there exists $C > 0$, such that for all $1 \leq r \leq \min(l,s)$,

$$h \| u_0 - u_{0h} \|_{H^1} + \| u_0 - u_{0h} \|_{L^2} \leq C h^{r+1}.$$

♣

For the time discretization, we partition the interval $(0,T)$ by using the following points

$$0 = t_0 \leq t_1 \leq \cdots \leq t_M = T$$

with $t_n = n\tau$, for $n = 0,\ldots,M$ and $\tau = T/M$. For a continuous mapping $u : (0,T) \longrightarrow L^2(\Omega)$, we define $u^n = u(.,t_n)$ for $n = 0,\ldots,M$. For a given sequence $(u^n)_{n=0,\ldots,M}$ in $L^2(\Omega)$, we define its difference quotient as $\partial_\tau u^n = (u^n - u^{n-1})/\tau$. The step is chosen sufficiently small to guarantee both the time accuracy and convergence of the solution.

With the above notation and the discretization of (9.2) and (9.1) by the composite trapezoidal rule in time, we formulate the finite approximation of the problems (9.2) and (9.1) as follows: find a saddle point in $U_{\mathrm{ad},h} \times V_{\mathrm{ad},h}$ of

$$\begin{aligned} J_h(\phi_h, \psi_h) = & \frac{\tau}{2} \sum_{n=0}^{M} \theta_n \int_\Omega \mid \mathcal{C}(u_h^n - u_{\mathrm{obs}}^n) \mid^2 \mathrm{d}x \\ & + \frac{\mu}{2} \int_\Omega \mid u_h^M - v_{\mathrm{obs}} \mid^2 \mathrm{d}x \\ & + \frac{\tau}{2} \sum_{n=0}^{M} \theta_n \int_\Omega \alpha \mid \phi_h^n \mid^2 \mathrm{d}x - \frac{\tau}{2} \sum_{n=0}^{M} \theta_n \int_\Omega \beta \mid \psi_h^n \mid^2 \mathrm{d}x, \end{aligned} \tag{9.25}$$

with $u_h^n \in V_h$ satisfying $(\forall v_h \in V_h)$

$$\begin{aligned} & \int_\Omega \partial_\tau u_h^n v_h \mathrm{d}x + a(u_h^n, v_h) + \int_\Omega K(\mathbf{V}_h^n, u_h^n) v_h \mathrm{d}x + \int_\Omega F_h^n(u_h^n) v_h \mathrm{d}x \\ & \qquad = \int_\Omega B_1 \phi_h^n v_h \mathrm{d}x + \int_\Omega B_2 \psi_h^n v_h \mathrm{d}x, \\ & u_h^0 = u_0 \quad \text{on } \Omega, \end{aligned} \tag{9.26}$$

for $n = 1, 2, \ldots, M$. Here $\theta_0 = \theta_M = 1/2$ and $\theta_n = 1$ for $n \neq 0$ and $n \neq M$.

In order to solve the discretized finite element minimization of $J_h(\phi_h^n, \psi_h^n)$ over $U_{\mathrm{ad},h} \times V_{\mathrm{ad},h}$, we need to calculate the partial derivatives of $J_h(\phi_h^n, \psi_h^n)$ at direction p_h^n and q_h^n. According to (9.7) we have that the evaluation of the derivative of $J(\phi_h^n, \psi_h^n)$ is

$$\begin{aligned} \frac{\partial J}{\partial \phi}(\phi_h^n, \psi_h^n).p_h^n &= \tau \sum_{n=0}^{M} \theta_n \int_\Omega (\alpha \phi_h^n + B_1^* \tilde{u}_h^n) p_h^n \mathrm{d}x, \\ \frac{\partial J}{\partial \psi}(\phi_h^n, \psi_h^n).q_h^n &= \tau \sum_{n=0}^{M} \theta_n \int_\Omega (-\beta \psi_h^n + B_2^* \tilde{u}_h^n) q_h^n \mathrm{d}x. \end{aligned} \tag{9.27}$$

where $\tilde{u}_h^n$ is the solution of following discrete adjoint problem (we use the discrete backward Euler approximation of time)

$$\begin{aligned} & - \int_\Omega \partial_\tau \tilde{u}_h^{n+1} v_h \mathrm{d}x + a(\tilde{u}_h^n, v_h) + \int_\Omega (G_h^n(u_h^n))^* \tilde{u}_h^n v_h \mathrm{d}x \\ & \qquad + \int_\Omega K^*(V_h^n, \tilde{u}_h^n) v_h \mathrm{d}x = \int_\Omega \mathcal{C}^* \mathcal{C}(u_h^n - u_{\mathrm{obs}}^n) v_h \mathrm{d}x, \\ & \tilde{u}_h^M = \mu(u_h^M - v_{\mathrm{obs}}) \quad \text{on } \Omega, \end{aligned} \tag{9.28}$$

for $n = 0, 1, \ldots, M-1$.

9.3.2 Discrete Gradient Algorithm

According to the previous discrete formula (9.27) we can now present the following gradient scheme in order to solve the discrete minimization problem (9.25)–(9.26).

For $k = 1, \ldots$ (the iteration index) we denote by $X_{h_k} = (\phi_{h,k}, \psi_{h,k})$ the numerical approximation of the control-disturbance at the kth iteration of the algorithm.

1. Initialization: $(\phi_{h,0}, \psi_{h,0})$ (given).
2. Compute the primal problem (9.26) with the source term $(\phi_{h,k}, \psi_{h,k})$, gives $u^n_{h,k}$, for $n = 1, 2, \ldots, M$.
3. Compute the adjoint problem (9.28) (based on $(\phi_{h,k}, \psi_{h,k}, u_{h,k})$), gives $\tilde{u}^n_{h,k}$, for $n = M-1, M-2, \ldots, 0$.
4. Compute the components of the gradient of J at $(\phi_{h,k}, \psi_{h,k})$, corresponding to all the basic functions $\{\phi_m\}$ from V_h

$$(\mathbf{GJD}) \begin{cases} c_{k,h} = \alpha\phi_{k,h} + B_1^*\tilde{u}_{k,h}, \\ d_{k,h} = -\beta\psi_{k,h} + B_2^*\tilde{u}_{k,h}. \end{cases}$$

5. Find $\lambda_{h,k} > 0$ and $\delta_{h,k}$ such that

$$J_h(\phi_{h,k} - \lambda c_{h,k}, \psi_{h,k}) \text{ is minimized over all } \lambda > 0,$$
$$J_h(\phi_{h,k}, \psi_{h,k} + \delta d_{h,k}) \text{ is maximized over all } \delta > 0.$$

6. Compute $(\phi_{h,k+1}, \psi_{h,k+1})$:

$$\begin{cases} \phi_{h,k+1} := \phi_{h,k} - \lambda_{h,k} c_{h,k}, \\ \psi_{h,k+1} := \psi_{h,k} + \delta_{h,k} d_{h,k}. \end{cases}$$

7. IF the gradient is sufficiently small (convergence) THEN end; ELSE set $k := k+1$ and GOTO 2. The discrete approximation of the optimal solution (ϕ^*, ψ^*, u^*) is $(\phi_{h,k}, \psi_{h,k}, u_{h,k})$.

For convenience, we denote the approximation of the optimal solution by the discrete gradient method, with the objective function J_h and initial guess $(\phi_{h,0}, \psi_{h,0})$, by

$$(\phi^*_{h,0}, \psi^*_{h,0}, u^*_h) = Grad_h(J_h, \phi_{h,0}, \psi_{h,0}).$$

Remark 9.5. (*i*) By using the same discretization as in the gradient method we can obtain easily the conjugate gradient method in order to solve the discrete saddle point problem (9.25)–(9.26) by using the discrete formula (9.27).
(*ii*) In order to accelerate the convergence of the non-linear gradient method, we propose in this section the non-linear multi-grid method (NMGM). Because in the iterative methods the frequency components of the residual error are obtained most rapidly on the grid corresponding to them, and the multi-grid method has the ability to use this characteristic in an effective way and thus to exploit many grids in order to converge rapidly. ◇

9.3.3 Multi-grid Gradient Method

Next we propose a non-linear multi-grid gradient method (NMGM) as in Yamamoto and Zou [305] for solving the non-linear minimization problem (9.25)-(9.26) (for more details on the multi-grid methods see, *e.g.*, Hackbusch [148], Wesseling [299] and the references therein). The basic element of NMGM is the well-known defect correction principle (see Bohmer and Steller [55]). The defect correction scheme starts with fine grid and iterates a few times using an iterative method (called pre-smoothing) and then goes to a coarser grid to solve the residual equation to achieve some coarse correction for the approximate solution obtained on the fine grid (called coarse-grid correction), again applying the same iterative method, a few iterations, for the residual equation (called post-smoothing).

In order to develop the NMGM we introduce a set of regular partitions $(\mathcal{T}_h^j)_{j=0,N}$, with $\mathcal{T}_h^{j+1}$ being the refinement of $\mathcal{T}_h^j$ and the spaces $(V_h^j)_{j=0,N}$ that are the continuous piecewise linear finite element spaces defined on $(\mathcal{T}_h^j)_{j=0,N}$, such that $V_h^0 \subset V_h^1 \subset \ldots \subset V_h^j \subset V_h^{j+1} \subset \ldots \subset V_h^{N-1} \subset V_h^N = V_h$.

Our goal is to solve the discrete saddle point problem (9.25)–(9.26), which is defined on the finest space V_h, by using the auxiliary spaces V_h^j, for $j = 0, \ldots N-1$. For this, we introduce some more notations. For each partition $\mathcal{T}_h^j$, we divide the time interval $[0, T]$ into M_j subintervals using the following points: $0 = t_j^0 < t_j^1 \ldots t_j^k < t_j^{k+1} \ldots t_j^{M_j-1} < t_j^{M_j} = T$, with $t_j^k = k\tau_j$ and $\tau_j = T/M_j$. Similar to $U_{\mathrm{ad},h}$ and $V_{\mathrm{ad},h}$, we define constrained subsets $U_{\mathrm{ad},h}^j$ and $V_{\mathrm{ad},h}^j$, respectively.

(I) *First step of NMGM:* For the initialization step of the NMGM to be introduced below, we solve the saddle point problem on each coarse space V_h^j as follows: find a saddle point in $U_{\mathrm{ad},h}^j \times V_{\mathrm{ad},h}^j$ of

$$\begin{aligned} J_{h_j}^0(\phi_{h_j}, \psi_{h_j}) = {} & \frac{\tau_j}{2} \sum_{n=0}^{M_j} \theta_n \int_\Omega | \mathcal{C}(u_{h_j}^n - u_{\mathrm{obs}}^n) |^2 \, dx \\ & + \frac{\mu}{2} \int_\Omega | u_{h_j}^{M_j} - v_{\mathrm{obs}} |^2 \, dx \\ & + \frac{\tau_j}{2} \sum_{n=0}^{M_j} \theta_n \int_\Omega \alpha \, | \phi_{h_j}^n |^2 \, dx \\ & - \frac{\tau_j}{2} \sum_{n=0}^{M} \theta_n \int_\Omega \beta \, | \psi_{h_j}^n |^2 \, dx, \end{aligned} \tag{9.29}$$

where $u_{h_j}^n \in V_h^j$ solves (9.26) on V_h^j, with the forcing (ϕ_{h_j}, ψ_{h_j}), $\theta_0 = \theta_{M_j} = 1/2$ and $\theta_n = 1$ if $n \neq 0, M_j$.

(II) *Second step of NMGM:* We introduce the coarse grid correction, for which we define the following functional:

$$
\begin{aligned}
J_{h_j}^0(\phi_h+\phi_{h_j},\psi_h+\psi_{h_j}) = {} & \frac{\tau_j}{2}\sum_{n=0}^{M_j}\theta_n\int_\Omega |\,\mathcal{C}(U_{h_j}^n-u_{\mathrm{obs}}^n)\,|^2\,\mathrm{d}x \\
& +\frac{\mu}{2}\int_\Omega |\,U_{h_j}^{M_j}-v_{\mathrm{obs}}\,|^2\,\mathrm{d}x \\
& +\frac{\tau_j}{2}\sum_{n=0}^{M_j}\theta_n\int_\Omega \alpha\,|\,\phi_h+\phi_{h_j}^n\,|^2\,\mathrm{d}x \\
& -\frac{\tau_j}{2}\sum_{n=0}^{M}\theta_n\int_\Omega \beta\,|\,\psi_h+\psi_{h_j}^n\,|^2\,\mathrm{d}x,
\end{aligned}
\tag{9.30}
$$

where $U_{h_j}^n \in V_h^j$ solves $(\forall v_{h_j} \in V_h^j)$

$$
\begin{aligned}
& \int_\Omega \partial_{\tau_j}U_{h_j}^n v_{h_j}\,\mathrm{d}x + a(U_{h_j}^n, v_{h_j}) + \int_\Omega K(\mathbf{V}_{h_j}^n, U_{h_j}^n)v_{h_j}\,\mathrm{d}x \\
& \qquad + \int_\Omega F_{h_j}^n(U_{h_j}^n)v_{h_j}\,\mathrm{d}x = \int_\Omega B_1(\phi_h^n+\phi_{h_j}^n)v_{h_j}\,\mathrm{d}x \\
& \qquad\qquad + \int_\Omega B_2(\psi_h^n+\psi_{h_j}^n)v_{h_j}\,\mathrm{d}x,
\end{aligned}
\tag{9.31}
$$

$$
U_{h_j}^0 = u_0 \quad \text{on } \Omega.
$$

(III) *Third step of NMGM:* Let us introduce the adjoint solution on each coarse space V_h^j and the evaluation of the partial derivatives of the functional $J_{h_j}(\phi_h+\phi_{h_j},\psi_h+\psi_{h_j})$ at direction $p_{h_j} \in U_{\mathrm{ad},h}^j$ and $q_{h_j} \in V_{\mathrm{ad},h}^j$, respectively:

$$
\begin{aligned}
& \frac{\partial J_{h_j}^0}{\partial \phi}(\phi_h+\phi_{h_j},\psi_h+\psi_{h_j}).p_{h_j} \\
& \qquad = \tau_j\sum_{n=0}^{M_j}\theta_n\int_\Omega (\alpha(\phi_h^n+\phi_{h_j}^n)+B_1^*\tilde{\mathbf{U}}_{h_j}^n)p_{h_j}^n\,\mathrm{d}x, \\
& \frac{\partial J_{h_j}^0}{\partial \psi}(\phi_h+\phi_{h_j},\psi_h+\psi_{h_j}).q_{h_j} \\
& \qquad = \tau_j\sum_{n=0}^{M_j}\theta_n\int_\Omega (-\beta(\psi_h^n+\psi_{h_j}^n)+B_2^*\tilde{\mathbf{U}}_{h_j}^n)q_{h_j}^n\,\mathrm{d}x,
\end{aligned}
\tag{9.32}
$$

where $\tilde{\mathbf{U}}_{h_j}^n$ is the solution of the following discrete adjoint problem $(\forall v_{h_j} \in V_h^j)$

$$
\begin{aligned}
& -\int_\Omega \partial_{\tau_j}\tilde{\mathbf{U}}_{h_j}^{n+1}v_{h_j}\,\mathrm{d}x + a(\tilde{\mathbf{U}}_{h_j}^n, v_{h_j}) + \int_\Omega (G_{h_j}^n(U_{h_j}^n))^*\tilde{\mathbf{U}}_{h_j}^n v_{h_j}\,\mathrm{d}x \\
& \qquad + \int_\Omega K^*(V_{h_j}^n, \tilde{\mathbf{U}}_{h_j}^n)v_{h_j}\,\mathrm{d}x = \int_\Omega \mathcal{C}^*\mathcal{C}(U_{h_j}^n-u_{\mathrm{obs}}^n)v_{h_j}\,\mathrm{d}x, \\
& \tilde{\mathbf{U}}_{h_j}^{M_j} = \mu(U_{h_j}^{M_j}-v_{\mathrm{obs}}) \quad \text{on } \Omega,
\end{aligned}
\tag{9.33}
$$

for $n = 0, 1, \ldots, M_j - 1$.

(IV) *Fourth step of NMGM:* The gradient method is used to solve the saddle point problem: find a saddle point of $J^0_{h_j}(\phi^n_h+\phi^n_{h_j}, \psi^n_h+\psi^n_{h_j})$, over all $(\phi^n_{h_j}, \psi^n_{h_j}) \in U^j_{\mathrm{ad},h} \times V^j_{\mathrm{ad},h}$ such that $(\phi^n_h + \phi^n_{h_j}, \psi^n_h + \psi^n_{h_j}) \in U^j_{\mathrm{ad},h} \times V^j_{\mathrm{ad},h}$.

Gradient algorithm: For $k = 1, \ldots$ (the iteration index) we denote by the pair $(\phi_{h_j,k}, \psi_{h_j,k})$ the numerical approximation of the control-disturbance at the kth iteration of the algorithm.

1. Initialization: $(\phi_{h_j,0}, \psi_{h_j,0})$ (given).
2. Compute the primal problem (9.31) with the source term $(\phi_{h_j,k}+\phi_h, \psi_{h_j,k}+\psi_h)$, gives $U^n_{h_j,k}$, for $n = 1, 2, \ldots, M_j$.
3. Compute the adjoint problem (9.33) (based on $(\phi_{h_j,k}+\phi_h, \psi_{h_j,k}+\psi_h, U_{h_j,k})$), gives $\tilde{\mathbf{U}}^n_{h,k}$, for $n = M_j-1, M_j-2, \ldots, 0$.
4. Compute the components of the gradient of J at $(\phi_{h_j,k} + \phi_h, \psi_{h_j,k} + \phi_h)$, corresponding to all the basic functions $\{\phi_m\}$ from V^j_h

$$(GJD) \begin{cases} c_{k,h} = \alpha(\phi_{h_j,k} + \phi_h) + B_1^* \tilde{\mathbf{U}}_{h_j,k}, \\ d_{k,h} = -\beta(\psi_{h_j,k} + \psi_h) + B_2^* \tilde{\mathbf{U}}_{h_j,k}. \end{cases}$$

5. Find $\lambda_{h_j,k} > 0$ and $\delta_{h_j,k}$ such that

$$J^0_{h_j}((\phi_{h_j,k} + \phi_h) - \lambda c_{h_j,k}, \psi_{h_j,k} + \psi_h) \text{ is minimized over all } \lambda > 0,$$

$$J^0_{h_j}(\phi_{h_j,k} + \phi_h, (\psi_{h_j,k} + \psi_h) + \delta d_{h_j,k}) \text{ is maximized over all } \delta > 0.$$

6. Compute $(\phi_{h_j,k+1}, \psi_{h_j,k+1})$:

$$\begin{cases} \phi_{h_j,k+1} := \phi_{h_j,k} - \lambda_{h_j,k} c_{h_j,k}, \\ \psi_{h_j,k+1} := \psi_{h_j,k} + \delta_{h_j,k} d_{h_j,k}. \end{cases}$$

7. IF the gradient is sufficiently small (convergence) THEN end; ELSE set $k := k+1$ and GOTO 2. The discrete approximation of the optimal solution $(\phi^*_{h_j}, \psi^*_{h_j}, U^*_{h_j})$ is $(\phi_{h_j,k}, \psi_{h_j,k}, U_{h_j,k})$.

Here, we denote the approximation of the optimal coarse correction of $X_h = (\phi_h, \psi_h)$ by the gradient method, with the objective function J_{h_j} and initial guess $X_{h_j,0}$, by

$$(\phi^*_{h_j}, \psi^*_{h_j}, U^*_{h_j}) = Grad_{h_j}(J_{h_j}, \phi_h, \psi_h, X_{h_j,0}).$$

We can now formulate the algorithm of the non-linear multi-grid gradient method for solving the finite element optimization problem (9.25)-(9.26).

The NMGM algorithm: Let $(\phi_{h_0}^{(0)}, \psi_{h_0}^{(0)}) \in U_{\mathrm{ad},h}^0 \times V_{\mathrm{ad},h}^0$ be a given initial guess (on the coarse finite element V_h^0).

(I) *Initialization of the coarse grid:*
  1. For $j = 0, \ldots, N-1$, do:
     IF $j \neq 0$, calculate $\phi_{h_j}^{(0)} = \Pi_{j-1,j}\phi_{h_{j-1}}^*$ and $\psi_{h_j}^{(0)} = \Pi_{j-1,j}\psi_{h_{j-1}}^*$,
     Compute $(\phi_{h_j}^*, \psi_{h_j}^*) = Grad_h(J_{h_j}^0, \phi_{h_j}^{(0)}, \psi_{h_j}^{(0)})$
     END
  2. Compute $\phi_h^{(0)} = \Pi_{N-1,N}\phi_{h_{N-1}}^*$ and $\psi_h^{(0)} = \Pi_{N-1,N}\psi_{h_{N-1}}^*$,
     where the operator $\Pi_{j-1,j}$ can be any interpolation from V_h^{j-1} onto V_h^j.

(II) *Smoothing and coarse grid correction: For* $k = 0, \ldots$ *(the iteration index):*
  3. Set $\tilde{X}_h^{(0)} = (\tilde{\phi}_h^{(0)}, \tilde{\psi}_h^{(0)}) := (\phi_h^{(0)}, \psi_h^{(0)})$
     For $j = N, \ldots, 0$, do:
     IF $j \neq N$, calculate $X_h^{(0)} = (\phi_h^{(0)}, \psi_h^{(0)}) = (\phi_h^*, \psi_h^*) = X_h^*$,
     Compute $X_h^* = (\phi_h^*, \psi_h^*) = Grad_{h_j}(J_{h_j}, X_h^{(0)}, 0)$
     END
  4. IF $\| X_h^* - \tilde{X}_h^{(0)} \|$ is sufficiently small, stop
     ELSE, set $X_h^{(0)} := X_h^*$, $k := k+1$ and GOTO 3.

Remark 9.6. We can consider, naturally, for the operator $\Pi_{j-1,j}$ the finite element injection of V_h^{j-1} into V_h^j. For the starting point $(\phi_{h_j}^0, \psi_{h_j}^0)$ we can take the value zero. ♢

Remark 9.7. In order to numerically solve the robust control problems (by taking into account the nature of the gradients), the reader can also adapt other numerical methods that exist in the literature for solving optimal control problems, in the context of the methods developed in this book. ♢

Part III

Applications in the Biological and Physical Sciences: Modeling and Stabilization

"As any human activity needs goals, mathematical research needs problems."

David Hilbert

"Nature not only suggests to us problems, she suggests their solution."

Henri Poincaré

"The miracle of the appropriateness of the language of mathematics for the formulation of the laws of physics is a wonderful gift which we neither understand nor deserve."

Eugene Wigner

We address three different topics in this part.

The first is that of material sciences. We first study the phenomenon of vortex structure in the phase transitions taking place in superconductor films with variable thickness. The objective of this chapter is to control the motion of vortices by taking into account the influence of fluctuations and noises in data (for example, thermal fluctuations and material impurities). The next chapter is about the microstructures (dendrite) that appear during isothermal solidification of a binary alloy. The goal is to predict and stabilize the microstructure dynamics (in this case, as in the previous application, the thermal fluctuations and material impurities also affect considerably solidification microstructure dynamics).

The second topic is that of oceanic currents, which have an influence on the climate system. Oceanic current movements play a key role in climate regulation. In the tropical zone, the oceanic circulation is characterized by steady zonal currents and by long waves propagating westward along the equator. These equatorial waves can be connected with strong vertical currents which are very sensitive to small changes in temperature (see, for example, the *El Niño* phenomenon). Our purpose is motivated by the robust regulation of the deviation of circulation from the mean circulation by taking into account the worst-case noise caused by a small variation of the surface temperature.

The third topic is that of biological systems. First, we analyze the transport of thermal energy in living tissues. A very important application of bioheat transfer models is in thermal therapy, especially in clinical cancer hyperthermia therapy (the goal is to destroy the pathological tissues, by applying heat to these tissues, with minimal damage to the surrounding tissues). The thermal conductivity of living tissues is a very complex process which uses different phenomenogical mechanisms, including conduction, convection, radiation and metabolism evaporation. Moreover, the studied model takes into account blood flow and extra–cellular water which affect considerably both the heat transfer in the tissues and the tissues thermal properties. Our contribution is to control and stabilize the online temperature. The last chapter deals with some general equations called *parabolic diffusive equations with multiple time-varying delays*. These systems describe resource management problems in which the objective is the stabilization of uncertain biological species. The applications are very varied: we can cite, for example, fishery resource systems, wildlife damage management, harvesting species, and so on. Dynamical population models govern diffusive biological species with logistic growth terms and time delays (such as birth rate, finished period of gestation, *etc.*).

Finally, we present other models which combine biology and fluid mechanics, namely micropolar systems (animal blood), and fluid mechanics and materials, namely semiconductors melts.

10

Vortex Dynamics in Superconductors and Ginzburg–Landau-type Models

In this chapter we formulate and study robust control problems for a two-dimensional time-dependent Ginzburg–Landau (*TDGL*) type model with Robin boundary conditions on the phase-field parameter, which describes the phase transitions taking place in superconductor films with variable thickness.

It is well known that thermal fluctuations and material impurities affect considerably the motion of vortices in superconductors. These effects are modeled by variants of the Ginzburg–Landau model containing either additive or multiplicative noise.

The objective is then to control the motion of vortices by taking into account the influence of fluctuations and noises in data. First, a variant of TDGL model (*MTDGL*) is introduced and analyzed. Second, we introduce the perturbation problem of the non-linear governing coupled system of equations MTDGL (the deviation from the desired target). The existence and uniqueness of the solution of the perturbation are proved as well as stability under mild assumptions. Afterwards the robust control problems are formulated in the case when the control is in the external magnetic field and in the case when the control is in the initial condition of the vector potential. We show the existence of an optimal solution, and we also find the necessary conditions for a saddle point optimality.

This work is a generalization of the recent research developed by Belmiloudi in [45].

10.1 Introduction

The aim of this contribution is the study of robust control problems to describe the phenomenon of vortex structure in the superconducting phase transitions, using the time-dependent Ginzburg–Landau (TDGL) complex superconductivity model. This model was derived by Gor'kov and Eliashberg in [136] from the microscopic BCS (Bardeen–Cooper–Schieffer) theory (see Bardeen *et al.* [23]) for a superconductor with paramagnetic impurities. It involves the real vector-valued potential U for the total magnetic field and a complex phase-field variable ϕ so that $|\phi|^2 = \phi\overline{\phi}$ ($\overline{\phi}$ is the complex conjugate of ϕ) gives the relative density of the superconducting charge carriers (Cooper pairs of electrons) which varies between 0 in the normal phase and 1 in the superconducting phase. The need for ϕ to be complex is associated with the macroscopic quantum nature of superconductivity. Here we are interested with the response of a superconducting material to an applied magnetic field under isothermal conditions below its critical temperature T_c (the transition from normally conducting to superconducting is usually associated with critical temperature).

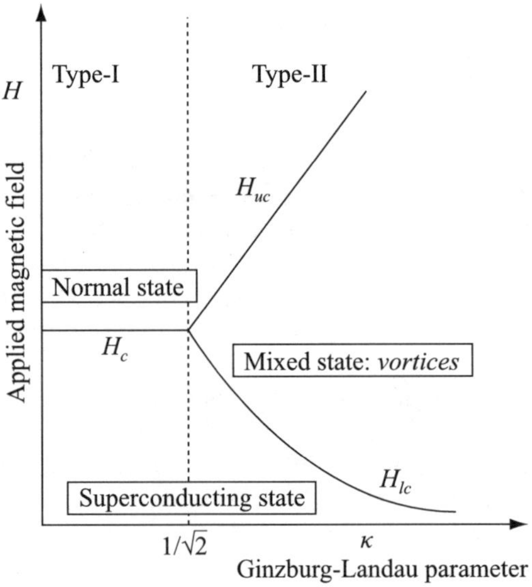

Figure 10.1. The reaction of a superconducting material to the applied magnetic field.

Most applications of superconducting materials involve type-II superconductors in high magnetic fields. It is known that for the type-II superconductors, there is a critical magnetic field which splits into a lower critical field H_{lc} and an upper critical field H_{uc} (cf. Figure 10.1). For the magnetic fields below H_{lc} the material is in the superconducting state and for magnetic field above H_{uc} the material is in the normal state. For the magnetic fields between H_{lc} and H_{uc} the material is in the mixed state. This mixed state is described by physicists as follows: around some isolated points (called vortices, which are most commonly arranged in a hexagonal arrangement, see Abrikosov [2]) inside the material, the superconducting property is destroyed and the magnetic field becomes stronger in the nearby regions surrounding these vortices. While, elsewhere, the superconducting property is still dominant and the magnetic field is excluded. Moreover, the motion of the vortices depends highly on the magnetization processes of the material (this is the result of the "Lorentz force"[1] on the magnetic flux line carried by the vortex due to the transport current, for example). The motion of the vortices is undesirable, because this motion dissipate energy and leads to an electric field. Therefore, it is very interesting to study the applied magnetic in order to prevent their motion. In practice, attempts are made to pin down vortices to particular locations in the material. Pinning down vortices inside superconductors is achieved by the presence of any form of inhomogeneity (for example, point defects, impurities or a variation in the thickness of the sample of superconducting material, see Du and Gunzburger [106]). A popular way of modeling the effect of pinning in the Ginzburg–Landau framework is to allow the equilibrium density of superconducting electrons to be a function of position (see, *e.g.*, Rubinstein [253]). In order to take into account the effect of inhomogeneities in superconducting thin films having variable thickness, we consider the following two-dimensional time-dependent and non-linear system defined on Ω:[2]

$$\begin{aligned}
&\eta\rho\frac{\partial\phi}{\partial t} - \mathrm{i}\eta\kappa\mathrm{div}(\rho U)\phi + b(U).(\rho b(U))(\phi) + \rho G(\phi) = 0 \quad \text{on } \mathcal{Q},\\
&\rho\frac{\partial U}{\partial t} + \mathrm{curl}(\rho\mathrm{curl}(U)) - \nabla(\mathrm{div}(\rho U))\\
&\qquad\qquad + \rho\mathcal{R}(b(U)(\phi)\overline{\phi}) = \mathrm{curl}(\rho H) \quad \text{on } \mathcal{Q},\\
&\text{subject to the Robin-type boundary conditions}\\
&\frac{1}{\kappa^2}\frac{\partial\phi}{\partial\mathbf{n}} = \mu\phi,\ U.\mathbf{n} = 0,\ \mathrm{curl}(U) = H \quad \text{on } \Sigma,\\
&\text{and the initial conditions}\\
&\phi(0) = \phi_0, U(0) = U_0 \quad \text{on } \Omega,
\end{aligned} \tag{10.1}$$

where $\mathcal{Q} = \Omega \times (0,T)$, $\Sigma = \partial\Omega \times (0,T)$ and the operator b (the covariant derivative) and the function G are defined by

$$\begin{aligned}
&b(U) = (\frac{\mathrm{i}}{\kappa}\nabla + U), \quad \overline{b}(U) = (-\frac{\mathrm{i}}{\kappa}\nabla + U),\\
&G(z) = (\mid z\mid^2 - \vartheta)z.
\end{aligned} \tag{10.2}$$

The equilibrium density of superconducting electron is denoted by the function $0 < \vartheta(x) < 1$ (spatially dependent). The function $\vartheta(x)$ can be thought of as measuring the quality of the superconductor. The domain Ω is an open bounded domain in $\mathbb{R}^2$ with Lipschitz boundary $\partial\Omega$, $\mathbf{n}$ is the unit normal to the surface of the superconductor $\Gamma = \partial\Omega$ and μ is an arbitrary real number (the boundary condition is appropriate for the superconductor interface with

---

[1] This force is more important than the anchorage forces of vortices, and causes the displacement of the vortex.

[2] This problem is corresponding to three-dimensional time-dependent model defined on $\Omega \times (-\epsilon\rho(x), \epsilon\rho(x))$, when ϵ tends to zero.

vacuum (or insulator) if $\mu = 0$ and for the superconductor interface with normal metal if $\mu \neq 0$). $\mathcal{R}(.)$ (resp. $\mathcal{I}(.)$) denotes the real part (resp. the imaginary part) of the quantity in (.) and curl denote the curl operators defined by (on the (x, y)-plane)

$$\begin{aligned} \text{curl}(\phi) &= (\frac{\partial \phi}{\partial y}, -\frac{\partial \phi}{\partial x})^T \ (\phi \text{ is a scalar}), \\ \text{curl}(U) &= \frac{\partial u_2}{\partial x} - \frac{\partial u_1}{\partial y} \ (U = (u_1, u_2) \text{ is a vector}). \end{aligned} \tag{10.3}$$

$\text{curl}(U)$ is the induced magnetic field, $\mathbf{J} = \text{curl}(\rho \text{curl}(U))$ is the current, H is the applied magnetic field, η is the non–dimensional diffusivity and $\rho > 0$ is a smooth function characterizing the vertical shape of the superconducting film and satisfying the following hypothesis:

$$\rho \in C^1(\overline{\Omega}) \text{ such that } \rho_0 \leq \rho \leq \rho_1 \text{ and } \mid \nabla \rho \mid \leq \rho_{\text{d}}, \tag{10.4}$$

where $(\rho_0, \rho_1, \rho_{\text{d}})$ are non-negative constants.

The positive constant κ is the Ginzburg–Landau parameter with $\kappa = \hbar/\ell$, where ℓ is a coherence length describing the size of thermodynamic fluctuations in the superconducting phase, and $\hbar$ is the London penetration depth describing the depth to which an external magnetic field can penetrate the superconductor. The parameter κ determines the type of superconducting material: $\kappa < 1/\sqrt{2}$ describes type-I superconductors, $\kappa > 1/\sqrt{2}$ describes type-II superconductors.

Various problems associated with the Ginzburg–Landau models in superconductivity have been studied recently (the literature on this model is vast, see, *e.g.*, Bertuel *et al.* [51], Chapman *et al.* [72, 73], Chen *et al.* [79, 80], Coskun and Kwong [88], Deckelnick *et al.* [97], Du *et al.* [107, 106, 105], Gropp *et al.* [141], Sandier and Serfaty [258], Tang and Wang [279] and the references therein). For the optimal control problems associated with the TDGL models, we can mention Chen and Hoffmann [81] in which the authors studied the control of the vortices in superconducting films through the external magnetic field. These works are applicable only to highly idealized physical situations that do not take into account factors such as inhomogeneities and thermal fluctuations. For example, it is well known that thermal fluctuations and material defects play an important role in the pinning of vortex in type-II superconductors. Moreover, recently, it was experimentally observed that as the temperature approaches the transition temperature and the effects of thermal fluctuations increase, the vortex lattice melts and moves towards a vortex–liquid state (see, *e.g.*, Ling *et al.* [200]). For the numerical analysis of vortexex in thermal equilibrium, by taking into account fluctuation, see, *e.g.*, Deang *et al.* [96], Sasik *et al.* [259] and the references therein.

It is then clear that it, in order to study the stability, dynamics and other properties of the vortex state, is very important to take into account the effects of thermal fluctuation and material defects on the vortex dynamics in type-II superconductors. In this study we consider, for thermal fluctuation, only the additive noise Ginzburg–Landau model, which is a simple modification of the Ginzburg–Landau model (10.1). More precisely, we introduce into the right-hand side of the first equation of (10.1) a complex-valued field in time and space. In this way, thermal fluctuations are modeled by deterministic Langevin-type dynamics. Thus, the additive deterministic noise Ginzburg–Landau model is given by

$$\begin{aligned}
&\eta\rho\frac{\partial\phi}{\partial t} - i\eta\kappa\mathrm{div}(\rho U)\phi + b(U).(\rho b(U))(\phi) + \rho G(\phi) = \rho\beta \quad \text{on } \mathcal{Q},\\
&\rho\frac{\partial U}{\partial t} + \mathrm{curl}(\rho\mathrm{curl}(U)) - \nabla(\mathrm{div}(\rho U))\\
&\qquad\qquad + \rho\mathcal{R}(b(U)(\phi)\overline{\phi}) = \mathrm{curl}(\rho H) \quad \text{on } \mathcal{Q},\\
&\text{subject to the Robin-type boundary conditions}\\
&\frac{1}{\kappa^2}\frac{\partial\phi}{\partial\mathbf{n}} = \mu\phi,\; U.\mathbf{n} = 0,\; \mathrm{curl}(U) = H \quad \text{on } \Sigma,\\
&\text{and the initial conditions}\\
&\phi(0) = \phi_0, U(0) = U_0 \quad \text{on } \Omega,
\end{aligned} \tag{10.5}$$

where β is a deterministic complex-valued field in time and space which is temperature dependent (it is depending on ratio between the temperature and the critical transition temperature T_c).[3] The choice of a deterministic noise is motivated by the fact that, if we consider the individual motion of a molecule from a microscopic point of view, the Brownian motion is a deterministic motion (see Shimizu and Morioka [269] and Shimizu and Yaghi [270]).

Here, we consider robust control problems, for the modified TDGL models (MTDGL) with Robin boundary conditions on the phase-field variable, which describes the phase transitions taking place in superconductor films, in order to take into account the influence of data noise. Indeed, such perturbations (noise) have the effect of impeding the ability of the material to become superconducting.

10.1.1 Assumptions and Notation

We denote by $V_n = \{U \in H^1(\Omega);\; U.\mathbf{n} = 0 \;\text{ on } \Omega\}$ and V_n' the dual of V_n. We denote by $<,>_{V_n',V_n}$ the duality product between V_n' and V_n. For any pair of real number $r, s \geq 0$, we introduce the Sobolev space $H^{r,s}(\mathcal{Q})$ defined by $H^{r,s}(\mathcal{Q}) = L^2(0,T;H^r(\Omega)) \cap H^s(0,T;L^2(\Omega))$, which is a Hilbert space normed by

$$\left(\int_0^T \| v \|^2_{H^r(\Omega)}\,\mathrm{d}t + \| v \|^2_{H^s(0,T;L^2(\Omega))}\right)^{1/2},$$

[3] β is said to be the chaotic-type force.

where $H^s(0,T;L^2(\Omega))$ denotes the Sobolev space of order s of functions defined on $(0,T)$ and taking values in $L^2(\Omega)$.

Remark 10.1. According to Girault and Raviart[134], we have the following embedding inequality on V_n:

$$\| U \|_{H^1}^2 \leq C(\| U \|_{L^2}^2 + \| \operatorname{div}(U) \|_{L^2}^2 + \| \operatorname{curl}(U) \|_{L^2}^2), \forall U \in V_n. \tag{10.6}$$

◊

Since the unknown function ϕ is complex valued, for the mathematical setting, we are naturally led here to introduce complex valued spaces (*this is the only occasion in this book in which such spaces appear*).

If X denotes some Banach space of real-valued functions, the corresponding space (its complexified space) of complex-valued functions will be denoted by $\mathcal{X}$ and the corresponding space of vector-valued functions, each of components belonging to X, will be denoted by $\mathbf{X}$, and we use $\| \,.\, \|_X$ to denote the norms of spaces X, $\mathbf{X}$ or $\mathcal{X}$. For example $\mathcal{L}^2(\Omega)$ is the complexified space of $L^2(\Omega)$ and $\mathbf{L}^2(\Omega)$ is the corresponding space of vector-valued functions, of $L^2(\Omega)$.

Remark 10.2. The norm on L^2 spaces is denoted here by $\| \,.\, \|_{L^2(\Omega)}$ or $\| \,.\, \|_{L^2}$, to avoid any confusion with the modulus $| \,.\, |$, which is frequently used. ◊

We can now introduce the following spaces

$$\begin{aligned}
&\mathbf{W}_n = L^2(0,T;V_n) \cap H^1(0,T;V_n'),\\
&\mathbf{E}_n = L^2(0,T;V_n) \cap L^\infty(0,T;\mathbf{L}^2(\Omega)),\\
&\mathcal{W} = L^2(0,T;\mathcal{H}^1(\Omega)) \cap H^1(0,T;(\mathcal{H}^1)'(\Omega)),\\
&\mathcal{E} = L^2(0,T;\mathcal{H}^1(\Omega)) \cap L^\infty(0,T;\mathcal{L}^2(\Omega)),\\
&\mathcal{L}^2_\infty(D) = \{\phi \in \mathcal{L}^2(D);\ |\,\phi\,| \text{ is bounded a.e. in } D\},
\end{aligned} \tag{10.7}$$

where $D = \Omega$ or $\mathcal{Q}$.

Remark 10.3. (*i*) $\mathcal{W}$ and $\mathbf{W}_n$ are continuously embedded into $\mathcal{C}([0,T];\mathcal{L}^2(\Omega))$ and $\mathcal{C}([0,T];\mathbf{L}^2(\Omega))$, respectively (see Lemma 6.6).
(*ii*) Although $\mathcal{L}^2_\infty$ is a subset of $\mathcal{L}^\infty$, we have described this space using the standard norm of the space $\mathcal{L}^2$. ◊

The weak formulation associated with problem (10.5) is then to find $(\phi,U) \in \mathcal{W} \times \mathbf{W}_n$ such that (a.e. $t \in (0,T)$)

$$\begin{aligned}
&\eta \int_\Omega \rho \frac{\partial \phi}{\partial t} q \mathrm{d}x - \mathrm{i}\eta\kappa \int_\Omega \operatorname{div}(\rho U)\phi q \mathrm{d}x + \int_\Omega \rho b(U)(\phi)\bar{b}(U)(q)\mathrm{d}x\\
&\qquad -\mathrm{i}\mu \int_\Gamma \rho\phi q \mathrm{d}\Gamma + \int_\Omega \rho G(\phi) q \mathrm{d}x = \int_\Omega \rho\beta q \mathrm{d}x \quad \forall q \in \mathcal{H}^1(\Omega),\\
&\int_\Omega \rho \frac{\partial U}{\partial t} v \mathrm{d}x + \int_\Omega \rho \operatorname{curl}(U)\operatorname{curl}(v)\mathrm{d}x + \int_\Omega \operatorname{div}(\rho U)\operatorname{div}(v)\mathrm{d}x\\
&\qquad + \int_\Omega \rho \mathcal{R}(b(U)(\phi)\overline{\phi}) v \mathrm{d}x = \int_\Omega \rho H \operatorname{curl}(v)\mathrm{d}x \quad \forall v \in V_n,\\
&\phi(0) = \phi_0, U(0) = U_0 \quad \text{on } \Omega.
\end{aligned} \tag{10.8}$$

10.1.2 Preliminary Results

Lemma 10.4. *(i) For all $z \in \mathbb{C}$, we have that $G(z)\overline{z} \in \mathbb{R}$.*
(ii) For all $(z_1, z_2) \in \mathbb{C}^2$, we have that

$$G(z_1) - G(z_2) = (| z_1 |^2 + | z_2 |^2 - \vartheta)(z_1 - z_2) + z_1 z_2 (\overline{z_1 - z_2}).$$

Lemma 10.5. *For (φ, u) and (ψ, v) sufficiently regular, we have:*

(i) $b(u)(\varphi) = b(v)(\varphi) + (u - v)\varphi$

(ii) $b(u)(\varphi) - b(u)(\psi) = b(u)(\varphi - \psi)$

(iii) $b(u)(\varphi).\overline{b(u)(\varphi)} = \frac{1}{\kappa^2} | \nabla\varphi |^2 + | u |^2 | \varphi |^2 - \frac{2}{\kappa}\mathcal{I}(\overline{\varphi}\nabla\varphi).u.$

The proof of the previous lemmas are immediate. □

Lemma 10.6. *For (u, v, w, X) sufficiently regular, we have:*

(i) $\| u \|_{H^1} \| v \|_{L^4} \| X \|_{L^4} \leq C_1 \| u \|_{H^1}^2 \| v \|_{L^2} + \delta \| \nabla X \|_{L^2}^2 + C_2(\| v \|_{H^1} + \| v \|_{H^1}^2) \| X \|_{L^2}^2,$

with δ chosen suitably at each situation

(ii) $\| u \|_{L^4} \| v \|_{L^4} \| w \|_{L^4} \| X \|_{L^4} \leq C_1 \| u \|_{L^4}^2 \| v \|_{L^4}^2 + \gamma \| \nabla X \|_{L^2}^2 + C_2(\| w \|_{L^4}^4 + \| w \|_{L^4}^2) \| X \|_{L^2}^2,$

with γ chosen suitably at each situation.

Proof. For the prove of this lemma, see Belmiloudi [45]. □

10.2 Existence and Uniqueness of the Solution of the MTDGL Model

The following results concern the existence and uniqueness of the solution of the modified Ginzburg–Landau model with Robin-type boundary conditions on the phase-field parameter (10.5).

Theorem 10.7. *For any function $\vartheta \in L^2(\Omega)$ such that $0 < M_1 \leq \vartheta \leq M_2 < 1$ a.e. in Ω, $H \in L^2(\mathcal{Q})$, and $(\phi_0, U_0, \beta) \in \mathcal{L}^2_\infty(\Omega) \times \mathbf{L}^2(\Omega) \times \mathcal{L}^2_\infty(\mathcal{Q})$ satisfying $| \phi_0 | \leq 1$ a.e. in Ω and $| \beta | < 1 - M_2$ a.e. in $\mathcal{Q}$, there exists a unique solution $(\phi, U) \in (\mathcal{W} \cap \mathcal{L}^\infty(\mathcal{Q})) \times \mathbf{W}_n$ of (10.8) satisfying $| \phi | \leq 1$ a.e. in $\mathcal{Q}$. Moreover the following estimate holds:*

$$\| \phi \|_{W}^2 + \| U \|_{W_n}^2 \leq C(\| H \|_{L^2(\mathcal{Q})}^2 + \| \phi_0 \|_{L^2(\Omega)}^2 + \| U_0 \|_{L^2(\Omega)}^2 + \| \beta \|_{L^2(\mathcal{Q})}^2).$$

Proof. The proof of this theorem is obtained by using the same technique as in Belmiloudi [45] (see also Hoffmann *et al.* [79, 81]). The existence of weak solutions and the estimate $| \phi | \leq 1$ are obtained by constructing approximate solutions of System (10.8) by the semidiscretized time approximation technique, by using the maximum principle (see Remark 10.8) and by employing standard convergence arguments. The uniqueness of the solution and the estimate given in the theorem can be obtained by the standard energy estimate arguments and Gronwall's formula. So, we omit the details. □

Remark 10.8. (*i*) The estimate $|\phi| \leq 1$ can be obtained by using the maximum principle as follows: by choosing $q = \overline{\phi}(|\phi|^2 - 1)^+ = \overline{\phi} r^+$ in the first part of (10.8) and by taking the real part we have ($\Omega^+ = \{x \in \Omega : |\phi|^2 - 1 > 0\}$):

$$\frac{\eta\rho_0}{2}\frac{\mathrm{d}\,\| r^+ \|_{L^2}^2}{\mathrm{d}t} + \int_{\Omega^+} \rho\, |\phi|^2\, (|\phi|^2 - \vartheta) r^+ \mathrm{d}x \leq \int_{\Omega^+} \rho\, |\beta|\, |\phi|\, r^+ \mathrm{d}x,$$

and then

$$\frac{\eta\rho_0}{2}\frac{\mathrm{d}\,\| r^+ \|_{L^2}^2}{\mathrm{d}t} + \int_{\Omega} \rho\, |\phi|^2\, (r^+)^2 \mathrm{d}x \leq \int_{\Omega^+} \rho(|\beta| - (1-\vartheta)\,|\phi|)\,|\phi|\, r^+ \mathrm{d}x.$$

Since $|\beta| \leq 1 - \|\vartheta\|_{L^\infty}$ and $\|\vartheta\|_{L^\infty} < 1$ then $|\beta| - (1-\vartheta)\,|\phi| \leq 0$ on Ω^+. Consequently, $\mathrm{d}\,\| r^+ \|_{L^2}^2 / \mathrm{d}t \leq 0$. By integrating with respect to time and by using the fact that $(|\phi_0|^2 - 1)^+ = 0$ (since $|\phi_0| \leq 1$) we can deduce that $|\phi| \leq 1$ a.e. in $\mathcal{Q}$.

(ii) Throughout the chapter, we suppose that the hypotheses of Theorem 10.7 are satisfied, to ensure that the solution of problem (10.5), is in $(\mathcal{W} \cap \mathcal{L}^\infty(\mathcal{Q})) \times \mathbf{W}_n$. ♢

10.3 The Perturbation Problem

10.3.1 Formulation of the Perturbation Problem

In the following, the solution (ϕ, U) of problem (10.5) will be treated as the target function. We are then interested in the robust regulation of the deviation of the problem from the desired target (ϕ, U). We analyze the full non-linear equation which models large perturbations (φ, u) to the target (ϕ, U), *i.e.*, we assume that (ϕ, U) satisfies the problem (10.5) with the data (ϕ_0, U_0, H, ξ) and $(\phi + \varphi, U + u)$ satisfies the problem (10.5) with the data $(\phi_0 + \varphi_0, U_0 + u_0, h + H, \xi + \beta)$. Hence, we consider the following system (for (ϕ, U) given satisfying the regularity of Theorem 10.7):

$$\begin{aligned}
&\eta\rho\frac{\partial\varphi}{\partial t} - \mathrm{i}\eta\kappa\mathrm{div}(\rho(u+U))\varphi - \mathrm{i}\eta\kappa\mathrm{div}(\rho u)\phi \\
&\qquad -b(U).(\rho b(U))(\phi) + b(u+U).(\rho b(u+U))(\varphi+\phi) \\
&\qquad +\rho(G(\phi+\varphi) - G(\phi)) = \rho\xi \quad \text{on } \mathcal{Q}, \qquad\qquad (10.9) \\
&\rho\frac{\partial u}{\partial t} + \mathrm{curl}(\rho\mathrm{curl}(u)) - \nabla(\mathrm{div}(\rho u)) \\
&\qquad +\rho\mathcal{R}(b(u+U)(\varphi+\phi)(\overline{\varphi}+\overline{\phi}) - b(U)(\phi)\overline{\phi}) = \mathrm{curl}(\rho h) \quad \text{on } \mathcal{Q},
\end{aligned}$$

subject to the Robin-type boundary conditions

$$\frac{1}{\kappa^2}\frac{\partial\varphi}{\partial\mathbf{n}} = \mu\varphi,\ u.\mathbf{n} = 0,\ \mathrm{curl}(u) = h \ \text{ on } \Sigma, \tag{10.10}$$

and the initial conditions

$$\varphi(0) = \varphi_0, u(0) = u_0 \ \text{ on } \Omega.$$

If we set $F(\varphi) = G(\varphi+\phi) - G(\phi)$, $B(u) = b(U+u)$ then (10.9) and (10.10) are reduced to

$$\begin{aligned}
&\eta\rho\frac{\partial\varphi}{\partial t} - \mathrm{i}\eta\kappa\mathrm{div}(\rho(u+U))\varphi - \mathrm{i}\eta\kappa\mathrm{div}(\rho u)\phi + \rho F(\varphi) \\
&\qquad +B(u).(\rho B(u))(\varphi+\phi) = B(0).(\rho B(0))(\phi) + \rho\xi \ \text{ on } \mathcal{Q}, \\
&\rho\frac{\partial u}{\partial t} + \mathrm{curl}(\rho\mathrm{curl}(u)) - \nabla(\mathrm{div}(\rho u)) + \rho\mathcal{R}(B(u)(\varphi+\phi)(\overline{\varphi+\phi})) \\
&\qquad = \rho\mathcal{R}(B(0)(\phi)\overline{\phi}) + \mathrm{curl}(\rho h) \ \text{ on } \mathcal{Q},
\end{aligned} \tag{10.11}$$

subject to the Robin-type boundary conditions

$$\frac{1}{\kappa^2}\frac{\partial\varphi}{\partial\mathbf{n}} = \mu\varphi,\ u.\mathbf{n} = 0,\ \mathrm{curl}(u) = h \ \text{ on } \Sigma,$$

and the initial conditions

$$\varphi(0) = \varphi_0, u(0) = u_0 \ \text{ on } \Omega.$$

Now we give the weak formulation associated with problem (10.11). Multiplying the first part of (10.11) by $q \in \mathcal{H}^1(\Omega)$ and the second part by $v \in V_n$ and integrating over Ω this gives (according to the third part of (10.11)) the weak formulation (a.e. $t \in (0,T)$)

$$\begin{aligned}
&\eta\int_\Omega \rho\frac{\partial\varphi}{\partial t}q\mathrm{d}x - \mathrm{i}\eta\kappa\int_\Omega \mathrm{div}(\rho(u+U))\varphi q\mathrm{d}x - \mathrm{i}\eta\kappa\int_\Omega \mathrm{div}(\rho u)\phi q\mathrm{d}x \\
&\qquad -\mathrm{i}\mu\int_\Gamma \rho\varphi q\mathrm{d}\Gamma + \int_\Omega (\rho B(u)(\varphi+\phi)\overline{B}(u)(q)\mathrm{d}x \\
&\qquad + \int_\Omega \rho F(\varphi)q\mathrm{d}x = \int_\Omega \rho B(0)(\phi)\overline{B}(0)(q)\mathrm{d}x + \int_\Omega \rho\xi q\mathrm{d}x, \\
&\int_\Omega \rho\frac{\partial u}{\partial t}v\mathrm{d}x + \int_\Omega \rho\mathrm{curl}(u)\mathrm{curl}(v)\mathrm{d}x + \int_\Omega \mathrm{div}(\rho u)\mathrm{div}(v)\mathrm{d}x \\
&\qquad + \int_\Omega \rho\mathcal{R}((\overline{\varphi}+\overline{\phi})B(u)(\varphi+\phi))v\mathrm{d}x \\
&\qquad = \int_\Omega \rho\mathcal{R}(\overline{\phi}B(0)(\phi))v\mathrm{d}x + \int_\Omega \rho h\mathrm{curl}(v)\mathrm{d}x, \\
&(\varphi(0), u(0)) = (\varphi_0, u_0).
\end{aligned} \tag{10.12}$$

10.3.2 Existence and Stability Results

Now we show the existence and uniqueness of the solution to problem (10.12), and give some Lipschitz continuity results.

Theorem 10.9. *(i) For any* $(\varphi_0, u_0) \in \mathcal{L}^2_\infty(\Omega) \times \mathbf{L}^2(\Omega)$, $\xi \in \mathcal{L}^2_\infty(\mathcal{Q})$ *and* $h \in L^2(\mathcal{Q})$, *there exists a unique* $(\varphi, u) \in (\mathcal{W} \cap \mathcal{L}^\infty(\mathcal{Q})) \times \mathbf{W}_n$ *solution of (10.12). Moreover, the following estimate holds:*

$$\| \varphi \|^2_{\mathcal{W}} + \| u \|^2_{\mathbf{W}_n} \leq C(\| h \|^2_{L^2(\mathcal{Q})} + \| \varphi_0 \|^2_{\mathcal{L}^2(\Omega)} + \| u_0 \|^2_{\mathbf{L}^2(\Omega)} + \| \xi \|^2_{\mathcal{L}^2(\mathcal{Q})}).$$

(ii) Let $(u_{01}, \varphi_{01}, h_1, \xi_1)$ *and* $(u_{02}, \varphi_{02}, h_2, \xi_2)$ *be functions of the space* $\mathbf{L}^2(\Omega) \times \mathcal{L}^2_\infty(\Omega) \times L^2(\mathcal{Q}) \times \mathcal{L}^2_\infty(\mathcal{Q})$. *If* $(u_1, \varphi_1) \in \mathbf{W}_n \times (\mathcal{W} \cap \mathcal{L}^\infty(\mathcal{Q}))$ *(respectively* $(u_2, \varphi_2) \in \mathbf{W}_n \times (\mathcal{W} \cap \mathcal{L}^\infty(\mathcal{Q}))$*) is the solution of (10.12) with the given data* $(\varphi_{01}, u_{01}, h_1, \xi_1)$ *(respectively* $(\varphi_{02}, u_{02}, h_1, \xi_2)$*), then the following Lipshitz continuity result holds:*

$$\| \varphi \|^2_{\mathcal{W}} + \| u \|^2_{\mathbf{W}_n} \leq C(\| \varphi_0 \|^2_{\mathcal{L}^2} + \| u_0 \|^2_{\mathbf{L}^2} + \| h \|^2_{L^2(\mathcal{Q})} + \| \xi \|^2_{\mathcal{L}^2(\mathcal{Q})}),$$

where $\varphi = \varphi_1 - \varphi_2$, $u = u_1 - u_2$, $\varphi_0 = \varphi_{01} - \varphi_{02}$, $u_0 = u_{01} - u_{02}$, $h = h_1 - h_2$ *and* $\xi = \xi_1 - \xi_2$.

Proof. The proof of this result can be obtained by using a technique that is similar to the one used in Belmiloudi [45]. So, we omit the details. □

We are now going to study the differentiability of the operator solution of problem (10.12).

10.4 Differentiability of the Operator Solution

Before proceeding to the investigation of the F-differentiability of the function $\mathcal{F} : (\varphi_0, u_0, h, \xi) \longrightarrow (\varphi, u)$, which maps the source term $(\varphi_0, u_0, h, \xi) \in \mathcal{L}^2_\infty(\Omega) \times \mathbf{L}^2(\Omega) \times L^2(\mathcal{Q}) \times \mathcal{L}^2_\infty(\mathcal{Q})$ of problem (10.12) into the corresponding solution $(\varphi, u) \in \mathcal{E} \times \mathbf{E}_n$, we study, for $(\psi_0, w_0, \lambda, k)$ be given data, the following problem $(\mathcal{P}_I)$:

find $(\psi, w) \in \mathcal{E} \times \mathbf{E}_n$ such that $(\forall (q, v) \in \mathcal{H}^1(\Omega) \times V_n$ and a.e. $t \in (0, T))$:

$$\begin{aligned}
&\eta \int_\Omega \rho \frac{\partial \psi}{\partial t} q \mathrm{d}x - \mathrm{i}\eta\kappa \int_\Omega \mathrm{div}(\rho U_1)\psi q \mathrm{d}x - \mathrm{i}\eta\kappa \int_\Omega \mathrm{div}(\rho w)\phi_1 q \mathrm{d}x \\
&\quad -\mathrm{i}\mu \int_\Gamma \rho\psi q \mathrm{d}\Gamma + \int_\Omega \rho(B(u)(\psi) + \phi_1 w)\overline{B}(u)(q)\mathrm{d}x \\
&\quad + \int_\Omega \rho B(u)(\phi_1) w q \mathrm{d}x + \int_\Omega \rho((2 \mid \phi_1 \mid^2 - \vartheta)\psi + \phi_1^2 \overline{\psi}) q \mathrm{d}x \\
&\quad = \int_\Omega \rho\lambda q \mathrm{d}x,
\end{aligned}$$

$$\int_\Omega \rho \frac{\partial w}{\partial t} v \mathrm{d}x + \int_\Omega \rho \mathrm{curl}(w)\mathrm{curl}(v)\mathrm{d}x + \int_\Omega \mathrm{div}(\rho w)\mathrm{div}(v)\mathrm{d}x$$
$$+ \int_\Omega \rho \mathcal{R}(\overline{\psi} B(u)(\phi_1)) v \mathrm{d}x + \int_\Omega \rho(\mathcal{R}(\overline{\phi_1} B(u)(\psi)) + w \mid \phi_1 \mid^2) v \mathrm{d}x$$
$$= \int_\Omega \rho k \mathrm{curl}(v)\mathrm{d}x,$$
$$(\psi(0), w(0)) = (\psi_0, w_0),$$

where $U_1 = U + u$ and $\phi_1 = \phi + \varphi$.

Remark 10.10. The problem $(\mathcal{P}_I)$ is the weak formulation of the problem:

$$\begin{aligned}
&\eta\rho \frac{\partial \psi}{\partial t} - \mathrm{i}\eta\kappa(\mathrm{div}(\rho U_1)\psi + \mathrm{div}(\rho w)\phi_1) \\
&\quad + B(u).(\rho(B(u)(\psi) + \phi_1 w)) + \rho B(u)(\phi_1) w \\
&\quad + \rho((2 \mid \phi_1 \mid^2 - \vartheta)\psi + \phi_1^2 \overline{\psi}) = \rho\lambda \ \text{ on } \mathcal{Q}, \\
&\rho \frac{\partial w}{\partial t} + \mathrm{curl}(\rho \mathrm{curl}(w)) + \nabla(\mathrm{div}(\rho w)) + \rho \mathcal{R}(\overline{\psi} B(u)(\phi_1)) \\
&\quad + \rho(\mathcal{R}(\overline{\phi_1} B(u)(\psi)) + w \mid \phi_1 \mid^2) = \mathrm{curl}(\rho k) \ \text{ on } \mathcal{Q},
\end{aligned} \tag{10.13}$$

subject to the Robin-type boundary conditions

$$\frac{1}{\kappa^2}\frac{\partial \psi}{\partial \mathbf{n}} = \mu\psi, \ w.\mathbf{n} = 0, \ \mathrm{curl}(w) = k \ \text{ on } \Sigma,$$

and the initial conditions

$$(\psi(0), w(0)) = (\psi_0, w_0) \ \text{ on } \Omega.$$

Theorem 10.11. *If* (u, φ) *and* (U, ϕ) *are in* $\mathbf{W}_n \times (\mathcal{W} \cap \mathcal{L}^\infty(\mathcal{Q}))$, *the following results hold:*

(*i*) *For any* $(\psi_0, w_0, k, \lambda) \in \mathcal{L}^2_\infty(\Omega) \times \mathbf{L}^2(\Omega) \times L^2(\mathcal{Q}) \times \mathcal{L}^2_\infty(\mathcal{Q})$, *there exists a unique couple of functions* $(\psi, w) \in \mathcal{E} \times \mathbf{E}_n$, *solution of problem* $(\mathcal{P}_I)$, *such that*

$$\begin{aligned}
&\| \psi \|^2_{\mathcal{E}} + \| w \|^2_{\mathbf{E}_n} \\
&\quad \leq C_e(\| \psi_0 \|^2_{L^2} + \| w_0 \|^2_{L^2} + \| k \|^2_{L^2(\mathcal{Q})} + \| \lambda \|^2_{L^2(\mathcal{Q})}).
\end{aligned} \tag{10.14}$$

(*ii*) *Let* $(\psi_{0i}, w_{0i}, k_i, \lambda_i)$, $i = 1, 2$ *be in* $\mathcal{L}^2_\infty(\Omega) \times \mathbf{L}^2(\Omega) \times L^2(\mathcal{Q}) \times \mathcal{L}^2_\infty(\mathcal{Q})$. *If* (ψ_i, w_i) *is the solution of* $(\mathcal{P}_I)$, *corresponding to data* $(\psi_{0i}, w_{0i}, k_i, \lambda_i)$ *for* $i = 1, 2$, *respectively, then*

$$\begin{aligned}
&\| \psi \|^2_{\mathcal{E}} + \| w \|^2_{\mathbf{E}_n} \\
&\quad \leq C_e(\| \psi_0 \|^2_{L^2} + \| w_0 \|^2_{L^2} + \| k \|^2_{L^2(\mathcal{Q})} + \| \lambda \|^2_{L^2(\mathcal{Q})})
\end{aligned} \tag{10.15}$$

where $w = w_1 - w_2$, $\psi = \psi_1 - \psi_2$, $w_0 = w_{01} - w_{02}$, $\psi_0 = \psi_{01} - \psi_{02}$, $k = k_1 - k_2$ *and* $\lambda = \lambda_1 - \lambda_2$.

Proof. The existence, uniqueness and Lipschitz continuity results of problem $(\mathcal{P}_I)$ are obtained in the same way as used to prove Theorem 10.9 and by using the regularity of (U_1, ϕ_1). For more details see Belmiloudi [45]. □

We are now going to study the F-differentiability of the operator solution $\mathcal{F}$. For simplicity, we denote by X the data $X := (\varphi_0, u_0, h, \xi)$, by $\mathcal{Z}$ the space $\mathcal{Z} := \mathcal{L}^2(\Omega) \times \mathbf{L}^2(\Omega) \times L^2(\mathcal{Q}) \times \mathcal{L}^2(\mathcal{Q})$ and by $\mathcal{Z}_\infty$ the space $\mathcal{Z}_\infty := \mathcal{L}^2_\infty(\Omega) \times \mathbf{L}^2(\Omega) \times L^2(\mathcal{Q}) \times \mathcal{L}^2_\infty(\mathcal{Q})$. The space $\mathcal{Z}$ and $\mathcal{Z}_\infty$ are equipped with the norm $\| X \|_{\mathcal{Z}} = (\| \varphi_0 \|^2_{L^2} + \| u_0 \|^2_{L^2} + \| h \|^2_{L^2(\mathcal{Q})} + \| \xi \|^2_{L^2(\mathcal{Q})})^{1/2}$ for all $X = (\varphi_0, u_0, h, \xi)$ in $\mathcal{Z}$ (or in $\mathcal{Z}_\infty$).

Theorem 10.12. *Let $X = (\varphi_0, u_0, h, \xi)$ and $Y = (p_0, q_0, z, \pi)$ be in $\mathcal{Z}_\infty$ with $\mathcal{F}(X)$ and $\mathcal{F}(X+Y)$ being the corresponding solutions of (10.12). Then*

$$\| \mathcal{F}(X) - \mathcal{F}(X+Y) - \mathcal{F}'(X).Y \|_{\mathcal{E}\times\mathbf{E}_n} \leq C \| Y \|^{3/2}_{\mathcal{Z}}, \tag{10.16}$$

where $\mathcal{F}'(X) : \mathcal{Z}_\infty \longrightarrow \mathcal{E}\times\mathbf{E}_n$ is a linear operator such that $(\psi, w) = \mathcal{F}'_\varphi(X).Y$ is the solution of the problem $(\mathcal{P}_I)$ with the initial condition $(\psi, w)(t=0) = (p_0, q_0)$ and the forcing $(k, \lambda) = (z, \pi)$.

Moreover, for all $X_i = (\varphi_{0i}, u_{0i}, h_i, \xi_i) \in \mathcal{Z}_\infty$, for $i = 1, 2$, we have the following estimate:

$$\begin{aligned} &\| \mathcal{F}'(X_1).Y - \mathcal{F}'(X_2).Y \|^2_{\mathcal{E}\times\mathbf{E}_n} \\ &\qquad \leq C_e(\| Y \|_{\mathcal{Z}} \| X \|^2_{\mathcal{Z}} + \| Y \|^2_{\mathcal{Z}} \| X \|_{\mathcal{Z}}), \end{aligned} \tag{10.17}$$

where $\varphi_0 = \varphi_{01} - \varphi_{02}$, $u_0 = u_{01} - u_{02}$, $h = h_1 - h_2$, $\xi = \xi_1 - \xi_2$ and $X = X_1 - X_2 = (\varphi_0, u_0, h, \xi)$.

Proof. The proof of this theorem can be obtained by using the same technique as used to prove the results of Theorem 4.3 of Belmiloudi [45]. So, we omit the details. □

10.5 Robust Control Problems

The objective of the robust control problem is to find the best admissible control in the presence of the worst disturbance which maximally spoils the control objective. We formulate the problem for two situations: first, where the control is in the external magnetic field and, second, where the control is in the initial condition u_0 (data assimilation).

10.5.1 Control in the External Magnetic Field

In this section, we consider two situations: first, where the worst disturbance is in the chaotic-type force ξ and, second, where the disturbance is in the external magnetic field h.

Distributed Disturbance in the Chaotic-Type Force

We suppose now that the control is in the external magnetic field h and the disturbance is in the force ξ, that is, $h = g$ ($g \in L^2(\mathcal{Q})$) and $\xi = f$ ($f \in \mathcal{L}^2_\infty(\mathcal{Q})$). Therefore, the function (φ, u) is assumed to be related to the disturbance f and control g through the problem (10.12) ($\forall (q, v) \in \mathcal{H}^1(\Omega) \times V_n$ and a.e. $t \in (0,T)$):

$$\begin{aligned}
&\eta \int_\Omega \rho \frac{\partial \varphi}{\partial t} q \mathrm{d}x - \mathrm{i}\eta\kappa \int_\Omega \mathrm{div}(\rho(u+U))\varphi q \mathrm{d}x - \mathrm{i}\eta\kappa \int_\Omega \mathrm{div}(\rho u)\phi q \mathrm{d}x \\
&\quad -\mathrm{i}\mu \int_\Gamma \rho\varphi q \mathrm{d}\Gamma + \int_\Omega (\rho B(u)(\varphi+\phi)\overline{B}(u)(q)\mathrm{d}x \\
&\quad + \int_\Omega \rho F(\varphi) q \mathrm{d}x = \int_\Omega \rho B(0)(\phi)\overline{B}(0)(q)\mathrm{d}x + \int_\Omega \rho f q \mathrm{d}x, \\
&\int_\Omega \rho \frac{\partial u}{\partial t} v \mathrm{d}x + \int_\Omega \rho \mathrm{curl}(u)\mathrm{curl}(v)\mathrm{d}x + \int_\Omega \mathrm{div}(\rho u)\mathrm{div}(v)\mathrm{d}x \\
&\quad + \int_\Omega \rho \mathcal{R}((\overline{\varphi} + \overline{\phi})B(u)(\varphi+\phi))v \mathrm{d}x \\
&\quad = \int_\Omega \rho \mathcal{R}(\overline{\phi} B(0)(\phi))v \mathrm{d}x + \int_\Omega \rho g \mathrm{curl}(v)\mathrm{d}x, \\
&(\varphi(0), u(0)) = (\varphi_0, u_0).
\end{aligned} \tag{10.18}$$

To obtain the regularity of Theorem 10.9, we assume that (φ_0, u_0) is in $\mathcal{L}^2_\infty(\Omega) \times \mathbf{L}^2(\Omega)$. Let $\mathcal{P} : (g, f) \longrightarrow (\varphi, u) = \mathcal{P}(g, f)$ be the map: $L^2(\mathcal{Q}) \times \mathcal{L}^2_\infty(\mathcal{Q}) \longrightarrow \mathcal{E} \times \mathbf{E}_n$ defined by (10.18) and introducing the cost function defined by

$$\begin{aligned}
J(g, f) = \frac{a}{4} \| | \varphi |^2 - \Lambda \|^2_{L^2(\mathcal{Q})} + \frac{b}{2} \| u - u_{\mathrm{obs}} \|^2_{L^2(\mathcal{Q})} \\
+ \frac{\alpha}{2} \| g \|^2_{L^2(\mathcal{Q})} - \frac{\gamma}{2} \| f \|^2_{L^2(\mathcal{Q})},
\end{aligned} \tag{10.19}$$

where a, b, α, γ are fixed such that $\alpha, \gamma > 0$,[4] $a, b \geq 0$ and $a + b > 0$. The functions $u_{\mathrm{obs}} \in \mathbf{L}^2(\mathcal{Q})$ and $\Lambda \in L^\infty(\mathcal{Q})$ are given and represent the observation.

Let $\mathcal{K} = \mathcal{K}_1 \times \mathcal{K}_2$ such that $\mathcal{K}_1$ and $\mathcal{K}_2$ are given non-empty, closed, convex, bounded subsets of $L^2(\mathcal{Q})$ and $\mathcal{L}^2_\infty(\mathcal{Q})$, respectively. We want to minimize the functional J with respect to g and maximize J with respect to f, *i.e.*, to study the following problem $(\mathcal{MP}_1)$:

find an admissible control $g^* \in \mathcal{K}_1$ and a disturbance $f^* \in \mathcal{K}_2$ such that: (g^*, f^*) is a saddle point of the functional J on $\mathcal{K}$, subject to system (10.18).

---

[4] The parameters α can be interpreted as the price of the control to the engineer and the parameter γ as the price of the disturbance to the nature.

Proposition 10.13. *The function $\mathcal{P}$ is continuously F-differentiable from $L^2(\mathcal{Q})\times\mathcal{L}^2_\infty(\mathcal{Q})$ to $\mathcal{E}\times\mathbf{E}_n$ with the derivative $\mathcal{P}'(g,f): Y=(\beta_1,\beta_2)\longrightarrow(\psi,w)$ given by the linear problem $(\mathcal{P}_{F1})$ ($\forall(q,v)\in\mathcal{H}^1(\Omega)\times V_n$ and a.e. $t\in(0,T)$):*

$$
\begin{aligned}
&\eta\int_\Omega\rho\frac{\partial\psi}{\partial t}q\mathrm{d}x-\mathrm{i}\eta\kappa\int_\Omega\mathrm{div}(\rho U_1)\psi q\mathrm{d}x-\mathrm{i}\eta\kappa\int_\Omega\mathrm{div}(\rho w)\phi_1 q\mathrm{d}x\\
&\qquad-\mathrm{i}\mu\int_\Gamma\rho\psi q\mathrm{d}\Gamma+\int_\Omega\rho(B(u)(\psi)+\phi_1 w)\overline{B}(u)(q)\mathrm{d}x\\
&\qquad+\int_\Omega\rho B(u)(\phi_1)wq\mathrm{d}x+\int_\Omega\rho((2\mid\phi_1\mid^2-\vartheta)\psi+\phi_1^2\overline{\psi})q\mathrm{d}x\\
&\qquad=\int_\Omega\rho\beta_2 q\mathrm{d}x,\\
&\int_\Omega\rho\frac{\partial w}{\partial t}v\mathrm{d}x+\int_\Omega\rho\mathrm{curl}(w)\mathrm{curl}(v)\mathrm{d}x+\int_\Omega\mathrm{div}(\rho w)\mathrm{div}(v)\mathrm{d}x\\
&\qquad+\int_\Omega\rho\mathcal{R}(\overline{\psi}B(u)(\phi_1))v\mathrm{d}x+\int_\Omega\rho(\mathcal{R}(\overline{\phi_1}B(u)(\psi))+w\mid\phi_1\mid^2)v\mathrm{d}x\\
&\qquad=\int_\Omega\rho\beta_1\mathrm{curl}(v)\mathrm{d}x,\\
&(\psi(0),w(0))=(0,0),
\end{aligned}
$$

where $(U_1,\phi_1)=(u+U,\varphi+\phi)$.

Moreover, for all $(f_i,g_i)\in L^2(\mathcal{Q})\times\mathcal{L}^2_\infty(\mathcal{Q})$, for $i=1,2$, we have the following estimates:

(i) $\parallel\mathcal{P}'(g_1,f_1)\parallel_{\mathcal{L}(L^2(\mathcal{Q})\times L^2(\mathcal{Q}),\mathcal{E}\times\mathbf{E}_n)}\leq C_e$

(ii) $\parallel\mathcal{P}'(g_1,f_1)Y-\mathcal{P}'(g_2,f_2)Y\parallel^2_{\mathcal{E}\times\mathbf{E}_n}$

$$
\begin{aligned}
&\leq C_e(\parallel X\parallel_{L^2(\mathcal{Q})\times L^2(\mathcal{Q})}\parallel Y\parallel^2_{L^2(\mathcal{Q})\times L^2(\mathcal{Q})}\\
&\qquad+\parallel X\parallel^2_{L^2(\mathcal{Q})\times L^2(\mathcal{Q})}\parallel Y\parallel_{L^2(\mathcal{Q})\times L^2(\mathcal{Q})}),
\end{aligned}
$$

where $f=f_1-f_2$, $g=g_1-g_2$ and $X=(g,f)$.

Proof. The proof of this proposition is the consequence of the result of Theorem 10.12. Here we omit the details. □

Proposition 10.14. *The mapping $\mathcal{P}$ defined by (10.18) is continuous from the weak topology of $L^2(\mathcal{Q})\times\mathcal{L}^2_\infty(\mathcal{Q})$ to the strong topology of $\mathcal{L}^2(\mathcal{Q})\times\mathbf{L}^2(\mathcal{Q})$.*

Proof. Let $\mathbf{f}=(g,f)$ be given in $L^2(\mathcal{Q})\times\mathcal{L}^2_\infty(\mathcal{Q})$ and let be a sequence $\mathbf{f}_k=(g_k,f_k)$ such that $\mathbf{f}_k$ is weakly convergent in $L^2(\mathcal{Q})\times\mathcal{L}^2(\mathcal{Q})$ to $\mathbf{f}$.

Set $(\varphi,u)=\mathcal{P}(g,f)$ and $(\varphi_k,u_k)=\mathcal{P}(g_k,f_k)$. Since $\mathbf{f}_k\rightharpoonup\mathbf{f}$ weakly in $L^2(\mathcal{Q})\times\mathcal{L}^2(\mathcal{Q})$ then $\mathbf{f}_k$ is uniformly bounded in $L^2(\mathcal{Q})\times\mathcal{L}^2(\mathcal{Q})$. In view of Theorem 10.9, we can deduce that the sequence (φ_k,u_k) is uniformly bounded in $\mathcal{E}\times\mathbf{E}_n$. Therefore, we can extract from $(\mathbf{f}_k,\varphi_k,u_k)$ a subsequence also denoted by $(\mathbf{f}_k,\varphi_k,u_k)$ and such that

$$
\begin{aligned}
&(g_k, f_k) \rightharpoonup (g, f) \text{ weakly in } L^2(\mathcal{Q}) \times \mathcal{L}^2(\mathcal{Q}),\\
&(\varphi_k, u_k) \rightharpoonup (\tilde{\varphi}, \tilde{u}) \text{ weakly in } \mathcal{E} \times \mathbf{E}_n,\\
&(\varphi_k, u_k) \longrightarrow (\tilde{\varphi}, \tilde{u}) \text{ strongly in } \mathcal{L}^2(\mathcal{Q}) \times \mathbf{L}^2(\mathcal{Q}),\\
&\varphi_k \rightharpoonup \tilde{\varphi} \text{ weakly in } \mathcal{L}^2(\Sigma).
\end{aligned} \tag{10.20}
$$

We can easily prove that $(\tilde{\varphi}, \tilde{u}) = \mathcal{P}(g, f)$ and according to the uniqueness of the solution of (10.18), we then have$\tilde{\varphi} = \varphi$ and $\tilde{u} = u$. □

Theorem 10.15. *For α and γ sufficiently large (i.e., there exists (α_l, γ_l) such that $\alpha \geq \alpha_l$ and $\gamma \geq \gamma_l$) there exists $(g^*, f^*) \in \mathcal{K}$ and $(\varphi^*, u^*) \in \mathcal{E} \times \mathbf{E}_n$ such that (g^*, f^*) is defined by $(\mathcal{MP}_1)$ and $(\varphi^*, u^*) = \mathcal{P}(g^*, f^*)$ is a solution of (10.18).*

Proof. Let P_f be the map: $g \longrightarrow J(g, f)$ and Q_g be the map: $f \longrightarrow J(g, f)$. To obtain the existence of the robust control problem we prove that P_f is convex and lower semi-continuous for all $f \in \mathcal{K}_2$, and Q_g is concave and upper semi-continuous for all $g \in \mathcal{K}_1$ and we use minimax duality theorems for infinite dimensions presented in Chapter 5.

First, we prove, for α and γ sufficiently large, the convexity of the map P_f and the concavity of the map Q_g. In order to prove the convexity, it is sufficient to show that for all $(g_1, g_2) \in \mathcal{K}_1$ we have $(P'_f(g_1) - P'_f(g_2)).g \geq 0$, where $g = g_1 - g_2$ (because P_f is G-differentiable). According to the definition of J, we have that

$$
\begin{aligned}
&(P'_f(g_1) - P'_f(g_2)).g\\
&= \alpha \parallel g \parallel^2_{L^2(\mathcal{Q})} + a \iint_{\mathcal{Q}} \mathcal{R}((\mid \varphi_1 \mid^2 - \mid \varphi_2 \mid^2)\psi_2\overline{\varphi}_2)\mathrm{d}x\mathrm{d}t\\
&\quad + a \iint_{\mathcal{Q}} \mathcal{R}((\mid \varphi_1 \mid^2 - \Lambda)(\varphi_1 - \varphi_2)\overline{\psi}_1)\mathrm{d}x\mathrm{d}t\\
&\quad + b \iint_{\mathcal{Q}} (u_1 - u_2)w_2\mathrm{d}x\mathrm{d}t + b \iint_{\mathcal{Q}} (u_1 - u_{\mathrm{obs}})(w_1 - w_2)\mathrm{d}x\mathrm{d}t\\
&\quad + a \iint_{\mathcal{Q}} \mathcal{R}((\mid \varphi_1 \mid^2 - \Lambda)(\psi_1 - \psi_2)\overline{\varphi}_2)\mathrm{d}x\mathrm{d}t,
\end{aligned} \tag{10.21}
$$

where $(\varphi_i, u_i) = \mathcal{P}(g_i, f)$ and $(\psi_i, w_i) = \mathcal{P}'(g_i, f).(g, 0)$ (solution of problem $(\mathcal{P}_{F1})$), for $i = 1, 2$.

According to Theorem 10.9 and Proposition 10.13 we have

$$
\begin{aligned}
a &\iint_{\mathcal{Q}} (\mathcal{R}((\mid \varphi_1 \mid^2 - \mid \varphi_2 \mid^2)\psi_2\overline{\varphi}_2)\mathrm{d}x\mathrm{d}t\\
&\qquad + a \iint_{\mathcal{Q}} \mathcal{R}((\mid \varphi_1 \mid^2 - \Lambda)(\varphi_1 - \varphi_2)\overline{\psi}_1))\mathrm{d}x\mathrm{d}t\\
&\qquad + b \iint_{\mathcal{Q}} (u_1 - u_2)w_2\mathrm{d}x\mathrm{d}t\\
&\qquad \leq c_1 \parallel \varphi \parallel_{L^2(\mathcal{Q})} (\parallel \psi_1 \parallel_{L^2(\mathcal{Q})} + \parallel \psi_2 \parallel_{L^2(\mathcal{Q})})\\
&\qquad + c_2 \parallel u \parallel_{L^2(\mathcal{Q})} \parallel w_2 \parallel_{L^2(\mathcal{Q})} \leq C_0 \parallel g \parallel^2_{L^2(\mathcal{Q})}
\end{aligned} \tag{10.22}
$$

and

$$
\begin{aligned}
a \iint_{\mathcal{Q}} \mathcal{R}((\mid \varphi_1 \mid^2 - \Lambda)((\psi_1 - \psi_2)\overline{\varphi}_2)\mathrm{d}x\mathrm{d}t \\
+ b \iint_{\mathcal{Q}} (u_1 - u_{\mathrm{obs}})(w_1 - w_2)\mathrm{d}x\mathrm{d}t \\
\leq c_3 \, \|\mid \varphi_1 \mid^2 - \Lambda \|_{L^2(\mathcal{Q})} \| \psi \|_{L^2(\mathcal{Q})} \\
+ c_4 \| u_1 - u_{\mathrm{obs}} \|_{L^2(\mathcal{Q})} \| w \|_{L^2(\mathcal{Q})} \\
\leq C_1 \| g \|^{3/2}_{L^2(\mathcal{Q})},
\end{aligned}
\tag{10.23}
$$

where $u = u_1 - u_2$, $\varphi = \varphi_1 - \varphi_2$, $w = w_1 - w_2$ and $\psi = \psi_1 - \psi_2$.

From (10.21), (10.22) and (10.23) we deduce that, for $\alpha \geq \alpha_l$ such that $\alpha_l > C_0$ and $(\alpha_l - C_0) \min_{g \in \mathcal{K}_1} \| g \|^{1/2}_{L^2(\mathcal{Q})} = C_1$, we have $(P'_f(g_1) - P'_f(g_2)).g \geq 0$ and then the convexity of P_f. In the same way, we can find γ_l such that for $\gamma \geq \gamma_l$ we have the concavity of Q_g.

We shall prove now that P_f is lower semi-continuous for all $f \in \mathcal{K}_2$, and Q_g is upper semi-continuous for all $g \in \mathcal{K}_1$. Let g_k be a minimizing sequence of J, *i.e.*, $\liminf_k J(g_k, f) = \min_{g \in \mathcal{K}_1} J(g, f)$ $(\forall f \in \mathcal{K}_2)$. Then g_k is uniformly bounded in $\mathcal{K}_1$ and we can extract from g_k a subsequence also denoted by g_k such that $g_k \rightharpoonup g_f$ weakly in $\mathcal{K}_1$. By using Proposition 10.14 we have then

$$
\mathcal{P}(g_k, f) \longrightarrow (\varphi_f, u_f) \text{ strongly in } \mathcal{L}^2(\mathcal{Q}) \times \mathbf{L}^2(\mathcal{Q}). \tag{10.24}
$$

Therefore, since the norm is lower semi-continuous we have that the map $P_f : g \longrightarrow J(g, f)$ is lower semi-continuous for all $f \in \mathcal{K}_2$. By using the same technique we obtain that Q_g is upper semi-continuous for all $g \in \mathcal{K}_1$. □

In order to obtain the necessary optimality conditions which have been satisfied by the solution of the robust control problem, we introduce the following adjoint problem corresponding to the primal problem (10.18) (we denote by $(\varphi, u) = \mathcal{P}(g, f)$ and $(\phi_1, U_1) = (\varphi + \phi, u + U)$):

find $(P, Q) \in \mathcal{E} \times \mathbf{E}_n$ such that $(\forall (q, v) \in \mathcal{H}^1(\Omega) \times V_n$ and a.e. $t \in (0, T))$:

$$
\begin{aligned}
-\eta \int_\Omega \rho \frac{\partial P}{\partial t} q \mathrm{d}x - \mathrm{i}\eta\kappa \int_\Omega \mathrm{div}(\rho U_1) P q \mathrm{d}x - \mathrm{i}\mu \int_\Gamma \rho P q \mathrm{d}\Gamma \\
+ \int_\Omega \rho \overline{B}(u)(P) B(u)(q) \mathrm{d}x + \int_\Omega \rho \overline{B(u)(\phi_1)} Q q \mathrm{d}x \\
+ \int_\Omega (\frac{-\mathrm{i}}{\kappa} \mathrm{div}(\rho \overline{\phi_1} Q) + \rho U_1 \overline{\phi_1} Q) q \mathrm{d}x \\
+ \int_\Omega \rho((2 \mid \phi_1 \mid^2 - \vartheta) P + \overline{\phi_1}^2 \overline{P}) q \mathrm{d}x = a \int_\Omega (\mid \varphi \mid^2 - \Lambda) \overline{\varphi} q \mathrm{d}x,
\end{aligned}
\tag{10.25}
$$

$$
- \int_\Omega \rho \frac{\partial Q}{\partial t} v \mathrm{d}x + \int_\Omega \rho \mathrm{curl}(Q) \mathrm{curl}(v) \mathrm{d}x + \int_\Omega \mathrm{div}(Q) \mathrm{div}(\rho q) \mathrm{d}x
$$

$$+\int_\Omega \rho\mathcal{R}(\mathrm{i}\eta\kappa\nabla(\phi_1 P) + PB(u)(\phi_1) + \phi_1\overline{B}(u)(P))v\mathrm{d}x$$
$$+\int_\Omega \rho Q \mid \phi_1 \mid^2 v\mathrm{d}x = b\int_\Omega (u - u_{\mathrm{obs}})v\mathrm{d}x,$$
$$(P(T), Q(T)) = (0, 0).$$

Remark 10.16. (*i*) The adjoint problem (10.25) is a linear system. By reversing sense of time, *i.e.*, $t := T - t$, and by applying the same way as to obtain the result of Theorem 10.11 we obtain the existence and uniqueness of (P, Q).
(*ii*) The system (10.25) is the weak formulation of the following problem:

$$\begin{aligned}
&-\eta\rho\frac{\partial P}{\partial t} - \mathrm{i}\eta\kappa\mathrm{div}(\rho U_1)P + \overline{B}(u).(\rho\overline{B}(u)(P)) + \rho\overline{B(u)(\phi_1)}Q\\
&\qquad -\frac{\mathrm{i}}{\kappa}\mathrm{div}(\rho\overline{\phi_1}Q) + \rho U_1\overline{\phi_1}Q + \rho((2 \mid \phi_1 \mid^2 - \vartheta)P + \overline{\phi_1}^2\overline{P})\\
&\quad = a(\mid \varphi \mid^2 - \Lambda)\overline{\varphi} \ \text{ on } \mathcal{Q},\\
&-\rho\frac{\partial Q}{\partial t} + \mathrm{curl}(\rho\mathrm{curl}(Q)) - \rho\nabla(\mathrm{div}(Q)) + \rho Q \mid \phi_1 \mid^2\\
&\qquad + \rho\mathcal{R}(\mathrm{i}\eta\kappa\nabla(\phi_1 P) + PB(u)(\phi_1) + \phi_1\overline{B}(u)(P))\\
&\quad = b(u - u_{\mathrm{obs}}) \ \text{ on } \mathcal{Q},\\
&\frac{1}{\kappa^2}\frac{\partial P}{\partial \mathbf{n}} = \mu P, \ Q.\mathbf{n} = 0, \ \mathrm{curl}(Q) = 0 \ \text{ on } \Sigma,\\
&(P(T), Q(T)) = (0, 0).
\end{aligned} \tag{10.26}$$

◇

We can now give the first-order optimality conditions for the robust control problem $(\mathcal{MP}_1)$.

Theorem 10.17. *Under the assumptions of Theorem 10.15, the optimal solution* $(g^*, f^*, u^*, \varphi^*) \in \mathcal{K} \times \mathbf{E}_n \times \mathcal{E}$ *such that* (g^*, f^*) *is defined by* $(\mathcal{MP}_1)$ *and* $(\varphi^*, u^*) = \mathcal{P}(g^*, f^*)$ *solution of (10.18), satisfies*

$$\begin{aligned}
&\iint_\mathcal{Q} (\rho\mathrm{curl}(Q^*) + \alpha g^*)(g - g^*)\mathrm{d}x\mathrm{d}t \geq 0,\\
&\iint_\mathcal{Q} \mathcal{R}((\rho P^* - \gamma\overline{f^*})(f - f^*))\mathrm{d}x\mathrm{d}t \leq 0 \qquad \forall (g, f) \in \mathcal{K},
\end{aligned} \tag{10.27}$$

where (P^*, Q^*) *is the solution of the adjoint problem (10.25) (corresponding to the primal solution* (φ^*, u^*)*).*

Proof. The cost function J is a composition of F-differentiable maps then J is differentiable and we have $(\forall Y = (\beta_1, \beta_2) \in \mathcal{K}\ )$

$$\begin{aligned}
J'(g, f).Y = a&\iint_\mathcal{Q} \mathcal{R}((\mid \varphi \mid^2 - \Lambda)\overline{\varphi}\psi)\mathrm{d}x\mathrm{d}t + b\iint_\mathcal{Q} (u - u_{\mathrm{obs}})w\mathrm{d}x\mathrm{d}t\\
&+\alpha\iint_\mathcal{Q} g\beta_1\mathrm{d}x\mathrm{d}t - \gamma\iint_\mathcal{Q} \mathcal{R}(\overline{f}\beta_2)\mathrm{d}x\mathrm{d}t,
\end{aligned} \tag{10.28}$$

where $(\psi, w) = \mathcal{P}'(g, f).Y$ is the solution of problem $(\mathcal{P}_{F1})$.

By taking $(q, v) = (P, Q)$ in $(\mathcal{P}_{F1})$, using Green's formula and integrating with respect to time, we obtain (according to the initial conditions of $(\mathcal{P}_{F1})$)

$$
\begin{aligned}
-\eta \iint_Q \rho \frac{\partial P}{\partial t} \psi \mathrm{d}x\mathrm{d}t + \eta \int_\Omega \rho P(T)\psi(T)\mathrm{d}x - \mathrm{i}\eta\kappa \iint_Q \mathrm{div}(\rho U_1) P\psi \mathrm{d}x\mathrm{d}t \\
+\mathrm{i}\eta\kappa \iint_Q \rho w \nabla(\phi_1 P)\mathrm{d}x\mathrm{d}t - \mathrm{i}\mu \iint_\Sigma \rho P\psi \mathrm{d}\Gamma\mathrm{d}t \\
+ \iint_Q \rho \overline{B}(u)(P)B(u)(\psi)\mathrm{d}x\mathrm{d}t + \iint_Q \rho\phi_1 w\overline{B}(u)(P)\mathrm{d}x\mathrm{d}t \\
+ \iint_Q \rho((2 \mid \phi_1 \mid^2 - \vartheta)P\psi + \phi_1^2 P\overline{\psi})\mathrm{d}x\mathrm{d}t \\
+ \iint_Q \rho B(u)(\phi_1) w P \mathrm{d}x\mathrm{d}t = \iint_Q \rho\beta_2 P \mathrm{d}x\mathrm{d}t, \qquad (10.29) \\
- \iint_Q \rho \frac{\partial Q}{\partial t} w \mathrm{d}x\mathrm{d}t + \int_\Omega \rho Q(T)w(T)\mathrm{d}x + \iint_Q \rho \mathrm{curl}(Q)\mathrm{curl}(w)\mathrm{d}x\mathrm{d}t \\
+ \iint_Q \mathrm{div}(Q)\mathrm{div}(\rho w)\mathrm{d}x\mathrm{d}t + \iint_Q \rho\mathcal{R}(\overline{\psi}B(u)(\phi_1))Q\mathrm{d}x\mathrm{d}t \\
+ \iint_Q \rho(\mathcal{R}(\overline{\phi_1}B(u)(\psi))Q + w \mid \phi_1 \mid^2 Q)\mathrm{d}x\mathrm{d}t \\
= \iint_Q \rho\beta_1 \mathrm{curl}(Q)\mathrm{d}x\mathrm{d}t.
\end{aligned}
$$

Since (P, Q) is a solution of (10.25), with null final conditions, we have that

$$
\begin{aligned}
\mathrm{i}\eta\kappa \iint_Q \rho w\nabla(\phi_1 P)\mathrm{d}x\mathrm{d}t + \iint_Q \rho\phi_1 w \overline{B}(u)(P)\mathrm{d}x\mathrm{d}t \\
+ \iint_Q \rho B(u)(\phi_1) w P\mathrm{d}x\mathrm{d}t + \iint_Q \rho(\phi_1^2 P\overline{\psi} - \overline{\phi_1}^2 \overline{P}\psi)\mathrm{d}x\mathrm{d}t \\
- \iint_Q \rho\overline{B(u)(\phi_1)}Q\psi\mathrm{d}x\mathrm{d}t - \iint_Q (\frac{-\mathrm{i}}{\kappa}\mathrm{div}(\rho\overline{\phi_1}Q) + \rho U_1\overline{\phi_1}Q)\psi\mathrm{d}x\mathrm{d}t \\
+a \iint_Q (\mid \varphi \mid^2 - \Lambda)\overline{\varphi}\psi\mathrm{d}x\mathrm{d}t = \iint_Q \rho\beta_2 P\mathrm{d}x\mathrm{d}t, \qquad (10.30) \\
\iint_Q \rho\mathcal{R}(\overline{\psi}B(u)(\phi_1) + \overline{\phi_1}B(u)(\psi))Q\mathrm{d}x\mathrm{d}t + b \iint_Q (u - u_{\mathrm{obs}})w\mathrm{d}x\mathrm{d}t \\
- \iint_Q \rho\mathcal{R}(\mathrm{i}\eta\kappa\nabla(\phi_1 P) + PB(u)(\phi_1) + \phi_1\overline{B}(u)(P))w\mathrm{d}x\mathrm{d}t \\
= \iint_Q \rho\beta_1\mathrm{curl}(Q)\mathrm{d}x\mathrm{d}t.
\end{aligned}
$$

By adding the real part of the first part of (10.30) and the second part of (10.30), we obtain

$$
\begin{aligned}
a \iint_{\mathcal{Q}} \mathcal{R}((| \varphi |^2 - \Lambda)\overline{\varphi}\psi)\mathrm{d}x\mathrm{d}t + b \iint_{\mathcal{Q}} (u - u_{\mathrm{obs}}) w \mathrm{d}x\mathrm{d}t \\
= \iint_{\mathcal{Q}} \rho\beta_1 \mathrm{curl}(Q)\mathrm{d}x\mathrm{d}t + \iint_{\mathcal{Q}} \rho\mathcal{R}(\beta_2 P)\mathrm{d}x\mathrm{d}t,
\end{aligned}
\tag{10.31}
$$

since $\int_{\Omega}(\frac{-\mathrm{i}}{\kappa}\mathrm{div}(\rho\overline{\phi_1}Q)+\rho U_1\overline{\phi_1}Q)\psi\mathrm{d}x = \int_{\Omega}\rho\overline{\phi_1}B(u)(\psi)Q\mathrm{d}x$ (because $Q.\mathbf{n}=0$),

$\mathcal{R}(\phi_1^2 P\overline{\psi} - \overline{\phi_1}^2\overline{P}\psi) = 0$ and $\mathcal{R}(-\overline{B(u)(\phi_1)}Q\psi + B(u)(\phi_1)Q\overline{\psi}) = 0$.

According to the expression of $J'(g,f).Y$ we can deduce that

$$
J'(g,f).Y = \iint_{\mathcal{Q}} (\rho\mathrm{curl}(Q) + \alpha g)\beta_1 \mathrm{d}x\mathrm{d}t + \iint_{\mathcal{Q}} \mathcal{R}((\rho P - \gamma\overline{f})\beta_2)\mathrm{d}x\mathrm{d}t. \tag{10.32}
$$

Since (f^*, g^*) is an optimal solution we have

$$
\frac{\partial J}{\partial g}(g^*, f^*).(g - g^*) \geq 0, \frac{\partial J}{\partial f}(g^*, f^*).(f - f^*) \leq 0 \quad \forall (g,f) \in \mathcal{K} \tag{10.33}
$$

and then

$$
\begin{aligned}
&\iint_{\mathcal{Q}} (\rho\mathrm{curl}(Q^*) + \alpha g^*)(g - g^*)\mathrm{d}x\mathrm{d}t \geq 0, \\
&\iint_{\mathcal{Q}} \mathcal{R}((\rho P^* - \gamma\overline{f^*})(f - f^*))\mathrm{d}x\mathrm{d}t \leq 0 \quad \forall (g,f) \in \mathcal{K}.
\end{aligned}
\tag{10.34}
$$

This completes the proof. □

Distributed Disturbance in the External Magnetic Field

In this section, the external magnetic field h is assumed to be decomposed into disturbance $f \in L^2(\mathcal{Q})$ and the control $g \in L^2(\mathcal{Q})$, *i.e.*, $h = f + g \in L^2(\mathcal{Q})$. So, the function (φ, u) is assumed to be related to the disturbance f and control g through the problem (10.12) ($\forall (q,v) \in \mathcal{H}^1(\Omega) \times V_n$ and a.e. $t \in (0,T)$)

$$
\begin{aligned}
&\eta \int_{\Omega} \rho \frac{\partial \varphi}{\partial t} q\mathrm{d}x - \mathrm{i}\eta\kappa \int_{\Omega} \mathrm{div}(\rho(u+U))\varphi q\mathrm{d}x - \mathrm{i}\eta\kappa \int_{\Omega} \mathrm{div}(\rho u)\phi q\mathrm{d}x \\
&\qquad -\mathrm{i}\mu \int_{\Gamma} \rho\varphi q\mathrm{d}\Gamma + \int_{\Omega} (\rho B(u)(\varphi + \phi)\overline{B}(u)(q)\mathrm{d}x \\
&\qquad + \int_{\Omega} \rho F(\varphi) q\mathrm{d}x = \int_{\Omega} \rho B(0)(\phi)\overline{B}(0)(q)\mathrm{d}x + \int_{\Omega} \rho\xi q\mathrm{d}x, \\
&\int_{\Omega} \rho \frac{\partial u}{\partial t} v\mathrm{d}x + \int_{\Omega} \rho\mathrm{curl}(u)\mathrm{curl}(v)\mathrm{d}x + \int_{\Omega} \mathrm{div}(\rho u)\mathrm{div}(v)\mathrm{d}x \\
&\qquad + \int_{\Omega} \rho\mathcal{R}((\overline{\varphi} + \overline{\phi})B(u)(\varphi + \phi))v\mathrm{d}x \\
&\qquad = \int_{\Omega} \rho\mathcal{R}(\overline{\phi}B(0)(\phi))v\mathrm{d}x + \int_{\Omega} \rho(f+g)\mathrm{curl}(v)\mathrm{d}x, \\
&(\varphi(0), u(0)) = (\varphi_0, u_0).
\end{aligned}
\tag{10.35}
$$

To obtain the regularity of Theorem 10.9, we suppose that $(\varphi_0, u_0) \in \mathcal{L}^2_\infty(\Omega) \times \mathbf{L}^2(\Omega)$ and $\xi \in \mathcal{L}^2_\infty(\mathcal{Q})$. Let $\mathcal{P} : (g, f) \longrightarrow (\varphi, u) = \mathcal{P}(g, f)$ be the map: $(L^2(\mathcal{Q}))^2 \longrightarrow \mathcal{E} \times \mathbf{E}_n$ defined by (10.35) and introducing the cost function defined by

$$J(g, f) = \frac{a}{4} \parallel\mid \varphi \mid^2 - \Lambda \parallel^2_{L^2(\mathcal{Q})} + \frac{b}{2} \parallel u - u_{\text{obs}} \parallel^2_{L^2(\mathcal{Q})} + \frac{\alpha}{2} \parallel g \parallel^2_{L^2(\mathcal{Q})} - \frac{\gamma}{2} \parallel f \parallel^2_{L^2(\mathcal{Q})}, \tag{10.36}$$

where α, β, a, b are fixed parameters such that $\alpha, \beta > 0$, $a, b \geq 0$ and $a + b > 0$. The pair of functions $(u_{\text{obs}}, \Lambda)$ is in $\mathbf{L}^2(\mathcal{Q}) \times L^\infty(\mathcal{Q})$ and represents the observation.

In this section we study the following robust control problem $(\mathcal{MP}_2)$:

find a saddle point (g^*, f^*) of the functional J on $\mathcal{K}$, subject to (10.35),

where $\mathcal{K} = \mathcal{K}_1 \times \mathcal{K}_2$ such that $\mathcal{K}_1$ and $\mathcal{K}_2$ are non-empty, closed, convex, bounded subsets of $L^2(\mathcal{Q})$. The proof of the following propositions and existence theorem is similar to that of Propositions 10.13 and 10.14 and Theorem 10.15. Therefore, we will omit the details.

Proposition 10.18. *The function $\mathcal{P}$ is continuously F-differentiable from $(L^2(\mathcal{Q}))^2$ to $\mathcal{E} \times \mathbf{E}_n$ with the derivative $\mathcal{P}'(g, f) : Y = (\beta_1, \beta_2) \longrightarrow (\psi, w)$ given by the linear problem $(\mathcal{P}_{F2})$ ($\forall (q, v) \in \mathcal{H}^1(\Omega) \times V_n$ and a.e. $t \in (0, T)$)*

$$\begin{aligned}
&\eta \int_\Omega \rho \frac{\partial \psi}{\partial t} q \mathrm{d}x - \mathrm{i}\eta\kappa \int_\Omega \operatorname{div}(\rho U_1) \psi q \mathrm{d}x - \mathrm{i}\eta\kappa \int_\Omega \operatorname{div}(\rho w) \phi_1 q \mathrm{d}x \\
&\quad - \mathrm{i}\mu \int_\Gamma \rho \psi q \mathrm{d}\Gamma + \int_\Omega \rho (B(u)(\psi) + \phi_1 w) \overline{B}(u)(q) \mathrm{d}x \\
&\quad + \int_\Omega \rho B(u)(\phi_1) w q \mathrm{d}x + \int_\Omega \rho ((2 \mid \phi_1 \mid^2 - \vartheta) \psi + \phi_1^2 \overline{\psi}) q \mathrm{d}x = 0, \\
&\int_\Omega \rho \frac{\partial w}{\partial t} u \mathrm{d}x + \int_\Omega \rho \operatorname{curl}(w) \operatorname{curl}(v) \mathrm{d}x + \int_\Omega \operatorname{div}(\rho w) \operatorname{div}(v) \mathrm{d}x \\
&\quad + \int_\Omega \rho \mathcal{R}(\overline{\psi} B(u)(\phi_1)) v \mathrm{d}x + \int_\Omega \rho (\mathcal{R}(\overline{\phi_1} B(u)(\psi)) + w \mid \phi_1 \mid^2) v \mathrm{d}x \\
&\quad = \int_\Omega \rho (\beta_1 + \beta_2) \operatorname{curl}(v) \mathrm{d}x, \\
&(\psi(0), w(0)) = (0, 0),
\end{aligned} \tag{10.37}$$

where $(U_1, \phi_1) = (u + U, \varphi + \phi)$.

Moreover, we have the estimates ($\forall (g_i, f_i) \in (L^2(\mathcal{Q}))^2, i = 1, 2$):

(i) $\parallel \mathcal{P}'(g_1, f_1) \parallel_{\mathcal{L}((L^2(\mathcal{Q}))^2, \mathcal{E} \times \mathbf{E}_n)} \leq C_e$

(ii) $\parallel \mathcal{P}'(g_1, f_1) Y - \mathcal{P}'(g_2, f_2) Y \parallel^2_{\mathcal{E} \times \mathbf{E}_n}$

$$\leq C_e (\parallel X \parallel_{(L^2(\mathcal{Q}))^2} \parallel Y \parallel^2_{(L^2(\mathcal{Q}))^2} + \parallel X \parallel^2_{(L^2(\mathcal{Q}))^2} \parallel Y \parallel_{(L^2(\mathcal{Q}))^2}),$$

where $f = f_1 - f_2$, $g = g_1 - g_2$ and $X = (g, f)$. □

Proposition 10.19. *The map $\mathcal{P}$ defined by (10.35) is continuous from the weak topology of $(L^2(\mathcal{Q}))^2$ to the strong topology of $\mathcal{L}^2(\mathcal{Q}) \times L^2(\mathcal{Q})$.* □

Theorem 10.20. *For α and γ sufficiently large, there exists $(g^*, f^*) \in \mathcal{K}$ and $(u^*, \varphi^*) \in \mathbf{E}_n \times \mathcal{E}$ such that (g^*, f^*) is defined by $(\mathcal{MP}_2)$ and $(\varphi^*, u^*) = \mathcal{P}(g^*, f^*)$ is a solution of (10.35).* □

Now we establish necessary optimality conditions for the robust control problem $(\mathcal{MP}_2)$.

Theorem 10.21. *Under the assumptions of Theorem 10.20, the optimal solution $(g^*, f^*, u^*, \varphi^*) \in \mathcal{K} \times \mathbf{E}_n \times \mathcal{E}$ such that (g^*, f^*) is defined by $(\mathcal{MP}_2)$ and $(\varphi^*, u^*) = \mathcal{P}(g^*, f^*)$ is a solution of (10.35), satisfies*

$$\begin{aligned} &\iint_{\mathcal{Q}} (\rho \mathrm{curl}(Q^*) + \alpha g^*)(g - g^*) \mathrm{d}x\mathrm{d}t \geq 0, \\ &\iint_{\mathcal{Q}} (\rho \mathrm{curl}(Q^*) - \gamma f^*)(f - f^*) \mathrm{d}x\mathrm{d}t \leq 0 \quad \forall (g, f) \in \mathcal{K}, \end{aligned} \tag{10.38}$$

where (P^*, Q^*) is the solution of problem (10.25) (corresponding to (φ^*, u^*)).

Proof. The cost function J is a composition of F-differentiable maps then J is differentiable and we have ($\forall Y = (\beta_1, \beta_2) \in \mathcal{K}$)

$$\begin{aligned} J'(g,f).Y = a \iint_{\mathcal{Q}} \mathcal{R}((| \varphi |^2 - \Lambda)\overline{\varphi}\psi) \mathrm{d}x\mathrm{d}t + b \iint_{\mathcal{Q}} (u - u_{\mathrm{obs}}) w \mathrm{d}x\mathrm{d}t \\ + \alpha \iint_{\mathcal{Q}} g\beta_1 \mathrm{d}x\mathrm{d}t - \gamma \iint_{\mathcal{Q}} f\beta_2 \mathrm{d}x\mathrm{d}t, \end{aligned} \tag{10.39}$$

where $(\psi, w) = \mathcal{P}'(g, f).Y$ is the solution of problem $(\mathcal{P}_{F2})$.

By taking $(q, v) = (P, Q)$ in $(\mathcal{P}_{F2})$ and integrating with respect to time we obtain (according to the initial condition)

$$\begin{aligned} -\eta \iint_{\mathcal{Q}} \rho \frac{\partial P}{\partial t} \psi \mathrm{d}x\mathrm{d}t + \eta \int_{\Omega} \rho P(T)\psi(T) \mathrm{d}x - \mathrm{i}\eta\kappa \iint_{\mathcal{Q}} \mathrm{div}(\rho U_1) P\psi \mathrm{d}x\mathrm{d}t \\ + \mathrm{i}\eta\kappa \iint_{\mathcal{Q}} \rho w \nabla(\phi_1 P) \mathrm{d}x\mathrm{d}t - \mathrm{i}\mu \int\int_{\Sigma} \rho\psi P \mathrm{d}\Gamma\mathrm{d}t \\ + \iint_{\mathcal{Q}} \rho \overline{B}(u)(P) B(u)(\psi) \mathrm{d}x\mathrm{d}t \\ + \iint_{\mathcal{Q}} \rho\phi_1 w \overline{B}(u)(P) \mathrm{d}x\mathrm{d}t + \iint_{\mathcal{Q}} \rho B(u)(\phi_1) w P \mathrm{d}x\mathrm{d}t \\ + \iint_{\mathcal{Q}} \rho((2 | \varphi |^2 - \vartheta) P\psi + \varphi^2 P\overline{\psi}) \mathrm{d}x\mathrm{d}t = 0, \end{aligned}$$

$$\begin{aligned} -\iint_{\mathcal{Q}} \rho \frac{\partial Q}{\partial t} w \mathrm{d}x\mathrm{d}t + \int_{\Omega} \rho Q(T) w(T) \mathrm{d}x + \iint_{\mathcal{Q}} \rho \mathrm{curl}(Q) \mathrm{curl}(w) \mathrm{d}x\mathrm{d}t \\ + \iint_{\mathcal{Q}} \mathrm{div}(Q) \mathrm{div}(\rho w) \mathrm{d}x\mathrm{d}t + \iint_{\mathcal{Q}} \rho \mathcal{R}(\overline{\psi} B(u)(\phi_1)) Q \mathrm{d}x\mathrm{d}t \\ + \iint_{\mathcal{Q}} \rho(\mathcal{R}(\overline{\phi_1} B(u)(\psi)) Q + w | \phi_1 |^2 Q) \mathrm{d}x\mathrm{d}t \end{aligned}$$

$$= \iint_Q \rho(\beta_1 + \beta_2)\mathrm{curl}(Q)\mathrm{d}x\mathrm{d}t.$$

Since (P, Q) is a solution of (10.25) we have that

$$\begin{aligned}
\mathrm{i}\eta\kappa \iint_Q & \rho w \nabla(\phi_1 P)\mathrm{d}x\mathrm{d}t + \iint_Q \rho\phi_1 w \overline{B}(u)(P)\mathrm{d}x\mathrm{d}t \\
&+ \iint_Q \rho B(u)(\phi_1) w P \mathrm{d}x\mathrm{d}t + \iint_Q \rho(\phi_1^2 P\overline{\psi} - \overline{\phi_1}^2 \overline{P}\psi)\mathrm{d}x\mathrm{d}t \\
&- \iint_Q (\frac{-\mathrm{i}}{\kappa}\mathrm{div}(\rho\overline{\phi_1}Q) + \rho U_1 \overline{\phi_1} Q)\psi \mathrm{d}x\mathrm{d}t \\
&- \iint_Q \rho\overline{B(u)(\phi_1)} Q\psi \mathrm{d}x\mathrm{d}t + a \iint_Q (\mid \varphi \mid^2 - \Lambda)\overline{\varphi}\psi \mathrm{d}x\mathrm{d}t = 0, \qquad (10.40)\\
\iint_Q & \rho \mathcal{R}(\overline{\psi} B(u)(\phi_1))Q\mathrm{d}x\mathrm{d}t + \iint_Q \rho\mathcal{R}(\overline{\phi_1} B(u)(\psi))Q\mathrm{d}x\mathrm{d}t \\
&- \iint_Q \rho\mathcal{R}(\mathrm{i}\eta\kappa\nabla(\phi_1 P) + PB(u)(\phi_1) + \phi_1 \overline{B}(u)(P)) w \mathrm{d}x\mathrm{d}t \\
&+ b \iint_Q (u - u_{\mathrm{obs}}) w \mathrm{d}x\mathrm{d}t = \iint_Q \rho(\beta_1 + \beta_2)\mathrm{curl}(Q)\mathrm{d}x\mathrm{d}t.
\end{aligned}$$

By adding the real part of the first part of (10.40) and the second part of (10.40) we obtain (by using Green's formula)

$$\begin{aligned}
a \iint_Q \mathcal{R}((\mid \varphi \mid^2 - \Lambda)\overline{\varphi}\psi)\mathrm{d}x\mathrm{d}t &+ b \iint_Q (u - u_{\mathrm{obs}}) w \mathrm{d}x\mathrm{d}t \\
&= \iint_Q \rho(\beta_1 + \beta_2)\mathrm{curl}(Q)\mathrm{d}x\mathrm{d}t.
\end{aligned}$$

According to the expression of $J'(g, f).Y$ we can deduce that

$$J'(g, f).Y = \iint_Q (\rho\mathrm{curl}(Q) + \alpha g)\beta_1 \mathrm{d}x\mathrm{d}t + \iint_Q (\rho\mathrm{curl}(Q) - \gamma f)\beta_2 \mathrm{d}x\mathrm{d}t.$$

Since (g^*, f^*) is an optimal solution we then have

$$\begin{aligned}
&\iint_Q (\rho\mathrm{curl}(Q^*) + \alpha g^*)(g - g^*)\mathrm{d}x\mathrm{d}t \geq 0, \\
&\iint_Q (\rho\mathrm{curl}(Q^*) - \gamma f^*)(f - f^*)\mathrm{d}x\mathrm{d}t \leq 0 \quad \forall (g, f) \in \mathcal{K}.
\end{aligned} \qquad (10.41)$$

This completes the proof. □

10.5.2 Control in the Initial Condition of the Vector Potential

In this section, we formulate the problem in two situations: first, where the worst disturbance is in the chaotic-type force ξ and, second, where the disturbance is in the external magnetic field h.

Distributed Disturbance in the Chaotic-Type Term

We suppose that the control is in the initial condition u_0, *i.e.*, $u_0 = g$ ($g \in \mathbf{L}^2(\Omega)$) and the disturbance is in the force ξ, *i.e.*, $\xi = f$ ($f \in \mathcal{L}^2_\infty(\mathcal{Q})$). So the function (φ, u) is assumed to be related to the disturbance f and control g through the problem (10.12) ($\forall (q,v) \in \mathcal{H}^1(\Omega) \times V_n$ and a.e. $t \in (0,T)$):

$$
\begin{aligned}
&\eta \int_\Omega \rho \frac{\partial \varphi}{\partial t} q\mathrm{d}x - \mathrm{i}\eta\kappa \int_\Omega \mathrm{div}(\rho(u+U))\varphi q\mathrm{d}x - \mathrm{i}\eta\kappa \int_\Omega \mathrm{div}(\rho u)\phi q\mathrm{d}x \\
&\quad -\mathrm{i}\mu \int_\Gamma \rho\varphi q\mathrm{d}\Gamma + \int_\Omega (\rho B(u)(\varphi+\phi)\overline{B}(u)(q)\mathrm{d}x + \int_\Omega \rho F(\varphi) q\mathrm{d}x \\
&\quad = \int_\Omega \rho B(0)(\phi)\overline{B}(0)(q)\mathrm{d}x + \int_\Omega \rho f q\mathrm{d}x, \\
&\int_\Omega \rho \frac{\partial u}{\partial t} v\mathrm{d}x + \int_\Omega \rho \mathrm{curl}(u)\mathrm{curl}(v)\mathrm{d}x + \int_\Omega \mathrm{div}(\rho u)\mathrm{div}(v)\mathrm{d}x \\
&\quad + \int_\Omega \rho \mathcal{R}((\overline{\varphi}+\overline{\phi})B(u)(\varphi+\phi))v\mathrm{d}x \\
&\quad = \int_\Omega \rho \mathcal{R}(\overline{\phi}B(0)(\phi))v\mathrm{d}x + \int_\Omega \rho h \mathrm{curl}(v)\mathrm{d}x, \\
&(\varphi(0), u(0)) = (\varphi_0, g).
\end{aligned}
\tag{10.42}
$$

To obtain the regularity of Theorem 10.9, we suppose that $h \in L^2(\mathcal{Q})$ and $\varphi_0 \in \mathcal{L}^2_\infty(\Omega)$. Let $\mathcal{P} : (g,f) \longrightarrow (\varphi, u) = \mathcal{P}(g,f)$ be the map: $\mathbf{L}^2(\Omega) \times \mathcal{L}^2_\infty(\mathcal{Q}) \longrightarrow \mathcal{E} \times \mathbf{E}_n$ defined by (10.42) and the cost function is defined by

$$
\begin{aligned}
J(g,f) = &\frac{a}{4} \| | \varphi |^2 - \Lambda \|^2_{L^2(\mathcal{Q})} + \frac{b}{2} \| u - u_{\mathrm{obs}} \|^2_{L^2(\mathcal{Q})} \\
&+ \frac{\alpha}{2} \| g \|^2_{L^2} - \frac{\gamma}{2} \| f \|^2_{L^2(\mathcal{Q})},
\end{aligned}
\tag{10.43}
$$

where $\alpha, \gamma > 0$, $a, b \geq 0$ and $a + b > 0$. The functions $(u_{\mathrm{obs}}, \Lambda) \in \mathbf{L}^2(\mathcal{Q}) \times L^\infty(\mathcal{Q})$ are given.

In this section we study the following robust control problem $(\mathcal{MP}_3)$:

find a saddle point (g^*, f^*) of the functional J on $\mathcal{K}$, subject to (10.42),

where $\mathcal{K} = \mathcal{K}_1 \times \mathcal{K}_2$ such that $\mathcal{K}_1$ and $\mathcal{K}_2$ are non-empty, closed, convex, bounded subsets of $\mathbf{L}^2(\Omega)$ and $\mathcal{L}^2_\infty(\mathcal{Q})$, respectively.

The arguments of Section 10.5.1 extend directly to the present case without requiring further estimates. We have then the following results.

Proposition 10.22. *The function $\mathcal{P}$ is continuously F-differentiable from $\mathbf{L}^2(\Omega) \times \mathcal{L}^2_\infty(\mathcal{Q})$ to $\mathcal{E} \times \mathbf{E}_n$ with the derivative $\mathcal{P}'(g,f) : Y = (\beta_1, \beta_2) \longrightarrow (\psi, w)$ given by the linear problem $(\mathcal{P}_{F3})$ ($\forall (q,v) \in \mathcal{H}^1(\Omega) \times V_n$ and a.e. $t \in (0,T)$):*

$$
\begin{aligned}
&\eta \int_\Omega \rho \frac{\partial \psi}{\partial t} q\mathrm{d}x - \mathrm{i}\eta\kappa \int_\Omega \mathrm{div}(\rho U_1)\psi q\mathrm{d}x - \mathrm{i}\eta\kappa \int_\Omega \mathrm{div}(\rho w)\phi_1 q\mathrm{d}x \\
&\quad -\mathrm{i}\mu \int_\Gamma \rho\psi q\mathrm{d}\Gamma + \int_\Omega \rho(B(u)(\psi) + \phi_1 w)\overline{B}(u)(q)\mathrm{d}x
\end{aligned}
$$

$$+\int_\Omega \rho B(u)(\phi_1)wq\mathrm{d}x + \int_\Omega \rho((2\mid\varphi\mid^2-\vartheta)\psi+\varphi^2\overline{\psi})q\mathrm{d}x$$
$$=\int_\Omega \rho\beta_2 q\mathrm{d}x,$$
$$\int_\Omega \rho\frac{\partial w}{\partial t}v\mathrm{d}x + \int_\Omega \rho\mathrm{curl}(w)\mathrm{curl}(v)\mathrm{d}x + \int_\Omega \mathrm{div}(\rho w)\mathrm{div}(v)\mathrm{d}x$$
$$+\int_\Omega \rho\mathcal{R}(\overline{\psi}B(u)(\phi_1))v\mathrm{d}x + \int_\Omega \rho(\mathcal{R}(\overline{\phi_1}B(u)(\psi)) + w\mid\phi_1\mid^2)v\mathrm{d}x = 0,$$
$$(\psi(0), w(0)) = (0, \beta_1),$$

where $(U_1, \phi_1) = (u + U, \varphi + \phi)$.

Moreover, we have the following estimates, for all $(g_i, f_i) \in \mathbf{L}^2(\Omega) \times \mathcal{L}^2_\infty(\mathcal{Q})$, *for* $i = 1, 2$:

(*i*) $\| \mathcal{P}'(g_1, f_1) \|_{\mathcal{L}(\mathbf{L}^2(\Omega)\times\mathcal{L}^2(\mathcal{Q}),\mathcal{E}\times\mathbf{E}_n)} \leq C_e$

(*ii*) $\| \mathcal{P}'(g_1, f_1)Y - \mathcal{P}'(g_2, f_2)Y \|^2_{\mathcal{E}\times\mathbf{E}_n}$
$$\leq C_e(\| X \|_{L^2\times L^2(\mathcal{Q})} \| Y \|^2_{L^2\times L^2(\mathcal{Q})} + \| X \|^2_{L^2\times L^2(\mathcal{Q})} \| Y \|_{L^2\times L^2(\mathcal{Q})}),$$

where $f = f_1 - f_2$, $g = g_1 - g_2$ *and* $X = (g, f)$. □

Proposition 10.23. *The mapping* $\mathcal{P}$ *defined by (10.42) is continuous from the weak topology of* $\mathbf{L}^2(\Omega) \times \mathcal{L}^2_\infty(\mathcal{Q})$ *to the strong topology of* $\mathcal{L}^2(\mathcal{Q}) \times \mathbf{L}^2(\mathcal{Q})$. □

Theorem 10.24. *For* α *and* γ *sufficiently large (i.e., there exists* (α_l, γ_l) *such that* $\alpha \geq \alpha_l$ *and* $\gamma \geq \gamma_l$*) there exists* $(g^*, f^*) \in \mathcal{K}$ *and* $(\varphi^*, u^*) \in \mathcal{E} \times \mathbf{E}_n$ *such that* (g^*, f^*) *is defined by* $(\mathcal{MP}_3)$ *and* $(\varphi^*, u^*) = \mathcal{P}(g^*, f^*)$ *is a solution of (10.42).* □

Next we establish necessary optimality conditions for the robust control problem $(\mathcal{MP}_3)$.

Theorem 10.25. *Under the assumptions of Theorem 10.24, the optimal solution* $(g^*, f^*, u^*, \varphi^*) \in \mathcal{K} \times \mathbf{E}_n \times \mathcal{E}$ *such that* (g^*, f^*) *is defined by* $(\mathcal{MP}_3)$ *and* $(\varphi^*, u^*) = \mathcal{P}(g^*, f^*)$ *is a solution of (10.42), satisfies*

$$\begin{aligned}&\int_\Omega (\rho Q^*(0) + \alpha g^*)(g - g^*)\mathrm{d}x \geq 0,\\ &\iint_\mathcal{Q} \mathcal{R}((\rho P^* - \gamma\overline{f^*})(f - f^*))\mathrm{d}x\mathrm{d}t \leq 0 \quad \forall (g, f) \in \mathcal{K},\end{aligned} \tag{10.44}$$

where (P^*, Q^*) *is the solution of the adjoint problem (10.25) (corresponding to the primal solution* (φ^*, u^*)*).*

Proof. Since the cost function J is a composition of F-differentiable maps, then J is differentiable and we have $(\forall Y = (\beta_1, \beta_2) \in \mathcal{K})$

$$\begin{aligned}J'(g,f).Y &= a\iint_Q \mathcal{R}((|\varphi|^2-\Lambda)\overline{\varphi}\psi)\mathrm{d}x\mathrm{d}t + b\iint_Q (u-u_{\mathrm{obs}})w\mathrm{d}x\mathrm{d}t \\ &+\alpha\int_\Omega g\beta_1\mathrm{d}x - \gamma\iint_Q \mathcal{R}(\overline{f}\beta_2)\mathrm{d}x\mathrm{d}t,\end{aligned} \tag{10.45}$$

where $(\psi,w)=\mathcal{P}'(g,f).Y$ is the solution of problem $(\mathcal{P}_{F3})$.

By taking $(q,v)=(P,Q)$ in $(\mathcal{P}_{F3})$ and integrating with respect to time we obtain (according to the homogeneous boundary conditions and initial condition)

$$\begin{aligned}&-\eta\iint_Q \rho\frac{\partial P}{\partial t}\psi\mathrm{d}x\mathrm{d}t + \int_\Omega \eta\rho P(T)\psi(T)\mathrm{d}x \\ &\quad -\mathrm{i}\eta\kappa\iint_Q \mathrm{div}(\rho U_1)P\psi\mathrm{d}x\mathrm{d}t + \mathrm{i}\eta\kappa\iint_Q \rho w\nabla(\phi_1 P)\mathrm{d}x\mathrm{d}t \\ &\quad -\mathrm{i}\mu\iint_\Sigma \rho\psi P\mathrm{d}\Gamma\mathrm{d}t + \iint_Q \rho\overline{B}(u)(P)B(u)(\psi)\mathrm{d}x\mathrm{d}t \\ &\quad +\iint_Q \rho\phi_1 w\overline{B}(u)(P)\mathrm{d}x\mathrm{d}t + \iint_Q \rho B(u)(\phi_1)wP\mathrm{d}x\mathrm{d}t \\ &\quad +\iint_Q \rho((2|\phi_1|^2-\vartheta)P\psi + \phi_1^2P\overline{\psi})\mathrm{d}x\mathrm{d}t = \iint_Q \rho\beta_2P\mathrm{d}x\mathrm{d}t, \\ &-\iint_Q \rho\frac{\partial Q}{\partial t}w\mathrm{d}x\mathrm{d}t + \int_\Omega \rho Q(T)w(T)\mathrm{d}x - \int_\Omega \rho Q(0)\beta_1\mathrm{d}x \\ &\quad +\iint_Q \rho\,\mathrm{curl}(Q)\mathrm{curl}(w)\mathrm{d}x\mathrm{d}t \\ &\quad +\iint_Q \mathrm{div}(Q)\mathrm{div}(\rho w)\mathrm{d}x\mathrm{d}t + \iint_Q \rho\mathcal{R}(\overline{\psi}B(u)(\phi_1))Q\mathrm{d}x\mathrm{d}t \\ &\quad +\iint_Q \rho(\mathcal{R}(\overline{\phi_1}B(u)(\psi))Q + w|\phi_1|^2Q)\mathrm{d}x\mathrm{d}t = 0.\end{aligned}$$

Since (P,Q) is a solution of (10.25) we have that

$$\begin{aligned}&\mathrm{i}\eta\kappa\iint_Q \rho w\nabla(\phi_1P)\mathrm{d}x\mathrm{d}t + \iint_Q \rho\phi_1 w\overline{B}(u)(P)\mathrm{d}x\mathrm{d}t \\ &\quad +\iint_Q \rho B(u)(\phi_1)wP\mathrm{d}x\mathrm{d}t + \iint_Q \rho(\phi_1^2P\overline{\psi} - \overline{\phi_1}^2\overline{P}\psi)\mathrm{d}x\mathrm{d}t \\ &\quad -\iint_Q \rho\overline{B(u)(\phi_1)}Q\psi\mathrm{d}x\mathrm{d}t - \iint_Q (\frac{-\mathrm{i}}{\kappa}\mathrm{div}(\rho\overline{\phi_1}Q) + \rho U_1\overline{\phi_1}Q)\psi\mathrm{d}x\mathrm{d}t \\ &\quad +a\iint_Q (|\varphi|^2-\Lambda)\overline{\varphi}\psi\mathrm{d}x\mathrm{d}t = \iint_Q \rho P\beta_2\mathrm{d}x\mathrm{d}t, \\ &-\int_\Omega \rho Q(0)\beta_1\mathrm{d}x + \iint_Q \rho\mathcal{R}(\overline{\psi}B(u)(\phi_1) + \overline{\phi_1}B(u)(\psi))Q\mathrm{d}x\mathrm{d}t \\ &\quad -\iint_Q \rho\mathcal{R}(\mathrm{i}\eta\kappa\nabla(\phi_1P) + PB(u)(\phi_1) + \phi_1\overline{B}(u)(P))w\mathrm{d}x\mathrm{d}t \\ &\quad +b\iint_Q (u-u_{\mathrm{obs}})w\mathrm{d}x\mathrm{d}t = 0.\end{aligned} \tag{10.46}$$

By adding the real part of the first part of (10.46) and the second part of (10.46) we obtain

$$\begin{aligned} a\iint_{\mathcal{Q}} \mathcal{R}((\mid \varphi \mid^2 - \Lambda)\overline{\varphi}\psi)\mathrm{d}x\mathrm{d}t + b\iint_{\mathcal{Q}}(u - u_{\mathrm{obs}})w\mathrm{d}x\mathrm{d}t \\ = \int_{\Omega} \rho Q(0)\beta_1 \mathrm{d}x + \iint_{\mathcal{Q}} \rho\mathcal{R}(P\beta_2)\mathrm{d}x\mathrm{d}t. \end{aligned} \tag{10.47}$$

According to the expression of $J'(g, f).Y$ we can deduce that

$$J'(g, f).Y = \int_{\Omega}(\rho Q(0) + \alpha g)\beta_1 \mathrm{d}x + \iint_{\mathcal{Q}} \mathcal{R}((\rho P - \gamma\overline{f})\beta_2)\mathrm{d}x\mathrm{d}t.$$

Since (g^*, f^*) is an optimal solution we then have

$$\begin{aligned} &\int_{\Omega}(\rho Q^*(0) + \alpha g^*)(g - g^*)\mathrm{d}x \geq 0, \\ &\iint_{\mathcal{Q}} \mathcal{R}((\rho P^* - \gamma\overline{f^*})(f - f^*))\mathrm{d}x\mathrm{d}t \leq 0 \quad \forall (g, f) \in \mathcal{K}. \end{aligned} \tag{10.48}$$

This completes the proof. □

Distributed Disturbance in the External Magnetic Field

In this section, the disturbance is in the magnetic field h and the control is in the initial condition u_0, *i.e.*, $u_0 = g$ ($g \in \mathbf{L}^2(\Omega)$), $h = f$ ($f \in L^2(\mathcal{Q})$). So the function (φ, u) is assumed to be related to the disturbance f and control g through the problem (10.12)($\forall (q, v) \in \mathcal{H}^1(\Omega) \times V_n$ and a.e. $t \in (0, T)$):

$$\begin{aligned} &\eta \int_{\Omega} \rho \frac{\partial \varphi}{\partial t} q\mathrm{d}x - \mathrm{i}\eta\kappa \int_{\Omega} \mathrm{div}(\rho(u + U))\varphi q\mathrm{d}x - \mathrm{i}\eta\kappa \int_{\Omega} \mathrm{div}(\rho u)\phi q\mathrm{d}x \\ &\quad -\mathrm{i}\mu \int_{\Gamma} \rho\varphi q\mathrm{d}\Gamma + \int_{\Omega}(\rho B(u)(\varphi + \phi)\overline{B}(u)(q)\mathrm{d}x \\ &\quad + \int_{\Omega} \rho F(\varphi)q\mathrm{d}x = \int_{\Omega} \rho B(0)(\phi)\overline{B}(0)(q)\mathrm{d}x + \int_{\Omega} \rho\xi q\mathrm{d}x, \\ &\int_{\Omega} \rho \frac{\partial u}{\partial t} v\mathrm{d}x + \int_{\Omega} \rho \mathrm{curl}(u)\mathrm{curl}(v)\mathrm{d}x + \int_{\Omega} \mathrm{div}(\rho u)\mathrm{div}(v)\mathrm{d}x \\ &\quad + \int_{\Omega} \rho\mathcal{R}((\overline{\varphi} + \overline{\phi})B(u)(\varphi + \phi))v\mathrm{d}x \\ &\quad = \int_{\Omega} \rho\mathcal{R}(\overline{\phi}B(0)(\phi))v\mathrm{d}x + \int_{\Omega} \rho f\mathrm{curl}(v)\mathrm{d}x, \\ &(\varphi(0), u(0)) = (\varphi_0, g). \end{aligned} \tag{10.49}$$

To obtain the regularity of Theorem 10.9, we suppose that $\varphi_0 \in \mathcal{L}^2_\infty(\Omega)$ and $\xi \in \mathcal{L}^2_\infty(\mathcal{Q})$. Let $\mathcal{P} : (g, f) \longrightarrow (\varphi, u) = \mathcal{P}(g, f)$ be the map: $\mathbf{L}^2(\Omega) \times L^2(\mathcal{Q}) \longrightarrow \mathcal{E} \times \mathbf{E}_n$ defined by (10.49) and the cost function is defined by

$$J(g,f) = \frac{a}{4} \parallel\mid \varphi \mid^2 - \Lambda \parallel^2_{L^2(Q)} + \frac{b}{2} \parallel u - u_{\text{obs}} \parallel^2_{L^2(Q)} \\ + \frac{\alpha}{2} \parallel g \parallel^2_{L^2} - \frac{\gamma}{2} \parallel f \parallel^2_{L^2(Q)}, \tag{10.50}$$

where $\alpha, \gamma > 0$, $a, b \geq 0$ and $a + b > 0$. The functions $(u_{\text{obs}}, \Lambda) \in \mathbf{L}^2(Q) \times L^\infty(Q)$ are given.

In this section we study the following robust control problem $(\mathcal{MP}_4)$:

find a saddle point (g^*, f^*) of the functional J on $\mathcal{K}$, subject to (10.49),

where, $\mathcal{K} = \mathcal{K}_1 \times \mathcal{K}_2$ such that $\mathcal{K}_1$ and $\mathcal{K}_2$ are non-empty, closed, convex, bounded subsets of $\mathbf{L}^2(\Omega)$ and $L^2(Q)$, respectively.

The proof of the following propositions and existence theorem is obtained by using similar arguments as used in Section 10.5.1. Therefore, we will omit the details.

Proposition 10.26. *The function $\mathcal{P}$ is continuously F-differentiable from $\mathbf{L}^2(\Omega) \times L^2(Q)$ to $\mathcal{E} \times \mathbf{E}_n$ with the derivative $\mathcal{P}'(g,f) : Y = (\beta_1, \beta_2) \longrightarrow (\psi, w)$ given by the linear problem $(\mathcal{P}_{F4})$ ($\forall (q,v) \in \mathcal{H}^1(\Omega) \times V_n$ and a.e. $t \in (0,T)$):*

$$\eta \int_\Omega \rho \frac{\partial \psi}{\partial t} q \mathrm{d}x - \mathrm{i}\eta\kappa \int_\Omega \mathrm{div}(\rho U_1)\psi q \mathrm{d}x - \mathrm{i}\eta\kappa \int_\Omega \mathrm{div}(\rho w)\phi_1 q \mathrm{d}x \\ - \mathrm{i}\mu \int_\Gamma \rho\psi q \mathrm{d}\Gamma + \int_\Omega \rho(B(u)(\psi) + \phi_1 w)\overline{B}(u)(q)\mathrm{d}x \\ + \int_\Omega \rho B(u)(\phi_1) w q \mathrm{d}x + \int_\Omega \rho((2 \mid \phi_1 \mid^2 - \vartheta)\psi + \phi_1^2 \overline{\psi}) q \mathrm{d}x = 0,$$

$$\int_\Omega \rho \frac{\partial w}{\partial t} u \mathrm{d}x + \int_\Omega \rho \mathrm{curl}(w)\mathrm{curl}(v)\mathrm{d}x + \int_\Omega \mathrm{div}(\rho w)\mathrm{div}(v)\mathrm{d}x \\ + \int_\Omega \rho \mathcal{R}(\overline{\psi} B(u)(\phi_1)) v \mathrm{d}x + \int_\Omega \rho(\mathcal{R}(\overline{\phi_1} B(u)(\psi)) + w \mid \phi_1 \mid^2) v \mathrm{d}x \\ = \int_\Omega \rho\beta_2 \mathrm{curl}(v)\mathrm{d}x, \\ (\psi(0), w(0)) = (0, \beta_1),$$

where $(U_1, \phi_1) = (u + U, \varphi + \phi)$.

Moreover, we have the following estimates, for all $(g_i, f_i) \in \mathbf{L}^2(\Omega) \times L^2(Q)$, for $i = 1, 2$):

(i) $\parallel \mathcal{P}'(g_1, f_1) \parallel_{\mathcal{L}(\mathbf{L}^2 \times L^2(Q), \mathcal{E} \times \mathbf{E}_n)} \leq C_e$

(ii) $\parallel \mathcal{P}'(g_1, f_1)Y - \mathcal{P}'(f_2, g_2)Y \parallel^2_{\mathcal{E} \times \mathbf{E}_n}$

$$\leq C_e(\parallel X \parallel_{\mathbf{L}^2 \times L^2(Q)} \parallel Y \parallel^2_{\mathbf{L}^2 \times L^2(Q)} + \parallel X \parallel^2_{\mathbf{L}^2 \times L^2(Q)} \parallel Y \parallel_{\mathbf{L}^2 \times L^2(Q)}),$$

where $f = f_1 - f_2$, $g = g_1 - g_2$ and $X = (g, f)$.

Proposition 10.27. *The map $\mathcal{P}$ defined by (10.49) is continuous from the weak topology of $\mathbf{L}^2(\Omega) \times L^2(Q)$ to the strong topology of $\mathcal{L}^2(Q) \times \mathbf{L}^2(Q)$.* □

Theorem 10.28. *For α and γ sufficiently large, there exists $(g^*, f^*) \in \mathcal{K}$ and $(u^*, \varphi^*) \in \mathbf{E}_n \times \mathcal{E}$ such that (g^*, f^*) is defined by $(\mathcal{MP}_4)$ and $(\varphi^*, u^*) = \mathcal{P}(g^*, f^*)$ is a solution of (10.49).* □

Next we give necessary optimality conditions for the robust control problem $(\mathcal{MP}_4)$.

Theorem 10.29. *Under the assumptions of Theorem 10.28, the optimal solution $(g^*, f^*, u^*, \varphi^*) \in \mathcal{K} \times \mathbf{E}_n \times \mathcal{E}$ such that (g^*, f^*) is defined by $(\mathcal{MP}_4)$ and $(\varphi^*, u^*) = \mathcal{P}(f^*, g^*)$ is a solution of (10.49), satisfies*

$$\begin{aligned} &\int_\Omega (\rho Q^*(0) + \alpha g^*)(g - g^*)\mathrm{d}x \geq 0, \\ &\iint_Q (\rho \mathrm{curl}(Q^*) - \gamma f^*)(f - f^*))\mathrm{d}x\mathrm{d}t \leq 0 \quad \forall (g, f) \in \mathcal{K}, \end{aligned} \tag{10.51}$$

where (P^*, Q^*) is the solution of problem (10.25) (corresponding to (φ^*, u^*)).

Proof. The proof is similar to that of Theorem 10.25, so we omit the details. □

We finish this section with the following remark.

Remark 10.30. (i) We can consider other types of controls and disturbances, as the technique developed previously will be still valid. For example, in the case where the control is in the initial condition of the order parameter, *i.e.*, $g = \varphi_0$ and the distributed disturbance is in the chaotic-type force, *i.e.*, $\xi = f$, we can prove, for α and γ sufficiently large, the existence theorem of the robust control problem and obtain necessary optimality conditions for its solution using the same method as previously. In this case the cost functional is given by

$$J(g,f) = \frac{a}{4} \,\|| \varphi |^2 - \Lambda \|^2_{L^2(Q)} + \frac{b}{2} \| u - u_{\mathrm{obs}} \|^2_{L^2(Q)} + \frac{\alpha}{2} \| g \|^2_{L^2} - \frac{\gamma}{2} \| f \|^2_{L^2(Q)},$$

where $\alpha, \gamma > 0$, $a, b \geq 0$ and $a + b > 0$.

Let $\mathcal{K} = \mathcal{K}_1 \times \mathcal{K}_2$ such that $\mathcal{K}_1$ and $\mathcal{K}_2$ are non-empty, closed, convex, bounded subsets of $\mathcal{L}^2_\infty(\Omega)$ and $\mathcal{L}^2_\infty(Q)$, respectively.

For α and γ sufficiently large, there exists (g^*, f^*, φ, u) satisfying

$$\begin{aligned} \eta \int_\Omega \rho \frac{\partial \varphi}{\partial t} q\mathrm{d}x - \mathrm{i}\eta\kappa \int_\Omega \mathrm{div}(\rho(u + U))\varphi q\mathrm{d}x - \mathrm{i}\eta\kappa \int_\Omega \mathrm{div}(\rho u)\phi q\mathrm{d}x \\ -\mathrm{i}\mu \int_\Gamma \rho\varphi P\mathrm{d}\Gamma + \int_\Omega (\rho B(u)(\varphi + \phi)\overline{B}(u)(q)\mathrm{d}x \\ + \int_\Omega \rho F(\varphi) q\mathrm{d}x = \int_\Omega \rho B(0)(\phi)\overline{B}(0)(q)\mathrm{d}x + \int_\Omega \rho f^* q\mathrm{d}x, \end{aligned}$$

$$
\begin{aligned}
&\int_\Omega \rho \frac{\partial u}{\partial t} v \mathrm{d}x + \int_\Omega \rho \mathrm{curl}(u)\mathrm{curl}(v)\mathrm{d}x + \int_\Omega \mathrm{div}(\rho u)\mathrm{div}(v)\mathrm{d}x \\
&\quad + \int_\Omega \rho \mathcal{R}((\overline{\varphi}+\overline{\phi})B(u)(\varphi+\phi))v\mathrm{d}x \\
&\quad = \int_\Omega \rho \mathcal{R}(\overline{\phi}B(0)(\phi))v\mathrm{d}x\mathrm{d}t + \int_\Omega \rho h \mathrm{curl}(v)\mathrm{d}x, \\
&(\varphi(0), u(0)) = (g^*, u_0)
\end{aligned}
$$

and the inequality

$$
\begin{aligned}
&\int_\Omega \mathcal{R}((\eta\rho P^*(0) + \alpha\overline{g^*})(g-g^*))\mathrm{d}x \geq 0, \\
&\iint_Q \mathcal{R}((\rho P^* - \gamma\overline{f^*})(f-f^*))\mathrm{d}x \leq 0 \quad \forall (g,f) \in \mathcal{K},
\end{aligned}
$$

where (P^*, Q^*) is the solution of problem (10.25) associated with the solution (φ, u) corresponding to data (g^*, f^*).
(*ii*) We can consider also other types of observation, the technique developed previously will be still valid. For example, consider the following cost functional:

$$
\begin{aligned}
J(g,f) = \frac{a}{4} \| | \varphi + \phi |^2 - \tilde{\Lambda} \|^2_{L^2(Q)} + \frac{b}{2} \| u - u_{\mathrm{obs}} \|^2_{L^2(Q)} \\
+ \frac{\alpha}{2} \| g \|^2_{L^2} - \frac{\gamma}{2} \| f \|^2_{L^2(Q)},
\end{aligned}
$$

where $\tilde{\Lambda} = \Lambda + | \phi |^2$, $\alpha, \gamma > 0$, $a, b \geq 0$ and $a + b > 0$.

Let $\mathcal{K} = \mathcal{K}_1 \times \mathcal{K}_2$ such that $\mathcal{K}_1$ and $\mathcal{K}_2$ are non-empty, closed, convex, bounded subsets of $\mathcal{L}^2_\infty(\Omega)$ and $\mathcal{L}^2_\infty(Q)$, respectively.

For α and γ sufficiently large, there exists (g^*, f^*, φ, u) satisfying

$$
\begin{aligned}
&\eta \int_\Omega \rho \frac{\partial \varphi}{\partial t} q\mathrm{d}x - \mathrm{i}\eta\kappa \int_\Omega \mathrm{div}(\rho(u+U))\varphi q\mathrm{d}x - \mathrm{i}\eta\kappa \int_\Omega \mathrm{div}(\rho u)\phi q\mathrm{d}x \\
&\quad - \mathrm{i}\mu \int_\Gamma \rho\varphi P \mathrm{d}\Gamma + \int_\Omega (\rho B(u)(\varphi+\phi)\overline{B}(u)(q)\mathrm{d}x \\
&\quad + \int_\Omega \rho F(\varphi) q\mathrm{d}x = \int_\Omega \rho B(0)(\phi)\overline{B}(0)(q)\mathrm{d}x + \int_\Omega \rho f^* q\mathrm{d}x, \\
&\int_\Omega \rho \frac{\partial u}{\partial t} v\mathrm{d}x + \int_\Omega \rho \mathrm{curl}(u)\mathrm{curl}(v)\mathrm{d}x + \int_\Omega \mathrm{div}(\rho u)\mathrm{div}(v)\mathrm{d}x \\
&\quad + \int_\Omega \rho \mathcal{R}((\overline{\varphi}+\overline{\phi})B(u)(\varphi+\phi))v\mathrm{d}x \\
&\quad = \int_\Omega \rho \mathcal{R}(\overline{\phi}B(0)(\phi))v\mathrm{d}x\mathrm{d}t + \int_\Omega \rho h \mathrm{curl}(v)\mathrm{d}x, \\
&(\varphi(0), u(0)) = (g^*, u_0)
\end{aligned}
\tag{10.52}
$$

and the inequality

$$\int_{\Omega} \mathcal{R}((\eta\rho P^*(0) + \alpha\overline{g^*})(g - g^*))\mathrm{d}x \geq 0,$$

$$\iint_{Q} \mathcal{R}((\rho P^* - \gamma\overline{f^*})(f - f^*))\mathrm{d}x \leq 0 \quad \forall (g, f) \in \mathcal{K},$$

where (P^*, Q^*) is the solution of the following adjoint problem (associated with the solution (φ, u) of (10.52) corresponding to data (g^*, f^*)):

$$\begin{aligned}
&-\eta \int_{\Omega} \rho \frac{\partial P}{\partial t} q\mathrm{d}x - \mathrm{i}\eta\kappa \int_{\Omega} \mathrm{div}(\rho U_1) P q \mathrm{d}x - \mathrm{i}\mu \int_{\Gamma} \rho P q \mathrm{d}\Gamma \\
&\qquad + \int_{\Omega} \rho \overline{B}(u)(P) B(u)(q)\mathrm{d}x + \int_{\Omega} \rho \overline{B(u)(\phi_1)} Q q \mathrm{d}x \\
&\qquad + \int_{\Omega} (\frac{-\mathrm{i}}{\kappa} \mathrm{div}(\rho \overline{\phi_1} Q) + \rho U_1 \overline{\phi_1} Q) q \mathrm{d}x \\
&\qquad + \int_{\Omega} \rho((2 \mid \phi_1 \mid^2 - \vartheta) P + \overline{\phi_1}^2 \overline{P}) q \mathrm{d}x = a \int_{\Omega} (\mid \phi_1 \mid^2 - \tilde{\Lambda}) \overline{\phi_1} q \mathrm{d}x, \\
&- \int_{\Omega} \rho \frac{\partial Q}{\partial t} v \mathrm{d}x + \int_{\Omega} \rho \mathrm{curl}(Q) \mathrm{curl}(v) \mathrm{d}x + \int_{\Omega} \mathrm{div}(Q) \mathrm{div}(\rho q) \mathrm{d}x \\
&\qquad + \int_{\Omega} \rho \mathcal{R}(\mathrm{i}\eta\kappa \nabla(\phi_1 P) + P B(u)(\phi_1) + \phi_1 \overline{B}(u)(P)) v \mathrm{d}x \\
&\qquad + \int_{\Omega} \rho Q \mid \phi_1 \mid^2 v \mathrm{d}x = b \int_{\Omega} (u - u_{\mathrm{obs}}) v \mathrm{d}x, \\
&(P(T), Q(T)) = (0, 0)
\end{aligned}$$

where $(\phi_1, U_1) = (\varphi + \phi, u + U))$. ♢

Remark 10.31. We recall that, for the numerical resolution, we can combine the optimal necessary conditions (which also give the gradients of the cost function) and, for example, the gradient-iterative algorithm (see Chapter 9). ♢

11

Multi-scale Modeling of Alloy Solidification and Phase-field Model

In this chapter we formulate and study robust control problems for a two-dimensional, non-linear, time-dependent and solutal phase-field model of the Warren–Boettinger-type (*TDWB*), which describes the isothermal solidification of a binary alloy. This model contains two unknowns, the relative concentration and a non-conserved structural order parameter (which is said to be the *phase-field variable*) coming from thermodynamics. The phase-field theory is a direct consequence of the Cahn–Hilliard and Ginzburg–Landau type classical field theoretic approaches to phase boundaries.

The order parameter describes the phase of the underlying substance: the order parameter is close to 1 if the system is in a liquid phase and is close to 0 if it is in a solid phase.

It is well known that thermal fluctuations and material impurities affect considerably the solidification microstructure dynamics. These effects are modeled by variants of Warren–Boettinger model containing additive noise due to thermal fluctuations and by modification of some operators to take into account impurities.

The objective is the prediction and stabilization of microstructure dynamics by taking into account the influence of fluctuations and data noises. First, a variant of TDWB model (*MTDWB*) is introduced and analyzed. Second, we introduce the perturbation problem of the non-linear governing coupled system of the MTDWB equations (the deviation from the desired target). The existence and uniqueness of the solution of the perturbation are proved as well as their stability under mild assumptions. Afterwards, some robust control problems are formulated for the cases when the control is in the initial condition of the concentration field and when the worst disturbance is the noise due to thermal fluctuations or is in the initial condition of the phase-field parameter. We show the existence of an optimal solution, and we also find the necessary conditions for a saddle point optimality.

This work is a generalization of recent research developed by Belmiloudi in [40].

11.1 Introduction

The aim of this chapter is the study of robust control problems related to the isotropic models for the isothermal solidification[1] of a binary alloy (*i.e.*, a mixture of two elements A and B), using a solutal phase-field model. This model was derived by Warren and Boettinger [293] to model dendrite growth in highly supersaturated binary melts. It involves the relative concentration U of solute B in solvent A and a phase-field variable ϕ with thermodynamically consistent evolution equations. The concentration U and the parameter ϕ vary sharply but smoothly between 0 in the solid phase and 1 in the liquid phase, over a thin layer which separates the phases. Our physical system is supposed to be closed, and no phase exchange takes place across its boundary. Therefore, we can close our system by considering Neumann boundary conditions and some initial physical conditions which are given between 0 and 1.

The time evolution system of $\mathbf{U} = (U, \phi)$ is governed by partial differential equations of the form

$$\begin{aligned}
&\frac{\partial \phi}{\partial t} - \frac{\epsilon^2}{\tau}\Delta\phi = F_1(t,\phi) + UF_2(t,\phi) \quad \text{on } \mathcal{Q} = \Omega \times (0,T),\\
&\frac{\partial U}{\partial t} - \operatorname{div}(D_1(t,\phi)\nabla U + D_2(t,\mathbf{U})\nabla\phi) = 0 \quad \text{on } \mathcal{Q},\\
&\text{subject to the boundary conditions}\\
&\frac{\partial \phi}{\partial \mathbf{n}} = 0, \ \frac{\partial U}{\partial \mathbf{n}} = 0 \quad \text{on } \Sigma = \partial\Omega \times (0,T),\\
&\text{and the initial conditions}\\
&\phi(0) = \phi_0, \ U(0) = U_0 \quad \text{on } \Omega,
\end{aligned} \tag{11.1}$$

where Ω is an open bounded domain in $\mathbb{R}^m$, $m \leq 2$ with a smooth boundary $\partial\Omega$ of class C^∞ and $\mathbf{n}$ is the unit normal to $\partial\Omega$. The positive constants τ and ϵ^2 are proportional to the relaxation time and interface thickness, respectively. In the sequel, we denote by ν the positive constant ϵ^2/τ.

The functions $(F_i)_{i=1,2}$ and $(D_i)_{i=1,2}$ appearing in problem (11.1) have the following properties (see the sketch of the modeling below):

1. The functions F_i, for $i = 1, 2$, are regular, depending on the temperature function, and satisfy $F_i(., \phi = 0) = F_i(., \phi = 1) = 0$, for $i = 1, 2$.
2. The function D_1 is positive regular and bounded above and below by two positive constants.
3. The function D_2 is regular and satisfies $D_2(., U = 0, .) = D_2(., U = 1, .) = 0$.

Next we give a brief review of the modeling leading to problem (11.1). We start with the choice of the free energy[2] for the Warren–Boettinger model

[1] So that the temperature is assumed to be externally imposed on the system.

[2] Generally, the choice of a model is conditioned by the choose form of the free energy functional.

[293]. A free energy for a binary alloy, which must decrease during any process, can be written as

$$\begin{aligned}\mathcal{F}(t) &= \mathcal{F}(\mathcal{T}(.,t),\phi(.,t),U(.,t)) \\ &= \int_{\Omega}(\frac{1}{2} \mid \epsilon_\phi \nabla\phi(.,t) \mid^2 + f_{AB}(\mathcal{T}(.,t),\phi(.,t),U(.,t)))\mathrm{d}x,\end{aligned} \tag{11.2}$$

where $\mid \epsilon_\phi \nabla\phi(.,t) \mid^2$ is a potential of Ginzburg–Landau type, the parameter ϵ_ϕ is an energy scale for the order parameter interface and $\mathcal{T}$ is the temperature. The corresponding width of this interface is given by $W_\phi = \epsilon_\phi/\sqrt{d}$, where d denotes the energy barrier between liquid and solid phases and is proportional to the liquid–solid surface energy, which is assumed to be the same for both A and B materials. The parameter ϵ_ϕ is made isotropic (respectively anisotropic) by making it independent (respectively dependent) on the interface normal vector, which is given by $\nabla\phi/ \mid \nabla\phi \mid$. The function f_{AB} denotes the bulk free energy density of the A–B mixture. It is designed to reproduce the thermodynamic phase diagram of the alloy and is built from the free energy densities f_A and f_B of the pure element ($U = 0$ and $U = 1$, respectively) and a mixing energy term. The simple form of f_{AB} can be written as follows (see Warren *et al.* [293, 294]):

$$\begin{aligned}f_{AB}(\mathcal{T},\phi,U) &= (1-U)f_A(\mathcal{T},\phi) + Uf_B(\mathcal{T},\phi) + H(\mathcal{T},U), \text{ with} \\ H(\mathcal{T},U) &= \frac{R\mathcal{T}}{v_m}(U\ln(U) + (1-U)\ln(1-U)),\end{aligned} \tag{11.3}$$

where R is the Bolzman constant, v_m denotes the molar volume of solid A and B phases and the function H describes the energy of ideal mixing between A and B. By using some basic thermodynamical principles, we can obtain a general form for f_A and f_B (which are depending on the latent heats of fusion and melting temperature L_A, $\mathcal{T}_A$ and L_B, $\mathcal{T}_B$ of A and B, respectively) and then for f_{AB} at any given temperature $\mathcal{T}$. If we assume that at the melting temperature the functions f_A and f_B are of double-well potential type between the solid and liquid phases and if at any given temperature $\mathcal{T}$, the functions f_A and f_B have each only two minima for the variable ϕ in [0,1], a particular explicit form for f_{AB} can be given by

$$f_M(\mathcal{T},\phi) = C_M(\mathcal{T})g(\phi) - \frac{(\mathcal{T}-\mathcal{T}_M)D_M(\mathcal{T})}{\mathcal{T}_M}p(\phi) + E_M(\mathcal{T}), \text{ for } M = A \text{ or } B,$$

where the functions C_M, D_M and E_M are positive such that $E_M(\mathcal{T}_M) = 0$ and C_M is depending on W_ϕ. The function $g(\phi) = \phi^2(1-\phi)^2$ provides a double-well potential between the solid and liquid phases, while $p(\phi)$ is, for example, $p(\phi) = \phi^2(3-2\phi)$ or $p(\phi) = \phi^3(6\phi^2 - 15\phi + 10)$ (which interpolates between the solid and liquid entropy states).

Once the free energy function is defined, the equations of motion for the concentration and the phase-field parameter are obtained as follows:

$$
\begin{aligned}
\tau_\phi \frac{\partial \phi}{\partial t} &= -\frac{\partial \mathcal{F}}{\partial \phi}, \\
\frac{\partial U}{\partial t} &= \operatorname{div}(M_u(\mathcal{T}, \phi, U)\nabla(\frac{\partial \mathcal{F}}{\partial U})) \\
&= \operatorname{div}(M_u(\mathcal{T}, \phi, U)(\nabla(\frac{\partial f_{AB}}{\partial U}) + \frac{R\mathcal{T}}{v_m}\frac{1}{U(1-U)}\nabla U)),
\end{aligned} \tag{11.4}
$$

where τ_ϕ represents the interface kinetic attachment time scale (which can be dependent on the interface normal vector) and M_u is the solute mobility which is assumed to be in the form

$$
M_u(\mathcal{T}, \phi, U) = D_1(\mathcal{T}, \phi)U(1-U),
$$

where $D_1(\mathcal{T}, \phi) = D_{11}(\mathcal{T})D_{12}(\phi)$ such that D_{11} is a non-negative function, D_{12} is an increasing smooth function for the variable ϕ such that $D_{12}(0) > 0$,[3] $D_{12}(1) > 0$ and D_1 is bounded above and below by two non-negative constants.

According to the expression (11.3) of the free energy, we can easily prove that there exist functions F_1 and F_2 such that

$$
-\frac{\partial f_{AB}}{\partial \phi} = F_1(\mathcal{T}, \phi) + UF_2(\mathcal{T}, \phi). \tag{11.5}
$$

If we suppose, moreover, that the given temperature $\mathcal{T}$ is independent of variable space, we can deduce that

$$
\nabla(\frac{\partial f_{AB}}{\partial U}) = -F_2(\mathcal{T}, \phi)\nabla\phi. \tag{11.6}
$$

If we suppose now that $\tau_\phi = \tau$ and $\epsilon_\phi = \epsilon$ are independent on ϕ then, according to (11.5) and (11.6) and the expression of the free energy function, System (11.4) becomes

$$
\begin{aligned}
\tau\frac{\partial \phi}{\partial t} &= \operatorname{div}(\epsilon^2\nabla\phi) + F_1(\mathcal{T}, \phi) + UF_2(\mathcal{T}, \phi), \\
\frac{\partial U}{\partial t} &= \operatorname{div}(D_1(\mathcal{T}, \phi)\nabla U + D_2(\mathcal{T}, U, \phi)\nabla\phi),
\end{aligned} \tag{11.7}
$$

where $D_2(\mathcal{T}, U, \phi) = D_1(\mathcal{T}, \phi)U(1-U)F_2(\mathcal{T}, \phi)$.

If we suppose that τ and ϵ^2 are constants we can easily obtain System (11.1).

Remark 11.1. In the general case the parameter ϵ_ϕ is dependent on $\nabla\phi$. This variable is an anisotropic tensor and is introduced for the simulation of dentritic growth, and accounts for the existence of privileged directions for solidification due to microscopic crystal growth. ♢

[3] D_{12} is a function that interpolates between solid and liquid diffusivities.

The greatest difficulty in this type of problem is the multiscale nature (for length and time scales) of solidification microstructures with thermal conditions and the material parameters used in experiments, which influence considerably the physical material properties (see, *e.g.*, Askeland [12]). The interface between liquid and solid is a few nanometers thick and microstructural features are on the scale of tens to hundreds of microns. Moreover, the atomic attachment kinetics take a few picoseconds, whereas diffusion of heat and solute is of the order of seconds. Consequently, in order to take into account this multiscale nature, it is necessary to develop adapted mathematical and computational models, which help the engineer to complete the experiment as well as to define the pertinent material parameters to be analyzed.

In recent years the so-called phase-field formulation has emerged as a powerful computational approach to modeling and predicting the range of phase transitions and complex "dendritic" growth structures occurring during solidification. This approach has proved to be an emerging technology that complements experimental research.[4] Various problems associated with phase-field models have been studied to treat both pure materials and binary alloys, either from the theoretical or numerical point of view (see, *e.g.*, Brochet *et al.* [62, 63], Caginalp [66], Grujicic *et al.* [142], Laurencot [186], Rappaz and Scheid [249], Warren *et al.* [293, 294] and the references therein). For control problems associated with phase-field models, refer to Hoffman and Jiang [159] (for optimal control problems) and Belmiloudi and Yvon [41] (for robust control problems). In these the authors have studied the problem of describing the phase transitions of pure materials due to thermal effects, and they have developed non-linear parabolic systems for the phase field and temperature.

It is then clear that, in order to study the stability, dynamics and other properties of the microstructure state, it is necessary to take into account the effects of thermal fluctuation and material defects on the solidification dynamics, because for longer solidification times, the solid/liquid interface becomes unstable with respect to small perturbations caused by the introduction of noise and fluctuation terms. In this chapter, which generalizes the previous results of Belmiloudi [40], we consider, in order to take into account thermal fluctuations, an additive noise solidification model[5] which is a modification of the model (11.1). More precisely, we introduce into the right-hand side of the first equation of (11.1) a real-valued field in time and space and assume that the operators appearing in (11.1) are modified.[6] Thus, the modified system is

---

[4] Because the experimental solidification research in metals that is required to link microstructural characteristics with processing regimes, is often limited in its ability to observe real-time development of microstructure and the associated segregation patterns.

[5] The thermal fluctuations can be modeled, for example, by deterministic Langevin-type dynamics.

[6] The polynom functions and coefficients appearing in the modeling are dependent on the variable space x. For example, the function g defined previously can be replaced by $\tilde{g}(x, \phi) = \phi^2(\vartheta(x) - \phi)^2$ with $0 < \vartheta < 1$ in Ω.

given by

$$\begin{aligned}
&\frac{\partial\phi}{\partial t} - \nu\Delta\phi = \tilde{F}_1(x,t,\phi) + U\tilde{F}_2(x,t,\phi) + \lambda \ \text{ on } \mathcal{Q},\\
&\frac{\partial U}{\partial t} - \operatorname{div}(\tilde{D}_1(x,t,\phi)\nabla U + \tilde{D}_2(x,t,\mathbf{U})\nabla\phi) = 0 \ \text{ on } \mathcal{Q},\\
&\text{subject to the boundary conditions}\\
&\frac{\partial\phi}{\partial\mathbf{n}} = 0, \ \frac{\partial U}{\partial\mathbf{n}} = 0 \ \text{ on } \Sigma,\\
&\text{and the initial conditions}\\
&\phi(0) = \phi_0, \ U(0) = U_0 \ \text{ on } \Omega.
\end{aligned} \tag{11.8}$$

In the following, we omit the "~" on the operators appearing in problem (11.8).

11.1.1 Assumptions and Notations

We denote by $V = H^1(\Omega)$ and V' the dual of V. We denote by $\langle\ ,\ \rangle_{V',V}$ the duality product between V' and V. For any pair of real number $r, s \geq 0$, we introduce the Sobolev space $H^{r,s}(\mathcal{Q})$ defined by $H^{r,s}(\mathcal{Q}) = L^2(0,T;H^r(\Omega)) \cap H^s(0,T;L^2(\Omega))$, which is normed by (see Chapter 6)

$$\left(\int_0^T \| v \|^2_{H^r(\Omega)} \,\mathrm{d}t + \| v \|^2_{H^s(0,T;L^2(\Omega))}\right)^{1/2}.$$

We can now introduce the following spaces: $\mathcal{H}_i = L^\infty(0,T;H^{i-1}(\Omega))$, $\mathcal{V}_i = L^2(0,T;H^i(\Omega))$, $\mathcal{W}_i = \mathcal{H}_i \cap \mathcal{V}_i$, (for $i = 1,3$), $\mathcal{H}_i^1 = H^1(0,T;H^{i-2}(\Omega))$, (for $i = 2,3$), $\mathcal{H}_1^1 = H^1(0,T;V')$, $\mathcal{W}_i^1 = \mathcal{H}_i^1 \cap \mathcal{V}_i$, (for $i = 1,3$) and the space $H_0^2 = \{v \in H^2(\Omega) : \ \partial v/\partial\mathbf{n} = 0\}$.

Remark 11.2. (*i*) $\mathcal{W}_i^1$ is compactly embedded into $\mathcal{V}_{i-1}$, for $i = 1,3$ (see Chapter 6).
(*ii*) $\mathcal{W}_i^1 \subset \mathcal{C}([0,T];H^{i-1}(\Omega))$, for $i = 1,3$ (see Chapter 6). ◊

We state the following hypotheses for the operators $(F_i)_{i=1,2}$ and $(D_i)_{i=1,2}$:

(H1) F_1 and F_2 are Carathéodory functions from $\mathcal{Q}\times\mathbb{R}$ into $\mathbb{R}$ such that, for almost all $(x,t) \in \mathcal{Q}$, $F_1(x,t,.)$ and $F_2(x,t,.)$ are Lipschitz and bounded functions with
(*i*) $\mid F_i(x,t,r) \mid \leq M_1 \ \ \forall\, i = 1,2, \ \forall\, r \in \mathbb{R}$ and a.e. $(x,t) \in \mathcal{Q}$
(*ii*) $F'_{ix}.n = 0$, on Σ.

(H2) D_1 is a Carathéodory function from $\mathcal{Q}\times\mathbb{R}$ into $\mathbb{R}$ such that, for almost all $(x,t) \in \mathcal{Q}$, $D_1(x,t,.)$ is Lipschitz positive and bounded function with

$$0 < D_0 \leq D_1(x,t,r) \leq D_1 \ \forall\, r \in \mathbb{R} \text{ and a.e. } (x,t) \in \mathcal{Q}.$$

(H3) D_2 is a Carathéodory function from $\mathcal{Q}\times\mathbb{R}^2$ into $\mathbb{R}$ such that, for almost all $(x,t)\in\mathcal{Q}$, $D_2(x,t,.)$ is Lipschitz positive and bounded function with

$$0 < D_0 \le D_2(x,t,\mathbf{r}) \le M_2,\ \forall\, \mathbf{r}=(r,r')\in\mathbb{R}^2 \text{ and a.e. } (x,t)\in\mathcal{Q}.$$

(H4) F_i, for $i=1,2$, and D_1 are differentiable with F'_{ix}, $G_i=F'_{ir}$, D'_{1x} and $H_1=D'_{1r}$ Lipschitz continuous a.e. in $\mathcal{Q}$.

(H5) D_2 is differentiable with D'_{2x}, $H_2=D'_{2\mathbf{r}}$ Lipschitz continuous a.e. in $\mathcal{Q}$.

Remark 11.3. The functions $(F_i, i=1,2)$ and $(D_i, i=1,2)$ are depending on temperature. ◊

Remark 11.4. If u and ϕ are sufficiently regular and satisfy $\partial u/\partial\mathbf{n}=0$ and $\partial\phi/\partial\mathbf{n}=0$, then according to the second assumption of **(H1)**, we have

$$\nabla(F_1(x,t,\phi)).n = \nabla(uF_2(x,t,\phi)).n = 0.$$

◊

In the following we will use C to denote some positive constant which may be different at each occurrence.

11.1.2 Preliminary Results

Lemma 11.5. *(Elliptic estimate) Let $k\in\mathbb{N}$ and $v\in H^2(\Omega)$ satisfy $\Delta v\in H^k$ and $\partial v/\partial\mathbf{n}=0$ on $\partial\Omega$. Then $v\in H^{k+2}(\Omega)$ and we have the following estimate: there exists $C>0$ (independent of v) such that*

$$\| v \|_{H^{k+2}} \le C(\| \Delta v \|_{H^k} + \| v \|_{H^k}).$$

Proof. For the proof of this lemma see, for example, Lions and Magenes [204]. □

Lemma 11.6. *Let $\mathbf{u}_l=(u_l,\varphi_l)$ be a sequence converging toward $\mathbf{u}=(u,\varphi)$ in $\mathcal{W}_1^1$ weakly and in $L^2(\mathcal{Q})$ strongly. Then we have the following convergence results:*

(i) $F_i(.,\varphi_l)\longrightarrow F_i(.,\varphi)$,*(i=1,2) in* $L^p(\mathcal{Q})$ *strongly,* $\forall p\in[1,+\infty)$

(ii) $D_1(.,\varphi_l)\longrightarrow D_1(.,\varphi)$ *in* $L^p(\mathcal{Q})$ *strongly,* $\forall p\in[1,+\infty)$

(iii) $u_lF_2(.,\varphi_l)\longrightarrow uF_2(.,\varphi)$ *in* $L^q(\mathcal{Q})$ *strongly,* $\forall q\in[1,2)$

(iv) $D_1(.,\varphi_l)\nabla u_l \rightharpoonup D_1(.,\varphi)\nabla u$ *in* $L^q(\mathcal{Q})$ *weakly,* $\forall q\in[1,2)$

(v) $D_2(.,\mathbf{u}_l)\nabla\varphi_l \rightharpoonup D_2(.,\mathbf{u})\nabla\varphi$ *in* $L^q(\mathcal{Q})$ *weakly,* $\forall q\in[1,2)$.

Proof. For the proof of this lemma we can use a classical technique based on taking the difference between the sequence and its limit in the form of the sum of two terms such that the first uses the weak convergence result and the other uses the strong convergence result (for a similar result see, *e.g.*, Rappaz *et al.* [249]). The results *(i)*–*(iii)* are a simple consequence of the strong convergence in $L^2(\mathcal{Q})$ and the assumptions (**H1**) and (**H2**). The proof of the results (*iii*) is similar to the proof of (*iv*). The proof of (*iv*) is based on the fact that $I_l = \iint_{\mathcal{Q}} (D_2(x,t,\mathbf{u}_l)\nabla\varphi_l - D_2(x,t,\mathbf{u})\nabla\varphi)v\mathrm{d}x\mathrm{d}t$, for all $v \in L^{q^*}(\mathcal{Q})$, can be written as follows:

$$I_l = \iint_{\mathcal{Q}} D_2(x,t,\mathbf{u})(\nabla\varphi_l - \nabla\varphi)v\mathrm{d}x\mathrm{d}t \\ + \iint_{\mathcal{Q}} (D_2(x,t,\mathbf{u}_l) - D_2(x,t,\mathbf{u}))\nabla\varphi_l v\mathrm{d}x\mathrm{d}t \tag{11.9}$$

where the convergence to 0 of the first term is a consequence of the weak convergence in $\mathcal{W}_1^1$ and the convergence to 0 of the second term is a consequence of the strong convergence in $L^2(\mathcal{Q})$ and the assumption (**H3**). □

11.2 Existence, Uniqueness and a Maximum Principle

11.2.1 Existence and Uniqueness Results

The following results concern the existence, uniqueness and regularity of a solution for problem (11.8).

Proposition 11.7. *Let Assumptions* (**H1**)–(**H3**) *be fulfilled. For any element* $(\phi_0, U_0, \lambda) \in H_0^2(\Omega) \times H^1(\Omega) \times L^2(0,T;H^1(\Omega))$, *then there exists a unique global solution* (ϕ, U) *of (11.8) satisfying*

$$\phi \in \mathcal{W}_3^1 \subset \mathcal{C}([0,T];H^2) \text{ and } U \in \mathcal{W}_2^1 \subset \mathcal{C}([0,T];H^1).$$

Proof. The proof of this proposition can be obtained in the same way as used to prove the results of Section 11.3. So, we omit the details. □

The following results concern the maximum principle.

11.2.2 A Maximum Principle

Assume that function λ is in the space

$$\Lambda = \{\lambda \in L^2(\mathcal{Q}) : \ 0 \le \lambda(x,t) \le M \quad \text{a.e. } (x,t) \in \mathcal{Q}\}.$$

We establish now a maximum principle under extra assumptions on the non-linear terms. So, in addition to Hypothesis (**H1**)–(**H3**), we assume that the non-linear terms F_1, F_2 and D_2 satisfy the following assumptions (a.e. $(x,t) \in \mathcal{Q}$):

$$\begin{aligned} &F_2(x,t,.) = 0 \text{ in }]-\infty,0] \cup [1,+\infty[, \\ &F_1(x,t,.) = 0 \text{ in }]-\infty,0] \text{ and } F_1(x,t,.) = -M \text{ in } [1,+\infty[, \\ &D_2(x,t,.,r') = 0 \text{ in }]-\infty,0] \cup [1,+\infty[\ \text{ for all } r' \in \mathbb{R}. \end{aligned} \tag{11.10}$$

Then we have the following result.

Theorem 11.8. *Let assumptions* (**H1**)–(**H3**) *and (11.10) be fulfilled. Assume that the initial data* $(\phi_0, U_0) \in (L^2(\Omega))^2$ *and the forcing* $\lambda \in \Lambda$. *Then every weak solution* $(\phi, U) \in (L^2(0,T;H^1(\Omega)) \cap H^1(0,T;V'))^2$*, satisfies for all* $t \in (0,T)$

$$0 \le \phi(x,t) \le 1 \text{ and } 0 \le U(x,t) \le 1 \quad a.e.\ x \in \Omega.$$

Proof. Let us consider the following notations: $r^+ = \max(r,0)$, $r^- = (-r)^+$ and then $r = r^+ - r^-$.

First, we prove that if $\phi_0 \ge 0$ and $U_0 \ge 0$, a.e. in Ω then $\phi(.,t) \ge 0$ and $U(.,t) \ge 0$, for all $t \in (0,T)$ and a.e. in Ω. According to Chapter 3, we have that $\phi^- \in L^2(0,T;H^1(\Omega))$ with $\nabla\phi^- = -\nabla\phi$ if $\phi > 0$ and $\nabla\phi^- = 0$ otherwise, a.e. in $\mathcal{Q}$ and the same properties hold for U^-. Then, multiplying the first equation of System (11.8) by $-\phi^-$ and the second equation of System (11.8) by $-U^-$ we have (a.e. in $(0,T)$)

$$\frac{\mathrm{d} \parallel \phi^- \parallel_{L^2}^2}{2\mathrm{d}t} + \nu \parallel \nabla\phi^- \parallel_{L^2}^2 = -\int_\Omega (F_1(.,t,\phi) + UF_2(.,t,\phi))\phi^- \mathrm{d}x - \int_\Omega \lambda\phi^- \mathrm{d}x$$
$$\frac{\mathrm{d} \parallel U^- \parallel_{L^2}^2}{2\mathrm{d}t} + \int_\Omega D_1(.,t,\phi) \mid \nabla U^- \mid^2 \mathrm{d}x = \int_{\Omega^-} D_2(.,t,U,\phi)\nabla\phi\nabla U \mathrm{d}x,$$

where $\Omega^- = \{x \in \Omega : U(t,x) < 0\}$.

Using assumptions (11.10) on the non-linear terms F_1, F_2 and D_2, respectively, we have

$$(F_1(.,t,\phi) + UF_2(.,t,\phi))\phi^- = 0 \text{ and } D_2(.,t,U,\phi) = 0 \text{ if } U < 0, \quad \text{a.e. in } \mathcal{Q}.$$

Consequently, for a.e. in $(0,T)$ (since $0 \le \lambda$)

$$\begin{aligned} &\frac{\mathrm{d} \parallel \phi^- \parallel_{L^2}^2}{2\mathrm{d}t} + \nu \parallel \nabla\phi^- \parallel_{L^2}^2 \le 0, \\ &\frac{\mathrm{d} \parallel U^- \parallel_{L^2}^2}{2\mathrm{d}t} + \int_\Omega D_1(.,t,\phi) \mid \nabla U^- \mid^2 \mathrm{d}x = 0 \end{aligned} \tag{11.11}$$

By integration over $(0,t)$, for any $t \in (0,T)$, we can deduce that

$$\parallel \phi^- \parallel_{L^2}^2 \le \parallel \phi_0^- \parallel_{L^2}^2, \ \parallel U^- \parallel_{L^2}^2 \le \parallel U_0^- \parallel_{L^2}^2 .$$

Therefore, for all $t \in (0,T)$, $\phi^-(.,t) = U^-(.,t) = 0$ a.e. in Ω (since $\phi_0^- = U_0^- = 0$ *a.e.* in Ω).

Next, we prove that if $\phi_0 \le 1$ and $U_0 \le 1$, a.e. in Ω then $\phi(.,t) \le 1$ and $U(.,t) \le 1$, for all $t \in (0,T)$ and a.e. in Ω.

By multiplying the first equation of System (11.8) by $(\phi - 1)^+$ and the second equation of System (11.8) by $(U-1)^+$ we have (a.e. in $(0,T)$)

$$\begin{aligned} &\frac{\mathrm{d} \parallel (\phi - 1)^+ \parallel_{L^2}^2}{2\mathrm{d}t} + \nu \parallel \nabla(\phi - 1)^+ \parallel_{L^2}^2 \\ &= \int_\Omega ((F_1(.,t,\phi) + \lambda) + UF_2(.,t,\phi))(\phi - 1)^+ \mathrm{d}x \end{aligned}$$

and

$$\frac{\mathrm{d} \parallel (U-1)^+ \parallel_{L^2}^2}{2\mathrm{d}t} + \int_\Omega D_1(.,t,\phi) \mid \nabla (U-1)^+ \mid^2 \mathrm{d}x$$
$$= \int_{\Omega^+} D_2(.,t,U,\phi) \nabla\phi\nabla U \mathrm{d}x,$$

where $\Omega^+ = \{x \in \Omega : U(t,x) > 1\}$.

Using assumptions (11.10) on the non-linear terms F_1, F_2 and D_2, respectively, we have (since $M \geq \lambda$), a.e. in $\mathcal{Q}$

$$(F_1(.,t,\phi) + \lambda + UF_2(.,t,\phi))(\phi-1)^+ \leq 0 \text{ and } D_2(.,t,U,\phi) = 0 \text{ if } U > 1.$$

Consequently, for a.e. in $(0,T)$

$$\begin{aligned} &\frac{\mathrm{d} \parallel (\phi-1)^+ \parallel_{L^2}^2}{2\mathrm{d}t} + \nu \parallel \nabla(\phi-1)^+ \parallel_{L^2}^2 \leq 0, \\ &\frac{\mathrm{d} \parallel (U-1)^+ \parallel_{L^2}^2}{2\mathrm{d}t} + \int_\Omega D_1(.,t,\phi) \mid \nabla (U-1)^+ \mid^2 \mathrm{d}x = 0. \end{aligned} \tag{11.12}$$

By integration over $(0,t)$, for any $t \in (0,T)$, we can deduce that

$$\parallel (\phi-1)^+ \parallel_{L^2}^2 \leq \parallel (\phi_0-1)^+ \parallel_{L^2}^2, \ \parallel (U-1)^+ \parallel_{L^2}^2 \leq \parallel (U_0-1)^+ \parallel_{L^2}^2 .$$

Therefore, for all $t \in (0,T)$, $(\phi-1)^+(.,t) = (U-1)^+(.,t) = 0$ a.e. in Ω (since $(\phi_0-1)^+ = (U_0-1)^+ = 0$ a.e. in Ω).

This completes the proof. □

11.3 The Perturbation Problem

In the following, the solution $\mathbf{U} = (U,\phi)$ of problem (11.8) will be treated as the target function. We are then interested in the robust regulation of the deviation of the problem from the desired target $\mathbf{U}$. We analyze the full nonlinear equation which models large perturbations $\mathbf{u} = (u,\varphi)$ to the target $\mathbf{U}$, *i.e.*, we assume that $\mathbf{U}$ satisfies the problem (11.8) with the data (U_0,ϕ_0,λ) and $\mathbf{U}+\mathbf{u}$ satisfies the problem (11.8) with the data $(U_0+u_0,\phi_0+\varphi_0,\lambda+\xi)$. Hence, we consider the following system (for $\mathbf{U} = (U,\phi)$ given satisfying the regularity of Proposition 11.7):

$$\begin{aligned} &\frac{\partial\varphi}{\partial t} - \nu\Delta\varphi = (F_1(x,t,\varphi+\phi) - F_1(x,t,\phi)) + uF_2(x,t,\varphi+\phi) \\ &\qquad\qquad + U(F_2(x,t,\varphi+\phi) - F_2(x,t,\phi)) + \xi \ \text{ on } \mathcal{Q}, \\ &\frac{\partial u}{\partial t} - \mathrm{div}(D_1(x,t,\varphi+\phi)\nabla u + D_2(x,t,\mathbf{u}+\mathbf{U})\nabla\varphi) \\ &\qquad\qquad = \mathrm{div}((D_1(x,t,\varphi+\phi) - D_1(x,t,\phi))\nabla U) \\ &\qquad\qquad\quad + \mathrm{div}((D_2(x,t,\mathbf{u}+\mathbf{U}) - D_2(x,t,\mathbf{U}))\nabla\phi) \ \text{ on } \mathcal{Q}, \\ &\frac{\partial\varphi}{\partial\mathbf{n}} = 0, \ \frac{\partial u}{\partial\mathbf{n}} = 0 \ \text{ on } \Sigma, \\ &(\varphi(0), u(0)) = (\varphi_0, u_0) \ \text{ on } \Omega. \end{aligned} \tag{11.13}$$

If we set

$$\begin{aligned}&\tilde{F}_1(x,t,\varphi) = F_1(x,t,\varphi+\phi) - F_1(x,t,\phi),\\&\tilde{F}_2(x,t,\varphi) = F_2(x,t,\varphi+\phi),\ \tilde{D}_1(x,t,\varphi) = D_1(x,t,\varphi+\phi),\\&\tilde{D}_2(x,t,\mathbf{u}) = D_2(x,t,\mathbf{u}+\mathbf{U}),\end{aligned} \tag{11.14}$$

then System (11.13) reduces to

$$\begin{aligned}&\frac{\partial\varphi}{\partial t} - \nu\Delta\varphi = \tilde{F}_1(x,t,\varphi) + u\tilde{F}_2(x,t,\varphi)\\&\qquad\qquad + U(\tilde{F}_2(x,t,\varphi) - \tilde{F}_2(x,t,0)) + \xi \quad \text{on } \mathcal{Q},\\&\frac{\partial u}{\partial t} - \operatorname{div}(\tilde{D}_1(x,t,\varphi)\nabla u + \tilde{D}_2(x,t,\mathbf{u})\nabla\varphi)\\&\qquad = \operatorname{div}((\tilde{D}_1(x,t,\varphi) - \tilde{D}_1(x,t,0))\nabla U)\\&\qquad\quad + \operatorname{div}((\tilde{D}_2(x,t,\mathbf{u}) - \tilde{D}_2(x,t,0))\nabla\phi) \quad \text{on } \mathcal{Q},\\&\frac{\partial\varphi}{\partial\mathbf{n}} = 0,\ \frac{\partial u}{\partial\mathbf{n}} = 0 \quad \text{on } \Sigma,\\&(\varphi(0), u(0)) = (\varphi_0, u_0) \quad \text{on } \Omega.\end{aligned} \tag{11.15}$$

Remark 11.9. (*i*) We can easily verify that $(\tilde{F}_i, i = 1,2)$ and $(\tilde{D}_i, i = 1,2)$ satisfy the same hypotheses that $(F_i, i = 1,2)$ and $(D_i, i = 1,2)$, *i.e.*, (**H1**)–(**H5**).
(*ii*) For simplicity of future reference, we omit the "~" on $\tilde{F}$ and $\tilde{D}$ for the system (11.15). ◇

Now we give the weak formulation associated with the problem (11.15). Multiplying the first part of (11.15) by $v \in V$ and the second part by $q \in V$ and integrating over Ω this gives (according to the third part of (11.15)) the weak formulation (a.e. $t \in (0,T)$)

$$\begin{aligned}&\int_\Omega \frac{\partial\varphi}{\partial t} v\mathrm{d}x + \nu\int_\Omega \nabla\varphi.\nabla v\mathrm{d}x = \int_\Omega F_1(.,t,\varphi)v\mathrm{d}x + \int_\Omega uF_2(.,t,\varphi)v\mathrm{d}x\\&\qquad\qquad + \int_\Omega U(F_2(.,t,\varphi) - F_2(.,t,0))v\mathrm{d}x + \int_\Omega \xi v\mathrm{d}x,\\&\int_\Omega \frac{\partial u}{\partial t} q\mathrm{d}x + \int_\Omega D_1(.,t,\varphi)\nabla u.\nabla q\mathrm{d}x = -\int_\Omega D_2(.,t,\mathbf{u})\nabla\varphi.\nabla q\mathrm{d}x\\&\qquad\qquad - \int_\Omega (D_1(.,t,\varphi) - D_1(.,t,0))\nabla U\nabla q\mathrm{d}x\\&\qquad\qquad - \int_\Omega (D_2(.,t,\mathbf{u}) - D_2(.,t,0))\nabla\phi\nabla q\mathrm{d}x,\\&(\varphi(0), u(0)) = (\varphi_0, u_0).\end{aligned} \tag{11.16}$$

Before giving the existence theorem, we study the following lemma.

Lemma 11.10. *Let assumptions* (**H1**)-(**H3**) *be fulfilled. For* $\mathbf{u} = (u,\varphi)$ *be sufficiently regular we have (for all* $i = 1,2$*):*

(i) $\mid \nabla(F_i(x,t,\varphi)) \mid \leq C(1+ \mid \nabla\varphi \mid)$
(ii) $\mid \nabla(D_1(x,t,\varphi)) \mid \leq C(1+ \mid \nabla\varphi \mid)$
(iii) $\mid \nabla(D_2(x,t,\mathbf{u})) \mid \leq C(1+ \mid \nabla\varphi \mid + \mid \nabla u \mid)$.

Proof. Since $\nabla(F_i(x,t,\varphi)) = F'_{ix}(x,t,\varphi) + F'_{i\varphi}(x,t,\varphi)\nabla\varphi$, for $i = 1,2$, by using $F_i \in W^{1,\infty}$ we have the result (i). By using the same technique, we obtain the results (ii) and (iii). □

Now we show the existence, the regularity and the uniqueness of the solution for the problem (11.16).

Theorem 11.11. *Let assumptions* (**H1**)–(**H3**) *be fulfilled. The following results hold:*

(i) *For any* (φ_0, u_0) *in* $(L^2(\Omega))^2$ *and* ξ *in* $L^2(Q)$, *there exists a couple of functions* (φ, u) *in* $\mathcal{W}_1^1 \times \mathcal{W}_1^1$ *which is a solution of problem (11.16).*

(ii) *For any* (φ_0, u_0) *in* $H^1(\Omega) \times L^2(\Omega)$ *and* ξ *in* $L^2(Q)$, *the couple of functions* (φ, u) *which is a solution of problem (11.16) is in* $\mathcal{W}_2^1 \times \mathcal{W}_1^1$.

(iii) *For any* (φ_0, u_0) *in* $H_0^2(\Omega) \times H^1(\Omega)$ *and* ξ *in* $L^2(0,T;H^1(\Omega))$, *there exists a unique couple of functions* (φ, u) *in* $\mathcal{W}_3^1 \times \mathcal{W}_2^1$ *which is the unique solution of problem (11.16). Moreover, the Lipschitz continuity relation is satisfied, i.e., for any elements* $(\varphi_{01}, u_{01}, \xi_1)$ *and* $(\varphi_{02}, u_{02}, \xi_2)$ *in* $H_0^2 \times H^1 \times L^2(0,T;H^1(\Omega))$, *we have*

$$\begin{aligned} &\| \varphi_1 - \varphi_2 \|^2_{\mathcal{W}_2^1} + \| u_1 - u_2 \|^2_{\mathcal{W}_1^1} \\ &\quad \leq C(\| \varphi_{01} - \varphi_{02} \|^2_{H^2} \\ &\quad + \| u_{01} - u_{02} \|^2_{H^1} + \| \xi_1 - \xi_2 \|^2_{L^2(0,T;H^1)}), \end{aligned} \tag{11.17}$$

where (φ_1, u_1) *(respectively* (φ_2, u_2)*) is the solution of (11.15), which corresponds to the data* $(\varphi_{01}, u_{01}, \xi_1)$ *(respectively* $(\varphi_{02}, u_{02}, \xi_2)$*).*

Proof. The proof of this theorem can be obtained using the same technique as used in Belmiloudi [40]. For example, the existence and regularity results follow from the Faedo–Galerkin method and Lemmas 11.5, 11.6 and 11.10; according to the properties of the different forms appearing in the weak formulation, we obtain the *a priori* estimates necessary to prove the convergence of an approximate solution (u_m, φ_m) for different necessary topologies. So, we omit the details. □

Remark 11.12. The results of (i) and (ii) are valid for any bounded and regular domain $\Omega \subset \mathbb{R}^m$ with $m \geq 1$ and the results of (iii) are valid for any bounded and regular domain $\Omega \subset \mathbb{R}^m$ with $m \leq 3$. The restriction $m \leq 2$ does not occur in the proof. ◇

11.4 Differentiability of the Operator Solution

Before proceeding to the investigation of the F-differentiability of the function $\mathcal{F} : (\varphi_0, u_0, \xi) \longrightarrow \mathbf{u} = (u, \varphi)$, which maps the source term (φ_0, u_0, ξ) in $H_0^2 \times H^1 \times L^2(0,T;H^1(\Omega))$ of problem (11.15) into the corresponding solution

(u, φ) in $\mathcal{W}_2^1 \times \mathcal{W}_3^1$, we study the following problem $(\mathcal{P}_I)$:

$$
\begin{aligned}
&\text{find } \mathbf{w} = (w, \psi) \text{ such that,} \\
&\frac{\partial \psi}{\partial t} - \nu \Delta \psi = G_1(x,t,\varphi)\psi + wF_2(x,t,\varphi) \\
&\qquad\qquad + U_1 G_2(x,t,\varphi)\psi + h \quad \text{on } \mathcal{Q}, \\
&\frac{\partial w}{\partial t} - \operatorname{div}(D_1(x,t,\varphi)\nabla w + D_2(x,t,\mathbf{u})\nabla \psi) \\
&\qquad = \operatorname{div}(H_1(x,t,\varphi)\psi \nabla U_1 + (H_2(x,t,\mathbf{u}).\mathbf{w})\nabla \phi_1) \quad \text{on } \mathcal{Q}, \\
&\frac{\partial \psi}{\partial \mathbf{n}} = 0, \quad \frac{\partial w}{\partial \mathbf{n}} = 0 \quad \text{on } \Sigma, \\
&\psi(0) = \psi_0, w(0) = w_0 \quad \text{on } \Omega,
\end{aligned}
\tag{11.18}
$$

where $U_1 = U + u$, $\phi_1 = \phi + \varphi$, and $(G_i)_{i=1,2}$, $(H_i)_{i=1,2}$ are given in **(H4)**–**(H5)**.[7]

Theorem 11.13. *Let assumptions* **(H1)**–**(H5)** *be fulfilled. If* $\mathbf{u} = (u, \varphi)$ *and* (U, ϕ) *are in* $\mathcal{W}_2^1 \times \mathcal{W}_3^1$, *then the following results hold:*

(*i*) *For any element* (ψ_0, w_0, h) *in* $H^1(\Omega) \times L^2(\Omega) \times L^2(\mathcal{Q})$, *there exists a unique couple of functions* (ψ, w) *in* $\mathcal{W}_2^1 \times \mathcal{W}_1^1$ *solution of problem* $(\mathcal{P}_I)$, *such that*

$$\| \psi \|_{\mathcal{W}_2^1}^2 + \| w \|_{\mathcal{W}_1^1}^2 \leq C_e(\| \psi_0 \|_{H^1}^2 + \| w_0 \|_{L^2}^2 + \| h \|_{L^2(\mathcal{Q})}^2).$$

(*ii*) *Let* (ψ_{0i}, w_{0i}, h_i), *for* $i = 1, 2$, *be two elements from* $H^1(\Omega) \times L^2(\Omega) \times L^2(\mathcal{Q})$. *If* (ψ_i, w_i) *is the solution of* $(\mathcal{P}_I)$, *where the data is* (ψ_{0i}, w_{0i}, h_i), *for* $i = 1, 2$, *then*

$$
\begin{aligned}
&\| \psi_1 - \psi_2 \|_{\mathcal{W}_2^1}^2 + \| w_1 - w_2 \|_{\mathcal{W}_1^1}^2 \\
&\leq C_e(\| \psi_{01} - \psi_{02} \|_{H^1}^2 + \| w_{01} - w_{02} \|_{L^2}^2 + \| h_1 - h_2 \|_{L^2(\mathcal{Q})}^2).
\end{aligned}
\tag{11.19}
$$

(*iii*) *For any element* (ψ_0, w_0, h) *in* $H_0^2(\Omega) \times H^1(\Omega) \times L^2(0,T; H^1(\Omega))$, *the couple of functions* (ψ, w) *which is the unique solution of problem* $(\mathcal{P}_I)$ *is in* $\mathcal{W}_3^1 \times \mathcal{W}_2^1$.

Proof. The existence, uniqueness and Lipschitz continuity results of the problem $(\mathcal{P}_I)$ can be obtained in the same way as used to prove Theorem 11.11 and by using the regularity of (U_1, ϕ_1). For more details see Belmiloudi [40].□

We are now going to study the F-differentiability of the operator solution $\mathcal{F}$. For simplicity, we denote by X the data $X = (\varphi_0, u_0, \xi)$, by $\mathcal{Z}$ the space

---

[7] Where $H_2(x,t,\mathbf{u}).\mathbf{w} = \dfrac{\partial D_2}{\partial u}(x,t,\mathbf{u}).w + \dfrac{\partial D_2}{\partial \varphi}(x,t,\mathbf{u}).\psi$.

$\mathcal{Z} = H_0^2(\Omega) \times H^1(\Omega) \times L^2(0,T;H^1(\Omega))$. The space $\mathcal{Z}$ is equipped with the following norm

$$\| X \|_{\mathcal{Z}} = (\| \varphi_0 \|_{H^2}^2 + \| u_0 \|_{H^1}^2 + \| \xi \|_{L^2(0,T;H^1)}^2)^{1/2},$$

for all $X = (\varphi_0, u_0, \xi)$ in $\mathcal{Z}$.

Theorem 11.14. *Let $X = (\varphi_0, u_0, \xi)$ and $Y = (p_0, q_0, \pi)$ be in $\mathcal{Z}$ with $\mathcal{F}(X)$ and $\mathcal{F}(X+Y)$ being the corresponding solutions of (11.16). Then*

$$\| \mathcal{F}(X) - \mathcal{F}(X+Y) - \mathcal{F}'(X).Y \|_{\mathcal{W}_1 \times \mathcal{W}_2} \leq C \| Y \|_{\mathcal{Z}}^{3/2}, \tag{11.20}$$

where $\mathcal{F}'(X) : \mathcal{Z} \longrightarrow \mathcal{W}_1 \times \mathcal{W}_2$ is a linear operator such that $(w, \psi) = \mathcal{F}'(X).Y$ is the solution of the problem $(\mathcal{P}_I)$ with the initial condition $(\psi, w)(t=0) = (p_0, q_0)$ and the forcing $h = \pi$.

Moreover, for all $X_i = (\varphi_{0i}, u_{0i}, \xi_i) \in \mathcal{Z}$, for $i = 1, 2$, we have the following estimate:

$$\begin{aligned} &\| \mathcal{F}'(X_1).Y - \mathcal{F}'(X_2).Y \|_{\mathcal{W}_1 \times \mathcal{W}_2}^2 \\ &\qquad \leq C_e(\| Y \|_{\mathcal{Z}} \| X \|_{\mathcal{Z}}^2 + \| Y \|_{\mathcal{Z}}^2 \| X \|_{\mathcal{Z}}), \end{aligned} \tag{11.21}$$

where $X = X_1 - X_2 = (\varphi_{01} - \varphi_{02}, u_{01} - u_{02}, \xi_1 - \xi_2)$.

Proof. The proof of this theorem can be obtained by using the same technique as used to prove the results of Theorem 4.3 of Belmiloudi [40]. So, we omit the details. □

11.5 Robust Control Problems

The objective of the robust control problem is to find the best estimate of the initial state u_0 in the presence of the disturbance which maximally spoils the control objective. We formulate the problem in two situations: where the worst disturbance is in the force ξ (*i.e.*, if we take into account the temperature fluctuation) and where the disturbance is in the initial condition of the phase-field parameter φ_0.

11.5.1 Disturbance in the Forcing of the Phase-field Parameter

We suppose now that the control is in the initial condition of the relative concentration u_0 and the disturbance is in the force ξ, *i.e.*, $u_0 = B_1 g$ ($g \in L^2(\Omega)$) and $\xi = B_2 f$ ($f \in L^2(\mathcal{Q})$), where B_1 is a continuous and bounded operator from $L^2(\Omega)$ into $H^1(\Omega)$ and B_2 is a continuous and bounded operator from $L^2(\mathcal{Q})$ into $L^2(0,T;H^1(\Omega))$. Therefore, the function (φ, u) is assumed to be related to the disturbance f and the control g through the problem (11.15):

$$
\begin{aligned}
&\frac{\partial \varphi}{\partial t} - \nu \Delta \varphi = F_1(x,t,\varphi) + uF_2(x,t,\varphi) \\
&\qquad\qquad + U(F_2(x,t,\varphi) - F_2(x,t,0)) + B_2 f \quad \text{on } \mathcal{Q}, \\
&\frac{\partial u}{\partial t} - \operatorname{div}(D_1(x,t,\varphi)\nabla u + D_2(x,t,\mathbf{u})\nabla\varphi) \\
&\qquad\qquad = \operatorname{div}((D_1(x,t,\varphi) - D_1(x,t,0))\nabla U) \\
&\qquad\qquad\quad + \operatorname{div}((D_2(x,t,\mathbf{u}) - D_2(x,t,0))\nabla\phi) \quad \text{on } \mathcal{Q}, \\
&\frac{\partial \varphi}{\partial \mathbf{n}} = 0, \ \frac{\partial u}{\partial \mathbf{n}} = 0 \quad \text{on } \Sigma, \\
&(\varphi(0), u_0) = (\varphi_0, B_1 g) \quad \text{on } \Omega.
\end{aligned}
\tag{11.22}
$$

To obtain the regularity of Theorem 11.11, we suppose that $(g,f) \in L^2(\Omega) \times L^2(\mathcal{Q})$ and $(U,\phi) \in \mathcal{W}_2^1 \times \mathcal{W}_3^1$. Let $\mathcal{P} : (g,f) \longrightarrow (u,\varphi) = \mathcal{P}(g,f)$ be the map: $L^2(\Omega) \times L^2(\mathcal{Q}) \longrightarrow \mathcal{W}_2^1 \times \mathcal{W}_3^1$ defined by (11.22) and introducing the cost function defined by

$$
\begin{aligned}
J(g,f) = {}& \frac{a}{2} \parallel \varphi - \varphi_{\text{obs}} \parallel^2_{L^2(\mathcal{Q})} + \frac{b}{2} \parallel u - u_{\text{obs}} \parallel^2_{L^2(\mathcal{Q})} \\
& + \frac{\alpha}{2} \parallel g \parallel^2_{L^2} - \frac{\beta}{2} \parallel f \parallel^2_{L^2(\mathcal{Q})},
\end{aligned}
\tag{11.23}
$$

where a, b, α, β are fixed such that $\alpha, \beta > 0$, $a, b \geq 0$ and $a + b > 0$. The functions $u_{\text{obs}} \in L^2(\mathcal{Q})$ and $\varphi_{\text{obs}} \in L^2(\mathcal{Q})$ are given and represent the observation.

Let $\mathcal{K} = \mathcal{K}_1 \times \mathcal{K}_2$ such that $\mathcal{K}_1$ and $\mathcal{K}_2$ are (given) non-empty, closed, convex, bounded subsets of $L^2(\Omega)$ and $L^2(\mathcal{Q})$, respectively. We want to minimize the functional J with respect to g and maximize J with respect to f, *i.e.*, to find $(g^*, f^*) \in \mathcal{K}$ such that

$$
J(g^*, f) \leq J(g^*, f^*) \leq J(g, f^*), \forall (g,f) \in \mathcal{K}. \tag{11.24}
$$

Proposition 11.15. *The function $\mathcal{P}$ is continuously F-differentiable from $L^2(\Omega) \times L^2(\mathcal{Q})$ to $\mathcal{W}_1 \times \mathcal{W}_2$ with the derivative*

$$
\mathcal{P}'(g,f) : \mathbf{h} = (h_1, h_2) \longrightarrow \mathbf{w} = (w, \psi)
$$

given by the following linear problem $(\mathcal{P}_{\text{LP}})$

$$
\begin{aligned}
&\frac{\partial \psi}{\partial t} - \nu \Delta \psi = G_1(x,t,\varphi)\psi + U_1 G_2(x,t,\varphi)\psi \\
&\qquad\qquad + wF_2(x,t,\varphi) + B_2 h_2 \quad \textit{on } \mathcal{Q}, \\
&\frac{\partial w}{\partial t} - \operatorname{div}(D_1(x,t,\varphi)\nabla w + D_2(x,t,\mathbf{u})\nabla\psi) \\
&\qquad = \operatorname{div}(H_1(x,t,\varphi)\psi\nabla U_1 + (H_2(x,t,\mathbf{u}).\mathbf{w})\nabla\phi_1) \quad \textit{on } \mathcal{Q}, \\
&\frac{\partial \psi}{\partial \mathbf{n}} = 0, \ \frac{\partial w}{\partial \mathbf{n}} = 0 \quad \textit{on } \Sigma, \\
&(\psi(0), w(0)) = (0, B_1 h_1) \quad \textit{on } \Omega,
\end{aligned}
\tag{11.25}
$$

where $(U_1, \phi_1) = (u + U, \varphi + \phi)$.

Moreover, $\forall (g_i, f_i) \in L^2(\Omega) \times L^2(\mathcal{Q})$, *for* $i = 1, 2$, *we have the following estimates:*

(i) $\| \mathcal{P}'(g_1, f_1) \|_{\mathcal{L}(L^2(\Omega)\times L^2(\mathcal{Q}), \mathcal{W}_1 \times \mathcal{W}_2)} \leq C_e$

(ii) $\| \mathcal{P}'(g_1, f_1)\mathbf{h} - \mathcal{P}'(g_2, f_2)\mathbf{h} \|^2_{\mathcal{W}_1 \times \mathcal{W}_2}$

$$\leq C_e\{(\| g \|^2_{L^2} + \| f \|^2_{L^2(\mathcal{Q})})^{1/2}(\| h_1 \|^2_{L^2} + \| h_2 \|^2_{L^2(\mathcal{Q})})$$
$$+(\| g \|^2_{L^2} + \| f \|^2_{L^2(\mathcal{Q})})(\| h_1 \|^2_{L^2} + \| h_2 \|^2_{L^2(\mathcal{Q})})^{1/2}\},$$

where $f = f_1 - f_2$ *and* $g = g_1 - g_2$.

Proof. The proof of this proposition is a consequence of the nature of the operators $B_i, i = 1, 2$ and the result of Theorem 11.14. So, we omit the details. □

Proposition 11.16. *Let assumptions* **(H1)**–**(H3)** *be satisfied. Then the mapping* $\mathcal{P}$ *defined by (11.22) is continuous from the weak topology of* $L^2(\Omega) \times L^2(\mathcal{Q})$ *to the strong topology of* $(L^2(\mathcal{Q}))^2$.

Proof. Let $\mathbf{f} = (g, f)$ be given in $L^2(\Omega) \times L^2(\mathcal{Q})$ and let a sequence $\mathbf{f}_k = (g_k, f_k)$ be such that $\mathbf{f}_k$ is weakly convergent in $L^2(\Omega) \times L^2(\mathcal{Q})$ to $\mathbf{f}$.

Set $B\mathbf{f} = (B_1 g, B_2 f)$, $(u, \varphi) = \mathcal{P}(g, f)$, $B\mathbf{f}_k = (B_1 g_k, B_2 f_k)$ and $(u_k, \varphi_k) = \mathcal{P}(g_k, f_k)$. Since $\mathbf{f}_k \rightharpoonup \mathbf{f}$ weakly in $L^2(\Omega) \times L^2(\mathcal{Q})$ then $\mathbf{f}_k$ is uniformly bounded in $L^2(\Omega) \times L^2(\mathcal{Q})$ and then $B\mathbf{f}_k$ is uniformly bounded in $H^1(\Omega) \times L^2(0, T; H^1(\Omega))$. In view of Theorem 11.11, we can deduce that the sequence (u_k, φ_k) is uniformly bounded in $\mathcal{W}_2^1 \times \mathcal{W}_3^1$. Therefore, we can extract from $(\mathbf{f}_k, u_k, \varphi_k)$ a subsequence also denoted by $(\mathbf{f}_k, u_k, \varphi_k)$ and such that

$$\begin{aligned}
&(g_k, f_k) \rightharpoonup (g, f) \text{ weakly in } L^2(\Omega) \times L^2(\mathcal{Q}),\\
&\varphi_k \rightharpoonup \tilde{\varphi} \text{ weakly in } \mathcal{W}_3,\\
&u_k \rightharpoonup \tilde{u} \text{ weakly in } \mathcal{W}_2,\\
&\varphi_k \longrightarrow \tilde{\varphi} \text{ strongly in } L^2(\mathcal{Q}),\\
&u_k \longrightarrow \tilde{u} \text{ strongly in } L^2(\mathcal{Q}).
\end{aligned}$$

We can easily prove that $(\tilde{u}, \tilde{\varphi}) = \mathcal{P}(g, f)$ and, according to the uniqueness of the solution of (11.22), we then have$\tilde{\varphi} = \varphi$ and $\tilde{u} = u$. □

Theorem 11.17. *Let assumptions* **(H1)**–**(H5)** *be satisfied. Then for sufficiently large* α *and* β *(i.e., there exists* (α_l, β_l) *such that* $\alpha \geq \alpha_l$ *and* $\beta \geq \beta_l$*) there exists* $(g^*, f^*) \in \mathcal{K}$ *and* $(u^*, \varphi^*) \in \mathcal{W}_2^1 \times \mathcal{W}_3^1$ *such that* (g^*, f^*) *is defined by (11.24) and* $(u^*, \varphi^*) = \mathcal{P}(g^*, f^*)$ *is a solution of (11.22).*

Proof. Let P_f be the mapping: $g \longrightarrow J(g, f)$ and Q_g be the mapping: $f \longrightarrow J(g, f)$. To obtain the existence of the robust control problem we prove that P_f is convex and lower semi-continuous for all $f \in \mathcal{K}_2$, and Q_g is concave and upper semi-continuous for all $g \in \mathcal{K}_1$ and we use the minimax theorems in infinite dimensions presented in Chapter 5.

First, we prove for sufficiently large α and β, the convexity of the mapping P_f and the concavity of the mapping Q_g. In order to prove the convexity, it is sufficient to show that for all $(g_1, g_2) \in \mathcal{K}_1 \times \mathcal{K}_1$ the estimate

$$(P'_f(g_1) - P'_f(g_2)).g \geq 0$$

holds, where $g = g_1 - g_2$ (because P_f is G-differentiable). According to the definition of J, we have

$$\begin{aligned}(P'_f(g_1) - P'_f(g_2)).g &= a \iint_Q (\varphi_1 - \varphi_2)\psi_2 \mathrm{d}x\mathrm{d}t \\ &+b \iint_Q (u_1 - u_2)w_2 \mathrm{d}x\mathrm{d}t + a \iint_Q (\varphi_1 - \varphi_{\mathrm{obs}})(\psi_1 - \psi_2)\mathrm{d}x\mathrm{d}t \\ &+b \iint_Q (u_1 - u_{\mathrm{obs}})(w_1 - w_2)\mathrm{d}x\mathrm{d}t + \alpha \parallel g \parallel^2_{L^2},\end{aligned} \tag{11.26}$$

where $(u_i, \varphi_i) = \mathcal{P}(g_i, f)$, $(w_i, \psi_i) = \mathcal{P}'(g_i, f).(g, 0)$ (solution of problem $(\mathcal{P}_{\mathrm{LP}})$), for $i = 1, 2$).

According to Theorem 11.11 and Proposition 11.15 we have

$$\begin{aligned}&\mid a \iint_Q (\varphi_1 - \varphi_2)\psi_2 \mathrm{d}x\mathrm{d}t + b \iint_Q (u_1 - u_2)w_2 \mathrm{d}x\mathrm{d}t \mid \\ &\qquad \leq a \parallel \varphi_1 - \varphi_2 \parallel_{L^2(Q)} \parallel \psi_2 \parallel_{L^2(Q)} \\ &\qquad\quad +b \parallel u_1 - u_2 \parallel_{L^2(Q)} \parallel w_2 \parallel_{L^2(Q)} \\ &\qquad \leq C_0 \parallel g \parallel^2_{L^2}, \\ &\mid a \iint_Q (\varphi_1 - \varphi_{\mathrm{obs}})(\psi_1 - \psi_2)\mathrm{d}x\mathrm{d}t + b \iint_Q (u_1 - u_{\mathrm{obs}})(w_1 - w_2)\mathrm{d}x\mathrm{d}t \mid \\ &\qquad \leq a \parallel \varphi_1 - \varphi_{\mathrm{obs}} \parallel_{L^2(Q)} \parallel \psi_1 - \psi_2 \parallel_{L^2(Q)} \\ &\qquad\quad +b \parallel u_1 - u_{\mathrm{obs}} \parallel_{L^2(Q)} \parallel w_1 - w_2 \parallel_{L^2(Q)} \\ &\qquad \leq C_1 C_2(u_{\mathrm{obs}}, v_{\mathrm{obs}}) \parallel g \parallel^{3/2}_{L^2}.\end{aligned} \tag{11.27}$$

From (11.26) and (11.27) we can deduce that for $\alpha \geq \alpha_l$ such that $\alpha_l > C_0$ and

$$(\alpha_l - C_0) \min_{g \in \mathcal{K}_1} \parallel g \parallel^{1/2}_{L^2} = C_1 C_2$$

the estimate $(P'_f(g_1) - P'_f(g_2)).g \geq 0$ holds and, therefore, P_f is convex. In the same way, we can find β_l such that Q_g is concave for $\beta \geq \beta_l$.

We shall prove now that P_f is lower semi-continuous for all $f \in \mathcal{K}_2$, and Q_g is upper semi-continuous for all $g \in \mathcal{K}_1$. Let g_k be a minimizing sequence of J, *i.e.*,

$$\liminf_{k} J(g_k, f) = \min_{g\in\mathcal{K}_1} J(g, f) \ \forall f \in \mathcal{K}_2.$$

Then g_k is uniformly bounded in $\mathcal{K}_1$ and we can extract from g_k a subsequence also denoted by g_k such that $g_k \rightharpoonup g_f$ weakly in $\mathcal{K}_1$. By using Proposition 11.16, we then have

$$\mathcal{P}(g_k, f) \longrightarrow (u_f, \varphi_f) \text{ strongly in } L^2(\mathcal{Q}).$$

Therefore, since the norm is lower semi-continuous, we have that the mapping $P_f : g \longrightarrow J(g, f)$ is lower semi-continuous for all $f \in \mathcal{K}_2$. By using the same technique as above, we obtain that Q_g is upper semi-continuous for all $g \in \mathcal{K}_1$. □

In order to obtain the necessary optimality conditions which are satisfied by the solution of the robust control problem, we introduce the following adjoint problem corresponding to the primal problem (11.22) (we denote by $\mathbf{u} = (u, \varphi) = \mathcal{P}(g, f)$ and $(U_1, \phi_1) = (u + U, \varphi + \phi)$):

$$\begin{aligned}
&-\frac{\partial p}{\partial t} - \nu\Delta p - G_1(x,t,\varphi)p - U_1 G_2(x,t,\varphi)p \\
&\qquad +H_1(x,t,\varphi)\nabla q.\nabla U_1 - \operatorname{div}(D_2(x,t,\mathbf{u})\nabla q) \\
&\qquad +\frac{\partial D_2}{\partial \varphi}(x,t,\mathbf{u})\nabla q \nabla \phi_1 = a(\varphi - \varphi_{\text{obs}}) \ \text{ on } \mathcal{Q}, \\
&-\frac{\partial q}{\partial t} - \operatorname{div}(D_1(x,t,\varphi)\nabla q) + \frac{\partial D_2}{\partial u}(x,t,\mathbf{u})\nabla q \nabla \phi_1 \\
&\qquad -F_2(x,t,\varphi)p = b(u - u_{\text{obs}}) \ \text{ on } \mathcal{Q}, \\
&\frac{\partial p}{\partial \mathbf{n}} = 0, \ \frac{\partial q}{\partial \mathbf{n}} = 0 \ \text{ on } \Sigma, \\
&p(T) = 0, \ q(T) = 0 \ \text{ on } \Omega.
\end{aligned} \tag{11.28}$$

Remark 11.18. The adjoint problem (11.28) is a linear system. By reversing sense of time, *i.e.*, $t := T - t$, and by following the same procedure as used to derive the result of Theorem 11.13, we obtain the existence and uniqueness of (p, q). ♢

Now we can give the first-order optimality conditions for the robust control problem (11.24).

Theorem 11.19. *Under the assumptions of Theorem 11.17, the optimal solution* $(g^*, f^*, u^*, \varphi^*) \in \mathcal{K} \times \mathcal{W}_2^1 \times \mathcal{W}_3^1$*, such that* (g^*, f^*) *is defined by (11.24) and* $(u^*, \varphi^*) = \mathcal{P}(g^*, f^*)$ *is a solution of (11.22), satisfies*

$$\begin{aligned}
&\int_\Omega (B_1^* q^*(0) + \alpha g^*)(g - g^*)\mathrm{d}x \geq 0, \\
&\iint_\mathcal{Q} (B_2^* p^* - \beta f^*)(f - f^*)\mathrm{d}x\mathrm{d}t \leq 0 \ \text{ for all } (g, f) \in \mathcal{K},
\end{aligned} \tag{11.29}$$

where (p^*, q^*) *is the solution of the adjoint problem (11.28), corresponding to the primal solution* (u^*, φ^*).

Proof. The cost function J is a composition of F-differentiable mappings then J is F-differentiable and we have $(\forall \mathbf{h} = (h_1, h_2) \in \mathcal{K})$

$$\begin{aligned} J'(g,f).\mathbf{h} = a \iint_Q (\varphi - \varphi_{\text{obs}})\psi \mathrm{d}x\mathrm{d}t + b \iint_Q (u - u_{\text{obs}}) w \mathrm{d}x\mathrm{d}t \\ +\alpha \int_\Omega g h_1 \mathrm{d}x - \beta \iint_Q f h_2 \mathrm{d}x\mathrm{d}t, \end{aligned} \tag{11.30}$$

where $(w, \psi) = \mathcal{P}'(g,f).\mathbf{h}$ is the solution of problem $(\mathcal{P}_{\text{LP}})$.

Multiplying the first part of $(\mathcal{P}_{\text{LP}})$ by p and the second part by q, using Green's formula, and integrating with respect to time, we obtain (according to the homogeneous Neumann boundary conditions and the fact that $H_2(x,t,\mathbf{u}).\mathbf{w} = (\partial D_2/\partial u)(x,t,\mathbf{u})w + (\partial D_2/\partial \varphi)(x,t,\mathbf{u})\psi)$

$$\begin{aligned} &\iint_Q \Big(-\frac{\partial p}{\partial t} - \nu \Delta p - G_1(.,t,\varphi)p - U_1 G_2(.,t,\varphi)p\Big)\psi \mathrm{d}x\mathrm{d}t \\ &\quad = \iint_Q F_2(.,t,\varphi) p w \mathrm{d}x\mathrm{d}t + \iint_Q B_2^* p.h_2 \mathrm{d}x\mathrm{d}t - \int_\Omega p(T)\psi(T)\mathrm{d}x, \\ &\iint_Q \Big(-\frac{\partial q}{\partial t} + \frac{\partial D_2}{\partial u}(.,t,\mathbf{u})\nabla q.\nabla \phi_1 - \mathrm{div}(D_1(.,t,\varphi)\nabla q)\Big) w \mathrm{d}x\mathrm{d}t \\ &\quad = \int_\Omega B_1^* q(0).h_1 \mathrm{d}x - \int_\Omega q(T) w(T) \mathrm{d}x \\ &\qquad + \iint_Q \mathrm{div}(D_2(.,t,\mathbf{u})\nabla q)\psi \mathrm{d}x\mathrm{d}t \\ &\qquad - \iint_Q \Big(H_1(.,t,\varphi)\nabla q.\nabla U_1 + \frac{\partial D_2}{\partial \varphi}(.,t,\mathbf{u})\nabla q.\nabla \phi_1\Big)\psi \mathrm{d}x\mathrm{d}t. \end{aligned} \tag{11.31}$$

Since (p,q) satisfies System (11.28), with null final conditions, we have

$$\begin{aligned} &\iint_Q \Big(-H_1(.,t,\varphi)\nabla q.\nabla U_1 + \mathrm{div}(D_2(.,t,\mathbf{u})\nabla q) - \frac{\partial D_2}{\partial \varphi}(.,t,\mathbf{u})\nabla q \nabla \phi_1\Big)\psi \mathrm{d}x\mathrm{d}t \\ &+ \iint_Q a(\varphi - \varphi_{\text{obs}})\psi \mathrm{d}x\mathrm{d}t = \iint_Q (F_2(.,t,\varphi)p) w \mathrm{d}x\mathrm{d}t + \iint_Q B_2^* p.h_2 \mathrm{d}x\mathrm{d}t, \\ &\iint_Q (F_2(.,t,\varphi)p) w \mathrm{d}x\mathrm{d}t + \iint_Q b(u - u_{\text{obs}}) w \mathrm{d}x\mathrm{d}t = \int_\Omega B_1^* q(0).h_1 \mathrm{d}x \\ &+ \iint_Q \Big(-H_1(.,t,\varphi)\nabla q.\nabla U_1 + \mathrm{div}(D_2(.,t,\mathbf{u})\nabla q) - \frac{\partial D_2}{\partial \varphi}(.,t,\mathbf{u})\nabla q.\nabla \phi_1\Big)\psi \mathrm{d}x\mathrm{d}t \end{aligned}$$

and, therefore,

$$\begin{aligned} &\iint_Q b(u - u_{\text{obs}}) w \mathrm{d}x\mathrm{d}t + \iint_Q a(\varphi - \varphi_{\text{obs}})\psi \mathrm{d}x\mathrm{d}t \\ &\quad = \int_\Omega B_1^* q(0).h_1 \mathrm{d}x + \iint_Q B_2^* p.h_2 \mathrm{d}x\mathrm{d}t. \end{aligned} \tag{11.32}$$

According to the expression of $J'(g,f).\mathbf{h}$ we can deduce that

$$J'(g,f).\mathbf{h} = \int_{\Omega}(B_1^*q(0)+\alpha g).h_1\mathrm{d}x + \iint_{Q}(B_2^*p-\beta f).h_2\mathrm{d}x\mathrm{d}t.$$

Since (g^*, f^*) is a saddle point of J we have

$$\frac{\partial J}{\partial g}(g^*,f^*).(g-g^*) \geq 0 \text{ and } \frac{\partial J}{\partial f}(g^*,f^*).(f-f^*) \leq 0, \forall (g,f)\in\mathcal{K}$$

and then

$$\begin{aligned} &\int_{\Omega}(B_1^*q^*(0)+\alpha g^*)(g-g^*)\mathrm{d}x \geq 0, \\ &\iint_{Q}(B_2^*p^*-\beta f^*)(f-f^*)\mathrm{d}x\mathrm{d}t \leq 0 \quad \forall (g,f)\in\mathcal{K}, \end{aligned} \tag{11.33}$$

where (p^*, q^*) is the solution of the problem (11.28), corresponding to the primal solution $(u^*, \varphi^*) = \mathcal{P}(g^*, f^*)$.

This completes the proof. □

Remark 11.20. In the case where we take into account the final observation, we obtain the same results. In this case the cost functional can be given as the form

$$\begin{aligned} J(g,f) = &\frac{a_1}{2}\parallel \varphi-\varphi_{\mathrm{obs}}\parallel^2_{L^2(Q)} + \frac{a_2}{2}\parallel u-u_{\mathrm{obs}}\parallel^2_{L^2(Q)} \\ &+\frac{a_3}{2}\parallel \varphi(T)-\eta_{\mathrm{obs}}\parallel^2_{L^2} + \frac{a_4}{2}\parallel u(T)-v_{\mathrm{obs}}\parallel^2_{L^2} \\ &+\frac{\alpha}{2}\parallel g\parallel^2_{L^2} - \frac{\beta}{2}\parallel f\parallel^2_{L^2(Q)}, \end{aligned} \tag{11.34}$$

where a_i, for $i=1,4$, α and β are fixed and such that $\alpha,\beta>0$, $a_i\geq 0$, for $i=1,4$, and $\sum_{i=1,4} a_i > 0$.

The functions $(u_{\mathrm{obs}}, \varphi_{\mathrm{obs}}) \in L^2(Q)\times L^2(Q)$ and $(\eta_{\mathrm{obs}}, v_{\mathrm{obs}}) \in H^1(\Omega)\times L^2(\Omega)$ are given and represent the observation. We can prove also an existence theorem for the robust control problem and obtain the necessary optimality conditions for its solution using the same method.

Let $\mathcal{K} = \mathcal{K}_1 \times \mathcal{K}_2$ such that $\mathcal{K}_1$ and $\mathcal{K}_2$ are non-empty, closed, convex, bounded subsets of $L^2(\Omega)$ and $L^2(Q)$, respectively.

Let assumptions (**H1**)–(**H5**) be satisfied. Then for sufficiently large α and β, there exists (u,φ,p,q,g^*,f^*) satisfying

$$\begin{aligned} \frac{\partial\varphi}{\partial t} - \nu\Delta\varphi = &F_1(x,t,\varphi) + uF_2(x,t,\varphi) \\ &+U(F_2(x,t,\varphi)-F_2(x,t,0)) + B_2 f^* \quad \text{on } Q, \end{aligned}$$

$$\begin{aligned} \frac{\partial u}{\partial t} &- \mathrm{div}(D_1(x,t,\varphi)\nabla u + D_2(x,t,\mathbf{u})\nabla\varphi) \\ &= \mathrm{div}((D_1(x,t,\varphi)-D_1(x,t,0))\nabla U) \\ &\quad +\mathrm{div}((D_2(x,t,\mathbf{u})-D_2(x,t,0))\nabla\phi) \quad \text{on } Q, \end{aligned}$$

$$
\begin{aligned}
&-\frac{\partial p}{\partial t} - \nu \Delta p - G_1(x,t,\varphi)p - U_1 G_2(x,t,\varphi)p \\
&\qquad\qquad + H_1(x,t,\varphi)\nabla q . \nabla U_1 - \operatorname{div}(D_2(x,t,\mathbf{u})\nabla q) \\
&\qquad\qquad + \frac{\partial D_2}{\partial \varphi}(x,t,\mathbf{u})\nabla q \nabla \phi_1 = a_1(\varphi - \varphi_{\text{obs}}) \ \text{ on } \mathcal{Q}, \\
&-\frac{\partial q}{\partial t} - \operatorname{div}(D_1(x,t,\varphi)\nabla q) + \frac{\partial D_2}{\partial u}(x,t,\mathbf{u})\nabla q \nabla \phi_1 \\
&\qquad\qquad - F_2(x,t,\varphi)p = a_2(u - u_{\text{obs}}) \ \text{ on } \mathcal{Q}, \\
&\frac{\partial \varphi}{\partial \mathbf{n}} = 0, \ \frac{\partial u}{\partial \mathbf{n}} = 0, \ \frac{\partial p}{\partial \mathbf{n}} = 0, \ \frac{\partial q}{\partial \mathbf{n}} = 0 \ \text{ on } \Sigma, \\
&(\varphi(0), u_0) = (\varphi_0, B_1 g^*) \ \text{ on } \Omega, \\
&p(T) = a_3(\varphi(T) - \eta_{\text{obs}}), \ q(T) = a_4(u(T) - v_{\text{obs}}) \ \text{ on } \Omega
\end{aligned}
$$

and inequalities

$$
\begin{aligned}
&\int_\Omega (B_1^* q(0) + \alpha g^*)(g - g^*) \mathrm{d}x \geq 0, \\
&\iint_\mathcal{Q} (B_2^* p - \beta f^*)(f - f^*) \mathrm{d}x\mathrm{d}t \leq 0 \qquad \forall (g,f) \in \mathcal{K}.
\end{aligned}
$$

◇

11.5.2 Distributed Disturbance in the Initial Condition of the Phase-field Variable

In this section, the disturbance is in the initial condition of the phase-field variable φ_0 and the control is in the initial condition of the concentration u_0, *i.e.*, $\varphi_0 = B_2 f$ ($f \in L^2(\Omega)$), $u_0 = B_1 g$ ($g \in L^2(\Omega)$), where B_1 is a bounded operator from $L^2(\Omega)$ into $H^1(\Omega)$ and B_2 is a bounded operator from $L^2(\Omega)$ into $H_0^2(\Omega)$. So the function (φ, u) is assumed to be related to the disturbance f and control g through the problem (11.15):

$$
\begin{aligned}
&\frac{\partial \varphi}{\partial t} - \nu \Delta \varphi = F_1(x,t,\varphi) + u F_2(x,t,\varphi) \\
&\qquad\qquad + U(F_2(x,t,\varphi) - F_2(x,t,0)) \ \text{ on } \mathcal{Q}, \\
&\frac{\partial u}{\partial t} - \operatorname{div}(D_1(x,t,\varphi)\nabla u + D_2(x,t,\mathbf{u})\nabla \varphi) \\
&\qquad\qquad = \operatorname{div}((D_1(x,t,\varphi) - D_1(x,t,0))\nabla U) \\
&\qquad\qquad + \operatorname{div}((D_2(x,t,\mathbf{u}) - D_2(x,t,0))\nabla \phi) \ \text{ on } \mathcal{Q}, \\
&\frac{\partial \varphi}{\partial \mathbf{n}} = 0, \ \frac{\partial u}{\partial \mathbf{n}} = 0 \ \text{ on } \Sigma, \\
&\varphi(0) = B_2 f, \quad u(0) = B_1 g \ \text{ on } \Omega.
\end{aligned}
\tag{11.35}
$$

To obtain the regularity of Theorem 11.11, we suppose that $(g,f) \in L^2(\Omega)^2$ and $(U,\phi) \in \mathcal{W}_2^1 \times \mathcal{W}_3^1$. Let $\mathcal{P} : (g,f) \longrightarrow (u,\varphi) = \mathcal{P}(g,f)$ be the map:

$(L^2(\Omega))^2 \longrightarrow \mathcal{W}_2^1 \times \mathcal{W}_3^1$ defined by (11.35). The cost function and the saddle point problem that we want to study are the same as in the previous section, *i.e.*, the functional defined by (11.23) and the problem defined by (11.24).

The arguments of the previous sections extend directly to the present case without further estimates. We have then the following results.

Proposition 11.21. *The function $\mathcal{P}$ is continuously F-differentiable from $(L^2(\Omega))^2$ to $\mathcal{W}_1 \times \mathcal{W}_2$ where the derivative*

$$\mathcal{P}'(g, f) : \mathbf{h} = (h_1, h_2) \longrightarrow (w, \psi)$$

is a linear mapping from $(L^2(\Omega))^2$ into $\mathcal{W}_1 \times \mathcal{W}_2$ such that $\mathcal{P}'(g, f).\mathbf{h}$ satisfies the linear problem $(\mathcal{P}_{\text{LPB}})$

$$\frac{\partial \psi}{\partial t} - \nu \Delta \psi = G_1(x, t, \varphi)\psi + uG_2(x, t, \varphi)\psi$$
$$+wF_2(x, t, \varphi) + UG_2(x, t, \varphi)\psi \quad \text{on } \mathcal{Q},$$

$$\frac{\partial w}{\partial t} - \operatorname{div}(D_1(x, t, \varphi)\nabla w + D_2(x, t, \mathbf{u})\nabla \psi)$$
$$= \operatorname{div}(H_1(x, t, \varphi)\psi \nabla U_1 + (H_2(x, t, \mathbf{u}).\mathbf{w})\nabla \phi_1) \quad \text{on } \mathcal{Q},$$

$$\frac{\partial \psi}{\partial \mathbf{n}} = 0, \ \frac{\partial w}{\partial \mathbf{n}} = 0 \quad \text{on } \Sigma,$$
$$\psi(0) = B_2 h_2 \quad w(0) = B_1 h_1 \quad \text{on } \Omega,$$

where $(U_1, \phi_1) = (u + U, \varphi + \phi)$.

Moreover, we have the estimates (for all $(f_i, g_i) \in (L^2(\Omega))^2$, for $i = 1, 2$):

(i) $\| \mathcal{P}'(f_1, g_1) \|_{\mathcal{L}((L^2(\Omega))^2, \mathcal{W}_1 \times \mathcal{W}_2)} \leq C_e$

(ii) $\| \mathcal{P}'(f_1, g_1) - \mathcal{P}'(f_2, g_2) \|^2_{\mathcal{W}_1 \times \mathcal{W}_2}$

$$\leq C_e\{(\| f \|^2_{L^2} + \| g \|^2_{L^2})^{1/2}(\| h_1 \|^2_{L^2} + \| h_2 \|^2_{L^2})$$
$$+ (\| f \|^2_{L^2} + \| g \|^2_{L^2})(\| h_1 \|^2_{L^2} + \| h_2 \|^2_{L^2})^{1/2}\},$$

where $f = f_1 - f_2$ and $g = g_1 - g_2$. □

Proposition 11.22. *Let assumptions* **(H1)**–**(H5)** *be satisfied. Then the map $\mathcal{P}$ defined by (11.35) is continuous from the weak topology of $(L^2(\Omega))^2$ to the strong topology of $(L^2(\mathcal{Q}))^2$.* □

Theorem 11.23. *Let assumptions* **(H1)**–**(H5)** *be satisfied. Then, for α and β sufficiently large, there exist $(g^*, f^*) \in \mathcal{K}$ and $(u^*, \varphi^*) \in \mathcal{W}_2^1 \times \mathcal{W}_3^1$ such that (g^*, f^*) is defined by (11.24) and $(u^*, \varphi^*) = \mathcal{P}(g^*, f^*)$ is a solution of (11.35).* □

Theorem 11.24. *Let assumptions* **(H1)**–**(H5)** *be satisfied, α and β sufficiently large, $(g^*, f^*) \in \mathcal{K}$ and $(u^*, \varphi^*) \in \mathcal{W}_2^1 \times \mathcal{W}_3^1$ such that (g^*, f^*) is defined by (11.24) and $(u^*, \varphi^*) = \mathcal{P}(g^*, f^*)$ is a solution of (11.35). Then*

$$\begin{aligned}&\int_{\Omega}(B_1^* q^*(0) + \alpha g^*)(g - g^*) \geq 0,\\&\int_{\Omega}(B_2^* p^*(0) - \beta f^*)(f - f^*) \leq 0, \forall (g, f) \in \mathcal{K},\end{aligned} \tag{11.36}$$

where (p^*, q^*) is the solution of the adjoint problem (11.28), corresponding to the primal solution (u^*, φ^*). □

Remark 11.25. We can consider other types of controls and disturbances, the technique developed previously is still valid. For example in the case where the control is in the initial condition of the order parameter, *i.e.*, $g = \varphi_0$ and the distributed disturbance is in the forcing ξ, *i.e.*, $\xi = f$, we can prove, under the assumptions **(H1)**–**(H5)** and for α and γ sufficiently large, the existence theorem of the robust control problem and obtain necessary optimality conditions for its solution using the same method as previously. In this case the cost functional can be given as the form (if we take also into account the final observation)

$$\begin{aligned}J(g, f) = &\frac{a_1}{2} \| \varphi - \varphi_{\text{obs}} \|^2_{L^2(\mathcal{Q})} + \frac{a_2}{2} \| u - u_{\text{obs}} \|^2_{L^2(\mathcal{Q})}\\&+ \frac{a_3}{2} \| \varphi(T) - \eta_{\text{obs}} \|^2_{L^2} + \frac{a_4}{2} \| u(T) - v_{\text{obs}} \|^2_{L^2}\\&+ \frac{\alpha}{2} \| g \|^2_{L^2} - \frac{\beta}{2} \| f \|^2_{L^2(\mathcal{Q})},\end{aligned} \tag{11.37}$$

where a_i, for $i = 1, 4$, α and β are fixed and such that $\alpha, \beta > 0$, $a_i \geq 0$, for $i = 1, 4$, and $\sum_{i=1,4} a_i > 0$.

The functions $(u_{\text{obs}}, \varphi_{\text{obs}}) \in L^2(\mathcal{Q}) \times L^2(\mathcal{Q})$ and $(\eta_{\text{obs}}, v_{\text{obs}}) \in H^1(\Omega) \times L^2(\Omega)$ are given and represent the observation.

Let $\mathcal{K} = \mathcal{K}_1 \times \mathcal{K}_2$ such that $\mathcal{K}_1$ and $\mathcal{K}_2$ are non-empty, closed, convex, bounded subsets of $L^2(\Omega)$ and $L^2(\mathcal{Q})$, respectively.

For sufficiently large α and β, there exists $(u, \varphi, p, q, g^*, f^*)$ satisfying

$$\begin{aligned}&\frac{\partial \varphi}{\partial t} - \nu \Delta \varphi = F_1(x, t, \varphi) + u F_2(x, t, \varphi)\\&\qquad + U(F_2(x, t, \varphi) - F_2(x, t, 0)) + B_2 f^* \quad \text{on } \mathcal{Q},\\&\frac{\partial u}{\partial t} - \operatorname{div}(D_1(x, t, \varphi)\nabla u + D_2(x, t, \mathbf{u})\nabla \varphi)\\&\qquad = \operatorname{div}((D_1(x, t, \varphi) - D_1(x, t, 0))\nabla U)\\&\qquad + \operatorname{div}((D_2(x, t, \mathbf{u}) - D_2(x, t, 0))\nabla \phi) \quad \text{on } \mathcal{Q},\\&-\frac{\partial p}{\partial t} - \nu \Delta p - G_1(x, t, \varphi)p - U_1 G_2(x, t, \varphi)p\\&\qquad + H_1(x, t, \varphi)\nabla q . \nabla U_1 - \operatorname{div}(D_2(x, t, \mathbf{u})\nabla q)\\&\qquad + \frac{\partial D_2}{\partial \varphi}(x, t, \mathbf{u})\nabla q \nabla \phi_1 = a_1(\varphi - \varphi_{\text{obs}}) \quad \text{on } \mathcal{Q},\\&-\frac{\partial q}{\partial t} - \operatorname{div}(D_1(x, t, \varphi)\nabla q) + \frac{\partial D_2}{\partial u}(x, t, \mathbf{u})\nabla q \nabla \phi_1\\&\qquad - F_2(x, t, \varphi)p = a_2(u - u_{\text{obs}}) \quad \text{on } \mathcal{Q},\end{aligned}$$

$$\frac{\partial \varphi}{\partial \mathbf{n}} = 0, \ \frac{\partial u}{\partial \mathbf{n}} = 0, \ \frac{\partial p}{\partial \mathbf{n}} = 0, \ \frac{\partial q}{\partial \mathbf{n}} = 0 \ \text{ on } \Sigma,$$
$$(\varphi(0), u_0) = (B_1 g^*, u_0) \ \text{ on } \Omega,$$
$$p(T) = a_3(\varphi(T) - \eta_{\text{obs}}), \quad q(T) = a_4(u(T) - v_{\text{obs}}) \ \text{ on } \Omega$$

and inequalities

$$\int_\Omega (B_1^* p(0) + \alpha g^*)(g - g^*) \mathrm{d}x \geq 0,$$
$$\iint_Q (B_2^* p - \beta f^*)(f - f^*) \mathrm{d}x\mathrm{d}t \leq 0 \quad \forall (g, f) \in \mathcal{K}.$$

◇

Remark 11.26. (*i*) To obtain more realistic simulations of dendrites using the phase-field system, it is interesting to include the anisotropy in our studied system. The most widely used method to include this anisotropy for two-dimensional case is to assume that $W_\phi = \epsilon_\phi / \epsilon$ (given in (11.2)) depends on an angle θ. The angle θ is corresponding to the orientation of the normal to the interface with respect to the x-axis given by $\tan\theta = (\partial\phi/\partial y)/(\partial\phi/\partial x)$. According to Kobayashi [175] (see also Taylor and Cahn [280] for a discussion about the choice of the anisotropy formulation), the coefficient can be written as

$$\epsilon_\phi = \epsilon(1 + \epsilon_0 \cos(N_s \theta)),$$

where $N_s \in \mathbb{N}$ is corresponding to the number of branching directions and $\epsilon_0 \in [0, 1[$ is corresponding to the anisotropy amptitude and is assumed to be sufficiently small.

Consequently, the problem (11.8) can be reformulated, by taking into account the anisotropy, as follows:

$$\begin{aligned}
&\frac{\partial \phi}{\partial t} - \operatorname{div}(\mathcal{A}(\nabla\phi)\nabla\phi) = F_1(x,t,\phi) + U F_2(x,t,\phi) + \lambda \ \text{ on } \mathcal{Q},\\
&\frac{\partial U}{\partial t} - \operatorname{div}(D_1(x,t,\phi)\nabla U + D_2(x,t,\mathbf{U})\nabla\phi) = 0 \ \text{ on } \mathcal{Q},\\
&\text{subject to the boundary conditions}\\
&\mathcal{A}(\nabla\phi)\nabla\phi.\mathbf{n} = 0, \ \ (D_1(x,t,\phi)\nabla U + D_2(x,t,\mathbf{U})\nabla\phi).\mathbf{n} = 0 \ \text{ on } \Sigma,\\
&\text{and the initial conditions}\\
&\phi(0) = \phi_0, U(0) = U_0 \ \text{ on } \Omega,
\end{aligned} \tag{11.38}$$

where $\mathcal{A}$ is an anisotropy tensor.

If we assume some strong regularity for the non-linear operator $\mathcal{A}$, we can extend directly the arguments of the previous sections to the present case. For example, in order to have a well-posedeness problem (11.38), we can make the following assumptions:

(**A1**) There exists $\nu > 0$ such that

$$\nu \parallel v_1 - v_2 \parallel_{L^2}^2 \leq \int_\Omega (\mathcal{A}(v_1)v_1 - \mathcal{A}(v_2)v_2)(v_1 - v_2)\mathrm{d}x,$$

for all $(v_1, v_2) \in L^2(\Omega) \times L^2(\Omega)$.

(**A2**) There exists $M > 0$ such that

$$\parallel \mathcal{A}(v_1)v_1 - \mathcal{A}(v_2)v_2 \parallel_{L^2} \leq M \parallel v_1 - v_2 \parallel_{L^2},$$

for all $(v_1, v_2) \in L^2(\Omega) \times L^2(\Omega)$.

(*ii*) It is clear that we can treat, by using the same technique developed in this chapter and more generally in this book, (at least from the "formal" viewpoint), different general physical models concerning the solidification process, for example the problem presented by Granasy *et al.* [137, 138, 139] and Warren *et al.* [293, 294, 295]. ◇

Remark 11.27. As indicated in Remark 10.31, for numerical resolution of the robust control problems, the reader is referred to Chapter 9. ◇

12

Large-scale Ocean in the Climate System

In this chapter, we study robust control problems arising from oceanic currents. The equations are non-linear, time-dependent and coupled, and are of Navier–Stokes type for the velocity and pressure, and of transport-diffusion type for the temperature and the salinity, with Robin-type boundary conditions. The objective is the prediction and robust regulation of the circulation from the mean circulation by taking into account the worst disturbance caused by small variations of the surface temperature.

First, the mathematical models are introduced and the existence, uniqueness and regularity results of the solution of the state systems are studied. The asymptotic behavior is also considered. Afterwards, robust control problems are formulated. The controls and disturbances are of Robin type and act on a part of the boundary during a time T. The existence of a robust control in the admissible sets is proved and, finally, first-order necessary conditions of optimality are obtained. The problem is considered first for a system with Boussinesq approximation and second for a system with hydrostatic approximation with vertical viscosity.

12.1 Introduction and Formulation of the Problem

12.1.1 Motivation

There are two types of currents whose main causes are solar radiation (the warming of the surface by downwelling radiation coming from the atmosphere), wind and gravity. The Earth unevenly receives solar energy, which is not the same in the polar region as in the equatorial region. Consequently, the intertropical zone receives so much more energy than the rest of the planet, which creates a thermal imbalance. This imbalance is the origin of the two types of oceanic currents (surface currents and deep currents), and, indeed, puts in motion the atmosphere and oceanic currents which trying to thermally readjust the whole system. These movements are influenced by a force caused by the rotation of the Earth, called the Coriolis force. Moreover, it generates winds, which are the main causes of surface currents (which affect approximately 10% of the water of the oceans and are generally limited to the top 400 meters of the oceans). The location of these surface currents changes significantly with the seasons; this phenomenon is particularly sensitive for the equatorial currents.

The thermal imbalance leads also to differences in temperature depending on latitude[1] (between the various layers of the ocean). This temperature difference leads to a difference in salinity and thus in density, creating deep oceanic currents (the difference in density is a function of the difference in temperature and salinity). Consequently, the surface currents and deep currents are interconnected. Thanks to the heat capacity of water (its thermal inertia is much larger than that of the air), the ocean tempers the seasonal temperature changes of the air masses, which otherwise would be much more important. Thus currents of the warm surface layers can warm a region's climate. In contrast, the rise of cold water from the sea depths to the surface moderates water temperatures in the equatorial region.[2] Therefore, the surface temperature variation plays an important role in oceanic current movements, which play a key role in the regulation of the climate, and ensure that a heat transport from the equator to the poles is as important as the atmosphere. We can not neglect the fluctuation caused by this variation (even if the variation is small) and consequently the analysis of its influence is very important.

The phenomenon that we want to model occurs in the tropical Atlantic and Pacific oceans: the circulation there is characterized by steady zonal currents and by long waves propagating westward along the equator and superposed on the mean currents. The equatorial waves are connected with strong vertical currents, called convective currents, which are very sensitive to small changes in temperature and induce "upwellings" or "downwellings." These phenomena modify the properties of the sea water near the surface –decreasing or increasing the temperature, supply of plankton *etc.*– and are therefore of great importance for climate, fishing *etc.* This climate variability can be illustrated, for example, by the *El Niño* phenomenon, which occurs every two to seven years. *El Niño* warming begins with a modest temperature elevation (2 or 3°C) of surface waters in the Pacific ocean along the equatorial coast of South America. The *El Niño* current then cuts the cold and deep upwelling (it results in a reduction of food, and, consequently, causes fish populations to decline). This phenomenon is accompanied with a sharp increase in rainfall and flooding. In contrast, around Australia and Asia, the high pressures and cold water temperature decrease the precipitation, and then cause severe drought, which often lead to a series of fires.

[1] The deep waters sink in the ocean basins located at high latitudes, where temperatures are low enough so that the density increases.

[2] But this movement is still poorly understood because it is difficult to measure directly.

These equatorial waves have been evidenced from "in situ" observations, and more recently from altimetric measurements[3] and data assimilation in connection with control theory, see Belmiloudi *et al.* [27, 28, 30, 31, 33, 34, 35, 36, 37], in which the authors analyze the process of these tropical instability waves. The equations of the large-scale systems of the ocean are derived from Navier–Stokes equations but take into account oceanographical assumptions such as Boussinesq approximation, *i.e.*, the density variations ρ are neglected in the system except in the buoyancy term (on the other hand, density is assumed to be constant and equal to a mean value ρ_{av} in the equations describing the horizontal motion) and in the equation of state.

Of course, there is a vast literature concerning the control theory and inverse problems in connection with fluid mechanics problems. The reader is referred, for example, to Blayo *et al.* [53], Gejadze *et al.* [129, 130], Gunzburger *et al.* [144, 145], Marchuk [217, 218], Parmuzin *et al.* [235], Sritharan *et al.* [119, 120, 273], Wunsch [303] and the references therein. For robust control analysis of Navier–Stokes equations in the Riccati operator, see Barbu and Srithran [22], in which the authors assume appropriate detectability and stability constraints on the system; while for a study using a similar approach as developed in this book, see Bewley *et al.* [52].

12.1.2 Primitive Equations and Study Domain

We consider a time interval $(0, T)$ and an oceanic domain Ω extending on both sides of the equator (10°S–10°N), and of constant depth H (for example $H = 3000$ m). The curvature of the Earth is neglected. The vertical extension of Ω $(-H \leq z \leq 0)$ corresponds to a part of the physical domain where the variability of the mean current is large. The studied model includes the following unknown functions: the velocity $\mathbf{u}$, the pressure p, the temperature $\mathcal{T}$ and the salinity $\mathcal{S}$.

Taking the explicit equation of state given by Washington and Parkinson [296] into consideration: $\rho = \delta_0 - \delta_{\text{T}}\mathcal{T} + \delta_{\text{S}}\mathcal{S}$, the total Boussinesq equations of the ocean are as follows:

$$
\begin{aligned}
&\frac{\partial \mathbf{u}}{\partial t} + (\mathbf{u}.\nabla)\mathbf{u} + \mathbf{F} \wedge \mathbf{u} - \operatorname{div}(\nu_1 \nabla \mathbf{u}) + \frac{1}{\rho_{\text{av}}}\nabla p = \frac{\rho}{\rho_{\text{av}}}\mathbf{G} \text{ on } \mathcal{Q},\\
&\operatorname{div}(\mathbf{u}) = 0 \text{ on } \mathcal{Q},\\
&\frac{\partial \mathcal{T}}{\partial t} + (\mathbf{u}.\nabla)\mathcal{T} - \operatorname{div}(\nu_2 \nabla \mathcal{T}) = 0 \text{ on } \mathcal{Q},\\
&\frac{\partial \mathcal{S}}{\partial t} + (\mathbf{u}.\nabla)\mathcal{S} - \operatorname{div}(\nu_3 \nabla \mathcal{S}) = 0 \text{ on } \mathcal{Q},\\
&\rho = \delta_0 - \delta_{\text{T}}\mathcal{T} + \delta_{\text{S}}\mathcal{S},
\end{aligned}
\tag{12.1}
$$

[3] Altimetric measurements give the distance between the satellite and the sea surface.

where $\mathcal{Q} = \Omega \times (0,T)$, $\mathbf{G} = (0,0,-g)$ is the gravity force, (x,y,z) are the cartesian coordinates: x,y are measured in the horizontal plane of the undisturbed sea-surface (x towards the east, y towards the north) and z is vertically ascendant. The constants δ_{T} and δ_{S} are expansion coefficients.

The function $\mathbf{F} \wedge \mathbf{u} = (0,0,2\omega \sin\phi) \wedge \mathbf{u}$ is the Coriolis acceleration, ω the rotation rate of Earth, ϕ the latitude (this acceleration becomes dominant for large-scale flows (see, *e.g.*, Pedlosky [236]), because the relative acceleration becomes small when the scale of the motion is large). Afterwards, we will suppose that ρ_{av} is equal to 1 (ρ_{av}=1 g cm^{-3}).

The sensitivity of ocean general circulation models to the parametrization of the vertical turbulent diffusion has been known for a long time and has been proved by Bryan [65]. The turbulent flux is usually modeled by the dissipative terms and linked to large-scale oceans by using the mixing coefficients: $\nu_{1\mathrm{h}}$, $\nu_{2\mathrm{h}}$ and $\nu_{3\mathrm{h}}$ (resp. $\nu_{1\mathrm{v}}$, $\nu_{2\mathrm{v}}$ and $\nu_{3\mathrm{v}}$) denote the coefficients of horizontal (resp. vertical) eddy viscosity and diffusivity. Moreover, the turbulent mixing is high above the thermocline and low in stratified regions. According to the parametrization described in Philander *et al.* [239, 240], the coefficients $\nu_{1\mathrm{v}}$, $\nu_{2\mathrm{v}}$ and $\nu_{3\mathrm{v}}$ are variable, positive and bounded functions. They are deduced from the given mean circulation. The coefficients $\nu_{1\mathrm{h}}$, $\nu_{2\mathrm{h}}$ and $\nu_{3\mathrm{h}}$ are constant and positive.

The total circulation $(\mathbf{u}, p, \mathcal{T}, \mathcal{S})$ is the sum of the mean circulation $(\mathbf{u_0}, p_0, \mathcal{T}_0, \mathcal{S}_0)$, which is given, and a variability (a perturbation) $(\tilde{\mathbf{u}}, \tilde{p}, \tilde{\mathcal{T}}, \tilde{\mathcal{S}})$, which is corresponding to small deviations from the target flow-temperature-salinity $(\mathbf{u_0}, p_0, \mathcal{T}_0, \mathcal{S}_0)$. This expansion is justified in tropical region: the steady mean circulation $(\mathbf{u_0}, p_0, \mathcal{T}_0, \mathcal{S}_0)$ is known for each tropical season; the variability $(\tilde{\mathbf{u}}, \tilde{p}, \tilde{\mathcal{T}}, \tilde{\mathcal{S}})$ is made of westward propagating waves. We assume that, at initial time ($t = 0$) the flow in the domain Ω is the mean circulation $(\mathbf{u_0}, p_0, \mathcal{T}_0, \mathcal{S}_0)$ which satisfies steady-state Equations (12.1), *i.e.*, $(\mathbf{u_0}, p_0, \mathcal{T}_0, \mathcal{S}_0)$ satisfies the system (since $\rho_{\mathrm{av}} = 1$)

$$
\begin{aligned}
&-\nu_{1\mathrm{h}} \Delta_2 \mathbf{u}_0 - \frac{\partial}{\partial z}\Big(\nu_{1\mathrm{v}} \frac{\partial \mathbf{u}_0}{\partial z}\Big) + (\mathbf{u}_0.\nabla)\mathbf{u}_0 \\
&\qquad\qquad\qquad\qquad + \mathbf{F} \wedge \mathbf{u}_0 + \nabla p_0 = \rho_0 \mathbf{G} \quad \text{on } \Omega, \\
&\mathrm{div}(\mathbf{u}_0) = 0 \quad \text{on } \Omega, \\
&-\nu_{2\mathrm{h}} \Delta_2 \mathcal{T}_0 - \frac{\partial}{\partial z}\Big(\nu_{2\mathrm{v}} \frac{\partial \mathcal{T}_0}{\partial z}\Big) + (\mathbf{u}_0.\nabla)\mathcal{T}_0 = 0 \quad \text{on } \Omega, \\
&-\nu_{3\mathrm{h}} \Delta_2 \mathcal{S}_0 - \frac{\partial}{\partial z}\Big(\nu_{3\mathrm{v}} \frac{\partial \mathcal{S}_0}{\partial z}\Big) + (\mathbf{u}_0.\nabla)\mathcal{S}_0 = 0 \quad \text{on } \Omega, \\
&\rho_0 = \delta_0 - \delta_{\mathrm{T}} \mathcal{T}_0 + \delta_{\mathrm{S}} \mathcal{S}_0.
\end{aligned}
\tag{12.2}
$$

The full non-linear system which models large perturbation $(\tilde{u}, \tilde{p}, \tilde{\mathcal{T}}, \tilde{\mathcal{S}})$ to the target $(\mathbf{u_0}, p_0, \mathcal{T}_0, \mathcal{S}_0)$ can be deduced from (12.1) and (12.2):

$$\frac{\partial \tilde{u}}{\partial t} + (\tilde{u}.\nabla)\mathbf{u_0} + (\mathbf{u_0}.\nabla)\tilde{u} + (\tilde{u}.\nabla)\tilde{u} + \mathbf{F} \wedge \tilde{u}$$
$$-\nu_{1\mathrm{h}}\Delta_2\tilde{u} - \frac{\partial}{\partial z}(\nu_{1\mathrm{v}}\frac{\partial \tilde{u}}{\partial z}) + \nabla\tilde{p} = (-\delta_{\mathrm{T}}\tilde{\mathcal{T}} + \delta_{\mathrm{S}}\tilde{\mathcal{S}})\mathbf{G} \quad \text{on } \mathcal{Q},$$
$$\mathrm{div}(\tilde{u}) = 0 \quad \text{on } \mathcal{Q},$$
$$\frac{\partial \tilde{\mathcal{T}}}{\partial t} + (\tilde{u}.\nabla)\mathcal{T}_0 + (\mathbf{u_0}.\nabla)\tilde{\mathcal{T}} + (\tilde{u}.\nabla)\tilde{\mathcal{T}} \tag{12.3}$$
$$-\nu_{2\mathrm{h}}\Delta_2\tilde{\mathcal{T}} - \frac{\partial}{\partial z}(\nu_{2\mathrm{v}}\frac{\partial \tilde{\mathcal{T}}}{\partial z}) = 0 \quad \text{on } \mathcal{Q},$$
$$\frac{\partial \tilde{\mathcal{S}}}{\partial t} + (\tilde{u}.\nabla)\mathcal{S}_0 + (\mathbf{u_0}.\nabla)\tilde{\mathcal{S}} + (\tilde{u}.\nabla)\tilde{\mathcal{S}}$$
$$-\nu_{3\mathrm{h}}\Delta_2\tilde{\mathcal{S}} - \frac{\partial}{\partial z}(\nu_{3\mathrm{v}}\frac{\partial \tilde{\mathcal{S}}}{\partial z}) = 0 \quad \text{on } \mathcal{Q},$$

where $\Delta_2. = \partial^2./\partial x^2 + \partial^2./\partial y^2$.

Nota bene: The notation "~" used for the perturbation of the mean circulation will now be omitted.

The flow domain Ω can be defined as $\Omega =]0, Lx[\times]-Ly, Ly[\times]-H, 0[$. Γ denotes its boundary: $\Gamma = \Gamma_0 \cup \Gamma_5 \cup \Gamma_1 \cup \Gamma_2 \cup \Gamma_3 \cup \Gamma_4$ where Γ_0 denotes the surface ($z = 0$), Γ_5 the bottom ($z = -H$), Γ_1 and Γ_2 the eastern and western boundaries, Γ_3 and Γ_4 the northern and southern boundaries.

We note that the vertical extension $-H \leq z \leq 0$ of the domain Ω doesn't correspond to the physical sea water but only to the layer where the variability is computed. We assume that for depths greater than H the perturbation is negligible. The perturbation of the current is made of zonal propagating waves. Therefore, we can choose the zonal extension of Ω as the greatest wavelength, and impose periodic conditions on the eastern and western boundaries. To take into account the phenomena we want to describe, we set mixed boundary conditions:

- the flow is periodic in the x-direction:

  $\mathbf{u}\,|_{\Gamma_1} = \mathbf{u}\,|_{\Gamma_2}$, $\mathcal{T}\,|_{\Gamma_1} = \mathcal{T}\,|_{\Gamma_2}$ and $\mathcal{S}\,|_{\Gamma_1} = \mathcal{S}\,|_{\Gamma_2}$,
- on the northern and southern boundaries we impose a sliding condition for the flow and an homogeneous Dirichlet condition for the temperature-salinity: (12.4)

  $$v = \mathcal{T} = \mathcal{S} = 0, \frac{\partial u}{\partial y} = 0 \text{ and } \frac{\partial w}{\partial y} = 0 \quad \text{on } \Gamma_3 \text{ and } \Gamma_4,$$
- the perturbation vanishes at $z = -H$:

  $\mathbf{u} = \mathbf{0}$ and $\mathcal{T} = \mathcal{S} = 0$ on Γ_5,

where $\mathbf{u} = (u, v, w)$.

In the surface Γ_0 the systems (12.3) with (12.4) are supplemented with the following Robin-type boundary conditions

$$\begin{aligned}
&\nu_{1\mathrm{v}}\frac{\partial u}{\partial z} = f_1, \nu_{1\mathrm{v}}\frac{\partial v}{\partial z} = f_2, w = 0, \\
&\nu_{2\mathrm{v}}\frac{\partial \mathcal{T}}{\partial z} + \delta\mathcal{T} = \varpi, \\
&\nu_{3\mathrm{v}}\frac{\partial \mathcal{S}}{\partial z} = 0,
\end{aligned} \tag{12.5}$$

where $\mathbf{u} = (u, v, w)$, δ is a positive constant related with the turbulent heating on the surface of the ocean, $\mathbf{f} = (f_1, f_2)$ is the variability of the wind stress on the surface of the ocean and ϖ is the disturbance of the heat stress on the surface of the ocean.

We assume that, at initial time $t = 0$, the mean circulation is not disturbed. Therefore, the initial condition is:

$$(\mathbf{u}, \mathcal{T}, \mathcal{S})(t = 0) = 0. \tag{12.6}$$

Remark 12.1. The perturbation of the mean circulation is driven by the perturbation of the surface stress, which depends, for example, on the perturbation of the velocity of the atmosphere. If we suppose that Γ_0 is the upper surface of the ocean, the perturbation of the vertical velocity w (with $w = \mathbf{u}.\mathbf{n}$) vanishes on Γ_0. This condition is classically named the impermeability condition on the sea surface. Moreover, the perturbation is not computed in the thin surface layer $0 \leq z \leq \xi$, where $z = \xi$ is the free sea surface observed by the satellite. ♢

Our purpose is motivated by the robust regulation of the deviation of circulation from the mean circulation $(\mathbf{u_0}, \mathcal{T}_0, \mathcal{S}_0, , p_0)$, by taking into account the worst disturbance caused by the small variation of the surface temperature. The variability of the surface stress $\mathbf{f}$ acts as the forcing of the perturbation and is unknown, and the variability of the surface temperature ϖ acts as the worst disturbance (generated by the warming of surface water). We take $\mathbf{f}$ as the control and ϖ as the disturbance in our robust control problem.

12.2 The Perturbation Problem

In this section we develop some results of the existence, uniqueness and regularity of the solutions for the non-linear and three-dimensional cases. These results will still be valid for the non-linear and two-dimensional, and for the linear and three-dimensional case. After introducing some notations and preliminary results useful in the linear as well as in the non-linear situation (in two or three-dimensional cases), we give the weak (or variational) formulations of the problem and the main existence, uniqueness and regularity results.

12.2.1 Preliminary Results and Weak Formulations

In order to study the problem (12.3) with the boundary conditions (12.4) and (12.5), and the initial condition (12.6), we introduce the following functional spaces:

$$
\begin{aligned}
&\mathcal{V}_0 = \{\mathbf{v} \in (H^1(\Omega))^3 : \mathrm{div}(\mathbf{v}) = 0, \mathbf{v}.\mathbf{n} = 0 \ \text{ on } \Gamma\},\\
&\mathcal{H}_1 = \{\mathbf{v} \in (L^2(\Omega))^3 : \mathrm{div}(\mathbf{v}) = 0, \mathbf{v}.\mathbf{n} = 0 \ \text{ on } \Gamma_0 \cup \Gamma_3 \cup \Gamma_4 \cup \Gamma_5,\\
&\qquad\qquad \mathbf{v}.\mathbf{n}|_{\Gamma_1} = -\mathbf{v}.\mathbf{n}|_{\Gamma_2}\},\\
&\mathcal{W}_1 = \{\mathbf{v} \in (H^1(\Omega))^3 : \mathbf{v} = \mathbf{0} \text{ on } \Gamma_5, \mathbf{v}.\mathbf{n} = 0 \ \text{ on } \Gamma_0 \cup \Gamma_3 \cup \Gamma_4,\\
&\qquad\qquad \mathbf{v}|_{\Gamma_1} = \mathbf{v}|_{\Gamma_2}\},\\
&\mathcal{V}_1 = \{\mathbf{v} \in \mathcal{W}_1 : \mathrm{div}(\mathbf{v}) = 0\},\\
&\mathcal{H}_2 = L^2(\Omega), \ \ \mathcal{V}_2 = \{\phi \in H^1(\Omega) : \phi = 0 \ \text{ on } \Gamma_3 \cup \Gamma_4 \cup \Gamma_5, \phi|_{\Gamma_1} = \phi|_{\Gamma_2}\},\\
&\mathcal{H} = \mathcal{H}_1 \times \mathcal{H}_2 \times \mathcal{H}_2, \qquad \mathcal{V} = \mathcal{V}_1 \times \mathcal{V}_2 \times \mathcal{V}_2,
\end{aligned}
$$

where $\mathbf{n}$ is the unit outward vector normal to Γ.

Nota bene: In the following, if X denotes some Banach space of real-valued functions, the corresponding space of vector-valued functions, each of components belonging to X, will be denoted, for simplicity, by the same notation X, and we use $\| \,.\, \|_X$ to denote the norms of spaces of real-valued functions or of vector-valued functions X. For example $L^2(\Omega)$ (respectively $L^2(\Gamma_0)$) is the corresponding space of vector-valued functions of $L^2(\Omega)$ (respectively $L^2(\Gamma_0)$), *etc.*

We use the following notations: $| \,.\, |_{1,\Omega}$ and $\| \,.\, \|_{1,\Omega}$ denote, respectively the semi-norm and the norm on $H^1(\Omega)$. They are equivalent on $\mathcal{W}_1$, $\mathcal{V}_1$ and $\mathcal{V}_2$, and we set $\| \mathbf{u} \| = | \mathbf{u} |_{1,\Omega} = \| \mathbf{u} \|_{\mathcal{W}_1} = \| \mathbf{u} \|_{\mathcal{V}_1}$, $\| \phi \| = | \phi |_{1,\Omega} = \| \phi \|_{\mathcal{V}_2}$. We denote by $| \mathbf{u} |$ and $| \phi |$ the norm in $L^2(\Omega)$, and by $| \,.\, |_{\Gamma_0}$ and $(.,.)_{\Gamma_0}$ the norm and the scalar product in $L^2(\Gamma_0)$.

Remark 12.2. (*i*) If $\mathbf{v} \in H(\mathrm{div}; \Omega)$, then $\mathbf{v}.\mathbf{n}$ makes sense on the boundary Γ (see Chapter 3).
(*ii*) If $\mathbf{v} \in \mathcal{H}_1$ (respectively $\mathbf{v} \in \mathcal{V}_1$), then $\mathbf{v}$ can be extended as a free divergent, x-periodic function, which belongs to the space L^2_{loc} (resp. H^1_{loc}).
(*iii*) If $\phi \in \mathcal{V}_2$, then ϕ can be extended as a x-periodic function, which belongs to the space H^1_{loc}. ♢

We now define the following forms:

$$
\begin{aligned}
&a_1(\mathbf{u}, \mathbf{v}) = \nu_{1\mathrm{h}}(\nabla_2 \mathbf{u}, \nabla_2 \mathbf{v}) + (\nu_{1\mathrm{v}} \frac{\partial \mathbf{u}}{\partial z}, \frac{\partial \mathbf{v}}{\partial z}), \qquad \text{with } \nabla_2 \mathbf{u} = (\frac{\partial \mathbf{u}}{\partial x}, \frac{\partial \mathbf{u}}{\partial y}),\\
&a_2(\mathcal{T}, \phi) = \nu_{2\mathrm{h}}(\nabla_2 \mathcal{T}, \nabla_2 \phi) + (\nu_{2\mathrm{v}} \frac{\partial \mathcal{T}}{\partial z}, \frac{\partial \phi}{\partial z}) + \delta(\mathcal{T}, \phi)_{\Gamma_0},\\
&a_3(\mathcal{S}, \psi) = \nu_{3\mathrm{h}}(\nabla_2 \mathcal{S}, \nabla_2 \psi) + (\nu_{3\mathrm{v}} \frac{\partial \mathcal{S}}{\partial z}, \frac{\partial \psi}{\partial z}), \qquad d(\mathbf{u}, \mathbf{v}) = (\mathbf{F} \wedge \mathbf{u}, \mathbf{v}),\\
&b_1(\mathbf{u}, \mathbf{v}, \mathbf{w}) = ((\mathbf{u}.\nabla)\mathbf{v}, \mathbf{w}), \quad b_2(\mathbf{u}, \phi, \psi) = ((\mathbf{u}.\nabla)\phi, \psi),\\
&l_0(\mathbf{u}, \mathbf{v}) = ((\mathbf{u}_0.\nabla)\mathbf{u}, \mathbf{v}) + ((\mathbf{u}.\nabla)\mathbf{u}_0, \mathbf{v}) + d(\mathbf{u}, \mathbf{v}),
\end{aligned}
$$

$$l_1(\mathcal{T},\mathcal{S},\mathbf{v}) = -((-\delta_{\mathrm{T}}\mathcal{T}+\delta_{\mathrm{S}}\mathcal{S})\mathbf{G},\mathbf{v}),$$
$$l_2(\mathbf{u},\mathcal{T},\phi) = ((\mathbf{u_0}.\nabla)\mathcal{T},\phi)+((\mathbf{u}.\nabla)\mathcal{T}_0,\phi),$$
$$l_3(\mathbf{u},\mathcal{S},\psi) = ((\mathbf{u_0}.\nabla)\mathcal{S},\psi)+((\mathbf{u}.\nabla)\mathcal{S}_0,\psi),$$
$$c(\mathbf{v},p) = -(\mathrm{div}(\mathbf{v}),p).$$

We now recall some properties of the previous operators, which can be found in Belmiloudi [35].

Lemma 12.3. *For the operators d, $(a_i)_{i=1,3}$ and $(b_i)_{i=1,2}$ the following properties hold:*

(i) a_1 (respectively a_i,for $i=2,3$) is a bilinear continuous and coercive form on $\mathcal{W}_1\times\mathcal{W}_1$ (respectively on $\mathcal{V}_2\times\mathcal{V}_2$), where the coercivity constants are denoted by α_i, for $i=1,3$
(ii) d is a bilinear continuous form on $\mathcal{W}_1\times\mathcal{W}_1$
(iii) b_1 (respectively b_2) is a trilinear continuous form on $\mathcal{W}_1\times\mathcal{W}_1\times\mathcal{W}_1$ (respectively $\mathcal{W}_1\times\mathcal{V}_2\times\mathcal{V}_2$). □

By using the interpolation theory and Gagliardo–Nirenberg inequalities (see Chapter 3), we can easily prove the following classical estimates (see, for example, Belmiloudi [32] and Temam [281]).

Lemma 12.4. *There exists a positive constant C (depending on Ω) such that the following estimates hold:*

$$\begin{aligned}
&| b_1(\mathbf{u},\mathbf{v},\mathbf{w}) |\leq C \parallel \mathbf{u} \parallel_{L^2(\Omega)}^{1/4}\parallel \mathbf{u} \parallel^{3/4}\parallel \mathbf{v} \parallel \parallel \mathbf{w} \parallel_{L^2(\Omega)}^{1/4}\parallel \mathbf{w} \parallel^{3/4}\\
&\qquad\qquad \textit{for all } \mathbf{u},\mathbf{v},\mathbf{w}\in\mathcal{W}_1,\\
&| b_1(\mathbf{u},\mathbf{v},\mathbf{w}) |\leq C \parallel \mathbf{u} \parallel\parallel \mathbf{v} \parallel^{1/2}\parallel \mathbf{v} \parallel_{H^2(\Omega)}^{1/2} \parallel \mathbf{w} \parallel_{L^2(\Omega)}\\
&\qquad\qquad \textit{for all } \mathbf{u}\in\mathcal{W}_1,\mathbf{v}\in H^2(\Omega)\cap\mathcal{W}_1,\mathbf{w}\in\mathcal{H}_1,\\
&| b_1(\mathbf{u},\mathbf{v},\mathbf{w}) |\leq C \parallel \mathbf{u} \parallel^{1/2}\parallel \mathbf{u} \parallel_{H^2(\Omega)}^{1/2}\parallel \mathbf{u} \parallel \parallel \mathbf{w} \parallel_{L^2(\Omega))}\\
&\qquad\qquad \textit{for all } \mathbf{u}\in H^2(\Omega)\cap\mathcal{W}_1,\mathbf{v}\in\mathcal{W}_1,\mathbf{w}\in\mathcal{H}_1,\\
&| b_2(\mathbf{u},\psi,\phi) |\leq C \parallel \mathbf{u} \parallel_{L^2(\Omega)}^{1/4}\parallel \mathbf{u} \parallel^{3/4}\parallel \psi \parallel \parallel \phi \parallel_{L^2(\Omega)}^{1/4}\parallel \phi \parallel^{3/4}\\
&\qquad\qquad \textit{for all } \mathbf{u}\in\mathcal{W}_1,\psi,\phi\in\mathcal{V}_2,\\
&| b_2(\mathbf{u},\psi,\phi) |\leq C \parallel \mathbf{u} \parallel\parallel \psi \parallel^{1/2}\parallel \psi \parallel_{H^2(\Omega)}^{1/2} \parallel \phi \parallel_{L^2(\Omega)}\\
&\qquad\qquad \textit{for all } \mathbf{u}\in\mathcal{W}_1,\psi\in H^2(\Omega)\cap\mathcal{V}_2,\phi\in\mathcal{H}_2,\\
&| b_2(\mathbf{u},\psi,\phi) |\leq C \parallel \mathbf{u} \parallel^{1/2}\parallel \mathbf{u} \parallel_{H^2(\Omega)}^{1/2}\parallel \psi \parallel \parallel \phi \parallel_{L^2(\Omega))}\\
&\qquad\qquad \textit{for all } \mathbf{u}\in H^2(\Omega)\cap\mathcal{W}_1,\psi\in\mathcal{V}_2,\phi\in\mathcal{H}_2.
\end{aligned}\tag{12.7}$$
□

In order to introduce the weak formulations associated with the perturbation problems (12.3)–(12.6), we give the following lemma.

Lemma 12.5. *Suppose that the function $\mathbf{f}=(f_1,f_2)$ is in $L^2(0,T;L^2(\Gamma_0))$, the function ϖ is in $L^2(0,T;L^2(\Gamma_0))$ and that $(\mathbf{u},p,\mathcal{T},\mathcal{S})$ is sufficiently regular and satisfies the boundary conditions (12.4) and (12.5), then the following relations hold:*

(*i*) *For all* $\mathbf{v} \in \mathcal{W}_1$,

$$-\nu_{1\mathrm{h}}(\Delta_2\mathbf{u},\mathbf{v})-(\frac{\partial}{\partial z}(\nu_{1\mathrm{v}}\frac{\partial \mathbf{u}}{\partial z}),\mathbf{v})+(\nabla p,\mathbf{v}) = a_1(\mathbf{u},\mathbf{v})+c(\mathbf{v},p)-(\mathbf{f}_{\mathrm{T}},\mathbf{v})_{\Gamma_0},$$

where $\mathbf{f}_{\mathrm{T}} = (\mathbf{f}, 0)$. *Moreover, if* $\mathbf{v} \in \mathcal{V}_1$, *then*

$$-\nu_{1\mathrm{h}}(\Delta_2\mathbf{u},\mathbf{v}) - (\frac{\partial}{\partial z}(\nu_{1\mathrm{v}}\frac{\partial \mathbf{u}}{\partial z}),\mathbf{v}) + (\nabla p,\mathbf{v}) = a_1(\mathbf{u},\mathbf{v}) - (\mathbf{f}_{\mathrm{T}},\mathbf{v})_{\Gamma_0}.$$

(*ii*) *For all* $\phi \in \mathcal{V}_2$,

$$-\nu_{2\mathrm{h}}(\Delta_2\mathcal{T},\phi) - (\frac{\partial}{\partial z}(\nu_{2\mathrm{v}}\frac{\partial \mathcal{T}}{\partial z}),\phi) = a_2(\mathcal{T},\phi) - (\varpi,\phi)_{\Gamma_0}.$$

(*iii*) *For all* $\psi \in \mathcal{V}_2$,

$$-\nu_{3\mathrm{h}}(\Delta_2\mathcal{S},\psi) - (\frac{\partial}{\partial z}(\nu_{3\mathrm{v}}\frac{\partial \mathcal{S}}{\partial z}),\psi) = a_3(\mathcal{S},\psi).$$

(*iv*) *For all* $\mathbf{u},\mathbf{v}$ *in* $\mathcal{W}_1$,

$$d(\mathbf{u},\mathbf{v}) = -d(\mathbf{v},\mathbf{u}).$$

(*v*) *For all* $\mathbf{u}$ *in* $\mathcal{V}_1$ *and* $\mathbf{v},\mathbf{w}$ *in* $\mathcal{W}_1$,

$$b_1(\mathbf{u},\mathbf{v},\mathbf{v}) = 0 \quad \textit{and} \quad b_1(\mathbf{u},\mathbf{v},\mathbf{w}) = -b_1(\mathbf{u},\mathbf{w},\mathbf{v}).$$

(*vi*) *For all* $\mathbf{u}$ *in* $\mathcal{V}_1$ *and* ϕ,ψ *in* $\mathcal{V}_2$,

$$b_2(\mathbf{u},\phi,\phi) = 0 \quad \textit{and} \quad b_2(\mathbf{u},\phi,\psi) = -b_2(\mathbf{u},\psi,\phi).$$

(*vii*) *Since* $\mathbf{u_0} \in \mathcal{V}_0$ *we have, for all* $\mathbf{v},\mathbf{w}$ *in* $\mathcal{W}_1$ *and* ϕ,ψ *in* $\mathcal{V}_2$, *the following identities:*

$$b_1(\mathbf{u_0},\mathbf{v},\mathbf{v}) = 0,\ b_2(\mathbf{u_0},\phi,\phi) = 0,$$
$$b_1(\mathbf{u_0},\mathbf{v},\mathbf{w}) = -b_1(\mathbf{u_0},\mathbf{w},\mathbf{v}),$$
$$b_2(\mathbf{u_0},\phi,\psi) = -b_2(\mathbf{u_0},\psi,\phi).$$

Proof. Results (*i*)–(*iv*) are deduced from the definition of the spaces $\mathcal{W}_1, \mathcal{V}_1$ and $\mathcal{V}_2$, and from the boundary conditions satisfied by $(\mathbf{u},\mathcal{T},\mathcal{S})$ on Γ. The orthogonality identities (*v*) and (*vi*) are consequences of $\mathrm{div}(\mathbf{u}) = 0$ and Green's formula. Finally, result (*vii*) is a direct consequence of (*v*)–(*vi*).

This completes the proof. □

According to Lemma 12.5, the problems (12.3)–(12.6) satisfied by the perturbation $(\mathbf{u},p,\mathcal{T},\mathcal{S})$ of the mean flow admits the two following equivalent weak formulations:

find $(\mathbf{u},\mathcal{T},\mathcal{S}) \in L^2(0,T;\mathcal{V})$ such that, for all (v,ϕ,ψ) in $\mathcal{V}$, a.e. in $(0,T)$

$$
\begin{aligned}
&(\frac{\partial \mathbf{u}}{\partial t}, \mathbf{v}) + a_1(\mathbf{u}, \mathbf{v}) + l_0(\mathbf{u}, \mathbf{v}) + l_1(\mathcal{T}, \mathcal{S}, \mathbf{v}) + b_1(\mathbf{u}, \mathbf{u}, \mathbf{v}) \\
&\qquad = (\mathbf{f}_{\mathrm{T}}, \mathbf{v})_{\Gamma_0}, \\
&(\frac{\partial \mathcal{T}}{\partial t}, \phi) + a_2(\mathcal{T}, \phi) + l_2(\mathbf{u}, \mathcal{T}, \phi) + b_2(\mathbf{u}, \mathcal{T}, \phi) = (\varpi, \phi)_{\Gamma_0}, \\
&(\frac{\partial \mathcal{S}}{\partial t}, \psi) + a_3(\mathcal{S}, \psi) + l_3(\mathbf{u}, \mathcal{S}, \psi) + b_2(\mathbf{u}, \mathcal{S}, \psi) = 0, \\
&(\mathbf{u}, \mathcal{T}, \mathcal{S})(t = 0) = 0
\end{aligned}
\tag{12.8}
$$

and

find $(\mathbf{u}, \mathcal{T}, \mathcal{S}, p) \in L^2(0, T; \mathcal{W}_1 \times \mathcal{V}_2 \times \mathcal{V}_2) \times \mathcal{D}'(\mathcal{Q})$ such that, for all (v, ϕ, ψ, q) in $\mathcal{W}_1 \times \mathcal{V}_2 \times \mathcal{V}_2 \times L^2(\Omega)$, a.e. in $(0, T)$

$$
\begin{aligned}
&(\frac{\partial \mathbf{u}}{\partial t}, \mathbf{v}) + a_1(\mathbf{u}, \mathbf{v}) + l_0(\mathbf{u}, \mathbf{v}) + l_1(\mathcal{T}, \mathcal{S}, \mathbf{v}) + b_1(\mathbf{u}, \mathbf{u}, \mathbf{v}) \\
&\qquad + c(\mathbf{v}, p) = (\mathbf{f}_{\mathrm{T}}, \mathbf{v})_{\Gamma_0}, \\
&(\frac{\partial \mathcal{T}}{\partial t}, \phi) + a_2(\mathcal{T}, \phi) + l_2(\mathbf{u}, \mathcal{T}, \phi) + b_2(\mathbf{u}, \mathcal{T}, \phi) = (\varpi, \phi)_{\Gamma_0}, \\
&(\frac{\partial \mathcal{S}}{\partial t}, \psi) + a_3(\mathcal{S}, \psi) + l_3(\mathbf{u}, \mathcal{S}, \psi) + b_2(\mathbf{u}, \mathcal{S}, \psi) = 0, \\
&c(\mathbf{u}, q) = 0, \\
&(\mathbf{u}, \mathcal{T}, \mathcal{S})(t = 0) = 0,
\end{aligned}
\tag{12.9}
$$

where $\mathbf{f}_{\mathrm{T}}$ is the vector $(\mathbf{f}, 0)$.

We end this section by the following lemma which shows the convergence results by applying the limit in the non-linear terms b_1 and b_2 (necessary to prove the existence and some regularity results).

Lemma 12.6. *If the sequence* $(\mathbf{u}_k, \psi_k)_k$ *converges towards* $(\mathbf{u}, \psi)$ *in the space* $L^2(0, T; \mathcal{V}_1 \times \mathcal{V}_2)$ *weakly and in* $L^2(0, T; \mathcal{H}_1 \times \mathcal{H}_2)$ *strongly, then for any vector function* $(\lambda(t), \mathbf{v}, \phi)$ *such that* $\lambda \in \mathcal{C}^1(0, T)$ *and* $(\mathbf{v}, \phi) \in \mathcal{V}_1 \times \mathcal{V}_2$ *we have:*

(i) $\int_0^T b_1(\mathbf{u}_k(.,t), \mathbf{u}_k(.,t), \lambda(t)\mathbf{v})dt$ *converges to* $\int_0^T b_1(\mathbf{u}(.,t), \mathbf{u}(.,t), \lambda(t)\mathbf{v})dt$

(ii) $\int_0^T b_2(\mathbf{u}_k(.,t), \psi_k(.,t), \lambda(t)\phi)dt$ *converges to* $\int_0^T b_2(\mathbf{u}(.,t), \psi(.,t), \lambda(t)\phi)dt.$

Proof. *(i)* By using Lemma 12.5 we can deduce that

$$
\int_0^T b_1(\mathbf{u}_k(.,t), \mathbf{u}_k(.,t), \lambda(t)\mathbf{v})dt = -\int_0^T b_1(\mathbf{u}_k(.,t), \lambda(t)\mathbf{v}, \mathbf{u}_k(.,t))dt.
$$

According to the assumptions of the lemma, the following convergence result holds:

$$\int_0^T b_1(\mathbf{u}_k(.,t), \lambda(t)\mathbf{v}, \mathbf{u}_k(.,t))\mathrm{d}t \longrightarrow \int_0^T b_1(\mathbf{u}(.,t), \lambda(t)\mathbf{v}, \mathbf{u}(.,t))\mathrm{d}t.$$

Consequently (by using again Lemma 12.5),

$$\int_0^T b_1(\mathbf{u}_k(.,t), \mathbf{u}_k(.,t), \lambda(t)\mathbf{v})\mathrm{d}t \longrightarrow \int_0^T b_1(\mathbf{u}(.,t), \mathbf{u}(.,t), \lambda(t)\mathbf{v})\mathrm{d}t$$

and then result (i) follows.

By using the same arguments as in (i), we obtain result (ii). □

12.2.2 Existence, Uniqueness and Regularity of the Solution

Theorem 12.7. *Let* $\mathbf{u_0}, \mathcal{T}_0, \mathcal{S}_0$ *and* $(\mathbf{f}, \varpi)$ *be given such that* $(\mathcal{T}_0, \mathcal{S}_0) \in H^1(\Omega)$, $\mathbf{u_0} \in \mathcal{V}_0$ *and* $(\mathbf{f}, \varpi) \in L^2(0,T; L^2(\Gamma_0))$. *Then there exists a solution* $(\mathbf{u}, \mathcal{T}, \mathcal{S})$ *of the problem (12.8) satisfying*

$$X = (\mathbf{u}, \mathcal{T}, \mathcal{S}) \in L^2(0,T;\mathcal{V}) \cap \mathcal{C}([0,T];\mathcal{H}) \text{ and } \frac{\partial X}{\partial t} \in L^2(0,T;\mathcal{V}').$$

Proof. The proof of this theorem is similar to the one used to obtain Theorem 2.1 in Belmiloudi [35]. The existence results are obtained by constructing approximate solutions of system (12.8) by the classical Faedo–Galerkin method, by using Lemma 12.3 and the estimates given in Lemma 12.4, and applying the limit. For applying the limit we use the convergence results of Lemma 12.6 and the compactness arguments. So, we omit the details. □

From now on we assume the following regularity for the surface-stress

$$\begin{aligned} \mathbf{f} \in U_f &= \{\mathbf{f} \in L^2(0,T;H^1(\Gamma_0)) : \frac{\partial \mathbf{f}}{\partial t} \in L^2(0,T;L^2(\Gamma_0))\}, \\ \varpi \in U_\varpi &= \{\varpi \in L^2(0,T;H^1(\Gamma_0)) : \frac{\partial \varpi}{\partial t} \in L^2(0,T;L^2(\Gamma_0))\}. \end{aligned} \tag{12.10}$$

The regularity (12.10) implies that $(\mathbf{f}, \varpi) \in \mathcal{C}([0,T]; L^2(\Gamma_0))$, a.e. on $[0,T]$ (see Lemma 6.6). Moreover, since the data $(\mathbf{f}, \varpi)$ must be consistent with the initial and boundary conditions imposed to the velocity $\mathbf{u}$ and temperature $\mathcal{T}$ on the open set $\Gamma_0 =]0, L_x[\times] - L_y, L_y[\times\{0\}$, we have to impose the following compatibility conditions

$$\begin{aligned} &(\mathbf{f}, \varpi)(t=0) = 0 \text{ on } \Gamma_0, \\ &f_2 = 0, \ \frac{\partial f_1}{\partial y} = 0, \ \varpi = 0 \text{ on } \gamma_3 \cup \gamma_4, \\ &(\mathbf{f}, \varpi)|_{\gamma_1} = (\mathbf{f}, \varpi)|_{\gamma_2}, \end{aligned} \tag{12.11}$$

where $\gamma = \cup_{i=1,4}\gamma_i$ denotes the boundary of Γ_0, with γ_1 and γ_2 the western and eastern boundaries, and γ_3 and γ_4 the southern and northern boundaries.

Theorem 12.8. *Assume that the data* $(\mathbf{f}, \varpi)$ *satisfies the conditions (12.10) and* $(\mathbf{f}, \varpi)(0) = 0$. *Then there exists* $R_\alpha > 0$ *such that if* $\| X_0 \| \leq R_\alpha$,[4] *where* $X_0 = (\mathbf{u}_0, \mathcal{T}_0, \mathcal{S}_0)$, *and if* $(\mathbf{f}, \varpi)$ *or the final time* T *is small, there exists a unique solution* $X = (\mathbf{u}, \mathcal{T}, \mathcal{S})$ *of the problem (12.8) such that*

$$\frac{\partial X}{\partial t} \in L^2(0, T; \mathcal{V}) \cap L^\infty(0, T; \mathcal{H}) \text{ and } X \in L^\infty(0, T; \mathcal{V}).$$

Proof. The proof of this theorem is similar to the one used to obtain Theorem 2.2 in Belmiloudi [35]. So, we omit the details (the proof is based on the classical Faedo–Galerkin method, Lemmas 12.3 and 12.6, the estimates given in Lemma 12.4 and the result of Theorem 12.7). □

Remark 12.9. (*i*) The existence, uniqueness and regularity given, for the non-linear and three-dimensional case, in Theorem 12.8 are still valid for the non-linear and two-dimensional case, and for the linear and three-dimensional case (see Belmiloudi [31]) *with no smallness assumptions on the data or on the final time.*
(*ii*) In order to obtain the uniqueness and the regularity of the solution $(\mathbf{u}, \mathcal{T}, \mathcal{S})$ of the non-linear and three-dimensional problem (12.8), the variability of the wind stress and the disturbance of heat stress on the surface of the ocean must be small enough. Physically, these conditions are not too restrictive, because they do not concern the total wind stress and heat stress on the surface, but concern only the variability of the wind stress and the small disturbance of the heat stress. ◊

The surface pressure is of great interest for oceanographic interpretation. To define it, it is necessary to have some regularity $H^\epsilon(\Omega)$, for $\epsilon > 0$. The following theorem shows that the pressure is in $H^1(\Omega)$.

Theorem 12.10. *Under the assumptions of Theorem 12.8, assume moreover that* X_0 *is in* $(H^2(\Omega) \cap \mathcal{V}_0) \times H^2(\Omega) \times H^2(\Omega)$, *and that the data* $(\mathbf{f}, \varpi)$ *satisfies assumptions (12.10) and (12.11), then the solution* $(\mathbf{u}, \mathcal{T}, \mathcal{S}, p) = (X, p)$ *of problem (12.9) is such that*

$$X \in L^2(0, T; H^2(\Omega)), \;\; p \in L^2(0, T; H^1(\Omega)).$$

Proof. The shape of the oceanic domain Ω, and especially the presence of corners, prevents us from applying a standard theorem of regularity. We use an extension method, with even-odd reflection, to prove this result. The idea is the following: by using the classical arguments (see, *e.g.*, Agmon *et al.* [5]) we can obtain the regularity on the open set Ω except near the corners. For the regularity in the corners we proceed in two steps. First, since the solution is periodic in the x-direction, the open set Ω can be extended from $x = -L_x$ to $x = 2L_x$ and the regularity near the western and eastern boundaries Γ_1 and Γ_2 is automatically obtained; second we define $(\tilde{\mathbf{u}}, \tilde{p}, \tilde{\mathcal{T}}, \tilde{\mathcal{S}})$,

[4] R_α depends on the coefficients $\nu_{i\mathrm{h}}, \nu_{i\mathrm{v}}$, for $i = 1, 3$.

extension of $(\mathbf{u}, p, \mathcal{T}, \mathcal{S})$ in $\tilde{\Omega} =]0, L_x[\times] - 2L_y, 2L_y[\times] - H, 0[$ such that $(\tilde{\mathbf{u}}, \tilde{p}, \tilde{\mathcal{T}}, \tilde{\mathcal{S}}) = (\mathbf{u}, p, \mathcal{T}, \mathcal{S})$ on Ω and $(\tilde{\mathbf{u}}, \tilde{p}, \tilde{\mathcal{T}}, \tilde{\mathcal{S}})$ verifies the equations and boundary conditions similar to (12.3),(12.4) and (12.5). Then we can apply the regularity result for $(\tilde{\mathbf{u}}, \tilde{p}, \tilde{\mathcal{T}}, \tilde{\mathcal{S}})$ on $\tilde{\Omega}$, except near the corners of $\tilde{\Omega}$. This implies the regularity of $(\mathbf{u}, p, \mathcal{T}, \mathcal{S})$ near the corners of Ω. For more details the reader can be referred to Belmiloudi *et al.* [31, 28]. So, we omit the details.

□

Remark 12.11. (*i*) Since the pressure is in $H^1(\Omega)$, we can now define the trace of p on the boundary Γ. The pressure p is defined regardless of any time-dependent function. This function can be fixed for example by setting the condition

$$\frac{1}{\text{mes}\Gamma_0} \int_{\Gamma_0} p \mathrm{d}\Gamma = p_{\mathrm{d}} \quad \text{where } p_{\mathrm{d}} \text{ is given.} \tag{12.12}$$

(*ii*) According to the first equation of System (12.8) and Green's formula, we can deduce that $p|_{\Gamma_1} = p|_{\Gamma_2}$.

(*iii*) According to Theorems 12.8 and 12.10, we have that

$$\begin{aligned} &X \in L^2(0,T; H^2(\Omega)) \cap L^\infty(0,T;\mathcal{V}),\\ &\frac{\partial X}{\partial t} \in L^2(0,T;\mathcal{V}) \cap L^\infty(0,T;\mathcal{H}),\\ &p \in L^2(0,T;H^1(\Omega)) \end{aligned} \tag{12.13}$$

and then, according to Lemma 6.6, that

$$X \in \mathcal{C}([0,T];\mathcal{V}) \cap H^{2,1}(\mathcal{Q}) \text{ and } \frac{\partial X}{\partial t} \in \mathcal{C}([0,T];\mathcal{H}).$$

◇

Proposition 12.12. *Under the assumptions of Theorem 12.8, assume moreover that X_0 is in $(H^2(\Omega) \cap \mathcal{V}_0) \times H^2(\Omega) \times H^2(\Omega)$. Let $(\mathbf{f}_1, \varpi_1)$ and $(\mathbf{f}_2, \varpi_2)$ be two functions in $U_f \times U_\varpi$ such that (12.11) holds. If $(\mathbf{u}_i, \mathcal{T}_i, \mathcal{S}_i, p_i) = (X_i, p_i)$ is the solution of problem (12.9), corresponding to the forcing $(\mathbf{f}_i, \varpi_i)$, for $i = 1, 2$, then the following estimate holds:*

$$\| X \|^2_{H^{2,1}(\mathcal{Q})} + \| p \|^2_{L^2(0,T;H^1(\Omega))} \leq C(\| \mathbf{f} \|^2_{U_f} + \| \varpi \|^2_{U_\varpi}), \tag{12.14}$$

where $X = X_1 - X_2$, $p = p_1 - p_2$, $\mathbf{f} = \mathbf{f}_1 - \mathbf{f}_2$ and $\varpi = \varpi_1 - \varpi_2$.

Proof. By using a similar technique as used in Belmiloudi [31] (see also, for the more general case, Agmon *et al.* [5]) we can obtain the desired estimate. So, we omit the details. □

Remark 12.13. (*i*) In a more general case for the desired target $(X_0, p_0) = (\mathbf{u_0}, \mathcal{T}_0, \mathcal{S}_0, p_0)$, *i.e.*, if we assume that the target (X_0, p_0) satisfies the complete evolution Equations (12.1) and such that

$$X_0 \in \mathcal{C}([0,T];\mathcal{V}) \cap H^{2,1}(\mathcal{Q}), \quad \frac{\partial X_0}{\partial t} \in \mathcal{C}([0,T];\mathcal{H}) \text{ and } p_0 \in L^2(0,T;H^1(\Omega)),$$

then the previous arguments and results (in particular the results of Theorem 12.10 and Proposition 12.14) extend directly to the present situation without further estimates.

(ii) We can consider other domains Ω and other boundary conditions, the previous arguments can be extended to these new situations. ◊

The following theorem shows more regularity for the pressure, which can be useful in the case of the use of the pressure, for example, as observation in robust control problems.

Theorem 12.14. *Under the assumptions of Theorem 12.10, we have that* $\partial p/\partial t \in L^2(\mathcal{Q})$.

Proof. Since $X = (\mathbf{u}, \mathcal{T}, \mathcal{S})$ satisfies $X \in L^2(0,T;H^2(\Omega)) \cap L^\infty(0,T;\mathcal{V})$, we can deduce that $((\partial\mathbf{u}/\partial t)\nabla)(\mathbf{u}+\mathbf{u_0})$, $((\mathbf{u}+\mathbf{u_0})\nabla)(\partial\mathbf{u}/\partial t)$ and $(-\delta_{\rm T}(\partial\mathcal{T}/\partial t)+\delta_{\rm S}(\partial\mathcal{S}/\partial t))\mathbf{G}$ are in $L^2(0,T;(H^1(\Omega))')$. Consequently, by differentiating the first equation of System (12.8) and by using the regularity of $\partial\mathbf{f}/\partial t$ (*i.e.*, $\partial\mathbf{f}/\partial t \in L^2(0,T;L^2(\Gamma_0))$,[5] we conclude that $\partial^2\mathbf{u}/\partial t^2 \in L^2(0,T;(H^1(\Omega))')$.

Finally, the first equation of (12.3) written in $L^2(0,T;(H^1(\Omega))')$ can be differentiated in $\mathcal{D}'(0,T;(H^1(\Omega))')$ and we obtain that

$$\nabla(\frac{\partial p}{\partial t}) = -\frac{\partial^2\mathbf{u}}{\partial t^2} - ((\mathbf{u}+\mathbf{u_0}).\nabla)\frac{\partial\mathbf{u}}{\partial t} - (\frac{\partial\mathbf{u}}{\partial t}.\nabla)(\mathbf{u}+\mathbf{u_0}) - \mathbf{F}\wedge\frac{\partial\mathbf{u}}{\partial t}$$
$$+\nu_{1\rm h}\Delta_2(\frac{\partial\mathbf{u}}{\partial t}) + \frac{\partial}{\partial z}(\nu_{1\rm v}\frac{\partial}{\partial z}(\frac{\partial\mathbf{u}}{\partial t})))(-\delta_{\rm T}\tilde{\mathcal{T}} + \delta_{\rm S}\tilde{\mathcal{S}})\mathbf{G},$$

in $\mathcal{D}'(0,T;(H^1(\Omega))')$.

The regularity of the right-hand side obtained before, yields $\nabla(\partial p/\partial t) \in L^2(0,T;(H^1(\Omega))')$ and then the result of the theorem.

This completes the proof. □

We finish this section by the following comments, which gives some asymptotic behavior of the perturbation under extra assumptions.

12.2.3 Comments on the Asymptotic Behavior

In order to give the stability result, we assume that there exist sufficiently regular functions $(\mathbf{g}, \varrho)$ on $\mathcal{Q}$ such that[6]

$$(\mathbf{f},\mathbf{v})_{\Gamma_0} = (\mathbf{g},\mathbf{v}), \ (\varpi,\mathbf{v})_{\Gamma_0} = (\varrho,\phi) \quad \text{for all } (\mathbf{v},\phi) \in \mathcal{V}_1 \times \mathcal{V}_2,$$

[5] The function $\mathbf{v} \longrightarrow (\partial\mathbf{f}/\partial t, \mathbf{v})_{\Gamma_0}$ is a linear continuous application from $\mathcal{V}_1$ into $\mathbb{R}$, which implies that there exists $\mathbf{g} \in \mathcal{V}_1'$ such that $(\partial\mathbf{f}/\partial t,\mathbf{v})_{\Gamma_0} = \langle\mathbf{g},\mathbf{v}\rangle_{\mathcal{V}_1',\mathcal{V}_1}$.

[6] $\mathbf{g}$ and ϱ can represent the wind stress and the heat stress, respectively, which can be considered as body forces.

the current $(\mathbf{u_0}, \mathcal{T}_0, \mathcal{S}_0)$ is given and real-valued, the functional spaces $\mathcal{V}_{i,i=1,2}$, $\mathcal{H}_{i,i=1,2}$, $\mathcal{V}$, $\mathcal{H}$ and all Sobolev spaces, will be considered as complex spaces and finally we introduce the following forms:

$$h_{1\mathrm{T}}(\phi, \mathbf{v}) = (-\delta_{\mathrm{T}}\phi\mathbf{G}, \mathbf{v}), \quad h_{1\mathrm{S}}(\psi, \mathbf{v}) = (\delta_{\mathrm{S}}\psi\mathbf{G}, \mathbf{v}),$$
$$h_{2\mathrm{T}}(\mathbf{v}, \phi) = b_2(\mathbf{v}, \mathcal{T}_0, \psi), \quad h_{2\mathrm{S}}(\mathbf{v}, \psi) = b_2(\mathbf{v}, \mathcal{S}_0, \psi),$$
$$\tilde{a}_1(\mathbf{u}, \mathbf{v}) = a_1(\mathbf{u}, \mathbf{v}) + l_0(\mathbf{u}, \mathbf{v}),$$
$$\tilde{a}_2(\mathcal{T}, \phi) = a_2(\mathcal{T}, \phi) + ((\mathbf{u}_0\nabla)\mathcal{T}, \phi), \quad \tilde{a}_3(\mathcal{S}, \psi) = a_3(\mathcal{S}, \psi) + ((\mathbf{u}_0\nabla)\mathcal{S}, \psi),$$
$$\tilde{a}(X, Y) = \tilde{a}_1(\mathbf{u}, \mathbf{v}) + \tilde{a}_2(\mathcal{T}, \phi) + \tilde{a}_3(\mathcal{S}, \psi), \text{ with } X = (\mathbf{u}, \mathcal{T}, \mathcal{S}), Y = (\mathbf{v}, \phi, \psi).$$

Now we formulate the problem (12.8) in the abstract form useful for the mathematical analysis. For this let us consider the operators $\tilde{A}_i$ associated with the forms $\tilde{a}_i$, for $i = 1, 3$, respectively and A_i the operators associated with the forms a_i for $i = 1, 3$, respectively. By noting P the well-known Leray–Hopf projector which is the orthogonal projector of $L^2(\Omega)$ onto the divergence-free space $\mathcal{H}_1$, we have then

$$\tilde{A}_1\mathbf{u} = A_1\mathbf{u} + L_1\mathbf{u}, \tilde{A}_2\mathcal{T} = A_2\mathcal{T} + L_2\mathcal{T} \text{ and } \tilde{A}_3\mathcal{S} = A_3\mathcal{S} + L_3\mathcal{S},$$

where

$$L_1\mathbf{u} = P((\mathbf{u_0}\nabla)\mathbf{u} + (\mathbf{u}\nabla)\mathbf{u_0} + \mathbf{F} \wedge \mathbf{u}), L_2\mathcal{T} = (\mathbf{u_0}\nabla)\mathcal{T} \text{ and } L_3\mathcal{S} = (\mathbf{u_0}\nabla)\mathcal{S}.$$

Let $H_{i\mathrm{T}}$ and $H_{i\mathrm{S}}$ be the operators associated with the forms $h_{i\mathrm{T}}$ and $h_{i\mathrm{S}}$, for $i = 1, 2$, respectively.

$$\text{Let } \mathcal{A} = \begin{pmatrix} A_1 & 0 & 0 \\ 0 & A_2 & 0 \\ 0 & 0 & A_3 \end{pmatrix}, \mathcal{C} = \begin{pmatrix} L_1 & H_{1\mathrm{T}} & H_{1\mathrm{S}} \\ H_{2\mathrm{T}} & L_2 & 0 \\ H_{2\mathrm{S}} & 0 & L_3 \end{pmatrix}, \tilde{\mathcal{A}} = \begin{pmatrix} \tilde{A}_1 & H_{1\mathrm{T}} & H_{1\mathrm{S}} \\ H_{2\mathrm{T}} & \tilde{A}_2 & 0 \\ H_{2\mathrm{S}} & 0 & \tilde{A}_3 \end{pmatrix}.$$

The domain of the operator $\tilde{A}_1$ is defined by: $D_1 = H^2(\Omega) \cap \mathcal{V}_1$ and the domain of the operator $\tilde{A}_2$ (respectively $\tilde{A}_3$) is defined by: $D_2 = H^2(\Omega) \cap \mathcal{V}_2$. Consequently, the domain of the operator $\tilde{\mathcal{A}}$ is defined by: $\mathcal{D} = D_1 \times D_2 \times D_2$. We can easily verify that System (12.8), can be written as the following Cauchy differential form:

$$\begin{aligned} &\frac{\mathrm{d}X}{\mathrm{d}t} + \tilde{\mathcal{A}}X = \tilde{\mathbf{g}}, \\ &X(t=0) = 0, \end{aligned} \tag{12.15}$$

where $X = (\mathbf{u}, \mathcal{T}, \mathcal{S})$, $\tilde{\mathbf{g}} = (\mathbf{g}_1, g_2, g_3)$, $\mathbf{g}_1 = \mathbf{g} - (\mathbf{u}\nabla)\mathbf{u}$, $g_2 = \varrho - (\mathbf{u}\nabla)\mathcal{T}$ and $g_3 = -(\mathbf{u}\nabla)\mathcal{S}$.

By using similar techniques as used in Belmiloudi [32], which are based on a method introduced by Prodi [245], the following asymptotic result holds:

If the wind stress $\mathbf{f}$ *and the heat stress* ϖ *act only during a finite time and are small enough, and if the eigenvalues* $\lambda_{\tilde{\mathcal{A}}}$ *of the operator* $\tilde{\mathcal{A}}$, *which are situated inside a parabola of the complex plane, have positive real parts, then, the perturbation* $X = (\mathbf{u}, \mathcal{T}, \mathcal{S})$ *tends to 0 as time tends to* $+\infty$ *and* X *is a strong solution on* $[0, +\infty[$ *of the differential problem (12.15).*

The assumptions of the previous result depend on the initial situation $(\mathbf{u}_0, \mathcal{T}_0, \mathcal{S}_0)$ and on the eddy viscosity and diffusivity (which can be deduced from the characteristics of the initial situation $(\mathbf{u}_0, \mathcal{T}_0, \mathcal{S}_0)$, see Philander and Pacanowski [238]).

12.3 Robust Control Problems

In the control framework, our objective is to find the best estimate of the variability of the wind stress $\mathbf{f}$ in the presence of the worst disturbance in the heat stress ϖ which maximally spoils the control objective. We assume then the control is in the variability of the wind stress and the disturbance is in heat stress in the context of non-cooperative game discussed in Chapter 8. Thus, we write $\mathbf{f}$ and ϖ as

$$\mathbf{f} = \mathcal{B}_1\xi, \quad \mathbf{f}_{\mathrm{T}} = \mathcal{B}_{1\mathrm{T}}\xi = (\mathcal{B}_1\xi, 0), \quad \varpi = \mathcal{B}_2\eta, \tag{12.16}$$

where $\mathcal{B}_1$ (respectively $\mathcal{B}_2$) is taken here as a given linear continuous and bounded operator from $L^2(\Sigma_0)$ (respectively $L^2(\Sigma_0)$) into U_f (respectively U_ϖ), with $\Sigma_0 = \Gamma_0 \times (0, T)$.

The function $(\mathbf{u}, \mathcal{T}, \mathcal{S})$ is assumed to be related to the disturbance η and control ξ through the problem (12.8):

$$\begin{aligned}
&(\frac{\partial \mathbf{u}}{\partial t}, \mathbf{v}) + a_1(\mathbf{u}, \mathbf{v}) + l_0(\mathbf{u}, \mathbf{v}) + l_1(\mathcal{T}, \mathcal{S}, \mathbf{v}) + b_1(\mathbf{u}, \mathbf{u}, \mathbf{v}) \\
&\qquad\qquad = (\mathcal{B}_{1\mathrm{T}}\xi, \mathbf{v})_{\Gamma_0}, \\
&(\frac{\partial \mathcal{T}}{\partial t}, \phi) + a_2(\mathcal{T}, \phi) + l_2(\mathbf{u}, \mathcal{T}, \phi) + b_2(\mathbf{u}, \mathcal{T}, \phi) = (\mathcal{B}_2\eta, \phi)_{\Gamma_0}, \\
&(\frac{\partial \mathcal{S}}{\partial t}, \psi) + a_3(\mathcal{S}, \psi) + l_3(\mathbf{u}, \mathcal{S}, \psi) + b_2(\mathbf{u}, \mathcal{S}, \psi) = 0, \\
&(\mathbf{u}, \mathcal{T}, \mathcal{S})(t = 0) = 0,
\end{aligned} \tag{12.17}$$

for all $(\mathbf{v}, \phi, \psi)$ in $\mathcal{V}$ and a.e. in $(0, T)$.

The cost functional considered in the present work is of the form

$$\begin{aligned}
J(\xi, \eta) = &\frac{1}{2}\iint_Q |\, \mathcal{C}(\mathbf{u} - \mathbf{u}_{\mathrm{obs}}) \,|^2 \, dxdt \\
&+ \frac{\alpha}{2}\iint_{\Sigma_0} |\, \xi \,|^2 \, dxdt - \frac{\beta}{2}\iint_{\Sigma_0} |\, \eta \,|^2 \, dxdt
\end{aligned} \tag{12.18}$$

where $\alpha, \beta > 0$ are fixed, $\mathbf{u}_{\mathrm{obs}} \in L^2(0, T; \mathcal{V}_1)$ is the observation and $\mathcal{C}$ is unbounded, linear operator on $L^2(\Omega)$ satisfying the condition (7.91) (here $\mathcal{V}_1$ plays the role of $\mathcal{D}$). In particular, we can consider the following two cases:

(*i*) $\mathcal{C}(.) = a * Id(.)$ which corresponds to the regulation of turbulent kinetic energy (*ii*) $\mathcal{C}(.) = a * \nabla \times \; . = a * \text{curl}(.)$ which corresponds to the regulation of the vorticity.

The robust control problem is then to minimize the functional J with respect to ξ and maximize J with respect to η, *i.e.*,

$$\begin{aligned} &\text{find } (\xi^*, \eta^*) \in \mathcal{U}_{\text{ad}} \times \mathcal{V}_{\text{ad}} \text{ such that} \\ &J(\xi^*, \eta) \leq J(\xi^*, \eta^*) \leq J(\xi, \eta^*), \forall (\xi, \eta) \in \mathcal{U}_{\text{ad}} \times \mathcal{V}_{\text{ad}}, \end{aligned} \tag{12.19}$$

where $\mathcal{U}_{\text{ad}}$ (respectively $\mathcal{V}_{\text{ad}}$) is a given non-empty, closed, convex and bounded subsets of vector space $L^2(\Sigma_0)$ (respectively scalar space $L^2(\Sigma_0)$).

We are now going to show the differentiability result of the operator solution of problem (12.17) $\mathcal{F} : (\xi, \eta) \longrightarrow X = (\mathbf{u}, \mathcal{T}, \mathcal{S})$, which maps the source term $(\xi, \eta) \in L^2(\Sigma_0) \times L^2(\Sigma_0)$ into the corresponding solution $X \in \mathcal{Z}$, where

$$\mathcal{Z} = \{Y \in L^2(0, T; \mathcal{V}) \cap \mathcal{C}([0, T]; \mathcal{H}) : \frac{\partial Y}{\partial t} \in L^2(0, T; \mathcal{V}')\}.$$

12.3.1 Differentiability of the Operator Solution

Proposition 12.15. *Under the assumptions of Proposition 12.12, the operator solution $\mathcal{F}$ is continuously F-differentiable from $L^2(\Sigma_0) \times L^2(\Sigma_0)$ to $\mathcal{Z}$ where the derivative $\mathcal{F}'(\xi, \eta) : (\mathbf{h}, \kappa) \longrightarrow (\mathbf{w}, \theta, \vartheta)$ is a linear mapping from $L^2(\Sigma_0) \times L^2(\Sigma_0)$ to $\mathcal{Z}$ such that $\mathcal{F}'(\xi, \eta).(\mathbf{h}, \kappa) = (\mathbf{w}, \theta, \vartheta)$ satisfies, a.e. in $(0, T)$ and for all $(\mathbf{v}, \phi, \psi) \in \mathcal{V}$*

$$\begin{aligned} &(\frac{\partial \mathbf{w}}{\partial t}, \mathbf{v}) + a_1(\mathbf{w}, \mathbf{v}) + d(\mathbf{w}, \mathbf{v}) + l_1(\theta, \vartheta, \mathbf{v}) + b_1(\mathbf{u}_1, \mathbf{w}, \mathbf{v}) \\ &\qquad\qquad + b_1(\mathbf{w}, \mathbf{u}_1, \mathbf{v}) = (\mathcal{B}_{1\mathrm{T}}\mathbf{h}, \mathbf{v})_{\Gamma_0}, \\ &(\frac{\partial \theta}{\partial t}, \phi) + a_2(\theta, \phi) + b_2(\mathbf{w}, \mathcal{T}_1, \phi) + b_2(\mathbf{u}_1, \theta, \phi) = (\mathcal{B}_2\kappa, \phi)_{\Gamma_0}, \\ &(\frac{\partial \vartheta}{\partial t}, \psi) + a_3(\vartheta, \psi) + b_2(\mathbf{w}, \mathcal{S}_1, \psi) + b_2(\mathbf{u}_1, \vartheta, \psi) = 0, \\ &(\mathbf{w}, \theta, \vartheta)(t = 0) = 0, \end{aligned} \tag{12.20}$$

where $(\mathbf{u}, \mathcal{T}, \mathcal{S}) = \mathcal{F}(\xi, \eta)$ is the solution of (12.17), $\mathbf{u}_1 = \mathbf{u} + \mathbf{u}_0$, $\mathcal{T}_1 = \mathcal{T} + \mathcal{T}_0$ and $\mathcal{S}_1 = \mathcal{S} + \mathcal{S}_0$.

Moreover, for any $(\xi_i, \eta_i) \in L^2(\Sigma_0) \times L^2(\Sigma_0)$, for $i = 1, 2$, we have the following estimates:

(*i*) $\| \mathcal{F}'(\xi_1, \eta_1) \|_{\mathcal{L}(L^2(\Sigma) \times L^2(\Sigma); \mathcal{Z})} \leq C$

(*ii*) $\| \mathcal{F}'(\xi_1, \eta_1) - \mathcal{F}'(\xi_2, \eta_2) \|_{\mathcal{L}(L^2(\Sigma) \times L^2(\Sigma); \mathcal{Z})} \leq C(\| \xi \|^2_{L^2(\Sigma)} + \| \eta \|^2_{L^2(\Sigma)})$,

where $\xi = \xi_1 - \xi_2$ and $\eta = \eta_1 - \eta_2$.

Proof. The existence of the F-derivative and its characterization by (12.20) as well as estimates (i) and (ii) can be obtained by using a technique similar to that employed in Section 8.6 and by using the estimates given in Lemma 12.4, the results of Lemma 12.3 and the regularity of the solution X. So, we omit the details. □

Remark 12.16. (i) A similar result concerning the existence of the F-derivatives for Navier–Stokes system can be found, *e.g.*, in Abergel and Temam [1].
(ii) System (12.20) is the weak formulation of the following system (on $\mathcal{Q}$):

$$\begin{aligned}
&\text{find } (\mathbf{w}, \pi, \theta, \vartheta) \text{ such that}\\
&\frac{\partial \mathbf{w}}{\partial t} + (\mathbf{w}.\nabla)\mathbf{u}_1 + (\mathbf{u}_1.\nabla)\mathbf{w} + \mathbf{F}\wedge\mathbf{w}\\
&\qquad -\nu_{1\mathrm{h}}\Delta_2\mathbf{w} - \frac{\partial}{\partial z}(\nu_{1\mathrm{v}}\frac{\partial \mathbf{w}}{\partial z}) + \nabla\pi = (-\delta_{\mathrm{T}}\theta + \delta_{\mathrm{S}}\vartheta)\mathbf{G},\\
&\mathrm{div}(\mathbf{w}) = 0,\\
&\frac{\partial \theta}{\partial t} + (\mathbf{w}.\nabla)\mathcal{T}_1 + (\mathbf{u}_1.\nabla)\theta - \nu_{2\mathrm{h}}\Delta_2\theta - \frac{\partial}{\partial z}(\nu_{2\mathrm{v}}\frac{\partial \theta}{\partial z}) = 0,\\
&\frac{\partial \vartheta}{\partial t} + (\mathbf{w}.\nabla)\mathcal{S}_1 + (\mathbf{u}_1.\nabla)\vartheta - \nu_{3\mathrm{h}}\Delta_2\vartheta - \frac{\partial}{\partial z}(\nu_{3\mathrm{v}}\frac{\partial \vartheta}{\partial z}) = 0,\\
&\text{with the initial condition}\\
&(\mathbf{w}, \theta, \vartheta)(t=0) = 0,\\
&\text{subject to the boundary conditions (12.4) and (12.5) where}\\
&\mathcal{B}_{1\mathrm{T}}\mathbf{h} \text{ (respectively } \mathcal{B}_2\kappa) \text{ plays the role of } \mathbf{f} \text{ (respectively } \varpi),
\end{aligned} \tag{12.21}$$

with $\mathbf{u}_1 = \mathbf{u} + \mathbf{u}_0$, $\mathcal{T}_1 = \mathcal{T} + \mathcal{T}_0$ and $\mathcal{S}_1 = \mathcal{S} + \mathcal{S}_0$. ◇

Concerning the system of Equations (12.21) (and then (12.20)) we have the following result.

Theorem 12.17. *Let* (ξ, η), $(\mathbf{h}, \kappa) \in L^2(\Sigma_0) \times L^2(\Sigma_0)$ *and* $(\mathbf{u}, \mathcal{T}, \mathcal{S}) = \mathcal{F}(\xi, \eta)$ *be the solution of the non-linear problem (12.17). Then there exists a unique* $Y = (\mathbf{w}, \theta, \vartheta) \in \mathcal{Z}$ *and* $\pi \in \mathcal{D}'(\mathcal{Q})$ *(up to the addition of a distribution in* $(0,T)$*) solution of the linear system (12.20). Moreover, if* $(\mathbf{h}_1, \kappa_1)$ *and* $(\mathbf{h}_2, \kappa_2)$ *are two functions in* $L^2(\Sigma_0) \times L^2(\Sigma_0)$ *and* $(\mathbf{w}_i, \theta_i, \vartheta_i, \pi_i) = (Y_i, \pi_i)$, *for i=1,2, are weak solutions of problem (12.20), corresponding to the forcing* $(\mathbf{h}_i, \kappa_i)$, *respectively, the following estimate holds.*

$$\| Y \|_{\mathcal{Z}}^2 \leq C(\| \mathbf{h} \|_{L^2(\Sigma)}^2 + \| \kappa \|_{L^2(\Sigma)}^2) \tag{12.22}$$

where $Y = Y_1 - Y_2$, $\mathbf{h} = \mathbf{h}_1 - \mathbf{h}_2$ *and* $\kappa = \kappa_1 - \kappa_2$.

Proof. The existence of solutions $Y = (\mathbf{w}, \theta, \vartheta) \in \mathcal{Z}$ of the linear system (12.20) can be proved by using the classical Faedo–Galerkin method as in the existence of the problem (12.8). The uniqueness of solutions can be obtained by the standard energy estimate arguments. For the proof of the estimate (12.22) we take $(\mathbf{v}, \phi, \psi) - (\mathbf{w}, \theta, \vartheta)$ as the test function in (12.20), integrate over $(0, t)$, for all t in $(0, T)$ and use the classical technique to obtain some energy inequalities. The existence of the pressure can be obtained by using the standard technique to those for the Navier–Stokes equations (see, *e.g.*, Temam [281]). Therefore we omit the details. □

We are now going to study the existence of a solution for the robust control problem.

12.3.2 Existence of an Optimal Solution

Theorem 12.18. *For α and β sufficiently large (i.e., there exists (α_l, β_l) such that $\alpha \geq \alpha_l$ and $\beta \geq \beta_l$) there exist $(\xi^*, \eta^*) \in \mathcal{U} = \mathcal{U}_{\rm ad} \times \mathcal{V}_{\rm ad}$ and $X^* = (\mathbf{u}^*, \mathcal{T}^*, \mathcal{S}^*) \in H^{2,1}(\mathcal{Q}) \cap L^\infty(0, T; \mathcal{V})$ such that (ξ^*, η^*) satisfies (12.19) and $(\mathbf{u}^*, \mathcal{T}^*, \mathcal{S}^*) = \mathcal{F}(\xi^*, \eta^*)$ is a solution of the primal problem (12.17).*

Proof. Let P_η be the map: $\xi \longrightarrow J(\xi, \eta)$ and Q_ξ be the map: $\eta \longrightarrow J(\xi, \eta)$. To obtain the existence of the robust control problem we prove that P_η is convex and lower semi-continuous for all $\eta \in \mathcal{V}_{\rm ad}$, and Q_ξ is concave and upper semi-continuous for all $\xi \in \mathcal{U}_{\rm ad}$ and we use the minimax duality theorems in infinite dimensions presented in Chapter 5.

First we prove, for α and β sufficiently large, the convexity of the map P_η and the concavity of the map Q_ξ. In order to prove the convexity, it is sufficient to show that for all $(\xi_1, \xi_2) \in \mathcal{U}_{\rm ad}$ we have
$(P'_\eta(\xi_1) - P'_\eta(\xi_2)).\xi \geq 0$, where $\xi = \xi_1 - \xi_2$ (because P_η is G-differentiable). According to the definition of J, we have that

$$
\begin{aligned}
&(P'_\eta(\xi_1) - P'_\eta(\xi_2)).\xi \\
&\quad = \alpha \parallel \xi \parallel^2_{L^2(\Sigma)} + \int_0^T (\mathcal{C}(\mathbf{u}_1 - \mathbf{u}_2), \mathcal{C}\mathbf{w}_2)_{L^2(\Omega)} \mathrm{d}t \\
&\qquad + \int_0^T (\mathcal{C}(\mathbf{u}_1 - \mathbf{u}_{\rm obs}), \mathcal{C}(\mathbf{w}_1 - \mathbf{w}_2))_{L^2(\Omega)} \mathrm{d}t,
\end{aligned}
\tag{12.23}
$$

where $(\mathbf{u}_i, \mathcal{T}_i, \mathcal{S}_i) = \mathcal{F}(\xi_i, \eta)$ and $(\eta_i, w_i) = (\partial \mathcal{F}/\partial \xi)(\xi_i, \eta).\xi - \mathcal{F}'(\xi_i, \eta).(\xi, 0)$, the solution of the problem (12.20), for $i = 1, 2$.

According to Proposition 12.12, Proposition 12.15 and the relation (7.91), satisfied by the operator $\mathcal{C}$, we can deduce that there exists a constant $C_0 > 0$ such that[7]

[7] The constant C_0 depends on the given data: the steady mean circulation X_0, the final time T, the domain Ω, the observation $\mathbf{u}_{\rm obs}$ and the coefficients of horizontal and vertical eddy viscosity and diffusivity $\nu_{i\rm h}$, $\nu_{i\rm v}$, for $i = 1, 3$.

$$\int_0^T (\mathcal{C}(\mathbf{u}_1 - \mathbf{u}_2), \mathcal{C}\mathbf{w}_2)_{L^2(\Omega)} \mathrm{d}t \\ + \int_0^T (\mathcal{C}(\mathbf{u}_1 - \mathbf{u}_{\mathrm{obs}}), \mathcal{C}(\mathbf{w}_1 - \mathbf{w}_2))_{L^2(\Omega)} \mathrm{d}t \leq C_0 \parallel \xi \parallel^2_{L^2(\Sigma)} . \tag{12.24}$$

From (12.23) and (12.24) we deduce that for $\alpha \geq \alpha_l$ such that $\alpha_l \geq C_0$ we have

$$(P'_\eta(\xi_1) - P'_\eta(\xi_2)).\xi \geq (\alpha - C_0) \parallel \xi \parallel^2_{L^2(\Sigma)} \geq (\alpha_l - C_0) \parallel \xi \parallel^2_{L^2(\Sigma)} \geq 0$$

and then the convexity of P_η. In the same way, we can find β_l such that for $\beta \geq \beta_l$ we have the concavity of Q_ξ.

We shall now prove that P_η is lower semi-continuous for all $\eta \in \mathcal{V}_{\mathrm{ad}}$, and Q_ξ is upper semi-continuous for all $\xi \in \mathcal{U}_{\mathrm{ad}}$. Let ξ_k be a minimizing sequence of J, *i.e.*, $\liminf_k J(\xi_k, \eta) = \min_{\xi \in \mathcal{U}_{\mathrm{ad}}} J(\xi, \eta)$ ($\forall \eta \in \mathcal{V}_{\mathrm{ad}}$). Then ξ_k is uniformly bounded in $\mathcal{U}_{\mathrm{ad}}$ and we can extract from ξ_k a subsequence also denoted by ξ_k such that $\xi_k \rightharpoonup \xi_\eta$ weakly in $\mathcal{U}_{\mathrm{ad}}$. Let us consider $(\mathbf{u}_k, \mathcal{T}_k, \mathcal{S}_k) = \mathcal{F}(\xi_k, \eta)$ which is the solution of (12.17), *i.e.*,

$$\begin{aligned}
&(\frac{\partial \mathbf{u}_k}{\partial t}, \mathbf{v}) + a_1(\mathbf{u}_k, \mathbf{v}) + l_0(\mathbf{u}_k, \mathbf{v}) + l_1(\mathcal{T}_k, \mathcal{S}_k, \mathbf{v}) \\
&\qquad\qquad + b_1(\mathbf{u}_k, \mathbf{u}_k, \mathbf{v}) = (\mathcal{B}_{1\mathrm{T}}\xi_k, \mathbf{v})_{\Gamma_0}, \\
&(\frac{\partial \mathcal{T}_k}{\partial t}, \phi) + a_2(\mathcal{T}_k, \phi) + l_2(\mathbf{u}_k, \mathcal{T}_k, \phi) + b_2(\mathbf{u}_k, \mathcal{T}_k, \phi) = (\mathcal{B}_2\eta, \phi)_{\Gamma_0}, \\
&(\frac{\partial \mathcal{S}_k}{\partial t}, \psi) + a_3(\mathcal{S}_k, \psi) + l_3(\mathbf{u}_k, \mathcal{S}_k, \psi) + b_2(\mathbf{u}_k, \mathcal{S}_k, \psi) = 0, \\
&(\mathbf{u}_k, \mathcal{T}_k, \mathcal{S}_k)(t = 0) = 0.
\end{aligned} \tag{12.25}$$

By taking $(\mathbf{v}, \phi, \psi) = (\mathbf{u}_k, \mathcal{T}_k, \mathcal{S}_k)$ as the test function in (12.25), integrating over $(0, t)$, for all t in $(0, T)$ and using the classical technique we obtain that the sequence $X_k = (\mathbf{u}_k, \mathcal{T}_k, \mathcal{S}_k)$ is uniformly bounded in $L^\infty(0, T; \mathcal{H}) \cap L^2(0, T; \mathcal{V})$ and $\partial X_k / \partial t$ is uniformly bounded in $L^2(0, T; \mathcal{V}')$. This result, the compactness results and the fact that ξ_k is uniformly bounded in $\mathcal{U}_{\mathrm{ad}}$ makes it possible to extract from (X_k, ξ_k) a subsequence also denoted by (X_k, ξ_k) such that

$$\begin{aligned}
&\xi_k \rightharpoonup \xi_\eta \text{weakly in } \mathcal{U}_{\mathrm{ad}}, \\
&X_k \rightharpoonup X_\eta = (\mathbf{u}_\eta, \mathcal{T}_\eta, \mathcal{S}_\eta) \text{ weakly in } L^2(0, T; \mathcal{V}), \\
&X_k \longrightarrow X_\eta \text{ strongly in } L^2(0, T; \mathcal{H}).
\end{aligned} \tag{12.26}$$

Using the same techniques as to prove the existence of a solution to problem (12.8), we may apply the limit in (12.25) to conclude that $(\mathbf{u}_\eta, \mathcal{T}_\eta, \mathcal{S}_\eta) = \mathcal{F}(\xi_\eta, \eta)$ satisfies the problem (12.17). Therefore, since the norm is lower semi-continuous we have that the mapping $P_\eta : \xi \longrightarrow J(\xi, \eta)$ is lower semi-continuous for all $\eta \in \mathcal{V}_{\mathrm{ad}}$. By using the same technique we obtain that Q_ξ is upper semi-continuous for all $\xi \in \mathcal{U}_{\mathrm{ad}}$.

This completes the proof. □

We next wish to show the appropriate first-order necessary conditions for the robust control problem.

12.3.3 Optimality Conditions

In order to obtain the necessary optimality conditions which have been satisfied by the solution of the robust control problem, we introduce the following adjoint problem corresponding to the primal problem (12.17):

$$
\begin{aligned}
&\text{find } (\mathbf{R}, D, P) \in \mathcal{Z} \text{ such that, for all } (\mathbf{v}, \phi, \psi) \in \mathcal{V} \text{ and a.e. in } (0,T),\\
&-(\frac{\partial \mathbf{R}}{\partial t}, \mathbf{v}) + a_1(\mathbf{R}, \mathbf{v}) + b_1(\mathbf{v}, \mathbf{u}_1, \mathbf{R}) - b_1(\mathbf{u}_1, \mathbf{R}, \mathbf{v}) - d(\mathbf{R}, \mathbf{v})\\
&\qquad\qquad + b_2(\mathbf{v}, \mathcal{T}_1, D) + b_2(\mathbf{v}, \mathcal{S}_1, P) = (\mathcal{C}^*\mathcal{C}(\mathbf{u} - \mathbf{u}_{\text{obs}}), \mathbf{v})_{L^2}\\
&-(\frac{\partial D}{\partial t}, \phi) + a_2(D, \phi) - b_2(\mathbf{u}_1, D, \phi) + (-\delta_{\text{T}}\mathbf{G}.\mathbf{R}, \phi)_{L^2} = 0,\\
&-(\frac{\partial P}{\partial t}, \psi) + a_3(P, \psi) - b_2(\mathbf{u}_1, P, \psi) + (\delta_{\text{S}}\mathbf{G}.\mathbf{R}, \psi)_{L^2} = 0,\\
&(\mathbf{R}, D, P)(t = T) = 0,
\end{aligned}
\tag{12.27}
$$

where $(\mathbf{u}, \mathcal{T}, \mathcal{S}) = \mathcal{F}(\xi, \eta)$, $(\xi, \eta) \in L^2(\Sigma_0) \times L^2(\Sigma_0)$, with $\mathbf{u}_1 = \mathbf{u} + \mathbf{u}_0$, $\mathcal{T}_1 = \mathcal{T} + \mathcal{T}_0$ and $\mathcal{S}_1 = \mathcal{S} + \mathcal{S}_0$.

Remark 12.19. (*i*) The adjoint problem (12.27) is a linear system. By reversing sense of time, *i.e.*, $t := T - t$, and by applying the same way as to obtain the result of Theorem 12.17 we obtain the existence and uniqueness of the solution $(\mathbf{R}, D, P)$ in $\mathcal{Z}$.
(*ii*) The system (12.27) is the weak formulation of the following system:

$$
\begin{aligned}
&\text{find } (\mathbf{R}, \Pi, D, P) \text{ such that}\\
&-\frac{\partial \mathbf{R}}{\partial t} - (\mathbf{u}_1.\nabla)\mathbf{R} + (\nabla \mathbf{R})^t \mathbf{u}_1 - \mathbf{F} \wedge \mathbf{R} + (\nabla \mathcal{T}_1)^t D + (\nabla \mathcal{S}_1)^t P\\
&\qquad\qquad -\nu_{1\text{h}}\Delta_2 \mathbf{R} - \frac{\partial}{\partial z}(\nu_{1\text{v}}\frac{\partial \mathbf{R}}{\partial z}) + \nabla \Pi = \mathcal{C}^*\mathcal{C}(\mathbf{u} - \mathbf{u}_{\text{obs}}),\\
&\operatorname{div}(\mathbf{R}) = 0,\\
&-\frac{\partial D}{\partial t} - (\mathbf{u}_1\nabla)D - \nu_{2\text{h}}\Delta_2 D - \frac{\partial}{\partial z}(\nu_{2\text{v}}\frac{\partial D}{\partial z}) = \delta_{\text{T}}\mathbf{G}.\mathbf{R},\\
&-\frac{\partial P}{\partial t} - (\mathbf{u}_1\nabla)P - \nu_{3\text{h}}\Delta_2 P - \frac{\partial}{\partial z}(\nu_{3\text{v}}\frac{\partial P}{\partial z}) = -\delta_{\text{S}}\mathbf{G}.\mathbf{R},\\
&\text{with the initial condition}\\
&(\mathbf{R}, D, P)(t = T) = 0,\\
&\text{subject to the boundary conditions (12.4) and (12.5)}\\
&\text{where the vector 0 (respectively the scalar 0) plays the role}\\
&\text{of } \mathbf{f} \text{ (respectively } \varpi),
\end{aligned}
\tag{12.28}
$$

with $\mathbf{u}_1 = \mathbf{u} + \mathbf{u_0}$, $\mathcal{T}_1 = \mathcal{T} + \mathcal{T}_0$ and $\mathcal{S}_1 = \mathcal{S} + \mathcal{S}_0$. ♢

We can now give the first-order optimality conditions for the robust control problem (12.19).

Theorem 12.20. *Under the assumptions of Theorem 12.18, the optimal solutions* $(\xi^*, \eta^*) \in \mathcal{U}$ *and* $X^* = (\mathbf{u}^*, \mathcal{T}^*, \mathcal{S}^*) \in H^{2,1}(\mathcal{Q}) \cap L^\infty(0,T;\mathcal{V})$ *such that* (ξ^*, η^*) *satisfies (12.19) and* $(\mathbf{u}^*, \mathcal{T}^*, \mathcal{S}^*) = \mathcal{F}(\xi^*, \eta^*)$ *is a solution of the primal (12.17), satisfy*

$$\begin{aligned} &\iint_{\Sigma_0} (\alpha\xi^* + \mathcal{B}_{1\mathrm{T}}^*\mathbf{R}^*).(\xi - \xi^*)\mathrm{d}\Gamma\mathrm{d}t \geq 0, \\ &\iint_{\Sigma_0} (-\beta\eta^* + \mathcal{B}_2^* D^*)(\eta - \eta^*)\mathrm{d}\Gamma\mathrm{d}t \leq 0 \quad \forall(\xi,\eta) \in \mathcal{U}, \end{aligned} \tag{12.29}$$

where $(\mathbf{R}^*, D^*, P^*)$ *is the adjoint solution of the problem (12.27) corresponding to the primal solution* $\mathcal{F}(\xi^*, \eta^*)$.

Proof. The cost function J is a composition of F-differentiable mappings then J is differentiable and we have $(\forall(\mathbf{h},\kappa) \in \mathcal{U}\ )$

$$\begin{aligned} J'(\xi,\eta).(\mathbf{h},\kappa) = &\iint_{\mathcal{Q}} \mathcal{C}^*\mathcal{C}(\mathbf{u} - \mathbf{u}_{\mathrm{obs}})\mathbf{w}\mathrm{d}x\mathrm{d}t \\ &+\alpha \iint_{\Sigma_0} \xi.\mathbf{h}\mathrm{d}\Gamma\mathrm{d}t - \beta \iint_{\Sigma_0} \kappa\eta\mathrm{d}\Gamma\mathrm{d}t, \end{aligned} \tag{12.30}$$

where $(\mathbf{w}, \theta, \vartheta) = \mathcal{F}'(\xi,\eta).(\mathbf{h},\kappa)$ is the solution of problem (12.20).

By taking $(\mathbf{v}, \phi, \psi) = (\mathbf{R}, D, P)$ in (12.20), and integrating with respect to time, we obtain (according to the initial conditions of (12.20))

$$\begin{aligned} &\int_0^T \left[(-\frac{\partial \mathbf{R}}{\partial t}, \mathbf{w}) + a_1(\mathbf{R},\mathbf{w}) - d(\mathbf{R},\mathbf{w}) - b_1(\mathbf{u}_1, \mathbf{R}, \mathbf{w}) + b_1(\mathbf{w}, \mathbf{u}_1, \mathbf{R}) \right] dt \\ &\qquad = -(\mathbf{R}(T), \mathbf{w}(T))_{L^2} - \int_0^T l_1(\theta, \vartheta, \mathbf{R})\mathrm{d}t + \int_0^T (\mathcal{B}_{1\mathrm{T}}^*\mathbf{R}, \mathbf{h})_{\Gamma_0}\mathrm{d}t, \\ &\int_0^T \left[(-\frac{\partial D}{\partial t}, \theta) + a_2(D,\theta) - b_2(\mathbf{u}_1, D, \theta) \right] \mathrm{d}t \\ &\qquad = -(D(T), \theta(T))_{L^2} - \int_0^T b_2(\mathbf{w}, \mathcal{T}_1, D)\mathrm{d}t + \int_0^T (\mathcal{B}_2^* D, \kappa)_{\Gamma_0}\mathrm{d}t, \\ &\int_0^T \left[(-\frac{\partial P}{\partial t}, \vartheta) + a_3(P,\vartheta) - b_2(\mathbf{u}_1, P, \vartheta) \right] \mathrm{d}t \\ &\qquad = -(P(T), \vartheta(T))_{L^2} - \int_0^T b_2(\mathbf{w}, \mathcal{S}_1, P)\mathrm{d}t. \end{aligned}$$

Since $(\mathbf{R}, D, P)$ is a solution of (12.27), with null final conditions, we can deduce that (since $l_1(\theta, \vartheta, \mathbf{R}) = ((-\delta_{\mathrm{T}}\theta + \delta_{\mathrm{S}}\vartheta)\mathbf{G}, \mathbf{R})_{L^2}$)

$$
\begin{aligned}
&\int_0^T b_2(\mathbf{w}, \mathcal{T}_1, D)\mathrm{d}t + \int_0^T b_2(\mathbf{w}, \mathcal{S}_1, P)\mathrm{d}t - \int_0^T (\mathcal{C}^*\mathcal{C}(\mathbf{u} - \mathbf{u}_{\mathrm{obs}}), \mathbf{w})_{L^2}\mathrm{d}t \\
&\qquad = \int_0^T l_1(\theta, \vartheta, \mathbf{R})\mathrm{d}t - \int_0^T (\mathcal{B}_{1\mathrm{T}}^*\mathbf{R}, \mathbf{h})_{\Gamma_0}\mathrm{d}t, \\
&- \int_0^T (\delta_{\mathrm{T}}\theta\mathbf{G}, \mathbf{R})_{L^2}\mathrm{d}t = \int_0^T b_2(\mathbf{w}, \mathcal{T}_1, D)\mathrm{d}t - \int_0^T (\mathcal{B}_2^* D, \kappa)_{\Gamma_0}\mathrm{d}t, \\
&\int_0^T (\delta_{\mathrm{S}}\vartheta\mathbf{G}, \mathbf{R})_{L^2}\mathrm{d}t = \int_0^T b_2(\mathbf{w}, \mathcal{S}_1, P)\mathrm{d}t.
\end{aligned}
\tag{12.31}
$$

By adding the first, the second and the third part of (12.31), we obtain that

$$
\int_0^T (\mathcal{C}^*\mathcal{C}(\mathbf{u} - \mathbf{u}_{\mathrm{obs}}), \mathbf{w})_{L^2}\mathrm{d}t = \int_0^T (\mathcal{B}_{1\mathrm{T}}^*\mathbf{R}, \mathbf{h})_{\Gamma_0}\mathrm{d}t + \int_0^T (\mathcal{B}_2^* D, \kappa)_{\Gamma_0}\mathrm{d}t. \tag{12.32}
$$

According to the expression of $J'(\xi, \eta).(\mathbf{h}, \kappa)$ we can deduce that

$$
J'(\xi, \eta).(\mathbf{h}, \kappa) = \iint_{\Sigma_0} (\alpha\xi + \mathcal{B}_{1\mathrm{T}}^*\mathbf{R}).\mathbf{h}\mathrm{d}\Gamma\mathrm{d}t - \iint_{\Sigma_0} (\beta\eta - \mathcal{B}_2^* D)\kappa\mathrm{d}\Gamma\mathrm{d}t, \tag{12.33}
$$

Since (ξ^*, η^*) is an optimal solution we have

$$
\frac{\partial J}{\partial \xi}(\xi^*, \eta^*).(\xi - \xi^*) \geq 0, \frac{\partial J}{\partial \eta}(\xi^*, \eta^*).(\eta - \eta^*) \leq 0 \quad \forall (\xi, \eta) \in \mathcal{U} \tag{12.34}
$$

and then

$$
\begin{aligned}
&\iint_{\Sigma_0} (\alpha\xi^* + \mathcal{B}_{1\mathrm{T}}^*\mathbf{R}^*).(\xi - \xi^*)\mathrm{d}\Gamma\mathrm{d}t \geq 0, \\
&\iint_{\Sigma_0} (-\beta\eta^* + \mathcal{B}_2^* D^*)(\eta - \eta^*)\mathrm{d}\Gamma\mathrm{d}t \leq 0 \qquad \forall (\xi, \eta) \in \mathcal{U},
\end{aligned}
\tag{12.35}
$$

where $(\mathbf{R}^*, D^*, P^*)$ is the adjoint solution corresponding to the primal solution $\mathcal{F}(\xi^*, \eta^*)$.

This completes the proof. □

We end this section by the following results, which correspond to the case where we take into account the final observation. In this case the cost functional can be given by

$$
\begin{aligned}
J(\xi, \eta) = &\frac{1}{2} \iint_{Q} | \mathcal{C}(\mathbf{u} - \mathbf{u}_{\mathrm{obs}}) |^2 \,\mathrm{d}x\mathrm{d}t \\
&+ \frac{1}{2} \int_{\Omega} | \mathcal{C}_1(\mathbf{u}(., T) - \mathbf{v}_{\mathrm{obs}}) |^2 \,\mathrm{d}x \\
&+ \frac{\alpha}{2} \iint_{\Sigma_0} | \xi |^2 \,\mathrm{d}\Gamma\mathrm{d}t - \frac{\beta}{2} \iint_{\Sigma_0} | \eta |^2 \,\mathrm{d}\Gamma\mathrm{d}t
\end{aligned}
\tag{12.36}
$$

where $\alpha, \beta > 0$ are fixed, $(\mathbf{u}_{\text{obs}}, \mathbf{v}_{\text{obs}}) \in L^2(0,T;\mathcal{V}_1) \times \mathcal{V}_1$ is the observation and the operators $\mathcal{C}$ and $\mathcal{C}_1$ are unbounded and linear operators on $L^2(\Omega)$ satisfying the condition (7.91) (here $\mathcal{V}_1$ plays the role of $\mathcal{D}$).

The arguments of the previous results extend directly to the present case without further estimates. Therefore, we have the following results.

Theorem 12.21. *Under the assumptions of Theorem 12.18, the robust control problem (12.19) admits a solution $(\xi^*, \eta^*) \in \mathcal{U}$, which is characterized by the following necessary optimality conditions:*

$$\begin{aligned} &\iint_{\Sigma_0} (\alpha\xi^* + \mathcal{B}_{1\mathrm{T}}^*\mathbf{R}^*).(\xi - \xi^*)\mathrm{d}\Gamma \mathrm{d}t \geq 0, \\ &\iint_{\Sigma_0} (-\beta\eta^* + \mathcal{B}_2^* D^*)(\eta - \eta^*)\mathrm{d}\Gamma \mathrm{d}t \leq 0 \qquad \forall(\xi,\eta) \in \mathcal{U}, \end{aligned} \tag{12.37}$$

where $(\mathbf{R}^*, D^*, P^*)$ is the unique solution in $\mathcal{Z}$ of the following adjoint problem (corresponding to the unique solution $(\mathbf{u}^*, \mathcal{T}^*, \mathcal{S}^*) = \mathcal{F}(\xi^*, \eta^*) \in H^{2,1}(\mathcal{Q}) \cap L^\infty(0,T;\mathcal{V})$ of the primal problem (12.17)):

$$\begin{aligned} -(\frac{\partial \mathbf{R}^*}{\partial t}, \mathbf{v}) &+ a_1(\mathbf{R}^*, \mathbf{v}) + b_1(\mathbf{v}, \mathbf{u}_1, \mathbf{R}^*) - b_1(\mathbf{u}_1, \mathbf{R}^*, \mathbf{v}) \\ &\qquad\qquad + b_2(\mathbf{v}, \mathcal{T}_1, D^*) + b_2(\mathbf{v}, \mathcal{S}_1, P^*) \\ &\qquad\qquad - d(\mathbf{R}^*, \mathbf{v}) = (\mathcal{C}^*\mathcal{C}(\mathbf{u}^* - \mathbf{u}_{\text{obs}}), \mathbf{v})_{L^2} \\ -(\frac{\partial D^*}{\partial t}, \phi) &+ a_2(D^*, \phi) - b_2(\mathbf{u}_1, D^*, \phi) + (-\delta_{\mathrm{T}}\mathbf{G}.\mathbf{R}^*, \phi)_{L^2} = 0, \\ -(\frac{\partial P^*}{\partial t}, \psi) &+ a_3(P^*, \psi) - b_2(\mathbf{u}_1, P^*, \psi) + (\delta_{\mathrm{S}}\mathbf{G}.\mathbf{R}^*, \psi)_{L^2} = 0, \\ (\mathbf{R}^*, D^*, P^*)(t = T) &= \mathcal{C}_1^*\mathcal{C}_1(\mathbf{u}^*(.,T) - \mathbf{v}_{\text{obs}}), \end{aligned} \tag{12.38}$$

for all $(\mathbf{v}, \phi, \psi) \in \mathcal{V}$ and a.e. in $(0,T)$, with $\mathbf{u}_1 = \mathbf{u} + \mathbf{u}_0$, $\mathcal{T}_1 = \mathcal{T} + \mathcal{T}_0$ and $\mathcal{S}_1 = \mathcal{S} + \mathcal{S}_0$. □

12.4 Primitive Ocean Equations with Vertical Viscosity

In this section we focus our attention on different physical assumptions. In the equatorial ocean the horizontal scale is much bigger than the vertical one.[8] Therefore, we can simplify the total Boussinesq equations of the ocean (12.1) by neglecting some terms. However, we must take into account the viscosity, which plays a very important role in the dynamics of the ocean in such region. The velocity solution is now denoted by $\mathbf{u} = (\mathbf{u}', w) = (u, v, w)$. Therefore, we can replace the equation describing the vertical motion by the following

[8] The value of the ratio between the vertical and horizontal scales of the oceanic domain is approximately 10^{-3} for the physical problems considered here.

equation called the *hydrostatic approximation with vertical viscosity*[9] (see Lions *et al.* [205]):

$$-\nu_{1\mathrm{h}}\Delta w - \frac{\partial}{\partial z}(\nu_{1\mathrm{v}}\frac{\partial w}{\partial z}) + \frac{1}{\rho_{\mathrm{av}}}\frac{\partial p}{\partial z} + \frac{\rho}{\rho_{\mathrm{av}}}g = 0. \tag{12.39}$$

Applying these physical assumptions to (12.1), we obtain the following equations (since $\rho_{\mathrm{av}} = 1$):

$$\begin{aligned}
&\frac{\partial \mathbf{u}'}{\partial t} + (\mathbf{u}'.\nabla_2)\mathbf{u}' + w\frac{\partial \mathbf{u}'}{\partial z} - \nu_{1\mathrm{h}}\Delta\mathbf{u}' - \frac{\partial}{\partial z}(\nu_{1\mathrm{v}}\frac{\partial \mathbf{u}'}{\partial z}) \\
&\qquad\qquad +\mathbf{F}\wedge_2 \mathbf{u}' + \nabla_2 p = 0, \\
&-\nu_{1\mathrm{h}}\Delta_2 w - \frac{\partial}{\partial z}(\nu_{1\mathrm{v}}\frac{\partial w}{\partial z}) + \frac{\partial p}{\partial z} + \rho g = 0, \\
&\mathrm{div}_2(\mathbf{u}') + \frac{\partial w}{\partial z} = 0, \\
&\frac{\partial \mathcal{T}}{\partial t} + (\mathbf{u}'.\nabla_2)\mathcal{T} + w\frac{\partial \mathcal{T}}{\partial z} - \nu_{2\mathrm{h}}\Delta_2\mathcal{T} - \frac{\partial}{\partial z}(\nu_{2\mathrm{v}}\frac{\partial \mathcal{T}}{\partial z}) = 0, \\
&\frac{\partial \mathcal{S}}{\partial t} + (\mathbf{u}'.\nabla_2)\mathcal{S} + w\frac{\partial \mathcal{S}}{\partial z} - \nu_{2\mathrm{h}}\Delta_2\mathcal{S} - \frac{\partial}{\partial z}(\nu_{2\mathrm{v}}\frac{\partial \mathcal{S}}{\partial z}) = 0, \\
&\rho = \delta_0 - \delta_{\mathrm{T}}\mathcal{T} + \delta_{\mathrm{S}}\mathcal{S},
\end{aligned} \tag{12.40}$$

with $\mathbf{F}\wedge_2 \mathbf{u}' = \mathbf{F}\wedge(\mathbf{u}',0)$, $\nabla_2. = (\partial./\partial x, \partial./\partial y)$ and $\mathrm{div}_2(\mathbf{u}') = \nabla_2.\mathbf{u}'$.

Nota bene: The notation " $'$ " used for the separation between the vertical and the horizontal velocity will now be omitted.

We now split up the circulation $(\mathbf{u}, w, \mathcal{T}, \mathcal{S}, p)$ into a given mean value $(\mathbf{u_0}, w_0, \mathcal{T}_0, \mathcal{S}_0, p_0)$ and a variability $(\tilde{\mathbf{u}}, \tilde{w}, \tilde{\mathcal{T}}, \tilde{\mathcal{S}}, \tilde{p})$ that will be computed by the model. The mean circulation $(\mathbf{u_0}, w_0, \mathcal{T}_0, \mathcal{S}_0, p_0)$ has to satisfy the steady state Equations (12.40).

The full non-linear system which models large perturbation $(\tilde{\mathbf{u}}, \tilde{w}, \tilde{\mathcal{T}}, \tilde{\mathcal{S}}, \tilde{p})$ to the target $(\mathbf{u_0}, w_0, \mathcal{T}_0, \mathcal{S}_0, p_0)$ can be deduced from (12.40). It can be written as

$$\begin{aligned}
&\frac{\partial \tilde{\mathbf{u}}}{\partial t} + (\tilde{\mathbf{u}}.\nabla_2)\tilde{\mathbf{u}} + \tilde{w}\frac{\partial \tilde{\mathbf{u}}}{\partial z} + (\mathbf{u_0}.\nabla_2)\tilde{\mathbf{u}} + w_0\frac{\partial \tilde{\mathbf{u}}}{\partial z} + (\tilde{\mathbf{u}}.\nabla_2)\mathbf{u_0} \\
&\qquad +\tilde{w}\frac{\partial \mathbf{u_0}}{\partial z} + \mathbf{F}\wedge_2 \tilde{\mathbf{u}} - \nu_{1\mathrm{h}}\Delta_2\tilde{\mathbf{u}} - \frac{\partial}{\partial z}(\nu_{1\mathrm{v}}\frac{\partial \tilde{\mathbf{u}}}{\partial z}) + \nabla\tilde{p} = 0, \\
&-\nu_{1\mathrm{h}}\Delta_2\tilde{w} - \frac{\partial}{\partial z}(\nu_{1\mathrm{v}}\frac{\partial \tilde{w}}{\partial z}) + \frac{\partial \tilde{w}}{\partial z} + \frac{\partial \tilde{p}}{\partial z} + g(-\delta_{\mathrm{T}}\tilde{\mathcal{T}} + \delta_{\mathrm{S}}\tilde{\mathcal{S}}) = 0, \\
&\frac{\partial \tilde{\mathcal{T}}}{\partial t} + (\tilde{\mathbf{u}}.\nabla_2)\tilde{\mathcal{T}} + \tilde{w}\frac{\partial \tilde{\mathcal{T}}}{\partial z} + (\tilde{\mathbf{u}}.\nabla_2)\mathcal{T}_0 + \tilde{w}\frac{\partial \mathcal{T}_0}{\partial z} + (\mathbf{u_0}.\nabla_2)\tilde{\mathcal{T}}
\end{aligned}$$

[9] In contrast with the classical hydrostatic approximation, where the vertical motion is replaced by the hydrostatic pressure equation: $\partial p/\partial z = -\rho g$.

$$
\begin{aligned}
&\qquad +w_0\frac{\partial \tilde{\mathcal{T}}}{\partial z} - \nu_{2\mathrm{h}}\Delta_2\tilde{\mathcal{T}} - \frac{\partial}{\partial z}(\nu_{2\mathrm{v}}\frac{\partial \tilde{\mathcal{T}}}{\partial z}) = 0,\\
&\frac{\partial \tilde{\mathcal{S}}}{\partial t} + (\tilde{\mathbf{u}}.\nabla_2)\tilde{\mathcal{S}} + \tilde{w}\frac{\partial \tilde{\mathcal{S}}}{\partial z} + (\tilde{\mathbf{u}}.\nabla_2)\mathcal{S}_0 + \tilde{w}\frac{\partial \mathcal{S}_0}{\partial z} + (\mathbf{u_0}.\nabla_2)\tilde{\mathcal{S}}\\
&\qquad +w_0\frac{\partial \tilde{\mathcal{S}}}{\partial z} - \nu_{2\mathrm{h}}\Delta_2\tilde{\mathcal{S}} - \frac{\partial}{\partial z}(\nu_{2\mathrm{v}}\frac{\partial \tilde{\mathcal{S}}}{\partial z}) = 0,\\
&\mathrm{div}_2(\tilde{\mathbf{u}}) + \frac{\partial \tilde{w}}{\partial z} = 0.
\end{aligned}
\tag{12.41}
$$

Nota bene: The notation "~" used for the perturbation of the mean flow will now be omitted.

To take into account the phenomena we want to describe, we assume that the circulation $(\mathbf{u}, w, \mathcal{T}, \mathcal{S}, p)$ satisfies the boundary conditions (12.4) and (12.5), and the initial conditions (12.6).

Equations (12.41) are not of a Cauchy–Kovalevsky type with respect to all variables because of the continuity equation and of the hydrostatic equation with vertical viscosity (12.39). The method proposed by Lions *et al.* in [205, 206] is the integration of these two diagnostic equations with respect to the vertical variable, and then we obtain another formulation of the primitive equations, which is a three dimensional evolution system. Therefore we can use Faedo–Galerkin method to solve the weak formulations.

By integrating the continuity equation with respect to z and taking the boundary conditions for w into account, we obtain

$$
\begin{aligned}
&w(x, y, z; t) = W(\mathbf{u})(x, y, z; t) = -\mathrm{div}\left(\int_{-H}^{z} \mathbf{u}(x, y, s; t)\mathrm{d}s\right),\\
&-\mathrm{div}\left(\int_{-H}^{0} \mathbf{u}(., ., z; .)dz\right) = 0
\end{aligned}
\tag{12.42}
$$

By integrating the hydrostatic equation with vertical viscosity in (12.41), we obtain

$$
p = P + g\int_{z}^{0}(-\delta_{\mathrm{T}}\mathcal{T}(s) + \delta_{\mathrm{S}}\mathcal{S}(s))\mathrm{d}s + \int_{z}^{0} L(\mathbf{u})(s)\mathrm{d}s \tag{12.43}
$$

where $L(\mathbf{u}) = -\nu_{1\mathrm{h}}\Delta W(\mathbf{u}) - (\partial/\partial z)\,(\nu_{1\mathrm{v}}(\partial W(\mathbf{u})/\partial z))$ and P is the pressure of the sea water on the surface of the ocean.

Then by using (12.42) and (12.43) we obtain the following reformulation of Equations (12.41)

$$
\begin{aligned}
&\frac{\partial \mathbf{u}}{\partial t} + (\mathbf{u}.\nabla_2)\mathbf{u} + W(\mathbf{u})\frac{\partial \mathbf{u}}{\partial z} + (\mathbf{u_0}.\nabla_2)\mathbf{u} + w_0\frac{\partial \mathbf{u}}{\partial z} + (\mathbf{u}.\nabla_2)\mathbf{u_0}\\
&\qquad +W(\mathbf{u})\frac{\partial \mathbf{u_0}}{\partial z} + \mathbf{F}\wedge_2 \mathbf{u} - \nu_{1\mathrm{h}}\Delta_2\mathbf{u} - \frac{\partial}{\partial z}(\nu_{1\mathrm{v}}\frac{\partial \mathbf{u}}{\partial z})
\end{aligned}
$$

$$
\begin{aligned}
&+\nabla P + g\nabla(-\delta_{\mathrm{T}}R(\mathcal{T}) + \delta_{\mathrm{S}}R(\mathcal{S})) + \nabla\mathcal{L}(\mathbf{u}) = 0,\\
&\frac{\partial \mathcal{T}}{\partial t} + (\mathbf{u}.\nabla_2)\mathcal{T} + W(\mathbf{u})\frac{\partial \mathcal{T}}{\partial z} + (\mathbf{u}.\nabla_2)\mathcal{T}_0 + W(\mathbf{u})\frac{\partial \mathcal{T}_0}{\partial z} + (\mathbf{u_0}.\nabla_2)\mathcal{T}\\
&\qquad +w_0\frac{\partial \mathcal{T}}{\partial z} - \nu_{2\mathrm{h}}\Delta_2\mathcal{T} - \frac{\partial}{\partial z}(\nu_{2\mathrm{v}}\frac{\partial \mathcal{T}}{\partial z}) = 0,\\
&\frac{\partial \mathcal{S}}{\partial t} + (\mathbf{u}.\nabla_2)\mathcal{S} + W(\mathbf{u})\frac{\partial \mathcal{S}}{\partial z} + (\mathbf{u}.\nabla_2)\mathcal{S}_0 + W(\mathbf{u})\frac{\partial \mathcal{S}_0}{\partial z} + (\mathbf{u_0}.\nabla_2)\mathcal{S}\\
&\qquad +w_0\frac{\partial \mathcal{S}}{\partial z} - \nu_{2\mathrm{h}}\Delta_2\mathcal{S} - \frac{\partial}{\partial z}(\nu_{2\mathrm{v}}\frac{\partial \mathcal{S}}{\partial z}) = 0,\\
&\mathrm{div}\left(\int_{-H}^{0}\mathbf{u}(z)dz\right) = 0,
\end{aligned}
\tag{12.44}
$$

where

$$
R(\phi) = \int_z^0 \phi(s)\mathrm{d}s, \quad \mathcal{L}(\mathbf{u}) = \int_z^0 L(\mathbf{u})(s)\mathrm{d}s.
$$

Remark 12.22. The above system is three-dimensional, but the unknown function P is only a function of x, y and time t. ♢

For obtaining the variational formulation of System (12.44) with Conditions (12.4), (12.5) and (12.6), we use the following spaces:

$$
\begin{aligned}
\mathcal{H}_1 &= \{\mathbf{v} \in (L^2(\Omega))^2 : (\mathbf{v},0).\mathbf{n} = 0 \ \text{ on } \Gamma_0 \cup \Gamma_3 \cup \Gamma_4 \cup \Gamma_5,\\
&\qquad (\mathbf{v},0).\mathbf{n}|_{\Gamma_1} = -(\mathbf{v},0).\mathbf{n}|_{\Gamma_2}\},\\
\mathcal{W}_1 &= \{\mathbf{v} \in (H^1(\Omega))^2 : W(\mathbf{v}) = -\int_{-H}^{z}\mathrm{div}(\mathbf{v})\mathrm{d}s \in H^1(\Omega), \ \mathbf{v} = \mathbf{0} \ \text{ on } \Gamma_5,\\
&\qquad (\mathbf{v},0).\mathbf{n} = 0 \ \text{ on } \Gamma_0 \cup \Gamma_3 \cup \Gamma_4, \mathbf{v}|_{\Gamma_1} = \mathbf{v}|_{\Gamma_2}\},\\
\mathcal{V}_1 &= \{\mathbf{v} \in \mathcal{W}_1 : \int_{-H}^{0}\mathrm{div}(\mathbf{v})dz = 0\},\\
\mathcal{H}_2 &= L^2(\Omega), \quad \mathcal{V}_2 = \{\phi \in H^1(\Omega) : \phi = 0 \ \text{ on } \Gamma_3 \cup \Gamma_4 \cup \Gamma_5, \phi|_{\Gamma_1} = \phi|_{\Gamma_2}\},\\
\mathcal{H} &= \mathcal{H}_1 \times \mathcal{H}_2 \times \mathcal{H}_2, \qquad \mathcal{V} = \mathcal{V}_1 \times \mathcal{V}_2 \times \mathcal{V}_2,
\end{aligned}
$$

where $\mathbf{n}$ is the unit outward vector normal to Γ.

Remark 12.23. (*i*) The spaces $\mathcal{W}_1$ and $\mathcal{V}_1$ are equipped with the norm

$$
\| \mathbf{u} \|_W^2 = \| \mathbf{u} \|^2 + \| \int_{-H}^{z}\mathrm{div}(\mathbf{u})\mathrm{d}s \|^2 \qquad \text{for all } \mathbf{u} \in \mathcal{W}_1 \text{ (respectively } \mathcal{V}_1)
$$

and $\mathcal{V}_2$ is equipped with the norm $\| \phi \|$, for all $\phi \in \mathcal{V}_2$.
(*ii*) The space $\mathcal{V}$ is equipped with the norm

$$
\| X \|_W^2 = \| \mathbf{u} \|_W^2 + \| \mathcal{T} \|^2 + \| \mathcal{S} \|^2 \quad \text{for all } X = (\mathbf{u},\mathcal{T},\mathcal{S}) \in \mathcal{V}.
$$

♢

We now define the following forms:

$$
\begin{aligned}
&a_{11}(\mathbf{u},\mathbf{v}) = \nu_{1\mathrm{h}}(\nabla_2\mathbf{u},\nabla_2\mathbf{v}) + (\nu_{1\mathrm{v}}\frac{\partial\mathbf{u}}{\partial z},\frac{\partial\mathbf{v}}{\partial z}),\\
&a_{12}(\mathbf{u},\mathbf{v}) = \nu_{1\mathrm{h}}(\nabla_2 W(\mathbf{u}),\nabla_2 W(\mathbf{v})) + (\nu_{1\mathrm{v}}\mathrm{div}_2(\mathbf{u}),\mathrm{div}_2(\mathbf{v})),\\
&a_1(\mathbf{u},\mathbf{v}) = a_{11}(\mathbf{u},\mathbf{v}) + a_{12}(\mathbf{u},\mathbf{v}),\\
&a_2(\mathcal{T},\phi) = \nu_{2\mathrm{h}}(\nabla_2\mathcal{T},\nabla_2\phi) + (\nu_{2\mathrm{v}}\frac{\partial\mathcal{T}}{\partial z},\frac{\partial\phi}{\partial z}),\\
&a_3(\mathcal{S},\psi) = \nu_{3\mathrm{h}}(\nabla_2\mathcal{S},\nabla_2\psi) + (\nu_{3\mathrm{v}}\frac{\partial\mathcal{S}}{\partial z},\frac{\partial\psi}{\partial z}),\\
&d(\mathbf{u},\mathbf{v}) = (\mathbf{F}\wedge_2\mathbf{u},\mathbf{v}),\\
&b_1(\mathbf{u},\mathbf{v},\mathbf{w}) = ((\mathbf{u}.\nabla_2)\mathbf{v},\mathbf{w}) + (W(\mathbf{u})\frac{\partial\mathbf{v}}{\partial z},\mathbf{w}),\\
&b_2(\mathbf{u},\phi,\psi) = ((\mathbf{u}.\nabla_2)\phi,\psi) + (W(\mathbf{u})\frac{\partial\phi}{\partial z},\psi),\\
&l_0(\mathbf{u},\mathbf{v}) = b_1(\mathbf{u},\mathbf{u_0},\mathbf{v}) + ((\mathbf{u_0}.\nabla_2)\mathbf{u},\mathbf{v}) + (w_0\frac{\partial\mathbf{u}}{\partial z},\mathbf{v}) + d(\mathbf{u},\mathbf{v}),\\
&l_1(\mathcal{T},\mathcal{S},\mathbf{v}) = g(-\delta_{\mathrm{T}}\mathcal{T} + \delta_{\mathrm{S}}\mathcal{S},\int_{-H}^{z}\mathrm{div}(\mathbf{v})\mathrm{d}s),\\
&l_2(\mathbf{u},\mathcal{T},\phi) = b_2(\mathbf{u},\mathcal{T}_0,\phi) + ((\mathbf{u_0}.\nabla_2)\mathcal{T},\phi) + (w_0\frac{\partial\mathcal{T}}{\partial z},\phi),\\
&l_3(\mathbf{u},\mathcal{T},\psi) = b_2(\mathbf{u},\mathcal{S}_0,\psi) + ((\mathbf{u_0}.\nabla_2)\mathcal{S},\psi) + (w_0\frac{\partial\mathcal{S}}{\partial z},\psi).
\end{aligned}
$$

In order to introduce the weak formulations associated with the perturbation problem (12.44), (12.4), (12.5), (12.6), we give the following lemmas, which can be found in Belmiloudi [32, 36].

Lemma 12.24. *For all* $(\mathbf{u},\mathbf{v})$ *in* $(\mathcal{W}_1\cap H^2(\Omega))^2$*, the following relation holds:*

$$
\begin{aligned}
(\nabla\mathcal{L}(\mathbf{u}),\mathbf{v}) = \nu_{1\mathrm{h}}(\nabla(\int_{-H}^{z}\mathrm{div}(\mathbf{u})\mathrm{d}s),\nabla(\int_{-H}^{z}\mathrm{div}(\mathbf{v})\mathrm{d}s)) + (\nu_{1\mathrm{v}}\mathrm{div}(\mathbf{u}),\mathrm{div}(\mathbf{v}))\\
- (\nu_{1\mathrm{v}}\mathrm{div}(\mathbf{u}),\int_{-H}^{0}\mathrm{div}(\mathbf{v})\mathrm{d}z)_{\Gamma_0}.
\end{aligned}
$$

Moreover, if $\mathbf{v}\in\mathcal{V}_1$*, the previous relation becomes*

$$
\begin{aligned}
(\nabla\mathcal{L}(\mathbf{u}),\mathbf{v}) = \nu_{1\mathrm{h}}(\nabla(\int_{-H}^{z}\mathrm{div}(\mathbf{u})\mathrm{d}z'),\nabla(\int_{-H}^{z}\mathrm{div}(\mathbf{v})\mathrm{d}z'))\\
+ (\nu_{1\mathrm{v}}\mathrm{div}(\mathbf{u}),\mathrm{div}(\mathbf{v})) = a_{12}(\mathbf{u},\mathbf{v}).
\end{aligned}
$$

□

Lemma 12.25. *Suppose that* $\mathbf{f} = (f_1,f_2)\in L^2(\Gamma_0)$*,* $\varpi\in L^2(\Gamma_0)$ *and* $(\mathbf{u},\mathcal{T},\mathcal{S},P)$ *is sufficiently regular and satisfies the boundary condition (12.4), (12.5). Then the following relations hold:*

$(i)\quad -\nu_{1\mathrm{h}}(\Delta_2\mathbf{u},\mathbf{v}) - (\frac{\partial}{\partial z}(\nu_{1\mathrm{v}}\frac{\partial \mathbf{u}}{\partial z}),\mathbf{v}) = a_{11}(\mathbf{u},\mathbf{v}) - (\mathbf{f},\mathbf{v})_{\Gamma_0},$ *for all* $\mathbf{v} \in \mathcal{W}_1$.

$(ii)\quad -\nu_{2\mathrm{h}}(\Delta_2\mathcal{T},\phi) - (\frac{\partial}{\partial z}(\nu_{2\mathrm{v}}\frac{\partial \mathcal{T}}{\partial z}),\phi) = a_2(\mathcal{T},\phi) - (\varpi,\phi)_{\Gamma_0},$ *for all* $\phi \in \mathcal{V}_2$.

$(iii)\ -\nu_{2\mathrm{h}}(\Delta_2\mathcal{S},\psi) - (\frac{\partial}{\partial z}(\nu_{2\mathrm{v}}\frac{\partial \mathcal{S}}{\partial z}),\psi) = a_3(\mathcal{S},\psi),$ *for all* $\psi \in \mathcal{V}_2$.

$(iv)\quad (\nabla P,\mathbf{v}) = -(P,\int_{-H}^{0}\mathrm{div}(\mathbf{v})\mathrm{d}z)_{\Gamma_0},$ *for all* $\mathbf{v} \in \mathcal{W}_1$.

$(v)\quad (\nabla P,\mathbf{v}) = 0,$ *for all* $\mathbf{v} \in \mathcal{V}_1$.

$$(vi)\quad (\nabla(\int_z^0 \phi(s)\mathrm{d}s),\mathbf{v}) = -(\int_z^0 \phi(s)\mathrm{d}s, \mathrm{div}(\mathbf{v})).$$
$$= (\phi, \int_{-H}^{z}\mathrm{div}(\mathbf{v})\mathrm{d}s),\ \text{for all } \mathbf{v} \in \mathcal{W}_1,\ \phi \in \mathcal{V}_2. \square$$

We now recall some properties of the previous operators.

Proposition 12.26. *The following properties hold:*

(i) a_1 *(respectively* a_i, $i = 2, 3$*) is a bilinear continuous and coercive form on* $\mathcal{W}_1 \times \mathcal{W}_1$ *(respectively on* $\mathcal{V}_2 \times \mathcal{V}_2$*).*

(ii) d *is a bilinear continuous form on* $\mathcal{W}_1 \times \mathcal{W}_1$.

(iii) *The operator* W *is linear and satisfies:* $\| W(\mathbf{u}) \|_{L^2(\Omega)} \leq C \| \mathbf{u} \|$.

$$(iv)\quad |(\phi, \int_{-H}^{z}\mathrm{div}(\mathbf{v})\mathrm{d}z')| \leq c \| \phi \|_{L^2(\Omega)} \| \mathbf{v} \|,$$
$$\leq c \| \mathbf{v} \|_{L^2(\Omega)} \| \phi \|,\ \text{for all } (\mathbf{v},\phi) \in \mathcal{W}_1 \times \mathcal{V}_2.$$

Proof. The proof of this proposition may be found in Belmiloudi [32, 36]. □

Remark 12.27. The trilinear continuous forms $b1$ and b_2 satisfy the same estimates and relations as in Lemmas 12.4 and 12.5. ♢

According to Lemma 12.25, the problem (12.44) with the boundary conditions (12.4), (12.5) and the initial conditions (12.6) satisfied by the perturbation $(\mathbf{u}, p, \mathcal{T}, \mathcal{S})$ of the mean flow admits the following weak formulation:

find $(\mathbf{u}, \mathcal{T}, \mathcal{S}) \in L^2(0, T; \mathcal{V})$ such that, for all $(\mathbf{v}, \phi, \psi)$ in $\mathcal{V}$, a.e. in $(0, T)$

$$\begin{aligned}
&(\frac{\partial \mathbf{u}}{\partial t},\mathbf{v}) + a_1(\mathbf{u},\mathbf{v}) + l_1(\mathcal{T},\mathcal{S},\mathbf{v}) + b_1(\mathbf{u},\mathbf{u},\mathbf{v}) \\
&\qquad + l_0(\mathbf{u},\mathbf{v}) = (\mathbf{f},\mathbf{v})_{\Gamma_0}, \\
&(\frac{\partial \mathcal{T}}{\partial t},\phi) + a_2(\mathcal{T},\phi) + l_2(\mathbf{u},\mathcal{T},\phi) + b_2(\mathbf{u},\mathcal{T},\phi) = (\varpi,\phi)_{\Gamma_0}, \\
&(\frac{\partial \mathcal{S}}{\partial t},\psi) + a_3(\mathcal{S},\psi) + l_3(\mathbf{u},\mathcal{S},\psi) + b_2(\mathbf{u},\mathcal{S},\psi) = 0, \\
&(\mathbf{u},\mathcal{T},\mathcal{S})(t=0) = 0.
\end{aligned} \tag{12.45}$$

From the properties of the forms $(a_i)_{i=1,3}$, $(b_i)_{i=1,2}$ and $(l_i)_{i=0,3}$ (Proposition 12.26, Lemmas 12.24 and 12.25 and Lemma 12.4), we derive the following existence and uniqueness results (see Belmiloudi [32, 36] for more details).

Proposition 12.28. *Under the assumptions of Theorem 12.8, assume moreover, that* $X_0 = (\mathbf{u}_0, w_0, \mathcal{T}_0, \mathcal{S}_0)$ *is in* $(H^2(\Omega) \cap \mathcal{V}_0) \times H^2(\Omega) \times H^2(\Omega)$. *The problem (12.45) admits a unique solution* $(\mathbf{u}, \mathcal{T}, \mathcal{S}) \in H^{2,1}(\mathcal{Q}) \cap \mathcal{C}([0,T]; \mathcal{V})$. *Moreover, if* $(\mathbf{f}_1, \varpi_1)$ *and* $(\mathbf{f}_2, \varpi_2)$ *are two functions in* $U_f \times U_\varpi$ *such that (12.11) holds, and* $X_i = (\mathbf{u}_i, \mathcal{T}_i, \mathcal{S}_i)$ *is the solution of problem (12.45), corresponding to the forcing* $(\mathbf{f}_i, \varpi_i)$, *for* $i = 1, 2$, *then the following estimate holds:*

$$\| X \|^2_{H^{2,1}(\mathcal{Q})} \leq C(\| \mathbf{f} \|^2_{U_f} + \| \varpi \|^2_{U_\varpi}), \tag{12.46}$$

where $X = X_1 - X_2$, $\mathbf{f} = \mathbf{f}_1 - \mathbf{f}_2$ *and* $\varpi = \varpi_1 - \varpi_2$. □

For the robust control problem, we state similar problems as in Section 12.3. More precisely, we assume that the control is in the variability of the wind stress and the disturbance is in heat stress in the context of the non-cooperative game discussed in Chapter 8. Thus, we write $\mathbf{f}$ and ϖ as in (12.16). The function $(\mathbf{u}, \mathcal{T}, \mathcal{S})$ is then assumed to be related to the disturbance η and control ξ through the problem (12.45):

$$\begin{aligned}
(\frac{\partial \mathbf{u}}{\partial t}, \mathbf{v}) + a_1(\mathbf{u}, \mathbf{v}) \quad & + l_1(\mathcal{T}, \mathcal{S}, \mathbf{v}) + b_1(\mathbf{u}, \mathbf{u}, \mathbf{v}) \\
& + l_0(\mathbf{u}, \mathbf{v}) = (\mathcal{B}_1 \xi, \mathbf{v})_{\Gamma_0}, \\
(\frac{\partial \mathcal{T}}{\partial t}, \phi) + a_2(\mathcal{T}, \phi) \quad & + l_2(\mathbf{u}, \mathcal{T}, \phi) + b_2(\mathbf{u}, \mathcal{T}, \phi) = (\mathcal{B}_2 \eta, \phi)_{\Gamma_0}, \\
(\frac{\partial \mathcal{S}}{\partial t}, \psi) + a_3(\mathcal{S}, \psi) \quad & + l_3(\mathbf{u}, \mathcal{S}, \psi) + b_2(\mathbf{u}, \mathcal{S}, \psi) = 0, \\
(\mathbf{u}, \mathcal{T}, \mathcal{S})(t = 0) = 0. &
\end{aligned} \tag{12.47}$$

for all $(\mathbf{v}, \phi, \psi)$ in $\mathcal{V}$ and a.e. in $(0, T)$.

The cost functional considered here is the same as in (12.18), *i.e.*, of the form

$$\begin{aligned}
J(\xi, \eta) = & \frac{1}{2} \iint_{\mathcal{Q}} | \mathcal{C}(\mathbf{u} - \mathbf{u}_{\text{obs}}) |^2 \, dx dt \\
& + \frac{\alpha}{2} \iint_{\Sigma_0} | \xi |^2 \, d\Gamma dt - \frac{\beta}{2} \iint_{\Sigma_0} | \eta |^2 \, d\Gamma dt,
\end{aligned}$$

where $\alpha, \beta > 0$ are fixed, $\mathbf{u}_{\text{obs}} \in L^2(0, T; \mathcal{V}_1)$ is the observation and $\mathcal{C}$ is an unbounded, linear operator on $L^2(\Omega)$ satisfying the condition (7.91).

The arguments of Section 12.3 extend directly to the present case. So, we omit the details.

Therefore, we have the following existence and first-order optimality condition results.

Theorem 12.29. *Under the assumptions of Theorem 12.18, the robust control problem (12.19) admits a solution $(\xi^*, \eta^*) \in \mathcal{U}$, which is charaterized by the following necessary optimality conditions:*

$$\begin{aligned} &\iint_{\Sigma_0} (\alpha\xi^* + \mathcal{B}_1^*\mathbf{R}^*).(\xi - \xi^*)\mathrm{d}\Gamma\mathrm{d}t \geq 0, \\ &\iint_{\Sigma_0} (-\beta\eta^* + \mathcal{B}_2^* D^*)(\eta - \eta^*)\mathrm{d}\Gamma\mathrm{d}t \leq 0 \qquad \forall(\xi, \eta) \in \mathcal{U}, \end{aligned} \tag{12.48}$$

where $(\mathbf{R}^*, D^*, P^*)$ is the unique solution in $\mathcal{Z}$ of the following adjoint problem (corresponding to the unique solution $(\mathbf{u}^*, \mathcal{T}^*, \mathcal{S}^*) = \mathcal{F}(\xi^*, \eta^*) \in H^{2,1}(\mathcal{Q}) \cap L^\infty(0,T;\mathcal{V})$ of the primal problem (12.47)): [10]

$$\begin{aligned} &-(\frac{\partial \mathbf{R}^*}{\partial t}, \mathbf{v}) + a_1(\mathbf{R}^*, \mathbf{v}) + b_1(\mathbf{v}, \mathbf{u}_1, \mathbf{R}^*) - b_1(\mathbf{u}_1, \mathbf{R}^*, \mathbf{v}) \\ &\qquad\qquad -d(\mathbf{R}^*, \mathbf{v}) + b_2(\mathbf{v}, \mathcal{T}_1, D^*) \\ &\qquad\qquad +b_2(\mathbf{v}, \mathcal{S}_1, P^*) = (\mathcal{C}^*\mathcal{C}(\mathbf{u}^* - \mathbf{u}_{\mathrm{obs}}), \mathbf{v})_{L^2}, \\ &-(\frac{\partial D^*}{\partial t}, \phi) + a_2(D^*, \phi) - b_2(\mathbf{u}_1, D^*, \phi) - (g\delta_{\mathrm{T}} W(\mathbf{R}^*), \phi)_{L^2} = 0, \\ &-(\frac{\partial P^*}{\partial t}, \psi) + a_3(P^*, \psi) - b_2(\mathbf{u}_1, P^*, \psi) + (g\delta_{\mathrm{S}} W(\mathbf{R}^*), \psi)_{L^2} = 0, \\ &(\mathbf{R}^*, D^*, P^*)(t = T) = 0, \end{aligned} \tag{12.49}$$

for all $(\mathbf{v}, \phi, \psi) \in \mathcal{V}$ and a.e. in $(0,T)$, with $\mathbf{u}_1 = \mathbf{u} + \mathbf{u_0}$, $\mathcal{T}_1 = \mathcal{T} + \mathcal{T}_0$ and $\mathcal{S}_1 = \mathcal{S} + \mathcal{S}_0$. □

Remark 12.30. As indicated in Remark 10.31, for numerical resolution of the robust control problems, the reader is referred to Chapter 9. ◇

Remark 12.31. It is clear that we can treat, by using the techniques developed in this chapter and more generally in this book, other physical models concerning the ocean, atmosphere or coupled ocean–atmosphere, for example the problems presented by Lions *et al.* [205, 206, 207] and Belmiloudi [37]. ◇

[10] Where $\mathcal{Z} = \{Y \in L^2(0,T;\mathcal{V}) \cap \mathcal{C}([0,T];\mathcal{H}) : \partial Y/\partial t \in L^2(0,T;\mathcal{V}')\}$.

13

Heat Transfer Laws on Temperature Distribution in Biological Tissues

This chapter considers non-linear robust control problems governed by some generalized transient bioheat transfer type models in biological systems with directional blood flow and Robin boundary conditions. The model equation depends on the blood perfusion rate, the heat transfer parameter, the distributed energy source terms and the heat flux due to the evaporation, which affect the effects of thermal and physical properties on the transient temperature of biological tissues. The result can be very beneficial for thermal diagnostics and treatments in medical practices, for example for laser surgery, and photo- and thermotherapy for regional hyperthermia, oftenly used in the treatment of cancer. First, the mathematical models are introduced and the existence, uniqueness and regularity of the solution of the state equation are proved as well as the stability and maximum principle under extra assumptions. Second, the control problems are formulated for different situations in order to control and stabilize the desired online temperature. An optimal solution is proven to exist and finally the necessary optimality conditions are given.

Our problem incorporates the effect of blood flow in the heat transfer equation in a way that captures the directionality of the blood flow and incorporates the convection features of the heat transfer between blood and solid tissue.

13.1 Introduction

13.1.1 Motivation and Statement of the Problem

The mathematical problem studied in this chapter is derived from the modeling of the transport of thermal energy in living tissues. The evaluation of thermal conductivities in living tissues is a very complex process which uses different phenomenological mechanisms including conduction, convection, radiation, metabolism, evaporation and others. Moreover, blood flow and extracellular water affect considerably both the heat transfer in the tissues and the tissues thermal properties. The bioheat transfer process in tissues is dependent on the behavior of blood perfusion along the vascular system. The first model, taking account of the blood perfusion, was introduced by Pennes [237] in 1948. The model is based on the classical thermal diffusion system, by incorporating the effects of metabolism and blood perfusion. The Pennes model has been adapted by many biologists for the analysis of various heat transfer phenomena in a living body. Others, after evaluating the Pennes model in specific situations, have concluded that many of the hypotheses (which are foundational to the model) are not valid. They then modified and generalized the model for these systems, see, *e.g.*, Arkin *et al.* [11], Chato [75], Chen and Holmes [77], Hirst [158], Valvano *et al.* [287] and Weinbaum and Jiji [298] (see also, *e.g.*, Charney [74] for a review on the mathematical modeling of the influence of blood perfusion).

The goal of this chapter is to use the robust control technique in order to analyze the effects of the fluctuations and the uncertainties on the transient temperature of biological tissues. To treat the system of motion in living body, we consider the following transient bioheat transfer type model introduced by Belmiloudi [48]:[1]

$$\frac{\partial U}{\partial t} = \operatorname{div}(\kappa(x)\nabla U) - p(U - U_{\mathrm{a}}) - K(\varpi, U) + F(x,t,U) + f \quad \text{on } \mathcal{Q},$$

subject to the heat flux boundary condition

$$(\kappa\nabla U).\mathbf{n} = -q(U - U_{\mathrm{b}}) + g \quad \text{on } \Sigma = \partial\Omega \times (0,T), \tag{13.1}$$

and the initial condition

$$U(0) = U_0 \quad \text{on } \Omega,$$

under the pointwise constraints

$$a_1 \le p \le a_2 \quad \text{a.e. in } \mathcal{Q}, \tag{13.2}$$

where the cylinder $\mathcal{Q} = \Omega \times (0,T)$, the state function U is the temperature distribution, $T > 0$ is a fixed constant (a given final time), the body Ω is an open bounded domain in $\mathbb{R}^m$, $m \le 3$ with a smooth boundary $\Gamma = \partial\Omega$ which is sufficiently regular, and Ω is totally on one side of Γ, $\mathbf{n}$ is the unit outward normal to Γ and a_i, for $i = 1, 2$, are given positive constants. The quantity p is the blood perfusion rate and q describes the heat transfer coefficient. The heat capacity is assumed to be constant and thermal conductivity of tissue κ is assumed to be variable and satisfies $\nu \ge \kappa = \sigma^2 \ge \mu > 0$ (where ν and μ are two positive constants). The second term on the right-hand side of the state Equation (13.1) describes the heat transport between the tissue and microcirculatory blood perfusion, the third term K corresponds, for example, to the directional convective mechanism of heat transfer due to blood flow, the fourth term F is the body heating function which describes the physical

[1] To simplify the presentation we have neglected the radiative terms.

properties of material (depending for example on the thermal absorptivity).[2] The source term f describes a distributed energy source which can be generated through a variety of sources, such as focused ultrasound, radio-frequency, microwave, resistive heating, laser beams and others (depending on the difference between the energy generated by the metabolic processes and the heat exchanged between, for example, the electrode and the tissue). The first term on the right-hand side of the boundary condition in (13.1) describes the convective component and the second term g is the heat flux due to evaporation. The function U_{a} is the blood temperature, and the function U_{b} is the bolus temperature, and they are assumed to be in $L^{\infty}(\mathcal{Q})$ and $L^{\infty}(\Sigma)$, respectively. The initial value U_0 is assumed to be in $L^2(\Omega)$.

13.1.2 Thermal Damage Calculations

After obtaining the temperature distribution, we can calculate the accumulation of thermal damage, which is associated with injury to tissue. For this, we can use the well known Arrhennuis damage integral formulation (see, *e.g.*, Tropea and Lee [285]):

$$D(x, \tau_{\mathrm{exp}}) = \ln\left(\frac{C(0)}{C(\tau_{\mathrm{exp}})}\right) = A \int_0^{\tau_{\mathrm{exp}}} \exp\left(\frac{-E}{RU(x,t)}\right) \mathrm{d}t, \qquad (13.3)$$

where D is the non-dimensional degree of tissue injury, C is the concentration of living cells, τ_{exp} is the duration of the exposure, A is the molecular collision frequency (s^{-1}), E is the denaturation activation energy ($\mathrm{J\,mol}^{-1}$) and R is the universal gas constant, equal to $8.314\,\mathrm{J\,mol}^{-1}\,\mathrm{K}^{-1}$. The parameters A and E are dependent on the type of tissue, and the cumulative damage can be interpreted as the fraction of hypothetical indicator molecules that are denatured.

Example 13.1. (Examples of models)

1. For the Pennes model, the operator K is null and the operator F is constant.
2. In the Chen–Holmes model [77] (the model has been formulated after the analyzing of blood vessel thermal equilibration length), the model incorporates the effect of blood flow in the heat transfer equation in a way that captures the directionality of the blood flow and incorporates the convection features of the heat transfer between blood and solid tissue,[3] see Figure 13.1. The operator K is then the transport operator $\varpi.\nabla$ and it is

---

[2] The introduction of the non-linear term F in the bioheat system is very important, because the physical properties of material have power law dependence on temperature.

[3] The blood-perfused tumor tissue volume, including blood flow in microvascular bed with the blood flow direction, contains many vessels and can be regarded as a porous medium consisting of a tumor tissue (a solid domain) fully filled with blood (a liquid domain).

corresponding to a directional convective term due to the net flux of the equilibrated blood. The operator $F = F(x,t)$ is independent of the temperature. The conductivity term is the sum of two terms: $\kappa = \kappa_t + \kappa_p$, where $\mathrm{div}(\kappa_p \nabla T)$ is corresponding to the enhancement of thermal conductivity in tissue due to the flow of blood within thermally significant blood vessels and the term $\mathrm{div}(\kappa_t \nabla T)$ is similar to the Pennes model. Moreover, if we assume that the velocity field ϖ is known and satisfies $\varpi \in L^\infty(Q)$ and $\mathrm{div}(\varpi) = 0$, so we have easily the estimate given in hypothesis (**H2**) (given below) where $\gamma_v = \| \varpi \|_{L^\infty(Q)}$, and that $K^* = -K$.
3. In biological modeling, the non-linear operator F can be chosen as a polynomial functions of the temperature. ♣

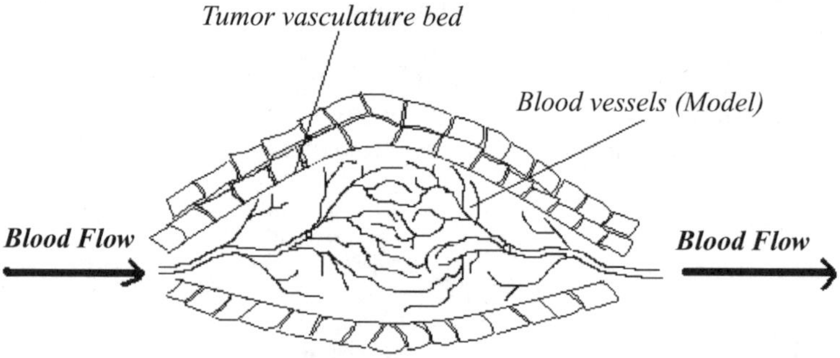

Figure 13.1. Relationship between tumor vascular and blood flow direction.

13.1.3 Background and Motivation

The mathematical modeling of cancer treatments (chemotherapy, thermotherapy, *etc.*) is an highly challenging area of applied mathematics. Recently, a large number of studies and research related to the cancer treatments, in particular by chemotherapy or thermotherapy, have been the object of numerous developments.

Various problems associated with cancer chemotherapy (drug treatment) have been intensively studied (see, *e.g.*, De Pillis *et al.* [99], Liu and Freedman [209], Mishra *et al.* [224], and the references therein). In order to control the drug dosage administered, many control problems have been studied (see, *e.g.*, Alamir and Chareyron [7], Belmiloudi [47], Fister and Panetta [124], Jackson and Byrne [168], Kimmel and Swierniak [174], Ledzewicz and Sachttler [189, 190], and the references therein).

As an alternative to the traditional surgical treatment or to enhance the effect of conventional chemotherapy, various problems associated with localized thermal therapy have been intensively studied, see *e.g.*, Deuflhard and Seebass [100], Hill and Pincombe [156], Liu *et al.* [208], Marchant and Lui

[216], Martin *et al.* [219], Pincombe and Smyth [242], Seip and Ebbini [266], Shih *et al.* [268], Sturesson and Engels [278], Tropea and Lee [285], and the references therein. In order to improve the treatments, several approaches have been proposed recently to control the temperature during thermal therapy. We can mention, *e.g.*, Belmiloudi [46, 48], Bohm *et al.* [54], Hutchinson *et al.* [163], Kohler *et al.* [176], Kowalski and Jin [181], Malinen *et al.* [215], Ganzler *et al.* [126], Vanne and Hynynen [289], and the references therein. The essential aspect of these contributions has been the numerical analysis, MRI-based optimization techniques and mathematical analysis.

An important application of all bioheat transfer models in interdisciplinary research areas, joining the mathematical, biological and medical fields, is the analysis of the temperature field which develops in living tissue when heat is applied to the tissue, especially in clinical cancer therapy hyperthermia and in accidental heating injury, such as burns (in hyperthermia, tissue is heated to enhance the effect of an accompanying radio or chemotherapy). Indeed, thermal therapy (performed with a laser, focused ultrasound or microwaves) gives the possibility of destroying the pathological tissues with minimal damage to the surrounding tissues. Moreover, due to the self-regulating capability of the biological tissue, the blood perfusion depends on the evolution of the temperature and vary significantly between different patients, and between different therapy sessions (for the same patient). Consequently, in order to have optimal thermal diagnostics and maximize the benefit of the therapy to the patient, it is necessary to study the value of the perfusion parameter.

The new feature introduced in this chapter concerns the study of robust control problems for a generalized non-linear evolutive bioheat transfer system, where the goal is to stabilize and control the desired online temperature control m provided by magnetic resonance imaging (MRI) measurements (MRI is a new efficient tool in medicine in order to control surgery and treatments). The online temperature can be modeled by the relation $\mathfrak{m} \approx \gamma U + \delta P$, where the non-negative functions δ and γ satisfy $\gamma \approx \delta$ in muscle and $\gamma \leq \delta$ in fat.

13.1.4 Assumptions and Notations

First, we state the main hypotheses on the body heating coefficient F and the operator K:

(H1) F is a Carathéodory function from $\mathcal{Q} \times \mathbb{R}$ into $\mathbb{R}$ such that

(*i*) for almost all $(x,t) \in \mathcal{Q}$, $F(x,t,.)$ is Lipschitz and bounded function with

$$| F(x,t,r) | \leq M_1, \ \forall \ r \in \mathbb{R} \text{ and a.e. in } \mathcal{Q},$$

(*ii*) F is differentiable and the partial derivatives $F'_x(.,r) = (\partial F/\partial x)(.,r)$ and $G = F'_r(.,r) = (\partial F/\partial r)(.,r)$ are Lipschitz continuous in $\mathcal{Q}$ for all $r \in \mathbb{R}$, and are globally bounded in $\mathcal{Q} \times \mathbb{R}$.

(H2) For all sufficiently regular functions ϖ, $K(\varpi, .)$ is a linear operator and satisfies the following estimate: there exists a constant $\gamma_v \geq 0$ such that $\| K(\varpi, v) \|_{L^2(\Omega)} \leq \gamma_v \| v \|_{H^1(\Omega)}, \forall v \in H^1(\Omega)$.[4]
We denote by K^* the adjoint of K, *i.e.*,

$$\langle K^*(\varpi, u), v\rangle = \langle K(\varpi, v), u\rangle, \forall (u, v) \in (H^1(\Omega))^2.$$

We can now introduce the following spaces:

$$\mathcal{H} = L^\infty(0, T; L^2(\Omega)), \quad V = H^1(\Omega), \quad \mathcal{V} = L^2(0, T; V),$$
$$\mathcal{W} = \{w \in L^2(0, T; V) : \frac{\partial w}{\partial t} \in L^2(0, T; V')\}$$

and the set of the admissible functions describing the constraints

$$C_{\text{ad}} = \{p \in L^2(\mathcal{Q}) : a_1 \leq p \leq a_2 \text{ a.e. in } \mathcal{Q}\},$$

where a_i, for $i = 1, 2$, are two constants.

In the rest of this application, we may omit the variables x and t in their functions without ambiguity.

Lemma 13.2. *Let u_l be a sequence converging toward u in $\mathcal{W}$ weakly and in $L^2(\mathcal{Q})$ strongly we have $F(., u_l) \longrightarrow F(., u)$ in $L^p(\mathcal{Q})$ strongly $\forall p \in [1, +\infty)$.*

Proof. We prove easily the result by using the convergence result of u_l and assumption (i) of (**H1**). □

13.2 The State System

In this section we present the existence, uniqueness and regularity of the solution of the problem (13.1) and obtain stability and maximum principle results.

13.2.1 Existence and Stability Results

A function $U \in \mathcal{W}$ is a weak solution of system (13.1) provided, for all $v \in V$ and a.e. $t \in (0, T)$,

$$\begin{aligned} \langle\frac{\partial U}{\partial t}, v\rangle &+ \int_\Omega \kappa \nabla U.\nabla v \mathrm{d}x + \int_\Omega K(\varpi, U) v \mathrm{d}x \\ &+ \int_\Gamma q(U - U_{\text{b}}) v \mathrm{d}\Gamma + \int_\Omega p(U - U_{\text{a}}) v \mathrm{d}x \\ &= \int_\Omega F(x, t, U) v \mathrm{d}x + \int_\Omega f v \mathrm{d}x + \int_\Gamma g v \mathrm{d}\Gamma, \\ U(0) &= U_0 \quad \text{on } \Omega, \end{aligned} \tag{13.4}$$

where $\langle ., .\rangle$ denotes the duality between V' and V.

[4] The constant γ_v depends on the norm of the given function ϖ.

Theorem 13.3. *Let assumptions* **(H1)**–**(H2)** *be fulfilled:*

(*i*) *Let the initial condition* U_0 *be given in* $L^2(\Omega)$ *and the source term* (p, q, f, g) *be in* $C_{\mathrm{ad}} \times L^\infty(\Sigma) \times L^2(Q) \times L^2(\Sigma)$. *Then there exists a unique solution* U *in* $\mathcal{W} \cap \mathcal{H}$ *of problem (13.1).*

(*ii*) *Let* $q \in L^\infty(\Sigma)$ *and let* (p_i, f_i, g_i, U_{0i}), *for* $i = 1, 2$, *be two functions of* $C_{\mathrm{ad}} \times L^2(Q) \times L^2(\Sigma) \times L^2(\Omega)$. *If* $U_i \in \mathcal{W} \cap \mathcal{H}$ *is the solution of (13.1) corresponding to data* (p_i, f_i, g_i, U_{0i}), *for* $i = 1, 2$, *then, a.e.* $t \in (0, T)$

$$\| U \|^2_{\mathcal{W}\cap\mathcal{H}} \leq c(\| p \|^2_{L^2(Q)} + \| f \|^2_{L^2(Q)} + \| g \|^2_{L^2(\Sigma)} + \| U_0 \|^2_{L^2(\Omega)}), \quad (13.5)$$

where $U = U_1 - U_2$, $p = p_1 - p_2$, $f = f_1 - f_2$, $g = g_1 - g_2$ *and* $U_0 = U_{01} - U_{02}$.

Proof. To prove the existence of a solution, we use the Faedo–Galerkin method.

Set $(v_1, \ldots, v_l, \ldots)$ a total and free sequence in V. Set $u_l = \sum_{i=1}^{l} g_{il}(t) v_i$ (the functions g_{il} are scalar functions defined on $[0, T]$) as an approximation of the solution of problem (13.4), verifying the following problem

$$\begin{aligned}
&\langle \frac{\partial u_l}{\partial t}, v_k \rangle + \int_\Omega \kappa \nabla u_l . \nabla v_k \mathrm{d}x + \int_\Omega K(\varpi, u_l) v_k \mathrm{d}x \\
&\qquad + \int_\Gamma q(u_l - U_{\mathrm{b}}) v_k \mathrm{d}\Gamma + \int_\Omega p(u_l - U_{\mathrm{a}}) v_k \mathrm{d}x \\
&\quad = \int_\Omega F(x, t, u_l) v_k \mathrm{d}x + \int_\Omega f v_k \mathrm{d}x + \int_\Gamma g v_k \mathrm{d}\Gamma, \\
&u_l(0) = P_l U_0 \quad \text{on } \Omega,
\end{aligned} \qquad (13.6)$$

where P_l is the L^2-projector onto the space spanned by $v_1, \ldots, v_l$ (the initial data U_0 satisfies then $\| P_l U_0 \|_{L^2(\Omega)} \leq \| U_0 \|_{L^2(\Omega)}$ and $P_l U_0 \longrightarrow U_0$ in $L^2(\Omega)$ as $l \longrightarrow \infty$).

We can deduce from (13.6) that

$$\begin{aligned}
&\frac{1}{2}\frac{\mathrm{d}}{\mathrm{d}t} \| u_l \|^2_{L^2} + \int_\Omega \kappa \mid \nabla u_l \mid^2 \mathrm{d}x + \int_\Omega K(\varpi, u_l) u_l \mathrm{d}x + \int_\Gamma q(u_l - U_{\mathrm{b}}) u_l \mathrm{d}\Gamma \\
&\quad + \int_\Omega p(u_l - U_{\mathrm{a}}) u_l \mathrm{d}x = \int_\Omega F(x, t, u_l) u_l \mathrm{d}x + \int_\Omega f u_l \mathrm{d}x + \int_\Gamma g u_l \mathrm{d}\Gamma
\end{aligned}$$

and then

$$\begin{aligned}
&\frac{1}{2}\frac{\mathrm{d}}{\mathrm{d}t} \| u_l \|^2_{L^2} + \int_\Omega \kappa \mid \nabla u_l \mid^2 \mathrm{d}x + \int_\Gamma q \mid u_l \mid^2 \mathrm{d}\Gamma + \int_\Omega p \mid u_l \mid^2 \mathrm{d}x \\
&\quad = - \int_\Omega K(\varpi, u_l) u_l \mathrm{d}x + \int_\Gamma q U_{\mathrm{b}} u_l \mathrm{d}\Gamma \\
&\qquad + \int_\Omega p U_{\mathrm{a}} u_l \mathrm{d}x + \int_\Omega F(x, t, u_l) u_l \mathrm{d}x + \int_\Omega f u_l \mathrm{d}x + \int_\Gamma g u_l \mathrm{d}\Gamma.
\end{aligned}$$

According to the positivity of the function q, the boundedness of the functions p, U_{a}, U_{b} and assumptions (**H1**)–(**H2**) we can deduce by using Young's inequality that

$$\frac{\mathrm{d}}{\mathrm{d}t} \| u_l \|_{L^2}^2 + \mu \| \nabla u_l \|_{L^2}^2 \le c_1 \| u_l \|_{L^2}^2 + c_2 (\| p \|_{L^2}^2 + \| f \|_{L^2}^2 + \| g \|_{L^2(\Gamma)}^2),$$

which implies that u_l is bounded in $L^2(0,T;V) \cap L^\infty(0,T;L^2(\Omega))$.

Prove now that $\partial u_l/\partial t \in L^2(0,T;V')$. According to (13.6) we have that, for all v in V,

$$\begin{aligned}\langle \frac{\partial u_l}{\partial t}, v\rangle = &- \int_\Omega \kappa \nabla u_l . \nabla v \mathrm{d}x - \int_\Omega K(\varpi, u_l) v \mathrm{d}x \\ &- \int_\Gamma q(u_l - U_{\mathrm{b}}) v \mathrm{d}\Gamma - \int_\Omega p(u_l - U_{\mathrm{a}}) v \mathrm{d}x \\ &+ \int_\Omega F(x,t,u_l) v \mathrm{d}x + \int_\Omega f v \mathrm{d}x + \int_\Omega g v \mathrm{d}\Gamma.\end{aligned}$$

Thanks to the boundedness of p, q, U_{a} and U_{b} we then have

$$| \langle \frac{\partial u_l}{\partial t}, v\rangle | \le c_1 \| u_l \|_{H^1} \| v \|_{H^1} + c_2 (\| f \|_{L^2} + \| g \|_{L^2(\Gamma)}) \| v \|_{H^1} .$$

Applying this result (according to Young's formula) we can deduce that

$$\| \frac{\partial u_l}{\partial t} \|_{V'} \le c_1 \| u_l \|_{H^1} + c_2 (\| f \|_{L^2} + \| g \|_{L^2(\Gamma)}). \tag{13.7}$$

Since u_l is bounded in $L^\infty(0,T;L^2(\Omega)) \cap L^2(0,T;H^1(\Omega))$, we can deduce that $\partial u_l/\partial t$ is bounded in $L^2(0,T;V')$.

According to Lemma 6.6 the injection of $\mathcal{W}$ into $L^2(0,T;L^2(\Omega))$ is compact. So this result makes it possible to extract from u_l a subsequence also denoted by u_l and such that

$$\begin{aligned} &u_l \longrightarrow U \text{ strongly in } L^2(0,T;L^2(\Omega)), \\ &u_l \rightharpoonup U \text{ weakly in } L^2(0,T;H^1(\Omega)). \end{aligned} \tag{13.8}$$

By using the previous results we can prove easily that U is the solution of problem (13.1) and verifies the regularity $U \in L^2(0,T;V) \cap L^\infty(0,T;L^2(\Omega))$ and $\partial U/\partial t \in L^2(0,T;V')$.

Now we prove the uniqueness and the stability results given in (*ii*). To obtain the uniqueness result, we suppose that there exist two solutions U_1, U_2 of (13.1). Then $U = U_1 - U_2$ is a solution of the following problem:

$$\begin{aligned} &\frac{\partial U}{\partial t} - \operatorname{div}(\kappa \nabla U) + K(\varpi, U) + pU = F(.,U_1) - F(.,U_2) \quad \text{on } \mathcal{Q}, \\ &(\kappa \nabla U).\mathbf{n} = -qU \quad \text{on } \Sigma, \\ &U(0) = 0 \quad \text{on } \Omega. \end{aligned}$$

Multiplying the previous system by U and integrating over $\Omega \times (0,t)$, for $t \in (0,T)$ (by using Green's formula), gives

$$\frac{1}{2} \| U(t) \|_{L^2}^2 + \int_0^t \| \sigma \nabla U \|_{L^2}^2 \,\mathrm{d}s + \int_0^t\!\!\int_\Omega p \mid U \mid^2 \mathrm{d}x\mathrm{d}s + \int_0^t\!\!\int_\Gamma q \mid U \mid^2 \mathrm{d}\Gamma\mathrm{d}s$$
$$= \int_0^t\!\!\int_\Omega (F(x,s,U_1) - F(x,s,U_2))U\mathrm{d}x\mathrm{d}s - \int_0^t\!\!\int_\Omega K(\varpi,U)U\mathrm{d}x\mathrm{d}s.$$

By using the hypotheses (**H1**)–(**H2**), the regularity of p and q, and the positivity of q we have that

$$\| U \|_{L^2}^2 (t) + \mu \int_0^t \| \nabla U \|_{L^2}^2 \,\mathrm{d}s \leq c_1 \int_0^t \| U \|_{L^2}^2 \,\mathrm{d}s.$$

According to Gronwall's formula (since $U(0) = 0$) we prove easily that $U_1 = U_2$ and then the uniqueness result follows.

To prove the estimate given in (*ii*), we set $p = p_1 - p_2$, $f = f_1 - f_2$, $g = g_1 - g_2$, $U_0 = U_{01} - U_{02}$ and $U = U_1 - U_2$. Then U is the solution of

$$\frac{\partial U}{\partial t} - \mathrm{div}(\kappa \nabla U) + p(U_2 - U_\mathrm{a}) + p_1 U + K(\varpi, U)$$
$$= F(.,U_1) - F(.,U_2) + f \quad \text{on } \mathcal{Q},$$
$$(\kappa \nabla U).\mathbf{n} = -qU + g \quad \text{on } \Sigma,$$
$$U(0) = U_0 \quad \text{on } \Omega.$$

Multiplying the previous system by U and integrating over $\Omega \times (0,t)$, for $t \in (0,T)$ (by using Green's formula), gives

$$\frac{1}{2} \| U(t) \|_{L^2}^2 + \int_0^t \| \sigma \nabla U \|_{L^2}^2 \,\mathrm{d}s + \int_0^t\!\!\int_\Omega p_1 \mid U \mid^2 \mathrm{d}x\mathrm{d}s$$
$$+ \int_0^t\!\!\int_\Gamma q \mid U \mid^2 \mathrm{d}\Gamma\mathrm{d}s = - \int_0^t\!\!\int_\Omega K(\varpi,U)U\mathrm{d}x\mathrm{d}s$$
$$- \int_0^t\!\!\int_\Omega p(u_2 - U_\mathrm{a})U\mathrm{d}x\mathrm{d}s$$
$$+ \int_0^t\!\!\int_\Omega (F(x,s,U_1) - F(x,s,U_2))U\mathrm{d}x\mathrm{d}s$$
$$+ \frac{1}{2} \| U_0 \|_{L^2}^2 + \int_0^t\!\!\int_\Omega fU\mathrm{d}x\mathrm{d}s + \int_0^t\!\!\int_\Gamma gU\mathrm{d}\Gamma\mathrm{d}s.$$

According to the regularity of U_i, p_i, f_i, g_i (for $i = 1,2$), q, U_a and U_b, and to the hypotheses (**H1**)–(**H2**) we have, by using Gronwall's formula,

$$\| U \|_{\mathcal{V}\cap\mathcal{H}}^2 \leq c(\| p \|_{L^2(\mathcal{Q})}^2 + \| f \|_{L^2(\mathcal{Q})}^2 + \| g \|_{L^2(\Sigma)}^2 + \| U_0 \|_{L^2(\Omega)}^2). \quad (13.9)$$

By using similar arguments as used to obtain the estimates (13.7) and (13.9) we can deduce that

$$\begin{aligned} &\| \frac{\partial U}{\partial t} \|^2_{L^2(0,T;V')} \\ &\qquad \le c(\| p \|^2_{L^2(\mathcal{Q})} + \| f \|^2_{L^2(\mathcal{Q})} + \| g \|^2_{L^2(\Sigma)} + \| U_0 \|^2_{L^2(\Omega)}). \end{aligned} \tag{13.10}$$

This completes the proof. □

13.2.2 A Maximum Principle

In this section we establish a maximum principle under extra assumptions on the data. In addition to (**H1**)–(**H2**), we suppose, for a couple of constants $(u_{\mathrm{f}}, u_{\mathrm{s}})$ such that $0 \le u_{\mathrm{f}} < u_{\mathrm{s}}$, the following assumptions

(**H3**) $F(x,t,.) = 0$ in $]-\infty, u_{\mathrm{f}}] \cup [u_{\mathrm{s}}, +\infty[$, for almost all $(x,t) \in \mathcal{Q}$.
(**H4**) $K(\varpi, s) = 0$, for all constant $s \in \mathbb{R}$ and

$$\int_\Omega K(\varpi, v) v \mathrm{d}x \ge 0 \quad \text{for all } v \in H^1(\Omega).$$

(**H5**) $u_{\mathrm{f}} \le U_{\mathrm{a}} \le u_{\mathrm{s}}$ and $u_{\mathrm{f}} \le U_{\mathrm{b}} \le u_{\mathrm{s}}$ for all in $\mathcal{Q}$ and in Σ, respectively.

Then we have the following theorem.

Theorem 13.4. *Let* (**H1**)–(**H5**) *be fulfilled. Suppose that the initial data* U_0 *is such that* $u_{\mathrm{f}} \le U_0 \le u_{\mathrm{s}}$, *a.e. in* Ω *and* f, g *are positive functions. Then, the weak solution* $U \in \mathcal{W} \cap \mathcal{H}$ *of (13.1) satisfies, for all* $t \in (0,T)$, $u_{\mathrm{f}} \le U(.,t) \le u_{\mathrm{s}}$ *a.e. in* Ω.

Proof. Let us consider the following notations: $r^+ = \max(r, 0)$, $r^- = (-r)^+$ and then $r = r^+ - r^-$.

We prove now that if $U_0 \ge u_{\mathrm{f}}$, a.e. in Ω then $U(.,t) \ge u_{\mathrm{f}}$, for all $t \in [0,T]$ and a.e. in Ω. According to Chapter 3, we have that $(U - u_{\mathrm{f}})^- \in L^2(0,T; H^1(\Omega))$ with $\nabla(U-u_{\mathrm{f}})^- = -\nabla(U - u_{\mathrm{f}})$ if $(U - u_{\mathrm{f}}) > 0$ and $\nabla(U - u_{\mathrm{f}})^- = 0$ otherwise, a.e. in $\mathcal{Q}$. Then, taking $v = -(U - u_{\mathrm{f}})^-$ in Equation (13.4), we have (a.e. in $(0,T)$)

$$\begin{aligned} &\frac{\mathrm{d}}{2\mathrm{d}t} \| (U - u_{\mathrm{f}})^- \|^2_{L^2} + \int_\Omega \kappa \mid \nabla (U - u_{\mathrm{f}})^- \mid^2 \mathrm{d}x \\ &\qquad + \int_\Omega K(\varpi, (U - u_{\mathrm{f}})^-)(U - u_{\mathrm{f}})^- \mathrm{d}x \\ &\qquad + \int_\Gamma q(U_{\mathrm{b}} - u_{\mathrm{f}})(U - u_{\mathrm{f}})^- \mathrm{d}\Gamma + \int_\Gamma q((U - u_{\mathrm{f}})^-)^2 \mathrm{d}\Gamma \\ &\qquad + \int_\Omega p(U_{\mathrm{a}} - u_{\mathrm{f}})(U - u_{\mathrm{f}})^- \mathrm{d}x + \int_\Omega p((U - u_{\mathrm{f}})^-)^2 \mathrm{d}x \\ &\qquad\quad = - \int_\Omega F(x,t,U)(U - u_{\mathrm{f}})^- \mathrm{d}x - \int_\Omega f (U - u_{\mathrm{f}})^- \mathrm{d}x \\ &\qquad\quad\quad - \int_\Gamma g(U - u_{\mathrm{f}})^- \mathrm{d}\Gamma. \end{aligned}$$

According to hypothesis (**H3**), and the assumptions $U_{\rm a} \geq u_{\rm f}$ and $U_{\rm b} \geq u_{\rm f}$, we can deduce that

$$\frac{\rm d}{2{\rm d}t} \parallel (U - u_{\rm f})^- \parallel_{L^2}^2 \leq 0$$

and then, for all time $t \in (0, T)$ (by integrating with respect to time),

$$\parallel (U - u_{\rm f})^- \parallel_{L^2}^2 (t) \leq \parallel (U_0 - u_{\rm f})^- \parallel_{L^2}^2 .$$

Using the assumption $U_0 \geq u_{\rm f}$ we can deduce that $U(t, .) \geq u_{\rm f}$ for all $t \in (0, T)$ and a.e. in Ω.

To prove that, for all $t \in (0, T)$, $U(., t) \leq u_{\rm s}$ a.e. in Ω, we take $v = (U - u_{\rm s})^+$ in Equation (13.4) and use the same technique as before.

This completes the proof. □

13.3 The Perturbation Problem

In this section we formulate the perturbation problem and present the existence, uniqueness and regularity results of the perturbation solution.

13.3.1 Formulation of the Perturbation Problem

In the following, the solution U of problem (13.1) will be treated as the target function. We are then interested in the robust regulation of the deviation of the problem from the desired target U.

We analyze the full non-linear equation which models large perturbations u to the target U, *i.e.*, we assume that U satisfies the problem (13.1) with the data $(U_0, P, f, g, U_{\rm a}, U_{\rm b})$ and $U + u$ satisfies problem (13.1) with the data $(U_0 + u_0, P + p, f + \xi, g + \eta, U_{\rm a} + u_{\rm a}, U_{\rm b} + u_{\rm b})$.

Hence we consider the following system (for a given U satisfying the regularity of Theorem 13.3):

$$\begin{aligned}
&\frac{\partial u}{\partial t} = \mathrm{div}(\kappa \nabla u) - P(u - u_{\rm a}) - p(u - u_{\rm a} + v_{\rm a}) \\
&\qquad -K(\varpi, u) + (F(., u + U) - F(., U)) + \xi \quad \text{on } Q, \\
&\text{subject to the heat flux boundary condition} \\
&(\kappa \nabla u).\mathbf{n} = -q(u - u_{\rm b}) + \eta \quad \text{on } \Sigma, \\
&\text{and the initial condition} \\
&u(0) = u_0 \quad \text{on } \Omega,
\end{aligned} \tag{13.11}$$

where $v_{\rm a} = U - U_{\rm a}$.

If we set $\tilde{F}(., u) = F(., u + U) - F(., U)$ then (13.11) is reduced to the following system:

$$\frac{\partial u}{\partial t} = \operatorname{div}(\kappa \nabla u) - P(u - u_{\mathrm{a}}) - p(u - u_{\mathrm{a}} + v_{\mathrm{a}})$$
$$-K(\varpi, u) + \tilde{F}(x, t, u) + \xi \quad \text{on } \mathcal{Q},$$

subject to the heat flux boundary condition (13.12)

$$(\kappa \nabla u).\mathbf{n} = -q(u - u_{\mathrm{b}}) + \eta \quad \text{on } \Sigma,$$

and the initial condition

$$u(0) = u_0 \quad \text{on } \Omega.$$

Remark 13.5. (*i*) We can easily verify that $\tilde{F}$ satisfies the same hypotheses that F, *i.e.*, (**H1**)–(**H2**).
(*ii*) For simplicity of future reference, we omit the "~" on $\tilde{F}$. ◇

Next we give the weak formulation associated with the problem (13.12). Multiplying the first part of (13.12) by $v \in V$ and integrating over Ω this gives (according to the second part of (13.12)) the weak formulation (a.e. $t \in (0, T)$)

$$\begin{aligned} \langle \frac{\partial u}{\partial t}, v\rangle + \int_\Omega \kappa \nabla u.\nabla v \mathrm{d}x + \int_\Omega K(\varpi, u) v \mathrm{d}x + \int_\Gamma q(u - u_{\mathrm{b}}) v \mathrm{d}\Gamma \\ + \int_\Omega P(u - u_{\mathrm{a}}) v \mathrm{d}x + \int_\Omega p(u - u_{\mathrm{a}} + v_{\mathrm{a}}) v \mathrm{d}x \\ = \int_\Omega F(., u) v \mathrm{d}x + \int_\Omega \xi v \mathrm{d}x + \int_\Gamma \eta v \mathrm{d}\Gamma, \end{aligned} \tag{13.13}$$
$$u(0) = u_0 \quad \text{on } \Omega.$$

13.3.2 Existence and Stability Results

Now we show the existence and uniqueness of the solution to the problem (13.13), and give some Lipschitz continuity results.

Theorem 13.6. *Let assumptions* (**H1**)–(**H2**) *be fulfilled:*

(*i*) *Let the initial condition* u_0 *be given in* $L^2(\Omega)$ *and the source term* (p, ξ, η) *be in* $L^\infty(\mathcal{Q}) \times L^2(\mathcal{Q}) \times L^2(\Sigma)$. *Then there exists a unique solution* u *in* $\mathcal{W} \cap \mathcal{H}$ *of (13.12).*

(*ii*) *Let* $(p_i, \xi_i, \eta_i, u_{0i})$, *for* $i = 1, 2$, *be two functions of* $L^\infty(\mathcal{Q}) \times L^2(\mathcal{Q}) \times L^2(\Sigma)$. *If* $u_i \in \mathcal{W} \cap \mathcal{H}$ *is the solution of (13.12) corresponding to data* $(p_i, \xi_i, \eta_i, u_{0i})$, $i = 1, 2$, *then the following Lipshitz continuity result holds:*

$$\| u \|^2_{\mathcal{W}\cap\mathcal{H}} \leq c(\| p \|^2_{L^2(\mathcal{Q})} + \| \xi \|^2_{L^2(\mathcal{Q})} + \| \eta \|^2_{L^2(\Sigma)} + \| u_0 \|^2_{L^2(\Omega)}), \tag{13.14}$$

where $u = u_1 - u_2$, $p = p_1 - p_2$, $\xi = \xi_1 - \xi_2$, $\eta = \eta_1 - \eta_2$ *and* $u_0 = u_{01} - u_{02}$.

Proof. The proof of this result can be obtained by using a similar technique as in the proof of Theorem 13.3. So, we omit the details. □

13.4 Robust Control Problems

In this section we formulate the robust control problem and study the existence and necessary optimality conditions for an optimal solution.

13.4.1 Formulation of the Control Problem and Differentiability

Our problem, in this section, is to find the best admissible perfusion function in the presence of the worst disturbance in the distributed energy sources. We then suppose that the control is in the perfusion function p and the disturbance is in the force ξ, that is, $p = \varphi$ ($\varphi \in L^\infty(\mathcal{Q})$) and $\xi = \psi$ ($\psi \in L^2(\mathcal{Q})$). Therefore, the function u is assumed to be related to the disturbance ψ and control ϕ through the problem

$$\left.\begin{aligned} &\frac{\partial u}{\partial t} = \operatorname{div}(\kappa \nabla u) - P(u - u_{\mathrm{a}}) - \varphi(u - u_{\mathrm{a}} + v_{\mathrm{a}}) \\ &\qquad -K(\varpi, u) + F(.,u) + \psi \quad \text{on } \mathcal{Q}, \\ &\text{subject to the heat flux boundary condition} \\ &(\kappa \nabla u).\mathbf{n} = -q(u - u_{\mathrm{b}}) + \eta \quad \text{on } \Sigma, \\ &\text{and the initial condition} \\ &u(0) = u_0 \quad \text{on } \Omega, \end{aligned}\right. \tag{13.15}$$

under the pointwise constraint

$$\tau_1 \leq \varphi \leq \tau_2 \qquad \text{a.e. in } \mathcal{Q}, \tag{13.16}$$

where $[\tau_1, \tau_2]$ contains 0 and, $u_0 \in L^2(\Omega)$ and $\eta \in L^2(\Sigma)$ are assumed to be given.

Let $U_{\mathrm{ad}} = \{\varphi \in L^2(\mathcal{Q}) : \tau_1 \leq \varphi \leq \tau_2 \quad \text{a.e. in } \mathcal{Q}\} \subset L^\infty(\mathcal{Q})$. Our control problem is then,

$$\left.\begin{aligned} &\text{find } (\varphi, \psi) \in U_{\mathrm{ad}} \times V_{\mathrm{ad}} \text{ such that the following cost functional :} \\ &J(\varphi, \psi) = \frac{1}{2} \parallel \gamma u + \delta \varphi - \mathfrak{m}_{\mathrm{obs}} \parallel^2_{L^2(\mathcal{Q})} \\ &\qquad + \frac{\alpha}{2} \parallel \varphi \parallel^2_{L^2(\mathcal{Q})} - \frac{\beta}{2} \parallel \psi \parallel^2_{L^2(\mathcal{Q})}, \\ &\text{is minimized with respect to } \varphi \text{ and maximized with respect to } \psi \\ &\text{subject to the problem (13.15),} \end{aligned}\right. \tag{13.17}$$

where V_{ad} is a convex, closed, non-empty and bounded subset of $L^2(\mathcal{Q})$, the function $\mathfrak{m}_{\mathrm{obs}}$ is in $L^2(\mathcal{Q})$, the parameters α and β are fixed constants and the functions γ, δ are positive with space-time dependent entries such that

$$0 < \gamma_1 \leq \gamma \leq \gamma_2, \quad 0 < \delta_1 \leq \delta \leq \delta_2 \quad \text{on } \overline{\mathcal{Q}}.$$

We are now going to study the differentiability of the operator solution of problem (13.15).

Before proceeding to investigation of the differentiability of the function $\mathcal{F} : (\varphi, \psi) \longrightarrow u$, which maps the source term $(\varphi, \psi) \in L^\infty(\mathcal{Q}) \times L^2(\mathcal{Q})$ of problem (13.15) into the corresponding solution $u \in \mathcal{W} \cap \mathcal{H}$, we study, for $(h, \vartheta, \kappa, w_0)$ given data, the following problem:

find $w \in \mathcal{W} \cap \mathcal{H}$ such that, for a.e. $t \in (0,T)$)

$$
\begin{aligned}
&\frac{\partial w}{\partial t} = \operatorname{div}(\kappa \nabla w) - (P + \varphi) w - h(u - u_{\mathrm{a}} + v_{\mathrm{a}}) \\
&\qquad -K(\varpi, w) + G(., u) w + \vartheta \quad \text{on } \mathcal{Q}, \\
&\text{subject to the heat flux boundary condition} \\
&(\kappa \nabla w).\mathbf{n} = -qw + \theta \quad \text{on } \Sigma, \\
&\text{and the initial condition} \\
&w(0) = w_0 \quad \text{on } \Omega,
\end{aligned}
\tag{13.18}
$$

where $u = \mathcal{F}(\varphi, \psi)$ and G is the partial derivative function of F.

Theorem 13.7. *If u is in $\mathcal{W} \cap \mathcal{H}$, the following results hold:*

(i) For any $(h, \vartheta, \theta, w_0) \in L^\infty(\mathcal{Q}) \times L^2(\mathcal{Q}) \times L^2(\Sigma) \times L^2(\Omega)$, there exists a unique function $w \in \mathcal{W} \cap \mathcal{H}$, solution of problem (13.18), such that

$$
\begin{aligned}
\| w \|^2_{\mathcal{W} \cap \mathcal{H}} \leq C_e(&\| h \|^2_{L^2(\mathcal{Q})} + \| \vartheta \|^2_{L^2(\mathcal{Q})} \\
&+ \| \theta \|^2_{L^2(\Sigma)} + \| w_0 \|^2_{L^2(\Omega)}).
\end{aligned}
\tag{13.19}
$$

(ii) Let $(h_i, \vartheta_i, \theta_i, w_{0i})$, for $i = 1, 2$ be in $L^\infty(\mathcal{Q}) \times L^2(\mathcal{Q}) \times L^2(\Sigma) \times L^2(\Omega)$. If w_i is the solution of (13.18), corresponding to data $(h_i, \vartheta_i, \theta_i, w_{0i})$ for $i = 1, 2$, respectively, then

$$
\begin{aligned}
\| w \|^2_{\mathcal{W} \cap \mathcal{H}} \leq C_e(&\| h \|^2_{L^2(\mathcal{Q})} + \| \vartheta \|^2_{L^2(\mathcal{Q})} \\
&+ \| \theta \|^2_{L^2(\Sigma)} + \| w_0 \|^2_{L^2(\Omega)}),
\end{aligned}
\tag{13.20}
$$

where $w = w_1 - w_2$, $h = h_1 - h_2$, $w_0 = w_{01} - w_{02}$, $\vartheta = \vartheta_1 - \vartheta_2$ and $\theta = \theta_1 - \theta_2$.

Proof. The existence, uniqueness and Lipschitz continuity results of problem (13.18) are obtained in the same way as used to prove Theorem 13.6 and by using the regularity of u. So, we omit the details. □

We are now going to study the differentiability of the operator solution $\mathcal{F}$.

Theorem 13.8. *Let assumptions* (**H1**)–(**H2**) *be fulfilled, the initial condition u_0 be in $L^2(\Omega)$ and the perturbation of the heat flux g be in $L^2(\Sigma)$. Then the function $\mathcal{F}$ is differentiable with respect to $X = (\varphi, \psi) \in U_{\mathrm{ad}} \times V_{\mathrm{ad}}$ where* *its derivative* $\mathcal{F}'(X) : L^\infty(\mathcal{Q}) \times L^2(\mathcal{Q}) \longrightarrow \mathcal{H} \cap \mathcal{V}$ *is a linear operator such that* $w = \mathcal{F}'(X)\,Y$ *is the unique solution of the problem (13.18), with data* $Y = (h, \vartheta)$*, the given initial condition* $w(t = 0) = 0$ *and the boundary forcing* $\theta = 0$*, i.e.,*

$$\begin{aligned}
&\frac{\partial w}{\partial t} = \operatorname{div}(\kappa \nabla w) - (P + \varphi)w - h(u - u_{\mathrm{a}} + v_{\mathrm{a}}) \\
&\qquad -K(\varpi, w) + G(.,u)w + \vartheta \quad \text{on } \mathcal{Q}, \\
&\text{subject to the heat flux boundary condition} \\
&(\kappa \nabla w).\mathbf{n} = -qw \quad \text{on } \Sigma, \\
&\text{and the initial condition} \\
&w(0) = 0 \quad \text{on } \Omega.
\end{aligned} \tag{13.21}$$

Moreover, we have the estimate (13.20), i.e., for all w_i *is the solution of (13.18), with data* $(h_i, \vartheta_i) \in L^\infty(\mathcal{Q}) \times L^2(\mathcal{Q})$*, the given initial condition* $w(t = 0) = 0$ *and the boundary forcing* $\theta = 0$*, for* $i = 1, 2$*, respectively, the following estimate holds:*

$$\| w \|^2_{\mathcal{H}\cap\mathcal{V}} \leq C_e(\| h \|^2_{L^2(\mathcal{Q})} + \| \vartheta \|^2_{L^2(\mathcal{Q})}), \tag{13.22}$$

where $w = w_1 - w_2$, $h = h_1 - h_2$ *and* $\vartheta = \vartheta_1 - \vartheta_2$.

Moreover, for all $X_i = (\varphi_i, \psi_i) \in U_{\mathrm{ad}} \times V_{\mathrm{ad}}$*, for* $i = 1, 2$*, we have the following estimate, for all* $Y = (h, \vartheta) \in L^\infty(\mathcal{Q}) \times V_{\mathrm{ad}}$*:*

$$\begin{aligned}
&\| \mathcal{F}'(X_1).Y - \mathcal{F}'(X_2).Y \|^2_{\mathcal{H}\cap\mathcal{V}} \\
&\qquad \leq C_e(\| Y \|_{L^2(\mathcal{Q})\times L^2(\mathcal{Q})} \| X \|^2_{L^2(\mathcal{Q})\times L^2(\mathcal{Q})} \\
&\qquad\qquad + \| Y \|^2_{L^2(\mathcal{Q})\times L^2(\mathcal{Q})} \| X \|_{L^2(\mathcal{Q})\times L^2(\mathcal{Q})}),
\end{aligned} \tag{13.23}$$

where $\varphi = \varphi_1 - \varphi_2$, $\psi = \psi_1 - \psi_2$ *and* $X = X_1 - X_2 = (\varphi, \psi)$.

Proof. The proof of this theorem can be obtained by using the same technique as used to prove the results of Proposition 8.13, by taking into account the properties of the non-linear operator F and by using the following expression (obtained by a simple manipulation):[5]

$$F(., v) - F(., u) = \int_0^1 G(., u + s(v - u))(v - u)\mathrm{d}s$$

(for the differentiability of $\mathcal{F}$ see also Belmiloudi [48], in which the studied model take into account the non-linear radiative terms in boundary). So, we omit the details. □

In the next subsection we study the existence of an optimal solution.

[5] The operator F is similar to F_1 in Section 11.3.

13.4.2 Existence of an Optimal Solution

Theorem 13.9. *For α and β sufficiently large, there exists $(\varphi^*, \psi^*) \in U_{\rm ad} \times V_{\rm ad}$ and $u^* \in \mathcal{W} \cap \mathcal{H}$ such that (φ^*, ψ^*) is the optimal solution of (13.17) and $u^* = \mathcal{F}(\varphi^*, \psi^*)$ is the solution of (13.15).*

Proof. Let P_ψ be the mapping: $\varphi \longrightarrow J(\varphi, \psi)$ and Q_φ be the mapping: $\psi \longrightarrow J(\varphi, \psi)$. To obtain the existence of the robust control problem, we prove first that P_ψ is convex and lower semi-continuous for all $\psi \in V_{\rm ad}$, second that Q_φ is concave and upper semi-continuous for all $\varphi \in U_{\rm ad}$ and, finally, we use the minimax theorems in infinite dimensions presented in Chapter 5.

In order to prove the convexity, it is sufficient to show that for $(\varphi_1, \varphi_2) \in U_{\rm ad} \times U_{\rm ad}$, we have $(P'_\psi(\varphi_1) - P'_\psi(\varphi_2)).\varphi \geq 0$, where $\varphi = \varphi_1 - \varphi_2$.

From the expression of G-differentiable cost functional J (a composition of G-differentiable mappings), it follows that P_ψ is G-differentiable and, for $i = 1, 2$,

$$\begin{aligned} P'_\psi(\varphi_i).\varphi = & \iint_Q (\gamma u_i + \delta\varphi_i - \mathfrak{m}_{\rm obs})(\gamma w_i + \delta\varphi) {\rm d}x{\rm d}t \\ & + \alpha \iint_Q \varphi_i \varphi {\rm d}x{\rm d}t, \end{aligned}$$

where $u_i = \mathcal{F}(\varphi_i, \psi)$ and $w_i = \mathcal{F}'(\varphi_i, \psi).(\varphi, 0)$.

Consequently,

$$\begin{aligned} & (P'_\psi(\varphi_1) - P'_\psi(\varphi_2)).\varphi \\ & = \alpha \parallel \varphi \parallel^2_{L^2(Q)} + \iint_Q \delta^2 \mid \varphi \mid^2 {\rm d}x{\rm d}t + \iint_Q \gamma\delta u \varphi {\rm d}x{\rm d}t \\ & \quad + \iint_Q (\gamma u + \delta\varphi)\gamma w_1 {\rm d}x{\rm d}t + \iint_Q (\gamma u_2 + \delta\varphi_2 - \mathfrak{m}_{\rm obs})\gamma w {\rm d}x{\rm d}t, \end{aligned} \tag{13.24}$$

where $u = u_1 - u_2$ and $w = w_1 - w_2$. The estimates (13.23), (13.22) and (13.14) imply that

$$\begin{aligned} & \iint_Q \gamma\delta u\varphi {\rm d}x{\rm d}t + \iint_Q (\gamma u + \delta\varphi)\gamma w_1 {\rm d}x{\rm d}t \leq C_0 \parallel \varphi \parallel^2_{L^2(Q)}, \\ & \iint_Q (\gamma u_2 + \delta\varphi_2 - \mathfrak{m}_{\rm obs})\gamma w {\rm d}x{\rm d}t \leq C_1 \parallel \varphi \parallel^{3/2}_{L^2(Q)} . \end{aligned} \tag{13.25}$$

From (13.24) and the previous results (13.25), we can deduce that there exists a constant $\alpha_l > 0$ such that, for $\alpha \geq \alpha_l$, we have

$$(P'_\psi(\varphi_1) - P'_\psi(\varphi_2)).\varphi \geq (\delta^2 + \alpha_l - C_0) \parallel \varphi \parallel^2_{L^2(Q)} - C_1 \parallel \varphi \parallel^{3/2}_{L^2(Q)} \geq 0$$

and then the convexity of P_ξ is established.

In the same way, we can find $\beta_l > 0$ such that, for $\beta \geq \beta_l$, Q_φ is concave.

We prove now that P_ψ (respectively Q_φ) is lower (respectively upper) semi-continuous for all $\psi \in V_{\rm ad}$ (respectively $\varphi \in U_{\rm ad}$).

Let $\varphi_k \in U_{\rm ad}$ be a minimizing sequence of P_ψ, *i.e.*,

$$\liminf_{k\longrightarrow\infty} J(\varphi_k,\psi) = \inf_{\varphi\in L^2(\mathcal{Q})} J(\varphi,\psi).$$

Then, according to the nature of the cost function J, we can deduce that φ_k is uniformly bounded in $U_{\rm ad}$ and we can extract from φ_k a subsequence also denoted by φ_k such that $\varphi_k \rightharpoonup \varphi_\xi$ weakly in $U_{\rm ad}$. Therefore, by using the same technique as to obtain the estimate (13.14), the function $u_k = \mathcal{F}(\varphi_k,\psi)$ is uniformly bounded in $\mathcal{W}\cap\mathcal{H}$. Moreover, according now to Lemma 6.6, the injection of $\mathcal{W}$ into $L^2(\mathcal{Q})$ is compact. Consequently, these results make it possible to extract from u_k a subsequence also denoted by u_k such that

$$\begin{aligned} &u_k \rightharpoonup u_\psi \text{ weakly in } L^2(0,T;V),\\ &u_k \longrightarrow u_\psi \text{ strongly in } L^2(\mathcal{Q}),\\ &\varphi_k \rightharpoonup \varphi_\psi \text{ weakly in } L^2(\mathcal{Q}) \text{ and } \varphi_\psi \in U_{\rm ad}. \end{aligned} \tag{13.26}$$

Prove now that $u_k\varphi_k \longrightarrow u_\psi\varphi_\psi$ weakly in $L^2(\mathcal{Q})$. Since $u_k\varphi_k - u_\psi\varphi_\psi = (u_k-u_\psi)\varphi_k + u_\psi(\varphi_k-\varphi_\psi)$, then according to the first and second parts of (13.26), we then obtain the result. It is easy to prove that u_ψ is a solution of (13.15) with data (φ_ψ,ψ) and according to the uniqueness of the solution of the problem (13.15), we then have $u_\psi = \mathcal{F}(\varphi_\psi,\psi)$.

Since the norm is lower semi-continuous, therefore we have that the map P_ψ is lower semi-continuous for all $\psi\in V_{\rm ad}$. By applying similar argument as in the proof of the previous result we obtain that Q_φ is upper semi-continuous for all $\varphi\in U_{\rm ad}$.

This completes the proof. □

We next wish to show the appropriate first-order necessary conditions (optimality conditions) of the saddle point problem (13.17).

13.4.3 Optimality Conditions

Theorem 13.10. *Let assumptions* **(H1)**–**(H2)** *be fulfilled,* $(\varphi^*,\psi^*,u^*)\in U_{\rm ad}\times V_{\rm ad}\times(\mathcal{W}\cap\mathcal{H})$ *be an optimal solution such that* (φ^*,ψ^*) *is defined by (13.17) and* $u^* = \mathcal{F}(\varphi^*,\psi^*)$ *is a solution of (13.15). Then, for* $(h,\vartheta)\in U_{\rm ad}\times V_{\rm ad}$,

$$\begin{aligned} &\iint_{\mathcal{Q}}((u^*-u_{\rm a}+v_{\rm a})\tilde{u}^*+\alpha\varphi^*)(h-\varphi^*){\rm d}x{\rm d}t\\ &\qquad\qquad+\iint_{\mathcal{Q}}(\delta(\gamma u^*+\delta\varphi^*-\mathfrak{m}_{\rm obs}))(h-\varphi^*){\rm d}x{\rm d}t\geq 0, \qquad (13.27)\\ &\iint_{\mathcal{Q}}(-\tilde{u}^*-\beta\psi^*)(\vartheta-\psi^*){\rm d}x{\rm d}t\leq 0. \end{aligned}$$

Moreover, the gradient of J *at* (φ^*,ψ^*) *in direction* (h,ϑ) *is given by*

$$J'(\varphi^*,\psi^*).(h,\vartheta) = \iint_{\mathcal{Q}} ((u^* - u_{\rm a} + v_{\rm a})\tilde{u}^* + \alpha\varphi^* + \delta(\gamma u^* + \delta\varphi^* - \mathfrak{m}_{\rm obs}))h \mathrm{d}x\mathrm{d}t$$
$$+ \iint_{\mathcal{Q}} (-\tilde{u}^* - \beta\psi^*)\vartheta \mathrm{d}x\mathrm{d}t.$$

Otherwise (in the weak sense),

$$\frac{\partial J}{\partial \varphi}(\varphi^*,\psi^*) = (u^* - u_{\rm a} + v_{\rm a})\tilde{u}^* + \alpha\varphi^* + \delta(\gamma u^* + \delta\varphi^* - \mathfrak{m}_{\rm obs}),$$
$$\frac{\partial J}{\partial \psi}(\varphi^*,\psi^*) = -\tilde{u}^* - \beta\psi^*,$$

where $\tilde{u}^*$ is the solution of the adjoint problem (13.29) (given below), corresponding to the primal solution u^*.

Proof. Let $(\varphi^*,\psi^*) \in U_{\rm ad} \times V_{\rm ad}$ be a saddle point of J, *i.e.*, the corresponding solution of problem (13.17) and $u^* = \mathcal{F}(\varphi^*,\psi^*)$ be the solution of problem (13.1). From Theorem 13.8 we know that $\mathcal{F}$ is differentiable. Therefore

$$\begin{aligned} J'(\varphi,\psi).(h,\vartheta) &= \frac{\mathrm{d}}{\mathrm{d}\lambda} J(\varphi + \lambda h, \psi + \lambda\vartheta)|_{\lambda=0} \\ &= \iint_{\mathcal{Q}} (\gamma u + \delta\varphi - \mathfrak{m}_{\rm obs})(\gamma w + \delta h)\mathrm{d}x\mathrm{d}t \\ &\quad + \alpha \iint_{\mathcal{Q}} \varphi h \mathrm{d}x\mathrm{d}t - \beta \iint_{\mathcal{Q}} \psi\vartheta \mathrm{d}x\mathrm{d}t, \end{aligned} \tag{13.28}$$

where $w = \mathcal{F}'(\varphi,\psi).(h,\vartheta)$ is the solution of problem (13.21).

In order to simplify (13.28), we introduce the following adjoint problem corresponding to the primal problem (13.15) (we denote by $u = \mathcal{F}(\varphi,\psi)$):

$$\left\{ \begin{aligned} &-\frac{\partial \tilde{u}}{\partial t} - \mathrm{div}(\kappa\nabla\tilde{u}) + \gamma(\gamma u + \delta\varphi - \mathfrak{m}_{\rm obs}) \\ &\qquad + K^*(\varpi,\tilde{u}) + (P + \varphi)\tilde{u} = (G(.,u))^*\tilde{u} \text{ on } \mathcal{Q}, \\ &\text{subject to the boundary condition} \\ &(\kappa\nabla\tilde{u}).\mathbf{n} = -q\tilde{u} \text{ on } \Sigma, \\ &\text{and the final condition} \\ &\tilde{u}(T) = 0 \text{ on } \Omega. \end{aligned} \right. \tag{13.29}$$

To prove the existence of a unique solution $\tilde{u} \in \mathcal{W} \cap \mathcal{H}$, we change the variables of problem (13.29) by reversing the sense of time, *i.e.*, $t := T - t$, and we apply a similar argument as used in the proof of Theorem 13.7.

Due to (13.21) and (13.29) (multiplying (13.21) by the solution of (13.29) and integrating with respect to space and time) and using Green's formula we obtain according to the second part of (13.21) and of (13.29) that

$$\int_\Omega \tilde{u}(T)w(T)\mathrm{d}x - \int_\Omega \tilde{u}(0)w(0)\mathrm{d}x + \int_0^T\int_\Omega h(u-u_\mathrm{a}+v_\mathrm{a})\tilde{u}\mathrm{d}x\mathrm{d}r$$
$$- \int_0^T\int_\Omega \vartheta\tilde{u}\mathrm{d}x\mathrm{d}r = \int_0^T\int_\Omega \gamma(\gamma u+\delta\varphi-\mathfrak{m}_\mathrm{obs})w\mathrm{d}x\mathrm{d}r.$$

Since $\tilde{u}(T)=0$ and $w(0)=0$, then

$$\iint_Q h(u-u_\mathrm{a}+v_\mathrm{a})\tilde{u}\mathrm{d}x\mathrm{d}t - \iint_Q \vartheta\tilde{u}\mathrm{d}x\mathrm{d}t = \iint_Q \gamma(\gamma u+\delta\varphi-\mathfrak{m}_\mathrm{obs})w\mathrm{d}x\mathrm{d}t.$$

According to the expression of $J'(\varphi,\psi)$ we can deduce that

$$\begin{aligned} J'(\varphi,\psi).(h,\vartheta) = &\iint_Q ((u-u_\mathrm{a}+v_\mathrm{a})\tilde{u}+\alpha\varphi)h\mathrm{d}x\mathrm{d}t \\ &+ \iint_Q \delta(\gamma u+\delta\varphi-\mathfrak{m}_\mathrm{obs})h\mathrm{d}x\mathrm{d}t \\ &+ \iint_Q (-\tilde{u}-\beta\psi)\vartheta\mathrm{d}x\mathrm{d}t. \end{aligned}$$

Since (φ^*,ψ^*) is a saddle point we can deduce easily the optimality conditions (13.27) given in this theorem.

This completes the proof. □

Remark 13.11. We can also prove by using the sign of the variation h (depending on the size of φ^*) that

$$\varphi^* = \max\left(\tau_1, \min\left(-\frac{\delta(\gamma u^*-\mathfrak{m}_\mathrm{obs})+(u^*-u_\mathrm{a}+v_\mathrm{a})\tilde{u}^*}{\alpha+\delta^2}, \tau_2\right)\right).$$

13.5 Other Situations

13.5.1 Data Assimilation

If we want to take into account the initial condition disturbance $\psi = u_0$ (data assimilation), we obtain the same results as in Section 13.4. In this case the cost functional can be given by

$$J(\varphi,\psi) = \frac{1}{2}\parallel \gamma u+\delta\varphi-\mathfrak{m}_\mathrm{obs}\parallel^2_{L^2(Q)} + \frac{\alpha}{2}\parallel\varphi\parallel^2_{L^2(Q)} - \frac{\beta}{2}\parallel\psi\parallel^2_{L^2(\Omega)}.$$

We can also prove an existence theorem of the control problem and obtain necessary optimality conditions for its solution using the same method as in Section 13.4. Let now $\mathcal{K} = U_\mathrm{ad}\times V_\mathrm{ad}$ such that V_ad is a non-empty closed convex and bounded subset of $L^2(\Omega)$ and U_ad is similar to previous section. Then, for α and β sufficiently large, there exist $(\varphi^*,\psi^*)\in\mathcal{K}$ and $u^*\in\mathcal{W}\cap\mathcal{H}$ that satisfy

$$\begin{aligned}
\frac{\partial u^*}{\partial t} &= \operatorname{div}(\kappa \nabla u^*) - P(u^* - u_{\mathrm{a}}) \\
&\qquad -\varphi^*(u^* - u_{\mathrm{a}} + v_{\mathrm{a}}) - K(\varpi, u^*) + F(., u^*) + \xi \quad \text{on } \mathcal{Q}, \\
(\kappa \nabla u^*).\mathbf{n} &= -q(u - u_{\mathrm{b}}) + \eta \quad \text{on } \Sigma, \\
u^*(0) &= \psi^* \quad \text{on } \Omega
\end{aligned} \tag{13.30}$$

and, for $(h, \vartheta) \in U_{\mathrm{ad}} \times V_{\mathrm{ad}}$, we have the optimality conditions

$$\iint_{\mathcal{Q}} ((u^* - u_{\mathrm{a}} + v_{\mathrm{a}})\tilde{u}^* + \alpha\varphi^* + \delta(\gamma u^* + \delta\varphi^* - \mathfrak{m}_{\mathrm{obs}}))(h - \varphi^*)\mathrm{d}x\mathrm{d}t \geq 0,$$

$$\int_{\Omega} (-\tilde{u}^*(0) - \beta\psi^*)(\vartheta - \psi^*)\mathrm{d}x \leq 0,$$

where $\tilde{u}^*$ is the solution of the adjoint problem (13.29) corresponding to the solution u^*.

13.5.2 Boundary Disturbance

If we want to take into account the heat flux disturbance (due to evaporation) $\psi = \eta$, we obtain the same results as in Section 13.4. In this case the cost functional can be given by

$$J(\varphi, \psi) = \frac{1}{2} \| \gamma u + \delta\varphi - \mathfrak{m}_{\mathrm{obs}} \|^2_{L^2(\mathcal{Q})} + \frac{\alpha}{2} \| \varphi \|^2_{L^2(\mathcal{Q})} - \frac{\beta}{2} \| \psi \|^2_{L^2(\Sigma)} .$$

We can also prove an existence theorem of the control problem and obtain necessary optimality conditions for its solution by using the same method as in Section 13.4. Let now $\mathcal{K} = U_{\mathrm{ad}} \times V_{\mathrm{ad}}$ such that V_{ad} is a non-empty closed convex and bounded subset of $L^2(\Sigma)$. Then, for α and β sufficiently large, there exist $(\varphi^*, \psi^*) \in \mathcal{K}$ and $u^* \in \mathcal{W} \cap \mathcal{H}$ that satisfy

$$\begin{aligned}
\frac{\partial u^*}{\partial t} &= \operatorname{div}(\kappa \nabla u^*) - P(u^* - u_{\mathrm{a}}) \\
&\qquad -\varphi^*(u^* - u_{\mathrm{a}} + v_{\mathrm{a}}) - K(\varpi, u^*) + F(., u^*) + \xi \quad \text{on } \mathcal{Q}, \\
(\kappa \nabla u^*).\mathbf{n} &= -q(u - u_{\mathrm{b}}) + \psi^* \quad \text{on } \Sigma, \\
u^*(0) &= u_0 \quad \text{on } \Omega
\end{aligned} \tag{13.31}$$

and, for $(h, \vartheta) \in U_{\mathrm{ad}} \times V_{\mathrm{ad}}$, we have the optimality conditions

$$\iint_{\mathcal{Q}} ((u^* - u_{\mathrm{a}} + v_{\mathrm{a}})\tilde{u}^* + \alpha\varphi^* + \delta(\gamma u^* + \delta\varphi^* - \mathfrak{m}_{\mathrm{obs}}))(h - \varphi^*)\mathrm{d}x\mathrm{d}t \geq 0,$$

$$\iint_{\Sigma} (-\tilde{u}^* - \beta\psi^*)(\vartheta - \psi^*)\mathrm{d}\Gamma\mathrm{d}t \leq 0,$$

where $\tilde{u}^*$ is the solution of the adjoint problem (13.29) corresponding to the solution u^*.

13.5.3 Finite Number of Measurments

In many biological situations, we can measure u in only some points in space time domain. Let now some points be in $\Omega \times (0,T)$ where we assume that we can measure u. Let $x_i \in \Omega, i = 1, \dots l$ such that $x_i \neq x_j$ if $i \neq j$, $0 < t_1 < t_2 < \dots < T$, and assume that we measure quantities ϵ_{ij} which are meant to be the tolerance uncertainty of the function $M = \gamma u + \delta p$ at point (x_i, t_j), for $i = 1, \dots, l$ and $j = 1, \dots, N$. Let $(\Omega_i)_{i=1,l}$ be a sequence of disjoint small balls in Ω such that $x_i \in \Omega_i$, $\forall i = 1, \dots, l$. Let $(I_j)_{j=1,N}$ be also a sequence of disjoint intervals in $(0,T)$ such that $t_j \in I_j$, $\forall j = 1, \dots, N$. We denote the average operator over the domain $\mathcal{Q}_{ij} = \Omega_i \times I_j$ (for $i = 1, \dots, l$, $j = 1, \dots, N$) by

$$< v >_{ij} = \frac{1}{\mid \mathcal{Q}_{ij} \mid} \int_{\mathcal{Q}_{ij}} v(x,t) \mathrm{d}x\mathrm{d}t$$

and propose the following cost function:

$$J(\varphi, \psi) = \frac{1}{2} \sum_{i=1,l} \sum_{j=1,N} \mid < M >_{ij} - \epsilon_{ij} \mid^2 + \frac{\alpha}{2} \parallel \varphi \parallel^2_{L^2(\mathcal{Q})} - \frac{\beta}{2} \parallel \psi \parallel^2_{L^2(\mathcal{Q})} .$$

Let χ_D be the usual characteristic function of a domain D, *i.e.*, $\chi_D = 1$ on D, and 0 outside of D. Let $\mathcal{S} : L^2(\mathcal{Q}) \longrightarrow L^2(\mathcal{Q})$ be defined by $(\forall v \in L^2(\mathcal{Q}))$

$$\mathcal{S}(v) = \sum_{i=1,l} \sum_{j=1,N} \frac{1}{\mid \mathcal{Q}_{ij} \mid} \chi_{\mathcal{Q}_{ij}} < v >_{ij}$$

and let m be the following value:

$$m = \sum_{i=1,l} \sum_{j=1,N} \frac{1}{\mid \mathcal{Q}_{ij} \mid} \chi_{\mathcal{Q}_{ij}} \epsilon_{ij}.$$

By using the same technique as used in the proof of the previous results of Section 13.4, we can prove the existence theorem of the control problem and obtain necessary optimality conditions. Then, for α and β sufficiently large, there exist $(\varphi^*, \psi^*) \in U_{\mathrm{ad}} \times V_{\mathrm{ad}}$ and $u^* \in \mathcal{W}$ such that $u^* = \mathcal{F}(\varphi^*, \psi^*)$ is a solution of (13.11) and, for $(h, \vartheta) \in U_{\mathrm{ad}} \times V_{\mathrm{ad}}$,

$$\iint_{\mathcal{Q}} ((u^* - u_{\mathrm{a}})\tilde{u}^* + \alpha \varphi^* + \delta(\mathcal{S}(M^*) - m))(h - \varphi^*) \mathrm{d}x\mathrm{d}t \geq 0,$$

$$\iint_{\mathcal{Q}} (-\tilde{u}^* - \beta \psi^*)(\vartheta - \psi^*) \mathrm{d}x\mathrm{d}t \leq 0,$$

where $M^* = \gamma u^* + \delta \varphi^*$ and $\tilde{u}^*$ is the solution of the following adjoint problem (corresponding to the solution u^*):

$$\begin{aligned} &-\frac{\partial \tilde{u}^*}{\partial t} - \mathrm{div}(\kappa \nabla \tilde{u}^*) + \gamma(\mathcal{S}(M^*) - m) \\ &\qquad + K^*(\varpi, \tilde{u}^*) + (P + \varphi^*)\tilde{u}^* = (G(., u^*))^* \tilde{u}^* \quad \text{on } \mathcal{Q}, \\ &(\kappa \nabla \tilde{u}^*).\mathbf{n} = -q\tilde{u}^* \quad \text{on } \Sigma, \\ &\tilde{u}^*(T) = 0 \quad \text{on } \Omega. \end{aligned}$$

Remark 13.12. We obtain, by using the same techniques, similar results if we want to take into account the disturbance in the initial condition. ◇

13.5.4 Union of a Finite Number of Subdomains

Suppose now that the body is made of different types of tissue which occupy finitely many disjointed subdomains Ω_i, $i = 1, \ldots, N_D$, of Ω, such that $\overline{\Omega} = \bigcup_{i=1,N_D} \overline{\Omega}_i$. Moreover, we assume that the perfusion acts continuously according to the temperature in each domain Ω_i and discontinuously at the tissue boundaries. We propose the following cost function:

$$J(\varphi, \psi) = \frac{1}{2} \parallel \gamma u + \delta\varphi - \mathfrak{m}_{\text{obs}} \parallel^2_{L^2(Q)} + \frac{\alpha}{2} \parallel \varphi \parallel^2_{\tilde{U}_{\text{ad}}} - \frac{\beta}{2} \parallel \psi \parallel^2_{L^2(\Omega)},$$

where $(\varphi, \psi) \in \tilde{U}_{\text{ad}} \times V_{\text{ad}}$, $\tilde{U}_{\text{ad}} = L^2(0,T;\mathcal{R}) \cap U_{\text{ad}}$ and $\mathcal{R}$ is the Hilbert space

$$\mathcal{R} = \{q \in L^2(\Omega) : q|_{\Omega_i} \in H^1(\Omega_i), \text{ for } i = 1, \ldots, N_D\}$$

equipped with the following norm

$$\parallel q \parallel_{\mathcal{R}} = \left(\sum_{i=1,N_D} (\alpha_1 \parallel q \parallel^2_{L^2(\Omega_i)} + \alpha_2 \parallel \nabla q \parallel^2_{L^2(\Omega_i)}) \right)^{1/2},$$

with the fixed constants $\alpha_i > 0$ for $i = 1, 2$, and $\parallel \varphi \parallel_{\tilde{U}_{\text{ad}}} = \parallel \varphi \parallel_{L^2(0,T;\mathcal{R})}$.

Let $\Lambda : L^2(\Omega) \longrightarrow \mathcal{R}$ be a linear operator such that, for all $v \in L^2(\Omega)$, the function $\pi = \Lambda(v)$ is the solution of

$$\begin{aligned} -\alpha_2 \Delta\pi_i + \alpha_1 \pi_i &= v|_{\Omega_i} \quad \text{on } \Omega_i, \\ (\nabla\pi_i).\mathbf{n} &= 0 \quad \text{on } \partial\Omega_i, \end{aligned} \tag{13.32}$$

where $\pi_i = \pi$, a.e. in Ω_i and $v|_{\Omega_i}$ denotes the restriction of v on the subdomain Ω_i, for $i = 1, \ldots, N_D$.

By using the same technique as in the proof of the results of Section 13.4, we can prove the existence of the control problem and obtain the necessary optimality conditions. More precisely, for α and β sufficiently large, there exists an optimal solution $(\varphi^*, \psi^*, u^*) \in \tilde{U}_{\text{ad}} \times V_{\text{ad}} \times \mathcal{W}$ such that (φ^*, ψ^*) is defined by (13.17), $u^* = \mathcal{F}(\varphi^*, \psi^*)$ is the solution of (13.1) and, for $(h, \vartheta) \in \tilde{U}_{\text{ad}} \times V_{\text{ad}}$

$$\int_0^T \langle \Lambda((u^* - u_{\text{a}} + v_{\text{a}})\tilde{u}^* + \delta(M^* - \mathfrak{m}_{\text{obs}})) + \alpha\varphi^*, h - \varphi^* \rangle_{\mathcal{R}} \text{d}t \geq 0,$$

$$\iint_Q (-\tilde{u}^* - \beta\psi^*)(\vartheta - \psi^*) \leq 0,$$

where $M^* = \gamma u^* + \delta\varphi^*$ and $\tilde{u}^*$ is the solution of the adjoint problem (13.29) (corresponding to the solution u^*).

Remark 13.13. (*i*) We can also combine the different situations presented in this section and obtain the existence of the control problem and necessary optimality conditions without major difficulties.
(*ii*) It is very interesting to study the case where the disturbance is in the bolus temperature, because during, for example, hyperthermia with prolongated laser induction heat treatment, the exchange mechanisms at the body–air interface play a very important role on the total tissue temperature distribution and, moreover, small fluctuations of the bolus temperature can affect considerably the resulting temperature distribution and thus the treatment.
(*iii*) As indicated in Remark 10.31, for numerical resolution of robust control problems, the reader is referred to Chapter 9. ♢

14

Lotka–Volterra-type Systems with Logistic Time-varying Delays

In this chapter, we consider a resource management problem in which the management objectives are the stabilization of uncertain biological species. Such situations may arise in many different fields of resource economics and wildlife management. The stabilizing management consists of robust control strategies for a class of systems governed by parabolic equations governing diffusive biological species with logistic growth terms and multiple time-varying delays. First, we prove the existence, uniqueness and regularity results for these parabolic systems. Then, we introduce the perturbation problem and analyze its well-posedness. Second, we formulate the control problem in different situations. The existence and condition of uniqueness of the optimal solution are derived and first-order necessary conditions of optimality, characterizing the optimal solution, are obtained.

14.1 Introduction and Mathematical Setting

14.1.1 Motivation

The aim of this chapter is the study of optimal control problem of a biological species, whose logistic growth is governed by a degenerate parabolic diffusive equation with multiple time-varying delays. Various problems associated with biology and economics behind diffusion population models with logistic growth terms have been studied extensively in the literature since few last years (see, *e.g.*,, Clark [87], Cushing [91], Edelstein [110], Murray [227], Waltman [291], and the references therein). For the optimal control and PDEs (partial diferential equations) see, *e.g.*, Fife [122]. We can mention also Leung and Stojanovic [197] in which the authors present optimal results for uniformly elliptic PDEs with logistic growth terms; He *et al.* [151, 152] in which the authors study optimal harvesting control of a periodic parabolic system with maximization of profit; Lenhart *et al.* [192, 193, 194] in which the authors treat a wildlife management problem in the case of constant diffusion and in the case of degenerate parabolic equation; Belmiloudi [42, 43] in which the author considers a wildlife management problem for time-varying delay non-linear and degenerate parabolic equations with logistic growth terms; and Belmiloudi [44] in which the author treats a minimax optimal control problem, for two parabolic competition systems with logistic growth terms, in order to minimize damage and trapping costs of the first species, and to maximize the difference between economic revenue and cost of the second species.

In this chapter, we consider resource management problems of some time-varying delay non-linear parabolic equations governing diffusive biological species with logistic growth terms. The management objectives are the stabilization of uncertain biological species in order to take into account the influence of the population at earlier times on the regulatory effect. The incorporation of multiple time-varying delays is motivated by the fact that the birth rate does not act instantaneously, since there is a time delay to take into account the time to reach maturity, the finished period of gestation, *etc.*

14.1.2 Studied Equations

In this contribution we consider non-linear parabolic partial differential delay equations of the form

$$\begin{aligned} &\frac{\partial U}{\partial t} + FU = U(d_0 - d\,U - \sum_{i=1,n} d_i U(.,t-e_i(t)) - P) \text{ on } \mathcal{Q}, \\ &U = H_0 \text{ on } \mathcal{Q}_0 = \Omega \times [-\delta(0), 0), \\ &U(.,0) = U_0 \text{ on } \Omega, \end{aligned} \tag{14.1}$$

subject to homogeneous Neumann boundary conditions (no-flux boundary conditions). The cylinder $\mathcal{Q}$ is $\mathcal{Q} = \Omega \times (0,T)$, the domain Ω (the habitat sits) is a bounded subset of $\mathbb{R}^m$, $m \geq 1$, where its boundary Γ is sufficiently regular and Ω is totally on one side of Γ. The given final time $T > 0$ (the planning period) is a fixed constant. The function $U(x,t)$ is the species concentration and the function $P(x,t)$ denotes the trapping rate of the species or the harvesting effort on the species. The function d_0 is the growth parameter and d is the crowding effect. The functions $(e_i, i = 1,n)$ are sufficiently regular functions representing multiple time-varying delays and $\delta(0) = \max_{i=1,n}(e_i(0))$. The function U_0 gives the initial condition of a given population and the function H_0 is the initial function on $\mathcal{Q}_0$ of this given population. The operator F has the form

$$Fv = -\sum_{i,j=1}^{m} \frac{\partial}{\partial x_i}\left(a_{ij}(x)\frac{\partial v}{\partial x_j}\right). \tag{14.2}$$

We state the following hypotheses for the operator F:

($\mathbf{A_1}$) The functions $a_{ij} : \overline{\Omega} \longrightarrow \mathbb{R}$ are $\mathcal{C}^1$ and the matrix $A(.)=(a_{ij}(.))_{1\leq i,j\leq m}$ is symmetric and satisfies the ellipticity conditions (for all $x \in \overline{\Omega}$ and $(y_i)_{i=1,m} \in \mathbb{R}^m$):

$$\mu \sum_{i=1}^{m} y_i^2 \geq \sum_{i,j=1}^{m} a_{ij}(x) y_i y_j \geq \nu \sum_{i=1}^{m} y_i^2, \text{ with } \nu > 0 \text{ and } \mu > 0 \tag{14.3}$$

and there exists a $\mathcal{C}^1$ function $\sigma : \overline{\Omega} \longrightarrow \mathbb{R}^{m\times m}$, such that

$$A(x) = \sigma^*(x)\sigma(x), \ x \in \overline{\Omega}, \tag{14.4}$$

where σ^* is the dual of the matrix σ.

According to the assumption ($\mathbf{A_1}$) we can now express the boundary conditions satisfied by the problem (14.1). Problem (14.1) is then subject to

$$\begin{aligned}&\text{homogeneous Neumann boundary conditions}\\ &\frac{\partial U}{\partial n_a} = (\sigma^*\sigma\nabla U).\mathbf{n} = 0 \ \text{ on } \Sigma = \Gamma \times (0,T),\end{aligned} \tag{14.5}$$

with $\mathbf{n}$ being the outward normal to Γ.

We make the following assumptions (similar to Section 8.7):

($\mathbf{A_2}$) The functions $(d_i)_{i=1,n}$, d_0 and $d : \overline{Q} \longrightarrow \mathbb{R}$ are non-negative and $\mathcal{C}^\infty$ functions on $\overline{Q}$.

($\mathbf{A_3}$) The functions $r_i : t \in [0,T) \longrightarrow r_i(t) = t - e_i(t)$ are strictly increasing functions and the functions e_i are $\mathcal{C}^1$ non-negative on $[0,T)$.

Remark 14.1. According to the assumptions ($\mathbf{A_3}$), the inverse functions f_i of r_i exist, for all i and the properties given in Remark 8.52 hold. ♢

We recall now the notations used in Section 8.7: we take the following subdivision $s_{-1} = -\delta(0)$, $s_0 = 0$ and $\forall j \in \mathbb{N} - \{0\}, s_j = \min_{i=1,n}(f_i(s_{j-1}))$, the time intervals $I_j =]s_{-1}, s_j[$ and the cylinders $Q_j = \Omega \times I_j$ for $j \geq 0$.

Next, let the Hilbert space $H_\sigma(\Omega)$ be the completion of $\mathcal{C}^\infty(\overline{\Omega})$, under norm

$$\| v \|_{H_\sigma} = \left(\| v \|^2_{L^2(\Omega)} + \| \sigma\nabla v \|^2_{L^2(\Omega)} \right)^{1/2}, \tag{14.6}$$

which is equivalent to the standard $H^1(\Omega)$-norm $\| v \|_{H^1(\Omega)}$. We denote its dual by $H^*_\sigma(\Omega)$. Then the embeddings $H_\sigma(\Omega) \subset L^2(\Omega) \subset H^*_\sigma(\Omega)$ are continuous.

Now we introduce the following sets:

$$\mathcal{U} = \{v \in L^\infty(\Omega) : 0 < v \ \text{ in } \Omega\}, \quad \mathcal{B} = \{f \in L^\infty(Q) : 0 < f \ \text{ in } Q\},$$
$$L^\infty_+(D) = \{h \in L^\infty(D) : 0 \leq h \ \text{ in } D\},$$

where D is a bounded subset of $\Omega \times (-\infty, +\infty)$ and spaces:

$$\mathcal{H}(\mathcal{Q}) = L^\infty(0,T;L^2(\Omega)), \qquad \mathcal{V}_\sigma(\mathcal{Q}) = L^2(0,T;H_\sigma(\Omega)),$$
$$\mathcal{W}_\sigma(\mathcal{Q}) = \mathcal{V}_\sigma(\mathcal{Q}) \cap H^1(0,T;H^*_\sigma(\Omega)).$$

According to Lemma 6.6, the space $\mathcal{W}_\sigma$ is continuously embedded in $\mathcal{C}([0,T];L^2(\Omega))$.

14.2 Existence and Uniqueness of the Solution

In this section, we prove the existence and uniqueness of the solution of the state equation. First, we state the following theorem.

Theorem 14.2. *Let assumptions* $(\mathbf{A_1})$–$(\mathbf{A_3})$ *hold. Then, for any* $U_{\text{int}} \in \mathcal{U}$, $g \in L^\infty_+(\mathcal{Q}_{1,2})$ $(\mathcal{Q}_{1,2} = \Omega \times (T_1,T_2))$ *and* $P \in \mathcal{B}$ *such that*

$$0 < \| P \|_{L^\infty(\mathcal{Q})} + \| g \|_{L^\infty(\mathcal{Q}_{1,2})} \leq \inf_{\mathcal{Q}}(d_0), \tag{14.7}$$

the problem

$$\begin{aligned} &\frac{\partial V}{\partial t} + FV = V(d_0 - g - d\,V - P) \quad \text{on } \mathcal{Q}_{1,2},\\ &V(.,T_1) = U_{\text{int}} \text{ on } \Omega,\\ &\text{subject to homogeneous Neumann boundary conditions,} \end{aligned} \tag{14.8}$$

has a unique non-negative solution V *in* $\mathcal{W}_\sigma(\mathcal{Q}_{1,2}) \cap L^\infty_+(\mathcal{Q}_{1,2})$, *satisfying the estimate, for a.e.* $(x,t) \in \mathcal{Q}_{1,2}$,

$$\frac{1}{2}U_{\text{int}}(x)\mathrm{e}^{-\delta t} \leq V(x,t) \leq C_0 = \max\left(\frac{\| d_0 \|_\infty}{\inf_{\mathcal{Q}}(d)}, \| U_{\text{int}} \|_{L^\infty(\Omega)}\right), \tag{14.9}$$

with $\delta = C_0 \| d \|_\infty + \| g \|_{L^\infty(\mathcal{Q}_{1,2})} + \| P \|_{L^\infty(\mathcal{Q})} - \inf_{\mathcal{Q}}(d_0)$. *Moreover, we have the following a priori estimate:*

$$\| V \|^2_{\mathcal{W}_\sigma(\mathcal{Q}_{1,2})} \leq C(\| U_{\text{int}} \|^2_{L^2(\Omega)} + \| P \|^2_{L^2(\mathcal{Q})} + \| g \|^2_{L^2(\mathcal{Q}_{1,2})}). \tag{14.10}$$

Proof. By using the classical technique (by the construction of a monotone decreasing sequence of solutions convergent towards the solution of (14.8)) and the maximum principle (according to hypothesis (14.7)), the problem (14.8) admits a unique solution V such that

$$0 \leq V \leq C_0 = \max\left(\frac{\| d_0 \|_\infty}{\inf_{\mathcal{Q}}(d)}, \| U_{\text{int}} \|_{L^\infty(\Omega)}\right).$$

Here, we sketch only the proof of the existence.
Let $C_1 > 0$ be a constant such that

$$C_1 d_0 - dC_1^2 \geq 0 \quad \text{and} \quad C_1 \geq \| U_{\text{int}} \|_{L^2(\Omega)} \tag{14.11}$$

and let C_2 such that

$$\sup_{\mathcal{Q}_{1,2}}(2dC_1) + \sup_{\mathcal{Q}_{1,2}}(P + g - d_0) < C_2. \tag{14.12}$$

We define the initial iterate by

$$V_0 = C_1 \tag{14.13}$$

and we obtain inductively V_k, $k = 1, 2, \ldots$, as the unique non-negative solution in $\mathcal{W}_\sigma(\mathcal{Q}_{1,2})$ of the problem:

$$\begin{aligned} &\frac{\partial V_k}{\partial t} + FV_k + C_2\, V_k \\ &\qquad = V_{k-1}(d_0 - g - d\, V_{k-1} - P) + C_2\, V_{k-1} \quad \text{on } \mathcal{Q}_{1,2}, \\ &V_k(., T_1) = U_{\text{int}} \quad \text{on } \Omega, \end{aligned} \tag{14.14}$$

subject to homogeneous Neumann boundary conditions.

By the choice of C_2 in (14.12), the right-hand side of the first part of (14.14) is an increasing function of V_{k-1} for $0 \leq V_{k-1} \leq V_0$. By comparing the right-hand side of (14.14) for V_k and V_{k-1}, we obtain that

$$V_k \leq V_{k-1} \leq V_0, \quad \text{for } k = 2, 3, \ldots.$$

Consequently, using the regularity of different data appearing in (14.14) and the uniform boundedness of the right-hand side of (14.14), we can easily obtain that the solution V_k is uniformly bounded in $\mathcal{W}_\sigma(\mathcal{Q}_{1,2})$. By using the classical argument, we can deduce the weak convergence of V_k in $\mathcal{W}_\sigma(\mathcal{Q}_{1,2})$ and its strong convergence in $L^2(\mathcal{Q})$ to V where V is a solution of (14.8) and satisfies the estimate (14.10) given in this theorem.

In order to obtain (14.9), we introduce the following problem (see, *e.g.*, Chipot [82] for the existence of such solution)

$$\begin{aligned} &FW \leq 0 \quad \text{on } \Omega, \\ &\frac{U_{\text{int}}}{2} \leq W \leq U_{\text{int}} \quad \text{on } \Omega, \end{aligned}$$

subject to the same boundary conditions as V

and by using the maximum principle we obtain that

$$V \geq \mathrm{e}^{-\delta t} W \geq \frac{\mathrm{e}^{-\delta t}}{2} U_{\text{int}} \text{ in } \mathcal{Q}_{1,2},$$

where $\delta = C_0 \| d \|_\infty + \| g \|_\infty + \| P \|_\infty - \inf_{\mathcal{Q}}(d_0)$. □

Remark 14.3. By taking into account the regularity of the right-hand side and the estimate (14.10), we can deduce that the solution U is in $H^{2,1}(\mathcal{Q}_{1,2})$ and satisfies an estimate similar to (14.10) in $H^{2,1}(\mathcal{Q}_{1,2})$ by mean of parabolic estimates. ◇

We can now prove the main result of this section.

Theorem 14.4. *Let assumptions* $(\mathbf{A_1})$–$(\mathbf{A_3})$ *hold:*

(i) Assume that $U_0 \in \mathcal{U}$, $H_0 \in L^\infty_+(\mathcal{Q}_0)$ *and* $P \in \mathcal{B}$ *such that*

$$\| P \|_{L^\infty(\mathcal{Q})} \leq \inf_{\mathcal{Q}}(d_0) - D_\infty \max(C_0, \| H_0 \|_{L^\infty(\mathcal{Q}_0)}), \tag{14.15}$$

where

$$C_0 = \max\left(\frac{\| d_0 \|_\infty}{\inf_{\mathcal{Q}}(d)}, \| U_0 \|_{L^\infty(\Omega)}\right) \text{ and } D_\infty = \sum_{i=1,n} \| d_i \|_\infty .$$

Then there exists a unique solution $U \in \mathcal{W}_\sigma(\mathcal{Q}) \cap L^\infty_+(\mathcal{Q})$, *of the state Equation (14.1) that satisfies*

$$0 < U(x,t) \leq \max(C_0, \| H_0 \|_{L^\infty(\mathcal{Q}_0)}), \ a.e. \ (x,t) \in \mathcal{Q}.$$

Moreover, we have the following a priori estimate:

$$\| U \|^2_{\mathcal{W}_\sigma(\mathcal{Q})} \leq C(\| U_0 \|^2_{L^2(\Omega)} + \| P \|^2_{L^2(\mathcal{Q})} + \| H_0 \|^2_{L^2(\mathcal{Q}_0)}). \tag{14.16}$$

(ii) Let $(U_{0,i}, H_{0,i}, P_i)$, *for* $i = 1, 2$ *be two smooth data of* $\mathcal{U} \times L^\infty_+(\mathcal{Q}_0) \times \mathcal{B}$. *If* U_1 *(respectively* U_2*) is the solution of (14.1) corresponding to data* $(U_{0,1}, H_{0,1}, P_1)$ *(respectively* $(U_{0,2}, H_{0,2}, P_2)$*) then*

$$\| U \|^2_{\mathcal{W}_\sigma(\mathcal{Q})} \leq C(\| U_0 \|^2_{L^2(\Omega)} + \| P \|^2_{L^2(\mathcal{Q})} + \| H_0 \|^2_{L^2(\mathcal{Q}_0)}), \tag{14.17}$$

where $U = U_1 - U_2$, $P = P_1 - P_2$, $U_0 = U_{0,1} - U_{0,2}$ *and* $H_0 = H_{0,1} - H_{0,2}$.

Proof. In order to prove the existence of a unique solution on $\mathcal{Q}$, we first establish the existence of a unique solution on $\mathcal{Q}_j, j \geq 1$ and obtain some estimates, like in Section 8.7.

We solve the problem on $\mathcal{Q}_1$ and obtain the existence of a unique solution on $\mathcal{Q}_1$. Then, the existence of a unique solution on $\mathcal{Q}_2$ is proved by using the solution on $\mathcal{Q}_1$ to generate the initial data at s_1. This advancing process is repeated for $\mathcal{Q}_3, \mathcal{Q}_4, \ldots$ until the final set is reached. Hereafter, the solution on $\mathcal{Q}_j$ will be denoted by $\tilde{U}_j$ for $j = 1, \ldots$.

We shall now introduce the following problems $(\mathcal{P}_j)$ for $j \in \mathbb{N} - \{0\}$:

$$\frac{\partial V_j}{\partial t} + FV_j = V_j(d_0 - g_j - d\, V_j - P) \ \text{ on } \Omega \times (s_{j-1}, s_j),$$
$$V_j(., s_{j-1}) = \tilde{U}_{j-1}(., s_{j-1}) \ \text{ on } \Omega,$$

subject to homogeneous Neumann boundary conditions,

where $g_j = \sum_{i=1,n} d_i \tilde{U}_{j-1}(.,r_i)$, $\tilde{U}_{j-1} \in \mathcal{W}_\sigma(\mathcal{Q}_{j-1}) \cap \mathcal{C}([s_{-1}, s_{j-1}]; L^2(\Omega))$ and $\tilde{U}_{j-1}(., s_{j-1}) \in \mathcal{U}$. We take $U_0(., s_0) = U_0$ and for any i, $\tilde{U}_0(., r_i) = H_0(., r_i)$ on $\Omega \times (s_0, s_1)$.

Since $P \in \mathcal{B}$, $(d_i, i = 0, n) \in \mathcal{C}^\infty(\overline{\mathcal{Q}})$ and according to the remark 14.1, we can use the result of Theorem 14.2 and we obtain that the problem $(\mathcal{P}_j)$ admits a unique solution $V_j \in \mathcal{W}_\sigma(\Omega \times]s_{j-1}, s_j[) \cap \mathcal{C}([s_{s-1}, s_j]; L^2(\Omega))$, such that

$$0 < V_j \le \max\left(\frac{\| d_0 \|_\infty}{\inf_{\mathcal{Q}}(d)}, \| \tilde{U}_{j-1}(., s_{j-1}) \|_{L^\infty(\Omega)}\right),$$

where $\| g_j \|_{L^\infty(\Omega \times (s_{j-2}, s_{j-1}))} + \| P \|_{L^\infty(\mathcal{Q})} - \inf_{\mathcal{Q}}(d_0) \le 0$. Then we can extend the result to the cylinder set $\mathcal{Q}_{j+1}$ by taking $\tilde{U}_j = \tilde{U}_{j-1}$ on $\mathcal{Q}_{j-1}$ and $\tilde{U}_j = V_j$ on $\Omega \times (s_{j-1}, s_j)$.

We can now prove the existence and uniqueness result of the problem (14.1).

We observe that, for $j = 1$, we have $g_1 = \sum_{i=1,n} d_i \tilde{U}_0(., r_i) = \sum_{i=1,n} d_i H_0(., r_i)$ is in $L^\infty_+(\Omega \times (s_{-1}, s_0))$ and $\tilde{U}_0(., s_0) = U_0$ is in $\mathcal{U}$. Then, by using Theorem 14.2, the problem $(\mathcal{P}_1)$ admits a unique solution V_1 such that $V_1 \in \mathcal{W}_\sigma(\Omega \times]s_0, s_1[) \cap \mathcal{C}([s_0, s_1]; L^2(\Omega))$, and we obtain the solution $\tilde{U}_1$ such that

$$0 < \tilde{U}_1 \le \max(C_0, \| H_0 \|_{L^\infty(\mathcal{Q}_0)})$$

and, according to (14.15),

$$\| g_2 \|_{L^\infty(\Omega \times (s_0, s_1))} + \| P \|_{L^\infty(\mathcal{Q})} - \inf_{\mathcal{Q}}(d_0) \le 0.$$

We shall now inject $\tilde{U}_1$ in the problem $(\mathcal{P}_2)$ and by using the same approach, we obtain the existence and uniqueness of $V_2 \in \mathcal{W}_\sigma(\Omega \times]s_1, s_2[) \cap \mathcal{C}([s_1, s_2]; L^2(\Omega))$, the solution of $(\mathcal{P}_2)$.

We can now iterate the process for any domain $\mathcal{Q}_j$, for $j \ge 1$ and obtain the existence and uniqueness of $V_j \in \mathcal{W}_\sigma(\Omega \times]s_{j-1}, s_j[) \cap \mathcal{C}([s_{j-1}, s_j]; L^2(\Omega))$, solution of $(\mathcal{P}_j)$. We deduce then the existence and uniqueness of the solution $U \in \mathcal{W}_\sigma(\mathcal{Q}) \cap \mathcal{C}([0, T]; L^2(\Omega))$ of (14.1) such that $U|_{\mathcal{Q}_j} = \tilde{U}_j, j \ge 1$ and

$$0 < \tilde{U}_j \le \max(C_0, \| H_0 \|_{L^\infty(\mathcal{Q}_0)}).$$

The second part of result (i) of the theorem can be obtained in the same way as to obtain the estimate (ii). We are now going to prove the estimate (ii).

Let us consider two smooth data $(U_{0,i}, H_{0,i}, P_i)$, for $i = 1, 2$, and the corresponding solution U_i of (14.1), respectively. Set $U_0 = U_{0,1} - U_{0,2}$, $H_0 = H_{0,1} - H_{0,2}$, $P = P_1 - P_2$ and $U = U_1 - U_2$. Then, U is a solution of the system:

$$\begin{aligned}
&\frac{\partial U}{\partial t} + FU = U(d_0 - G(U_1, P_1)) - U_2 G(U, P) \;\; \text{on } \mathcal{Q}, \\
&U = H_0 \;\; \text{on } \mathcal{Q}_0, \\
&U(.,0) = U_0 \;\; \text{on } \Omega, \\
&\text{subject to homogeneous Neumann boundary conditions,}
\end{aligned} \tag{14.18}$$

where $G(v, q) = d\, v + \sum_{i=1,n} d_i v(., r_i) + q$.

Multiplying (14.18) by U and integrating over $\Omega \times (0, t)$ (by using Green's formula), gives

$$\begin{aligned}
\| U(t) \|_{L^2}^2 + 2 \int_0^t \| \sigma \nabla U \|_{L^2}^2 \,\mathrm{d}s = 2 \int_0^t \int_\Omega U(d_0 - G(U_1, P_1)) U \mathrm{d}x\mathrm{d}s \\
-2 \int_0^t \int_\Omega U_2 G(U, P) U \mathrm{d}x\mathrm{d}s + \| U_0 \|_{L^2}^2 .
\end{aligned} \tag{14.19}$$

We shall now estimate the term $A = \int_0^t \int_\Omega d_i U(., r_i) U U_2 \mathrm{d}x\mathrm{d}s$. Since U_2 and d_i are in $L^\infty(\mathcal{Q})$ then

$$A \leq c_1 \left(\int_0^t \| U(., r_i) \|_{L^2(\Omega)}^2 \,\mathrm{d}s \right)^{1/2} \left(\int_0^t \| U \|_{L^2(\Omega)}^2 \,\mathrm{d}s \right)^{1/2}.$$

Putting $a = r_i(s)$, we then have $s = f_i(a)$ and $\mathrm{d}s = f_i'(a)\mathrm{d}a$. So

$$\int_0^t \| U(., r_i) \|_{L^2(\Omega)}^2 \,\mathrm{d}s = \int_{-e_i(0)}^{t - e_i(t)} \| U \|_{L^2}^2 \, f_i'(a)\mathrm{d}a.$$

Since $\forall i = 1, n$, $-\delta(0) \leq -e_i(0)$, and $U = H_0$ on $\mathcal{Q}_0$, we can deduce that

$$\int_0^t \| U(., r_i) \|_{L^2}^2 \,\mathrm{d}s \leq c_2 \left(\| H_0 \|_{L^2(\mathcal{Q}_0)}^2 + \int_0^t \| U \|_{L^2}^2 \,\mathrm{d}s \right)$$

and then

$$A = \int_0^t \int_\Omega d_i U(., r_i) U U_2 \mathrm{d}x\mathrm{d}s \leq c_3 \int_0^t \| U \|_{L^2}^2 \,\mathrm{d}s + c_4 \| H_0 \|_{L^2(\mathcal{Q}_0)}^2 . \tag{14.20}$$

According to (14.20), the relation (14.19) becomes, since $(d_i)_{i=0,n}$, P_1, P_2, U_1, U_2 and d are in $L^\infty(\mathcal{Q})$

$$\begin{aligned}
&\| U(., t) \|_{L^2}^2 + \int_0^t \| \sigma \nabla U \|_{L^2}^2 \,\mathrm{d}s \\
&\leq c_5 \left(\| H_0 \|_{L^2(\mathcal{Q}_0)}^2 + \| U_0 \|_{L^2}^2 + \| P \|_{L^2(\mathcal{Q})}^2 \right) + c_6 \int_0^t \| U \|_{L^2}^2 \,\mathrm{d}s.
\end{aligned} \tag{14.21}$$

By using Gronwall's formula we can deduce that

$$\| U \|^2_{\mathcal{H}(\mathcal{Q})} + \| U \|^2_{\mathcal{V}_\sigma(\mathcal{Q})} \leq C(\| H_0 \|^2_{L^2(\mathcal{Q}_0)} + \| U_0 \|^2_{L^2} + \| P \|^2_{L^2(\mathcal{Q})}). \quad (14.22)$$

Using this result and the equation (14.1) we prove easily that U satisfies the following estimate

$$\| U \|^2_{H^1(0,T;H^*_\sigma(\Omega))} \leq C(\| H_0 \|^2_{L^2(\mathcal{Q}_0)} + \| U_0 \|^2_{L^2(\Omega)} + \| P \|^2_{L^2(\mathcal{Q})}). \quad (14.23)$$

According to (14.22) and (14.23) we can deduce the result (ii) of the theorem. □

14.3 The Perturbation Problem

In the following, the concentration U will be treated as the desired target concentration of species. We are then interested in the robust regulation of the deviation of the problem from this concentration U. We analyze the full non-linear equation which models large perturbations u to the target U, *i.e.*, we assume that U satisfies the problem (14.1) with the data (P, H_0, U_0) and the perturbed concentration of species $U + u$ satisfies the problem (14.1) with the data $(P+p, H_0+h_0, U_0+u_0)$. Hence, we consider the system with multiple time-varying delays

$$\begin{aligned} &\frac{\partial u}{\partial t} + Fu = u(d_0 - du - \sum_{i=1,n} d_i u(.,r_i) - p - \tilde{G}) \\ &\qquad\qquad -U(du + \sum_{i=1,n} d_i u(.,r_i) + p) \ \text{ on } \mathcal{Q}, \\ &u = h_0 \ \text{ on } \mathcal{Q}_0, \\ &u(.,0) = u_0 \ \text{ on } \Omega, \end{aligned} \quad (14.24)$$

subject to homogeneous Neumann boundary conditions,

where $\tilde{G} = dU + \sum_{i=1,n} d_i U(.,r_i) + P \in L^\infty_+(\mathcal{Q})$.

Theorem 14.5. *Let assumptions* $(\mathbf{A_1})$–$(\mathbf{A_3})$ *hold:*

(i) *Assume that* $u_0 \in \mathcal{U}$, $h_0 \in L^\infty_+(\mathcal{Q}_0)$ *and* $p \in \mathcal{B}$. *Then there exists* $M > 0$ *such that, if*

$$\| p \|_{L^\infty(\mathcal{Q})} \leq M,$$

the perturbation problem (14.24) admits a unique solution $u \in \mathcal{W}_\sigma(\mathcal{Q}) \cap L^\infty_+(\mathcal{Q})$ *and the following a priori estimate holds:*

$$\| u \|^2_{\mathcal{W}_\sigma(\mathcal{Q})} \leq C(\| u_0 \|^2_{L^2} + \| p \|^2_{L^2(\mathcal{Q})} + \| h_0 \|^2_{L^2(\mathcal{Q}_0)}). \quad (14.25)$$

(ii) *Let* $(u_{0,i}, h_{0,i}, p_i), i = 1,2$ *be two smooth data of* $\mathcal{U} \times L^\infty_+(\mathcal{Q}_0) \times \mathcal{B}$ *such that* $p_i \leq M$, *for* $i = 1,2$. *If* u_1 *(respectively* u_2*) is the solution of (14.24) corresponding to data* $(u_{0,1}, h_{0,1}, p_1)$ *(respectively* $(u_{0,2}, h_{0,2}, p_2)$*) then*

$$\| u \|^2_{\mathcal{W}_\sigma(\mathcal{Q})} \leq C(\| u_0 \|^2_{L^2} + \| p \|^2_{L^2(\mathcal{Q})} + \| h_0 \|^2_{L^2(\mathcal{Q}_0)}), \tag{14.26}$$

where $u = u_1 - u_2$, $p = p_1 - p_2$, $h_0 = h_{0,1} - h_{0,2}$ *and* $u_0 = u_{0,1} - u_{0,2}$.

Proof. The existence, uniqueness and Lipschitz continuity results of the problem (14.24) are obtained in the same way as used to prove Theorem 14.4 and by using the regularity of U. So, we omit the details. □

14.4 Robust Control Problems

In this section we formulate the robust control problem and study the existence and necessary optimality conditions for an optimal solution.

14.4.1 Formulation of the Control Problem and Differentiability

Our problem, in this section, is to find the best admissible parameter function p in the presence of the worst disturbance in the initial condition. We then suppose that the control is in the parameter function p and the disturbance is in the initial condition u_0, that is, $p = \varphi$ ($\varphi \in L^\infty(\mathcal{Q})$) and $u_0 = \psi$ ($\psi \in L^\infty(\Omega)$). So the function u is assumed to be related to the disturbance ψ and control ϕ through the problem

$$\begin{aligned}
&\frac{\partial u}{\partial t} + Fu = u(d_0 - du - \sum_{i=1,n} d_i u(.,r_i) - \varphi - \tilde{G}) \\
&\qquad\qquad -U(du + \sum_{i=1,n} d_i u(.,r_i) + \varphi) \quad \text{on } \mathcal{Q}, \\
&u = h_0 \quad \text{on } \mathcal{Q}_0, \\
&u(.,0) = \psi \quad \text{on } \Omega, \\
&\text{subject to homogeneous Neumann boundary conditions,}
\end{aligned} \tag{14.27}$$

under the pointwise constraints

$$\begin{aligned}
&0 \leq \tau_1 \leq \varphi \leq \tau_2 \leq M \quad \text{a.e. in } \mathcal{Q}, \\
&0 \leq \pi_1 \leq \psi \leq \pi_2 \quad \text{a.e. in } \Omega.
\end{aligned} \tag{14.28}$$

Let us consider the following admissible constraint spaces:

$$\begin{aligned}
U_{\text{ad}} &= \{\varphi \in L^2(\mathcal{Q}) : 0 \leq \tau_1 \leq \varphi \leq \tau_2 \quad \text{a.e. in } \mathcal{Q}\}, \\
V_{\text{ad}} &= \{\psi \in L^2(\Omega) : 0 \leq \pi_1 \leq \psi \leq \pi_2 \quad \text{a.e. in } \Omega\}.
\end{aligned}$$

Let $\mathcal{F} : (\varphi, \psi) \longrightarrow u = \mathcal{F}(\varphi, \psi)$ be the map: $U_{\text{ad}} \times V_{\text{ad}} \longrightarrow \mathcal{W}_\sigma(\mathcal{Q})$ defined by (14.27). Our control problem is then find $(\varphi, \psi) \in U_{\text{ad}} \times V_{\text{ad}}$ such that the cost functional

$$J(\varphi, \psi) = \frac{1}{2} \parallel u - u_{\text{obs}} \parallel^2_{L^2(\mathcal{Q})} + \frac{\alpha}{2} \parallel \varphi \parallel^2_{L^2(\mathcal{Q})} - \frac{\beta}{2} \parallel \psi \parallel^2_{L^2(\Omega)} \tag{14.29}$$

is minimized with respect to φ and maximized with respect to ψ subject to the problem (14.27),

where α, β are fixed with constraints $\alpha > 0$ and $\beta > 0$.

We are now going to study the differentiability of the operator solution of the problem (14.27).

Before proceeding with investigation of the differentiability of the function $\mathcal{F} : (\varphi, \psi) \longrightarrow u$, which maps the source term $(\varphi, \psi) \in L^\infty_+(\mathcal{Q}) \times L^\infty_+(\Omega)$ of the problem (14.27) into the corresponding solution $u \in \mathcal{W}_\sigma(\mathcal{Q}) \cap L^\infty_+(\mathcal{Q})$, we study, for (ξ, η_0, ϑ) given data, the following problem:

find $w \in \mathcal{W}_\sigma(\mathcal{Q})$ such that

$$\begin{aligned} \frac{\partial w}{\partial t} + Fw = w(d_0 - du - \sum_{i=1,n} d_i u(.,r_i) - \varphi - \tilde{G}) \\ -U_1(dw + \sum_{i=1,n} d_i w(.,r_i) + \xi) \ \text{ on } \mathcal{Q}, \end{aligned} \tag{14.30}$$

$w = \eta_0$ on $\mathcal{Q}_0$,

$w(.,0) = \vartheta$ on Ω,

subject to homogeneous Neumann boundary conditions,

where $u = \mathcal{F}(\varphi, \psi)$ and $U_1 = u + U$.

Theorem 14.6. *If u and U are in $\mathcal{W}_\sigma(\mathcal{Q}) \cap L^\infty_+(\mathcal{Q})$, the following results hold:*

(i) For any $(\xi, \eta_0, \vartheta) \in L^\infty(\mathcal{Q}) \times L^\infty(\mathcal{Q}_0) \times L^\infty(\Omega)$, there exists a unique function $w \in \mathcal{W}_\sigma(\mathcal{Q})$, solution of the problem (14.30), such that

$$\parallel w \parallel^2_{\mathcal{W}_\sigma(\mathcal{Q})} \leq C_e(\parallel \xi \parallel^2_{L^2(\mathcal{Q})} + \parallel \eta_0 \parallel^2_{L^2(\mathcal{Q}_0)} + \parallel \vartheta \parallel^2_{L^2(\Omega)}). \tag{14.31}$$

(ii) Let $(\xi_i, \eta_{0i}, \vartheta_i)$, for $i = 1, 2$ be in $L^\infty(\mathcal{Q}) \times L^\infty(\mathcal{Q}_0) \times L^\infty(\Omega)$. If w_i is the solution of (14.30), corresponding to data $(\xi_i, \eta_0, \vartheta_i)$, for $i = 1, 2$, rcspcctively, then

$$\parallel w \parallel^2_{\mathcal{W}_\sigma(\mathcal{Q})} \leq C_e(\parallel \xi \parallel^2_{L^2(\mathcal{Q})} + \parallel \eta_0 \parallel^2_{L^2(\mathcal{Q}_0)} + \parallel \vartheta \parallel^2_{L^2(\Omega)}), \tag{14.32}$$

where $w = w_1 - w_2$, $\xi = \xi_1 - \xi_2$, $\eta_0 = \eta_{01} - \eta_{02}$, $\vartheta = \vartheta_1 - \vartheta_2$.

Proof. The existence, uniqueness and Lipschitz continuity results of problem (14.30) are obtained in the same way as used to prove Theorem 14.4 and by using the regularity of u and U. So, we omit the details. □

We shall now study the differentiability of the operator solution $\mathcal{F}$.

Theorem 14.7. *Let assumptions* $(\mathbf{A_1})$–$(\mathbf{A_3})$ *be fulfilled and the perturbation of the initial function* h_0 *be in* $L^\infty_+(\mathcal{Q}_0)$. *Then the following results hold:*

(*i*) *The function* $\mathcal{F}$ *is differentiable with respect to* $X = (\varphi, \psi) \in U_{\text{ad}} \times V_{\text{ad}}$ *and its derivative* $\mathcal{F}'(X) : L^\infty(\mathcal{Q}) \times L^\infty(\Omega) \longrightarrow \mathcal{W}_\sigma(\mathcal{Q})$ *is a linear operator such that* $w = \mathcal{F}'(X).Y$ *is the unique solution of the problem (14.30), with data* $Y = (\xi, \vartheta)$ *and the given initial function* $\eta_0 = 0$, *i.e.,*

$$
\begin{aligned}
&\frac{\partial w}{\partial t} + Fw = w(d_0 - du - \sum_{i=1,n} d_i u(.,r_i) - \varphi - \tilde{G}) \\
&\qquad\qquad -U_1(dw + \sum_{i=1,n} d_i w(.,r_i) + \xi) \quad \text{on } \mathcal{Q}, \\
&w = 0 \quad \text{on } \mathcal{Q}_0, \\
&w(.,0) = \vartheta \quad \text{on } \Omega,
\end{aligned}
\tag{14.33}
$$

subject to homogeneous Neumann boundary conditions.

Moreover, the Lipschitz continuous result (14.32) holds, i.e., for all w_i *solution of (14.33), with data* $(\xi_i, \vartheta_i) \in L^\infty(\mathcal{Q}) \times L^\infty(\mathcal{Q})$, *for* $i = 1, 2$, *respectively, we have the estimate*

$$
\| w \|^2_{\mathcal{W}_\sigma(\mathcal{Q})} \leq C_e(\| \xi \|^2_{L^2(\mathcal{Q})} + \| \vartheta \|^2_{L^2(\Omega)}),
\tag{14.34}
$$

where $w = w_1 - w_2$, $\xi = \xi_1 - \xi_2$ *and* $\vartheta = \vartheta_1 - \vartheta_2$.

(*ii*) *For all* $X_i = (\varphi_i, \psi_i) \in U_{\text{ad}} \times V_{\text{ad}}$, *for* $i = 1, 2$, *we have the following estimate, for all* $Y = (h, \vartheta) \in L^\infty(\mathcal{Q}) \times L^\infty(\Omega)$:

$$
\begin{aligned}
&\| \mathcal{F}'(X_1).Y - \mathcal{F}'(X_2).Y \|^2_{\mathcal{W}_\sigma(\mathcal{Q})} \\
&\qquad \leq C_e(\| Y \|_{L^2(\mathcal{Q})\times L^2(\Omega)} \| X \|^2_{L^2(\mathcal{Q})\times L^2(\Omega)} \\
&\qquad\qquad + \| Y \|^2_{L^2(\mathcal{Q})\times L^2(\Omega)} \| X \|_{L^2(\mathcal{Q})\times L^2(\Omega)}),
\end{aligned}
\tag{14.35}
$$

where $\varphi = \varphi_1 - \varphi_2$, $\psi = \psi_1 - \psi_2$ *and* $X = X_1 - X_2 = (\varphi, \psi)$.

Proof. The proof of this theorem is obtained by using a similar technique as used to prove the results of Proposition 8.13[1] (for the differentiability of $\mathcal{F}$ see also Belmiloudi [38], in which the author considers a bioeconomic model). So, we omit the details. □

Next, we will study the existence of an optimal solution.

14.4.2 Existence of an Optimal Solution

Theorem 14.8. *For* α *and* β *sufficiently large, there exists an optimal solution* $(\varphi^*, \psi^*, u^*) \in U_{\text{ad}} \times V_{\text{ad}} \times \mathcal{W}_\sigma(\mathcal{Q})$ *such that* (φ^*, ψ^*) *is a solution of (14.29) and* $u^* = \mathcal{F}(\varphi^*, \psi^*)$ *is the solution of (14.27).*

[1] See also Belmiloudi [44], in which minimax problems are considered for periodic competing systems.

Proof. Let P_ψ be the mapping: $\varphi \longrightarrow J(\varphi,\psi)$ and Q_φ be the mapping: $\psi \longrightarrow J(\varphi,\psi)$ In order to obtain the existence of the robust control problem, we prove first that P_ψ is convex and lower semi-continuous for all $\psi \in V_{\rm ad}$, second that Q_φ is concave and upper semi-continuous for all $\varphi \in U_{\rm ad}$ and, finally, we use the minimax theorems in infinite dimensions presented in Chapter 5.

In order to prove the convexity, it is sufficient to show that for $(\varphi_1,\varphi_2) \in U_{\rm ad} \times U_{\rm ad}$, we have $(P'_\psi(\varphi_1) - P'_\psi(\varphi_2)).\varphi \geq 0$, where $\varphi = \varphi_1 - \varphi_2$.

From the expression of G-differentiable cost functional J (a composition of G-differentiable mappings), it follows that P_ψ is G-differentiable and, for $i = 1, 2$,

$$P'_\psi(\varphi_i).\varphi = \iint_Q (u_i - u_{\rm obs}) w_i {\rm d}x{\rm d}t + \alpha \iint_Q \varphi_i \varphi {\rm d}x{\rm d}t,$$

where $u_i = \mathcal{F}(\varphi_i,\psi)$ and $w_i = \mathcal{F}'(\varphi_i,\psi).(\varphi,0)$.

Consequently,

$$\begin{aligned}(P'_\psi(\varphi_1) - P'_\psi(\varphi_2)).\varphi = \alpha \parallel \varphi \parallel^2_{L^2(Q)} + \iint_Q u w_1 {\rm d}x{\rm d}t \\ + \iint_Q (u_2 - u_{\rm obs}) w {\rm d}x{\rm d}t,\end{aligned} \tag{14.36}$$

where $u = u_1 - u_2$ and $w = w_1 - w_2$. The estimates (14.35), (14.34) and (14.26) imply that

$$\begin{aligned}&\iint_Q u w_1 {\rm d}x{\rm d}t \leq C_0 \parallel \varphi \parallel^2_{L^2(Q)}, \\ &\iint_Q (u_2 - u_{\rm obs}) w {\rm d}x{\rm d}t \leq C_1 \parallel \varphi \parallel^{3/2}_{L^2(Q)} .\end{aligned} \tag{14.37}$$

From (14.36) and the previous results (14.37), we can deduce that there exists a constant $\alpha_l > 0$ such that for $\alpha \geq \alpha_l$ we have

$$(P'_\psi(\varphi_1) - P'_\psi(\varphi_2)).\varphi \geq (\alpha_l - C_0) \parallel \varphi \parallel^2_{L^2(Q)} - C_1 \parallel \varphi \parallel^{3/2}_{L^2(Q)} \geq 0$$

and then the convexity of P_ξ is established.

In the same way, we can find $\beta_l > 0$ such that for $\beta \geq \beta_l$, Q_θ is concave.

We shall now prove that P_ψ (respectively Q_φ) is lower (respectively upper) semi-continuous for all $\psi \in V_{\rm ad}$ (respectively $\varphi \in U_{\rm ad}$).

Let $\varphi_k \in U_{\rm ad}$ be a minimizing sequence of P_ψ, *i.e.*,

$$\liminf_{k \longrightarrow \infty} J(\varphi_k,\psi) = \inf_{\varphi \in U_{\rm ad}} J(\varphi,\psi).$$

Then, according to the nature of the cost function J, we can deduce that φ_k is uniformly bounded in $U_{\rm ad}$ and we can extract from φ_k a subsequence also denoted by φ_k such that $\varphi_k \rightharpoonup \varphi_\xi$ weakly in $U_{\rm ad}$. Therefore, by using the same technique as to obtain the estimate (14.16), $u_k = \mathcal{F}(\varphi_k,\psi)$ is uniformly bounded in $\mathcal{W}_\sigma(\mathcal{Q})$. Moreover, according now to Lemma 6.6, the injection of $\mathcal{W}_\sigma(\mathcal{Q})$ into $L^2(\mathcal{Q})$ is compact. Consequently, these results make it possible to extract from u_k a subsequence also denoted by u_k such that

$$\begin{aligned} &u_k \rightharpoonup u_\psi \text{ weakly in } \mathcal{W}_\sigma(\mathcal{Q}),\\ &u_k \longrightarrow u_\psi \text{ strongly in } L^2(\mathcal{Q}),\\ &\varphi_k \rightharpoonup \varphi_\psi \text{ weakly in } L^2(\mathcal{Q}) \text{ and } \varphi_\psi \in U_{\text{ad}}. \end{aligned} \tag{14.38}$$

We shall now prove that $u_k\varphi_k \longrightarrow u_\psi\varphi_\psi$ weakly in $L^2(\mathcal{Q})$. Since $u_k\varphi_k - u_\psi\varphi_\psi = (u_k - u_\psi)\varphi_k + u_\psi(\varphi_k - \varphi_\xi)$, and according to the first and second parts of (14.38), we then obtain the result. Is is easy to prove that u_ψ is a solution of (14.27) with data (φ_ψ, ψ) and according to the uniqueness of the solution of the problem (14.27), we then have $u_\psi = \mathcal{F}(\varphi_\psi, \psi)$.

Since the norm is lower semi-continuous, therefore we have that the map P_ψ is lower semi-continuous for all $\psi \in V_{\text{ad}}$. By applying similar arguments as in the proof of the previous result we obtain that Q_φ is upper semi-continuous for all $\varphi \in U_{\text{ad}}$.

This completes the proof. □

We next wish to show the appropriate first-order necessary conditions (optimality conditions) of the saddle point problem (14.29).

14.4.3 Optimality Conditions

For simplicity we suppose that $e_i(T) = \delta(T)$ for all $i = 1, n$, where $\delta(T)$ is a given non-negative constant. In order to characterize the optimal control, we introduce the following adjoint problem corresponding to the primal problem (14.27) (we denote by $u = \mathcal{F}(\varphi, \psi)$ the solution of the problem (14.27) where the forcing is (φ, ψ)):

$$\begin{aligned} -\frac{\partial \tilde{u}}{\partial t} + F\tilde{u} &= (d_0 - d(U_1 + u) - \sum_{i=1,n} d_i u(., r_i) - \varphi - \tilde{G})\tilde{u}\\ &\quad - \sum_{i=1,n} d_i(., f_i)\tilde{u}(., f_i)U_1(., f_i)f_i' + (u - u_{\text{obs}}) \text{ on } \tilde{\mathcal{Q}},\\ -\frac{\partial \tilde{u}}{\partial t} + F\tilde{u} &= (d_0 - d(U_1 + u) - \sum_{i=1,n} d_i u(., r_i) - \varphi - \tilde{G})\tilde{u}\\ &\quad +(u - u_{\text{obs}}) \text{ on } \mathcal{Q}_T,\\ \tilde{u}(., T) &= 0 \text{ on } \Omega, \end{aligned} \tag{14.39}$$

subject to homogeneous Neumann boundary conditions,

where $\tilde{\mathcal{Q}} = \Omega \times (0, T - \delta(T))$ and $\mathcal{Q}_T = \Omega \times (T - \delta(T), T)$.

Proposition 14.9. *Let assumptions* $(\mathbf{A_1})$–$(\mathbf{A_3})$ *hold and let* u *be in* $\mathcal{W}_\sigma(\mathcal{Q}) \cap L^\infty(\mathcal{Q})$. *Then the problem (14.39) has a unique solution in* $\mathcal{W}_\sigma(\mathcal{Q}) \cap L^\infty(\mathcal{Q})$ *with* $\tilde{u}$ *satisfying*

$$\| \tilde{u} \|_{\mathcal{W}_\sigma(\mathcal{Q})} \leq C \| u - u_{\text{obs}} \|_{L^2(\mathcal{Q})} .$$

Proof. In order to prove the existence of a unique solution $\tilde{u}$, we change the variables of problem (14.39) by reversing sense of time, *i.e.*, $t := T - t$ and we apply the same approach (a constructive method) as to obtain the existence and uniqueness result of the problem (14.39), given in Theorem 14.4. So, we omit the details.

Next, we prove the estimate given in the proposition. Multiplying (14.39) by $\tilde{u}$ and integrating over $(t,T) \times \Omega$, we obtain (since $\tilde{u}(T) = 0$)

$$\begin{aligned} &\| \tilde{u}(.,t) \|_{L^2}^2 + 2 \int_t^T \| \sigma \nabla \tilde{u} \|_{L^2}^2 \, \mathrm{d}s \\ &\quad \leq c_1 \int_t^T \| \tilde{u} \|_{L^2}^2 \, \mathrm{d}s + \int_t^T \| u - u_{\text{obs}} \|_{L^2}^2 \, \mathrm{d}s \\ &\quad + 2 \sum_{i=1,n} \int_{\min(t,T-\delta(T))}^{T-\delta(T)} \int_\Omega d_i(.,f_i) U_1(.,f_i) \tilde{u}(.,f_i) f_i' \tilde{u} \mathrm{d}x \mathrm{d}s. \end{aligned} \tag{14.40}$$

We shall now estimate

$$I_T = 2 \int_{\min(t,T-\delta(T))}^{T-\delta(T)} \int_\Omega d_i(.,f_i) U_1(.,f_i) \tilde{u}(.,f_i) f_i' \tilde{u} \mathrm{d}x \mathrm{d}s.$$

According to the regularity of d_i, f_i and U_1 we then obtain

$$I_T \leq C_i \Big(\int_{\min(t,T-\delta(T))}^{T-\delta(T)} \| \tilde{u}(.,f_i) \|_{L^2}^2 \, f_i' \mathrm{d}s \Big)^{1/2} \Big(\int_t^T \| \tilde{u} \|_{L^2}^2 \, \mathrm{d}s \Big)^{1/2},$$

where $C_i = c_2 \| d_i \|_\infty (\| f_i' \|_\infty)^{1/2}$.

If we set $a = f_i(s)$, we have $\mathrm{d}a = f_i'(s)\mathrm{d}s$ and then, according to $r_i(T) = T - \delta(T)$ and $r_i(t) \leq t$,

$$I_T \leq C_i \int_t^T \| \tilde{u} \|_{L^2}^2 \, \mathrm{d}s. \tag{14.41}$$

According to (14.41) the estimate (14.40) becomes

$$\| \tilde{u}(.,t) \|_{L^2}^2 + 2 \int_t^T \| \sigma \nabla \tilde{u} \|_{L^2}^2 \, \mathrm{d}s \leq c_3 \int_t^T \| \tilde{u} \|_{L^2}^2 \, \mathrm{d}s + \int_t^T \| u - u_{\text{obs}} \|_{L^2}^2 \, \mathrm{d}s.$$

By using Gronwall's formula we then have

$$\| \tilde{u} \|_{\mathcal{H}(Q)}^2 + \| \tilde{u} \|_{\mathcal{V}_\sigma(Q)}^2 \leq C \| u - u_{\text{obs}} \|_{L^2(Q)}^2 . \tag{14.42}$$

Using this result and Equation (14.39) we prove easily that $\tilde{u}$ satisfies

$$\| \tilde{u} \|_{H^1(0,T;H_\sigma^*(Q))}^2 \leq C \| u - u_{\text{obs}} \|_{L^2(Q)}^2 . \tag{14.43}$$

Applying (14.42) and (14.43), we obtain the second result of the proposition. □

We can now give the optimality system for the saddle point problem (14.29).

Theorem 14.10. *Let assumptions* $(\mathbf{A_1})$–$(\mathbf{A_3})$ *hold, and* $(\varphi^*, \psi^*, u^*) \in U_{\text{ad}} \times V_{\text{ad}} \times (\mathcal{W}_\sigma(\mathcal{Q}) \cap L^\infty_+(\mathcal{Q}))$ *such that* (φ^*, ψ^*) *is defined by (14.29) and* $u^* = \mathcal{F}(\varphi^*, \psi^*)$ *is a solution of (14.27). Then, for all* $(\varphi, \psi) \in U_{\text{ad}} \times V_{\text{ad}}$,

$$\begin{aligned} &\iint_{\mathcal{Q}} (\alpha\varphi^* - \tilde{u}^* U_1^*)(\varphi - \varphi^*)\mathrm{d}x\mathrm{d}t \geq 0, \\ &\int_\Omega (\tilde{u}^*(.,0) - \beta\psi^*)(\psi - \psi^*)\mathrm{d}x \leq 0, \end{aligned} \tag{14.44}$$

where $U_1^* = u^* + U$ *and* $\tilde{u}^*$ *is the solution of the adjoint problem (14.39), corresponding to the primal solution* u^**. Otherwise,*

$$\begin{aligned} \varphi^* &= \min\left(\tau_2, \max\left(\frac{\tilde{u}^* U_1^*}{\alpha}, \tau_1\right)\right), \\ \psi^* &= \min\left(\pi_2, \max\left(\frac{\tilde{u}^*(.,0)}{\beta}, \pi_1\right)\right). \end{aligned} \tag{14.45}$$

Moreover, the gradient of the functional J*, in the weak sense, at point* (φ^*, ψ^*) *is*

$$\frac{\partial J}{\partial \varphi}(\varphi^*, \psi^*) = \alpha\varphi^* - \tilde{u}^* U_1^* \quad \textit{and} \quad \frac{\partial J}{\partial \varphi}(\varphi^*, \psi^*) = \tilde{u}^*(.,0) - \beta\psi^*.$$

Proof. The cost functional J is a composition of differentiable maps then J is differentiable and, for all $(\xi, \vartheta) \in L^\infty(\mathcal{Q}) \times L^\infty(\Omega)$ such that $(\varphi + \epsilon\xi, \psi + \epsilon\vartheta) \in U_{\text{ad}} \times V_{\text{ad}}$ for ϵ small, we have

$$\begin{aligned} J'(\varphi, \psi).(\xi, \vartheta) &= \lim_{\epsilon \longrightarrow 0} \frac{1}{\epsilon}(J(\varphi + \epsilon\xi, \psi + \epsilon\vartheta) - J(\varphi, \psi)) \\ &= \iint_{\mathcal{Q}} (u - u_{\text{obs}}) w \mathrm{d}x\mathrm{d}t + \alpha \iint_{\mathcal{Q}} \varphi\xi \mathrm{d}x\mathrm{d}t \\ &\quad -\beta \iint_{\mathcal{Q}} \psi\vartheta \mathrm{d}x\mathrm{d}t, \end{aligned} \tag{14.46}$$

where w is the solution of (14.33).

Multiplying (14.30) by $\tilde{u}$, integrating over $\mathcal{Q}$ and integrating by parts with respect to time t, we obtain

$$\begin{aligned} &\int_0^T \int_\Omega (-\frac{\partial \tilde{u}}{\partial t} + F\tilde{u} - \tilde{u}(d_0 - d(U_1 + u) - \sum_{i=1,n} d_i u(., r_i) - \varphi - \tilde{G})) w \mathrm{d}x\mathrm{d}t \\ &\quad = -\int_0^T \int_\Omega U_1 \xi \tilde{u} \mathrm{d}x\mathrm{d}t - \sum_{i=1,n} \int_0^T \int_\Omega d_i w(., r_i) U_1 \tilde{u} \mathrm{d}x\mathrm{d}t \\ &\quad - \int_\Omega w(., T)\tilde{u}(., T)\mathrm{d}x + \int_\Omega \vartheta\tilde{u}(., 0)\mathrm{d}x. \end{aligned}$$

Next, we shall calculate the term $A = \int_0^T \int_\Omega d_i w(.,r_i) U_1 \tilde{u} \mathrm{d}x\mathrm{d}t$. Let $s = r_i(t)$, then $t = f_i(s)$ and $\mathrm{d}t = f_i'(s)\mathrm{d}s$. Therefore,

$$A = \int_{-e_i(0)}^{T-e_i(T)} \int_\Omega d_i(.,f_i(s))\tilde{u}(.,f_i(s))U_1(.,f_i(s))f_i'(s)w(.,s)\mathrm{d}x\mathrm{d}s.$$

Since $e_i(T) = \delta(T), \forall i = 1, n$ and according to the second part of (14.30) we have

$$A = \int_0^{T-\delta(T)} \int_\Omega d_i(.,f_i(s))\tilde{u}(.,f_i(s))U_1(.,f_i(s))f_i'(s)w(.,s)\mathrm{d}x\mathrm{d}s. \tag{14.47}$$

Since $\tilde{u}$ is a solution of the adjoint problem (14.39) and according to (14.47), we obtain that

$$\iint_Q (u - u_{\mathrm{obs}})w\mathrm{d}x\mathrm{d}t = -\iint_Q \xi\tilde{u}U_1\mathrm{d}x\mathrm{d}t + \int_\Omega \vartheta\tilde{u}(.,0)\mathrm{d}x. \tag{14.48}$$

Applying (14.48) and according to the expression of $J'(\varphi,\psi)$ given by (14.46) we then have

$$J'(\varphi,\psi).(\xi,\vartheta) = \iint_Q (\alpha\varphi - \tilde{u}U_1)\xi\mathrm{d}x\mathrm{d}t + \int_\Omega (\tilde{u}(.,0) - \beta\psi)\vartheta\mathrm{d}x. \tag{14.49}$$

Since (φ^*,ψ^*) is an optimal solution we have

$$\frac{\partial J}{\partial\varphi}(\varphi^*,\psi^*).\xi \geq 0 \text{ and } \frac{\partial J}{\partial\psi}(\varphi^*,\psi^*).\vartheta \leq 0 \quad \text{for all } (\xi,\vartheta) \in L^\infty(Q) \times L^\infty(\Omega)$$

and so we obtain, for all $(\xi,\vartheta) \in L^\infty(Q) \times L^\infty(\Omega)$,

$$\begin{aligned} &\iint_Q (\alpha\varphi^* - \tilde{u}^*U_1^*)\xi\mathrm{d}x\mathrm{d}t \geq 0, \\ &\int_\Omega (\tilde{u}^*(.,0) - \beta\psi^*)\vartheta\mathrm{d}x \leq 0, \end{aligned} \tag{14.50}$$

where $U_1^* = u^* + U$.

By using a classical control argument concerning the sign of the variations ξ and ϑ (depending on the size of φ^* and ψ^* respectively), we obtain that

$$\begin{aligned} \varphi^* &= \min\left(\tau_2, \max\left(\frac{\tilde{u}^*U_1^*}{\alpha}, \tau_1\right)\right), \\ \psi^* &= \min\left(\pi_2, \max\left(\frac{\tilde{u}^*(.,0)}{\beta}, \pi_1\right)\right). \end{aligned}$$

This completes the proof. □

Remark 14.11. (*i*) By similar arguments as used in Belmiloudi [42, 43], the uniqueness of the optimal solution can be proved under, for example, the condition that the final time is small enough or the parameters α and β are large enough.

(*ii*) We can treat, by using the same technique developed in this chapter, other dynamical population models. For example, coupled cooperative diffusion systems of various populations (for a well-posedness problem, without delay, see *e.g.*, Jia and Feng [169]).

(*iii*) We can also consider the situation of heterogeneous environment with homogeneous Dirichlet[2] or Neumann boundary conditions. In this case, the diffusion can be degenerated in some areas and we can assume that the diffusion operator F given by (14.2) may be degenerate and satisfies the following (see, *e.g.*, Belmiloudi [42, 43] and Lenhart and Yong [193]):

The functions $a_{ij} : \overline{\Omega} \longrightarrow \mathrm{IR}$ are $\mathcal{C}^1$ and the matrix $A(.) = (a_{ij}(.))_{1\leq i,j\leq m}$ is symmetric positive semidefinite in $\overline{\Omega}$. Moreover,

$$\text{meas}\{x \in \Omega \text{ such that } \det(A(x)) = 0\} = 0, \tag{14.51}$$

and there exists a $\mathcal{C}^1$ function $\sigma : \overline{\Omega} \longrightarrow \mathrm{IR}^{m\times m}$, such that

$$A(x) = \sigma^*(x)\sigma(x), \; x \in \overline{\Omega}, \tag{14.52}$$

where σ^* is the dual of the matrix σ.

For the existence result, we can use the vanishing viscosity method (see, *e.g.*, Crandall *et al.* [90]) in the adequate spaces and for the robust control problem we can use a similar technique to the one developed in this chapter. ◊

14.5 Other Situations

14.5.1 Disturbance in the Parameter Function *p*

If we assume that the control and the disturbance are in the parameter p, *i.e.*, $p = \varphi + \psi$, we obtain the same results as in Section 14.4. In this case, the cost functional can be given by

$$J(\varphi, \psi) = \frac{1}{2} \parallel u - u_{\text{obs}} \parallel^2_{L^2(Q)} + \frac{\alpha}{2} \parallel \varphi \parallel^2_{L^2(Q)} - \frac{\beta}{2} \parallel \pi_2 - \psi \parallel^2_{L^2(Q)} .$$

We can also prove an existence theorem of the saddle point problem and obtain necessary optimality conditions for its solution by using the same method as in Section 14.4. Let now

[2] The given species does not live on the edge of the region.

$$U_{\text{ad}} = \{\varphi \in L^2(\mathcal{Q}) : \tau_1 \le \varphi \le \tau_2 \text{ a.e. in } \mathcal{Q}\},$$
$$V_{\text{ad}} = \{\psi \in L^2(\mathcal{Q}) : 0 \le \psi \le \pi_2 \text{ a.e. in } \mathcal{Q}\}.$$

Then, for α and β sufficiently large, there exist $(\varphi^*, \psi^*) \in U_{\text{ad}} \times V_{\text{ad}}$ and $u^* \in \mathcal{W}_\sigma(\mathcal{Q}) \cap L^\infty(\mathcal{Q})$ satisfying

$$\begin{aligned} &\frac{\partial u^*}{\partial t} + Fu^* = u^*(d_0 - du^* - \sum_{i=1,n} d_i u^*(.,r_i) - \varphi^* - \psi^* - \tilde{G}) \\ &\qquad -U(du^* + \sum_{i=1,n} d_i u^*(.,r_i) + \varphi^* + \psi^*) \text{ on } \mathcal{Q}, \\ &u^* = h_0 \text{ on } \mathcal{Q}_0, \\ &u^*(.,0) = u_0 \text{ on } \Omega, \end{aligned} \tag{14.53}$$

subject to homogeneous Neumann boundary conditions,

and for all $(\varphi, \psi) \in U_{\text{ad}} \times V_{\text{ad}}$, we have the optimality conditions

$$\iint_{\mathcal{Q}} (\alpha\varphi^* - \tilde{u}^* U_1^*)(\varphi - \varphi^*) dx dt \ge 0,$$
$$\iint_{\mathcal{Q}} (-\tilde{u}^* U_1^* - \beta(-\pi_2 + \psi^*))(\psi - \psi^*) dx dt \le 0,$$

otherwise,

$$\varphi^* = \min\left(\tau_2, \max\left(\frac{\tilde{u}^* U_1^*}{\alpha}, \tau_1\right)\right),$$
$$\psi^* = \min\left(\pi_2, \max\left(\pi_2 - \frac{\tilde{u}^* U_1^*}{\beta}, 0\right)\right),$$

where u^* is the solution of the problem (14.53) with the forcing (φ^*, ψ^*), $U_1^* = u^* + U$ and $\tilde{u}^*$ is the solution of the following adjoint problem corresponding to the primal solution u^*

$$\begin{aligned} &-\frac{\partial \tilde{u}^*}{\partial t} + F\tilde{u}^* = (d_0 - dU_1^* - \sum_{i=1,n} d_i u^*(.,r_i) - \varphi^* - \psi^* - \tilde{G})\tilde{u}^* \\ &\qquad - \sum_{i=1,n} d_i(.,f_i)\tilde{u}^*(.,f_i)U_1(.,f_i)f_i' + (u^* - u_{\text{obs}}) \text{ on } \tilde{\mathcal{Q}}, \\ &-\frac{\partial \tilde{u}^*}{\partial t} + F\tilde{u}^* = (d_0 - dU_1^* - \sum_{i=1,n} d_i u^*(.,r_i) - \varphi^* - \psi^* - \tilde{G})\tilde{u}^* \\ &\qquad +(u^* - u_{\text{obs}}) \text{ on } \mathcal{Q}_T, \\ &\tilde{u}^*(.,T) = 0 \text{ on } \Omega, \end{aligned} \tag{14.54}$$

subject to homogeneous Neumann boundary conditions,

with $\tilde{\mathcal{Q}} = \Omega \times (0, T - \delta(T))$ and $\mathcal{Q}_T = \Omega \times (T - \delta(T), T)$.

Moreover, the gradient of the functional J, in the weak sense, at point (φ^*, ψ^*) is

$$\frac{\partial J}{\partial \varphi}(\varphi^*, \psi^*) = \alpha\varphi^* - \tilde{u}^* U_1^* \text{ and } \frac{\partial J}{\partial \varphi}(\varphi^*, \psi^*) = -\tilde{u}^* U_1^* - \beta(-\pi_2 + \psi^*).$$

14.5.2 Remarks on Boundary Control and Habitat Hostility

If we want to take into account the influence of the hostility of the boundary environment, we can use a similar consideration as in Lenhart at al. [194]. More precisely, the hostility of the boundary environment can be represented by some parameter η, multiplying the state variable in the Robin boundary condition. For a given η, the corresponding concentration of species U satisfies the following system delay system:

$$\begin{aligned}
&\frac{\partial U}{\partial t} + FU = U(d_0 - dU - \sum_{i=1,n} d_i U(.,r_i) - P) \quad \text{on } \mathcal{Q}, \\
&U = H_0 \quad \text{on } \mathcal{Q}_0 = \Omega \times [-\delta(0), 0), \\
&U(.,0) = U_0 \quad \text{on } \Omega, \\
&(\sigma^* \sigma \nabla U).\mathbf{n} = -\eta U \quad \text{on } \Sigma = \Gamma \times (0,T).
\end{aligned} \tag{14.55}$$

The solution U will be treated as the target function and we analyze the non-linear equation which models large perturbations u to the target U, *i.e.*, we assume that U satisfies the problem (14.55) with the data (P, H_0, U_0, η) and $U+u$ satisfies the problem (14.55) with the data $(P+p, H_0+h_0, U_0+u_0, \eta+\xi)$. Hence, we consider the system with multiple time-varying delays

$$\begin{aligned}
&\frac{\partial u}{\partial t} + Fu = u(d_0 - du - \sum_{i=1,n} d_i u(.,r_i) - p - \tilde{G}) \\
&\qquad\qquad -U(du + \sum_{i=1,n} d_i u(.,r_i) + p) \quad \text{on } \mathcal{Q}, \\
&u = h_0 \quad \text{on } \mathcal{Q}_0, \\
&u(.,0) = u_0 \quad \text{on } \Omega, \\
&(\sigma^* \sigma \nabla u).\mathbf{n} = -\xi(U+u) - \eta u \quad \text{on } \Sigma,
\end{aligned} \tag{14.56}$$

where $\tilde{G} = dU + \sum_{i=1,n} d_i U(.,r_i) + P \in L^\infty_+(\mathcal{Q})$.

Our problem, in this section, is to find the best admissible parameter function p in the presence of the worst disturbance in the hostility boundary parameter. We then suppose that the control is in the parameter function p and the disturbance is in the boundary parameter η, *i.e.*, $p = \varphi$ $(\varphi \in L^\infty(\mathcal{Q}))$ and $\xi = \psi$ $(\psi \in L^\infty(\Sigma))$. So the function u is assumed to be related to the disturbance ψ and control φ through the problem

$$\begin{aligned}
&\frac{\partial u}{\partial t} + Fu = u(d_0 - du - \sum_{i=1,n} d_i u(.,r_i) - \varphi - \tilde{G}) \\
&\qquad\qquad -U(du + \sum_{i=1,n} d_i u(.,r_i) + \varphi) \quad \text{on } \mathcal{Q}, \\
&u = h_0 \quad \text{on } \mathcal{Q}_0, \\
&u(.,0) = u_0 \quad \text{on } \Omega, \\
&(\sigma^* \sigma \nabla u).\mathbf{n} = -\psi(U+u) - \eta u \quad \text{on } \Sigma,
\end{aligned} \tag{14.57}$$

under the pointwise constraints

$$\begin{aligned} &0 \leq \tau_1 \leq \varphi \leq \tau_2 \quad \text{a.e. in } \mathcal{Q}, \\ &0 \leq \psi \leq \pi_2 \quad \text{a.e. in } \Sigma. \end{aligned} \tag{14.58}$$

We obtain the same results as in Section 14.4. In this case, the cost functional can be given by

$$J(\varphi, \psi) = \frac{1}{2} \parallel u - u_{\text{obs}} \parallel^2_{L^2(\mathcal{Q})} + \frac{\alpha}{2} \parallel \varphi \parallel^2_{L^2(\mathcal{Q})} - \frac{\beta}{2} \parallel \pi_2 - \psi \parallel^2_{L^2(\Sigma)} .$$

We can also prove an existence theorem of the control problem and obtain necessary optimality conditions for its solution using the same method as in Section 14.4. Let now

$$\begin{aligned} U_{\text{ad}} &= \{\varphi \in L^2(\mathcal{Q}) : 0 \leq \tau_1 \leq \varphi \leq \tau_2 \ \text{ a.e. in } \mathcal{Q}\}, \\ V_{\text{ad}} &= \{\psi \in L^2(\Sigma) : 0 \leq \psi \leq \pi_2 \ \text{ a.e. in } \Sigma\}. \end{aligned}$$

Then, for α and β sufficiently large, there exists an optimal solution (φ^*, ψ^*, u^*) in $U_{\text{ad}} \times V_{\text{ad}} \times (\mathcal{W}_\sigma \cap L^\infty(\mathcal{Q}))$ such that u^* is a solution of the problem (14.57), with the forcing (φ^*, ψ^*) and, for all $(\varphi, \psi) \in U_{\text{ad}} \times V_{\text{ad}}$, the following optimality conditions hold:

$$\begin{aligned} &\iint_{\mathcal{Q}} (\alpha\varphi^* - \tilde{u}^* U_1^*)(\varphi - \varphi^*) \mathrm{d}x\mathrm{d}t \geq 0, \\ &\iint_{\Sigma} (-\tilde{u}^* U_1^* - \beta(-\pi_2 + \psi^*))(\psi - \psi^*) \mathrm{d}\Gamma\mathrm{d}t \leq 0, \end{aligned}$$

otherwise,

$$\begin{aligned} \varphi^* &= \min\left(\tau_2, \max\left(\frac{\tilde{u}^* U_1^*}{\alpha}, \tau_1\right)\right), \\ \psi^* &= \min\left(\pi_2, \max\left(\pi_2 - \frac{\tilde{u}^* U_1^*}{\beta}|_\Sigma, 0\right)\right), \end{aligned}$$

where $u^* = \mathcal{F}(\varphi^*, \psi^*)$ is the solution of the problem (14.57) with the forcing (φ^*, ψ^*), $U_1^* = u^* + U$ and $\tilde{u}^*$ is the solution of the following adjoint problem corresponding to the primal solution u^*

$$\begin{aligned} &-\frac{\partial \tilde{u}^*}{\partial t} + F\tilde{u}^* = (d_0 - dU_1^* - \sum_{i=1,n} d_i u^*(.,r_i) - \varphi^* - \tilde{G})\tilde{u}^* \\ &\qquad\qquad - \sum_{i=1,n} d_i(.,f_i)\tilde{u}^*(.,f_i)U_1(.,f_i)f_i' + (u^* - u_{\text{obs}}) \ \text{ on } \tilde{\mathcal{Q}}, \\ &-\frac{\partial \tilde{u}^*}{\partial t} + F\tilde{u}^* = (d_0 - dU_1^* - \sum_{i=1,n} d_i u^*(.,r_i) - \varphi^* - \tilde{G})\tilde{u}^* \\ &\qquad\qquad +(u^* - u_{\text{obs}}) \ \text{ on } \mathcal{Q}_T, \\ &\tilde{u}^*(.,T) = 0 \ \text{ on } \Omega, \\ &(\sigma^* \sigma \nabla u).\mathbf{n} = -(\eta + \psi^*)\tilde{u}^* \ \text{ on } \Sigma, \end{aligned} \tag{14.59}$$

with $\tilde{\mathcal{Q}} = \Omega \times (0, T - \delta(T))$ and $\mathcal{Q}_T = \Omega \times (T - \delta(T), T)$.

Moreover, the gradient of the functional J, in the weak sense, at point (φ^*, ψ^*) is

$$\frac{\partial J}{\partial \varphi}(\varphi^*, \psi^*) = \alpha\varphi^* - \tilde{u}^* U_1^* \text{ and } \frac{\partial J}{\partial \varphi}(\varphi^*, \psi^*) = -\tilde{u}^* U_1^*|_\Sigma - \beta(-\pi_2 + \psi^*).$$

15

Other Systems

The methods developed in this book can be applied to various physical, biological and chemical systems. We will not detail all the possible applications and will only quote two very interesting systems arising in micropolar fluids and semiconductor melts, namely the motion of animal blood, which is described by using the model of micropolar fluids, and Czochralski growth configurations and semiconductor melts in zone-melting. The main theorems of this book can be extended to these situations.

15.1 Micropolar Fluids and Blood Pressure

15.1.1 Introduction and Mathematical Setting

A micropolar fluid is a viscous, non-Newtonian fluid with local microstructure, which contains suspensions of rigid particles. A mathematical description of such fluids, which cannot be described, rheologically, by classical Navier–Stokes systems (especially when the diameter of the domain of flow becomes small), was first given by Eringen [115]. Animal blood, liquid crystals (with dumbbell type molecules), polymeric fluids, and certain colloidal fluids, whose fluid elements exhibit microrotations and complex biological structures, are examples of fluids modeled by micropolar fluid theory (see Eringen [116] and Popel *et al.* [243]).

Various problems associated with the micropolar fluid model have been studied recently (see, *e.g.*, Calmelet and Rosenhaus [67], Lukaszewicz [210], Yamaguchi [304]). For the optimal control problems, see Stavre [274], in which the author controls the blood pressure.

In the motion of blood, a problem of physical and biological interest is to control and stabilize the blood pressure. The goal of our study is to stabilize a desired blood pressure, by taking into account disturbances in the external forces. For this we consider the following constitutive system for two-dimensional incompressible time-dependent micropolar fluids given by Eringen [115]:

$$
\begin{aligned}
&\frac{\partial \mathbf{U}}{\partial t} + (\mathbf{U}.\nabla)\mathbf{U} - (\mu+\chi)\Delta\mathbf{U} + \nabla P - \chi\text{curl}(\varpi) = \mathbf{f} \quad \text{on } \mathcal{Q},\\
&\text{div}(\mathbf{U}) = 0 \quad \text{on } \mathcal{Q},\\
&\xi\frac{\partial \varpi}{\partial t} + \xi(\mathbf{U}.\nabla)\varpi - \nu\Delta\varpi + 2\chi\varpi - \chi\text{curl}(\mathbf{U}) = g \quad \text{on } \mathcal{Q},\\
&\text{subject to homegenous Dirichlet boundary conditions}\\
&\mathbf{U} = 0, \quad \varpi = 0 \quad \text{on } \Sigma,\\
&\text{and the null initial conditions}\\
&\mathbf{U}(.,0) = 0, \quad \varpi(.,0) = 0 \quad \text{on } \Omega,
\end{aligned}
\tag{15.1}
$$

where Ω is the domain occupied by the blood with the boundary Γ, T is the final time, $\mathcal{Q} = \Omega\times(0,T)$, $\Sigma = \Gamma\times(0,T)$ denotes the space-time cylinders and the parameters χ, μ, ξ and ν are positive given constants associated with the properties of the material. Furthermore, the functions $\mathbf{f}$ and g are the external fields, and the functions $\mathbf{U}$, ϖ and P are the velocity, the microrotation (which describes the skew-symmetric gyration tensor in the two-dimensional case) and the blood pressure, respectively.

We now introduce the following spaces:

$$
\begin{aligned}
V &= \{\mathbf{v} \in H_0^1(\Omega) : \text{div}(\mathbf{v}) = 0\},\\
H &= \{\mathbf{v} \in L^2(\Omega) : \text{div}(\mathbf{v}) = 0, \quad \mathbf{v}.\mathbf{n} = 0 \quad \text{on } \Gamma\},\\
L_0^2(\Omega) &= \{p \in L^2(\Omega) : \int_\Omega p\text{d}x = 0\}.
\end{aligned}
$$

By using a similar technique as used in Chapter 12, we can prove the following:

For given $g \in L^2(\mathcal{Q})$ and $\mathbf{f} \in L^2(0,T;H)$, there exists a unique solution $(\mathbf{U},\varpi,P)$ of the problem (15.1) such that

$$
\begin{aligned}
&\mathbf{U} \in H^{2,1}(\mathcal{Q}) \cap \mathcal{C}([0,T];V),\\
&\varpi \in H^{2,1}(\mathcal{Q}) \cap \mathcal{C}([0,T];H_0^1(\Omega)),\\
&P \in L^2(0,T;H^1(\Omega) \cap L_0^2(\Omega)).
\end{aligned}
$$

Moreover, if $\mathbf{f}$ is in $\mathcal{R}(\mathcal{Q})$, where

$$
\mathcal{R}(\mathcal{Q}) = \{\mathbf{f} \in L^2(0,T;H) : \frac{\partial \mathbf{f}}{\partial t} \in L^2(0,T;H^{-1}(\Omega)), \mathbf{f}(.,0) \in H\},
$$

then the pressure satisfies

$$
\frac{\partial P}{\partial t} \in L^2(\mathcal{Q}).
$$

15.1.2 Fluctuation and Robust Regulation of the Blood Pressure

In order to study the robust regulation of deviation of the blood pressure from the desired target P, we introduce the perturbation problem, which models small fluctuations $(\mathbf{u}, \omega, p, \tilde{\mathbf{f}}, \tilde{g})$ to the target micropolar flow $(\mathbf{U}, \varpi, P, \mathbf{f}, g)$ with Dirichlet boundary condition and null initial condition (we assume that $(\mathbf{u} + \mathbf{U}, \omega + \varpi, p + P, \tilde{\mathbf{f}} + \mathbf{f}, \tilde{g} + g)$ is also a solution of (15.1)). Then the perturbation $(\mathbf{u}, \omega, p, \tilde{\mathbf{f}}, \tilde{g})$ satisfies the following micropolar type system:

$$
\begin{aligned}
&\frac{\partial \mathbf{u}}{\partial t} - (\mu + \chi)\Delta \mathbf{u} + (\mathbf{u}.\nabla)\mathbf{u} + (\mathbf{u}.\nabla)\mathbf{U} + (\mathbf{U}.\nabla)\mathbf{u} \\
&\qquad\qquad + \nabla p - \chi \mathrm{curl}(\omega) = \tilde{\mathbf{f}} \text{ on } \mathcal{Q}, \\
&\mathrm{div}(\mathbf{u}) = 0 \text{ on } \mathcal{Q}, \\
&\xi \frac{\partial \omega}{\partial t} - \nu \Delta \omega + \xi(\mathbf{u}.\nabla)\omega + \xi(\mathbf{U}.\nabla)\omega + \xi(\mathbf{u}.\nabla)\varpi \\
&\qquad\qquad + 2\chi\omega - \chi \mathrm{curl}(\mathbf{u}) = \tilde{g} \text{ on } \mathcal{Q}, \\
&\text{subject to homegenous Dirichlet boundary conditions} \\
&\mathbf{u} = 0, \quad \omega = 0 \text{ on } \Sigma, \\
&\text{and the null initial conditions} \\
&\mathbf{u}(.,0) = 0, \quad \omega(.,0) = 0 \text{ on } \Omega,
\end{aligned}
\tag{15.2}
$$

where the regularity required on $\tilde{\mathbf{f}}$, $\tilde{g}$ and $(\mathbf{U}, \varpi)$ are

$$
(\mathbf{U}, \varpi) \in H^{2,1}(\mathcal{Q}) \cap \mathcal{C}([0,T]; H_0^1(\Omega)), \ \tilde{g} \in L^2(\mathcal{Q}), \ \tilde{\mathbf{f}} \in \mathcal{R}(\mathcal{Q}). \tag{15.3}
$$

By again using a similar technique as used in Chapter 12 and according to the regularity (15.3), we can prove the following:

There exists a unique solution $(\mathbf{u}, \omega, p)$ of the problem (15.2) such that

$$
\begin{aligned}
&\mathbf{U} \in H^{2,1}(\mathcal{Q}) \cap \mathcal{C}([0,T]; V), \\
&\varpi \in H^{2,1}(\mathcal{Q}) \cap \mathcal{C}([0,T]; H_0^1(\Omega)), \\
&p \in \mathcal{P}(\mathcal{Q}) \subset \mathcal{C}([0,T]; L^2(\Omega)) \text{ (see Lemma 6.6)},
\end{aligned}
\tag{15.4}
$$

where

$$
\mathcal{P}(\mathcal{Q}) = \{p \in L^2(0,T; H^1(\Omega) \cap L_0^2(\Omega)) : \frac{\partial p}{\partial t} \in L^2(\mathcal{Q})\}.
$$

Our purpose is to stabilize a desired blood pressure: for this we assume that the external field $\tilde{g}$ is decomposed into a disturbance $\psi \in L^2(\mathcal{Q})$ and a control $\varphi \in L^2(\mathcal{Q})$. Thus, we write $\tilde{g}$ as

$$
\tilde{g} = B_1\varphi + B_2\psi,
$$

where B_1 and B_2 are taken as given bounded operators on $L^2(\Omega)$.

The function $(\mathbf{u}, \omega, p)$ is assumed to be related to the disturbance ψ and control φ through the problem (15.2):

$$\begin{aligned}
&\frac{\partial \mathbf{u}}{\partial t} - (\mu + \chi)\Delta\mathbf{u} + (\mathbf{u}.\nabla)\mathbf{u} + (\mathbf{u}.\nabla)\mathbf{U} + (\mathbf{U}.\nabla)\mathbf{u} \\
&\qquad\qquad + \nabla p - \chi \mathrm{curl}(\omega) = \tilde{\mathbf{f}} \ \text{ on } \mathcal{Q}, \\
&\mathrm{div}(\mathbf{u}) = 0 \ \text{ on } \mathcal{Q}, \\
&\xi\frac{\partial \omega}{\partial t} - \nu\Delta\omega + \xi(\mathbf{u}.\nabla)\omega + \xi(\mathbf{U}.\nabla)\omega + \xi(\mathbf{u}.\nabla)\varpi \\
&\qquad\qquad + 2\chi\omega - \chi\mathrm{curl}(\mathbf{u}) = B_1\varphi + B_2\psi \ \text{ on } \mathcal{Q},
\end{aligned} \tag{15.5}$$

subject to homegenous Dirichlet boundary conditions

$\mathbf{u} = 0, \quad \omega = 0 \ \text{ on } \Sigma,$

and the null initial conditions

$\mathbf{u}(.,0) = 0, \quad \omega(.,0) = 0 \ \text{ on } \Omega.$

The cost functional considered in the present work is of the form:

$$\begin{aligned}
J(\varphi, \psi) = &\frac{1}{2}\iint_{\mathcal{Q}} \mid p - p_{\mathrm{obs}} \mid^2 \mathrm{d}x\mathrm{d}t \\
&+ \frac{\alpha}{2}\iint_{\mathcal{Q}} \mid \varphi \mid^2 \mathrm{d}x\mathrm{d}t - \frac{\beta}{2}\iint_{\mathcal{Q}} \mid \psi \mid^2 \mathrm{d}x\mathrm{d}t,
\end{aligned} \tag{15.6}$$

where $\alpha, \beta > 0$ are fixed, and $p_{\mathrm{obs}} \in \mathcal{P}(\mathcal{Q})$ is the observation.

We shall consider the following robust control problem:

$$\begin{aligned}
&\text{find } (\varphi^*, \psi^*) \in U_{\mathrm{ad}} \times V_{\mathrm{ad}} \text{ such that} \\
&J(\varphi^*, \psi) \leq J(\varphi^*, \psi^*) \leq J(\varphi, \psi^*), \quad \forall (\varphi, \psi) \in U_{\mathrm{ad}} \times V_{\mathrm{ad}},
\end{aligned} \tag{15.7}$$

where U_{ad} and V_{ad} are given non-empty, closed convex and bounded subsets of $L^2(\mathcal{Q})$.

The arguments of Chapter 12 extend directly to the present chapter. So, we omit the details. Therefore, we have the following existence and first-optimality conditions results.

Theorem 15.1. *For α and β be sufficiently large, there exist $(\varphi^*, \psi^*) \in U_{\mathrm{ad}} \times V_{\mathrm{ad}}$ and $X^* = (\mathbf{u}^*, \omega^*, p^*) \in H^{2,1}(\mathcal{Q}) \cap \mathcal{C}([0,T]; H_0^1(\Omega))$ such that (φ^*, ψ^*) satisfies (15.7) and $(\mathbf{u}^*, \omega^*, p^*)$ is the solution of the primal problem (15.5) with data (φ^*, ψ^*). Moreover, the optimal solution (φ^*, ψ^*, X^*) is characterized by the following necessary optimality conditions, for all $(\varphi, \psi) \in U_{\mathrm{ad}} \times V_{\mathrm{ad}}$:*

$$\begin{aligned}
&\iint_{\mathcal{Q}} (\alpha\varphi^* + B_1^*\vartheta^*)(\varphi - \varphi^*)\mathrm{d}x\mathrm{d}t \geq 0, \\
&\iint_{\mathcal{Q}} (-\beta\psi^* + B_2^*\vartheta^*)(\psi - \psi^*)\mathrm{d}x\mathrm{d}t \leq 0,
\end{aligned} \tag{15.8}$$

where $(\mathbf{w}^*, \vartheta^*, \pi^*)$ *is the unique solution in* $W_u(\mathcal{Q}) \times W_r(\mathcal{Q}) \times \mathcal{D}'(\mathcal{Q})$ *(*π *is unique up to the addition of a distribution in* $(0,T)$*) of the following adjoint problem:*

$$-\frac{\partial \mathbf{w}}{\partial t} - (\mu+\chi)\Delta \mathbf{w} - (\mathbf{u}_1.\nabla)\mathbf{w} + (\nabla \mathbf{u}_1)^t \mathbf{w} - \xi \omega_1 \nabla \vartheta$$
$$+\nabla \pi - \chi \text{curl}(\vartheta) = 0 \quad on\ \mathcal{Q},$$
$$\text{div}(\mathbf{w}) = p - p_{\text{obs}} \quad on\ \mathcal{Q},$$
$$-\xi \frac{\partial \vartheta}{\partial t} - \nu \Delta \vartheta - \xi(\mathbf{u}_1.\nabla)\vartheta + 2\chi\vartheta - \chi \text{curl}(\mathbf{w}) = 0 \quad on\ \mathcal{Q}, \tag{15.9}$$

subject to homegenous Dirichlet boundary conditions

$$\mathbf{w} = 0, \quad \vartheta = 0 \quad on\ \Sigma,$$

and the final conditions

$$\mathbf{w}(.,T) = \mathbf{w}_0, \quad \vartheta(.,T) = 0 \quad on\ \Omega,$$

where $\mathbf{u}_1 = \mathbf{u} + \mathbf{U}$, $\omega_1 = \omega + \varpi$,

$$W_u(\mathcal{Q}) = \{\mathbf{w} \in L^2(0,T;H_0^1(\Omega)) \cap \mathcal{C}([0,T];L^2(\Omega)) : \frac{\partial \mathbf{w}}{\partial t} \in L^2(0,T;V')\},$$
$$W_r(\mathcal{Q}) = \{\vartheta \in L^2(0,T;H_0^1(\Omega)) : \frac{\partial \vartheta}{\partial t} \in L^2(0,T;H^{-1}(\Omega))\}$$

and $\mathbf{w}_0 \in L^2(\Omega)$ *is the unique solution of the following problem:*

$$\begin{aligned} -\text{div}(\mathbf{w}_0) &= p_{\text{obs}}(T) - p^*(T), \\ \mathbf{w}_0.\mathbf{n} &= 0 \quad on\ \Gamma. \end{aligned} \tag{15.10}$$

Remark 15.2. The proof of the existence and uniqueness of the solution of the problem (15.9) can be obtained by introducing a divergence-free non-linear system, which is equivalent to (15.9). For this we can consider, for a.e. in $(0,T)$, functions

$$\mathbf{w}_L = -\int_t^T \mathbf{v}(s)\text{d}s \in L^2(0,T;H_0^1(\Omega)),$$

with $\mathbf{w}_L \in L^2(0,T;H_0^1(\Omega))$ and $\mathbf{w}_T \in H_0^1(\Omega)$ such that $\mathbf{v}(t)$ and $\mathbf{w}_T$ are solutions, respectively, of the problems

$$\text{div}(\mathbf{v}) - \frac{\partial}{\partial t}(p_{\text{obs}} - p^*), \tag{15.11}$$

and

$$-\text{div}(\mathbf{w}_T) = p_{\text{obs}}(T) - p^*(T). \tag{15.12}$$

Set $\tilde{\mathbf{w}} = \mathbf{w} - \mathbf{w}_L - \mathbf{w}_T$, then $\text{div}(\tilde{\mathbf{w}}) = 0$, $\tilde{\mathbf{w}}(T) = \mathbf{w}_0 - \mathbf{w}_T$ and $\tilde{\mathbf{w}}(T).\mathbf{n} = 0$. The obtained system, satisfied by $(\tilde{\mathbf{w}}, \vartheta)$, is similar to (15.1) where the corresponding right-hand sides, $\mathbf{f}$ and g, depend on $\mathbf{w}_L$ and are in $L^2(0,T;V')$ and $L^2(\mathcal{Q})$, respectively. Therefore, by using similar technique as used to obtain the existence and the uniqueness of (15.1), we can prove the existence and uniqueness of $(\tilde{\mathbf{w}}, \vartheta)$ and then those of $(\mathbf{w}, \vartheta)$. ♢

15.2 Semiconductor Melt Flow in Crystal Growth

15.2.1 Introduction and Mathematical Setting

Liquid encapsulate Czochralski crystal growth is a major technique used to produce crystals, for example, GaAs and InP single crystals used in electronics and optoelectronics (see, *e.g.*, Neubert and Rudolph [229]). During the whole growth process, the behavior of the crystal depends considerably on the thermal regime and formation of the crystallization front geometry. Moreover, during experimental study of these Czochralski growth processes in axisymetric zone–melting devices, a transition from the two-dimensional incompressible turbulent flow regime (melt) to an unsteady three-dimensional behavior is observed.

The distribution temperature in the melt and at the crystal growth interface causes fluctuations in the microscopic growth rate of the crystal. These fluctuations (in the impurity segregation) decrease the material quality of the grown crystal. They are known to cause the so-called striations which transform into clusters of point defects. To make crystals striation-free, it is necessary to reduce the influence of the thermal convection in the melt. Therefore, in order to suppress or to reduce the convection and, in particular, thermal convection in the melt, during practical crystal production processes, magnetic fields are often applied by the physicists and engineers.

Various problems associated with the semiconductor melts model have been studied recently (see, *e.g.*, Choe [83], Fedoseyev and Alexander [121], Prasad *et al.* [244], Watanabe *et al.* [297]). For the optimal control problems, see Barwolff and Hinze [25] and Gunzburger *et al.* [146] in which the authors develop computational techniques and optimal strategy for the suppression of turbulent motions in the melt.

The goal of our study is to avoid such crystal defects by stabilizing the melt flow motion during the growth process. For this, we use a robust regulation method in order to maintain the desired state (the growth of crystal with desired properties) which represents a physically favourable situation, by taking into account the worst disturbances in the temperature flux on the wall of the crucible. The control function is in body forces due to a magnetic field.

The unsteady model of the melt flow motion as well as heat transfer in the crystal and crucibles is governed by the following incompressible viscous fluid with a Boussinesq approximation:

$$\begin{aligned}
&\frac{\partial \mathbf{U}}{\partial t} + (\mathbf{U}.\nabla)\mathbf{U} - \operatorname{div}(\nu_1 \nabla \mathbf{U}) + \frac{1}{\rho_{\rm av}}\nabla P + \xi \frac{\mathcal{T}}{\rho_{\rm av}}\mathbf{G} = F \quad \text{on } \mathcal{Q},\\
&\operatorname{div}(\mathbf{U}) = 0 \quad \text{on } \mathcal{Q},\\
&\frac{\partial \mathcal{T}}{\partial t} + (\mathbf{U}.\nabla)\mathcal{T} - \operatorname{div}(\nu_2 \nabla \mathcal{T}) = 0 \quad \text{on } \mathcal{Q},\\
&\text{with the initial conditions}\\
&\mathbf{U}(.,0) = \mathbf{u}_0, \quad \mathcal{T}(.,0) = \mathcal{T}_0 \quad \text{on } \Omega,
\end{aligned} \tag{15.13}$$

where $\Omega \subset \mathbb{R}^m$, for $m = 2$ or 3, is the flow domain with the boundary $\Gamma = \Gamma_c \cup \Gamma_d$, where Γ_c corresponds to the crucible walls and Γ_d corresponds to the solid–liquid interface. $\mathcal{Q} = \Omega \times (0, T)$ denotes the space-time cylinder, with T the final time, $\mathbf{G} = (0, 0, -g)$ is the gravity force, the density ρ_{av} is a constant mean value which is supposed to be equal to 1 and ξ is a given coefficient depending on the Prandtl–Rayleigh and thermal expansion coefficient. The function $\mathbf{U}$ is the velocity, P is the pressure, $\mathcal{T}$ is the temperature and F is a body force due to the magnetic field (the Lorentz force). The turbulent flux is usually modeled by the dissipative terms which correspond to Raynolds stress with the variable coefficient of eddy viscosity ν_1 and the variable coefficient of eddy diffusivity ν_2. The functions ν_1 and ν_2 are assumed to be positive and bounded functions above and below by non-negative constants in Ω.

On account of the phenomena we want to describe, the system (15.13) is supplied with the following mixed boundary conditions:

$$\begin{aligned} &\mathbf{U} = \mathbf{u}_B \quad \text{on } \Sigma, \\ &\nu_2 \frac{\partial \mathcal{T}}{\partial \mathbf{n}} + \gamma(\mathcal{T} - \mathcal{T}_e) = \tau \quad \text{on } \Sigma_c, \\ &\mathcal{T} = \mathcal{T}_{sl} \quad \text{on } \Sigma_d. \end{aligned} \tag{15.14}$$

where $\Sigma = \Gamma \times (0, T)$, $\Sigma_c = \Gamma_c \times (0, T)$ and $\Sigma_d = \Gamma_d \times (0, T)$. The function $\mathcal{T}_e$ is a given environmental temperature, γ denotes some physical constant, and $\mathbf{u}_B$ and $\mathcal{T}_{sl}$ are given boundary functions such that $\mathbf{u}_B.\mathbf{n} = 0$.

15.2.2 Fluctuation and Robust Regulation of the Melt Flow Motion

In order to study the robust regulation of deviation of the flow from the desired state, we introduce the perturbation problem, which models small fluctuations $(\mathbf{u}, \theta, p, \mathbf{f}, \eta)$ to the target flow $(\mathbf{U}, \mathcal{T}, P, F, \tau)$ with the same Dirichlet boundary conditions and initial conditions (we assume that $(\mathbf{u}+\mathbf{U}, \mathcal{T}+\theta, p+P, F+\mathbf{f}, \tau+\eta)$ is also a solution of (15.13) and (15.14)). Then the perturbation $(\mathbf{u}, \theta, p, \mathbf{f}, \eta)$ satisfies the following non-linear system:

$$\begin{aligned} &\frac{\partial \mathbf{u}}{\partial t} + (\mathbf{u}.\nabla)\mathbf{u} + (\mathbf{U}.\nabla)\mathbf{u} + (\mathbf{u}.\nabla)\mathbf{U} \\ &\qquad\qquad -\mathrm{div}(\nu_1 \nabla \mathbf{u}) + \nabla p + \xi\theta\mathbf{G} = \mathbf{f} \quad \text{on } \mathcal{Q}, \\ &\mathrm{div}(\mathbf{u}) = 0 \quad \text{on } \mathcal{Q}, \\ &\frac{\partial \theta}{\partial t} + (\mathbf{u}.\nabla)\theta + (\mathbf{u}.\nabla)\mathcal{T} + (\mathbf{U}.\nabla)\theta - \mathrm{div}(\nu_2 \nabla\theta) = 0 \quad \text{on } \mathcal{Q}, \\ &\text{with the initial conditions} \\ &\mathbf{u}(.,0) = 0, \quad \theta(.,0) = 0 \quad \text{on } \Omega, \\ &\text{and the boundary conditions} \\ &\mathbf{u} = 0 \quad \text{on } \Sigma, \\ &\nu_2 \frac{\partial \theta}{\partial \mathbf{n}} + \gamma\theta = \eta \quad \text{on } \Sigma_c, \quad \theta = 0 \quad \text{on } \Sigma_d. \end{aligned} \tag{15.15}$$

Assume that the functions $(\mathbf{U}, \mathcal{T}) \in H^{2,1}(\mathcal{Q}) \cap \mathcal{C}([0,T]; H^1(\Omega))$, $\mathbf{f} \in L^2(0,T; L^2(\Omega))$ and $\eta \in U_c = \{\eta \in L^2(0,T; H^1(\Gamma_c)) : \partial\eta/\partial t \in L^2(0,T; L^2(\Gamma_c))\}$ then, by using a similar technique as used in Chapter 12, we can prove (under the constraint of a small data for the non-linear 3D case) the existence and uniqueness of the solution $(\mathbf{u}, \theta) \in H^{2,1}(\mathcal{Q}) \cap \mathcal{C}([0,T]; H^1(\Omega))$ of the problem (15.15).

Our purpose is to stabilize the desired flow, and for this we assume that the control $\varphi \in L^2(\mathcal{Q})$ is in the external magnetic field and the disturbance $\psi \in L^2(\Sigma_c)$ is in the temperature flux on the wall of the crucible, *i.e.*,

$$\mathbf{f} = \mathcal{B}_1\varphi \quad \text{and} \quad \eta = \mathcal{B}_2\psi,$$

where $\mathcal{B}_1$ (respectively $\mathcal{B}_2$) is taken here as given linear continuous and bounded operator from $L^2(\mathcal{Q})$ (respectively $L^2(\Sigma_c)$) into $L^2(\mathcal{Q})$ (respectively U_c).

The function $(\mathbf{u}, \theta, p)$ is assumed to be related to the disturbance ψ and control φ through the problem (15.15):

$$\begin{aligned}
&\frac{\partial \mathbf{u}}{\partial t} + (\mathbf{u}.\nabla)\mathbf{u} + (\mathbf{U}.\nabla)\mathbf{u} + (\mathbf{u}.\nabla)\mathbf{U} \\
&\qquad\qquad -\mathrm{div}(\nu_1 \nabla \mathbf{u}) + \nabla p + \xi\theta\mathbf{G} = \mathcal{B}_1\varphi \quad \text{on } \mathcal{Q}, \\
&\mathrm{div}(\mathbf{u}) = 0 \quad \text{on } \mathcal{Q}, \\
&\frac{\partial \theta}{\partial t} + (\mathbf{u}.\nabla)\theta + (\mathbf{u}.\nabla)\mathcal{T} + (\mathbf{U}.\nabla)\theta - \mathrm{div}(\nu_2\nabla\theta) = 0 \quad \text{on } \mathcal{Q}, \\
&\text{with the initial conditions} \\
&\mathbf{u}(.,0) = 0, \quad \theta(.,0) = 0 \quad \text{on } \Omega, \\
&\text{and the boundary conditions} \\
&\mathbf{u} = 0 \quad \text{on } \Sigma, \\
&\nu_2 \frac{\partial\theta}{\partial \mathbf{n}} + \gamma\theta = \mathcal{B}_2\psi \quad \text{on } \Sigma_c, \quad \theta = 0 \quad \text{on } \Sigma_d.
\end{aligned} \tag{15.16}$$

The cost functional considered in the present work is of the form

$$\begin{aligned}
J(\varphi, \psi) = {} & \frac{1}{2} \iint_{\mathcal{Q}} |\, \mathcal{C}\mathbf{u} \,|^2 \,\mathrm{d}x\mathrm{d}t \\
& + \frac{\alpha}{2} \iint_{\mathcal{Q}} |\, \varphi \,|^2 \,\mathrm{d}x\mathrm{d}t - \frac{\beta}{2} \iint_{\Sigma_c} |\, \psi \,|^2 \,\mathrm{d}\Gamma\mathrm{d}t
\end{aligned} \tag{15.17}$$

where $\alpha, \beta > 0$ are fixed parameters and $\mathcal{C}$ is similar as in (12.18).

We shall consider the following robust control problem:

$$\begin{aligned}
&\text{find } (\varphi^*, \psi^*) \in \mathcal{U}_{ad} \times \mathcal{V}_{ad} \text{ such that} \\
&J(\varphi^*, \psi) \le J(\varphi^*, \psi^*) \le J(\varphi, \psi^*), \quad \forall (\varphi, \psi) \in \mathcal{U}_{ad} \times \mathcal{V}_{ad},
\end{aligned} \tag{15.18}$$

where $\mathcal{U}_{ad}$ (respectively $\mathcal{V}_{ad}$) is a given non-empty, closed convex and bounded subset of $L^2(\mathcal{Q})$ (respectively $L^2(\Sigma_c)$).

The arguments of Chapter 12 extend directly to the present work. So, we omit the details. Therefore, we have the following existence and first-optimality conditions results. More precisely, for α and β be sufficiently large, there exist $(\varphi^*, \psi^*) \in \mathcal{U}_{\rm ad} \times \mathcal{V}_{\rm ad}$ and $X^* = (\mathbf{u}^*, \mathcal{T}^*) \in H^{2,1}(\mathcal{Q}) \cap \mathcal{C}([0,T]; H^1(\Omega))$ such that (φ^*, ψ^*) satisfies (15.18) and $(\mathbf{u}^*, \mathcal{T}^*)$ is the solution of the primal problem (15.16) with data (φ^*, ψ^*). Moreover, the optimal solution (φ^*, ψ^*, X^*) is characterized by the following necessary optimality conditions, for all $(\varphi, \psi) \in \mathcal{U}_{\rm ad} \times \mathcal{V}_{\rm ad}$:

$$\begin{aligned} &\iint_{\mathcal{Q}} (\alpha\varphi^* + \mathcal{B}_1^* \mathbf{w}^*)(\varphi - \varphi^*) \mathrm{d}x\mathrm{d}t \geq 0, \\ &\iint_{\Sigma_{\rm c}} (-\beta\psi^* + \mathcal{B}_2^* \vartheta^*)(\psi - \psi^*) \mathrm{d}\Gamma \mathrm{d}t \leq 0, \end{aligned} \tag{15.19}$$

where $(\mathbf{w}^*, \vartheta^*, \pi^*)$ is the unique solution of the following adjoint problem:

$$\begin{aligned} &-\frac{\partial \mathbf{w}}{\partial t} - \mathrm{div}(\nu_1 \nabla \mathbf{w}) - (\mathbf{u}_1^*.\nabla)\mathbf{w} + (\nabla \mathbf{u}_1^*)^t \mathbf{w} \\ &\qquad\qquad + \theta_1^* \nabla \vartheta + \nabla \pi = \mathcal{C}^* \mathcal{C} \mathbf{u}^* \quad \text{on } \mathcal{Q}, \\ &\mathrm{div}(\mathbf{w}) = 0 \quad \text{on } \mathcal{Q}, \\ &-\frac{\partial \vartheta}{\partial t} - \mathrm{div}(\nu_2 \nabla \vartheta) - (\mathbf{u}_1^*.\nabla)\vartheta = \xi \mathbf{G}.\mathbf{w} \quad \text{on } \mathcal{Q}, \\ &\text{with the final conditions} \\ &\mathbf{w}(.,T) = 0, \quad \vartheta(.,T) = 0 \quad \text{on } \Omega, \\ &\text{and the boundary conditions} \\ &\mathbf{w} = 0 \quad \text{on } \Sigma, \\ &\nu_2 \frac{\partial \vartheta}{\partial \mathbf{n}} + \gamma \vartheta = 0 \quad \text{on } \Sigma_{\rm c}, \quad \vartheta = 0 \quad \text{on } \Sigma_{\rm d}. \end{aligned} \tag{15.20}$$

where $\mathbf{u}_1^* = \mathbf{u}^* + \mathbf{U}$ and $\theta_1^* = \theta^* + \mathcal{T}$.

Remark 15.3. It is clear that we can consider other control and disturbance functions and obtain the same results by using the same techniques. We can also consider other observations, for example, temperature gradients in the crystal in Czochralski crystal growth processes. ♢

References

1. Abergel, F., Temam, R. (1990) On some control problems in fluid mechanics. *Theoret. Comput. Fluid Dyn.*, 303–325.
2. Abrikosov, A.A. (1957) On the magnetic properties of superconductors of the second group. *Soviet. Phys. JETP*, 5:1174–1182.
3. Ackermann, J., Barlett, A., Kaesbauer, D., Sienel, W., Steinhauser, R. (1993) *Robust control Systems with uncertain physical parameters.* Communications and Control Engineering Series. Springer-Verlag, London.
4. Adams, R.A. (1975) *Sobolev spaces.* Academic, New-York.
5. Agmon, S., Douglis, A., Nirenberg, L. (1964) Estimates near the boundary for solutions of elliptic partial differential equations satisfying general boundary conditions II. *Comm. Pure Appl. Math.*, 17:35–92.
6. Ahmed, N.U., Xiang, X. (1997) Nonlinear uncertain systems and necessary conditions of optimality. *SIAM J. Control Optim.*, 35:1755–1772.
7. Alamir, M., Chareyron, S. (2007) State-constrained optimal control applied to cell-cycle-specific cancer chemotherapy. *Optimal Control Appl. Methods*, 28:175–190.
8. Alt, W. (1990) The Lagrange Newton method for infinite-dimentional optimization problems. *Numer. Funct. Anal. Optim.*, 11:201–224.
9. Alt, W. (1990) Parametric programming with applications to optimal control and sequential quadratic programming. *Bayreuth Math. Schriften*, 34:1–37.
10. Arada, N., Raymond, J.P. (1999) Minimax control of parabolic systems with state constraints. *SIAM J. Control Optim.*, 38:254–271.
11. Arkin, H., Holmes, K.R., Chen, M.M., Bottje, W.G. (1986) Thermal pulse decay method for simultaneous measurement of local thermal conductivity and blood perfusion: A theoretical analysis. *J. Biomech. Eng.*, 108:208–214.
12. Askeland, D. (2003) *The science of materials and engineering.* Thomson, London.
13. Asplund, E. (1967) Positivity of duality mappings. *Bull. Amer. Math. Soc.*, 2000–2003.
14. Asplund, E. (1967) Averaged norms. *Isreal J. Math.*, 5:227–233.
15. Auchmuty, G. (1983) Duality for non-convex variational principles. *J. Diff. Eq.*, 50:80–145.
16. Auchmuty, G. (1992) Minimax results and differential inclusions. In: Wiener, J., Hale, J., Weiner, J. (eds.), *Inter. Conf. on Theory and Appl. Diff. Eq.*, vol. 273, pp.6–10. Pitman Lecture Notes.

17. Auchmuty, G. (1997) Min-max problems for non-potential operator equations. *Contemprary Math.*, 209:19–28.
18. Baciotti, A. (1992) *Local stabilizability of nonlinear control systems.* Series on Advances in Mathematics and Applied Sciences. Word Scientific, Singapore.
19. Ball, J.M., Mardsen, J.E., Slemrod, M. (1982) Controllability for distributed bilinear systems. *SIAM J. Control Optim.*, 31:575-597.
20. Banks, R.B. (1994) *Growth and diffusion phenomena. Mathematical frameworks and applications.* Texts in Applied Mathematics, Springer-Verlag, Berlin.
21. Barbu, V., Precupanu, T. (1986) *Convexity and optimization in Banach spaces.* Reidel, Dordrecht.
22. Barbu, V., Sritharan, S.S. (1998) $\mathcal{H}_\infty$ control theory of fluid of fluid dynamics. *Proc. Royal Soc. London, Series A*, 454:3009–3033.
23. Bardeen, J., Cooper, L., Schrieffer, J. (1957) Theory of superconductivity. *Phys. Rev.*, 108:1175–1204.
24. Barron, E.N. (1990) The pontryagin maximum principle for minimax problems of optimal control. *Nonlinear Anal.*, 15:1155–1165.
25. Barwolff, G., Hinze, M. (2006) Optimization of semiconductor melts. *ZAMM. Z. Angew. Math. Mech.*, 86:423–437.
26. Başar, T., Bernhard, P. (1995) *H^∞-optimal control and related minimax design problems. A dynamic game approach.* Systems & Control: Foundations & Applications. Birkhuser, Boston.
27. Belmiloudi, A. (1996) Resolution of optimal control problem for a perturbation linearized Navier–Stokes type equations. In: *Application of Mathematics in Engineering and Business Sozopol*, Proceedings of the XXII Summer School, Sofia, pp. 39–51. The University of Sofia.
28. Belmiloudi, A., Brossier, F. (1997) A control method for assimilation of surface data in a linearized Navier–Stokes type problem related to oceanography. *SIAM J. Control Optim.*, 35:2183–2197.
29. Belmiloudi, A., Brossier, F. (1997) Regularity results for a Navier–Stokes type problem related to oceanography. *Acta. Appl. Math.*, 48:299–316.
30. Belmiloudi, A. (1999) A nonlinear optimal control problem for assimilation of surface data in a Navier–Stokes type equations related to oceanography. *Numer. Funct. Anal. Optim.*, 20:1–26.
31. Belmiloudi, A. (2000) Regularity results and optimal control problems for the perturbation of Boussinesq equations of the ocean. *Numer. Funct. Anal. Optim.*, 21:623–651.
32. Belmiloudi, A. (2000) Asymptotic behaviour of the perturbation of the primitive equations of the ocean with vertical viscosity. *Canadian Appl. Math. Quarterly*, 8:97–140.
33. Belmiloudi, A. (2001) Existence and characterization of an optimal control for the problem of long waves in a shallow-water model. *SIAM J. Control Optim.*, 39:1558–1584.
34. Belmiloudi, A. (2001) Studies in linear and nonlinear optimal control problems for viscous flows equations. In: Cheshankov, B.I., Todorov, M.D. (eds.), *Application of Mathematics in Engineering*, pp. 181–188. Heron, Sofia.
35. Belmiloudi, A. (2002) Robin-type boundary control problems for the nonlinear Boussinesq type equations. *J. Math. Anal. Appl.*, 21:428–456.

36. Belmiloudi, A. (2002) Mathematical analysis and optimal control problems for the perturbation of the primitive equations of the ocean with vertical viscosity. *J. Appl. Anal.*, 8:97–140.
37. Belmiloudi, A. (2003) Mathematical and computational methods of an optimal control for the primitive equations in a coupled atmosphere–ocean model. In: *IEEE MMAR, Optimisation of Infinite Dimensional Systems*, Miedzyzdroje, Poland, pp. 87–96. IEEE Control Society.
38. Belmiloudi, A. (2003) Robust control problems associated with time-varying delay nonlinear parabolic equations. *IMA J. Math. Control Inf.*, 20:305–334.
39. Belmiloudi, A. (2003) Nonlinear robust control problems of parabolic type equations with time-varying delays given in the integral form. *J. Dynam. Control Systems*, 9:469–512.
40. Belmiloudi, A. (2004) Robust and optimal control problems to a phase-field model for the solidification of a binary alloy with a constant temperature. *J. Dynam. Control Systems*, 10:453–499.
41. Belmiloudi, A., Yvon, J.P. (2005) Robust control of a non-isothermal solidification model. *WSEAS Trans. Systems*, 4:2291–2300.
42. Belmiloudi, A. (2005) Optimal control problems of nonlinear degenerate parabolic differential systems with logistic time-varying delays. *IMA J. Math. Control Inf.*, 22:88–108.
43. Belmiloudi, A. (2005) Nonlinear optimal control problems of degenerate parabolic equations with logistic time-varying delays of convolution type. *Nonlinear Analysis: Theory Meth. Appl.*, 63:1126–1152.
44. Belmiloudi, A. (2006) Minimax control problems of periodic competing parabolic systems with logistic growth terms. *Int. J. Control*, 79:150–161.
45. Belmiloudi, A. (2006) Robust control problems of vortex dynamics in superconducting films with Ginzburg–Landau complex systems. *Abstract and Applied Analysis*, 1–43.
46. Belmiloudi, A. (2006) On some control problems for heat transfer systems in perfused biological tissues. *WSEAS Trans. Systems*, 5:17–25.
47. Belmiloudi, A. (2007) Bilinear minimax control problems for a class of parabolic systems with applications to control of nuclear reactors. *J. Math. Anal. Appl.*, 327:620–642.
48. Belmiloudi, A. (2007) Analysis of the impact of nonlinear heat transfer laws on temperature distribution in irradiated biological tissues: mathematical models and optimal controls. *J. Dynam. Control Systems*, 108:217–254.
49. Berger, M. (1977) *Nonlinearity and functional analysis.* Academic, New York.
50. Bergounioux, M., Kunisch, K. (1997) Augmented Lagrangian techniques for elliptic state constrained optimal control problems. *SIAM J. Control Optim.*, 35:1524–1543.
51. Berthuel, F., Brezis, H., Helein, F. (1994) *Ginzburg–Landau vortices.* Progress in Nonlinear Differential Equations and their Applications. Birkhäuser, Boston.
52. Bewley, T.R., Temam, R., Ziane, M. (2000) A general framework for robust control in fluid mechanics. *Physica D*, 138:360-392.
53. Blayo, E., Blum, J., Verron, J. (1998) Variational assimilation of oceanographic data and reduction of the control dimension. In: *Equations aux Dérivées Partielles et Applications*, pp. 199–219. Gauthier-Villars, Paris.

54. Bohm, M., Kremer, J., Louis, A.K. (1993) Efficient algorithm for computing optimal control of antennas in hyperthermia. *Surveys Math. Indust.*, 3:233–251.
55. Bohmer, K., Steller, H.J. (1984) *Defect correction methods: Theory and applications.* Springer-Verlag, Austria.
56. Bourbaki, N. (1981) *Espaces vectoriels Topologiques.* Masson, Paris.
57. Bradley, M.E., Lenhart, S., Yong, J. (1999) Bilinear optimal control of the velocity term in a Kirchhoff plate equation. *J. Math. Anal. Appl.*, 238:451–467.
58. Brezis, H., Nirenberg, L., Stampacchia, G. (1972) A remark on Ky Fan's Minimax Principle. *Boll. Unione Mat. Ital.*, 4:293–300.
59. Brezis, H., Coron, J.M., Nirenberg, L. (1980) Free vibrations for a nonlinear wave equation and a theorem of P. Rabinowitz. *Comm. Pure Appl. Math.*, 33:667–689.
60. Brezis, H. (1983) *Analyse fonctionnelle: Théorie et applications.* Masson, Paris.
61. Brezis, H., Lieb, E.H. (1983) A relation between pointwise convergence of functions and convergence of functionals. *Proc. Amer. Math. Soc.*, 88:486–490.
62. Brochet, D., Chen, X., Hilhorst, D. (1993) Finite dimensional exponential attractor fo the phase-field model. *Appl. Anal.*, 49:197–212.
63. Brochet, D., Hilhorst, D., Novick-Cohen, A. (1996) Maximal attractor and inertial sets for a conserved phase field model. *Adv. Differ. Eq.*, 1:547–578.
64. Browder, F.E. (1966) *Problèmes non-linéaires.* Presses de l'Université de Montréal.
65. Bryan, F. (1987) Parameter sensitivity of primitive equation ocean general circulation models. *J. Phys. Oceanogr.*, 17:970–985.
66. Caginalp, G. (1986) An analysis of a phase field model of a free boundary. *Arch. Rat. Mech. Anal.*, 92:205–245.
67. Calmeled-Eluhu, C., Rosenhaus, V. (2001) Symmetries and solution of a micropolar fluid flow through a cylinder. *Acta Mechanica*, 147:59–72.
68. Cantoni, M.W., Glover, K. (2000) Gap-metric robustness analysis of linear periodically time-varying feedback systems. *SIAM J. Control Optim.*, 38:803–822.
69. Cartan, H. (1967) *Calcul différentiel, formes différentielles.* Hermann, Paris.
70. Chandrasekharan, P.C. (1996) *Robust Control of Linear Dynamical Systems.* Academic, New York.
71. Chang, K.C. (1993) *Infinite dimensional Morse theory and multiple solution problems.* Progress in Nonlinear Differential Equations and their Applications. Birkhäuser, Boston.
72. Chapman, S.J., Rubinstein, J., Schatzman, M. (1996) A mean-field model of superconducting vortices. *Europ. J. Appl. Math.*, 7:97–111.
73. Chapman, S.J. (2000) A hierarchy of models for type-II superconductors. *SIAM Rev.*, 42:555–598.
74. Charney, C.K., (1992) *Mathematical models of bioheat transfer. Adv. Heat Trans.*, 22:19–155.
75. Chato, J.C. (1980) Heat transfer to blood vessels. *J. Biomech. Eng.*, 102:110–118.
76. Chen, B. (1998) $\mathcal{H}^\infty$ *Control and its applications.* Lecture Notes in Control and Information Sciences. Springer-Verlag, Berlin.

77. Chen, M.M., Holmes, K.R. (1980) Microvascular contributions in tissue heat transfer. *Annals N.Y. Acad. Sci.*, 335:137–150.
78. Chen, M.M., Holmes, K.R. (1980) In vivo tissue thermal conductivity and local blood perfusion measured with heat pulse-decay method. *Adv. Bioeng.*, 113–115.
79. Chen, Z., Hoffmann, K.H., Liang, L. (1993) On a non-stationary Ginzburg–Landau model for superconductivity. *Math. Meth. Appl. Sci.*, 16:855–875.
80. Chen, Z., Hoffmann, K.H. (1995) Numerical studies of a non-stationary Ginzburg–Landau model for superconductivity. *Adv. Math. Sci. Appl.*, 5:363–389.
81. Chen, Z., Hoffmann, K.H. (1996) Optimal control of dynamical Ginzburg–Landau vortices in superconductivity. *Num. Funct. Anal. Optim.*, 17:241–258.
82. Chipot, M. (1984) *Variational inequalities and flow in porous media.* Springer-Verlag, New York.
83. Choe, K.S. (2004) Growth striations and impurity concentrations in HMCZ silicon crystals *J. Crystal Growth*, 262:35–39.
84. Christensen, G.S., Soliman, S.A., Nieva, R. (1990) *Optimal control of distributed nuclear reactors.* Plenum Press, New York.
85. Ciarlet, P.G. (1978) *The finite element method for elliptic problems.* North-Holland, Amsterdam.
86. Ciarlet, P.G. (1989) *Introduction to numerical linear algebra and optimization.* Cambridge, University Press, Cambridge.
87. Clark, C. (1990) *Mathematical bioeconomics: the optimal management of renewable resources*, 2nd ed. Wiley Interscience, New York.
88. Coskun, E., Kwong, M.K. (1997) Simulating vortex motion in superconducting films with the time-dependent Ginzburg–landau equations. *Nonlinearity*, 10:579–593.
89. Courant, R., Hilbert, D. (1953) *Methods of mathematical physics.* Interscience, London.
90. Crandall, M.G., Ishii, H., Lions, P.L. (1992) User's guide to viscosity solutions of second order partial differential equations. *Bull. Am. Math. Soc. New Ser.*, 27:1–67.
91. Cushing, J.M. (1977) *Integrodifferential equations and delay models in population dynamics.* Lecture Notes in Biomathematics, Springer-Verlag, Berlin.
92. Cushing, J.M. (1994) Structured population dynamics. In: Levin, S. (ed.), *Frontiers in Mathematical Biology*, Lecture Notes in Biomathematics, pp. 280–295. Springer-Verlag, Berlin.
93. Dale, W., Smith, M.C. (1993) Stabilizability and existence of system representations for discrete-time, time varying systems. *SIAM J. Control Optim.*, 31:1538–1557.
94. Dai, Y.H, Yuan, Y. (1999) A nonlinear conjugate gradient method with a strong global convergence property. *SIAM J. Optim.*, 10:177–182.
95. Dautray, R., Lions, J.L. (1984/1985) *Analyse mathématique et calcul numérique pour les sciences et les techniques.* Masson, Paris.
96. Deang, J., Du, Q., Gunzburger, M.D. (2002) Modeling and computation of random thermal fluctuations and material defects in the Ginzburg–Landau model of superconductivity. *J. Comput. Phys.*, 181:45–67.
97. Deckelnick, K., Elliott, C.M., Richardson, G. (1997) Long time asymptotics for forced curvature flow with applications to the motion of a superconducting vortex. *Nonlinearity*, 10:655–678.

98. Dennis, J.E., Schnabel, R.B. (1996) *Numerical methods for unconstrained optimization and nonlinear equations.* Classics in Applied Mathematics. SIAM, PA.
99. De Pillis, L.G., Gu, W., Radunskaya, A.E. (2006) Mixed immunotherapy and chemotherapy of tumors: modeling, applications and biological interpretations. *J. Theoret. Biol.*, 238:841–862.
100. Deuflhard, P., Seebass, M. (1998) Adaptive multilevel FEM as decisive tools in clinical cancer therapy hyperthermia. *Konrad-Zuse-Zentrum für Informationstechnik*, Berlin Takustr. 7, D-14195 Berlin.
101. Diekmann, O., Van Gils, S.A., Verduyn-Lunel, S.M., Walther, H.O. (1995) *Delay equations, functional-, complex and nonlinear analysis.* Applied Mathematical Sciences Series. Springer-Verlag, New York.
102. Diestel, J. (1975) *Geometry of Banach space-selected topics.* Lecture Notes in Mathematics. Springer-Verlag, Berlin.
103. Dietrich, H. (1988) Zur c-Konvexität und c-Subdifferenzierbarkeit von Funktionalen. *Optimization*, 19:355–371.
104. Dontchev, A.L. (1996) Uniform convergence of the newton method for Aubin continuous maps. *Serdica. Math. J.*, 22:385–398.
105. Du, Q., Gunzburger, M.D., Peterson, J.S. (1992) Analysis and approximation of the Ginzburg–Landau model of superconductivity. *SIAM Rev.*, 34:54–81.
106. Du, Q., Gunzburger, M.D. (1993) A model for superconducting thin films having variable thickness. *Physica D*, 69:215–231.
107. Du, Q. (1994) Global existence and uniqueness of solution of time-dependent Ginzburg–Landau model for superconductivity. *Appl. Anal.*, 52:1–17.
108. Dunford, N., Schwartz, J.T. (1958) *Linear operators.* Interscience, New York.
109. Duvaut, G., Lions, J.L. (1972) *Les inéquations en mécanique et en physique.* Dunod, Paris.
110. Edestein-Keshet, L. (1988) *Mathematical models in biology.* Random House, New York.
111. Ekeland, I. (1974) On the variational principle. *J. Math. Anal. Appl.*, 47:324–353.
112. Ekeland, I., Temam, R. (1976) *Convex analysis and variational problems.* North-Holland, Amsterdam.
113. Ekeland, I. (1990) *Convex methods and Hamiltonian Mechanics.* Springer-Verlag, Berlin.
114. Engl, H.W., Scherzer, O. (2000) Convergence rates results for iterative methods for solving nonlinear ill-posed problems. In: Colton, D., Engl, H.W., Louis, A., McLaughlin, J.R., Rundell, W. (eds.), *Survey on Solution Methods for Inverse Problems*, pp. 1–34. Springer-Verlag, Vienna.
115. Eringer, A.C. (1966) Theory of micropolar fluids. *J. Math. Mech.*, 16:1–18.
116. Eringer, A.C. (1978) Micropolar theory of liquid crystals. In: *Liquid Crystal and Ordered Fluids*, pp. 443–471. Plenum, New York.
117. Fan, K. (1961) A generalization of Tychonoff's fixed point theorem. *Math. Ann.*, 142:305–310.
118. Fan, K. (1972) A minimax inequality and its application. In: Shisha, O. (ed.), *Inequalities*, pp. 103–113. Academic, New York.
119. Fattorini, H.O., Sritharan, S.S. (1992) Existence of optimal controls for viscous flows problems. *Proc. Royal Soc. London, Series A*, 439:81–102.

120. Fattorini, H.O., Sritharan, S.S. (1994) Necessary and sufficient conditions for optimal controls in viscous flow. *Proc. Royal Soc. Edinburgh, Series A*, 124:211–251.
121. Fedoseyev, A.I., Alexander, J.I. (2000) Investigation of vibrational control of convective flows in Bridgman melt growth configurations. *J. Crystal Growth*, 211:34–42.
122. Fife, P. (1979) *Mathematical aspects of reacting and diffusing systems.* Lecture Notes in Biomathematics. Springer-Verlag, New York.
123. Figueiredo, D.G. (1989) *Lectures on Ekeland variational principle with applications and detours.* Tata Institute of Fundamental Research. Springer-Verlag, New York.
124. Fister, K.R., Panetta, J.C. (2003) Optimal control applied to competing chemotherapeutic cell-kill strategies. *SIAM J. Appl. Math.*, 63:1954–1971.
125. Foias, C., Ozbay, H., Taunnenbaum, A. (1996) *Robust control for infinite-dimensional systems. Frequency Domain Methods.* Lecture Notes in Control and Information Sciences. Springer-Verlag, Berlin.
126. Ganzler, T., Volkwein, S., Weiser, M. (2006) SQP methods for parameter identification problem arising in hyperthermia. *Optim. Meth. Soft.*, 21:869–887.
127. Gao, Y., Strang, G. (1989) Geometric nonlinearity: potential energy, complementary, and the gap function. *Quart. Appl. Math.*, 47:487–504.
128. Gao, Y. (2000) *Duality principles in nonconvex systems: theory, methods and applications.* Kluwer, Dordrecht.
129. Gejadze, I.Y., Copeland, G.J.M., Navon, I.M. (2006) Open boundary control problem for Navier–Stokes equations including a free surface: data assimilation. *Comput. Math. Appl.*, 52:1269–1288.
130. Gejadze, I.Y., Copeland, G.J.M. (2006) Open boundary control problem for Navier–Stokes equations including a free surface: adjoint sensitivity analysis. *Comput. Math. Appl.*, 52:1243–1268.
131. Ghoussoub, N. (1993) *Duality and perturbation methods in critical point theory.* Cambridge University Press, New York.
132. Gilbarg, D., Trudinger, N.S. (1983) *Elliptic partial differential equations of second order.* Springer-Verlag, Berlin.
133. Gill, P.E., Murray, W., Wright, M.H. (1981) *Practical Optimization.* Academic, San Diego.
134. Girault, V., Raviart, P.A (1986) *Finite element methods for Navier–stokes equations.* Springer-Verlag, Berlin.
135. Gopalsamy, K. (1992) *Stability and oscillations in delay differential equations of population dynamics.* Kluwer, Dordrecht.
136. Gor'kov, L.P., Eliashberg, G.M. (1968) Generalization of the Ginzburg–Landau equations for non-stationary problems in the case of alloys with paramagnetic impurties. *Soviet. Phy., JETP*, 27:328–334.
137. Granasy, L., Borzsonyi, T., Pusztai, T. (2002) Nucleation and bulk crystallization in binary phase field theory. *Phys. Rev. Lett.*, 88:206105_1–206105_4.
138. Granasy, L., Borzsonyi, T., Pusztai, T. (2002) Crystal nucleation and growth in binary phase-field theory. *J. Cryst. Growth.*, 237–239:1813–1817.
139. Granasy, L., Pusztai, T., Warren, J.A. (2004) Modelling polycrystalline solidification using phase field theory. *J. Phys.: Condens. Matter*, 16:1205–1235.
140. Green, M., Limebeer, D.J.N. (1995) *Linear robust control.* Prentice-Hall, Englewood Cliffs, NJ.

141. Gropp, W.D., Kaper, H.G., Leaf, G.K., Levine, D.M., Palumbo, M., Vinokur, V.M. (1996) Numerical simulation of vortex dynamics in type-II superconductors. *J. Comput. Phys.*, 123:254–266.
142. Grujicic, M., Cao, G., Miller, R.S. (2002) Computer modeling of evolution of dentrite microstructure in binary alloys during non-isothermal solidification. *J. Mater. Synth. Processing*, 10:191–203.
143. Gu, K., Kharitnov, V.L., Chen, J. (2003) *Stability and robust stability of time-delay systems.* Birkhauser, Boston.
144. Gunzburger, M.D., Hou, L.S., Svobodny, Th.P. (1991) Analysis and finite element approximation of optimal control problems for the stationary Navier–Stokes equations whith Dirichlet controls. *M2.A.N.*, 25:711–748.
145. Gunzburger, M.D. (ed.) (1995) *Flow control.* IMA, Mathematics and its Applications. Springer-Verlag, Berlin.
146. Gunzburger, M., Ozugurlu, E., Turner, J., Zhang, H. (2002) Controlling transport phenomena in the Czochralski crystal growth process. *J. Crystal Growth*, 234:47–62.
147. Gurtin, M.E (1981) *Introduction to continuum mechanics.* Academic, San Diego.
148. Hackbusch, W. (1985) *Multi-grid methods and applications.* Springer-Verlag, Berlin.
149. Haftka, R.T., Gurdal, G. (1992) *Elements of structural optimization.* Kluwer, Dordrecht.
150. Hale, J.K., Verduyn-Lunel, S.M. (1993) *Introduction to Functional Differential Equations.* Applied Math. Sciences, Springer-Verlag, New York.
151. He, F., Leung, A., Stojanovic, S. (1994) Periodic optimal control for competing parabolic Volterra–Lotka-type systems. *J. Comput. Appl. Math.*, 52:199–217.
152. He, F., A. Leung, A., Stojanovic, S. (1995) Periodic optimal control for parabolic Volterra–Lotka type equations. *Math. Methods Appl. Sci.*, 18:127–146.
153. Henrot, A., Pierre, M. (2005) *Variation et optimisation de formes. Une analyse géométrique.* Collection Mathématiques et Applications. Springer-Verlag, New York.
154. Hestenes, M.R. (1980) *Conjugate direction methods in optimization.* Springer-Verlag, New York.
155. Hewitt, E., Stromberg, K. (1982) *Real and abstract analysis.* Graduate texts in Mathematics, Springer-Verlag, New York.
156. Hill, J., Pincombe, A. (1992) Some similarity temperature profiles for the microwave heating of a half-space. *J. Austral. Math. Sco. Ser. B*, 33:290–320.
157. Hiriart-Urruty, J.B., Lemaréchal, C. (2001) *Fundamentals of convex analysis.* Springer-Verlag, Berlin.
158. Hirst, D.G. (1989) Tumor blood flow modification therapeutic benefit: is this approach ready for clinical application? In: Michael, B., Hance, M. (eds.), *GrayLaboratory 1989 Annual Report*, pp. 14–17. Cancer Research Campaign, London.
159. Hoffman, K.H., Jiang, L. (1992) Optimal control of a phase field model for solidification. *Num. Funct. Anal. Optim.*, 13:11–27.
160. Holmes, R. (1975) *Geometric functional analysis and its applications.* Springer-Verlag, New York.
161. Hrinca, L. (1997) An optimal control problem problem for the Lotka–Volterra system with delay. *Nonlinear Anal.*, 28:247–262.

162. Hu, C., Temam, R. (2001) Robust control of the Kuramoto–Sivashinsky equation. *Dyn. Cont. Impu. Syst.*, 8:315–338.
163. Hutchinson, E., Dahlch, M.A., Hynynen, K. (1998) The feasibility of MRI feedback control for intracavitary phased array hyperthermia treatments. *Int. J. Hyperthermia*, 14:39–56.
164. Ichikawa, A. (1982) Quadratic control of evolution equations with delays in control. *SIAM J. Control Optim.*, 20:645–668.
165. Isidori, A. (1995) *Nonlinear control systems*, 3rd ed., Springer-Verlag, London.
166. Ito, K., Kunisch, K. (1996) Augmented Lagrangian-SQP methods for nonlinear optimal control problems of tracking type. *SIAM J. Control Optim.*, 34:874–891.
167. Ito, K., Kunisch, K. (1996) Augmented Lagrangian-SQP methods in Hilbert spaces and application to control in the coefficients problems. *SIAM J. Control Optim.*, 6:96–125.
168. Jackson, T.L., Byrne, H.M. (2000) A mathematical model to study the effects of drug resistance and vascular on the response of solid tumours to chemotherapy. *Math. Biosci.*, 164:17–38.
169. Jia, C.H., Feng, D.X. (2007) An optimal control problem of a coupled nonlinear parabolic population system. *Acta Math. Appl. Sinica*, 23:377–388.
170. Kamppeter, T., Mertens, F., Moro, E., Sanchez, A., Bishop, A. (1999) Stochastic vortex dynamics in two-dimensional easy-plane ferromagnets: multiplicative versus additive noise. *Phys. Rev. B*, 59:11349–11357.
171. Kavian, O. (1993) *Introduction à la théorie des points critiques et applications aux problèmes elliptiques.* Springer-Verlag, Berlin.
172. Khaled, A.R.A., Vafai, K. (2003) The role of porous media in modeling flow and heat transfer in biological tissues. *Int. J. Heat. Mass transfer*, 46:4989–5003.
173. Khapalov, A. (2004) *Bilinear control for global controllability of the semilinear parabolic equations with superlinear terms*, 3rd ed., Lecture Notes in Pure and Applied Mathematics, Springer-Verlag, London.
174. Kimmel, M., Świerniak, A. (2004) Using control theory to make cancer chemotherapy beneficial from phase dependence and resistant to drug resistance. *Arch. Control Sci.*, 14:105–145.
175. Kobayashi, R. (1993) Modelling and numerical simulations of dentritic crystal growth. *Physica D*, 63:410–423.
176. Köhler, T., Maass, P., Wust, P., Seebass, M. (2001) A fast algorithm to find optimal controls of multiantenna applicators in regional hyperthermia. *Phys. Med. Biol.*, 46:2503–2514.
177. Kolmanovskii, V.B., Myshkis, A.D. (1992) *Applied theory of functional differential equations.* Kluwer, Dordrecht.
178. Kolmanovskii, V.B., Shaikhet, L.E. (1993) *Control of systems with aftereffect.* Translations of Mathematical Monographs. AMS, Providence, RI.
179. Kowalewski, A. (1993) Optimal control of parabolic systems with time-varying lags. *IMA J. Math. Control Inf.*, 10:113–129.
180. Kowalewski, A. (1998) Optimal control of distributed parabolic systems with multiple time-varying lags. *Int. J. Control*, 69:361–381.
181. Kowalski, M.E., Jin, J.M. (2004) Model-based optimization of phased arrays for electromagnetic hyperthermia. *IEEE Trans. Microwave Theory Tech.*, 52:1964–1977.

182. Krasnoselskii, M.A. (1964) *Topological methods in the theory of nonlinear integral equations.* Pergamon, London.
183. Kuang, Y. (1993) *Delay differential equations with applications in population dynamics.* Academic, Boston.
184. Ladyzenskaya, O.A., Solonnikov, V.A., Uralceva, N.N. (1968) *Linear and quasilinear equations parabolic type.* Translations of Mathematical Monographs. AMS, Providence, RI.
185. Landau, L.D., Lifschitz, E.M (1986) *Theory of Elasticity.* Butterworth, Oxford.
186. Laurencot, Ph. (1997) Weak solutions to a phase-field model with non-constant thermal conductivity. *Quart. Appl. Math.*, 4:739–760.
187. Lasiecka, I., Triggiani, R. (2000) *Control theory for partial differential equations: continuous and approximation theories.* Cambridge University Press, Cambridge.
188. Le Dimet, F.X., Shutyaev, V.P. (2000) On Newton methods in data assimilation. *Russ. J. Numer. Anal. Math. Model.*, 15:419–434.
189. Ledzewicz, U., Schttler, H. (2002) Optimal bang-bang controls for a two-compartment model in cancer chemotherapy. *J. Optim. Theory Appl.*, 114:609–637.
190. Ledzewicz, U., Schttler, H. (2006) Drug resistance in cancer chemotherapy as an optimal control problem. *Disc. Contin. Dyn. Syst. Ser. B*, 6:129–150.
191. Lee, C.S., Leitmann, G. (1988) Continuous feedback guaranteeing uniform ultimate boundedness for uncertain linear delay systems: an application to river pollution control. *Comput. Math. Appl.*, 16:929–938.
192. Lenhart, S.M., Bhat, M.G. (1992) Application of distributed parameter control model in wildlife damage management. *Math. Models Meth. Appl. Sci.*, 2:423–439.
193. Lenhart, S.M., Yong, J. (1995) Optimal control for degenerate parabolic equations with logistic Growth. *Nonlinear Analysis: Theory Meth. Appl.*, 25:681–698.
194. Lenhart, S.M., Liang, M., Protopopescu, V. (1999) Optimal control of boundary habitat hostility for interacting species. *Math. Meth. Appl. Sci.*, 22:1061–1077.
195. Lenhart, S.M., Liang, M. (2000) Bilinear optimal control for a wave equation with viscous damping. *Houston J. Math.*, 26:575–595.
196. Leung, A.W. (1989) *Systems of nonlinear partial differentil equations. Applications to biology and engeineering.* Mathematics and its Applications. Kluwer, Dordrecht.
197. Leung, A.W., Stojanovic, S. (1993) Optimal control for elliptic Volterra–Lotka type equations. *J. Math. Anal. Appl.*, 173:603–619.
198. Leung, A.W., Chen, G.S. (1999) Optimal control of multigroup neutron fission System. *Appl. Math. Optim.*, 40:39–60.
199. Li, X., Yong, J. (1995) *Optimal control theory for infinite dimensional systems.* Birkhäuser, Boston, MA.
200. Ling, X., Park, S., McClain, B., Choi, S., Dender, D., Lynn, J. (2001) Superheating and supercooling of vortex matter in Nb single crystal: direct evidence for a phase transition at the peak effect from neutron diffraction. *Phys. Rev. Lett.*, 86:712–715.
201. Lions, J.L. (1961) *Equations différentielles operationnelles.* Springer-Verlag, New York.

202. Lions, J.L. (1968) *Contrôle optimal de systèmes gouvernés par des équations aux dérivées partielles.* Dunod, Paris.
203. Lions, J.L. (1969) *Quelques méthodes de résolution des problèmes aux limites non linéaires.* Gauthier-Villars, Paris.
204. Lions, J.L., Magenes, E. (1972) *Nonhomogeneous boundary value problems and applications.* Springer-Verlag, New York.
205. Lions, J.L., Temam, R., Wang, S. (1992) New formulation of the primitive equations of atmosphere and application. *Nonlinearity*, 5:237–288.
206. Lions, J.L., Temam, R., Wang, S. (1992) On the equation of the large-scale ocean. *Nonlinearity*, 5:1007–1053.
207. Lions, J.L., Temam, R., Wang, S. (1995) Mathematical theory for the coupled atmosphere–ocean models. *J. Math. Pure. Appl.*, 74:105–163.
208. Liu, J., Zhu, L., Xu, X.L. (2000) Studies on three-dimensional temperature transients in the canine prostate during transurethal microwave thermal therapy. *ASME J. Biomech. Eng.*, 122:372–379.
209. Liu, W., Freedman, H.I. (2005) A mathematical model of vascular tumor treatment by chemotherapy. *Math. Comput. Modelling*, 42:1089–1112.
210. Lukaszewicz, G. (1999) *Micropolar fluids. Theory and applications. Modeling and simulation in science.* Engineering and technology. Birkhauser, Boston.
211. Lusternik, L., Schnirelman, L. (1934) *Méthodes topologiques dans les problèmes variationels.* Gauthier-Villars, Paris.
212. MacDonald, N. (1989) *Biological delay systems: linear stability theory.* Cambridge University Press, Cambridge.
213. Mahmoud, M.S. (2000) *Robust control and filtering for time-delay systems.* Control Engineering. Marcel Dekker, New York.
214. Malanowski, K.D. (2004) Convergence of the Lagrange–Newton method for optimal control problems. *Int. J. Appl. Math. Comput. Sci.*, 14:531–540.
215. Malinen, M., Duncan, S., Huttunen, T., Kaipio, J. (2006) Feedforward and feedback control of ultrasound surgery. *Appl. Numer. Math.* 56:55–79.
216. Marchant, T., Lui, B. (2001) On the heating of a two-dimensional slab in a microwave cavity: aperture effects. *ANZIAM J.*, 43:137–148.
217. Marchuk, G.I. (1986) *Mathematical models in environmental problems*, North Holland, Amsterdam.
218. Marchuk, G.I. (1995) *Adjoint equation and analysis of complex systems.* Kluwer, Dordrecht.
219. Martin, G.T., Bowman, H.F., Newman, W.H. (1992) Basic element method for computing the temperature field during hyperthermia therapy planning. *Adv. Bio. Heat Mass Transf.*, 23:75–80.
220. Mawhin, J., Willem, M. (1989) *Critical point theory and Hamiltonian systems.* Applied Mathematical Science. Springer-Verlag, New York.
221. Maz'ja, V.G. (1985) *Sobolev Spaces.* Springer-Verlag, New York.
222. McMillan, C., Triggiani, R. (1994) Min-max game theory and algebraic Riccati equations for boundary control problems with continuous input-solution map, part II: the general case. *Appl. Math. Optim.*, 29:1–64.
223. McMillan, C., Triggiani, R. (1994) Min-max game theory and algebraic Riccati equations for boundary control problems with analytic semigroups, part II: the general case. *Nonlinear Anal.* 22:431–465.
224. Mishra, S., Katiyar, V.K., Arora, V. (2007) Mathematical modeling of chemotherapy strategies in vascular tumor growth using nanoparticles. *Appl. Math. Comput.*, 189:1246–1254.

225. Mordukhovich, B., Zhang, B.S. (1994) Existence, approximation and suboptimality conditions for minimax control of heat transfer systems with state constraints. In: *Applied Mathematics*, pp. 251–270. Marcel Dekker Lecture Notes Pure.
226. Mordukhovich, B., Zhang, B.S. (1997) Minimax control of parabolic systems with Dirichlet boundary conditions and state constraints. *Appl. Math. Optim.*, 36:323–360.
227. Murray, J.D. (1993) *Mathematical Biology.* Springer-Verlag, New York.
228. Nashed, M.Z. (1971) Differentiability and related properties of nonlinear operators: some aspects of the role of differentials in nonlinear functional analysis. In: *Nonlinear Functional Analysis and Application*, pp. 103–309. Mathematics Research Center, The University of Wisconsin.
229. Neubert, M., Rudolph, P. (2001) Growth of semi-insulating GaAs crystals in low temperature gradients by using the vapour pressure controlled Czochralski method (VCz). *Prog. Cryst. Growth Charact. Mater.*, 43:119-185
230. Niculescu, S.I. (2001) *Delay effects on stability. A robust control approach.* Springer-Verlag, Heidelberg.
231. Oden, J.T., Reddy, J.N. (1983) *Variational methods in theoretical Mechanics.* Springer-Verlag, New York.
232. Paganini, F.G., Dullerud, G.E. (2000) *A Course in robust control theory: a convex approach.* Springer-Verlag, Heidelberg.
233. Palais, R.S. (1970) Critical point theory and the minimax principle. In: *Proceedings of Symposia in Pure Mathematics (Global Analysis)*, pp. 185–212. AMS Providence, RI.
234. Papageorgiou, N.S., Yannakakis, N. (2002) Minimax control of nonlinear evolution equations. *Appl. Math. Comput.*, 131:271–297.
235. Parmuzin, E.I., Shutyaev, V.P., Diansky, N.A. (2007) Numerical solution of a variational data assimilation problem for a 3D ocean thermohydrodynamics model with a nonlinear vertical heat exchange. *Russ. J. Numer. Anal. Math. Model.*, 22:177–198.
236. Pedlosky, J. (1979) *Geophysical fluid dynamics.* Springer-Verlag, New York.
237. Pennes, H.H. (1948) Analysis of tissue and arterial blood temperatures in the resting human forearm. *J. Appl. Phys.*, 1:93–122.
238. Philander, S.G.H., Pacanowski, R.C. (1980) The generation of equatorial currents. *J. Geophys. Res.*, 85:1123–1136.
239. Philander, S.G.H., Pacanowski, R.C. (1981) Parameterization of vertical mixing in numerical models of tropical oceans. *J. Phys. Oceanogr.*, 11:1443–1451.
240. Philander, S.G.H., Hurlin, W.J., Pacanowski, R.C. (1986) Properties of long equatorial waves in models of the seasonal cycle in the tropical Atlantic and Pacific oceans. *J. Geophys. Res.*, 91:14207-14211.
241. Petersen, I.R., Ugrinovskii, V.A., Savkin, A.V. (2000) *Robust control design using H^∞-methods.* Communications and Control Engineering Series. Springer-Verlag, Berlin.
242. Pincombe, A.H., Smyth, N.F. (1991) Microwave heating of materials with low conductivity. *Proc. R. Soc., A*, 433:479–498.
243. Popel, A.S., Regirer, S.A., Usick, P.I. (1974) A continuum model of blood flow. *Biorheology*, 11:427–437.
244. Prasad, V., Zhang, H., Anselmo, A.P. (1997) Transport phenomena in Czochralski crystal growth processes. In: *Advances in Heat Transfer*, vol. 30, pp. 313–435. Academic, New York.

245. Prodi, G. (1962) Theoremi di tipo locae per il sistema di Navier–Stokes a stabilita delle soluzionare. *Rc. semin. Mat. Univ. Padova*, 32:374–397.
246. Qu, Z. (1998) *Robust control of nonlinear uncertain systems.* Wiley Series in Nonlinear Science. Wiley, Chichester.
247. Rabinowitz, P. (1986) *Minimax methods in critical point theory and application to differential equations.* CBMS Regional Conference Series in Mathematics. AMS, Providence, RI.
248. Rachev, S.T., Rschendorf, L. (1998) *Mass transportation problems I and II.* Probability and its Applications. Springer-Verlag, New York.
249. Rappaz, J., Scheid, J.F. (2000) Existence of solution to a phase-field model for the isothermal solidification process of a binary alloy. *Math. Meth. Appl. Sci.*, 23:491–513.
250. Robinson, S.M. (1980) Strongly regular generalized equations. *Math. Oper. Res.*, 5:43–62.
251. Rockafellar, R.T. (1967) Duality and stability in extremum problems involving convex functions. *Pacific J. Math.*, 21:167–187.
252. Rockafellar, R.T. (1970) *Convex Analysis.* Princeton landmarks in mathematics. Princeton University Press, Princeton, NJ.
253. Rubinstein, J. (1995) On the equilibrium position of Ginzburg–Landau vortices. *Z. Angew. Math. Phys.*, 46:739–751.
254. Ruschendorf, L. (1995) Optimal solution of multivariate coupling problems. *Appl. Mathematicae*, 22:325–338.
255. Sachkov, Y.L. (1997) On positive orthant controllability of bilinear systems in small codimensions. *SIAM J. Control Optim.*, 35:29–35.
256. Sadek, I.S., Vedantham, R. (2000) Optimal control of distributed nuclear reactors with pointwise controllers. *Math. Comput. Model.*, 32:341–348.
257. SanchezPena, R.S., Sznaier, M. (1998) *Robust systems theory and applications.* Wiley, Chichester.
258. Sandier, E., Serfaty, S. (2003) Ginzburg–Landau minimizers near the first critical field have bounded vorticity. *Calc. Var. Partial Differ. Eq.*, 17:17–28.
259. Sasik, R., Bettencourt, L., Habib, S. (2000) Thermal vortex motion in a two-dimensional condensate. *Phys. Rev. B*, 62:1238–1246.
260. Scherzer, O. (1995) Convergence criteria of iterative methods based on Landweber iteration for solving nonlinear problems. *J. Math. Anal. Appl.*, 194:911–933.
261. Schwartz, J.T. (1969) *Nonlinear functional analysis.* Gordon-Breach, London.
262. Schwartz, L. (1957) *Théorie des distributions.* 2nd ed. Hermann, Paris.
263. Schwartz, L. (1957-1958) Distributions à valeurs vectorielles. *Ann. Inst. Fourier (Grenoble)*, 7:1-141 and 8:1-209.
264. Schwartz, L. (1973) *Théorie des distributions.* Hermann, Paris.
265. Seidman, T., Zhou, H.Z. (1982) Existence and uniqueness of optimal controls for a quasilinear parabolic equation. *SIAM J. Control Optim.*, 20:747–762.
266. Seip, R., Ebbini, E.S. (1995) Studies on three-dimensional temperature response to heating fields using diagnostic ultrasound. *IEEE Trans. Biomed. Eng.*, 42:828–839.
267. Sewell, M.J. (1987) *Maximum and minimum principles.* Cambridge University Press, Cambridge.
268. Shih, T.C., Kou, H.S., Lin, W.L. (2002) Effect of effective tissue conductivity on thermal dose distributions on living tissue with directional blood flow during thermal therapy. *Int. Comm. Heat Mass transfer*, 29:115–126.

269. Shimizu, T., Morioka, N. (1995) Stationary distribution of a nonlinear system driven by a chaotic force. *Physica A*, 218:390–402.
270. Shimizu, T., Yaghi, S. (2001) Comparison of an integral including a chaotic force with a stochastic integral. *J. Korean physical Society*, 38:521–524.
271. Simon, J. (1990) Nonhomogeneous viscous incompressible fluid: existence of velocity, density, and pressure. *SIAM J. Math. Anal.*, 21:1093–1117.
272. Sokolowski, J., Zolesio, J.P. (1992) *Introduction to shape optimization.* Springer-Verlag, Berlin.
273. Sritharan, S.S. (ed.) (1998) *Optimal control of viscous.* SIAM, Philadelphia, PA.
274. Stavre, R. (2002) The control of the pressure for a micropolar fluid. *Z. Angew. Math. Phys.*, 53:912–922.
275. Strang, G. (1986) *Introduction to applied mathematics.* Wellesley-Cambridge Press, MA.
276. Strauss, W.T. (1966) On continuity of functions with values in various Banach spaces. *Pacific J. Math.*, 19:543–551.
277. Struwe, M. (1996) *Variational methods: applications to nonlinear partial differential equations and Hamiltonian systems.* Springer-Verlag, New York.
278. Sturesson, C., Andersson-Engels, S. (1995) A mathematical model for predicting the temperature distribution in laser-induced hyperthermia. Experimental evaluation and applications. *Phys. Med. Biol.*, 40:2037–2052.
279. Tang, Q, Wang, S. (1995) Long time behavior of the Ginzburg–Landau superconductivity equations. *Appl. Math. Lett.*, 8:31–34.
280. Taylor, J.E., Cahn, J.W. (1998) Diffuse interface with sharp corners and facets: phase field model. *Physica D*, 112:381–411.
281. Temam, R. (1977) *Navier–Stokes equations.* North-Holland, Amsterdam.
282. Temam, R. (1984) *Navier–Stokes equations, theory and numerical analysis.* North-Holland, Amsterdam.
283. Trèves, F. (1967) *Topological vector spaces, distributions and kernels.* Pure and Applied Mathematics. Academic, London.
284. Troltzsch, F. (1999) On the Lagrange–Newton-SQP method for the optimal control of semilinear parabolic equations. *SIAM J. Control Optim.*, 38:294–312.
285. Tropea, B.I., Lee, R.C. (1992) Thermal injury kinetics in electrical trauma. *J. Biomech. Eng.*, 114:241–250.
286. Vainberg, M.M. (1964) *Variational methods for the study of the nonlinear operators.* Holden-Day, Inc., San Francisco.
287. Valvano, J.W. (1984) An isolated rat liver model for the evaluation of thermal techniques to measure perfusion. *J. Biomech. Eng.*, 106:187–191.
288. Van Keulen, B. (1993) *$\mathcal{H}_\infty$-control for disturbed parameter systems: a state-space approach.* Birkhäuser, Basel.
289. Vanne, Hynynen, A.K. (2003) MRI feedback temperature control for focused ultrasound surgery. *Phys. Med. Biol.*, 48:31–43.
290. Vinnicombe, G. (1999) *Uncertainty and Feedback: $\mathcal{H}_\infty$ loop-shaping and the ν-gap metric.* Imperial College Press, London.
291. Waltman, P. (1983) *Competition models in population biology.* SIAM, Philadelphia.
292. Wang, P.K.C. (1975) Optimal control of parabolic system with boundary conditions involving time delays. *SIAM J. Control*, 13:274–293.

293. Warren, J.A., Boettinger, W.J. (1995) Prediction of the dentritic growth and microsegregation patterns in a binary alloy using the phase-field model. *Acta Metall. Mater.*, 43:689–703.
294. Boettinger, W.J., Warren, J.A., Beckermann, C., Karma, A. (2002) Phase-field simulation of solidification. *Annu. Rev. Mater. Res.*, 32:163–194.
295. Warren, J.A., Kobayashi, R., Lobkovsky, A.E., Carter, W.C. (2003) Extending phase field models of solidification to polycrystalline materials. *Acta Mater.*, 51:6035–6058.
296. Washington, W.M., Parkinson, C.L. (1986) *An introduction to three-dimensional climate modeling.* Oxford University Press, Oxford.
297. Watanabe, M., Vizman, D., Friedrich, J., Mueller, G. (2006) Large modification of crystal-melt interface shape during Si crystal growth by using electromagnetic Czochralski method (EMCZ). *J. Crystal Growth*, 292:252–256.
298. Weinbaum, S., Jiji, L.M. (1985) A two simplified bioheat equation for the effect of blood flow on average tissue temperature. *J. Biomech. Eng.*, 107:131–139.
299. Wesseling, P. (1992) *An introduction to multigrid methods.* Pure and Applied Mathematics. Wiley, New York.
300. Whittle, P. (1990) *Risk-sensitive optimal control.* Wiley-Interscience Series in Systems and Optimization. Wiley, New York.
301. Willem, M. (1983) *Lecture notes on critical point theory.* Fundaçao Universidade de Brasilia.
302. Wissler, E.H. (1998) Pennes' 1948 paper revisited. *J. Appl. Phys.*, 85:35–41.
303. Wunsch, C. (1996) *The ocean circulation inverse problem.* Cambridge University Press, Cambridge.
304. Yamaguchi, N. (2005) Existence of global strong solution to the micropolar fluid system in a bounded domain. *Math. Meth. Appl. Sci.*, 28:1507–1526.
305. Yamamoto, M., Zou, J. (2001) Simultaneous reconstruction of the initial temperature and heat radiative coefficient. *Inverse Problems*, 17:1181–1202.
306. Yosida, K. (1974) *Functional analysis.* Die Grundlehren der mathematischen Wissenschaften, Band 123. Springer-Verlag, New York-Heidelberg.
307. Zames, G., Elsakkary, A.K. (1980) Unstable systems and feedback: the gap-metric. In: *Proceedings of the Allerton Conference*, pp. 380–385. University of Illinois, Monticello, IL.
308. Zeidler, E. (1985) *Nonlinear functional analysis and its applications III.* Springer-Verlag, New York.
309. Zhang, J., Chen, L., Che, X.D. (1999) Persistence and global stability for two species nonautonomous competition Lotka–Volterra patch-system with time delay. *Nonlinear Anal.*, 37:1019–1028.
310. Zhong, Q.C. (2006) *Robust control of time-delay systems.* Springer-Verlag, London.
311. Zhou, K., Doyle, J.C., Glover, K. (1996) *Robust and optimal control.* Prentice-Hall, Upper Saddle River, NJ.
312. Zhou, K., Doyle, J.C., (1998) *Essential of robust control.* Prentice-Hall, Upper Saddle River, NJ.
313. Zuazua, E., (2003) Remarks on the controllability of the Schrödinger equation. In: Bandrauk, A., Delfour, M.C., Le Bris, C. (eds.), *Quantum Control: Mathematical and Numerical Challenges*, pp. 181–199. CRM Proceedings Lecture Notes Series, vol. 33. AMS Publications, Providence, RI.

Index